Nonuniform Hyperbolicity

Dynamics of Systems with Nonzero Lyapunov Exponents

Designed to work as a reference and as a supplement to an advanced course on dynamical systems, this book presents a self-contained and comprehensive account of modern smooth ergodic theory. Among other things, this provides a rigorous mathematical foundation for the phenomenon known as deterministic chaos – the appearance of chaotic motions in pure deterministic dynamical systems. A sufficiently complete description of topological and ergodic properties of systems exhibiting deterministic chaos can be deduced from relatively weak requirements on their local behavior known as nonuniform hyperbolicity conditions.

Nonuniform hyperbolicity theory is an important part of the general theory of dynamical systems. Its core is the study of dynamical systems with nonzero Lyapunov exponents both conservative and dissipative, in addition to cocycles and group actions. The results of this theory are widely used in geometry (e.g., geodesic flows and Teichmüller flows), in rigidity theory, in the study of some partial differential equations (e.g., the Schrödinger equation), and in the theory of billiards, as well as in applications to physics, biology, engineering, and other fields.

Luis Barreira is a Professor of Mathematics at Instituto Superior Técnico in Lisbon. He obtained his Ph.D. from the Pennsylvania State University in 1996, under the guidance of Yakov Pesin, with whom he coauthored the book *Lyapunov Exponents and Smooth Ergodic Theory*. He has also written two surveys and more than 50 research papers.

Yakov Pesin is a Distinguished Professor of Mathematics at the Pennsylvania State University. He obtained his Ph.D. from the Gorky State University in 1979. He is the author of three books, *Dimension Theory in Dynamical Systems*, *Lectures on Partial Hyperbolicity and Stable Ergodicity*, and, with Luis Barreira, *Lyapunov Exponents and Smooth Ergodic Theory*, as well as six surveys and more than 70 research papers. He is an executive editor of the journal *Ergodic Theory and Dynamical Systems*.

ENCYCLOPEDIA OF MATHEMATICS AND ITS APPLICATIONS

Volume 115

The titles below, and earlier volumes in the series, are available from booksellers or from Cambridge University Press at www.cambridge.org.

58 R. Gardner *Geometric Topography*
80 O. Stormark *Lie's Structural Approach to PDE Systems*
81 C. F. Dunkl and Y. Xu *Orthogonal Polynomials of Several Variables*
82 J. P. Mayberry *The Foundations of Mathematics in the Theory of Sets*
83 C. Foias et al. *Navier-Stokes Equations and Turbulence*
84 B. Polster and G. Steinke *Geometries on Surfaces*
85 D. Kaminski and R. B. Paris *Asymptotics and Mellin–Barnes Integrals*
86 R. J. Mc Eliece *The Theory of Information and Coding, 2ed*
87 B. A. Magurn *An Algebraic Introduction to K-Theory*
88 T. Mora *Solving Polynomial Equation Systems I*
89 K. Bichteler *Stochastic Integration with Jumps*
90 M. Lothaire *Algebraic Combinatorics on Words*
91 A. A. Ivanov and S. V. Shpectorov *Geometry of Sporadic Groups II*
92 P. McMullen and E. Schulte *Abstract Regular Polytopes*
93 G. Gierz et al. *Continuous Lattices and Domains*
94 S. R. Finch *Mathematical Constants*
95 Y. Jabri *The Mountain Pass Theorem*
96 G. Gasper and M. Rahman *Basic Hypergeometric Series, 2ed*
97 M. C. Pedicchio and W. Tholen *Categorical Foundations*
98 M. Ismail *Classical and Quantum Orthogonal Polynomials in One Variable*
99 T. Mora *Solving Polynomial Equation Systems II*
100 E. Olivieri and M. E. Vares *Large Deviations and Metastability*
101 A. Kushner, V. Lychagin, and V. Roubtsov *Contact Geometry and Nonlinear Differential Equations*
102 L. Beineke and R. J. Wilson *Topics in Algebraic Graph Theory*
103 O. J. Staffans *Well-Posed Linear Systems*
104 J. Lewis, S. Lakshmivaraham, and S. Dhall *Dynamic Data Assimilation*
105 M. Lothaire *Applied Combinatorics on Words*
106 A. Markoe *Analytic Tomography*
107 P. A. Martin *Multiple Scattering*
108 R. A. Brualdi *Combinatorial Matrix Classes*

ENCYCLOPEDIA OF MATHEMATICS AND ITS APPLICATIONS

Nonuniform Hyperbolicity

Dynamics of Systems with Nonzero Lyapunov Exponents

LUIS BARREIRA
Instituto Superior Técnico, Lisbon

YAKOV PESIN
The Pennsylvania State University, University Park

CAMBRIDGE
UNIVERSITY PRESS

CAMBRIDGE
UNIVERSITY PRESS

32 Avenue of the Americas, New York NY 10013-2473, USA

Cambridge University Press is part of the University of Cambridge.

It furthers the University's mission by disseminating knowledge in the pursuit of education, learning and research at the highest international levels of excellence.

www.cambridge.org
Information on this title: www.cambridge.org/9780521832588

First published 2007

A catalogue record for this publication is available from the British Library

Library of Congress Cataloguing in Publication data

Barreira, Luis, 1968–
Nonuniform hyperbolicity: dynamics of systems with nonzero Lyapunov exponents / Luis Barreira and Yakov Pesin.
p. cm. – (Encyclopedia of mathematics and its applications ; 115)
Includes bibliographical references and index.
ISBN-13: 978-0-521-83258-8 (hardback)
ISBN-10: 0-521-83258-6 (hardback)
1. Lyapunov exponents. 2. Lyapunov stability. 3. Dynamics.
I. Pesin, Ya. B. II. Title. III. Series.
QA871.B365 2007
531′.11–dc22 2007013057

ISBN 978-0-521-83258-8 Hardback

To Clàudia

and

To Natasha, Irishka, and Lenochka

for their patience, encouragement, and inspiration

Contents

Part IV General Hyperbolic Measures

Preface

Writing this book was a long-term project that has taken several years, and at the early stages Anatole Katok's participation was crucial. He provided us with the text of his unpublished notes with Leonardo Mendoza that served as the basis for the first draft of Chapters 1–5 and parts of Chapters 6–7. He also fully participated in designing the content of the book in its present form.

In this book we present a self-contained and sufficiently complete description of the modern nonuniform hyperbolicity theory, that is, the theory of dynamical systems whose Lyapunov exponents are not zero. The reader will find all the core results of the theory as well as a good account of its recent developments.

The nonuniform hyperbolicity theory is rich in wonderful ideas and sophisticated techniques, which are widely used in many areas of dynamical systems as well as other areas of mathematics and beyond. The nonuniform hyperbolicity theory is very popular and finds a lot of applications outside mathematics – in physics, biology, engineering, and so on.

Despite (or should we say because of) a tremendous amount of research on the subject, there have been relatively few attempts to summarize and unify the results of the theory in a single manuscript or a survey (see the books [110, 139, 179], the surveys [18, 173, 175], and the lectures [19, 27]). This book is meant to cover this gap. It can be used as a reference book for the theory or as a supporting material for an advanced course on dynamical systems. During the long course of working on this book, we first produced its baby version [20] in which we described the core results of the theory and some principal examples, and then we wrote the survey [21] in which we presented the contemporary status of the theory.

Since the beginning of the 1970s, the nonuniform hyperbolicity theory has emerged as an independent discipline lying in the heart of the modern theory of dynamical systems. It studies both conservative (volume preserving) and dissipative systems, deterministic as well as random dynamical systems, and discrete and continuous-time systems, in addition to cocycles and group actions. The results of this theory have found their way into geometry (e.g., in the study of geodesic flows and Teichmüller flows), into rigidity theory, into the study of some partial differential equations (e.g., the Schrödinger equation and some reaction-diffusion equations), and into the theory of "chaotic" billiards.

Writing a book of such a scope can be deemed a daunting task and we therefore had to select topics so that personal taste, clearly biased toward our own interests, entered in our choices. As a result, some interesting topics are barely mentioned or not covered at all. In particular, we do not consider random dynamical systems, referring the reader to the books [9, 112, 129] and the survey [113], nor dynamical systems, with singularities (see the book [110]), in particular, leaving aside the rich theory of chaotic billiards. We restrict ourselves to the case of invertible dynamical systems and thus the theory of nonuniformly expanding maps is not discussed here (see the survey [132]) nor do we include one-dimensional chaotic maps (e.g., the logistic family, see [94]). We touch upon some recent results on Hénon-like attractors related to the study of Sinai–Ruelle–Bowen measures but we do not go deep into the theory of these attractors (see the survey [133]). We mention some results on hyperbolic group actions and refer the reader to [71] for a more complete account.

All the principal results of the nonuniform hyperbolicity theory are presented in the book with complete proofs, although some other results are included, without proofs, for the sake of completeness.

Most chapters of the book end with notes in which the reader can find some remarks of historical and bibliographical nature, comments on some related results, and references for further reading. In no way are these notes meant to present a significant account of the history of the subject or a sufficiently complete list of references.

Acknowledgments

While working on the book, several people helped us in various ways and it is our great pleasure to acknowledge their contributions.

As we mentioned above, it is impossible to overestimate the contribution of Anatole Katok, whom we heartily thank for his constant support and guidance.

We also thank Marlies Gerber for providing us with the text of Section 6.4.2 and Theorem 9.2.8 and for some useful remarks.

It is our pleasure to thank Anton Zorich for useful comments on Section 12.5.

We are grateful to Claudia Valls, who carefully read several chapters of the book and made many remarks that helped us clarify the presentation.

When the first draft of the book was ready, we sent it to several experts in the field, asking for their opinions and comments. We are grateful to Dima Dolgopyat, François Ledrappier, Mark Pollicott, and Federico Rodríguez Hertz for innumerable constructive suggestions that helped us improve the content and presentation of some results of the book as well as extend and enhance the bibliography.

Our special thanks go to Boris Hasselblatt, who thoroughly examined the draft and pointed out many places in the book that needed additional work. Reflecting on his comments, we introduced many changes improving the style and exposition of the material; we also added more informal discussions, hopefully making the book more reader friendly.

January 2007

Luis Barreira, Lisboa
Yakov Pesin, State College

Introduction

The goal of this book is to present smooth ergodic theory from a contemporary point of view. Among other things this theory provides a rigorous mathematical foundation for the phenomenon known as *deterministic chaos* – a term coined by Yorke – the appearance of highly irregular, unpredictable,"chaotic" motions in pure deterministic dynamical systems. The main idea beyond this phenomenon is that one can deduce a sufficiently complete description of topological and ergodic properties of the system from relatively weak requirements on its local behavior, known as *nonuniform hyperbolicity conditions*: the reason this theory is also called nonuniform hyperbolicity theory.

It originated in the seminal works of Lyapunov [134] and Perron [164] on stability of solutions of ordinary differential equations. To determine whether a given solution is stable one proceeds as follows. First, the equation is linearized along the solution and then the stability of the zero solution of the corresponding nonautonomous linear differential equation is examined. There are several methods (due to Hadamard [79], Perron [165], Fenichel [70], and Irwin [92]) aimed at exhibiting stability of solutions via certain information on the linear system. The approach by Lyapunov uses a special real-valued function on the space of solutions of the linear system known as the *Lyapunov exponent*. It measures in the logarithmic scale the rate of convergence of solutions so that the zero solution is asymptotically exponentially stable along any subspace where the Lyapunov exponent is negative.

The Lyapunov exponent is arguable the best way to characterize stability: the requirement that the Lyapunov exponent is negative is the weakest one that still guarantees that solutions of the linear system eventually decay exponentially to zero. The price to pay is that stability of the zero solution in this weak sense does not necessarily imply stability of the original solution of the nonlinear equation. The latter can be ensured under an additional and quite subtle requirement known as the *Lyapunov–Perron regularity*.

Verifying this requirement for a given solution may be a very difficult if not virtually impossible task, making verification more a principle than practical matter.

This could deem the whole approach useless if not for an important particular case when the differential equation is given by a vector field on a smooth compact Riemannian manifold. In this case, the celebrated Multiplicative Ergodic Theorem, also known as Oseledets' theorem, claims that a "typical" solution of the equation is Lyapunov–Perron regular, thus making the difficult task of checking the regularity requirement unnecessary. Here "typical" means that the statement holds for almost every trajectory with respect to a finite Borel measure invariant under the flow generated by the vector field.

A principal application of Oseledets' theorem in the context of smooth dynamical systems is that the Lyapunov exponent alone can be used to characterize stability of trajectories. Building upon this idea, in the beginning of 1970s Pesin introduced the class of systems whose Lyapunov exponent is nonzero along almost every trajectory with respect to some *smooth invariant measure* (i.e., a measure, which is equivalent to the Riemannian volume) and then he developed the stability theory (constructing local and global stable and unstable manifolds; see Section 7.5), as well as described their ergodic properties (including ergodicity, K- and Bernoulli properties; see Chapter 9). The collection of these results is known as Pesin's theory (see [18]). A crucial manifestation of this theory is the *formula for the entropy* connecting the measure-theoretic entropy of the system with its Lyapunov exponent (see Chapter 10). It should be pointed out that these results require that the system is of class of smoothness $C^{1+\alpha}$ for some $\alpha > 0$ and that they may indeed fail if the system is only of class C^1 (see Section 7.8).

Unlike classical uniformly hyperbolic systems (i.e., Anosov or more general axiom A systems) where contractions and expansions are *uniform everywhere* on a compact invariant set, Pesin's theory deals with systems satisfying the substantially weaker requirement that contractions and expansions occur *asymptotically almost everywhere* with respect to a smooth invariant measure. Because this requirement is weak, there are no topological obstructions for the existence of such systems on any phase space. Indeed, any smooth compact Riemannian manifold (of dimension ≥ 2 in the discrete-time case and of dimension ≥ 3 in the continuous-time case) admits a volume preserving system whose Lyapunov exponent is nonzero almost everywhere (see Sections 11.4 and 11.5). It is therefore remarkable that such a weak requirement ensures highly nontrivial ergodic and topological properties of the system.

A small perturbation of a diffeomorphism with nonzero Lyapunov exponents (in the C^r topology, $r > 1$) may not bear the same properties – the price to pay for the great generality of the nonuniformly hyperbolic theory. However, experts believe that nonuniformly hyperbolic *conservative* systems (i.e., systems preserving a smooth measure, in particular, volume preserving) are *typical* in some sense. This is reflected in the following conjectures: (We consider the case of systems with discrete time.)

1. Let f be a C^r, $r > 1$, volume preserving diffeomorphism of a smooth compact Riemannian manifold M. Assume that the Lyapunov exponent of f is nonzero along almost every trajectory of f. Then there exists a neighborhood $\mathcal{U}$

of f in the space of C^r volume preserving diffeomorphisms of M and a residual subset $\mathcal{A} \subset \mathcal{U}$ such that for every $g \in \mathcal{A}$ the Lyapunov exponent of g is nonzero along every orbit in a subset of positive volume.

2. Let f be a C^r, $r > 1$, volume preserving diffeomorphism of a smooth compact Riemannian manifold M. Then arbitrarily close to f in the space of C^r volume preserving diffeomorphisms of M, there exists a diffeomorphism g whose Lyapunov exponent is nonzero along every orbit in a subset of positive volume.

We stress that the assumption $r > 1$ is crucial as the conjectures fail if $r = 1$ due to a recent result of Bochi and Viana [28]. So far there has been little progress in solving these conjectures (see Section 11.7). On the positive side, crucial results on genericity of hyperbolic cocycles over dynamical systems have been recently obtained by Viana [221].

A persistent obstruction to nonuniform hyperbolicity is presence of elliptic behavior (see [232,233]). For example, for area preserving surface diffeomorphisms, as predicted by KAM theory, elliptic islands survive under small perturbations of the system. Numerical studies of such maps suggest that in this case elliptic islands coexist with what appears to be a "chaotic sea" – an ergodic component of positive area with nonzero Lyapunov exponents (see [135, 136]). In fact, one often considers a one-parameter family of area preserving surface diffeomorphisms, which starts from a completely integrable (nonchaotic) system and evolves eventually into a completely hyperbolic (chaotic) one demonstrating, for intermediate values of the parameter, the appearance of elliptic islands gradually giving way to a "chaotic sea". For billiard dynamical systems, coexistence of elliptic and hyperbolic behavior has been shown for the so-called "mushroom billiards" (see [41]). In the category of smooth maps, establishing coexistence is arguably one of the most difficult problems in the theory of dynamical systems. A simple but somewhat "artificial" example of coexistence was constructed in [183] (see also [130] and Section 6.6; for a more elaborate construction see [90]). Much more complicated examples where coexistence is expected are (1) the famous standard map (also known as the Chirikov–Taylor map; see [51] and [188, Section 8.5]) and (2) automorphisms of real $K3$ surfaces (see [151]).

The requirement that the Lyapunov exponent is nonzero along almost every trajectory with respect to an invariant Borel probability measure – such a measure is said to be *hyperbolic* – is equivalent to the fact that the system is nonuniformly hyperbolic. Thus nonuniform hyperbolicity can be viewed as presence of hyperbolic invariant measures leading to challenging problems of studying ergodic and topological properties of general (not necessarily smooth) hyperbolic measures as well as of constructing some *natural* hyperbolic measures.

A general hyperbolic measure does not have "good" ergodic properties. (Simply note that *any* invariant measure on a horseshoe is hyperbolic.) It is therefore quite remarkable that hyperbolic measures have abundance of topological properties whose study was initiated in the work of Katok [101] (see Chapters 14 and 15). For example, the set of hyperbolic periodic orbits is dense in the support of the measure. Surprisingly, general hyperbolic measures asymptotically have local product structure (similar to the one of Gibbs measures on horseshoes) and one can compute their Hausdorff dimension and entropy. The formula for the entropy of a general hyperbolic measure due to Ledrappier and L.-S. Young is a substantial generalization of the entropy formula for smooth hyperbolic measures but unlike the latter, it involves quite subtle characteristics of the measure other than the Lyapunov exponent.

Smooth measures form an important yet particular case of natural hyperbolic measures. The latter were introduced by Ledrappier as an extension to nonuniformly hyperbolic systems of the Sinai–Ruelle–Bowen (SRB) measures for classical uniformly hyperbolic attractors (see Chapter 13). These *generalized* SRB measures describe the limit distribution of the time averages of continuous functions along forward orbits for a set of initial points of positive Lebesgue measure in a small neighborhood of the attractor. According to a result by Ledrappier and Strelcyn, these measures can be characterized as being the only measures for which the entropy formula of Pesin holds. Ledrappier showed that the methods used in studying ergodic properties of smooth hyperbolic measures can be adjusted to describe ergodic properties of SRB measures.

Constructing SRB measures for nonuniformly hyperbolic systems is a difficult problem. Beyond uniform hyperbolicity, there are very few examples, of which best known are Hénon-like attractors, where existence of SRB measures was rigorously shown. L.-S. Young has introduced a class of dynamical systems with nonzero Lyapunov exponents, which admit the so-called Young's tower. For these systems, she established existence of SRB measures (see Section 13.3).

The recent theory of Hénon-like diffeomorphisms (see [25, 26, 219, 222, 223]) suggests the following approach to the genericity problem for nonuniformly hyperbolic *dissipative* systems: given a one-parameter family of C^2 diffeomorphisms f_a, $a \in [\alpha, \beta]$ with a trapping region R (i.e., R is an open set for which $\overline{f_a(R)} \subset R$ for any $a \in [\alpha, \beta]$), there exists a set $A \subset [\alpha, \beta]$ of positive Lebesgue measure such that for every $a \in A$, the diffeomorphism f_a possesses an SRB measure supported on the attractor $\Lambda_a = \bigcap_{n>0} f_a^n(R)$.

Evaluating Lyapunov exponents by a computer is a relatively easy procedure and in many models in science, the absence of zero exponents can be shown numerically. This is often viewed as a convincing evidence that the system under investigation exhibits chaotic behavior. In mathematics, several "artificial" examples of systems with nonzero exponents have been constructed (and the reader can find most of them in Chapter 6) and for some interesting "natural" dynamical systems (e.g., geodesic flows on nonpositively curved manifolds and Teichmüller geodesic flows; see Chapter 12) absence of zero exponents have been shown. In addition, various powerful methods have been developed (e.g., cone and Lyapunov function techniques; see Chapter 4) that allow one to verify whether a given dynamical system has some positive Lyapunov exponents.

Many results of the nonuniform hyperbolicity theory hold in greater generality than for actions of single dynamical systems and wherever possible we describe the theory with this view in mind. For example, the linear hyperbolicity theory (including the theory of Lyapunov exponents and its principal result – the Multiplicative Ergodic Theorem) is presented for linear cocycles over dynamical systems (or even over higher-rank Abelian actions), and the stable manifold theory is developed for sequences of diffeomorphisms. Even in the case of an action of a single dynamical system, we consider a more general case of nonuniform *partial* hyperbolicity where the requirement that the values of the Lyapunov exponent are *all* nonzero is replaced by a weaker one that *some* of the values of the Lyapunov exponent are nonzero. Such generalizations require some more complicated techniques and tools from various areas of mathematics to be used and thus make the exposition more complicated but they substantially broaden applications and show the great power of the nonuniform hyperbolicity theory.

Part I

Linear Theory

1

The Concept of Nonuniform Hyperbolicity

In this chapter we consider sequences of linear maps in the Euclidean space and we introduce the principal notions of Lyapunov exponents, Lyapunov–Perron regularity, normal bases, and so on. These concepts are used in a variety of settings of which the main one is the study of linearizations of a dynamical system along its orbits. Thus a sequence of linear maps can be thought of as the sequence of derivatives (differentials) of a smooth map along an orbit.

We stress that in the situations we consider there are no preferred coordinate systems. Accordingly, even though we often use matrix representations of linear maps, we only study properties that are independent of certain classes of coordinate changes. The most narrow class is that of orthogonal coordinate changes; in the smooth situation, this corresponds to fixing a Riemannian metric in the phase space. A broader class includes coordinate changes uniformly bounded from above and below; in the case of a smooth system on a compact space, this corresponds to an arbitrary choice of a smooth coordinate atlas.

As it turns out, of greatest importance for the theory developed in this book is still a broader class of tempered coordinate changes. This reflects the primary role that exponential behavior plays in our considerations. A tempered change allows arbitrarily large distortions if these distortions change with time with a subexponential rate. Thus it preserves not only the exponential character of the asymptotic behavior but also the actual rates of expansion and contraction in various directions.

In the next chapter these considerations will be extended to the case of cocycles over dynamical systems. The principal difference is in paying attention to the dependence on the base point that may be measurable, continuous, differentiable, and so on.

1.1 Motivation

Consider an invertible linear transformation A of an Euclidean space $\mathbb{R}^n$. Let $\lambda_1, \ldots, \lambda_n$ be its eigenvalues. The transformation A is said to be *hyperbolic* if

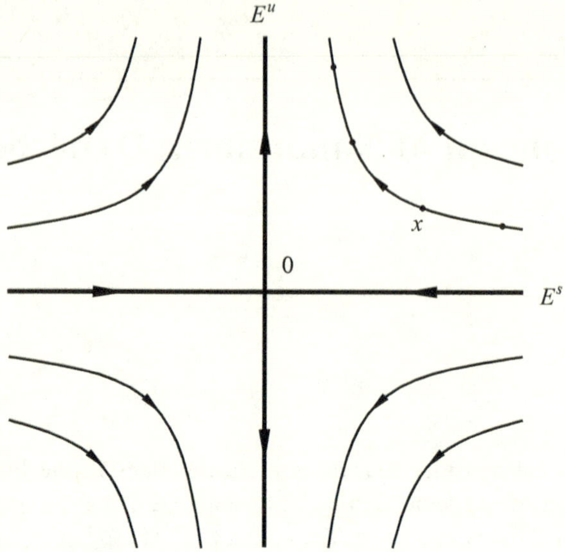

Figure 1.1. Hyperbolic fixed point of a linear map: the trajectory of a point x lies on the hyperbola passing through x.

$|\lambda_i| \neq 1$ for every i. If A is hyperbolic then there is a decomposition $\mathbb{R}^n = E^s \oplus E^u$ into *stable* and *unstable subspaces*, where

$$E^s = \mathbb{R}^n \cap \bigoplus_{i:|\lambda_i|<1} H_{\lambda_i} \quad \text{and} \quad E^u = \mathbb{R}^n \cap \bigoplus_{i:|\lambda_i|>1} H_{\lambda_i}.$$

Here H_{λ_i} is the root space

$$H_{\lambda_i} = \{v \in \mathbb{C}^n : (A - \lambda_i \mathrm{Id})^m v = 0 \text{ for some } m \in \mathbb{N}\}.$$

Set

$$\lambda = \max\left\{\max_{|\lambda_i|<1} |\lambda_i|, \max_{|\lambda_i|>1} |\lambda_i|^{-1}\right\} \in (0, 1).$$

Note that for any $\varepsilon > 0$, there exists $c = c(\varepsilon) > 0$ such that for every $m > 0$,

$$\|A^m v\| \leq c\lambda^m e^{\varepsilon m} \|v\| \quad \text{if } v \in E^s,$$

and

$$\|A^{-m} v\| \leq c\lambda^m e^{\varepsilon m} \|v\| \quad \text{if } v \in E^u.$$

Therefore the origin is a saddlelike point for A (see Fig. 1.1).

Notice that if A is diagonalizable, a similar estimate holds with $\delta = 0$ but if A has Jordan blocks then the growth in the corresponding root space has an extra subexponential factor.

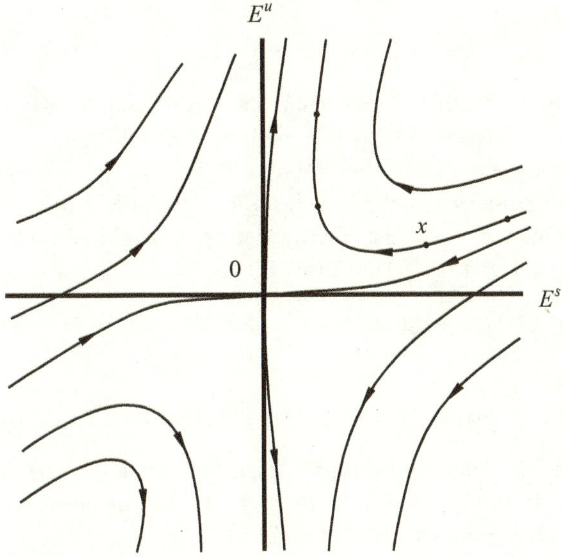

Figure 1.2. Hyperbolic fixed point of a diffeomorphism: near 0 the trajectory of a point x lies on a curve, which is the pre-image of a hyperbola under the conjugacy map h.

Let now f be a C^1 diffeomorphism of an open set U of $\mathbb{R}^n$ and $p \in U$ a *fixed point* for f, that is, $f(p) = p$. The point p is called *hyperbolic* if the linear transformation $A = d_p f$ is hyperbolic. By the Grobman–Hartman Theorem (see e.g., [104, Theorem 6.3.1]), there is a homeomorphism h defined in a small neighborhood $\widetilde{U} \subset U$ of p such that $h(f(x)) = A(h(x))$ for every $x \in \widetilde{U} \cap f^{-1}(\widetilde{U})$. This implies that the local orbit structure of a diffeomorphism in a small neighborhood of a hyperbolic fixed point resembles that of the linear transformation that is the differential (the linearization) of the map at that point (see Fig. 1.2).

By the Hadamard–Perron Theorem (see e.g., [104, Theorem 6.4.9]), the *stable set*

$$V^s(p) = \{x \in U : f^m(x) \in U \text{ for all } m > 0\}$$

and the *unstable set*

$$V^u(p) = \{x \in U : f^m(x) \in U \text{ for all } m < 0\}$$

are C^1 submanifolds passing through p such that $T_p V^s(p) = E^s$ and $T_p V^u(p) = E^u$. Moreover, similarly to the linear case above, given $\varepsilon > 0$, there exists $c = c(\varepsilon) > 0$ such that for every $m > 0$,

$$\|f^m(x) - p\| \leq c\lambda^m e^{\varepsilon m} \|x\| \quad \text{if } x \in V^s(p),$$

and

$$\|f^{-m}(x) - p\| \le c\lambda^m e^{\varepsilon m} \|x\| \quad \text{if } x \in V^u(p).$$

We call $V^s(p)$ and $V^u(p)$ the *local stable* and *unstable manifolds* at p.

This discussion extends directly to hyperbolic periodic points. A point $p \in U$ is *periodic* if for some $k \in \mathbb{N}$, we have $f^i(p) \in U$ for $i = 1, \ldots, k-1$ and $f^k(p) = p$. The periodic point p is called *hyperbolic* if the linear transformation $d_p f^k$ is hyperbolic. One can show that there exist open neighborhoods $U_0, \ldots, U_{k-1}$, respectively, of $p, \ldots, f^{k-1}(p)$ such that for each $i = 0, \ldots, k-1$, the sets

$$V^s(f^i(p)) = \{x \in U_i : f^m(x) \in U_{m+i} \text{ for all } m > 0\}$$

and

$$V^u(f^i(p)) = \{x \in U_i : f^m(x) \in U_{m+i} \text{ for all } m < 0\}$$

are C^1 submanifolds passing through $f^i(p)$. We call $V^s(f^i(p))$ and $V^u(f^i(p))$ the *local stable* and *unstable manifolds* at $f^i(p)$. Let us point out that if p is a hyperbolic periodic point of period k then for every $i = 1, \ldots, k-1$, the point $f^i(p)$ is a hyperbolic periodic point. In other words, the trajectory $\{f^i(p) : i = 0, \ldots, k-1\}$ is hyperbolic.

Now consider a nonperiodic point p of a diffeomorphism f. We would like to say that p is hyperbolic if the behavior of trajectories that start in a neighborhood of p resembles that of the trajectories in a neighborhood of a periodic hyperbolic point. More precisely, this means that one can construct local stable and unstable manifolds $V^s(p)$ and $V^u(p)$ at p such that every trajectory $f^m(v)$ with $v \in V^s(p)$ approaches $f^m(p)$ with an exponential rate and every trajectory $f^{-m}(v)$ with $v \in V^u(p)$ approaches $f^{-m}(p)$ with an exponential rate. In fact the orbit $\{f^m(x)\}_{m\in\mathbb{Z}}$ of a point $x \in V^s(p)$ may first diverge from the orbit $\{f^m(p)\}_{m\in\mathbb{Z}}$ until some time $T = T(x)$ before it starts converging to it with an exponential rate. If the rate of the exponential convergence does not depend on the point p in a compact f-invariant subset $\Lambda \subset U$, and the angle between local stable and unstable manifolds at p is uniformly bounded away from zero in p, then f is said to be *uniformly hyperbolic* on Λ. In this case the function $T(x)$ is uniformly bounded from above in x. Note that hyperbolicity refers to the whole orbit $\{f^m(p)\}_{m\in\mathbb{Z}}$.

In the general case, however, the rate of exponential convergence may vary from orbit to orbit. Moreover, the function $T(x)$ may not be bounded from above as x approaches p, forcing contraction estimates to deteriorate along the orbit. If this deterioration occurs with a subexponential rate (or a sufficiently small exponential rate) on a set Λ of orbits, then f is said to be *nonuniformly hyperbolic* on Λ. Observe that this set is invariant but not necessarily closed. The dynamics of f on the nonuniformly hyperbolic set is the main object of our study.

In view of the Grobman–Hartman Theorem, it is natural to try to introduce hyperbolicity in terms of the sequence of differentials $d_p f^m$. This leads us to the consideration of a sequence of linear transformations of an Euclidean space. It is important to keep in mind that unlike the case of a fixed or periodic point, we cannot fix coordinate systems in Euclidean space arbitrarily. For, by doing so one could simply reduce the sequence to the sequence of identity maps that is obviously meaningless. So we assume that a coordinate system is fixed and allow a certain kind of coordinate changes, which are discussed in the next section.

1.2 Basic Setting

1.2.1 Exponential Splitting and Nonuniform Hyperbolicity

Consider a (two-sided) sequence of matrices $\mathcal{A} = \{A_m\}_{m\in\mathbb{Z}} \subset GL(n, \mathbb{R})$, where $GL(n, \mathbb{R})$ is the group of invertible linear transformations of $\mathbb{R}^n$. For $m, k \in \mathbb{Z}$, set

$$\mathcal{A}_{m,k} = \begin{cases} A_{m-1} \cdots A_{k+1} A_k & \text{if } m > k \\ \text{Id} & \text{if } m = k \\ (A_m)^{-1} \cdots (A_{k-2})^{-1} (A_{k-1})^{-1} & \text{if } m < k \end{cases} \tag{1.1}$$

and write $\mathcal{A}_m = \mathcal{A}_{m,0}$. We also consider a sequence of inner products $\langle \cdot, \cdot \rangle_m$ on $\mathbb{R}^n$ and denote by $\|\cdot\|_m$ the norm and by $\sphericalangle(E_1, E_2)_m$ the angle between two subspaces $E_1, E_2 \subset \mathbb{R}^n$ induced by the inner product $\langle \cdot, \cdot \rangle_m$.

All the notions discussed below depend only on the choice of the inner products and thus, by definition, are independent of orthogonal coordinate changes C_m. In other words, we treat two sequences A_m and ${C_{m+1}}^{-1} A_m C_m$ as equivalent.

Let $0 < \lambda < \mu < \infty$ and $\varepsilon \geq 0$ be some numbers. Throughout the book we assume that $\lambda\, e^{\varepsilon} < \mu\, e^{-\varepsilon}$.

Definition 1.2.1. We say that the sequence of matrices $\mathcal{A}$ is *nonuniformly partially hyperbolic in the broad sense* (or admits a $(\lambda, \mu, \varepsilon)$-*splitting*) if the following conditions hold:

1. For each $m \in \mathbb{Z}$, there exists a decomposition $\mathbb{R}^n = E^1_m \oplus E^2_m$ such that $A_m E^1_m = E^1_{m+1}$ and $A_m E^2_m = E^2_{m+1}$.
2. There exist sequences of positive numbers $\{c_m\}_{m\in\mathbb{Z}}$ and $\{\gamma_m\}_{m\in\mathbb{Z}}$ such that
   (a) for $v \in E^1_k$, $k \in \mathbb{Z}$, and $m > 0$,
$$\|\mathcal{A}_{m+k,k} v\|_{m+k} \leq c_k \lambda^m\, e^{\varepsilon m}\, \|v\|_k; \tag{1.2}$$
   (b) for $v \in E^2_k$, $k \in \mathbb{Z}$, and $m < 0$,
$$\|\mathcal{A}_{m+k,k} v\|_{m+k} \leq c_k \mu^m\, e^{-\varepsilon m}\, \|v\|_k; \tag{1.3}$$
   (c) $\sphericalangle(E^1_m, E^2_m)_m \geq \gamma_m$ for $m \in \mathbb{Z}$;
   (d) $c_{m+k} \leq c_k\, e^{\varepsilon|m|}$ and $\gamma_{m+k} \geq \gamma_k\, e^{-\varepsilon|m|}$ for $m, k \in \mathbb{Z}$.

The requirement 2d is crucial and expresses the main feature of nonuniform hyperbolicity. It shows that the estimates (1.2) and (1.3) may deteriorate as $m \to \pm\infty$ but only with small exponential rate. In view of 2a–c, the contraction along E^1_m and the expansion along E^2_m are sufficiently strong to dominate the asymptotic behavior of the system.

Proposition 1.2.2. *The subspaces E^1_m and E^2_m are uniquely defined.*

Proof. For each $m \in \mathbb{Z}$, let $\mathbb{R}^n = E^1_m \oplus E^2_m$ and $\mathbb{R}^n = F^1_m \oplus F^2_m$ be two decompositions satisfying Conditions 1 and 2 given above. Given $v \in E^1_k$, write $v = v_1 + v_2$ with $v_1 \in F^1_k$ and $v_2 \in F^2_k$. For every $m > 0$,

$$\begin{aligned}\|\mathcal{A}_{m+k,k}v\|_{m+k} &\geq \|\mathcal{A}_{m+k,k}v_2\|_{m+k} - \|\mathcal{A}_{m+k,k}v_1\|_{m+k}\\ &\geq c^{-1}_{m+k}\mu^m\|v_2\|_k - c_k\lambda^m e^{\varepsilon m}\|v_1\|_k\\ &\geq c^{-1}_k(\mu\, e^{-\varepsilon})^m\|v_2\|_k - c_k\lambda^m e^{\varepsilon m}\|v_1\|_k.\end{aligned}$$

On the other hand, $\|\mathcal{A}_{m+k,k}v\|_{m+k} \leq c_k\lambda^m\|v\|_k$. Since $\lambda\, e^{\varepsilon} < \mu\, e^{-\varepsilon}$, this shows that v_2 must vanish. Therefore, $v \in F^1_k$ and $E^1_k \subset F^1_k$. By symmetry, $E^1_k = F^1_k$. A similar argument shows that $E^2_k = F^2_k$. Therefore the decomposition is unique. □

Note that either $\lambda\, e^{\varepsilon} < 1$ or $\mu\, e^{\varepsilon} > 1$. If $\lambda\, e^{\varepsilon} < 1$, the subspaces E^1_m are *stable* (or *contracting*) and if $\mu\, e^{-\varepsilon} > 1$, the subspaces E^2_m are *unstable* (or *expanding*). In this case we use the notation E^s_m and E^u_m, respectively.

Definition 1.2.3. We say that a sequence of matrices is *nonuniformly (completely) hyperbolic* if it admits a $(\lambda, \mu, \varepsilon)$-splitting and $\lambda\, e^{\varepsilon} < 1 < \mu\, e^{-\varepsilon}$.

We introduce a version of hyperbolicity that is much stronger than nonuniform hyperbolicity.

Definition 1.2.4. We say that the sequence of matrices $\mathcal{A}$ is *uniformly partially hyperbolic in the broad sense* (or admits a *uniform* $(\lambda, \mu, 0)$*-splitting* or simply a (λ, μ)*-splitting*) if

1. for each $m \in \mathbb{Z}$, there exists a decomposition $\mathbb{R}^n = E^1_m \oplus E^2_m$ such that $A_m E^1_m = E^1_{m+1}$ and $A_m E^2_m = E^2_{m+1}$;
2. there exist positive constants c and γ such that
   (a) for $v \in E^1_k$, $k \in \mathbb{Z}$, and $m > 0$,
   $$\|\mathcal{A}_{m+k,k}v\| \leq c\lambda^m\|v\|;$$
   (b) for $v \in E^2_k$, $k \in \mathbb{Z}$, and $m < 0$,
   $$\|\mathcal{A}_{m+k,k}v\| \leq c\mu^m\|v\|;$$
   (c) $\sphericalangle(E^1_m, E^2_m)_m \geq \gamma$ for $m \in \mathbb{Z}$.

1.2.2 Tempered Equivalence

The notions discussed above are invariant under a broader class of coordinate changes than orthogonal ones.

Definition 1.2.5. A sequence of linear transformations $\{C_m\}_{m\in\mathbb{Z}}$ in an Euclidean space is said to be *tempered* if

$$\lim_{m\to\pm\infty}\frac{1}{m}\log\|C_m\| = \lim_{m\to\pm\infty}\frac{1}{m}\log\|C_m{}^{-1}\| = 0.$$

Two sequences of linear transformations $\mathcal{A} = \{A_m\}_{m\in\mathbb{Z}}$ and $\mathcal{B} = \{B_m\}_{m\in\mathbb{Z}}$ are said to be *equivalent* (under tempered coordinate changes) if there exists a tempered sequence $\{C_m\}_{m\in\mathbb{Z}}$ such that

$$B_m = C_{m+1}{}^{-1}A_mC_m. \tag{1.4}$$

Proposition 1.2.6. *Given two equivalent sequences of linear transformations $\mathcal{A}$ and $\mathcal{B}$, if $\mathcal{A}$ is nonuniformly partially hyperbolic in the broad sense then so is $\mathcal{B}$.*

Proof. Assume that $\mathcal{A}$ admits a $(\lambda, \mu, \varepsilon)$-splitting with decompositions $\mathbb{R}^n = E_m^1 \oplus E_m^2$ and sequences of positive numbers $\{c_m\}_{m\in\mathbb{Z}}$ and $\{\gamma_m\}_{m\in\mathbb{Z}}$. By (1.4), for each sufficiently small $\delta > 0$, the sequence $\mathcal{B}$ admits the $(\lambda e^{\delta},\ \mu e^{-\delta}, \varepsilon + 2\delta)$-splitting with decompositions $\mathbb{R}^n = F_m^1 \oplus F_m^2$, where $F_m^1 = C_m{}^{-1}E_m^1$ and $F_m^2 = C_m{}^{-1}E_m^2$, and with sequences of positive numbers $\{c_m e^{2\delta|m|}\}_{m\in\mathbb{Z}}$ and $\{\gamma_m e^{-2\delta|m|}\}_{m\in\mathbb{Z}}$. □

Corollary 1.2.7. *Let $\mathcal{A}$ and $\mathcal{B}$ be two equivalent sequences of linear transformations. Then $\mathcal{A}$ is nonuniformly (partially) hyperbolic in the broad sense if and only if $\mathcal{B}$ is nonuniformly (partially) hyperbolic in the broad sense.*

1.2.3 The Continuous-Time Case

In the theory of ordinary differential equations and dynamical systems with continuous time (flows), one encounters the following counterpart of the situation described above.

Let $0 < \lambda < \mu < \infty$ and $\varepsilon \geq 0$. Assume that $\lambda e^{\varepsilon} < \mu e^{-\varepsilon}$. Consider a continuous family of linear transformations $\mathcal{A} = \{A_t\}_{t\in\mathbb{R}}$.

Definition 1.2.8. The family $\mathcal{A}$ is *nonuniformly partially hyperbolic in the broad sense* (or admits a $(\lambda, \mu, \varepsilon)$-*splitting*) if the following conditions hold:

1. For each $t \in \mathbb{R}$, there exists a decomposition $\mathbb{R}^n = E_t^1 \oplus E_t^2$ such that $\mathcal{A}_t E_0^1 = E_t^1$ and $\mathcal{A}_t E_0^2 = E_t^2$.
2. There exist sequences of positive numbers $\{c_t\}_{t\in\mathbb{R}}$, $\{d_t\}_{t\in\mathbb{R}}$ such that
   (a) if $v \in E_\tau^1$ and $\tau \in \mathbb{Z}$ then
   $$\|\mathcal{A}_{t+\tau}\mathcal{A}_\tau{}^{-1}v\|_{t+\tau} \leq c_\tau\lambda^t e^{\varepsilon t}\|v\|_\tau \text{ for every } t > 0;$$
   (b) if $v \in E_\tau^2$ and $\tau \in \mathbb{Z}$ then
   $$\|\mathcal{A}_{t+\tau}\mathcal{A}_\tau{}^{-1}v\|_{t+\tau} \leq c_\tau\mu^t e^{-\varepsilon t}\|v\|_\tau \text{ for every } t < 0;$$
   (c) $\sphericalangle(E_t^1, E_t^2)_t \geq \gamma_t$ for every $t \in \mathbb{R}$;
   (d) $c_{t+\tau} \leq c_\tau e^{\varepsilon|t|}$ and $\gamma_{t+\tau} \geq \gamma_\tau e^{-\varepsilon|t|}$ for every $t, \tau \in \mathbb{R}$.

Definition 1.2.9. We say that a family $\{\mathcal{A}_t\}_{t\in\mathbb{R}}$ of matrices is *nonuniformly (completely) hyperbolic* if it is nonuniformly partially hyperbolic in the broad sense and $\lambda e^{\varepsilon} < 1 < \mu e^{-\varepsilon}$.

Proposition 1.2.10. *The subspaces E_t^1 and E_t^2 are uniquely defined and continuous in t.*

Proof. The uniqueness property follows from Proposition 1.2.2. The continuous dependence of the decomposition in t follows from the continuity of the family $\mathcal{A}$. □

In the theory of ordinary differential equations, the families of linear transformations described above represent solutions of nonautonomous linear equations. In this setting there is a natural differentiability condition; that is, $\{\mathcal{A}_t\}_{t\in\mathbb{R}}$ depends smoothly on t. In this case we can define the *generator* of the family by

$$B_t = \frac{d\mathcal{A}_t}{dt}\mathcal{A}_t^{-1}.$$

The generator is the continuous counterpart of the elements A_m of the sequence in the discrete-time case. One can reconstruct the family from the differential equation $y' = B_t y$.

1.3 Lyapunov Exponents Associated to Sequences of Matrices

Establishing nonuniform hyperbolicity is a difficult problem. A powerful approach to this problem is based on characterizing nonuniform hyperbolicity in terms of the Lyapunov exponent – a quantity associated with a given sequence of matrices that we introduce and study in this section (see [20] for the abstract theory of Lyapunov exponents).

1.3.1 Definition of the Lyapunov Exponent

Let $\mathcal{A}^+ = \{A_m\}_{m\geq 0} \subset GL(n,\mathbb{R})$ be a one-sided sequence of matrices and let $\mathcal{A}_m = \mathcal{A}_{m,0}$ be given by (1.1). Consider the function $\chi^+\colon \mathbb{R}^n \to \mathbb{R}\cup\{-\infty\}$ defined by

$$\chi^+(v) = \chi^+(v,\mathcal{A}^+) = \overline{\lim_{m\to+\infty}} \frac{1}{m}\log\|\mathcal{A}_m v\|. \tag{1.5}$$

We make the convention $\log 0 = -\infty$.

Proposition 1.3.1. *The function χ^+ has the following properties:*

1. $\chi^+(\alpha v) = \chi^+(v)$ *for each* $v\in\mathbb{R}^n$ *and* $\alpha\in\mathbb{R}\setminus\{0\}$.
2. $\chi^+(v+w) \leq \max\{\chi^+(v),\chi^+(w)\}$ *for each* $v, w\in\mathbb{R}^n$.
3. $\chi^+(0) = -\infty$.

Proof. Properties 1 and 3 follow immediately from (1.5). Observe that for any $v, w \in \mathbb{R}^n$,

$$\|\mathcal{A}_m(v+w)\| \leq 2\max\{\|\mathcal{A}_m v\|, \|\mathcal{A}_m w\|\}$$

and that for any pair of sequences of real numbers $\{a_m\}_{m\geq 0}$, $\{b_m\}_{m\geq 0}$,

$$\varlimsup_{m\to+\infty} \max\{a_m, b_m\} = \max\left\{\varlimsup_{m\to+\infty} a_m, \varlimsup_{m\to+\infty} b_m\right\}.$$

Property 2 follows. □

We call $\chi^+(v)$ the *forward Lyapunov exponent of* v (*with respect to the sequence* $\mathcal{A}^+$).

Proposition 1.3.2. *There exist an integer* p, $1 \leq p \leq n$, *and real numbers* $\chi_1^+ < \cdots < \chi_p^+$ *such that the following statements hold:*

1. *The linear spaces* $E_i^+ = \{v \in \mathbb{R}^n : \chi^+(v) \leq \chi_i^+\}$ *form a filtration*
$$\{0\} = E_0^+ \subsetneqq E_1^+ \subsetneqq E_2^+ \subsetneqq \cdots \subsetneqq E_p^+ = \mathbb{R}^n. \tag{1.6}$$
2. *If* $v \in E_i^+ \setminus E_{i-1}^+$ *then* $\chi^+(v) = \chi_i^+$.

Proof. Let $v, w \in \mathbb{R}^n$ be such that $\chi^+(v) < \chi^+(w)$. We have

$$\chi^+(v+w) \leq \chi^+(w) = \chi^+(v+w-v) \leq \max\{\chi^+(v+w), \chi^+(v)\}.$$

It follows that if $\chi^+(v+w) < \chi^+(v)$ then $\chi^+(w) \leq \chi^+(v)$, which contradicts our assumption. Hence, $\chi^+(v+w) \geq \chi^+(v)$, and thus $\chi^+(v+w) = \chi^+(w)$. This implies that if for some $v_1, \ldots, v_m \in \mathbb{R}^n \setminus \{0\}$ the numbers $\chi^+(v_1), \ldots, \chi^+(v_m)$ are distinct, then the vectors $v_1, \ldots, v_m$ are linearly independent. It follows that the Lyapunov exponent for nonzero vectors can take no more than n distinct values. We denote them by $\chi_1^+ < \cdots < \chi_p^+$ for some $p \leq n$. It is easy to see that the sets $E_i^+ = \{v \in \mathbb{R}^n : \chi^+(v) \leq \chi_i^+\}$ are linear subspaces satisfying (1.6). For any $v \in E_i^+ \setminus E_{i-1}^+$, we have $\chi_{i-1}^+ < \chi^+(v) \leq \chi_i^+$. Since χ^+ attains no value strictly between χ_{i-1}^+ and χ_i^+, we obtain $\chi^+(v) = \chi_i^+$. □

The spaces E_i^+ form the *filtration* (1.6) of $\mathbb{R}^n$ associated with χ^+. We denote this filtration by $\mathcal{V}_{\chi^+}$. The number

$$k_i^+ = \dim E_i^+ - \dim E_{i-1}^+$$

is called the *multiplicity* of the value χ_i^+, and the collection of pairs

$$\operatorname{Sp} \chi^+ = \{(\chi_i^+, k_i^+) : i = 1, \ldots, p\}$$

the *Lyapunov spectrum* of χ^+.

In a similar way, given a sequence of matrices $\mathcal{A}^- = \{A_m\}_{m<0}$, we define the *backward Lyapunov exponent* (*with respect to the sequence* $\mathcal{A}^-$) $\chi^- : \mathbb{R}^n \to \mathbb{R} \cup \{-\infty\}$ by

$$\chi^-(v) = \chi^-(v, \mathcal{A}^-) = \varlimsup_{m\to-\infty} \frac{1}{|m|} \log\|\mathcal{A}_m v\|, \tag{1.7}$$

where $\mathcal{A}_m = (A_m)^{-1} \cdots (A_{-2})^{-1}(A_{-1})^{-1}$ for each $m < 0$. The function χ^- has basic properties similar to those of χ^+ (see Proposition 1.3.1), and χ^- attains finitely many distinct *values* $\chi_1^- > \cdots > \chi_{p^-}^-$ for some $1 \le p^- \le n$. The subspaces $E_i^- = \{v \in \mathbb{R}^n : \chi^-(v) \le \chi_i^-\}$ form the *filtration* $\mathcal{V}_{\chi^-}$ of $\mathbb{R}^n$ associated to χ^-,

$$\mathbb{R}^n = E_1^- \supsetneqq \cdots \supsetneqq E_{p^-}^- \supsetneqq E_{p^-+1}^- = \{0\}.$$

The number

$$k_i^- = \dim E_i^- - \dim E_{i+1}^-$$

is the *multiplicity* of the value χ_i^-, and the collection of pairs

$$\operatorname{Sp} \chi^- = \{(\chi_i^-, k_i^-): i = 1, \ldots, p^-\}$$

is the *Lyapunov spectrum* of χ^-.

In the case of a sequence of matrices $\mathcal{A} = \{A_m\}_{m \in \mathbb{Z}}$ generated by a single matrix A (i.e., $A_m = A$ for all $m \in \mathbb{Z}$), the Lyapunov spectrum is calculated as follows: take all the eigenvalues λ_i, $i = 1, \ldots, k$, with $|\lambda_i| = r$ for some number r. Then $\log r$ is a value of the Lyapunov exponent whose multiplicity is k.

Equation (1.5) implies that for every $\varepsilon > 0$, there exists a number $C_+ = C_+(v, \varepsilon) > 0$ such that for all $m \ge 0$,

$$\|\mathcal{A}_m v\| \le C_+ e^{(\chi^+(v)+\varepsilon)m} \|v\|. \tag{1.8}$$

Similarly, (1.7) implies that for every $\varepsilon > 0$, there exists a number $C_- = C_-(v, \varepsilon) > 0$ such that for all $m \ge 0$,

$$\|\mathcal{A}_{-m} v\| \le C_- e^{(\chi^-(v)+\varepsilon)m} \|v\|. \tag{1.9}$$

1.3.2 Forward and Backward Regularity

In order to explain the relation between Lyapunov exponents and nonuniform hyperbolicity, let us consider the case when the spectrum of the Lyapunov exponent χ^+ does not contain zero and that indeed, the following condition holds:

$$\text{there is a number } j,\ 1 \le j < p, \text{ such that } \chi_j < 0 < \chi_{j+1}. \tag{1.10}$$

Set $\lambda = \exp \chi_j < 1$ and $\mu = \exp \chi_{j+1} > 1$, and consider the subspaces E^1 and E^2 that consist of vectors whose Lyapunov exponents are respectively negative and positive. One can hope that the subspaces $E_m^1 = A_m E^1$ and $E_m^2 = A_m E^2$ generate a $(\lambda, \mu, \varepsilon)$-splitting of $\mathbb{R}^n$ for all sufficiently small ε. This may not be true, as the sequence may fail to satisfy the crucial property 2d in Definition 1.2.1 (see Example 1.3.9).

We impose a condition on the sequence of matrices that guarantees that the subspaces E_m^1 and E_m^2 indeed generate a $(\lambda, \mu, \varepsilon)$-splitting of $\mathbb{R}^n$. We call such a sequence regular. The regularity condition is expressed in terms of the asymptotic behavior of the sequence as time goes to $+\infty$ or $-\infty$.

Given vectors $v_1, \ldots, v_k \in \mathbb{R}^n$, we denote by $\operatorname{vol}(v_1, \ldots, v_k)$ the volume of the k-parallelepiped formed by $v_1, \ldots, v_k$. The *forward* and *backward k-dimensional*

Lyapunov exponents of the vectors $v_1, \dots, v_k$ are defined, respectively, by

$$\begin{aligned}\chi^+(v_1, \dots, v_k) &= \chi^+(v_1, \dots, v_k, \mathcal{A}^+) \\ &= \overline{\lim_{m\to+\infty}} \frac{1}{m} \log \operatorname{vol}(\mathcal{A}_m v_1, \dots, \mathcal{A}_m v_k),\end{aligned} \tag{1.11}$$

and

$$\begin{aligned}\chi^-(v_1, \dots, v_k) &= \chi^-(v_1, \dots, v_k, \mathcal{A}^-) \\ &= \overline{\lim_{m\to-\infty}} \frac{1}{|m|} \log \operatorname{vol}(\mathcal{A}_m v_1, \dots, \mathcal{A}_m v_k).\end{aligned}$$

Observe that these exponents depend only on the linear subspace generated by the vectors $v_1, \dots, v_k$. Since $\operatorname{vol}(v_1, \dots, v_k) \le \prod_{i=1}^k \|v_i\|$, we obtain

$$\chi^+(v_1, \dots, v_k) \le \sum_{i=1}^k \chi^+(v_i). \tag{1.12}$$

A similar inequality holds for the backward Lyapunov exponent.

Whether the inequality in (1.12) is strict depends on the choice of vectors and even for a "good" sequence of matrices (e.g., diagonal matrices), there is always a "bad" choice of a collection of vectors, which makes the inequality strict. For example, consider the sequence of matrices $A_m = \begin{pmatrix} 2 & 0 \\ 0 & 1/2 \end{pmatrix}$ and the vectors $v_1 = (1, 0)$ and $v_2 = (1, 1)$. We have $\chi^+(v_1) = \chi^+(v_2) = \log 2$. On the other hand, since $\det A_m = 1$, we have $\chi^+(v_1, v_2) = 0 < \chi^+(v_1) + \chi^+(v_2)$.

We wish to single out those sequences of matrices for which one has equality in (1.12) for a "representative" collection of vectors $v_1, \dots, v_n$. We call such sequences forward regular and show that they have many "nice" properties. We proceed with the formal description.

Definition 1.3.3. A sequence of matrices $\mathcal{A}^+$ is said to be *forward regular* if

$$\lim_{m\to+\infty} \frac{1}{m} \log|\det \mathcal{A}_m| = \sum_{i=1}^n \chi_i', \tag{1.13}$$

where $\chi_1', \dots, \chi_n'$ are the finite values of the exponent χ^+ counted with their multiplicities.

We stress that (1.13) includes two requirements: first, the limit of $\frac{1}{m} \log|\det \mathcal{A}_m|$ exists as $m \to \infty$ and second, this limit is equal to the sum of all finite values of χ^+. Moreover, by (1.12), the identity (1.13) is equivalent to

$$\underline{\lim_{m\to+\infty}} \frac{1}{m} \log|\det \mathcal{A}_m| \ge \sum_{i=1}^n \chi_i'.$$

If a sequence of matrices $\mathcal{A} = \{A_m\}_{m\in\mathbb{Z}}$ is such that $A_m = A$, where A is a constant matrix, then this sequence is forward regular.

It is easy to see that a sequence of matrices $\mathcal{A}^+$ is forward regular if and only if there exists a positive definite symmetric matrix Λ such that

$$\lim_{m\to\infty}\frac{1}{m}\log\|\mathcal{A}_m\Lambda^{-m}\|=0 \quad\text{and}\quad \lim_{m\to\infty}\frac{1}{m}\log\|\Lambda^m\mathcal{A}_m^{-1}\|=0. \tag{1.14}$$

Indeed, if Λ is a positive definite symmetric matrix satisfying (1.14), then (1.13) holds. On the other hand, let $\mathcal{A}=\{A_m\}_{m\in\mathbb{Z}}$ be a forward regular sequence of matrices. Consider the filtration (1.6) and let F_i be the orthogonal complement of E_{i-1} in E_i. Define a positive definite symmetric matrix Λ by requiring that $\Lambda v=e^{\chi_i}v$ for every $v\in F_i$, $1\le i\le p$. Clearly, the relation (1.14) holds.

Proposition 1.3.4. *Any sequence of linear transformations equivalent to a forward regular sequence is forward regular.*

Proof. Let $\mathcal{A}^+=\{A_m\}_{m\ge 0}$ and $\mathcal{B}^+=\{B_m\}_{m\ge 0}$ be two equivalent sequences of matrices. It is easy to verify that their Lyapunov spectra coincide. On the other hand, by (1.4), $\det\mathcal{B}_m=\det C_m^{-1}\det\mathcal{A}_m\det C_0$ and since C_m is tempered,

$$\begin{aligned}
&\underline{\lim}_{m\to+\infty}\frac{1}{m}\log|\det\mathcal{B}_m|\\
&\quad\ge \underline{\lim}_{m\to+\infty}\frac{1}{m}\log|\det C_m{}^{-1}|+\underline{\lim}_{m\to+\infty}\frac{1}{m}\log|\det\mathcal{A}_m|\\
&\quad\ge \underline{\lim}_{m\to+\infty}-\frac{n}{m}\log\|C_m\|+\underline{\lim}_{m\to+\infty}\frac{1}{m}\log|\det\mathcal{A}_m|\\
&\quad\ge \underline{\lim}_{m\to+\infty}\frac{1}{m}\log|\det\mathcal{A}_m|.
\end{aligned}$$

Note that the roles of $\mathcal{A}^+$ and $\mathcal{B}^+$ can be reversed and the desired statement follows. □

Definition 1.3.5. The Lyapunov exponent χ^+ is said to be *exact* if for any $1\le k\le n$ and any collection $\{v_1,\dots,v_k\}\subset\mathbb{R}^n$ of vectors,

$$\chi^+(v_1,\dots,v_k)=\lim_{m\to+\infty}\frac{1}{m}\log\operatorname{vol}(\mathcal{A}_mv_1,\dots,\mathcal{A}_mv_k); \tag{1.15}$$

that is, the upper limit in (1.11) is indeed a limit.

If the Lyapunov exponent is exact then, in particular, for $v\in\mathbb{R}^n$, one has

$$\chi^+(v)=\lim_{m\to+\infty}\frac{1}{m}\log\|\mathcal{A}_mv\|;$$

equivalently (compare to (1.8) and (1.9)): for every $\varepsilon>0$, there exists $C=C(v,\varepsilon)>0$ such that for $m\ge 0$,

$$C^{-1}e^{(\chi^+(v)-\varepsilon)m}\le\|\mathcal{A}_mv\|\le Ce^{(\chi^+(v)+\varepsilon)m}.$$

We introduce a special class of bases in $\mathbb{R}^n$ associated with the filtration generated by the Lyapunov exponent. These bases provide an important technical tool in studying forward (backward) regularity.

Definition 1.3.6. Given a filtration $\mathcal{V} = \{E_i : i = 0, \ldots, s\}$, we say that a basis $\{v_1, \ldots, v_n\}$ is *normal* with respect to $\mathcal{V}$ if for every $1 \le i \le s$, there exists a basis of E_i composed of vectors from $\{v_1, \ldots, v_n\}$. In other words, those vectors in the basis $\{v_1, \ldots, v_n\}$ that lie in E_i form a basis of E_i.

Given a filtration $\mathcal{V}$, there is always a normal basis with respect to $\mathcal{V}$. Indeed, there is a special procedure that allows one to obtain such a basis. Starting with a basis $\mathbf{v} = \{v_1, \ldots, v_n\}$ of $\mathbb{R}^n$, we define a sequence of bases $\mathbf{w}_i = \{w_{i1}, \ldots, w_{in}\}$, $\mathbf{w}_0 = \mathbf{v}$, which makes the sum $\sum_{j=1}^n \chi^+(w_{ij})$ decrease as i increases. Since χ^+ takes on only finitely many values, this process ends up after a finite number of steps. The resulting basis $\mathbf{w}$ is normal with respect to $\mathcal{V}$. To construct the basis $\mathbf{w}_1$, assume that there is a linear combination $w = \sum_{j=1}^n \alpha_j v_j$, with not all α_j being zero, such that

$$\chi^+(w) < \max\{\chi^+(v_j) : 1 \le j \le n \text{ and } \alpha_j \ne 0\}.$$

Choose a vector v_k for which

$$\chi^+(v_k) = \max\{\chi^+(v_j) : 1 \le j \le n \text{ and } \alpha_j \ne 0\}$$

and replace it with the vector w. Since $\alpha_k \ne 0$, the vectors $v_1, \ldots, v_{k-1}, w, v_{k+1}, \ldots, v_n$ are linearly independent and hence form a basis $\mathbf{w}_1$. Continuing in a similar fashion, we obtain a basis $\mathbf{w} = \{w_1, \ldots, w_n\}$ with the property that for any linear combination $w = \sum_{j=1}^n \alpha_j w_j$, with not all α_j being zero,

$$\chi^+(w) = \max\{\chi^+(w_j) : 1 \le j \le n \text{ and } \alpha_j \ne 0\}. \tag{1.16}$$

We claim that the basis $\mathbf{w}$ is normal with respect to $\mathcal{V}$. Otherwise, there exists i such that the number of vectors in the basis lying in the space E_i is strictly less than $\dim E_i$, and these vectors do not form a basis of E_i. Hence there is a vector $w \in E_i$, which is linearly independent of these vectors and $\chi^+(w) \le \chi_i$. On the other hand, we can write $w = \sum_{j=1}^n \alpha_j w_j$ with some $\alpha_k \ne 0$ such that a vector w_k lies outside of the space E_i. Since $\chi^+(w_k) > \chi_i$, this contradicts (1.16). We conclude that the basis $\mathbf{w}$ is normal.

We show that forward regularity implies the exactness of the forward Lyapunov exponent.

Theorem 1.3.7. *If the sequence of matrices $\mathcal{A}^+$ is forward regular, then the Lyapunov exponent χ^+ is exact.*

Proof. We may assume that the vectors $v_1, \ldots, v_n$ are linearly independent (otherwise the volumes vanish and the statement is trivial). Let $\{e_1, \ldots, e_n\}$ be the canonical basis of $\mathbb{R}^n$. For each k, there exists an orthogonal matrix R such that $Re_1, \ldots, Re_k$ span the same vector space as the vectors $v_1, \ldots, v_k$. Hence,

$$\chi^+(v_1, \ldots, v_k) = \chi^+(Re_1, \ldots, Re_k, \mathcal{A}^+) = \chi^+(e_1, \ldots, e_k, \widetilde{\mathcal{A}}^+),$$

where $\widetilde{A}_0 = A_0 R$ and $\widetilde{A}_m = A_m$ for any $m \geq 1$. Moreover,

$$\mathrm{vol}(\mathcal{A}_m v_1, \ldots, \mathcal{A}_m v_k) = \mathrm{vol}(\mathcal{A}_m Re_1, \ldots, \mathcal{A}_m Re_k).$$

Therefore it is sufficient to prove the theorem for the vectors $e_1, \ldots, e_n$.

As the sequence $\mathcal{A}^+$ is forward regular, the limit of $\frac{1}{m}\log|\det \mathcal{A}_m|$ exists as $m \to +\infty$. Hence the limit in (1.15) exists if and only if it exists when $\mathcal{A}_m$ is substituted by $\mathcal{A}_m/|\det \mathcal{A}_m|^{1/n}$. Taking into account that

$$|\det(\mathcal{A}_m/|\det \mathcal{A}_m|^{1/n})| = 1,$$

we may assume, without loss of generality, that $\det \mathcal{A}_m = 1$ for every $m \geq 1$.

Any matrix $\mathcal{A}_m$ can be written in the form $\mathcal{A}_m = R_m T_m$, where R_m is orthogonal and T_m is lower triangular (i.e., $T_m = (t_{ij})$ with $t_{ij} = 0$ for $i < j$). We obtain

$$\mathrm{vol}(\mathcal{A}_m e_1, \ldots, \mathcal{A}_m e_k) = \mathrm{vol}(T_m e_1, \ldots, T_m e_k) = \prod_{i=1}^{k} t_{im}, \tag{1.17}$$

where t_{im}, for $i = 1, \ldots, m$, are the diagonal entries of T_m.

Without loss of generality, we may assume that the standard basis $\{e_1, \ldots, e_n\}$ is ordered in such a way that $\chi^+(e_1) \geq \cdots \geq \chi^+(e_n)$.

Lemma 1.3.8. *There exists a normal basis $\{f_1, \ldots, f_n\}$ of $\mathbb{R}^n$ such that for each i, we have $f_i = e_i + g_i$ for some $g_i \in \mathrm{span}\{e_1, \ldots, e_{i-1}\}$.*

Proof of the lemma. Let $\chi_1^+ < \cdots < \chi_p^+$ be the distinct values of the Lyapunov exponent and let k_i, $i = 1, \ldots, p$, be their multiplicities. Note that

$$\chi^+(e_1) = \cdots = \chi^+(e_{k_1}) = \chi_p^+$$

(otherwise, the vectors $e_1, \ldots, e_n$ would not form a basis). We set $f_i = e_i$, $i = 1, \ldots, k_1$. If $\chi^+(e_{k_1+1}) < \chi_p^+$ then

$$\chi^+(e_{k_1+1}) = \cdots = \chi^+(e_{k_1+k_2}) = \chi_{p-1}^+$$

and we proceed with the vector $e_{k_1+k_2+1}$. Therefore, we assume that $\chi^+(e_{k_1+1}) = \chi_p^+$. Then there exists $w = \sum_{j=1}^{k_1+1} \alpha_j e_j$ for which $\chi^+(w) = \chi_{p-1}^+$. Clearly, $\alpha_{k_1+1} \neq 0$ and hence the vector $f_{k_1+1} = \alpha_{k_1+1}^{-1} w$ is of the form $f_{k_1+1} = e_{k_1+1} + g_{k_1+1}$ for some $g_{k_1+1} \in \mathrm{span}\{e_1, \ldots, e_{k_1}\}$. Also, $\chi^+(f_{k_1+1}) = \chi_{p-1}^+$. Continuing in the same way we construct the desired vectors f_i. □

Applying the lemma we find that $T_m f_i = t_{im} e_i + T_m g_i$. Since T_m is lower triangular, the vectors e_i and $T_m g_i$ are orthogonal. Thus, $\|T_m f_i\| \geq t_{im}$ and

$$\chi^+(f_i) \geq \varlimsup_{m\to+\infty} \frac{1}{m} \log t_{im} =: \lambda_i.$$

Since $\det \mathcal{A}_m = 1$ for every $m \geq 1$, one has $\prod_{i=1}^{n} t_{im} = 1$, and hence $\sum_{i=1}^{n} \lambda_i \geq 0$. Moreover, since the basis $\{f_1, \ldots, f_n\}$ is normal, we have

$$0 = \lim_{m\to+\infty} \frac{1}{m} \log|\det \mathcal{A}_m| = \sum_{i=1}^{n} \chi^+(f_i) \geq \sum_{i=1}^{n} \lambda_i \geq 0.$$

Hence, $\sum_{i=1}^{n} \lambda_i = 0$. Observe that $\log t_{jm} = -\sum_{i \neq j} \log t_{im}$ (since $\det \mathcal{A}_m = 1$). Therefore given $\varepsilon > 0$, we have that for all sufficiently large m,

$$\frac{1}{m} \log t_{jm} \geq -\sum_{i \neq j} (\lambda_i + \varepsilon) = -\sum_{i \neq j} \lambda_i - (n-1)\varepsilon = \lambda_j - (n-1)\varepsilon,$$

and hence

$$\lambda_j - (n-1)\varepsilon \leq \frac{1}{m} \log t_{jm} < \lambda_j + \varepsilon.$$

This implies that

$$\lim_{m \to +\infty} \frac{1}{m} \log t_{jm} = \lambda_j.$$

Since the number $\chi^+(v_1, \ldots, v_k)$ is defined by (1.11), the desired result follows from (1.17). □

Exactness of the Lyapunov exponent is a weaker property than forward regularity as the following simple example illustrates. It shows that the existence of the limit in (1.5) does *not* guarantee that the Lyapunov exponent χ^+ is forward regular.

Example 1.3.9. *Let $\mathcal{A}^+ = \{\mathcal{A}_m\}_{m \geq 0}$ be the sequence of matrices where $A_0 = \begin{pmatrix} 1 & 0 \\ 2 & 4 \end{pmatrix}$ and $A_m = \begin{pmatrix} 1 & 0 \\ -2^{m+1} & 4 \end{pmatrix}$ for each $m \geq 1$ so that $\mathcal{A}_m = \begin{pmatrix} 1 & 0 \\ 2^m & 4^m \end{pmatrix}$ for every $m \geq 1$. Given a vector $v = (a, b) \neq (0, 0)$, we have $\chi^+(v) = \log 2$ if $b = 0$ and $\chi^+(v) = \log 4$ if $b \neq 0$. Let $v_1 = (1, 0)$ and $v_2 = (0, 1)$; then $\chi^+(v_1) = \log 2$ and $\chi^+(v_2) = \log 4$. Since $\det \mathcal{A}_m = 4^m$, there exists the limit*

$$\chi^+(v_1, v_2) = \lim_{m \to +\infty} \frac{1}{m} \log |\det \mathcal{A}_m| = \log 4.$$

Therefore, χ^+ is exact. On the other hand,

$$\log 4 < \log 2 + \log 4 = \chi^+(v_1) + \chi^+(v_2)$$

and the sequence of matrices $\mathcal{A}^+$ is not forward regular.

In the one-dimensional case the situation is obviously different.

Proposition 1.3.10. *A sequence of numbers $\mathcal{A}^+ \subset GL(1, \mathbb{R}) = \mathbb{R} \setminus \{0\}$ is forward regular if and only if the Lyapunov exponent χ^+ is exact.*

We now establish an important characteristic property of forward regularity, which is quite useful in applications.

Theorem 1.3.11. *A sequence of matrices $\mathcal{A}^+$ is forward regular if and only if for any normal basis $\mathbf{v}$ of $\mathbb{R}^n$ with respect to the filtration $\mathcal{V}_{\chi^+}$ and any subset $K \subset \{1, \ldots, n\}$, the following statements hold:*

1.

$$\chi^+(\{v_i\}_{i \in K}) = \lim_{m \to +\infty} \frac{1}{m} \log \operatorname{vol}(\{\mathcal{A}_m v_i\}_{i \in K}) = \sum_{i \in K} \chi^+(v_i). \qquad (1.18)$$

2. *The angle* σ_m *between the subspaces* $\operatorname{span}\{\mathcal{A}_m v_i : i \in K\}$ *and* $\operatorname{span}\{\mathcal{A}_m v_i : i \notin K\}$ *satisfies*

$$\lim_{m\to+\infty} \frac{1}{m} \log \sin\sigma_m = 0.$$

Proof. If (1.18) holds then $\mathcal{A}^+$ is clearly forward regular.

Assume now that the sequence of matrices $\mathcal{A}^+$ is forward regular. Choose any normal basis $\{v_1, \dots, v_n\}$ of $\mathbb{R}^n$ and observe that

$$\operatorname{vol}(\mathcal{A}_m v_1, \dots, \mathcal{A}_m v_n) = \operatorname{vol}(\{\mathcal{A}_m v_i\}_{i\in K}) \operatorname{vol}(\{\mathcal{A}_m v_i\}_{i\notin K}) \sin\sigma_m. \tag{1.19}$$

By (1.12), we have

$$\chi^+(\{v_i\}_{i\in K}) \le \sum_{i\in K} \chi^+(v_i) \quad \text{and} \quad \chi^+(\{v_i\}_{i\notin K}) \le \sum_{i\notin K} \chi^+(v_i), \tag{1.20}$$

and hence

$$\chi^+(v_1, \dots, v_n) \le \sum_{i=1}^{n} \chi^+(v_i) + \underline{\lim}_{m\to+\infty} \frac{1}{m} \log \sin\sigma_m.$$

Since $\mathcal{A}^+$ is forward regular, we have

$$\sum_{i=1}^{n} \chi^+(v_i) = \chi^+(v_1, \dots, v_n),$$

and thus

$$\underline{\lim}_{m\to+\infty} \frac{1}{m} \log \sin\sigma_m \ge 0.$$

Since $\sin\sigma_m$ is bounded, we conclude that

$$\lim_{m\to+\infty} \frac{1}{m} \log \sin\sigma_m = 0.$$

Therefore using (1.19),

$$\begin{aligned} \sum_{i\in K} \chi^+(v_i) + \sum_{i\notin K} \chi^+(v_i) &= \chi^+(v_1, \dots, v_n) \\ &\le \chi^+(\{v_i\}_{i\in K}) + \chi^+(\{v_i\}_{i\notin K}). \end{aligned}$$

It follows from (1.20) that

$$\chi^+(\{v_i\}_{i\in K}) = \sum_{i\in K} \chi^+(v_i) \quad \text{and} \quad \chi^+(\{v_i\}_{i\notin K}) = \sum_{i\notin K} \chi^+(v_i).$$

This completes the proof of the theorem. □

A sequence of matrices $\mathcal{A}^- = \{A_m\}_{m<0}$ is said to be *backward regular* if

$$\lim_{m\to-\infty} \frac{1}{|m|} \log|\det \mathcal{A}_m| = \sum_{i=1}^{n} \widetilde{\chi}'_i,$$

where $\widetilde{\chi}'_1, \dots, \widetilde{\chi}'_n$ are the finite values of χ^- counted with their multiplicities.

We say that the Lyapunov exponent χ^- is *exact* if

$$\chi^-(v_1,\ldots,v_k) = \lim_{m\to-\infty} \frac{1}{|m|} \log \operatorname{vol}(\mathcal{A}_m v_1,\ldots,\mathcal{A}_m v_k)$$

for any $1 \le k \le n$ and every collection $\{v_1,\ldots,v_k\} \subset \mathbb{R}^n$.

Given a sequence $\mathcal{A}^- = \{A_m\}_{m<0}$, we construct a new sequence $\mathcal{B}^+ = \{B_m\}_{m\ge 0}$ by setting $B_m = (A_{-m-1})^{-1}$. It follows from (1.1) that the backward regularity of $\mathcal{A}^-$ is equivalent to the forward regularity of $\mathcal{B}^+$. This reduction allows us to translate any fact about forward regularity into a corresponding fact about backward regularity.

For example, the backward regularity of the sequence of matrices $\mathcal{A}^-$ implies that the Lyapunov exponent χ^- is exact. Moreover, if the Lyapunov exponent χ^- is exact (in particular, if it is backward regular) then

$$\chi^-(v) = \lim_{m\to-\infty} \frac{1}{|m|} \log\|\mathcal{A}_m v\|$$

for every $v \in \mathbb{R}^n$. This is equivalent to the following statement: for every $\varepsilon > 0$, there exists $C = C(v,\varepsilon) > 0$ such that if $m \ge 0$ then

$$C^{-1} e^{-(\chi^-(v)-\varepsilon)m} \le \|\mathcal{A}_{-m} v\| \le C e^{-(\chi^-(v)+\varepsilon)m}.$$

1.3.3 A Criterion of Forward Regularity for Triangular Matrices

Let $\mathcal{A}^+ = \{A_m\}_{m\ge 0}$ be a sequence of matrices. One can write each A_m in the form $A_m = R_m T_m$, where R_m is orthogonal and T_m is lower triangular. In general, the diagonal entries of T_m alone do *not* determine the values of the Lyapunov exponent associated to $\mathcal{A}^+$. Indeed, consider the sequence of matrices where $A_m = \left(\begin{smallmatrix} 0 & -1/2 \\ 2 & 0 \end{smallmatrix}\right)$ for each $m > 0$. It is easy to see that $\mathcal{A}_m = R_m T_m$, where $R_m = \left(\begin{smallmatrix} 0 & -1 \\ 1 & 0 \end{smallmatrix}\right)$ is orthogonal and $T_m = \left(\begin{smallmatrix} 2 & 0 \\ 0 & 1/2 \end{smallmatrix}\right)$ is diagonal. Since $\mathcal{A}_m{}^2 = -\mathrm{Id}$, we obtain $\chi^+(v,\mathcal{A}^+) = 0$ for every $v \ne 0$ whereas the values of the Lyapunov exponent for $\mathcal{T} = \{T_m\}_{m\ge 0}$ are equal to $\pm\log 2$.

However, in certain situations one can reduce the study of sequences of arbitrary matrices to the study of sequences of lower triangular matrices (see Section 3.4). Therefore, we begin by considering sequences of lower triangular matrices and present a useful criterion of regularity of Lyapunov exponents. This criterion will be used in Chapter 3 in the proof of the Multiplicative Ergodic Theorem 3.4.3 that is one of our central results. We write $\log^+ a = \max\{\log a, 0\}$ for a positive number a.

Theorem 1.3.12. *Let $\mathcal{A}^+ = \{(a_{ij}^m)\}_{m\ge 0} \subset GL(n,\mathbb{R})$ be a sequence of lower triangular matrices such that*

1. *the limit*

$$\lim_{m\to+\infty} \frac{1}{m} \sum_{k=0}^{m} \log|a_{ii}^k| =: \lambda_i, \quad i = 1,\ldots,n$$

exists and is finite;

2. *we have*

$$\varlimsup_{m\to+\infty}\frac{1}{m}\log^+|a_{ij}^m|=0,\quad i,j=1,\ldots,n.$$

Then the sequence $\mathcal{A}^+$ is forward regular, and the numbers λ_i are the values of the Lyapunov exponent χ^+ (counted with their multiplicities but possibly not ordered).

Proof. Before going into the detailed proof let us explain the main point. For the sake of this discussion let us count each exponent according to its multiplicity. Thus we have exactly n exponents. To verify (1.13), we will produce a basis $\{v_1,\ldots,v_n\}$, which is normal with respect to the standard filtration (i.e., related with the standard basis by an upper triangular coordinate change) and such that $\chi^+(v_i)=\lambda_i$.

If the exponents are ordered so that $\lambda_1\ge\lambda_2\ge\cdots\ge\lambda_n$ then the standard basis is in fact forward regular. To see this notice that while multiplying lower triangular matrices, one obtains a matrix whose off-diagonal entries contain a polynomially growing number of terms each of which can be estimated by the growth of the product of diagonal terms below.

However, if the exponents are not ordered that way then an element e_i of the standard basis will grow according to the maximal of the exponents λ_j for $j\ge i$. In order to produce the right growth, one has to compensate the growth caused by off-diagonal terms by subtracting from the vector e_i a certain linear combination of vectors e_j for which $\lambda_j>\lambda_i$. This can be done in a unique fashion. The detailed proof proceeds by induction in n.

For $n=1$, the result follows immediately from Condition 1. Proceeding by induction in n, we assume that the sequence of lower triangular matrices $A_m\in GL(n+1,\mathbb{R})$ satisfies Conditions 1 and 2. For each $m\ge1$, let A'_m be the triangular $n\times n$ matrix obtained from A_m by deleting its first row and first column. The matrices A'_m satisfy Conditions 1 and 2. Consider the Lyapunov exponent

$$\chi'(v)=\varlimsup_{m\to+\infty}\frac{1}{m}\log\|\mathcal{A}'_m v\|,$$

where $\mathcal{A}'_m=A'_{m-1}\cdots A'_0$, and the filtration

$$\{0\}=E_0\subsetneq E_1\subsetneq\cdots\subsetneq E_p=\mathbb{R}^n$$

associated with the Lyapunov exponent χ'. By the induction hypothesis, χ' is forward regular. Write $\mathbb{R}^n=\bigoplus_{i=1}^p H_i$ where $H_i=E_{i-1}{}^\perp\cap E_i$ and let $\lambda'_2\le\cdots\le\lambda'_{n+1}$ be the numbers $\lambda_2,\ldots,\lambda_{n+1}$ in Condition 1 written in nondecreasing order. Since $\mathcal{A}'_m$ is forward regular, Conditions 1 and 2 guarantee that for $i\ge2$ and $v\in H_i\setminus\{0\}$,

$$\lim_{m\to+\infty}\frac{1}{m}\log\|\mathcal{A}'_m v\|=\lambda'_i.\tag{1.21}$$

It remains only to show that λ_1 is a value of the Lyapunov exponent χ for some vector $v\in\mathbb{R}^{n+1}\setminus\bigoplus_{i=1}^p H_i$. Indeed, since $\mathcal{A}'_m$ is forward regular, we obtain that

$$\lim_{m\to+\infty}\frac{1}{m}\log|\det\mathcal{A}_m| = \lim_{m\to+\infty}\frac{1}{m}\sum_{k=0}^{m}\log|a_{11}^k|$$

$$+\lim_{m\to+\infty}\frac{1}{m}\log|\det\mathcal{A}'_m| = \sum_{i=1}^{n+1}\lambda_i,$$

and thus $\mathcal{A}_m$ is regular. By Theorem 1.3.11, we have

$$\lim_{m\to+\infty}\frac{1}{m}\log|\sin\sphericalangle(H_i^m, \widehat{H}_i^m)| = 0, \tag{1.22}$$

where

$$H_i^m = \mathcal{A}'_m H_i \quad \text{and} \quad \widehat{H}_i^m = \bigoplus_{j\neq i}\mathcal{A}'_m H_i.$$

Let $\{v_1^m, \ldots, v_n^m\}$ be a basis of $\mathbb{R}^n$ such that $\{v_{n_{i-1}+1}^m, \ldots, v_{n_i}^m\}$ is an orthonormal basis of H_i^m, where $n_i = \dim E_i$, $i = 1, \ldots, p$. We denote by C_m the coordinate change from the standard basis $\{e_0, \ldots, e_n\}$ of $\mathbb{R}^{n+1}$ to the basis $\{e_0, v_1^m, \ldots, v_n^m\}$. It follows from (1.22) that

$$\lim_{m\to+\infty}\frac{1}{m}\log\|C_m\| = \lim_{m\to+\infty}\frac{1}{m}\log\|C_m{}^{-1}\| = 0. \tag{1.23}$$

Consider the sequence of matrices $B_m = C_{m+1}{}^{-1}A_m C_m$. Note that for each i,

$$B_m\ \text{span}\{e_{n_{i-1}+1}, \ldots, e_{n_i}\} = \text{span}\{e_{n_{i-1}+1}, \ldots, e_{n_i}\}.$$

Hence the matrix B_m has the form

$$B_m = \begin{pmatrix} a_{11}^m & & & & \\ g_1^m & B_1^m & & & \\ g_2^m & 0 & B_2^m & & \\ \vdots & \vdots & \ddots & \ddots & \\ g_p^m & 0 & \cdots & 0 & B_p^m \end{pmatrix},$$

where each B_i^m is a $k_i \times k_i$ matrix and each $g_i^m \in \mathbb{R}^{k_i}$ is a column vector. Set $\mathcal{B}_m = B_{m-1}\cdots B_0$ and observe that

$$\mathcal{B}_m = C_m{}^{-1}\mathcal{A}_m C_0. \tag{1.24}$$

The following lemma establishes a crucial property of the column vector g_i^m.

Lemma 1.3.13. *For each $i = 1, \ldots, p$, we have*

$$\overline{\lim_{m\to+\infty}}\frac{1}{m}\log^+\|g_i^m\| = 0.$$

Proof of the lemma. Let $g^m \in \mathbb{R}^n$ be the column vector composed of the components of $g_1^m, \ldots, g_p^m$. We have

$$\|g^m\| \le \|C_{m+1}{}^{-1}\| \cdot \|a_{21}^m e_1 + \cdots + a_{n+1,1}^m e_n\| \cdot \|C_m\|.$$

Therefore using (1.23) and Condition 2, we obtain

$$\overline{\lim_{m\to+\infty}} \frac{1}{m} \log^+ \|g^m\| \le \overline{\lim_{m\to+\infty}} \frac{1}{m} \log^+ \max\{|a^m_{i1}| : 2 \le i \le n+1\} = 0. \tag{1.25}$$

Moreover, since g^m_i is the projection of g^m onto H^{m+1}_i along $\widehat{E}^{m+1}_i$, we have

$$\|g^m_i\| \le \|g^m\| / \sin \sphericalangle(H^{m+1}_i, \widehat{H}^{m+1}_i),$$

and hence

$$\frac{1}{m} \log\|g^m_i\| \le \frac{1}{m} \log\|g^m\| - \frac{1}{m} \log \sin \sphericalangle(H^{m+1}_i, \widehat{H}^{m+1}_i).$$

By (1.22) and (1.25),

$$0 \le \overline{\lim_{m\to+\infty}} \frac{1}{m} \log^+ \|g^m_i\| \le \overline{\lim_{m\to+\infty}} \frac{1}{m} \log^+ \|g^m\| = 0$$

and the lemma follows. □

We proceed with the proof of the theorem.

Case 1. $\lambda_1 \ge \lambda'_{n+1}$ or $\lambda_1 \ge \lambda_j$ for all $2 \le j \le n+1$. This is an easy situation since the off-diagonal elements will grow sufficiently slowly compared to the first element of the basis. Observe that

$$\mathcal{B}_m e_0 = \prod_{i=0}^{m-1} a^i_{11} e_0 + \sum_{j=1}^{p} \sum_{i=1}^{m-1} a^{i-1}_{11} \cdots a^0_{11} B^{m-1}_j \cdots B^{i+1}_j g^i_j. \tag{1.26}$$

Moreover, $\mathcal{B}_m v = B^{m-1}_j \cdots B^0_j v$ for each $v \in H_j$. Given j, $1 \le j \le p$, set

$$c_{mj}(v) = \frac{\|B^m_j \cdots B^0_j v\|}{\|B^{m-1}_j \cdots B^0_j v\|}, \quad v \in C_0{}^{-1} H_j \setminus \{0\}, \quad m \ge 0.$$

By (1.21), (1.23), and (1.24), we obtain for such v,

$$\begin{aligned} \lambda'_j &= \lim_{m\to+\infty} \frac{1}{m} \log\|\mathcal{A}'_m C_0 v\| = \lim_{m\to+\infty} \frac{1}{m} \log \frac{\|\mathcal{B}_m v\|}{\|v\|} \\ &= \lim_{m\to+\infty} \frac{1}{m} \log(c_{0j}(v) \cdots c_{m-1j}(v)) = \lim_{m\to+\infty} \frac{1}{m} \sum_{l=0}^{m-1} \log c_{lj}(v). \end{aligned} \tag{1.27}$$

Fix i, $0 \le i < m$, and set

$$K_{mij} = \log\|a^{i-1}_{11} \cdots a^0_{11} B^{m-1}_j \cdots B^{i+1}_j g^i_j\|.$$

Note that

$$\begin{aligned} K_{mij} &= \sum_{\ell=0}^{i-1} \log|a^\ell_{11}| + \log\|B^{m-1}_j \cdots B^{i+1}_j g^i_j\| \\ &= \sum_{\ell=0}^{i-1} \log|a^\ell_{11}| + \sum_{r=i+1}^{m-1} \log c_{rj}(v) + \log\|g^i_j\|, \end{aligned}$$

where $v = (B_j^i \cdots B_j^0)^{-1} g_j^i$. By Lemma 1.3.13, for each $\varepsilon > 0$, there exists i_0 independent of m such that $\log^+ \|g_j^i\| < \varepsilon i < \varepsilon m$ for every $i_0 \le i < m$. By Condition 1 and (1.27), we may assume (choosing a larger i_0 if necessary) that

$$(\lambda_1 - \varepsilon)i < \sum_{\ell=0}^{i-1} \log|a_{11}^\ell| < (\lambda_1 + \varepsilon)i$$

and

$$(\lambda_j' - \varepsilon)(m - i) < \sum_{r=i+1}^{m-1} \log c_{rj}(v) < (\lambda_j' + \varepsilon)(m - i).$$

Since $\lambda_1 \ge \lambda_j'$, it follows that

$$K_{mij} < (\lambda_1 + \varepsilon)i + (\lambda_j' + \varepsilon)(m - i) + \varepsilon m \le (\lambda_1 + 2\varepsilon)m.$$

By (1.26), we conclude that

$$\begin{aligned}
\exp[(\lambda_1 - \varepsilon)m] &\le \|\mathcal{B}_m e_0\| \\
&\le \exp[(\lambda_1 + \varepsilon)m] + \sum_{j=1}^{p} \sum_{i=0}^{m-1} \exp K_{mij} \\
&\le \exp[(\lambda_1 + \varepsilon)m] + pm\, \exp[(\lambda_1 + 2\varepsilon)m] \\
&\le (p + 1) \exp[(\lambda_1 + 3\varepsilon)m]
\end{aligned}$$

for all sufficiently large m for which $m \le \exp(\varepsilon m)$. It follows from (1.23) and (1.24) that

$$\chi'(e_0) = \lim_{m \to +\infty} \frac{1}{m} \log\|\mathcal{B}_m e_0\| = \lambda_1.$$

This completes the proof of the theorem in the case $\lambda_1 \ge \lambda_{n+1}'$.

Case 2. $\lambda_1 < \lambda_j'$ for some j such that $2 \le j \le n + 1$. Using again Lemma 1.3.13, given $\varepsilon > 0$, we find that $\|g_j^m\| < \exp(\varepsilon m)$. Furthermore,

$$\|(B_j^m \cdots B_j^0)^{-1} g_j^m\| \le \exp[(-\lambda_j' + \varepsilon)m]$$

for all sufficiently large m. Therefore,

$$\begin{aligned}
&\sum_{m=1}^{\infty} \|a_{11}^{m-1} \cdots a_{11}^0 (B_j^m \cdots B_j^0)^{-1} g_j^m\| \\
&\qquad \le \sum_{m=1}^{\infty} \exp[(\lambda_1 + \varepsilon)m + (-\lambda_j' + \varepsilon)m] \\
&\qquad \le \sum_{m=1}^{\infty} \exp[(\lambda_1 - \lambda_j' + 2\varepsilon)m] < +\infty
\end{aligned}$$

for all $m \geq 1$ and all sufficiently small $\varepsilon > 0$. This shows that the formula

$$h_j = -\sum_{m=1}^{\infty} a_{11}^{m-1} \cdots a_{11}^{0} (B_j^m \cdots B_j^0)^{-1} g_j^m$$

defines a vector in H_j. This is exactly where the compensation needed to correct the excessive growth takes place. Set

$$v = e_0 + \sum_{j:\lambda_1 < \lambda'_j} h_j$$

and denote by $\mathrm{P}_{H_j^m}$ the projection onto H_j^m along $\widehat{E}_j^m$. We have

$$\begin{aligned}
\mathrm{P}_{H_j^m} \mathcal{B}_m v &= \mathrm{P}_{H_j^m} \mathcal{B}_m (e_0 + h_j) \\
&= \sum_{i=1}^{m-1} a_{11}^{i-1} \cdots a_{11}^{0} B_j^{m-1} \cdots B_j^{i+1} g_j^i + \mathcal{B}_m h_j \\
&= \sum_{i=1}^{m-1} a_{11}^{i-1} \cdots a_{11}^{0} B_j^{m-1} \cdots B_j^{i+1} g_j^i + B_j^{m-1} \cdots B_j^0 h_j \\
&= -\sum_{i=m}^{\infty} a_{11}^{i-1} \cdots a_{11}^{0} (B_j^i \cdots B_j^m)^{-1} g_j^i.
\end{aligned}$$

Proceeding as before we obtain

$$\begin{aligned}
\|\mathrm{P}_{E_j^m} \mathcal{B}_m v\| &\leq \sum_{i=m}^{\infty} \|a_{11}^{i-1} \cdots a_{11}^{0} (B_j^i \cdots B_j^m)^{-1} g_j^i\| \\
&\leq \sum_{i=m}^{\infty} \exp[(\lambda_1 + \varepsilon)i + (-\lambda'_j + \varepsilon)(i - m) + \varepsilon i] \\
&\leq \exp[(\lambda_1 + \varepsilon)m] \sum_{i=m}^{\infty} \exp[(\lambda_1 - \lambda'_j + 2\varepsilon)(i - m) + \varepsilon i] \\
&= D(\varepsilon) \exp[(\lambda_1 + 2\varepsilon)m]
\end{aligned} \tag{1.28}$$

for all sufficiently large $m \geq 1$. Here

$$D(\varepsilon) = \sum_{i=1}^{\infty} \exp[(\lambda_1 - \lambda'_j + 3\varepsilon)i] < +\infty,$$

provided $\varepsilon > 0$ is chosen such that $\lambda_1 - \lambda'_j + 3\varepsilon < 0$ (which is always possible since $\lambda_1 < \lambda'_j$).

Denote by P_{e_0} the projection on e_0 along $\mathbb{R}^n = \bigoplus_{i=1}^{p} H_i$. Exploiting the special form of the matrix B_m described above, we obtain

$$\|\mathrm{P}_{e_0} \mathcal{B}_m v\| = \prod_{i=0}^{m-1} a_{11}^i.$$

By Theorem 1.3.11, for each j,

$$\lim_{m\to+\infty}\frac{1}{m}\log|\sin\sphericalangle(E_j^m,\widehat{E}_j^m)|=0.$$

Since e_0 is orthogonal to $\bigoplus_{i=1}^p H_i$, there exist $c_1>0$ and $c_2>0$ such that

$$c_1\|\mathcal{B}_m v\|\le\prod_{i=0}^{m-1}a_{11}^i+\sum_{j:\lambda_j>\lambda_1}\|\mathrm{P}_{E_j^m}\mathcal{B}_m v\|\le c_2\|\mathcal{B}_m v\|$$

for every $m\ge 1$. By (1.23), (1.28), and Condition 1, we have

$$\chi'(C_0{}^{-1}v)=\lim_{m\to+\infty}\frac{1}{m}\log\|\mathcal{B}_m v\|=\lambda_1.$$

The above discussion implies that the Lyapunov exponent χ' is exact with respect to the vector

$$v_{n+1}=e_0+C_0{}^{-1}\sum_{j:\lambda_1<\lambda'_j}h_j$$

and that $\chi'(v_{n+1})=\lambda_1$. This completes the proof of the theorem. □

An analogous criterion for forward regularity holds for sequences of upper triangular matrices.

Using the correspondence between forward and backward sequences of matrices, we immediately obtain the corresponding criterion for backward regularity.

Theorem 1.3.14. *Let $\mathcal{A}^-=\{(a_{ij}^m)\}_{m<0}\subset GL(n,\mathbb{R})$ be a sequence of lower triangular matrices such that*

1. *the limit*

$$\lim_{m\to-\infty}\frac{1}{|m|}\sum_{k=1}^{m}\log|a_{ii}^k|=:\lambda_i^-,\quad i=1,\dots,n$$

exists and is finite;

2. *we have*

$$\overline{\lim_{m\to-\infty}}\frac{1}{|m|}\log^+|a_{ij}^m|=0,\quad i,j=1,\dots,n.$$

Then the sequence $\mathcal{A}^-$ is backward regular, and the numbers λ_i^- are the values of the Lyapunov exponent χ^- (counted with multiplicities but possibly not ordered).

1.3.4 The Lyapunov–Perron Regularity

We finally introduce the crucial property of the Lyapunov–Perron regularity. It deals with a two-sided sequence of matrices $\mathcal{A}=\{A_m\}_{m\in\mathbb{Z}}$ in $GL(n,\mathbb{R})$. Set $\mathcal{A}^+=\{A_m\}_{m\ge0}$ and $\mathcal{A}^-=\{A_m\}_{m<0}$. Consider the forward and backward Lyapunov exponents χ^+ and χ^- of the sequence $\mathcal{A}$, that is, of the sequences $\mathcal{A}^+$ and

$\mathcal{A}^-$, respectively (see (1.5) and (1.7)). Denote by

$$\mathcal{V}_{\chi^+} = \{E_i^+ : i = 1, \ldots, p^+\} \quad \text{and} \quad \mathcal{V}_{\chi^-} = \{E_i^- : i = 1, \ldots, p^-\}$$

the filtrations of $\mathbb{R}^n$ associated with the Lyapunov exponents χ^+ and χ^-.

Definition 1.3.15. We say that the filtrations $\mathcal{V}_{\chi^+}$ and $\mathcal{V}_{\chi^-}$ are *coherent* if the following properties hold:

1. $p^+ = p^- =: p$.
2. There exists a decomposition

$$\mathbb{R}^n = \bigoplus_{j=1}^{p} H_j$$

into subspaces H_j such that for $i = 1, \ldots, p$,

$$E_i^+ = \bigoplus_{j=1}^{i} H_j \quad \text{and} \quad E_i^- = \bigoplus_{j=i}^{p} H_j.$$

(Note that necessarily $H_i = E_i^+ \cap E_i^-$.)

3. For $v \in H_i \setminus \{0\}$, $i = 1, \ldots, p$,

$$\lim_{m\to\pm\infty} \frac{1}{m} \log\|\mathcal{A}_m v\| = \chi_i^+ = -\chi_i^- =: \chi_i,$$

with uniform convergence on $\{v \in H_i : \|v\| = 1\}$.

Definition 1.3.16. We say that the sequence $\mathcal{A}$ is *Lyapunov–Perron regular* or simply *regular* if

1. $\mathcal{A}$ is simultaneously forward and backward regular (i.e., $\mathcal{A}^+$ is forward regular and $\mathcal{A}^-$ is backward regular);
2. the filtrations $\mathcal{V}_{\chi^+}$ and $\mathcal{V}_{\chi^-}$ are coherent.

Note that a sequence of matrices $\mathcal{A} = \{A_m\}_{m\in\mathbb{Z}}$ generated by a single matrix A (i.e., $A_m = A$ for all $m \in \mathbb{Z}$) is Lyapunov–Perron regular since

$$\sum_{i=1}^{p} \chi_i \dim H_i = \log|\det A|.$$

Proposition 1.3.17. *If a sequence of matrices $\mathcal{A}$ is a regular then*

1. *the exponents χ^+ and χ^- are exact;*
2. *for $i = 1, \ldots, p$, we have $\chi_i^- = -\chi_i$, $\dim H_i = k_i$, and*

$$\lim_{m\to\pm\infty} \frac{1}{m} \log|\det(\mathcal{A}_m | H_i)| = \chi_i k_i.$$

Proof. All statements immediately follow from the definitions and Theorems 1.3.7 and 1.3.11 applied to the set $K = \{n_{i+1}, \ldots, n_{i+k_i}\}$. ☐

Simultaneous forward and backward regularity of a sequence of matrices $\mathcal{A}$ is not sufficient for the Lyapunov–Perron regularity. Forward (respectively, backward) regularity does not depend on the backward (respectively, forward) behavior of $\mathcal{A}$, that is, for $m \le 0$ (respectively, $m \ge 0$). On the other hand, the Lyapunov–Perron regularity requires some compatibility between the forward and backward behavior, which is expressed in terms of the filtrations $\mathcal{V}_{\chi^+}$ and $\mathcal{V}_{\chi^-}$.

Example 1.3.18. *Let*

$$A_m = \begin{cases} \begin{pmatrix} 2 & 0 \\ 0 & 1/2 \end{pmatrix} & \text{if } m \ge 0 \\ \begin{pmatrix} 5/4 & -3/4 \\ -3/4 & 5/4 \end{pmatrix} & \text{if } m < 0 \end{cases}.$$

Note that

$$\begin{pmatrix} 2 & 0 \\ 0 & 1/2 \end{pmatrix} = R^{-1} \begin{pmatrix} 5/4 & -3/4 \\ -3/4 & 5/4 \end{pmatrix} R,$$

where R is the rotation around 0 by $\pi/4$. One can easily check that $\chi^+(1,0) = \chi^-(1,1) = \log 2$ and $\chi^+(0,1) = \chi^-(1,-1) = -\log 2$. Hence, $E_1^+ \neq E_1^-$, and thus $\mathcal{A}$ is not regular. On the other hand, since $\det A_m = 1$ *for every $m \in \mathbb{Z}$, we have*

$$\chi^+(v_1, v_2) = \chi^-(v_1, v_2) = \log 2 - \log 2 = 0,$$

and the exponents χ^+ and χ^- are exact. Therefore the sequence $\mathcal{A}$ is simultaneously forward and backward regular.

Consider a regular sequence of matrices $\mathcal{A}$ and assume that the spectrum of its Lyapunov exponent satisfies the following condition:

$$\text{there is a number } j,\ 1 \le j < p, \text{ such that } \chi_j < 0. \tag{1.29}$$

Theorem 1.3.19. *A sequence of matrices $\mathcal{A}$ satisfying* (1.29) *is nonuniformly partially hyperbolic in the broad sense. Moreover, it admits a $(\lambda, \mu, \varepsilon)$-splitting of $\mathbb{R}^n$ for every sufficiently small ε. In addition, if a regular sequence of matrices $\mathcal{A}$ satisfies Condition* (1.10) *then it is nonuniformly completely hyperbolic.*

Proof. The first statement is a particular case of Theorem 3.3.1 with

$$\lambda = \exp \chi_j, \quad \mu = \exp \chi_{j+1}, \quad E_m^1 = A_m E_j^+, \quad E_m^2 = A_m E_j^-.$$

The second statement follows from the first one. □

1.4 Notes

The Lyapunov exponents were introduced by Lyapunov [134] in his study of stability of solutions of differential equations. A comprehensive but somewhat outdated reference for the abstract theory of Lyapunov exponents as well as its applications to the theory of differential equations is the book [49] that is available only in Russian. A contemporary account of this theory is presented in [20].

The notion of forward regularity and regularity appeared in the work of Lyapunov [134] and Perron [164, 165] in connection with the study of stability of solutions of linear ordinary differential equations with nonconstant coefficients (see [20] for a detailed discussion). We follow their approach but adopt it for the discrete-time case. In the literature regularity is often called "Lyapunov regularity." However, Perron's contribution in studying regularity is essential and to acknowledge this fact, throughout the book we adopt the term "Lyapunov–Perron regularity".

The characterization of normal bases (see Theorem 1.3.7) was first given by Lyapunov in [134].

The main results of Section 1.3 – Theorems 1.3.11 and 1.3.12 – are due to Oseledets [162]. Our presentation follows the unpublished notes by Katok and Mendoza [107].

2

Lyapunov Exponents for Linear Extensions

A typical example of a sequence of matrices is the sequence of differentials of a smooth transformation along an orbit. More generally, one considers linear extensions of dynamical systems. If the corresponding vector bundles are trivial in a sense, such an extension can be defined by a cocycle, that is, a function from the phase space of the system to the group $GL(n, \mathbb{R})$. In this setting the sequence of matrices corresponds to the restriction of a cocycle to an orbit. Namely, let

$$\mathcal{A}(\ell, m) = A_{\ell+m-1} \cdots A_\ell = \mathcal{A}_{\ell+m,\ell},$$

where $\mathcal{A}_{\ell+m,\ell}$ is defined by (1.1). It follows immediately that

$$\mathcal{A}(\ell, m+k) = \mathcal{A}(\sigma^k(\ell), m)\mathcal{A}(\ell, k),$$

where $\sigma : \mathbb{Z} \to \mathbb{Z}$ is the shift; that is, $\sigma(\ell) = \ell + 1$. This property identifies the function $\mathcal{A}$ as a cocycle over the shift σ. One can now produce a general concept of a cocycle by interpreting the shift as an action of an invertible map on an orbit, and one can extend the notions of Lyapunov exponents, regularity, and so on to cocycles.

2.1 Cocycles over Dynamical Systems

2.1.1 Cocycles and Linear Extensions

In what follows we assume that X is a measure space, endowed with a σ-algebra of measurable subsets, and that $f \colon X \to X$ is an invertible measurable transformation. For most substantive statements we assume that f preserves a finite measure.

Definition 2.1.1. A function $\mathcal{A} \colon X \times \mathbb{Z} \to GL(n, \mathbb{R})$ is called a *linear multiplicative cocycle over* f or simply a *cocycle* if the following properties hold:

1. For each $x \in X$, we have $\mathcal{A}(x, 0) = \mathrm{Id}$, and given $m, k \in \mathbb{Z}$,
$$\mathcal{A}(x, m+k) = \mathcal{A}(f^k(x), m)\mathcal{A}(x, k). \tag{2.1}$$
2. For every $m \in \mathbb{Z}$, the function $\mathcal{A}(\cdot, m) \colon X \to GL(n, \mathbb{R})$ is measurable.

Note another property of the cocycle: $\mathcal{A}(f^{-m}(x), m)^{-1} = \mathcal{A}(x, -m)$.

Given a measurable function $A\colon X \to GL(n, \mathbb{R})$, we define a cocycle by

$$\mathcal{A}(x, m) = \begin{cases} A(f^{m-1}(x)) \cdots A(f(x))A(x) & \text{if } m > 0 \\ \mathrm{Id} & \text{if } m = 0. \\ A(f^m(x))^{-1} \cdots A(f^{-2}(x))^{-1} A(f^{-1}(x))^{-1} & \text{if } m < 0 \end{cases}$$

We call the map A the *generator* of the cocycle $\mathcal{A}$ and we say that the cocycle $\mathcal{A}$ is *generated* by the function A. Note that every cocycle $\mathcal{A}$ is generated by the function $A(\cdot) = \mathcal{A}(\cdot, 1)$.

A cocycle $\mathcal{A}$ over f induces a *linear extension* $F\colon X \times \mathbb{R}^n \to X \times \mathbb{R}^n$ of f to $X \times \mathbb{R}^n$, or a *linear skew product*, defined by

$$F(x, v) = (f(x), A(x)v). \tag{2.2}$$

In other words, the action of F on the fiber over x to the fiber over $f(x)$ is given by the linear map $A(x)$. If $\pi : X \times \mathbb{R}^n \to X$ is the projection defined by $\pi(x, v) = x$, then the diagram

$$\begin{array}{ccc} X \times \mathbb{R}^n & \xrightarrow{F} & X \times \mathbb{R}^n \\ {\scriptstyle \pi}\downarrow & & \downarrow{\scriptstyle \pi} \\ X & \xrightarrow{f} & X \end{array}$$

is commutative and for each $m \in \mathbb{Z}$, we obtain

$$F^m(x, v) = (f^m(x), \mathcal{A}(x, m)v).$$

Linear extensions are particular cases of bundle maps of measurable vector bundles that we now consider. Let E and X be measure spaces and $\pi : E \to X$ a measurable map. E is called a *measurable vector bundle* over X if there exists a countable collection of measurable subsets $Y_i \subset X$, which cover X, and measurable maps $\pi_i : X \times \mathbb{R}^n \to X$ with $\pi_i^{-1}(Y_i) = Y_i \times \mathbb{R}^n$. A bundle map $F\colon E \to E$ over a measurable transformation $f\colon X \to X$ is a measurable map, which makes the following diagram commutative:

$$\begin{array}{ccc} E & \xrightarrow{F} & E \\ {\scriptstyle \pi}\downarrow & & \downarrow{\scriptstyle \pi} \\ X & \xrightarrow{f} & X \end{array}.$$

The following proposition shows that from the measure theory point of view, every vector bundle over a compact metric space is trivial, and hence without loss of generality, one may always assume that $E = X \times \mathbb{R}^n$. In other words every bundle map of E is essentially a linear extension, provided that the base space X is a compact metric space.

Proposition 2.1.2. *If E is a measurable vector bundle over a compact metric space (X, ν), then there is a subset $Y \subset X$ such that $\nu(Y) = 1$ and $\pi^{-1}(Y)$ is (isomorphic to) a trivial vector bundle.*

Proof. Since $\bigcup_{r>0} \partial B(x,r)$ is an uncountable disjoint union, we may choose $r_x > 0$ such that $\nu(\partial B(x, r_x)) = 0$ and $\pi | B(x, r_x)$ is measurably isomorphic to $B(x, r_x) \times \mathbb{R}^n$. Since X is compact, there exists a finite cover $\{B(x_1, r_1), \ldots, B(x_k, r_k)\}$ of X such that $\nu(\partial B(x_i, r_i)) = 0$ and $\pi | B(x_i, r_i)$ is measurably isomorphic to $B(x_i, r_i) \times \mathbb{R}^n$ for each $i = 1, \ldots, k$. Setting $Y = X \setminus \bigcup_{i=1}^{k} \partial B(x_i, r_i)$ the desired result follows. □

2.1.2 Cohomology and Tempered Equivalence

Let $A: X \to GL(n, \mathbb{R})$ be the generator of a cocycle $\mathcal{A}$ over an invertible measurable transformation $f: X \to X$. The cocycle acts on the linear coordinate v_x on the fiber $\{x\} \times \mathbb{R}^n$ of $X \times \mathbb{R}^n$ by $v_{f(x)} = A(x)v_x$. Let $L(x) \in GL(n, \mathbb{R})$ be a linear coordinate change in each fiber, given by $u_x = L(x)v_x$ for each $x \in X$. We assume that the function $L: X \to GL(n, \mathbb{R})$ is measurable. Consider the function $B: X \to GL(n, \mathbb{R})$ for which $u_{f(x)} = B(x)u_x$. One can easily verify that

$$A(x) = L(f(x))^{-1} B(x) L(x)$$

and that B generates a new cocycle $\mathcal{B}$ over f. One can naturally think of the cocycles $\mathcal{A}$ and $\mathcal{B}$ as equivalent. However, since the function L is in general only measurable, without any additional assumption on L the measure-theoretic properties of the cocycles $\mathcal{A}$ and $\mathcal{B}$ can be very different. We now introduce a sufficiently general class of coordinate changes, which make the notion of equivalence productive.

Let $Y \subset X$ be an f-invariant nonempty measurable set.

Definition 2.1.3. A measurable function $L: X \to GL(n, \mathbb{R})$ is said to be *tempered on Y with respect to f* or simply *tempered on Y* if for every $x \in Y$, we have

$$\lim_{m \to \pm\infty} \frac{1}{m} \log \|L(f^m(x))\| = \lim_{m \to \pm\infty} \frac{1}{m} \log \|L(f^m(x))^{-1}\| = 0.$$

A cocycle over f is said to be *tempered on Y* if its generator is tempered on Y.

If the real functions $x \mapsto \|L(x)\|$ and $x \mapsto \|L(x)^{-1}\|$ are bounded or, more generally, have finite essential supremum, then the function L is tempered with respect to any invertible transformation $f: X \to X$ on any f-invariant nonempty measurable subset $Y \subset X$. The following statement provides a more general criterion for a function L to be tempered.

Proposition 2.1.4. *Let $f: X \to X$ be an invertible transformation preserving a probability measure ν and $L: X \to GL(n, \mathbb{R})$ a measurable function. If*

$$\log \|L\|, \log \|L^{-1}\| \in L^1(X, \nu),$$

then L is tempered on some set of full ν-measure.

Proof. We need the following lemma:

Lemma 2.1.5. *Let $f\colon X \to X$ be a (not necessarily invertible) transformation preserving a probability measure ν. If the function $\varphi\colon X \to \mathbb{R}$ is in $L^1(X, \nu)$, then*

$$\lim_{m\to+\infty} \frac{\varphi(f^m(x))}{m} = 0 \tag{2.3}$$

for ν-almost every $x \in X$.

Proof of the lemma. For each $n \geq 1$ and $\varepsilon > 0$, set

$$A_n^\varepsilon = \{x \in X : |\varphi(f^n(x))| > n\varepsilon\}.$$

Property (2.3) holds if and only if, given $\varepsilon > 0$, we have $|\varphi(f^n(x))| \leq n\varepsilon$ for all sufficiently large n, and hence if and only if for any $\varepsilon > 0$,

$$x \notin \overline{\lim_{n\to+\infty}} A_n^\varepsilon := \bigcap_{n=1}^{\infty} \bigcup_{k\geq n} A_k^\varepsilon.$$

Therefore Property (2.3) holds for ν-almost every $x \in X$ if and only if

$$\nu\left(\bigcap_{k\geq 1}(X \setminus \overline{\lim_{n\to+\infty}} A_n^{1/k})\right) = 1$$

and hence if and only if

$$\nu\left(\overline{\lim_{n\to+\infty}} A_n^\varepsilon\right) = 0 \tag{2.4}$$

for any $\varepsilon > 0$. We wish to prove that (2.4) holds. For each $n \geq 1$ and $\varepsilon > 0$, set

$$B_n^\varepsilon = \{x \in X : n\varepsilon < |\varphi(x)| \leq (n+1)\varepsilon\}.$$

Observe that $B_n^\varepsilon \cap B_m^\varepsilon = \varnothing$ for any $n \neq m$. Since $\varphi \in L^1(X, \nu)$ and ν is f-invariant, we obtain

$$\nu\left(\bigcup_{m\geq n} B_m^\varepsilon\right) = \nu(A_n^\varepsilon).$$

Furthermore,

$$\begin{aligned}
\infty > \int_X |\varphi|\, d\nu &\geq \sum_{n=1}^{\infty} \int_{B_n^\varepsilon} |\varphi|\, d\nu \geq \varepsilon \sum_{n=1}^{\infty} n\nu(B_n^\varepsilon) \\
&= \varepsilon \sum_{n=1}^{\infty} \sum_{m=1}^{n} \nu(B_n^\varepsilon) = \varepsilon \sum_{m=1}^{\infty} \sum_{n=m}^{\infty} \nu(B_n^\varepsilon) \\
&= \varepsilon \sum_{m=1}^{\infty} \nu\left(\bigcup_{n\geq m} B_k^\varepsilon\right) = \varepsilon \sum_{m=1}^{\infty} \nu(A_n^\varepsilon).
\end{aligned}$$

Therefore,

$$\nu\left(\overline{\lim_{n\to+\infty}} A_n^{\varepsilon}\right) \le \nu\left(\bigcup_{k\ge n} A_k^{\varepsilon}\right) \le \sum_{k=n}^{\infty} \nu(A_k^{\varepsilon}) \to 0$$

as $n \to +\infty$, and (2.4) follows. This completes the proof of the lemma. □

The desired statement follows by applying Lemma 2.1.5 to the functions $x \mapsto \log\|L(x)\|$ and $x \mapsto \log\|L(x)^{-1}\|$. □

Let $A, B\colon X \to GL(n, \mathbb{R})$ be the generators of cocycles $\mathcal{A}$ and $\mathcal{B}$ over an invertible measurable transformation f, respectively, and $Y \subset X$ a measurable subset.

Definition 2.1.6. The cocycles $\mathcal{A}$ and $\mathcal{B}$ are said to be *equivalent on* Y or *cohomologous on* Y if there exists a measurable function $L\colon X \to GL(n, \mathbb{R})$, which is tempered on Y, such that for every $x \in Y$, we have

$$A(x) = L(f(x))^{-1} B(x) L(x). \tag{2.5}$$

This is clearly an equivalence relation and if two cocycles $\mathcal{A}$ and $\mathcal{B}$ are equivalent, we write $\mathcal{A} \sim_Y \mathcal{B}$. Equation (2.5) is called *cohomological equation.*

It follows from (2.5) that for any $x \in Y$ and $m \in \mathbb{Z}$,

$$\mathcal{A}(x, m) = L(f^m(x))^{-1} \mathcal{B}(x, m) L(x). \tag{2.6}$$

Proposition 2.1.4 immediately implies the following result.

Corollary 2.1.7. *If* $L\colon X \to \mathbb{R}$ *is a measurable function such that* $\log\|L\|$, $\log\|L^{-1}\| \in L^1(X, \nu)$ *then any two cocycles* $\mathcal{A}$ *and* $\mathcal{B}$ *satisfying* (2.6) *are equivalent cocycles.*

We now consider the notion of equivalence for cocycles over different transformations. Let $f\colon X \to X$ and $g\colon Y \to Y$ be invertible measurable transformations. Assume that f and g are *measurably conjugated*; that is, $h \circ f = g \circ h$ for some invertible measurable transformation $h\colon X \to Y$. Let $\mathcal{A}$ be a cocycle over f and $\mathcal{B}$ a cocycle over g.

Definition 2.1.8. The cocycles $\mathcal{A}$ and $\mathcal{B}$ are said to be *equivalent* if there exists a measurable function $L\colon X \to GL(n, \mathbb{R})$, which is tempered on Y with respect to g, such that for every $x \in Y$, we have

$$A(h^{-1}(x)) = L(g(x))^{-1} B(x) L(x).$$

2.1.3 Examples and Basic Constructions

We describe various examples of measurable cocycles over dynamical systems. Perhaps the simplest one is provided by the cocycles generated by a single matrix. Starting from a given cocycle one can build other cocycles using some basic constructions in ergodic theory and algebra. Thus one obtains power cocycles, induced cocycles, and exterior power cocycles.

Let $\mathcal{A}$ be a measurable cocycle over a measurable transformation $f: X \to X$. We call a cocycle $\mathcal{A}$ *rigid* if it is equivalent to a cocycle whose generator A is a constant map. Rigid cocycles naturally arise in the classical Floquet Theory (where the dynamical system in the base is a periodic flow) and among smooth cocycles over translations on the torus with rotation vector satisfying a Diophantine condition (see [66, 119] and the references therein). In the setting of actions of groups other than $\mathbb{Z}$ and $\mathbb{R}$, rigid cocycles appear in the measurable setting for actions of higher-rank semisimple Lie groups and lattices in such groups (see [240]) and in the smooth setting for hyperbolic actions of higher-rank Abelian groups (see [108, 109]).

Let $\mathcal{A}$ be a cocycle over a measurable transformation f of a Lebesgue space X. For each $m \geq 1$, consider the transformation $f^m: X \to X$ and the measurable cocycle $\mathcal{A}^m$ over f^m with the generator

$$A^m(x) := A(f^{m-1}(x)) \cdots A(x).$$

The cocycle $\mathcal{A}^m$ is called the *mth power cocycle* of $\mathcal{A}$.

Let $f: X \to X$ be a measure preserving transformation of a Lebesgue space (X, ν) and $Y \subset X$ a measurable subset of positive ν-measure. By Poincaré's Recurrence Theorem the set $Z \subset Y$ of points $x \in Y$, such that $f^n(x) \in Y$ for infinitely many positive integers n, has measure $\nu(Z) = \nu(Y)$. We define the transformation $f_Y: Y \to Y$ (mod 0) as follows: for each $x \in Z$, set

$$k_Y(x) = \min\{k \geq 1 : f^k(x) \in Y\} \quad \text{and} \quad f_Y(x) = f^{k_Y(x)}(x).$$

One can easily verify that the function k_Y and the map f_Y are measurable on Z. We call k_Y the *(first) return time* to Y and f_Y the *(first) return map* or *induced transformation* on Y.

Lemma 2.1.9 (see, e.g., [52]).

1. *We have* $\int_Y k_Y \, d\nu = \nu(\bigcup_{n\geq 0} f^n(Y))$ *and* $k_Y \in L^1(X, \nu)$.
2. *The measure* ν *is invariant under* f_Y.

Since $k_Y \in L^1(X, \nu)$, by Birkhoff's Ergodic Theorem, the function

$$\tau_Y(x) = \lim_{k \to +\infty} \frac{1}{k} \sum_{i=0}^{k-1} k_Y(f_Y^i(x)) \tag{2.7}$$

is well defined for ν-almost all $x \in Y$ and $\tau_Y \in L^1(X, \nu)$.

If $\mathcal{A}$ is a measurable linear cocycle over f with generator A, we define the *induced cocycle* $\mathcal{A}_Y$ over f_Y to be the cocycle with the generator

$$A_Y(x) = A^{k_Y(x)}(x). \tag{2.8}$$

Let $\mathcal{A}$ be a cocycle over a measurable transformation f of a Lebesgue space X. For each $k = 1, \ldots, n$, consider the cocycle $\mathcal{A}^{\wedge k}: X \times \mathbb{Z} \to (GL(n, \mathbb{R}))^{\wedge k}$

defined by $\mathcal{A}^{\wedge k}(x, m) = \mathcal{A}(x, m)^{\wedge k}$ (see Section 3.1 for the definition of the exterior power). We call $\mathcal{A}^{\wedge k}$ the *k-fold exterior power cocycle* of $\mathcal{A}$.

2.2 Hyperbolicity of Cocycles

Let $\mathcal{A}$ be a cocycle over an invertible measurable transformation f of a measure space X with the generator $A\colon X \to GL(n, \mathbb{R})$. Note that for each $x \in X$, the cocycle $\mathcal{A}$ generates a sequence of matrices $\{A_m\}_{m\in\mathbb{Z}} = \{A(f^m(x))\}_{m\in\mathbb{Z}}$. Therefore every cocycle can be viewed as a collection of sequences of matrices, which are indexed by the trajectories of f. This allows us to extend the notion of hyperbolicity for sequences of matrices to cocycles.

2.2.1 Hyperbolic Cocycles

Consider a family of inner products $\langle\cdot,\cdot\rangle = \{\langle\cdot,\cdot\rangle_x : x \in X\}$ on $\mathbb{R}^n$. Given $x \in X$, we denote by $\|\cdot\|_x$ the norm and by $\sphericalangle(\cdot,\cdot)_x$ the angle induced by the inner product $\langle\cdot,\cdot\rangle_x$. In order to simplify the notation, we often write $\|\cdot\|$ and $\sphericalangle(\cdot,\cdot)$ omitting the reference point x at which they are computed. Let $Y \subset X$ be an f-invariant nonempty measurable subset. Let also $\lambda, \mu\colon Y \to (0, \infty)$, and $\varepsilon\colon Y \to [0, \varepsilon_0]$ with $\varepsilon_0 > 0$ be measurable functions satisfying

$$\lambda(f(x)) = \lambda(x), \quad \mu(f(x)) = \mu(x), \quad \varepsilon(f(x)) = \varepsilon(x),$$

(i.e., these functions are f-invariant) and

$$\lambda(x)e^{\varepsilon(x)} < \mu(x)e^{-\varepsilon(x)}. \tag{2.9}$$

Definition 2.2.1. We say that the cocycle $\mathcal{A}$ is *nonuniformly partially hyperbolic in the broad sense* (or admits a *$(\lambda(x), \mu(x), \varepsilon(x))$-splitting on* Y) if there exist measurable functions $C, K\colon Y \to (0, \infty)$ such that for every $x \in Y$, the sequence of matrices $\mathcal{A}(x) = \{\mathcal{A}(x, m)\}_{m\in\mathbb{Z}}$ admits a $(\lambda, \mu, \varepsilon)$-splitting with

$$\lambda = \lambda(x), \quad \mu = \mu(x), \quad \varepsilon = \varepsilon(x), \quad c_m = C(f^m(x)), \quad \gamma_m = K(f^m(x))$$

(see Section 1.2.1). In other words,

1. there exists a decomposition $\mathbb{R}^n = E^1(x) \oplus E^2(x)$, depending measurably on $x \in Y$, such that $A(x)E^1(x) = E^1(f(x))$ and $A(x)E^2(x) = E^2(f(x))$;
2. (a) for $v \in E^1(x)$ and $m > 0$,

$$\|\mathcal{A}(x, m)v\| \le C(x)\lambda(x)^m\, e^{\varepsilon(x)m}\,\|v\|; \tag{2.10}$$

   (b) for $v \in E^2(x)$ and $m < 0$,

$$\|\mathcal{A}(x, m)v\| \le C(x)\mu(x)^m\, e^{-\varepsilon(x)m}\,\|v\|; \tag{2.11}$$

   (c) $\sphericalangle(E^1(x), E^2(x)) \ge K(x)$;
   (d) for $m \in \mathbb{Z}$,

$$C(f^m(x)) \le C(x)e^{\varepsilon(x)|m|}, \quad K(f^m(x)) \ge K(x)e^{-\varepsilon(x)|m|}. \tag{2.12}$$

Note that the dimensions of E^1 and E^2 are measurable f-invariant functions and hence the set Y can be decomposed into finitely many disjoint invariant measurable subsets on which the dimensions of E^1 and E^2 are constant. In view of this observation, we may assume whenever necessary that the dimensions of the subspaces E^1 and E^2 are constant on the whole set Y.

Later in the book we will consider cocycles, which are continuous or even smooth. However, even for such cocycles, one should expect the functions λ, μ, ε, C, and K to be only measurable.

Proposition 2.2.2. *If the cocycle $\mathcal{A}\colon X \times \mathbb{Z} \to GL(n, \mathbb{R})$ over f is nonuniformly partially hyperbolic in the broad sense, then for every $x \in Y$, we have*

1. $\mathcal{A}(x, m)E^1(x) = E^1(f^m(x))$ *and* $\mathcal{A}(x, m)E^2(x) = E^2(f^m(x))$;
2. *for* $v \in E^1(x)$ *and* $m < 0$,
$$\|\mathcal{A}(x, m)v\| \geq C(f^m(x))^{-1}\lambda(x)^m\, e^{\varepsilon(x)m}\,\|v\|; \tag{2.13}$$
3. *for* $v \in E^2(x)$ *and* $m > 0$,
$$\|\mathcal{A}(x, m)v\| \geq C(f^m(x))^{-1}\mu(x)^m\, e^{-\varepsilon(x)m}\,\|v\|. \tag{2.14}$$

Proof. The first statement follows immediately from Condition 1 in Definition 2.2.1. Using (2.10) we obtain

$$\|v\| \geq C(f^{-m}(x))^{-1}\lambda(x)^{-m}\, e^{-\varepsilon(x)m}\,\|\mathcal{A}(f^{-m}(x), m)v\|$$

for every $v \in E^1(f^{-m}(x))$ and $m > 0$. Set $w = \mathcal{A}(x, -m)^{-1}v$. By Statement 1, we have $v \in E^1(f^{-m}(x))$ if and only if $w \in E^1(x)$. Hence,

$$\|\mathcal{A}(x, -m)w\| \geq C(f^{-m}(x))^{-1}\lambda(x)^{-m}\, e^{-\varepsilon(x)m}\,\|w\|$$

for every $w \in E^1(x)$ and $m > 0$. Statement 2 follows. The last statement can be proved in a similar fashion using (2.11). □

Definition 2.2.3. We say that the cocycle $\mathcal{A}$ is *nonuniformly (completely) hyperbolic on Y* if it is nonuniformly partially hyperbolic in the broad sense on Y and

$$\lambda(x)e^{\varepsilon(x)} < 1 < \mu(x)e^{-\varepsilon(x)}.$$

We now introduce the notion of uniform hyperbolicity. Let $0 < \lambda < \mu$ be some numbers and $Y \subset X$ a measurable subset. We stress that Y need not be invariant.

Definition 2.2.4. A cocycle $\mathcal{A}$ is said to be *uniformly hyperbolic* on a set Y if there exist $c > 0$ and $\gamma > 0$ such that for every $x \in Y$,

1. there exists a decomposition $\mathbb{R}^n = E^1(x) \oplus E^2(x)$, depending measurably on $x \in Y$, such that $A(x)E^1(x) = E^1(f(x))$ and $A(x)E^2(x) = E^2(f(x))$ whenever $f(x) \in Y$;
2. (a) for $v \in E^1(x)$ and $m > 0$,
$$\|\mathcal{A}(x, m)v\| \leq c\lambda^m\,\|v\|;$$

(b) for $v \in E^2(x)$ and $m < 0$,

$$\|\mathcal{A}(x, m)v\| \leq c\mu^m \|v\|;$$

(c) $\measuredangle(E^1(x), E^2(x)) \geq \gamma$.

For a cocycle, which is nonuniformly partially hyperbolic in the broad sense on Y, we set

$$C_Y := \sup_{x \in Y} C(x) \quad \text{and} \quad K_Y := \inf_{x \in Y} K(x).$$

Clearly, $C_Y \leq \infty$ and $K_Y \geq 0$.

Proposition 2.2.5. *Let $0 < \lambda < \mu < \infty$ be constants. A cocycle $\mathcal{A}$ is uniformly hyperbolic on an f-invariant subset Y if and only if it admits a $(\lambda, \mu, 0)$-splitting on Y for which $C_Y < \infty$ and $K_Y > 0$.*

Proof. Clearly, a uniformly hyperbolic cocycle $\mathcal{A}$ on Y admits a $(\lambda, \mu, 0)$-splitting on Y with $C(x, \delta) = c$ and $K(x, \delta) = \gamma$.

On the other hand, a cocycle admitting a $(\lambda, \mu, 0)$-splitting on Y with $C_Y < \infty$ and $K_Y > 0$ is uniformly hyperbolic on Y with $c = C_Y$ and $\gamma = K_Y$. □

One can show that if a cocycle admits a $(\lambda, \mu, 0)$-splitting on Y, then for every $x \in Y$ and $k \in \mathbb{Z}$,

$$C(f^k(x)) = C(x) \quad \text{and} \quad K(f^k(x)) = K(x).$$

The principal example of uniformly hyperbolic cocycles are the derivative cocycles generated by Anosov diffeomorphisms (see Sections 5.1 and 6.1.1 below) and more generally Axiom A diffeomorphisms. The principal examples of nonuniformly hyperbolic cocycles are cocycles with nonzero Lyapunov exponents (see Section 5.4).

We will show below that a nonuniformly hyperbolic cocycle on a set Y of full measure (with respect to an invariant measure) is, in fact, uniformly hyperbolic on a (noninvariant) set $Y_\delta \subset Y$ of measure at least $1 - \delta$ for arbitrarily small $\delta > 0$. This observation is crucial in studying topological and measure-theoretic properties of such cocycles. However, the "parameters" of uniform hyperbolicity, that is, the numbers c and γ may vary with δ approaching ∞ and 0, respectively. We will then show a crucial fact: this can occur only with a subexponential rate. We proceed with the formal description.

2.2.2 Regular Sets of Hyperbolic Cocycles

Let $\mathcal{A}$ be a cocycle over an invertible measurable transformation f of a measure space X and let λ, μ, and ε be positive numbers, satisfying

$$0 < \lambda e^{\varepsilon} < \mu e^{-\varepsilon}, \quad \lambda < 1. \tag{2.15}$$

Given an integer j, $1 \le j < n$, and $\ell \ge 1$, we denote by $\Lambda^{\ell}_{\lambda\mu\varepsilon j}$ the set of points $x \in X$ for which there exists a decomposition $\mathbb{R}^n = E^1_x \oplus E^2_x$ such that for every $k \in \mathbb{Z}$ and $m > 0$, the following properties hold:

1. $\dim E^1_x = j$ (and hence, $\dim E^2_x = n - j$).
2. If $v \in \mathcal{A}(x,k)E^1_x$ then

$$\|\mathcal{A}(f^k(x), m)v\| \le \ell\lambda^m \, e^{\varepsilon(m+|k|)}\|v\| \tag{2.16}$$

and

$$\|\mathcal{A}(f^k(x), -m)v\| \ge \ell^{-1}\lambda^{-m} \, e^{-\varepsilon(|k-m|+m)}\|v\|. \tag{2.17}$$

3. If $v \in \mathcal{A}(x,k)E^2_x$ then

$$\|\mathcal{A}(f^k(x), -m)v\| \le \ell\mu^{-m} \, e^{\varepsilon(m+|k|)}\|v\| \tag{2.18}$$

and

$$\|\mathcal{A}(f^k(x), m)v\| \ge \ell^{-1}\mu^{m} \, e^{-\varepsilon(|k+m|+m)}\|v\|. \tag{2.19}$$

4.

$$\sphericalangle(\mathcal{A}(x,k)E^1_x, \mathcal{A}(x,k)E^2_x) \ge \ell^{-1} \, e^{-\varepsilon|k|}. \tag{2.20}$$

Definition 2.2.6. The set $\Lambda^{\ell}_{\lambda\mu\varepsilon j}$ is called a *regular set* (or a *Pesin set*).

We also introduce the *level set*

$$\Lambda_{\lambda\mu\varepsilon j} := \bigcup_{\ell \ge 1} \Lambda^{\ell}_{\lambda\mu\varepsilon j}. \tag{2.21}$$

It is easy to see that these sets have the following properties.

Proposition 2.2.7. *The following statements hold:*

1. $\Lambda^{\ell}_{\lambda\mu\varepsilon j} \subset \Lambda^{\ell+1}_{\lambda\mu\varepsilon j}$*; that is, regular sets are nested.*
2. *If* $m \in \mathbb{Z}$ *then* $f^m(\Lambda^{\ell}_{\lambda\mu\varepsilon j}) \subset \Lambda^{\ell'}_{\lambda\mu\varepsilon j}$, *where* $\ell' = \ell \exp(|m|\varepsilon)$.
3. *The set* $\Lambda_{\lambda\mu\varepsilon j}$ *is* f*-invariant.*

By Property 2, if $\varepsilon < \log(1 + 1/\ell)$ then

$$\Lambda^{\ell-1}_{\lambda\mu\varepsilon j} \subset f(\Lambda^{\ell}_{\lambda\mu\varepsilon j}) \subset \Lambda^{\ell+1}_{\lambda\mu\varepsilon j} \quad \text{and} \quad \Lambda^{\ell-1}_{\lambda\mu\varepsilon j} \subset f^{-1}(\Lambda^{\ell}_{\lambda\mu\varepsilon j}) \subset \Lambda^{\ell+1}_{\lambda\mu\varepsilon j}.$$

Consider now a cocycle $\mathcal{A}$, which admits a $(\lambda(x), \mu(x), \varepsilon(x))$-splitting on a subset $Y \subset X$ (where $\lambda(x)$, $\mu(x)$, and $\varepsilon(x)$ are f-invariant measurable functions on Y). Given positive numbers λ, μ, and ε satisfying (2.15) and an integer j, $1 \le j < n$, consider the measurable set

$$Y_{\lambda\mu\varepsilon j} = \{x \in Y : \lambda(x) \le \lambda < \mu \le \mu(x), \ \varepsilon(x) \le \varepsilon, \ E^1(x) = j\}.$$

Clearly, $Y_{\lambda\mu\varepsilon j}$ is invariant under f and is not empty if the numbers λ, μ, ε, and j are chosen appropriately. For each integer $\ell \ge 1$, consider the measurable subset

$$Y_{\lambda\mu\varepsilon j}^{\ell} = \{x \in Y_{\lambda\mu\varepsilon j} : C(x) \le \ell,\ K(x) \ge \ell^{-1}\}.$$

We have $Y_{\lambda\mu\varepsilon j}^{\ell} \subset Y_{\lambda\mu\varepsilon j}^{\ell+1}$. Note that the set $Y_{\lambda\mu\varepsilon j}^{\ell}$ need not be f-invariant.

Proposition 2.2.8. *The following properties hold:*

1. $Y_{\lambda\mu\varepsilon j}^{\ell} \subset \Lambda_{\lambda\mu\varepsilon j}^{\ell}$ *for every* $\ell \ge 1$.
2. $E_x^1 = E^1(x)$ *and* $E_x^2 = E^2(x)$ *for every* $x \in Y$.

Proof. By (2.10), if $x \in Y_{\lambda\mu\varepsilon j}^{\ell}$, $v \in E^1(f^k(x))$, $k \in \mathbb{Z}$, and $m > 0$ then

$$\begin{aligned} \|\mathcal{A}(f^k(x), m)v\| &\le C(x)e^{\varepsilon(x)|k|}\lambda(x)^m\, e^{\varepsilon(x)m}\,\|v\| \\ &\le \ell\, e^{\varepsilon(m+|k|)}\lambda^m\,\|v\|, \end{aligned} \tag{2.22}$$

and by (2.13),

$$\begin{aligned} \|\mathcal{A}(f^k(x), -m)v\| &\ge C(f^{k-m}(x))^{-1}\lambda(x)^{-m}\, e^{-\varepsilon(x)m}\,\|v\| \\ &\ge \ell^{-1}\lambda^{-m}\, e^{-\varepsilon(|k-m|+m)}\,\|v\|. \end{aligned} \tag{2.23}$$

Hence, (2.16) and (2.17) hold. Similarly, using (2.11) and (2.14), one can show that (2.18) and (2.19) hold. Furthermore,

$$\sphericalangle(E^1(f^k(x)), E^2(f^k(x))) \ge K(x)e^{-\varepsilon(x)|k|} \ge \ell^{-1}\, e^{-\varepsilon|k|}.$$

This implies that $x \in \Lambda_{\lambda\mu\varepsilon j}^{\ell}$, and hence $Y_{\lambda\mu\varepsilon j}^{\ell} \subset \Lambda_{\lambda\mu\varepsilon j}^{\ell}$.

The inequalities (2.22) and (2.23) show that $E^1(x) \subset E_x^1$. Similarly, one can show that $E^2(x) \subset E_x^2$. Since

$$\dim E^1(x) + \dim E^2(x) = \dim E_x^1 + \dim E_x^2 = n,$$

we conclude that $E^1(x) = E_x^1$ and $E^2(x) = E_x^2$ for every $x \in Y_{\lambda\mu\varepsilon j}^{\ell}$. The desired result follows. □

It follows that every cocycle $\mathcal{A}$ admitting a $(\lambda(x), \mu(x), \varepsilon(x))$-splitting possesses the family of nonempty level sets $\{\Lambda_{\lambda\mu\varepsilon j} : \lambda, \mu, \varepsilon$ satisfy (2.15)$\}$ and for each λ, μ, and ε, the collection of nonempty regular sets $\{\Lambda_{\lambda\mu\varepsilon j}^{\ell} : \ell \ge 1\}$. Note that $\mathcal{A}$ is nonuniformly partially hyperbolic in the broad sense on each level set, as well as on the set $\Lambda = \bigcup \Lambda_{\lambda\mu\varepsilon j}$ (where the union is taken over all numbers λ, μ, and ε satisfying (2.15)) that can be viewed as an "extension" of the set Y. We stress that the rates of exponential contraction $\lambda(x)$ and of exponential contraction or expansion $\mu(x)$ are uniformly bounded away from 1 on each level set $\Lambda_{\lambda\mu\varepsilon j}$ but may be arbitrarily close to 1 on Λ. It is easy to see that there is a countable collection of level sets $\{\Lambda_{\lambda_i\mu_i\varepsilon_i j} : i \ge 1\}$ such that $\Lambda = \bigcup_{i=1} \Lambda_{\lambda_i\mu_i\varepsilon_i j}$. We say that this collection and the corresponding collection of the level sets $\{\Lambda_{\lambda_i\mu_i\varepsilon_i j}^{\ell} : \ell \ge 1\}$ are *associated with the cocycle* $\mathcal{A}$.

2.2.3 Cocycles over Topological Spaces

We consider a cocycle $\mathcal{A}$ over a topological space X and assume that $\mathcal{A}$ admits a $(\lambda(x), \mu(x), \varepsilon(x))$-splitting on a Borel subset $Y \subset X$.

Proposition 2.2.9. *Assume that the inner products $\langle \cdot, \cdot \rangle_x$ vary continuously with $x \in X$ and that $\mathcal{A}$ is continuous. Then the following statements hold:*

1. *The regular sets $\Lambda^\ell = \Lambda^\ell_{\lambda\mu\varepsilon j}$ are closed.*
2. *The subspaces E^1_x and E^2_x vary continuously with $x \in \Lambda^\ell$ (with respect to the distance in the Grassmannian bundle).*

Proof. The proof of both statements is a modification of the standard argument for Anosov maps. Fix $\ell \geq 1$ and consider a sequence of points $x_n \in \Lambda^\ell$, which converges to a point x. We have to show that $x \in \Lambda^\ell$ and that the subspaces $E^1_{x_n}$ and $E^2_{x_n}$ converge respectively to E^1_x and E^2_x. Passing to a subsequence if necessary we may assume that the dimension of $E^1_{x_n}$ is constant and hence so is the dimension of $E^2_{x_n}$. Passing again to a subsequence if necessary we may also assume that the subspaces $E^1_{x_n}$ converge to a subspace $F^1 \subset \mathbb{R}^n$ and that the subspaces $E^2_{x_n}$ converge to a subspace $F^2 \subset \mathbb{R}^n$. Clearly, the inequalities (2.16) and (2.17) hold for every $v \in \mathcal{A}(x, k)E^1_x$ with $k \in \mathbb{Z}$ and $m > 0$. Similarly, the inequalities (2.18) and (2.19) hold for every $v \in \mathcal{A}(x, k)E^2_x$ with $k \in \mathbb{Z}$ and $m > 0$, and the inequality (2.20) holds as well. Hence, $x \in \Lambda^\ell$, and $F^1 = E^1_x$ and $F^2 = E^2_x$. The desired result follows. □

It follows from this proposition that the dimensions of the subspaces E^1_x and E^2_x vary continuously with $x \in \Lambda^\ell$. Hence, they are locally constant but they may not be constant.

2.3 Lyapunov Exponents for Cocycles

In this section we extend the notion of Lyapunov exponents for cocycles over dynamical systems that is one of the main concepts of the book. We essentially generalize the definitions given in Chapter 1 for sequences of matrices to cocycles.

As we saw before, a cocycle $\mathcal{A}$ over an invertible transformation f of a measure space X generates a sequence of matrices $\{A(f^m(x))\}_{m\in\mathbb{Z}}$, where A is the generator. One can associate to each of these sequences of matrices a Lyapunov exponent (see Section 1.3).

However, one should carefully examine the dependence of the Lyapunov exponent when one moves from a sequence of matrices to another one (see Propositions 2.3.3–2.3.5) – this is what constitutes a substantial difference in studying cocycles over dynamical systems and sequences of matrices (as will be seen in Chapter 3). As in the case of sequences of matrices, the Lyapunov exponent is a main tool in studying nonuniform hyperbolicity for cocycles. We now proceed with the formal definition of the Lyapunov exponent for cocycles.

Definition 2.3.1. Given a point $(x, v) \in X \times \mathbb{R}^n$, define the *forward Lyapunov exponent of* (x, v) *(with respect to the cocycle* $\mathcal{A}$*)* by

$$\chi^+(x, v) = \chi^+(x, v, \mathcal{A}) = \overline{\lim_{m\to+\infty}} \frac{1}{m} \log\|\mathcal{A}(x, m)v\|. \tag{2.24}$$

We note that the number $\chi^+(x, v)$ does not depend on the inner product on $\mathbb{R}^n$. With the convention $\log 0 = -\infty$, we obtain $\chi^+(x, 0) = -\infty$ for every $x \in X$.

Observe that for each $x \in X$, the formula (2.24) produces the same function $\chi^+(x, \cdot)$ as the formula (1.5) for the sequence of matrices $\mathcal{A}_m = \mathcal{A}(x, m)$. We therefore obtain the following statement.

Proposition 2.3.2. *For every* $x \in X$*, the function* $\chi^+(x, \cdot)\colon \mathbb{R}^n \to \mathbb{R} \cup \{-\infty\}$ *is a Lyapunov exponent; that is, it has the properties described in Proposition 1.3.1.*

Fix $x \in X$. Applying Propositions 2.3.2 and 1.3.2, we conclude that there exist a positive integer $p^+(x) \le n$, a collection of numbers

$$\chi_1^+(x) < \chi_2^+(x) < \cdots < \chi_{p^+(x)}^+(x),$$

and a filtration $\mathcal{V}_x^+$ of linear subspaces

$$\{0\} = E_0^+(x) \subsetneqq E_1^+(x) \subsetneqq \cdots \subsetneqq E_{p^+(x)}^+(x) = \mathbb{R}^n, \tag{2.25}$$

such that

1. $E_i^+(x) = \{v \in \mathbb{R}^n : \chi^+(x, v) \le \chi_i^+(x)\}$;
2. $\chi^+(x, v) = \chi_i^+(x)$ for $v \in E_i^+(x) \setminus E_{i-1}^+(x)$.

The numbers $\chi_i^+(x)$ are called the *values* of the Lyapunov exponent χ^+ at x. We refer to the collection of linear spaces $E_i^+(x, \mathcal{A}) = E_i^+(x)$ as the *filtration* $\mathcal{V}_x^+$ of $\mathbb{R}^n$ associated to χ^+ at x. The number

$$k_i^+(x) = \dim E_i^+(x) - \dim E_{i-1}^+(x)$$

is called the *multiplicity* of the value $\chi_i^+(x)$. We also write

$$n_i^+(x) := \dim E_i^+(x) = \sum_{j=1}^{i} k_j^+(x).$$

We define the *Lyapunov spectrum* of χ^+ at x as the collection of pairs

$$\operatorname{Sp}_x^+ \mathcal{A} = \{(\chi_i^+(x), k_i^+(x))\colon i = 1, \dots, p^+(x)\}.$$

Observe that $k_i^+(f(x)) = k_i^+(x)$ for every $x \in X$ and $1 \le i \le p^+(x)$, and hence $\operatorname{Sp}_{f(x)}^+ \mathcal{A} = \operatorname{Sp}_x^+ \mathcal{A}$ for any $x \in X$.

Proposition 2.3.3. *The following properties hold:*

1. *The functions* χ^+ *and* p^+ *are Borel measurable.*
2. $\chi^+ \circ F = \chi^+$ *and* $p^+ \circ f = p^+$*, where* F *is the linear extension given by* (2.2).

3. *$A(x)E_i^+(x) = E_i^+(f(x))$ and $\chi_i^+(f(x)) = \chi_i^+(x)$ for every $x \in X$ and $1 \le i \le p^+(x)$.*

Proof. We have

$$(\chi^+ \circ F)(x, v) = \overline{\lim_{m\to+\infty}} \frac{1}{m} \log\|\mathcal{A}(f(x), m)A(x)v\|$$
$$= \overline{\lim_{m\to+\infty}} \frac{1}{m} \log\|\mathcal{A}(x, m+1)v\| = \chi^+(x, v),$$

and hence $\chi^+ \circ F = \chi^+$. Observe now that

$$A(x)E_i^+(x) = \{A(x)v \in \mathbb{R}^n : \chi^+(f(x), A(x)v) \le \chi_i^+(x)\}$$
$$= \{w \in \mathbb{R}^n : \chi^+(f(x), w) \le \chi_i^+(x)\}.$$

If $w \in A(x)E_i^+(x) \setminus A(x)E_{i-1}^+(x)$ then $\chi^+(f(x), w) = \chi_i^+(x)$. This implies that

1. $p^+(f(x)) = p^+(x)$;
2. $\chi_i^+(f(x)) = \chi_i^+(x)$ for every $1 \le i \le p^+(x)$;
3. $A(x)E_i^+(x) = E_i^+(f(x))$ for every $1 \le i \le p^+(x)$.

The Borel measurability of the function χ^+ follows immediately from (2.24). This also implies that the function p^+ is Borel measurable. □

We now discuss the measurability of the functions $\chi_i^+(x)$ and $E_i^+(x)$. Note that the former takes on values in the real line $\mathbb{R}$, while the latter attains values in the Grassmannian bundle of subspaces of $\mathbb{R}^n$. We introduce the values of the Lyapunov exponent χ^+ at a point x counted with their multiplicities

$$\chi_1'(x) \le \cdots \le \chi_n'(x)$$

and the spaces

$$E_1'(x) \subseteq \cdots \subseteq E_n'(x)$$

such that $\chi_i'(x) = \chi_j^+(x)$ and $E_i'(x) = E_j^+(x)$ for each $i = n_{j-1}^+(x) + 1, \ldots, n_j^+(x)$ and each $j = 1, \ldots, p^+(x)$ (where we assume that $n_0^+(x) = 0$). We note that for each $i = 1, \ldots, n$,

$$E_i'(x) = \{v \in \mathbb{R}^n : \chi^+(x, v) \le \chi_i'(x)\}. \tag{2.26}$$

Proposition 2.3.4. *The functions $\chi_i'(x)$ and subspaces $E_i'(x)$ are Borel measurable on x for $i = 1, \ldots, n$.*

Proof. Note that $\chi_i'(x) = \min \max\{\chi^+(x, v) : v \in R_i\}$, where the minimum is taken over all subspaces $R_i \subset \mathbb{R}^n$ such that $\dim R_i = i$. This implies that χ_i' is Borel measurable. The Borel measurability of the subspaces E_i' follows immediately from (2.26). □

Let ν be an ergodic f-invariant Borel measure. There exist numbers $p = p^\nu$, $\chi_i = \chi_i^\nu$, and $k_i = k_i^\nu$ for $i = 1, \ldots, p$ such that

$$p(x) = p, \quad \chi_i(x) = \chi_i, \quad k_i(x) = k_i$$

for ν-almost every x. The collection of pairs

$$\operatorname{Sp}\chi(\nu) = \{(\chi_i, k_i) : 1 \le i \le p\}$$

is called the *Lyapunov spectrum* of the measure ν.

For every $(x, v) \in X \times \mathbb{R}^n$, we set

$$\chi^-(x, v) = \chi^-(x, v, \mathcal{A}) = \overline{\lim_{m\to-\infty}} \frac{1}{|m|} \log\|\mathcal{A}(x, m)v\|.$$

We call $\chi^-(x, v)$ the *backward Lyapunov exponent of* (x, v) (with respect to the cocycle $\mathcal{A}$). One can show that for every $x \in X$, there exist a positive integer $p^-(x) \le n$, the *values*

$$\chi_1^-(x) > \chi_2^-(x) > \cdots > \chi_{p^-(x)}^-(x),$$

and the *filtration* $\mathcal{V}_x^-$ of $\mathbb{R}^n$ associated to χ^- at x,

$$\mathbb{R}^n = E_1^-(x) \supsetneq \cdots \supsetneq E_{p^-(x)}^-(x) \supsetneq E_{p^-(x)+1}^-(x) = \{0\}, \tag{2.27}$$

such that $E_i^-(x) = \{v \in \mathbb{R}^n : \chi^-(x, v) \le \chi_i^-(x)\}$. The number

$$k_i^-(x) = \dim E_i^-(x) - \dim E_{i+1}^-(x)$$

is called the *multiplicity* of the value $\chi_i^-(x)$. We define the *Lyapunov spectrum* of χ^- at x by

$$\operatorname{Sp}_x^- \mathcal{A} = \{(\chi_i^-(x), k_i^-(x)) \colon i = 1, \ldots, p^-(x)\}.$$

Proposition 2.3.5. *Let $Y \subset X$ be an f-invariant nonempty measurable subset. The following statements hold:*

1. *If $\mathcal{A}$ is a cocycle, which is nonuniformly partially hyperbolic in the broad sense on Y, then*
$$Y \subset \left\{x \in X : \chi^+(x, v) \ne 0 \textit{ for some } v \in \mathbb{R}^n\right\}.$$
2. *If $\mathcal{A}$ is a nonuniformly hyperbolic cocycle on Y, then*
$$Y \subset \left\{x \in X : \chi^+(x, v) \ne 0 \textit{ for all } v \in \mathbb{R}^n\right\}.$$

Proof. Without loss of generality, we assume that $\lambda(x)e^{\varepsilon(x)} < 1$ for every $x \in X$. If $v \in E^s(x)$ then by (2.10),

$$\chi^+(x, v) = \overline{\lim_{m\to+\infty}} \frac{1}{m} \log\|\mathcal{A}(x, m)v\| \le \log\left(\lambda(x)e^{\varepsilon(x)}\right) < 0.$$

Furthermore, whenever $\mu(x)e^{-\varepsilon(x)} > 1$ for $x \in X$, we obtain by (2.14), that for $v \in E^u(x)$,

$$\chi^+(x, v) \ge \underline{\lim_{m\to+\infty}} \frac{1}{m} \log\|\mathcal{A}(x, m)v\| \ge \log\left(\mu(x)e^{-\varepsilon(x)}\right) > 0.$$

These inequalities imply the desired statements. □

The Lyapunov exponents, its spectrum, and associated filtrations are invariant under a tempered coordinate change. Let Y be a measurable subset in a measure space X.

Proposition 2.3.6. *Let $\mathcal{A}$ and $\mathcal{B}$ be equivalent cocycles on Y over the measurable transformation $f\colon X \to X$ and let $L: X \to GL(n, \mathbb{R})$ be any measurable function satisfying* (2.5) *that is tempered on Y. For any $x \in Y$,*

1. *the forward and backward Lyapunov spectra at x coincide; that is,*
$$\mathrm{Sp}_x^+ \mathcal{A} = \mathrm{Sp}_x^+ \mathcal{B} \quad \textit{and} \quad \mathrm{Sp}_x^- \mathcal{A} = \mathrm{Sp}_x^- \mathcal{B};$$
2. *$L(x)$ preserves the forward and backward filtrations of $\mathcal{A}$ and $\mathcal{B}$; that is, for each $i = 1, \ldots, p^+(x)$,*
$$L(x)E_i^+(x, \mathcal{A}) = E_i^+(x, \mathcal{B}),$$
and for each $i = 1, \ldots, p^-(x)$,
$$L(x)E_i^-(x, \mathcal{A}) = E_i^-(x, \mathcal{B}).$$

Proof. Since the function L satisfies (2.6), for each $v \in \mathbb{R}^n$, we obtain

$$\varlimsup_{m\to\pm\infty} \frac{1}{|m|} \log\|\mathcal{A}(x, m)v\| \le \varlimsup_{m\to\pm\infty} \frac{1}{|m|} \log\|L(f^m(x))^{-1}\| + \varlimsup_{m\to\pm\infty} \frac{1}{|m|} \|\mathcal{B}(x, m)L(x)v\|.$$

As L is tempered, this implies that

$$\chi^+(x, v, \mathcal{A}) \le \chi^+(x, L(x)v, \mathcal{B}) \quad \text{and} \quad \chi^-(x, v, \mathcal{A}) \le \chi^-(x, L(x)v, \mathcal{B}).$$

Writing $\mathcal{B}(x, m)L(x) = L(f^m(x))\mathcal{A}(x, m)$, we conclude in a similar way that

$$\chi^+(x, L(x)v, \mathcal{B}) \le \chi^+(x, v, \mathcal{A}) \quad \text{and} \quad \chi^-(x, L(x)v, \mathcal{B}) \le \chi^-(x, v, \mathcal{A}).$$

The desired result now follows. □

Thus for a rigid cocycle the Lyapunov spectrum is the same as for the constant coefficient cocycle generated by a single matrix. Namely, the Lyapunov spectrum consists of logarithms of absolute values of the eigenvalues of the generator. The multiplicities are sums of the multiplicities of the corresponding eigenvalues.

2.4 Spaces of Cocycles

Let $\mathcal{A}$ be a cocycle over a measurable transformation f of a Lebesgue space X and $A\colon X \to GL(n, \mathbb{R})$ its generator. We shall consider the space of all such cocycles that can be identified with the space of all functions $A\colon X \to GL(n, \mathbb{R})$, and we endow it with various structures specified by the structure of the base transformation:

1. X is a measure space and μ an f-invariant Lebesgue measure in X; the space of cocycles is $L^p(X, GL(n,\mathbb{R}))$, $p \geq 1$.
2. X is a compact topological space and f a homeomorphism; the space of cocycles is $C(X, GL(n,\mathbb{R}))$.
3. X is a compact metric space and f a Hölder or Lipschitz homeomorphism; the space of cocycles is $C^\nu(X, GL(n,\mathbb{R}))$, $0 < \nu \leq 1$.
4. X is a smooth compact Riemannian manifold and f a diffeomorphism; the space of cocycles is $C^r(X, GL(n,\mathbb{R}))$, $r \geq 1$.

One can also consider cocycles with values in a subgroup $G \subset GL(n,\mathbb{R})$ (an important case is $G = SL(n,\mathbb{R})$) and introduce the corresponding spaces of cocycles.

It is an open problem whether the set of cocycles, whose associated Lyapunov exponent has all or some nonzero values, is residual (open and dense) in the corresponding space of cocycles. Some results have been obtained for cocycles with values in the group $SL(2,\mathbb{R})$ (see [11, 83, 237]). We present one of the most general results in this direction due to Viana [221] that provides a partial answer to the problem. We consider the case when X is a smooth compact Riemannian manifold and $f\colon X \to X$ a volume preserving C^r Anosov diffeomorphism (see Section 6.1.1 below for the definition of Anosov diffeomorphisms).

Theorem 2.4.1. *For every $\nu \in (0, r]$, there exists an open and dense subset $\mathfrak{U}$ of $C^\nu(X, SL(n,\mathbb{R}))$ such that every $A \in \mathfrak{U}$ has largest Lyapunov exponent $\chi(A, m) > 0$ where m is the volume.*

This result can be generalized to the case when f is a C^r diffeomorphism preserving a hyperbolic measure μ with local direct product structure (see Section 14.2 for the notion of measures with local direct product structure).

We briefly discuss the proof of the above theorem assuming that $n = 2$ and $\nu = 1$. Let us call a function $A\colon X \to SL(2,\mathbb{R})$ *bundle free* if given $n \geq 1$, there exists no Lipschitz continuous map $\psi\colon x \to \{v_1(x), \dots, v_n(x)\}$, $v_i(x) \in \mathbb{RP}^1$, which is invariant under the cocycle generated by A; that is, for any $x \in X$,

$$A(x)(\{v_1(x), \dots, v_n(x)\}) = \{v_1(f(x)), \dots, v_n(f(x))\}.$$

For $n = 1$, this means that the cocycle has no invariant Lipschitz subbundles.

Lemma 2.4.2. *Suppose that $A \in C^1(X, SL(2,\mathbb{R}))$ is bundle free and there is a periodic point $p \in X$ for f such that A is hyperbolic over the orbit of p. Then $\chi(A, m) > 0$.*

The condition that "there is a periodic point $p \in X$ for f such that A is hyperbolic over the orbit of p" is satisfied for an open and dense subset of $C^1(X, SL(2,\mathbb{R}))$.

To illustrate this statement consider the case when $X = S^1$, $f(x) = ka \pmod 1$ for some $k \geq 2$, and $A\colon S^1 \to SL(2, \mathbb{R})$ given by

$$A(x) = \begin{pmatrix} \beta(x) & 0 \\ 0 & 1/\beta(x) \end{pmatrix}$$

for some function $\beta(x)$ such that $\int_{S^1} \log \beta(x)\, dm(x) = 0$. It is easy to arrange that the set $\beta^{-1}(\mathrm{Id})$ is finite and does not contain $x = 0$. Then A is hyperbolic but $\chi(A, m) = 0$. Observe that A is *not* bundle free.

The statement of Theorem 2.4.1 is now an easy corollary of the following result.

Lemma 2.4.3. *Suppose that* $A \in C^1(X, SL(2, \mathbb{R}))$ *has* $\chi(A, m) = 0$. *Then for any* $\varepsilon > 0$, *there exists a function* $B \in C^1(X, SL(2, \mathbb{R}))$ *such that*

1. *B is ε-close to A in $C^1(X, SL(2, \mathbb{R}))$;*
2. *B is stably bundle free; that is, any sufficiently small perturbation of B in $C^1(X, SL(2, \mathbb{R}))$ is bundle free.*

3

Regularity of Cocycles

In this chapter we introduce the crucial concept of regularity for cocycles over dynamical systems and we present one of our principal results – the Multiplicative Ergodic Theorem. It asserts that regularity of cocycles is a typical property in the measure-theoretic sense. We discuss various versions of this theorem as well as some of its manifestations including the Reduction Theorems. The latter describe normal forms of cocycles according to their Lyapunov exponents. The regularity of cocycles will also be used to deduce nonuniform hyperbolicity from absence of zero Lyapunov exponents.

3.1 The Lyapunov–Perron regularity

Let $\mathcal{A}$ be a cocycle over an invertible measurable transformation f of a measurable space X. Given $x \in X$, the cocycle $\mathcal{A}$ generates the sequence of matrices $\{A_m\}_{m\in\mathbb{Z}} = \{A(f^m(x))\}_{m\in\mathbb{Z}}$.

Definition 3.1.1. We say that x is a *forward* (respectively, *backward*) regular point for $\mathcal{A}$ if the sequence of matrices $\{A(f^m(x))\}_{m\in\mathbb{Z}}$ is forward (respectively, backward) regular (see Section 1.3.2).

One can easily check that if x is a forward (respectively, backward) regular point for $\mathcal{A}$ then for every $m \in \mathbb{Z}$, the point $f^m(x)$ is also forward (respectively, backward) regular for $\mathcal{A}$. Furthermore, if $\mathcal{A}$ and $\mathcal{B}$ are equivalent cocycles on Y then the point $y \in Y$ is forward (respectively, backward) regular for $\mathcal{A}$ if and only if it is forward (respectively, backward) regular for $\mathcal{B}$.

In view of (1.14), forward regularity is equivalent to the following statement: there exists a positive definite symmetric matrix $\Lambda(x)$ such that

$$\begin{aligned}\lim_{m\to\infty} \frac{1}{m} \log \|\mathcal{A}(x,m)\Lambda(x)^{-m}\| &= 0,\\ \lim_{m\to\infty} \frac{1}{m} \log \|\Lambda(x)^m \mathcal{A}(x,m)^{-1}\| &= 0.\end{aligned} \tag{3.1}$$

There is an interesting geometric description of forward regularity [96]. Consider the symmetric space $Y = GL(n, \mathbb{R})/O(n, \mathbb{R})$ (where $O(n, \mathbb{R})$ is the group of orthogonal matrices) and a cocycle $\mathcal{A}$ with values in $O(n, \mathbb{R})$ over an ergodic transformation f of a measure space (X, μ). Let $A\colon X \to X$ be the generator. Fix a point $y \in O(n, \mathbb{R})$. For $g \in GL(n, \mathbb{R})$, let λ_i be the eigenvalues of $(gg^*)^{1/2}$, where g^* is the transpose of g. The distance in Y between y and gy is

$$\rho(y, gy) = \left(\sum_{i=1}^{n}(\log \lambda_i)^2\right)^{1/2}.$$

A geodesic in Y starting at y is of the form $\gamma(t) = e^{tH}y$, where H is a symmetric matrix. Then $\Lambda = e^H$ is a positive definite symmetric matrix. One can show that

$$\lim_{m\to\infty} \frac{1}{m}\rho(\Lambda^{-m}y, \mathcal{A}(x, m)^{-1}y) = 0$$

for almost every $x \in X$. In view of (3.1), this means that x is forward regular.

We give another useful characterization of forward (backward) regularity.

Proposition 3.1.2. *The following statements hold:*

1. *If* $\sup_{m>0} \|A(f^m(x))\| < \infty$ *and* $\chi^+(x, v_1, \ldots, v_k)$ *is exact for any choice of linearly independent vectors* $v_1, \ldots, v_k \in \mathbb{R}^n$ *and* $k = 1, \ldots, n$*, then* x *is forward regular.*
2. *If* $\sup_{m>0} \|A(f^{-m}(x))^{-1}\| < \infty$ *and* $\chi^-(x, v_1, \ldots, v_k)$ *is exact for any choice of linearly independent vectors* $v_1, \ldots, v_k \in \mathbb{R}^n$ *and* $k = 1, \ldots, n$*, then* x *is backward regular.*

Proof. For each m choose the orthogonal matrix R_m such that

$$T_m = R_{m+1}{}^{-1}A(f^m(x))R_m$$

is lower triangular. Note that $\chi^+(x, v_1, \ldots, v_k)$ is exact for any linearly independent vectors $v_1, \ldots, v_k \in \mathbb{R}^n$ and $k = 1, \ldots, n$ if and only if $\chi^+(x, v_1, \ldots, v_k, \{T_m\}_{m\in\mathbb{Z}})$ is exact. Therefore, it suffices to obtain the forward regularity of x with respect to the sequence $\{T_m\}_{m\in\mathbb{Z}}$. In particular,

$$\det\bigl(T_m \,|\, \operatorname{span}\{e_k, \ldots, e_n\}\bigr) = \prod_{i=k}^{n} t_{ii}^m,$$

where t_{ii}^m, $i = 1, \ldots, n$, are the diagonal entries of T_m with respect to the standard basis $\{e_1, \ldots, e_n\}$. Hence,

$$\begin{aligned} c_k := \chi^+\bigl(e_k, \ldots, e_n, \{T_m\}_{m\in\mathbb{Z}}\bigr) &= \overline{\lim_{m\to+\infty}} \frac{1}{m} \log \prod_{j=1}^{m}\prod_{i=k}^{n}|t_{ii}^j| \\ &= \lim_{m\to+\infty} \frac{1}{m} \sum_{j=1}^{m}\sum_{i=k}^{n} \log|t_{ii}^j|. \end{aligned}$$

Therefore, for each $k = 1, \ldots, n$, there exists the limit

$$\lim_{m\to+\infty} \frac{1}{m} \sum_{j=1}^{m} \log|t_{kk}^{j}| = c_k - c_{k+1}$$

and Condition 1 of Theorem 1.3.12 holds. Since the matrices R_m are orthogonal and $\sup_{m>0} \|A(f^m(x))\| < \infty$, we have $\sup_{m>0} \|T_m\| < \infty$ and Condition 2 of Theorem 1.3.12 holds. Hence, x is a forward regular point. The proof of Statement 2 can be obtained from Theorem 1.3.14 by similar arguments. □

Consider the families of filtrations $\mathcal{V}^+ = \{V_x^+\}_{x\in X}$ and $\mathcal{V}^- = \{V_x^-\}_{x\in X}$ of the cocycle $\mathcal{A}$. Recall that for each $x \in X$, $\mathcal{V}^+(x)$ and $\mathcal{V}^-(x)$ are filtrations associated with the Lyapunov exponents χ^+ and χ^- of the cocycle $\mathcal{A}$. For each $x \in X$, these filtrations determine filtrations V_x^+ and V_x^- of the Lyapunov exponents $\chi^+(x, \cdot)$ and $\chi^-(x, \cdot)$ for the sequence of matrices $\{A_m\}_{m\in\mathbb{Z}} = \{A(f^m(x))\}_{m\in\mathbb{Z}}$.

Definition 3.1.3. We say that the families of filtrations $\mathcal{V}^+$ and $\mathcal{V}^-$ are *coherent* at a point $x \in X$ if the filtrations V_x^+ and V_x^- are coherent with respect to the sequence of matrices $\{A(f^m(x))\}_{m\in\mathbb{Z}}$ (see Definition 1.3.15).

In other words, the families of filtrations $\mathcal{V}^+$ and $\mathcal{V}^-$ are coherent at a point $x \in X$ if the following properties hold:

1. $p^+(x) = p^-(x) =: p(x)$.
2. There exists a decomposition

$$\mathbb{R}^n = \bigoplus_{j=1}^{p(x)} H_j(x) \tag{3.2}$$

into subspaces $H_j(x)$ such that $A(x)H_j(x) = H_j(f(x))$ and

$$E_i^+(x) = \bigoplus_{j=1}^{i} H_j(x) \quad \text{and} \quad E_i^-(x) = \bigoplus_{j=i}^{p(x)} H_j(x). \tag{3.3}$$

3. For $v \in H_i(x) \setminus \{0\}$, $i = 1, \ldots, p(x)$,

$$\lim_{m\to\pm\infty} \frac{1}{m} \log\|\mathcal{A}(x, m)v\| = \chi_i^+(x) = -\chi_i^-(x) =: \chi_i(x), \tag{3.4}$$

with uniform convergence on $\{v \in H_i(x) : \|v\| = 1\}$.

Definition 3.1.4. We call the decomposition (3.2) the *Oseledets decomposition*, and we call the subspaces $H_i(x)$ the *Oseledets subspaces* at x.

We remark that Property 2 is equivalent to the following property: for $i = 1, \ldots, p(x)$, the spaces

$$H_i(x) = E_i^+(x) \cap E_i^-(x) \tag{3.5}$$

satisfy (3.2) (see (2.25) and (2.27)).

Definition 3.1.5. We say that a point $x \in X$ is *Lyapunov–Perron regular* or simply *regular* for $\mathcal{A}$ if the following conditions hold:

1. x is simultaneously forward and backward regular for $\mathcal{A}$.
2. The families of filtrations $\mathcal{V}^+$ and $\mathcal{V}^-$ are coherent at x.

One can easily see that the set of regular points is f-invariant. Condition 2 requires some degree of compatibility between forward and backward regularity. We remark that if $\mathcal{A}$ and $\mathcal{B}$ are equivalent cocycles on Y, then the point $y \in Y$ is regular for $\mathcal{A}$ if and only if it is regular for $\mathcal{B}$. We will show in Theorem 3.4.3 that under fairly general assumptions the set of regular points has full measure with respect to any invariant measure.

For each integer k such that $1 \le k \le n$, let $(\mathbb{R}^n)^{\wedge k}$ be the space of alternating k-linear forms on $\mathbb{R}^n$. For any linear transformation A of $\mathbb{R}^n$, the *k-fold exterior power* $A^{\wedge k}$ of A is the unique linear transformation $A^{\wedge k}$ of $(\mathbb{R}^n)^{\wedge k}$ such that

$$A^{\wedge k}(v_1 \wedge \cdots \wedge v_k) = Av_1 \wedge \cdots \wedge Av_k$$

for any $v_1, \ldots, v_k \in (\mathbb{R}^n)^{\wedge 1} \equiv \mathbb{R}^n$. One can define an inner product in $(\mathbb{R}^n)^{\wedge k}$ by requiring that for any $v_1 \wedge \cdots \wedge v_k, w_1 \wedge \cdots \wedge w_k \in (\mathbb{R}^n)^{\wedge k}$,

$$\langle v_1 \wedge \cdots \wedge v_k, w_1 \wedge \cdots \wedge w_k \rangle = \det B,$$

where $B = (b_{ij})$ is the $k \times k$ matrix with entries $b_{ij} = \langle v_i, w_j \rangle$ for each i and j. The induced norm satisfies the following properties:

1. $\|v_1 \wedge \cdots \wedge v_k\| \le \|v_1 \wedge \cdots \wedge v_\ell\| \cdot \|v_{\ell+1} \wedge \cdots \wedge v_k\|$ for any $\ell < k$.
2. For any linear transformations A, B of $\mathbb{R}^n$ and $1 \le k, \ell \le n$, the induced operator norm in $(\mathbb{R}^n)^{\wedge k}$ satisfies
   (a) $\|(AB)^{\wedge k}\| \le \|A^{\wedge k}\| \cdot \|B^{\wedge k}\|$;
   (b) $\|A^{\wedge(k+\ell)}\| \le \|A^{\wedge k}\| \cdot \|A^{\wedge \ell}\| \le \|A\|^{k+\ell}$;
   (c) $\|A^{\wedge k}\| = \prod_{j=1}^k d_j$, where $d_1 \ge d_2 \ge \cdots \ge d_n \ge 0$ are the eigenvalues of $(A^*A)^{1/2}$.

Proposition 3.1.6. *If $x \in X$ is a regular point for $\mathcal{A}$ then the following properties hold:*

1. *The exponents $\chi^+(x, \cdot)$ and $\chi^-(x, \cdot)$ are exact.*
2. $\dim H_i(x) = k_i^+(x) = k_i^-(x) =: k_i(x)$.
3. *For any vectors $v_1, \ldots, v_{k_i(x)} \in H_i(x)$ with $\operatorname{vol}(v_1, \ldots, v_{k_i(x)}) \neq 0$, we have*

$$\lim_{m \to \pm\infty} \frac{1}{m} \log \operatorname{vol}(\mathcal{A}(x, m)v_1, \ldots, \mathcal{A}(x, m)v_{k_i(x)}) = \chi_i(x)k_i(x).$$

   (Recall that $\operatorname{vol}(v_1, \ldots, v_n)$ is the volume of the parallelepiped formed by the vectors $v_1, \ldots, v_n$.)
4. *If $k = 1, \ldots, n$ then*

$$\lim_{m \to \pm\infty} \frac{1}{m} \log\|\mathcal{A}(x, m)^{\wedge k}\| = \sum_{j=1}^{k} \chi'_{n-j+1}(x).$$

Proof. Property 1 follows from Theorem 1.3.7. Properties 2–4 follow immediately from the definitions and Theorem 1.3.11 applied to the set $K = \{n^+_{i+1}(x), \dots, n^+_{i+k_i^+(x)}(x)\}$. □

Identifying the space $H_i(f^m(x))$ with $\mathbb{R}^{k_i(x)}$ for every $m \in \mathbb{Z}$, one can rewrite Property 3 in the following way: if $i = 1, \dots, p(x)$ then

$$\lim_{m\to\pm\infty} \frac{1}{m} \log|\det(\mathcal{A}(x, m)|H_i(x))| = \chi_i(x)k_i(x).$$

Proposition 3.1.7. *Assume that*

1. $\sup_{m>0} \|A(f^m(x))\| < \infty$ *and* $\sup_{m>0} \|A(f^{-m}(x))^{-1}\| < \infty$*;*
2. $\chi^+(x, v_1, \dots, v_k)$ *and* $\chi^-(x, v_1, \dots, v_k)$ *are exact for any linearly independent vectors* $v_1, \dots, v_k \in T_xM$ *and* $k = 1, \dots, n$*;*
3. $p^+(x) = p^-(x) =: p(x)$ *and* $\chi_i^+(x) = -\chi_i^-(x)$ *for* $i = 1, \dots, p(x)$*;*
4. $\bigoplus_{i=1}^{p(x)}(E_i^+(x) \cap E_i^-(x)) = \mathbb{R}^n$.

Then x *is regular.*

Proof. By Proposition 3.1.2 the point x is forward and backward regular. Setting $H_i(x) = E_i^+(x) \cap E_i^-(x)$ for $i = 1, \dots, p(x)$, one can easily see that the identities in (3.2)–(3.4) hold. Therefore the filtrations $\mathcal{V}_x^+$ and $\mathcal{V}_x^-$ are coherent at x, and x is a regular point. □

Remark 3.1.8. *It also follows from Theorem 1.3.11 that for every regular point* $x \in X$, $1 \le i, j \le p(x)$ *with* $i \ne j$, *and every distinct vectors* $v, w \in H_i(x)$, *we have*

$$\lim_{m\to\pm\infty} \frac{1}{m} \log \sin\sphericalangle(H_i(f^m(x)), H_j(f^m(x))) = 0;$$

that is, the angles between any two spaces $H_i(x)$ *and* $H_j(x)$ *can grow at most subexponentially along the orbit of* x, *and*

$$\lim_{m\to\pm\infty} \frac{1}{m} \log \sin\sphericalangle(\mathcal{A}(x, m)v, \mathcal{A}(x, m)w) = 0.$$

3.2 Lyapunov Exponents and Basic Constructions

Consider a cocycle $\mathcal{A}$ over an invertible measurable transformation f of a measure space X.

Proposition 3.2.1. *For every* $x \in X$ *and every* $v \in \mathbb{R}^n$, *if the exponent* $\chi^+(x, v, \mathcal{A})$ *is exact, then* $\chi^+(x, v, \mathcal{A}^m)$ *is exact and*

$$\chi^+(x, v, \mathcal{A}^m) = m\chi^+(x, v, \mathcal{A}).$$

Proof. This is an immediate consequence of the definition of the exponent $\chi^+(x, v, \mathcal{A}^m)$. □

It follows that if $x \in X$ is Lyapunov–Perron regular with respect to the cocycle $\mathcal{A}$ then so it is with respect to the cocycle $\mathcal{A}^m$. Moreover, the Oseledets decomposition at a regular point $x \in X$ for the cocycle $\mathcal{A}$ provides the Oseledets decomposition at x for the cocycle $\mathcal{A}^m$.

Proposition 3.2.2. *Let $\mathcal{A}$ be a measurable cocycle over a measurable transformation f preserving a finite Borel measure ν. If $Y \subset X$ is a measurable subset of positive ν-measure, then for ν-almost every $x \in Y$ and every $v \in \mathbb{R}^n$, if $\chi^+(x, v, \mathcal{A})$ is exact then $\chi^+(x, v, \mathcal{A}_Y)$ is exact and*

$$\chi^+(x, v, \mathcal{A}_Y) = \tau_Y(x)\chi^+(x, v, \mathcal{A}).$$

Proof. Let $\tau_k(x) = \sum_{i=0}^{k-1} k_Y(f_Y^i(x))$. Then $\mathcal{A}_Y(x, k) = \mathcal{A}(x, \tau_k(x))$. It follows from Birkhoff's Ergodic Theorem that for ν-almost every $x \in Y$,

$$\begin{aligned}\chi^+(x, v, \mathcal{A}_Y) &= \lim_{k\to+\infty} \frac{1}{k} \log\|\mathcal{A}_Y(x, k)v\| \\ &= \lim_{k\to+\infty} \frac{1}{k} \log\|\mathcal{A}(x, \tau_k(x))v\| \\ &= \lim_{k\to+\infty} \frac{\tau_k(x)}{k}\chi^+(x, v, \mathcal{A}) = \tau_Y(x)\chi^+(x, v, \mathcal{A}).\end{aligned}$$

The desired result follows. □

It follows that for ν-almost every point $x \in Y$, the point x is regular with respect to the cocycle $\mathcal{A}_Y$ if and only if it is regular with respect to the cocycle $\mathcal{A}$. Moreover, the Oseledets decomposition at x for $\mathcal{A}$ provides the Oseledets decomposition at x for $\mathcal{A}_Y$.

Proposition 3.2.3. *For every $x \in X$, $k = 1, \ldots, n$, and $v_1 \wedge \cdots \wedge v_k \in (\mathbb{R}^n)^{\wedge k}$, if the exponent $\chi^+(x, v_i, \mathcal{A})$ is exact for $i = 1, \ldots, k$, then $\chi^+(x, v_1 \wedge \cdots \wedge v_k, \mathcal{A}^{\wedge k})$ is exact and*

$$\chi^+(x, v_1 \wedge \cdots \wedge v_k, \mathcal{A}^{\wedge k}) = \sum_{i=1}^{k} \chi^+(x, v_i, \mathcal{A}).$$

Proof. This is an immediate consequence of the definition of the exponent $\chi^+(x, v, \mathcal{A}^{\wedge k})$. □

It follows that if $x \in X$ is Lyapunov–Perron regular with respect to the cocycle $\mathcal{A}$ then so it is with respect to the cocycle $\mathcal{A}^{\wedge k}$. Moreover, from the Oseledets decomposition $\bigoplus_{i=1}^{s(x)} H_i(x)$ at a regular point $x \in X$ for the cocycle $\mathcal{A}$, we obtain the Oseledets decomposition

$$\bigoplus_{i_1, \ldots, i_k} H_{i_1}(x)^{\wedge 1} \wedge \cdots \wedge H_{i_k}(x)^{\wedge 1}$$

of $(\mathbb{R}^n)^{\wedge k}$ at x for the cocycle $\mathcal{A}^{\wedge k}$.

3.3 Lyapunov Exponents and Hyperbolicity

Let $\mathcal{A}$ be a cocycle over an invertible measurable transformation f of a measure space X and $\mathcal{R}$ the set of its regular points. For $x \in \mathcal{R}$ let $\chi_i(x)$, $i = 1, \dots, p(x)$, be the values of the Lyapunov exponent at x. Consider the set

$$Y = \{x \in \mathcal{R} : \text{there exists } k = k(x),\ 1 \le k < p(x) \\ \text{such that } \chi_k(x) < 0\}. \tag{3.6}$$

Clearly, Y is invariant under f. The following result reveals a relation between nonuniform hyperbolicity of cocycles and nonvanishing of Lyapunov exponents. We assume that the function $k = k(x)$ is Borel.

Theorem 3.3.1. *The cocycle $\mathcal{A}$ is nonuniformly partially hyperbolic in the broad sense on the set Y in the sense of Definition 2.2.1 where $\lambda(x) = e^{\chi_k(x)}$, $\mu(x) = e^{\chi_{k+1}(x)}$, and the number ε_0 (and hence the function $\varepsilon(x)$) can be chosen arbitrarily small (with an appropriate choice of functions $C(x)$ and $K(x)$).*

Proof. We need the following lemma:

Lemma 3.3.2. *Let $Z \subset X$ be an f-invariant Borel set and $C(x, \varepsilon)$ a positive Borel function on $Z \times [0, \varepsilon_0)$, $0 < \varepsilon_0 < 1$, such that for every $\varepsilon_0 \ge \varepsilon > 0$, $x \in Z$, and $m \in \mathbb{Z}$,*

$$M_1(x, \varepsilon)e^{-\varepsilon|m|} \le C(f^m(x), \varepsilon) \le M_2(x, \varepsilon)e^{\varepsilon|m|},$$

where $M_1(x, \varepsilon)$ and $M_2(x, \varepsilon)$ are Borel functions. Then one can find positive Borel functions $B_1(x, \varepsilon)$ and $B_2(x, \varepsilon)$ such that

$$B_1(x, \varepsilon) \le C(x, \varepsilon) \le B_2(x, \varepsilon), \tag{3.7}$$

and for $m \in \mathbb{Z}$,

$$B_1(x, \varepsilon)e^{-2\varepsilon|m|} \le B_1(f^m(x), \varepsilon) \quad \text{and} \quad B_2(x, \varepsilon)e^{2\varepsilon|m|} \ge B_2(f^m(x), \varepsilon). \tag{3.8}$$

Proof of the lemma. It follows from the conditions of the lemma that there exists $m(x, \varepsilon) > 0$ such that if $m \in \mathbb{Z}$ and $|m| > m(x, \varepsilon)$ then

$$-2\varepsilon \le \frac{1}{|m|} \log C(f^m(x), \varepsilon) \le 2\varepsilon.$$

Set

$$B_1(x, \varepsilon) = \min_{-m(x,\varepsilon) \le i \le m(x,\varepsilon)} \left\{1, C(f^i(x), \varepsilon)e^{2\varepsilon|i|}\right\},$$

$$B_2(x, \varepsilon) = \max_{-m(x,\varepsilon) \le i \le m(x,\varepsilon)} \left\{1, C(f^i(x), \varepsilon)e^{-2\varepsilon|i|}\right\}.$$

The functions $B_1(x, \varepsilon)$ and $B_2(x, \varepsilon)$ are Borel measurable. Moreover, if $n \in \mathbb{Z}$ then

$$B_1(x, \varepsilon)e^{-2\varepsilon|n|} \le C(f^n(x), \varepsilon) \le B_2(x, \varepsilon)e^{2\varepsilon|n|}. \tag{3.9}$$

Furthermore, if $b_1 \le 1 \le b_2$ are such that

$$b_1 e^{-2\varepsilon|n|} \le C(f^n(x), \varepsilon) \tag{3.10}$$

and

$$b_2\, e^{2\varepsilon|n|} \geq C(f^n(x), \varepsilon) \tag{3.11}$$

for all $n \in \mathbb{Z}$ then $b_1 \leq B_1(x, \varepsilon)$ and $b_2 \geq B_1(x, \varepsilon)$. In other words,

$$\begin{aligned} B_1(x, \varepsilon) &= \sup\{b \leq 1 : \text{inequality (3.10) holds for all } n \in \mathbb{Z}\}, \\ B_2(x, \varepsilon) &= \inf\{b \geq 1 : \text{inequality (3.11) holds for all } n \in \mathbb{Z}\}. \end{aligned} \tag{3.12}$$

Inequalities (3.7) follow from (3.9) (with $n = 0$). We also have

$$C(f^{n+m}(x), \varepsilon) \leq B_2(x, \varepsilon)e^{2\varepsilon|n+m|} \leq B_2(x, \varepsilon)e^{2\varepsilon|n|+2\varepsilon|m|},$$

$$C(f^{n+m}(x), \varepsilon) \geq B_1(x, \varepsilon)e^{-2\varepsilon|n+m|} \geq B_1(x, \varepsilon)e^{-2\varepsilon|n|-2\varepsilon|m|},$$

Comparing these inequalities with (3.9) written at the point $f^m(x)$ and taking (3.12) into account we obtain (3.8). The proof of the lemma is complete. □

Let $h\colon Y \to \mathbb{R}$ be a Borel function. We call it *tempered* at the point x if

$$\lim_{m\to\pm\infty} \frac{1}{m} \log h(f^m(x)) = 0.$$

In other words the Lyapunov exponent of the function $h(x)$ is exact and is zero. It follows that the function $C(x, \varepsilon) = h(x)$ satisfies the conditions of Lemma 3.3.2 for all $\varepsilon > 0$.

Set $\lambda(x) = e^{\chi_k(x)}$, $\mu(x) = e^{\chi_{k+1}(x)}$, and then

$$E^1(x) = \bigoplus_{j=1}^{k} H_j(x) \quad \text{and} \quad E^2(x) = \bigoplus_{j=k+1}^{p(x)} H_j(x).$$

Let us also choose an f-invariant Borel function $\varepsilon(x)$ such that it satisfies (2.9). Consider the function $\gamma(x) = \sphericalangle(E^1(x), E^2(x))$ on Y. It follows from Proposition 3.1.6 (see Remark 3.1.8) that $\gamma(x)$ is tempered. Therefore applying Lemma 3.3.2, we conclude that the function $K(x) = B_1(x, \varepsilon(x)/2)$ satisfies (2.12).

We shall now construct the function $C(x)$.

Lemma 3.3.3. *There exists a positive Borel function $D(x, \varepsilon)$ (for $x \in Y$ and each sufficiently small $\varepsilon = \varepsilon(x) > 0$), such that if $m \in \mathbb{Z}$ and $1 \leq i \leq p$, then*

$$D(f^m(x), \varepsilon) \leq D(x, \varepsilon)^2\, e^{2\varepsilon|m|} \tag{3.13}$$

and

$$\|\mathcal{A}^n_{ix}\| \leq D(x, \varepsilon)\, e^{(\chi_i+\varepsilon)n} \quad \textit{and} \quad \|\mathcal{A}^{-n}_{ix}\| \geq D(x, \varepsilon)^{-1}\, e^{-(\chi_i+\varepsilon)n}$$

for any $n \geq 0$, where $\mathcal{A}^n_{ix} = \mathcal{A}(x, n)|H_i(x)$.

Proof of the lemma. Let $x \in Y$. By Proposition 3.1.6 there exists a number $n(x, \varepsilon) \in \mathbb{N}$ such that if $n \geq n(x, \varepsilon)$, then

$$\chi_i - \varepsilon \leq \frac{1}{n} \log \|\mathcal{A}^n_{ix}\| \leq \chi_i + \varepsilon \quad \text{and} \quad -\chi_i - \varepsilon \leq \frac{1}{n} \log \|\mathcal{A}^{-n}_{ix}\| \leq -\chi_i + \varepsilon,$$

and

$$-\chi_i - \varepsilon \le \frac{1}{n}\log\|\mathcal{B}^n_{ix}\| \le -\chi_i + \varepsilon \quad \text{and} \quad \chi_i - \varepsilon \le \frac{1}{n}\log\|\mathcal{B}^{-n}_{ix}\| \le \chi_i + \varepsilon.$$

Here $\mathcal{B}^n_{ix} = \mathcal{B}(x,n)|H_i^*(x)$, $H_i^*(x)$ is the dual space to $H_i(x)$, and

$$\mathcal{B}(x,n) = \begin{cases} (A(x)^*)^{-1}(A(f(x))^*)^{-1}\cdots(A(f^{n-1}(x))^*)^{-1} & \text{if } n > 0 \\ \text{Id} & \text{if } n = 0. \\ A(f^{-1}(x))^* A(f^{-2}(x))^* \cdots A(f^n(x))^* & \text{if } n < 0 \end{cases}$$

Set

$$D_1^+(x,\varepsilon) = \min_{1\le i\le s}\ \min_{0\le j\le n(x,\varepsilon)} \left\{1, \|\mathcal{A}^j_{ix}\|e^{(-\chi_i+\varepsilon)j}, \|\mathcal{B}^j_{ix}\|e^{(\chi_i+\varepsilon)j}\right\},$$

$$D_1^-(x,\varepsilon) = \min_{1\le i\le s}\ \min_{-n(x,\varepsilon)\le j\le 0} \left\{1, \|\mathcal{A}^j_{ix}\|e^{(-\chi_i-\varepsilon)j}, \|\mathcal{B}^j_{ix}\|e^{(\chi_i-\varepsilon)j}\right\},$$

$$D_2^+(x,\varepsilon) = \max_{1\le i\le s}\ \max_{0\le j\le n(x,\varepsilon)} \left\{1, \|\mathcal{A}^j_{ix}\|e^{(-\chi_i-\varepsilon)j}, \|\mathcal{B}^j_{ix}\|e^{(\chi_i-\varepsilon)j}\right\},$$

$$D_2^-(x,\varepsilon) = \max_{1\le i\le s}\ \max_{-n(x,\varepsilon)\le j\le 0} \left\{1, \|\mathcal{A}^j_{ix}\|e^{(-\chi_i+\varepsilon)j}, \|\mathcal{B}^j_{ix}\|e^{(\chi_i+\varepsilon)j}\right\},$$

and

$$\begin{aligned} D_1(x,\varepsilon) &= \min\{D_1^+(x,\varepsilon), D_1^-(x,\varepsilon)\}, \\ D_2(x,\varepsilon) &= \max\{D_2^+(x,\varepsilon), D_2^-(x,\varepsilon)\}, \\ D(x,\varepsilon) &= \max\{D_1(x,\varepsilon)^{-1}, D_2(x,\varepsilon)\}. \end{aligned}$$

The function $D(x,\varepsilon)$ is measurable, and for $n \ge 0$ and $1 \le i \le p$,

$$\begin{aligned} D(x,\varepsilon)^{-1}\, e^{(\chi_i-\varepsilon)n} &\le \|\mathcal{A}^n_{ix}\| \le D(x,\varepsilon)e^{(\chi_i+\varepsilon)n}, \\ D(x,\varepsilon)^{-1}\, e^{(-\chi_i-\varepsilon)n} &\le \|\mathcal{A}^{-n}_{ix}\| \le D(x,\varepsilon)e^{(-\chi_i+\varepsilon)n}, \\ D(x,\varepsilon)^{-1}\, e^{(-\chi_i-\varepsilon)n} &\le \|\mathcal{B}^n_{ix}\| \le D(x,\varepsilon)e^{(-\chi_i+\varepsilon)n}, \\ D(x,\varepsilon)^{-1}\, e^{(\chi_i-\varepsilon)n} &\le \|\mathcal{B}^{-n}_{ix}\| \le D(x,\varepsilon)e^{(\chi_i+\varepsilon)n}. \end{aligned} \tag{3.14}$$

Moreover, if $d \ge 1$ is a number for which the inequalities (3.14) hold for all $n \ge 0$ and $1 \le i \le p$ with $D(x,\varepsilon)$ replaced by d, then $d \ge D(x,\varepsilon)$. Therefore,

$$\begin{aligned} D(x,\varepsilon) = \inf\{d \ge 1 : &\text{ the inequalities (3.14) hold for all } n \ge 0 \\ &\text{and } 1 \le i \le p \text{ with } D(x,\varepsilon) \text{ replaced by } d\}. \end{aligned} \tag{3.15}$$

We wish to compare the values of the function $D(x,\varepsilon)$ at the points x and $f^m(x)$ for $m \in \mathbb{Z}$. We introduce the identification map $\tau_x : (\mathbb{R}^n)^* \to \mathbb{R}^n$ such that $\langle\tau_x(\varphi), v\rangle = \varphi(v)$ where $v \in \mathbb{R}^n$ and $\varphi \in (\mathbb{R}^n)^*$.

Let $\{v^n_k : k = 1,\dots,\ell\}$ be a basis of $H_i(f^n(x))$ and $\{w^n_k : k = 1,\dots,\ell\}$ the dual basis of $H_i^*(f^n(x))$. We have $\tau_{f^n(x)}(w^n_k) = v^n_k$. Denote by $A^i_{n,m}$ and $B^i_{n,m}$ the matrices corresponding to the linear maps $\mathcal{A}^n_{if^m(x)}$ and $\mathcal{B}^n_{if^m(x)}$ with respect to the above bases. We have that

$$A^i_{m,0}(B^i_{m,0})^* = \text{Id}$$

where $*$ stands for matrix transposition. Hence for every $n > 0$, the matrix corresponding to the map $\mathcal{A}^n_{if^m(x)}$ is

$$A^i_{n,m} = A^i_{n+m,0}(A^i_{m,0})^{-1} = A^i_{n+m,0}(B^i_{m,0})^*.$$

Therefore, in view of (3.14), we obtain that

1. if $n > 0$ then
$$\begin{aligned}\|\mathcal{A}^n_{if^m(x)}\| &\le D(x,\varepsilon)^2\, e^{(\chi_i+\varepsilon)(n+m)+(-\chi_i+\varepsilon)m}\\ &= D(x,\varepsilon)^2\, e^{2\varepsilon m}\, e^{(\chi_i+\varepsilon)n},\\ \|\mathcal{A}^n_{if^m(x)}\| &\ge D(x,\varepsilon)^{-2}\, e^{(\chi_i-\varepsilon)(n+m)+(-\chi_i-\varepsilon)m}\\ &= D(x,\varepsilon)^{-2}\, e^{-2\varepsilon m}\, e^{(\chi_i-\varepsilon)n};\end{aligned}$$
2. if $n > 0$ and $m - n \ge 0$ then
$$\begin{aligned}\|\mathcal{A}^{-n}_{if^m(x)}\| &\le D(x,\varepsilon)^2\, e^{(\chi_i+\varepsilon)(m-n)+(-\chi_i+\varepsilon)m}\\ &= D(x,\varepsilon)^2\, e^{2\varepsilon m}\, e^{(-\chi_i+\varepsilon)n},\\ \|\mathcal{A}^{-n}_{if^m(x)}\| &\ge D(x,\varepsilon)^{-2}\, e^{(\chi_i-\varepsilon)(m-n)+(-\chi_i-\varepsilon)m}\\ &= D(x,\varepsilon)^{-2}\, e^{-2\varepsilon m}\, e^{(-\chi_i-\varepsilon)n};\end{aligned}$$
3. if $n > 0$ and $n - m \ge 0$ then
$$\begin{aligned}\|\mathcal{A}^{-n}_{if^m(x)}\| &\le D(x,\varepsilon)^2\, e^{(\chi_i+\varepsilon)(n-m)+(-\chi_i+\varepsilon)m}\\ &= D(x,\varepsilon)^2\, e^{2\varepsilon m}\, e^{(-\chi_i+\varepsilon)n},\\ \|\mathcal{A}^{-n}_{if^m(x)}\| &\ge D(x,\varepsilon)^{-2}\, e^{(\chi_i-\varepsilon)(n-m)+(-\chi_i-\varepsilon)m}\\ &= D(x,\varepsilon)^{-2}\, e^{-2\varepsilon m}\, e^{(-\chi_i-\varepsilon)n}.\end{aligned}$$

Similar inequalities hold for the maps $\mathcal{B}^n_{if^m(x)}$ for each $n, m \in \mathbb{Z}$. Comparing this with the inequalities (3.14), applied to the point $f^m(x)$, and using (3.15), we conclude that if $m \ge 0$, then

$$D(f^m(x),\varepsilon) \le D(x,\varepsilon)^2\, e^{2\varepsilon m}. \tag{3.16}$$

Similar arguments show that if $m \le 0$, then

$$D(f^{-m}(x),\varepsilon) \le D(x,\varepsilon)^2\, e^{-2\varepsilon m}. \tag{3.17}$$

It follows from (3.16) and (3.17) that if $m \in \mathbb{Z}$, then

$$D(f^m(x),\varepsilon) \le D(x,\varepsilon)^2\, e^{2\varepsilon|m|}.$$

This completes the proof of the lemma. □

We proceed with the construction of the function $C(x)$. Replacing in (3.13) m by $-m$ and x by $f^m(x)$, we obtain

$$D(f^m(x),\varepsilon) \ge \sqrt{D(x,\varepsilon)}e^{-\varepsilon|m|}. \tag{3.18}$$

Consider two disjoint subsets $\sigma_1, \sigma_2 \subset [1, s] \cap \mathbb{N}$ and set

$$L_1(x) = \bigoplus_{i \in \sigma_1} H_i(x) \quad \text{and} \quad L_2(x) = \bigoplus_{i \in \sigma_2} H_i(x)$$

and $\gamma_{\sigma_1\sigma_2}(x) = \sphericalangle(L_1(x), L_2(x))$. By Theorem 3.4.3 the function $\gamma_{\sigma_1\sigma_2}$ is tempered and hence in view of Lemma 3.3.2 one can find a function $K_{\sigma_1\sigma_2}(x, \varepsilon) = B_1(x, \varepsilon/2)$ such that

$$\gamma_{\sigma_1\sigma_2}(x) \geq K_{\sigma_1\sigma_2}(x, \varepsilon).$$

Set

$$T(x, \varepsilon) = \min K_{\sigma_1\sigma_2}(x, \varepsilon),$$

where the minimum is taken over all pairs of disjoint subsets $\sigma_1, \sigma_2 \subset [1, s] \cap \mathbb{N}$.

Let $v \in E^1(x)$. Write $v = \sum_{i=1}^k v_i$ where $v_i \in H_i(x)$. We have

$$\|v\| \leq \sum_{i=1}^k \|v_i\| \leq LT^{-1}(x, \varepsilon)\|v\|,$$

where $L > 1$ is a constant. Let us set

$$C'(x, \varepsilon) = LD(x, \varepsilon)T(x, \varepsilon)^{-1}.$$

It follows from (3.13) and (3.18) that the function $C'(x, \varepsilon)$ satisfies the condition of Lemma 3.3.2 with

$$M_1(x, \varepsilon) = \frac{2}{\pi} L\sqrt{D(x, \varepsilon)} \quad \text{and} \quad M_2(x, \varepsilon) = LD(x, \varepsilon)^2 T(x, \varepsilon)^{-1}.$$

Therefore there exists a function $C_1(x, \varepsilon) \geq C'(x, \varepsilon)$ for which the statements of Lemma 3.3.2 hold.

Applying the above arguments to the inverse map f^{-1} and the subspace $E^2(x)$, one can construct a function $C_2(x, \varepsilon)$ for which the statements of Lemma 3.3.2 hold. The desired function $C(x)$ is defined by

$$C(x) = \max\{C_1(x, \varepsilon(x)/2), C_2(x, \varepsilon(x)/2)\}$$

and it satisfies (2.12). This completes the proof of the theorem. □

Given a cocycle $\mathcal{A}$ over an invertible measurable transformation f, consider the set $\mathcal{R}$ of regular points and the set

$$Z = \{x \in \mathcal{R} : \text{there exists } s(x),\ 1 \leq s(x) < p(x) \\ \text{such that } \chi_{s(x)}(x) < 0 < \chi_{s(x)+1}(x)\}. \tag{3.19}$$

Clearly, Z is invariant under f. The following result is an immediate corollary of Theorem 3.3.1.

Theorem 3.3.4. *The cocycle $\mathcal{A}$ is nonuniformly completely hyperbolic on the set Z in the sense of Definition 2.2.3 where $\lambda(x) = e^{\chi_{s(x)}(x)}$, $\mu(x) = e^{\chi_{s(x)+1}(x)}$, and the* *number ε_0 (and hence the function $\varepsilon(x)$) can be chosen arbitrarily small (with an appropriate choice of functions $C(x)$ and $K(x)$).*

3.4 The Multiplicative Ergodic Theorem

We present the Multiplicative Ergodic Theorem. It states that for every cocycle $\mathcal{A}$ over a dynamical system $f: X \to X$ preserving a Borel probability measure ν on X, almost every point $x \in X$ is regular for $\mathcal{A}$, provided that the generator A of the cocycle $\mathcal{A}$ satisfies the integrability condition

$$\log^+ \|A\|, \log^+ \|A^{-1}\| \in L^1(X, \nu), \tag{3.20}$$

where $\log^+ a = \max\{\log a, 0\}$. We notice that Property (3.20) holds for any cocycle $A: X \to GL(n, \mathbb{R})$ for which there is a positive constant c such that $\|A(x)^{\pm 1}\| \leq c$ for ν-almost all $x \in X$.

A crucial manifestation of Oseledets' theorem is a connection between dynamical systems with nonzero Lyapunov exponents and nonuniformly hyperbolic dynamical systems. Namely, a cocycle $\mathcal{A}$, whose Lyapunov exponents do not vanish almost everywhere, is nonuniformly hyperbolic on a set of full ν-measure. Furthermore, for every f-invariant measurable function $\varepsilon: X \to [0, +\infty)$, the cocycle $\mathcal{A}$ admits a $(\lambda, \mu, \varepsilon)$-splitting on a set of full measure.

3.4.1 One-Dimensional Cocycles and Birkhoff's Ergodic Theorem

We begin with the most simple cocycles, that is, the cocycles with values in $GL(1, \mathbb{R})$. Let $\mathcal{A}: X \times \mathbb{Z} \to \mathbb{R} \setminus \{0\}$ be such a cocycle and $A(x) = \mathcal{A}(x, 1)$ its generator. Notice that

$$\log|\mathcal{A}(x, m)| = \sum_{j=0}^{m-1} \log|A(f^j(x))|.$$

Applying Birkhoff's Ergodic Theorem we obtain the following result.

Theorem 3.4.1. *Let f be an invertible measure preserving transformation of a Lebesgue space (X, ν) and $\mathcal{A}: X \times \mathbb{Z} \to \mathbb{R} \setminus \{0\}$ a measurable cocycle over f such that*

$$\log|A| \in L^1(X, \nu). \tag{3.21}$$

Then there exists an f-invariant measurable function $\varphi: X \to \mathbb{R}$ such that

$$\varphi(x) = \lim_{m \to \pm\infty} \frac{1}{m} \log|\mathcal{A}(x, m)|$$

for ν-almost every $x \in X$. Moreover, $\varphi \in L^1(X, \nu)$, and

$$\int_X \varphi \, d\nu = \int_X \log|A| \, d\nu.$$

We remark that the assumption (3.21) holds if and only if $\log^+|A|$, $\log^-|A| \in L^1(X, \nu)$, where $\log^- a = \max\{-\log a, 0\}$, and hence if and only if (3.20) holds.

As a direct consequence of Theorem 3.4.1, we obtain the Multiplicative Ergodic Theorem for one-dimensional cocycles.

Theorem 3.4.2. *Let f be an invertible measure preserving transformation of a Lebesgue space (X, ν) and $\mathcal{A}\colon X \times \mathbb{Z} \to \mathbb{R} \setminus \{0\}$ a measurable cocycle over f satisfying* (3.21). *Then the set of regular points for A has full ν-measure.*

Proof. By Theorem 3.4.1, there exists a set $X' \subset X$ with $\nu(X') = \nu(X)$ such that if $x \in X'$ and $v \neq 0$ then the exponents $\chi^+(x, v)$ and $\chi^-(x, v)$ are exact, and $\chi^+(x, v) = -\chi^-(x, v) = \varphi(x)$. This implies that $p^+(x) = p^-(x) = 1$, $E_1^+(x) = E_1^-(x)$, and $\chi_1^+(x) = -\chi_1^-(x)$ for every $x \in X'$. Hence the filtrations $\mathcal{V}_x^+$ and $\mathcal{V}_x^-$ are coherent at every $x \in X'$. By Proposition 1.3.10 the points in X' are simultaneously forward and backward regular for A. The result follows. □

3.4.2 Oseledets' Proof of the Multiplicative Ergodic Theorem

In this section we present a proof of the Multiplicative Ergodic Theorem for measurable cocycles over measure preserving transformations, following Oseledets' original approach (see [162]).

Theorem 3.4.3 (Multiplicative Ergodic Theorem [162]). *Let f be an invertible measure preserving transformation of a Lebesgue space (X, ν) and $\mathcal{A}$ a measurable cocycle over f whose generator satisfies* (3.20). *Then the set of regular points for $\mathcal{A}$ has full ν-measure.*

Proof. We split the proof of the theorem into two steps. First, we show how to reduce the general case to the case of triangular cocycles and then we present the proof of the latter case.

Step I: Reduction to triangular cocycles.

We construct an extension of the transformation f,

$$F\colon X \times SO(n, \mathbb{R}) \to X \times SO(n, \mathbb{R}),$$

where $SO(n, \mathbb{R})$ is the group of orthogonal $n \times n$ matrices. Given $(x, U) \in X \times SO(n, \mathbb{R})$, one can apply the Gram-Schmidt orthogonalization procedure to the columns of the matrix $A(x)U$ and write

$$A(x)U = R(x, U)T(x, U), \tag{3.22}$$

where $R(x, U)$ is orthogonal and $T(x, U)$ is lower triangular (with positive entries on the diagonal). The two matrices $R(x, U)$ and $T(x, U)$ are uniquely defined and their entries are linear combinations of the entries of U. Set

$$F(x, U) = (f(x), R(x, U)).$$

Consider the projection $\pi : (x, U) \mapsto U$. By (3.22), we obtain

$$T(x, U) = ((\pi \circ F)(x, U))^{-1} A(x)\pi(x, U). \tag{3.23}$$

Let $\widetilde{\mathcal{A}}$ and $\mathcal{T}$ be two cocycles over F defined respectively by $\widetilde{A}(x, U) = A(x)$ and $\mathcal{T}(x, U) = T(x, U)$. Since $\|U\| = 1$ for every $U \in SO(n, \mathbb{R})$, it follows from (3.23) that the cocycles $\widetilde{\mathcal{A}}$ and $\mathcal{T}$ are equivalent on $X \times SO(n, \mathbb{R})$. Therefore a point $(x, U) \in X \times SO(n, \mathbb{R})$ is regular for $\widetilde{\mathcal{A}}$ if and only if it is regular for $\mathcal{T}$.

By the Representation Theorem for Lebesgue spaces, we may assume that X is a compact metric space and that $f : X \to X$ is continuous. Let $\mathcal{M}$ be the set of all Borel probability measures $\widetilde{\nu}$ on $X \times SO(n, \mathbb{R})$ that satisfy

$$\widetilde{\nu}(B \times SO(n, \mathbb{R})) = \nu(B) \tag{3.24}$$

for all measurable sets $B \subset X$. Then $\mathcal{M}$ is a compact convex subset of a locally convex topological vector space. The map $F_* : \mathcal{M} \to \mathcal{M}$ defined by

$$(F_*\widetilde{\nu})(B) = \widetilde{\nu}(F^{-1}(B))$$

is a bounded linear operator. By the Tychonoff Fixed Point Theorem, there exists a fixed point $\widetilde{\nu}_0 \in \mathcal{M}$ for the operator F_*, that is, a measure $\widetilde{\nu}_0$ such that $\widetilde{\nu}_0(F^{-1}(B)) = \widetilde{\nu}_0(B)$ for every measurable set $B \subset X \times SO(n, \mathbb{R})$. By (3.24), we conclude that the set of regular points for $\mathcal{A}$ has full ν-measure if and only if the set of regular points for $\widetilde{\mathcal{A}}$ has full $\widetilde{\nu}_0$-measure, and hence if and only if the set of regular points for $\mathcal{T}$ has full $\widetilde{\nu}_0$-measure.

Step II: Proof for triangular cocycles.

From now on we assume that $A(x) = (a_{ij}(x))$ is a lower triangular matrix (i.e., $a_{ij}(x) = 0$ if $i < j$). Write $A(x)^{-1} = (b_{ij}(x))$ and note that $b_{ii}(x) = 1/a_{ii}(x)$ for each i. By (3.20) we obtain $\log^+|a_{ij}|, \log^+|b_{ij}| \in L^1(X, \nu)$. It follows from Lemma 2.1.5 that

$$\lim_{m\to+\infty} \frac{1}{m} \log^+|a_{ij}(f^m(x))| = \lim_{m\to-\infty} \frac{1}{m} \log^+|b_{ij}(f^m(x))| = 0 \tag{3.25}$$

for every $1 \le i, j \le n$ and ν-almost every $x \in X$. Note that

$$|\log|a_{ii}|| = \log^+|a_{ii}| - \log^-|a_{ii}| = \log^+|a_{ii}| - \log^+|b_{ii}|. \tag{3.26}$$

By (3.20) and (3.26) we have $\log|a_{ii}| = -\log|b_{ii}| \in L^1(X, \nu)$. Birkhoff's Ergodic Theorem guarantees the existence of measurable functions $\lambda_i \in L^1(X, \nu)$ for $i = 1, \ldots, n$, such that

$$\lim_{m\to+\infty} \frac{1}{m} \sum_{k=0}^{m-1} \log|a_{ii}(f^k(x))| = \lim_{m\to-\infty} \frac{1}{m} \sum_{k=m}^{-1} \log|b_{ii}(f^k(x))| = \lambda_i(x) \tag{3.27}$$

for each $i = 1, \ldots, n$ and ν-almost every $x \in X$. Let $Y \subset X$ be the set of points $x \in X$ for which (3.25) and (3.27) hold. Notice that Y is a set of full ν-measure. We will show that Y consists of regular points for $\mathcal{A}$, and thus the set of regular points for $\mathcal{A}$ has full ν-measure.

By Theorems 1.3.12 and 1.3.14, the sequence $\{A(f^m(x))\}_{m\in\mathbb{Z}}$ is simultaneously forward and backward regular for every $x \in Y$. Moreover, the numbers $\lambda_i(x)$ are the forward Lyapunov exponents counted with their multiplicities (but possibly not ordered) and are the symmetric of the backward Lyapunov exponents counted with their multiplicities (but possibly not ordered). We conclude that $p^+(x) = p^-(x) =: p(x)$ and $\chi_i^-(x) = -\chi_i^+(x)$ for $i = 1, \ldots, p(x)$. We now show that the spaces $H_1(x), \ldots, H_{p(x)}(x)$ defined by (3.5) satisfy (3.2).

Lemma 3.4.4. *If $x \in Y$ and $v \in \mathbb{R}^n \setminus \{0\}$, then $\chi^+(x, v) + \chi^-(x, v) \geq 0$.*

Proof of the lemma. For every $v \in \mathbb{R}^n \setminus \{0\}$, there exists i such that $v = \alpha_i e_i + \cdots + \alpha_n e_n$ with $\alpha_i \neq 0$, where $\{e_1, \ldots, e_n\}$ is the standard basis of $\mathbb{R}^n$. For any $m > 0$, the projection of $\mathcal{A}(x, m)v$ over $\operatorname{span}\{e_i\}$ along its orthogonal complement is

$$\prod_{k=0}^{m-1} a_{ii}(f^k(x))\alpha_i e_i.$$

Thus,

$$\log\|\mathcal{A}(x, m)v\| \geq \sum_{k=0}^{m-1} \log|a_{ii}(f^k(x))| + \log|\alpha_i|,$$

and $\chi^+(x, v) \geq \lambda_i(x)$. In a similar way, for every $m < 0$,

$$\log\|\mathcal{A}(x, m)v\| \geq \sum_{k=-m}^{-1} \log|b_{ii}(f^k(x))| + \log|\alpha_i|,$$

and $\chi^-(x, v) \geq -\lambda_i(x)$. Hence, $\chi^+(x, v) + \chi^-(x, v) \geq 0$. □

Lemma 3.4.5. *If $x \in Y$ then $\bigoplus_{i=1}^{p(x)} H_i(x) = \mathbb{R}^n$.*

Proof of the lemma. Since

$$A(x)E_i^+(x) = E_i^+(f(x)) \quad \text{and} \quad A(x)E_i^-(x) = E_i^-(f(x))$$

(see (2.25) and (2.27)), we have $A(x)H_i(x) = H_i(f(x))$ for each i. Take $v \in E_i^+(x) \cap E_{i+1}^-(x)$. Then $\chi^+(x, v) \leq \chi_i^+(x)$ and $\chi^-(x, v) \leq -\chi_{i+1}^+(x)$. Hence,

$$\chi^+(x, v) + \chi^-(x, v) \leq \chi_i^+(x) - \chi_{i+1}^+(x) < 0,$$

and by Lemma 3.4.4 we obtain $v = 0$. Therefore, $E_i^+(x) \cap E_{i+1}^-(x) = \{0\}$. Moreover, for each $i < j$, we have

$$H_i(x) \cap H_j(x) \subset E_i^+(x) \cap E_j^-(x) \subset E_i^+(x) \cap E_{i+1}^-(x) = \{0\}.$$

Observe that

$$\begin{aligned}\dim H_i(x) &\geq \dim E_i^+(x) + \dim E_i^-(x) - n \\ &= (k_1(x) + \cdots + k_i(x)) + (k_i(x) + \cdots + k_{p(x)}(x)) - n = k_i(x).\end{aligned}$$

Therefore,

$$\dim \bigoplus_{i=1}^{p(x)} H_i(x) \geq k_1(x) + \cdots + k_{p(x)}(x) = n$$

and hence $\bigoplus_{i=1}^{p(x)} H_i(x) = \mathbb{R}^n$. This completes the proof of the lemma. □

It remains to show that the convergence in (3.4) is uniform on $C_i(x) = \{v \in H_i(x) : \|v\| = 1\}$.

Lemma 3.4.6. *If $x \in Y$ then*

$$\lim_{m\to\pm\infty} \frac{1}{|m|} \log \inf_v \|\mathcal{A}(x,m)v\| = \lim_{m\to\pm\infty} \frac{1}{|m|} \log \sup_v \|\mathcal{A}(x,m)v\| = \chi_i(x),$$

with the infimum and supremum taken over all $v \in C_i(x)$.

Proof of the lemma. Let $x \in Y$. Consider an orthonormal basis $\{e_1, \ldots, e_{k_i(x)}\}$ of $H_i(x)$. Let now

$$u_m = \sum_{j=1}^{k_i(x)} c_{m,j} e_j$$

be a vector in $C_i(x)$ at which $v \mapsto \|\mathcal{A}(x,m)v\|$ attains its minimum. Choose an integer $j(m)$ such that $|c_{m,j(m)}| = \max_j |c_{m,j}|$. Since

$$\sum_{j=1}^{k_i(x)} c_{m,j}{}^2 = 1,$$

we have $|c_{m,j(m)}| \geq 1/\sqrt{k_i(x)}$. We denote by $\rho_{m,j}$ and $\varphi_{m,j}$ respectively the distance and the angle between $\mathcal{A}(x,m)e_j$ and $\mathcal{A}(x,m)\operatorname{span}\{e_i : i \neq j\}$. Note that

$$\rho_{m,j} = \|\mathcal{A}(x,m)e_j\| \sin \varphi_{m,j}.$$

We have

$$\mathcal{A}(x,m)u_m = c_{m,j(m)} \left(\mathcal{A}(x,m)e_{j(m)} + \sum_{j\neq j(m)} \frac{c_{m,j}}{c_{m,j(m)}} \mathcal{A}(x,m)e_j \right),$$

and thus

$$\|\mathcal{A}(x,m)u_m\| \geq |c_{m,j(m)}|\rho_{m,j(m)} \geq \frac{1}{\sqrt{k_i(x)}} \|\mathcal{A}(x,m)e_{j(m)}\| \sin \varphi_{m,j(m)}.$$

Since x is forward and backward regular, it follows from Theorem 1.3.11 (and the analogous statement in the case of backward regularity) that

$$\lim_{m\to\pm\infty} \frac{1}{m} \log|\sin \varphi_{m,j}| = 0.$$

Since $j(m)$ can take on only a finite number of values, we obtain

$$
\begin{aligned}
&\varliminf_{m\to\pm\infty} \frac{1}{|m|} \log\|\mathcal{A}(x,m)u_m\| \\
&\geq \varliminf_{m\to\pm\infty} \frac{1}{|m|} \log\|\mathcal{A}(x,m)e_{j(m)}\| + \varliminf_{m\to\pm\infty} \frac{1}{|m|} \log|\sin\varphi_{m,j(m)}| \\
&\geq \min_j \varliminf_{m\to\pm\infty} \frac{1}{|m|} \log\|\mathcal{A}(x,m)e_j\| \\
&\quad + \min_j \varliminf_{m\to\pm\infty} \frac{1}{|m|} \log|\sin\varphi_{m,j}| = \chi_i(x),
\end{aligned}
\tag{3.28}
$$

where the minima are taken over all $j \in \{1,\dots,k_i(x)\}$. Choose a vector

$$v_m = \sum_{j=1}^{k_i(x)} d_{m,j}e_j$$

in $C_i(x)$ for which the function $v \mapsto \|\mathcal{A}(x,m)v\|$ attains its maximum. We have

$$\|\mathcal{A}(x,m)v_m\| \leq \sum_{j=1}^{k_i(x)} |d_{m,j}| \cdot \|\mathcal{A}(x,m)e_j\| \leq \sum_{j=1}^{k_i(x)} \|\mathcal{A}(x,m)e_j\|,$$

and hence

$$\varlimsup_{m\to\pm\infty} \frac{1}{|m|} \log\|\mathcal{A}(x,m)v_m\| \leq \chi_i(x). \tag{3.29}$$

The desired result follows now from (3.28) and (3.29). □

Lemmas 3.4.5 and 3.4.6 show that the filtrations $\mathcal{V}^+$ and $\mathcal{V}^-$ are coherent at every point $x \in Y$, and hence the set Y consists of regular points for $\mathcal{A}$. This implies that the set of regular points for $\mathcal{A}$ has full ν-measure. □

3.4.3 Lyapunov Exponents and Subadditive Ergodic Theorem

In the next section we present another proof of the Multiplicative Ergodic Theorem. We follow the approach of Raghunathan (see [187]) and exploits Kingman's Subadditive Ergodic Theorem. Here we present some preliminary information.

Let $f\colon X \to X$ be a measurable transformation.

Definition 3.4.7. A measurable function $\mathcal{B}\colon X \times \mathbb{Z} \to \mathbb{R} \setminus \{0\}$ is called a *subadditive cocycle over f* if for every $x \in X$,

1. $\mathcal{B}(x,0) = 1$;
2. $\mathcal{B}(x,m+k) \leq \mathcal{B}(f^k(x),m) + \mathcal{B}(x,k)$ for $m,k \in \mathbb{Z}$.

If $\mathcal{A}\colon X \times \mathbb{Z} \to GL(n,\mathbb{R})$ is a cocycle over f, then $\mathcal{B} = \log\|\mathcal{A}\|$ is a subadditive cocycle. Indeed, by (2.1), for $m,k \in \mathbb{Z}$ and $x \in X$,

$$\log\|\mathcal{A}(x,m+k)\| \leq \log\|\mathcal{A}(f^k(x),m)\| + \log\|\mathcal{A}(x,k)\|.$$

We need the following immediate corollary of Kingman's Subadditive Ergodic Theorem (see [114, Theorem 1.8]).

Theorem 3.4.8. *Let f be an invertible measure preserving transformation of a Lebesgue space (X, ν) and $\mathcal{A}$ a measurable cocycle over f whose generator satisfies* (3.20). *Then there exist f-invariant measurable functions $\varphi_+ \colon X \to \mathbb{R}$ and $\varphi_- \colon X \to \mathbb{R}$ such that*

$$\varphi_+(x) = \lim_{m\to+\infty} \frac{1}{m} \log\|\mathcal{A}(x,m)\| = -\lim_{m\to-\infty} \frac{1}{m} \log\|\mathcal{A}(x,m)^{-1}\|,$$

$$\varphi_-(x) = \lim_{m\to-\infty} \frac{1}{m} \log\|\mathcal{A}(x,m)\| = -\lim_{m\to+\infty} \frac{1}{m} \log\|\mathcal{A}(x,m)^{-1}\|,$$

for ν-almost every $x \in X$. Moreover $\varphi_+, \varphi_- \in L^1(X, \nu)$, and

$$\begin{aligned}\int_X \varphi_+ \, d\nu &= \lim_{m\to+\infty} \frac{1}{m} \int_X \log\|\mathcal{A}(x,m)\| \, d\nu(x) \\ &= -\lim_{m\to-\infty} \frac{1}{m} \int_X \log\|\mathcal{A}(x,m)^{-1}\| \, d\nu(x),\end{aligned}$$

$$\begin{aligned}\int_X \varphi_- \, d\nu &= \lim_{m\to-\infty} \frac{1}{m} \int_X \log\|\mathcal{A}(x,m)\| \, d\nu(x) \\ &= -\lim_{m\to+\infty} \frac{1}{m} \int_X \log\|\mathcal{A}(x,m)^{-1}\| \, d\nu(x).\end{aligned}$$

The following statement shows that the values of the Lyapunov exponent χ^+ are integrable functions provided that (3.20) holds.

Proposition 3.4.9. *Let f be an invertible measure preserving transformation of a Lebesgue space (X, ν) and $\mathcal{A}$ a measurable cocycle over f whose generator satisfies* (3.20). *For every $i = 1, \dots, n$, the function $x \mapsto \chi_i'(x)$ is in $L^1(X, \nu)$.*

Proof. If $v \in \mathbb{R}^n$ then

$$\|\mathcal{A}(x,m)^{-1}\|^{-1} \le \|\mathcal{A}(x,m)v\| \le \|\mathcal{A}(x,m)\|.$$

By Theorem 3.4.8, we obtain $\varphi_-(x) \le \chi_i'(x) \le \varphi_+(x)$ for μ-almost every $x \in X$, where φ_+ and φ_- are the functions in Theorem 3.4.8. Since $\varphi_+, \varphi_- \in L^1(X, \nu)$, we obtain the desired statement. □

Using similar arguments to those in the proof of the proposition, one can show that the values of the Lyapunov exponent χ^- are integrable functions provided that (3.20) holds.

3.4.4 Raghunathan's Proof of the Multiplicative Ergodic Theorem

Let $\mathcal{A}$ be a measurable cocycle over the transformation f. We present another proof of the Multiplicative Ergodic Theorem 3.4.3 that is an elaboration of the approach by Raghunathan [187].

For each $i = 1, \ldots, n$, consider the *i-fold exterior cocycle* of A, that is, the function $\mathcal{A}^{\wedge i} : X \times \mathbb{Z} \to GL(\binom{n}{i}, \mathbb{R})$ defined by $\mathcal{A}^{\wedge i}(x, m) = \mathcal{A}(x, m)^{\wedge i}$. Note that the function $\log\|\mathcal{A}^{\wedge i}\|$ is a subadditive cocycle. We refer the reader to Sections 2.1.3, 3.1, and 3.2 for more information on exterior power cocycles.

Theorem 3.4.10. *Let f be an invertible measure preserving transformation of a Lebesgue space (X, ν) and $\mathcal{A}$ a measurable cocycle over f whose generator satisfies* (3.20). *Then there exists a set $Y \subset X$ of full ν-measure such that for $x \in Y$,*

1. *x is a regular point for $\mathcal{A}$;*
2. *there exist matrices A_x^+ and A_x^- such that*
$$\lim_{m \to \pm\infty} (\mathcal{A}(x, m)^* \mathcal{A}(x, m))^{1/(2|m|)} = A_x^\pm;$$
3. *the distinct eigenvalues of the matrix A_x^+ are the numbers $e^{\chi_1(x)}, \ldots, e^{\chi_{p(x)}(x)}$;*
4. *the distinct eigenvalues of the matrix A_x^- are the numbers $e^{\chi_1(x)}, \ldots, e^{\chi_{p(x)}(x)}$.*

Proof. Since
$$\big|\log\|A(x)\|\big| = \log^+\|A(x)\| - \log^-\|A(x)\|,$$
it follows from (3.20) that $x \mapsto \log\|A(x)\|$ is ν-integrable. Furthermore, for each $i = 1, \ldots, n$, the function $(x, m) \mapsto \log\|\mathcal{A}(x, m)^{\wedge i}\|$ is a subadditive cocycle and
$$\int_X \big|\log\|A(x)^{\wedge i}\|\big|\, d\nu(x) \le i \int_X \big|\log\|A(x)\|\big|\, d\nu(x) < \infty.$$
Therefore by Lemma 2.1.5 and Kingman's Subadditive Ergodic Theorem, there is a set $Y \subset X$ of full ν-measure such that for every $x \in Y$ and $i = 1, \ldots, n$, there exist the limits
$$\lim_{m \to \pm\infty} \frac{1}{m} \log\|A(f^m(x))\| = 0, \tag{3.30}$$
$$\lim_{m \to +\infty} \frac{1}{m} \log\|\mathcal{A}(x, m)^{\wedge i}\| = -\lim_{m \to -\infty} \frac{1}{m} \log\|(\mathcal{A}(x, m)^{-1})^{\wedge i}\|, \tag{3.31}$$
and
$$\lim_{m \to -\infty} \frac{1}{m} \log\|\mathcal{A}(x, m)^{\wedge i}\| = -\lim_{m \to +\infty} \frac{1}{m} \log\|(\mathcal{A}(x, m)^{-1})^{\wedge i}\|. \tag{3.32}$$
Fix $x \in Y$ and let $\lambda_1(x, m) \le \cdots \le \lambda_n(x, m)$ be the eigenvalues of the matrix $(\mathcal{A}(x, m)^* \mathcal{A}(x, m))^{1/2}$. We have
$$\lim_{m \to \pm\infty} \frac{1}{m} \log\|\mathcal{A}(x, m)^{\wedge i}\| = \lim_{m \to \pm\infty} \frac{1}{m} \log \prod_{k=n-i+1}^{n} \lambda_k(x, m).$$
Therefore, there exist the limits
$$\lim_{m \to +\infty} \frac{1}{m} \log \lambda_i(x, m) \quad \text{and} \quad \lim_{m \to -\infty} \frac{1}{m} \log \lambda_i(x, m). \tag{3.33}$$

Let $\rho_1(x) < \cdots < \rho_{t(x)}(x)$ be the distinct values of the first limit in (3.33) and let $G_i(x, m)$ be the eigenspace of the matrix $(\mathcal{A}(x, m)^* \mathcal{A}(x, m))^{1/2}$ associated to the eigenvalue $\lambda_i(x, m)$. For $j = 1, \ldots, t(x)$, we write

$$F_j(x, m) = \bigoplus_i G_i(x, m),$$

where the sum is taken over all i for which

$$\lim_{m \to \pm\infty} \frac{1}{m} \log \lambda_i(x, m) \leq \rho_j(x).$$

By (3.30), given $\varepsilon > 0$, there exists $m(x, \varepsilon) \in \mathbb{N}$ such that if $m \geq m(x, \varepsilon)$ then $\|A(f^m(x))\| \leq e^{m\varepsilon}$. Let $v_m \in F_j(x, m)$ with $\|v_m\| = 1$. There exists $m'(x, \varepsilon) \geq m(x, \varepsilon)$ such that for $m \geq m'(x, \varepsilon)$,

$$\begin{aligned} \|\mathcal{A}(x, m+1)v_m\| &= \|A(f^m(x))\mathcal{A}(x, m)v_m\| \\ &\leq e^{m\varepsilon} \|\mathcal{A}(x, m)v_m\| \\ &\leq e^{(\rho_j(x)+2\varepsilon)m}. \end{aligned} \tag{3.34}$$

One can write the vector v_m in the form $v_m = \alpha_{m+1} v_{m+1} + u$, where $\alpha_{m+1} \geq 0$, $v_{m+1} \in F_j(x, m+1)$ with $\|v_{m+1}\| = 1$, and $u \in \mathbb{R}^n \setminus F_j(x, m+1)$. There exists $m''(x, \varepsilon) \geq m'(x, \varepsilon)$ such that for $m \geq m''(x, \varepsilon)$,

$$\|\mathcal{A}(x, m+1)v_m\| \geq \|\mathcal{A}(x, m+1)u\| \geq e^{(\rho_{j+1}(x)-\varepsilon)(m+1)} \|u\|. \tag{3.35}$$

By (3.34) and (3.35), we conclude that

$$\|u\| \leq e^{\varepsilon - \rho_{j+1}(x)} e^{(\rho_j(x)-\rho_{j+1}(x)+3\varepsilon)m}.$$

Since $\alpha_{m+1} v_{m+1} = v_m - u$, we obtain $1 - \|u\| \leq \alpha_{m+1} \leq 1 + \|u\|$. It follows that $|1 - \alpha_{m+1}| \leq \|u\|$ and

$$\begin{aligned} \|v_m - v_{m+1}\| &\leq \|v_m - \alpha_{m+1} v_{m+1}\| + \|(1 - \alpha_{m+1}) v_{m+1}\| \\ &= \|u\| + |1 - \alpha_{m+1}| \leq 2\|u\| \\ &\leq 2 e^{\varepsilon - \rho_{j+1}(x)} e^{(\rho_j(x)-\rho_{j+1}(x)+3\varepsilon)m}. \end{aligned}$$

One can obtain inductively a sequence of vectors $v_{m+k} \in F_j(x, m+k)$ with $\|v_{m+k}\| = 1$ such that $v_{m+k-1} - \alpha_{m+k} v_{m+k} \in \mathbb{R}^n \setminus F_j(x, m+k)$ for some $\alpha_{m+k} \geq 0$. This sequence satisfies

$$\|v_{m+k} - v_{m+k+1}\| \leq 2 e^{\varepsilon - \rho_{j+1}(x)} e^{(\rho_j(x)-\rho_{j+1}(x)+3\varepsilon)(m+k)}$$

for every sufficiently large m and all $k \geq 0$. Hence for $\ell > k \geq 0$,

$$\begin{aligned} \|v_{m+k} - v_{m+\ell+1}\| &\leq \sum_{i=k}^{\ell} \|v_{m+i} - v_{m+i+1}\| \\ &\leq 2 e^{\varepsilon - \rho_{j+1}(x)} \sum_{i=m+k}^{\infty} e^{(\rho_j(x)-\rho_{j+1}(x)+3\varepsilon)i} \\ &= C e^{(\rho_j(x)-\rho_{j+1}(x)+3\varepsilon)(m+k)}, \end{aligned}$$

where $C = 2\,e^{\varepsilon-\rho_{j+1}(x)}/(1 - e^{\rho_j(x)-\rho_{j+1}(x)+3\varepsilon})$ is independent of m. This means that $\{v_p\}_{p\geq m}$ is a Cauchy sequence. Note that for $j = t(x)$, we have $v_p = v_m$ for every $p \geq m$. Choose $\varepsilon > 0$ such that

$$\tau = \max\{\rho_j(x) - \rho_{j+1}(x) + 3\varepsilon : j = 1, \ldots, t(x) - 1\} < 0.$$

Let $\{v_m^1, \ldots, v_m^n\}$ be an orthonormal basis of $\mathbb{R}^n$ such that the collection of vectors $\{v_m^1, \ldots, v_m^{n_j(x)}\}$ is a basis of $F_j(x, m)$, where $n_j(x) = \dim F_j(x, m)$ and $j = 1, \ldots, t(x)$. For each $i = 1, \ldots, n$, we obtain a Cauchy sequence $\{v_p^i\}_{p\geq m}$ such that for $p \geq m$ and $k, \ell \geq 0$,

$$\|v_{p+k}^i - v_{p+\ell}^i\| \leq C\,e^{\tau(p+\max\{k,\ell\})} \quad \text{and} \quad \|v_p^i - v^i\| \leq C\,e^{\tau p},$$

where $v^i \in \mathbb{R}^n$ is the limit of the sequence $\{v_p^i\}_{p\geq m}$. Increasing m if necessary one can assume that $\{v_p^1, \ldots, v_p^n\}$, for each $p \geq m$, and $\{v^1, \ldots, v^n\}$ are bases of $\mathbb{R}^n$. This implies the convergence of the sequence $\{F_j(x, m)\}_{m\in\mathbb{N}}$. Let $F_j(x)$ denote its limit. Write $\mathcal{A}_j(x, p) = \mathcal{A}(x, p)|F_j(x, p)$. If $v^i \in F_j(x)$ then

$$\begin{aligned}
\|\mathcal{A}_j(x, p)v^i\| &\leq \|\mathcal{A}_j(x, p)v_p^i\| + \|\mathcal{A}_j(x, p)(v_p^i - v^i)\| \\
&\leq e^{(\rho_j(x)+\varepsilon)p} + e^{(\rho_j(x)+\varepsilon)p} C\,e^{\tau p} \\
&\leq 2C\,e^{(\rho_j(x)+\varepsilon)p},
\end{aligned}$$

and hence

$$\varlimsup_{p\to+\infty} \frac{1}{p} \log\|\mathcal{A}_j(x, p)|F_j(x)\| \leq \rho_j(x) + \varepsilon. \tag{3.36}$$

Let $\overline{v}_p^i$ be the projection of v_p^i over $F_{j-1}(x, p)$. If $v^i \notin F_{j-1}(x)$ then there exists $c > 0$ such that $\|v_p^i - \overline{v}_p^i\| \geq c$ for all sufficiently large p. Note that $v_p^i - \overline{v}_p^i \in F_j(x, p) \setminus F_{j-1}(x, p)$ and $\mathcal{A}_j(x, p)\overline{v}_p^i = \mathcal{A}_{j-1}(x, p)\overline{v}_p^i$. This implies that

$$\begin{aligned}
\|\mathcal{A}_j(x, p)v^i\| &\geq \|\mathcal{A}_j(x, p)(v_p^i - \overline{v}_p^i)\| \\
&\quad - \|\mathcal{A}_{j-1}(x, p)\overline{v}_p^i\| - \|\mathcal{A}_j(x, p)(v_p^i - v^i)\| \\
&\geq e^{(\rho_j(x)-\varepsilon)p} c - e^{(\rho_{j-1}(x)-\varepsilon)p} \|\overline{v}_p^i\| - e^{(\rho_j(x)+\varepsilon)p} C\,e^{\tau p}.
\end{aligned}$$

Note that $\{\overline{v}_p^i\}_{p\geq m}$ converges. Decreasing ε if necessary, we may assume that $\rho_j(x) - \varepsilon > \rho_j(x) + \varepsilon + \tau$, and hence

$$\varliminf_{p\to+\infty} \frac{1}{p} \log\|\mathcal{A}_j(x, p)v^i\| \geq \rho_j(x) - \varepsilon \tag{3.37}$$

whenever $v^i \notin F_{j-1}(x)$. Since ε is arbitrary, we conclude from (3.36) and (3.37) that

$$\lim_{p\to+\infty} \frac{1}{p} \log\|\mathcal{A}_j(x, p)v\| = \rho_j(x).$$

for every $v \in F_j(x) \setminus F_{j-1}(x)$. Therefore, $t(x) = p^+(x)$, $F_j(x) = E_j^+(x)$, and $\rho_j(x) = \chi_j^+(x)$ for $j = 1, \ldots, t(x)$. Without loss of generality, we may assume that the matrix S_p whose columns are $v_p^1, \dots, v_p^n$ is unitary, and hence $S_p{}^*(\mathcal{A}(x,p)^*\mathcal{A}(x,p))^{1/2}S_p$ is a diagonal matrix with entries $\lambda_1(x,p) \le \cdots \le \lambda_n(x,p)$ on the diagonal. Since $v_p^i \to v^i$ when $p \to \infty$, we conclude that

$$\lim_{p\to+\infty}(\mathcal{A}(x,p)^*\mathcal{A}(x,p))^{1/(2p)} = \lim_{p\to+\infty} S_p \begin{pmatrix} e^{\chi_1'(x)} & & 0 \\ & \ddots & \\ 0 & & e^{\chi_n'(x)} \end{pmatrix} S_p^*.$$

In particular,

$$\lim_{m\to+\infty}\frac{1}{m}\log|\det\mathcal{A}(x,m)| = \sum_{i=1}^n \chi_i'(x).$$

This shows that x is forward regular. In a similar way one can show that

$$\lim_{m\to-\infty}\frac{1}{|m|}\log\|\mathcal{A}(x,m)v\| = \chi_i'(x)$$

for every $v \in E_j^-(x) \setminus E_{j+1}^-(x)$ and that the following limit exists:

$$\lim_{m\to-\infty}(\mathcal{A}(x,m)^*\mathcal{A}(x,m))^{1/(2|m|)}.$$

Furthermore,

$$\lim_{m\to-\infty}\frac{1}{|m|}\log|\det\mathcal{A}(x,m)| = \sum_{i=1}^{p^-(x)} k_i^-(x)\chi_i^-(x),$$

and hence x is backward regular. We now show that the filtrations of the Lyapunov exponents χ^+ and χ^- are coherent at x. By (3.31) and (3.32),

$$\begin{aligned} 0 &\le \lim_{m\to+\infty}\frac{1}{m}\log\|\mathcal{A}(x,m)^{\wedge i}\| + \lim_{m\to-\infty}\frac{1}{|m|}\log\|\mathcal{A}(x,m)^{\wedge i}\| \\ &= \sum_{k=1}^i \chi_{n-k+1}'(x) + \sum_{k=1}^i \widetilde{\chi}_k'(x), \end{aligned} \tag{3.38}$$

where $\widetilde{\chi}_1'(x) \ge \cdots \ge \widetilde{\chi}_n'(x)$ are the values of the Lyapunov exponent χ^- counted with their multiplicities. For $i = n$,

$$(x,m) \mapsto \|\mathcal{A}(x,m)^{\wedge i}\| = |\det\mathcal{A}(x,m)|$$

is a cocycle with values in $GL(1,\mathbb{R})$. By Theorem 3.4.1, we conclude that

$$\begin{aligned} \sum_{k=1}^n(\chi_{n-k+1}'(x) + \widetilde{\chi}_k'(x)) &= \lim_{m\to+\infty}\frac{1}{m}\log|\det\mathcal{A}(x,m)| \\ &\quad + \lim_{m\to-\infty}\frac{1}{|m|}\log|\det\mathcal{A}(x,m)| = 0. \end{aligned}$$

By (3.38), for each i, we obtain

$$\sum_{k=1}^i \chi_{n-k+1}'(x) + \sum_{k=1}^i \widetilde{\chi}_k'(x) = 0.$$

Thus, $p^-(x) = p^+(x) =: p(x)$ and $\chi_i^-(x) = -\chi_i^+(x)$ for $i = 1, \ldots, p(x)$. In particular, it follows from (3.38) that

$$\lim_{m\to+\infty} \frac{1}{m} \log\|\mathcal{A}(x,m)^{\wedge i}\| + \lim_{m\to-\infty} \frac{1}{|m|} \log\|\mathcal{A}(x,m)^{\wedge i}\| = 0. \tag{3.39}$$

Set $H_i(x) = E_i^+(x) \cap E_i^-(x)$ for each $i = 1, \ldots, p(x)$. Proceeding as in the proof of Lemma 3.4.6, one can show that the convergence in (3.4) is uniform on $\{v \in H_i(x) : \|v\| = 1\}$. It remains to prove that $\bigoplus_{i=1}^{p(x)} H_i(x) = \mathbb{R}^n$. Let $v \in E_i^+(x) \cap E_{i+1}^-(x)$. Then

$$\begin{aligned}\lim_{m\to+\infty} \frac{1}{m} \log\|\mathcal{A}(x,m)v\| &+ \lim_{m\to-\infty} \frac{1}{|m|} \log\|\mathcal{A}(x,m)v\| \\ &\leq \chi_i^+(x) - \chi_{i+1}^+(x) < 0.\end{aligned} \tag{3.40}$$

By Theorem 3.4.8, (3.31), (3.32), and (3.39), we obtain

$$\begin{aligned}&\lim_{m\to+\infty} \frac{1}{m} \log\|\mathcal{A}(x,m)v\| + \lim_{m\to-\infty} \frac{1}{|m|} \log\|\mathcal{A}(x,m)v\| \\ &\geq \lim_{m\to+\infty} -\frac{1}{m} \log\|\mathcal{A}(x,m)^{-1}\| + \lim_{m\to-\infty} \frac{1}{m} \log\|\mathcal{A}(x,m)^{-1}\| \\ &= \lim_{m\to-\infty} \frac{1}{m} \log\|\mathcal{A}(x,m)\| - \lim_{m\to+\infty} \frac{1}{m} \log\|\mathcal{A}(x,m)v\| = 0.\end{aligned} \tag{3.41}$$

It follows from (3.40) and (3.41) that $E_i^+(x) \cap E_{i+1}^-(x) = \{0\}$. One can now proceed as in the proof of Lemma 3.4.5 to show that

$$\bigoplus_{i=1}^{p(x)} H_i(x) = \mathbb{R}^n.$$

Hence the filtrations of the Lyapunov exponents χ^+ and χ^- are coherent at x, and hence x is a regular point. This completes the proof of the theorem. □

3.5 Tempering Kernels and the Reduction Theorems

The goal of this section is to describe a "normal form" of a general measurable cocycle associated with its Lyapunov exponent. Let us begin with a simple particular case of a rigid cocycle $\mathcal{A}$ whose generator is a constant map; that is, $\mathcal{A}(x,m) = A^m$ for some matrix $A \in GL(n, \mathbb{R})$. It is easy to see that the cocycle $\mathcal{A}$ is equivalent to the rigid cocycle $\mathcal{B}$ whose generator is the Jordan block form of the matrix A. We consider $\mathcal{B}$ as the "normal form" of $\mathcal{A}$ and say that $\mathcal{A}$ is reduced to $\mathcal{B}$.

A general measurable cocycle $\mathcal{A}$ satisfying the integrability condition (3.20) is so to speak "weakly" rigid; that is, it can be reduced to a constant cocycle up to arbitrarily small error. We consider this constant cocycle as a "normal form" of $\mathcal{A}$. More precisely, we will show (see the Oseledets–Pesin Reduction Theorem 3.5.5 below) that given $\varepsilon > 0$, there exists a cocycle $\mathcal{A}_\varepsilon$, which is equivalent to $\mathcal{A}$ and has a block form such that the generator A^i_ε of each block satisfies

$$e^{\chi_i(x)-\varepsilon}\|v\| \le \|A^i_\varepsilon(x)v\| \le e^{\chi_i(x)+\varepsilon}\|v\|$$

for each regular point x and each $v \in H_i(x)$ (where $H_i(x)$ are the Oseledets subspaces at x; see (3.2)). We say that $\mathcal{A}_\varepsilon$ is the *reduced form* of $\mathcal{A}$.

3.5.1 Lyapunov Inner Products

Let $\mathcal{A}$ be a measurable cocycle over a measurable transformation $f\colon X \to X$. Consider a family of inner products $\langle\cdot,\cdot\rangle = \langle\cdot,\cdot\rangle_x$ on $\mathbb{R}^n$ for $x \in X$. We start with the following auxiliary result:

Proposition 3.5.1. *For each $\varepsilon > 0$, each regular point $x \in X$, and $i = 1, \dots, p(x)$, the formula*

$$\langle u, v\rangle'_{x,i} = \sum_{m\in\mathbb{Z}} \langle \mathcal{A}(x,m)u, \mathcal{A}(x,m)v\rangle e^{-2\chi_i(x)m-2\varepsilon|m|} \tag{3.42}$$

determines a scalar product on $H_i(x)$.

Proof. By (3.4), for each $\varepsilon > 0$, any regular point x, and any $v \in H_i(x) \setminus \{0\}$, there exists an integer $m_0 = m_0(x, v, \varepsilon) > 0$ such that for $|m| > m_0$,

$$\|\mathcal{A}(x,m)v\| \le e^{\chi_i(x)m+\varepsilon|m|/2}.$$

Therefore if $u, v \in H_i(x)$ then

$$\begin{aligned}\sum_{m\in\mathbb{Z}} &\left|\langle \mathcal{A}(x,m)u, \mathcal{A}(x,m)v\rangle\right| e^{-2\chi_i(x)m-2\varepsilon|m|} \\ &\le \sum_{|m|\le m_0} \left|\langle \mathcal{A}(x,m)u, \mathcal{A}(x,m)v\rangle\right| e^{-2\chi_i(x)m-2\varepsilon|m|} \\ &\quad + \sum_{|m|>m_0} e^{-\varepsilon|m|} < +\infty.\end{aligned}$$

This implies the desired result. □

Let $x \in X$ be a regular point. For a fixed $\varepsilon > 0$, we introduce a new inner product on $\mathbb{R}^n$ by

$$\langle u, v\rangle'_x = \sum_{i=1}^{p(x)} \langle u_i, v_i\rangle'_{x,i},$$

where u_i and v_i are the projections of the vectors u and v over $H_i(x)$ along $\bigoplus_{j\ne i} H_j(x)$.

Definition 3.5.2. We call $\langle\cdot,\cdot\rangle'_x$ the *Lyapunov inner product at x* and the corresponding norm $\|\cdot\|'_x$ the *Lyapunov norm at x*. The sequence of weights $\{e^{-2\chi_i(x)m-2\varepsilon|m|}\}_{m\in\mathbb{Z}}$ in (3.42) is called the *Pesin tempering kernel*. (This is a particular case of the tempering kernels that we will introduce in Section 3.5.3.)

We emphasize that the value of $\langle u, v\rangle'_x$ depends on the number ε. The following result is immediate:

Proposition 3.5.3. *Let $\langle\cdot,\cdot\rangle$ be a Lyapunov inner product at x. Then*

1. *it depends Borel measurably on x on the set of regular points in X;*
2. *for every regular point $x \in X$ and $i \neq j$, the spaces $H_i(x)$ and $H_j(x)$ are orthogonal with respect to the Lyapunov inner product.*

Definition 3.5.4. A coordinate change $C_\varepsilon\colon X \to GL(n,\mathbb{R})$ is called a *Lyapunov change of coordinates* if for each regular point $x \in X$ and $u, v \in \mathbb{R}^n$, it satisfies

$$\langle u, v\rangle_x = \langle C_\varepsilon(x)u, C_\varepsilon(x)v\rangle'_x. \tag{3.43}$$

We note that the formula (3.43) does not determine the function $C_\varepsilon(x)$ uniquely.

3.5.2 The Oseledets–Pesin Reduction Theorem

The following principal result describes "normal forms" of measurable cocycles.

Theorem 3.5.5 (Oseledets–Pesin Reduction Theorem). *Let $f\colon X \to X$ be an invertible measure preserving transformation of the Lebesgue space (X, ν) and $\mathcal{A}$ a measurable cocycle over f. Given $\varepsilon > 0$, if x is a regular point for $\mathcal{A}$ then*

1. *any Lyapunov change of coordinates C_ε sends the orthogonal decomposition $\bigoplus_{i=1}^{p(x)} \mathbb{R}^{k_i(x)}$ to the decomposition $\bigoplus_{i=1}^{p(x)} H_i(x)$ of $\mathbb{R}^n$;*
2. *the cocycle $A_\varepsilon(x) = C_\varepsilon(f(x))^{-1}A(x)C_\varepsilon(x)$ has the block form*

$$A_\varepsilon(x) = \begin{pmatrix} A_\varepsilon^1(x) & & \\ & \ddots & \\ & & A_\varepsilon^{p(x)}(x) \end{pmatrix}, \tag{3.44}$$

   where each block $A_\varepsilon^i(x)$ is a $k_i(x) \times k_i(x)$ matrix, and the entries are zero above and below the matrices $A_\varepsilon^i(x)$;
3. *each block $A_\varepsilon^i(x)$ satisfies*

$$e^{\chi_i(x)-\varepsilon} \le \|A_\varepsilon^i(x)^{-1}\|^{-1} \le \|A_\varepsilon^i(x)\| \le e^{\chi_i(x)+\varepsilon}; \tag{3.45}$$

4. *if the integrability condition (3.20) holds then the map C_ε is tempered ν-almost everywhere, and the spectra of $\mathcal{A}$ and $\mathcal{A}_\varepsilon$ coincide ν-almost everywhere.*

Proof. The first statement follows immediately from (3.43). For each i, we have

$$\begin{aligned} A_\varepsilon(x)\mathbb{R}^{k_i(x)} &= C_\varepsilon(f(x))^{-1}A(x)H_i(x) \\ &= C_\varepsilon(f(x))^{-1}H_i(f(x)) = \mathbb{R}^{k_i(f(x))} = \mathbb{R}^{k_i(x)}. \end{aligned}$$

Thus, $A_\varepsilon(x)$ has the block form (3.44).

Since $\mathcal{A}(x, m+1) = \mathcal{A}(f(x), m)A(x)$, for every $v \in H_i(x)$, we have that

$$\begin{aligned}(\|A(x)v\|'_{f(x)})^2 &= \sum_{m\in\mathbb{Z}} \|\mathcal{A}(f(x), m)A(x)v\|^2 e^{-2m\chi_i(x)-2\varepsilon|m|} \\ &= \sum_{m\in\mathbb{Z}} \|\mathcal{A}(x, m+1)v\|^2 e^{-2m\chi_i(x)-2\varepsilon|m|} \\ &= \sum_{k\in\mathbb{Z}} \|\mathcal{A}(x, k)v\|^2 e^{-2k\chi_i(x)-2\varepsilon|k|+\eta(x,\varepsilon)},\end{aligned}$$

where $\eta(x, \varepsilon) = 2\chi_i(x) + 2\varepsilon(|k| - |k-1|)$. Therefore,

$$e^{\chi_i(x)-\varepsilon}\|v\|'_x \le \|A(x)v\|'_{f(x)} \le e^{\chi_i(x)+\varepsilon}\|v\|'_x \tag{3.46}$$

for every $v \in H_i(x)$. By (3.43), for each $w \in \mathbb{R}^n$, we obtain $\|w\|_x = \|C_\varepsilon(x)w\|'_x$ and

$$\|A_\varepsilon(x)w\|_{f(x)} = \|C_\varepsilon(f(x))^{-1}A(x)C_\varepsilon(x)w\|_{f(x)} = \|A(x)C_\varepsilon(x)w\|'_{f(x)}.$$

Hence if $v = C_\varepsilon(x)w \in H_i(x)$, then

$$\|w\|_x = \|v\|'_x \quad \text{and} \quad \|A_\varepsilon(x)w\|_{f(x)} = \|A(x)v\|'_{f(x)}.$$

By (3.46), for $w \in \mathbb{R}^{k_i(x)}$,

$$e^{\chi_i(x)-\varepsilon}\|w\|_x \le \|A_\varepsilon(x)w\|_{f(x)} \le e^{\chi_i(x)+\varepsilon}\|w\|_x.$$

This implies the inequalities in (3.45).

We now proceed with the proof of the fourth statement. Since $\|w\|_x = \|C_\varepsilon(x)w\|'_x \ge \|C_\varepsilon(x)w\|_x$, we have

$$\|C_\varepsilon(x)\| \le 1 \quad \text{and} \quad \|C_\varepsilon(x)^{-1}\| \ge 1 \tag{3.47}$$

for every regular point $x \in X$. Observe now that

$$\mathcal{A}_\varepsilon(x, m) = C_\varepsilon(f^m(x))^{-1}\mathcal{A}(x, m)C_\varepsilon(x).$$

For each $N > 0$, consider the set

$$X_N = \{x \in X : x \text{ is regular}, \|C_\varepsilon(x)\| > N^{-1} \text{ and } \|C_\varepsilon(x)^{-1}\| < N\}.$$

By the Multiplicative Ergodic Theorem 3.4.3, $\nu(X_N) \to 1$ as $N \to \infty$. Let us fix $N > 0$ such that $\nu(X_N) > 0$. It follows from Poincaré's Recurrence Theorem that there exists a set $Y_N \subset X_N$ such that $\nu(X_N \setminus Y_N) = 0$, and both the forward and backward orbits of each point $y \in Y_N$ return infinitely many times to Y_N. Given $x \in Y_N$, let $m_k = m_k(x)$ be a sequence of positive integers such that $f^{m_k}(y) \in Y_N$ for all $k \ge 1$. For $i = 1, \ldots, p(x)$, we have

$$\begin{aligned}\|\mathcal{A}^i_\varepsilon(x, m_k)\| &\le \|C_\varepsilon(f^{m_k}(x))^{-1}\| \cdot \|\mathcal{A}(x, m_k)|H_i(x)\| \cdot \|C_\varepsilon(x)\| \\ &\le N\|\mathcal{A}(x, m_k)|H_i(x)\|,\end{aligned}$$

where $\mathcal{A}^i_\varepsilon(x, m)$ is the cocycle generated by the block $A^i_\varepsilon(x)$. Hence,

$$\overline{\lim_{m\to+\infty}} \frac{1}{m} \log\|\mathcal{A}^i_\varepsilon(x, m)\| \le \chi_i(x). \tag{3.48}$$

In a similar way, one can show that

$$\begin{aligned}\|\mathcal{A}(x, m_k)|H_i(x)\| &\le \|C_\varepsilon(f^{m_k}(x))\| \cdot \|\mathcal{A}^i_\varepsilon(x, m_k)\| \cdot \|C_\varepsilon(x)^{-1}\| \\ &\le N\|\mathcal{A}^i_\varepsilon(x, m_k)\|,\end{aligned}$$

and hence

$$\chi_i(x) \le \underline{\lim_{m\to+\infty}} \frac{1}{m} \log\|\mathcal{A}^i_\varepsilon(x, m)\|. \tag{3.49}$$

We conclude from (3.48) and (3.49) that for each $x \in Y_N$ and each $v \in \mathbb{R}^{k_i(x)} = C_\varepsilon(x)^{-1} H_i(x)$,

$$\lim_{m\to+\infty} \frac{1}{m} \log\|\mathcal{A}_\varepsilon(x, m)v\| = \chi_i(x).$$

In a similar way, one can prove that

$$\lim_{m\to-\infty} \frac{1}{|m|} \log\|\mathcal{A}_\varepsilon(x, m)v\| = \chi_i(x).$$

Since $\nu(Y_N) \to 1$ as $N \to \infty$, the spectra of $\mathcal{A}$ and $\mathcal{A}_\varepsilon$ coincide ν-almost everywhere.

We now show that the function $C_\varepsilon \colon X \to GL(n, \mathbb{R})$ is tempered. Observe that

$$\begin{aligned}\|C_\varepsilon(f^m(x))^{-1}\| &= \sup_{v\neq 0} \frac{\|\mathcal{A}_\varepsilon(x, m)C_\varepsilon(x)^{-1}\mathcal{A}(x, m)^{-1}v\|}{\|v\|} \\ &= \max_i \sup_{w\in H_i(x)} \frac{\|\mathcal{A}_\varepsilon(x, m)C_\varepsilon(x)^{-1}w\|}{\|\mathcal{A}(x, m)w\|}.\end{aligned}$$

Since $\|C_\varepsilon(x)^{-1}\| \ge 1$, using (3.48) we obtain that for ν-almost all regular points $x \in X$,

$$\begin{aligned}0 &\le \underline{\lim_{m\to\pm\infty}} \frac{1}{|m|} \log\|C_\varepsilon(f^m(x))^{-1}\| \\ &\le \overline{\lim_{m\to\pm\infty}} \frac{1}{|m|} \log\|C_\varepsilon(f^m(x))^{-1}\| \\ &= \max_i \sup_{w\in H_i(x)} \lim_{m\to\pm\infty} \frac{1}{|m|} \log \frac{\|\mathcal{A}_\varepsilon(x, m)C_\varepsilon(x)^{-1}w\|}{\|\mathcal{A}(x, m)w\|} = 0.\end{aligned}$$

This implies that for ν-almost every $x \in X$,

$$\lim_{m\to\pm\infty} \frac{1}{m} \log\|C_\varepsilon(f^m(x))^{-1}\| = 0.$$

In a similar way one can show that for ν-almost every $x \in X$,

$$\lim_{m\to\pm\infty} \frac{1}{m} \log\|C_\varepsilon(f^m(x))\| = 0.$$

This implies that C_ε is a tempered function ν-almost everywhere and completes the proof of the theorem. □

In dimension 2 there is a refined version of the Oseledets–Pesin Reduction Theorem 3.5.5 obtained by Thieullen [212]. We state it without proof. Let $f\colon X \to X$ be a measurable transformation preserving an ergodic probability measure ν. Given $r > 0$, we say that a function $\varphi\colon X \to \mathbb{R}$ is *cohomologous to zero mod* r if there is a measurable function $u\colon X \to \mathbb{R}$ such that $\varphi + u \circ f - u \in r\mathbb{Z}$ ν-almost everywhere.

Theorem 3.5.6. *Let $\mathcal{A}\colon X \times \mathbb{Z} \to SL(2,\mathbb{R})$ be a measurable cocycle over f whose generator A satisfies* (3.20). *Then there exists a map $P\colon X \to SL(2,\mathbb{R})$ such that $B(x) = P(f(x))^{-1}A(x)P(x)$ takes one of the following forms ν-almost everywhere:*

1. $B(x) = \begin{pmatrix} a(x) & 0 \\ 0 & 1/a(x) \end{pmatrix}$ *with* $\int_X \log|a|\,d\nu > 0$.
2. $B(x) = \begin{pmatrix} a(x) & b(x) \\ 0 & 1/a(x) \end{pmatrix}$ *with* $\int_X \log|a|\,d\nu = 0$ *and* $\log|b|$ *ν-integrable.*
3. $B(x) = \begin{pmatrix} \cos\theta(x) & -\sin\theta(x) \\ \sin\theta(x) & \cos\theta(x) \end{pmatrix}$ *with θ not cohomologous to zero mod π.*
4. $B(x) = R_E(x)\begin{pmatrix} a(x) & 0 \\ 0 & 1/a(x) \end{pmatrix}$ *with $\log|a|$ ν-integrable and the characteristic function of $E \subset X$ not cohomologous to zero mod 2, where $R_E(x)$ is the rotation by $\pi/2$ if $x \in E$ and $R_E(x) = \mathrm{Id}$ otherwise.*

The values of the Lyapunov exponent are $\pm \int_X \log|a|\,d\nu$ in the first case and zero in the other three. In all cases, $\|B(x)\| \le \|A(x)\|$ and thus the generator B satisfies (3.20).

3.5.3 A Tempering Kernel

Another important manifestation of the Oseledets–Pesin Reduction Theorem 3.5.5 is a crucial property of Lyapunov norms. It states that the function $x \mapsto \|v(x)\|'_x / \|v(x)\|_x$ is tempered on the set of regular points for every measurable vector field $X \ni x \mapsto v(x) \in \mathbb{R}^n \setminus \{0\}$. We recall that a positive function $K\colon X \to \mathbb{R}$ is said to be *tempered* on a set $Z \subset X$ if for any $x \in Z$,

$$\lim_{m\to\pm\infty} \frac{1}{m} \log K(f^m(x)) = 0. \tag{3.50}$$

We begin with a technical but crucial statement known as the Tempering Kernel Lemma.

Lemma 3.5.7. *Let $f: X \to X$ be a measurable transformation. If $K: X \to \mathbb{R}$ is a positive measurable function tempered on some subset $Z \subset X$, then for any $\varepsilon > 0$, there exists a positive measurable function $K_\varepsilon: Z \to \mathbb{R}$ such that $K(x) \le K_\varepsilon(x)$ and for $x \in Z$,*

$$e^{-\varepsilon} \le \frac{K_\varepsilon(f(x))}{K_\varepsilon(x)} \le e^{\varepsilon}. \tag{3.51}$$

Proof. By (3.50), there exists a measurable function $C(x, \varepsilon) > 0$ such that for every $m \in \mathbb{Z}$,

$$K(f^m(x)) \le C(x, \varepsilon) e^{\varepsilon|m|/2}. \tag{3.52}$$

Define

$$K_\varepsilon(x) = \sum_{m \in \mathbb{Z}} K(f^m(x)) e^{-\varepsilon|m|}.$$

Since K is positive, we have $K(x) \le K_\varepsilon(x)$. By (3.52),

$$K_\varepsilon(x) \le C(x, \varepsilon) \sum_{m \in \mathbb{Z}} e^{-\varepsilon|m|/2} < +\infty.$$

Moreover,

$$\begin{aligned} K_\varepsilon(f(x)) &= \sum_{m \in \mathbb{Z}} K(f^{m+1}(x)) e^{-\varepsilon|m|} \\ &= \sum_{k \in \mathbb{Z}} K(f^k(x)) e^{-\varepsilon|k|} e^{\varepsilon(|k|-|k-1|)}, \end{aligned}$$

and hence, $e^{-\varepsilon} \le K_\varepsilon(f(x))/K_\varepsilon(x) \le e^{\varepsilon}$. □

The function K_ε satisfying (3.51) is called a *tempering kernel*. Note that if f preserves a Lebesgue measure ν on the space X, then by Lemma 2.1.5, any positive function $K: X \to \mathbb{R}$ with $\log K \in L^1(X, \nu)$ satisfies (3.50).

Proposition 3.5.8. *Given $\varepsilon > 0$ there is a positive measurable function $K_\varepsilon: X \to \mathbb{R}$ such that if $x \in X$ is a regular point then*

1. $K_\varepsilon(x) e^{-\varepsilon|m|} \le K_\varepsilon(f^m(x)) \le K_\varepsilon(x) e^{\varepsilon|m|}$ *for every* $m \in \mathbb{Z}$*;*
2. $n^{-1/2}\|v\|_x \le \|v\|'_x \le K_\varepsilon(x)\|v\|_x$ *for every* $v \in \mathbb{R}^n$*;*
3. *for every measurable vector field $X \ni x \mapsto v(x) \in \mathbb{R}^n \setminus \{0\}$, the function $x \mapsto \|v(x)\|'_x / \|v(x)\|_x$ is tempered on the set of regular points.*

Proof. The first inequality in Statement 2 follows from the well-known inequality

$$n \sum_{i=1}^{n} a_i^2 \ge \left(\sum_{i=1}^{n} a_i \right)^2,$$

with equality if and only if $a_1 = \cdots = a_n$. Given a regular point $x \in X$ and $\varepsilon > 0$, let $C_\varepsilon(x)$ be the matrix defined by (3.43). Then $\|v\|'_x \le \|C_\varepsilon(x)^{-1}\| \cdot \|v\|_x$ for each $v \in \mathbb{R}^n$. By Theorem 3.5.5, the function C_ε is tempered. Thus, we can apply the

Tempering Kernel Lemma 3.5.7 to the positive function $K(x) = \|C_\varepsilon(x)^{-1}\|$, and we find a function K_ε satisfying the desired properties. □

3.5.4 Zimmer's Amenable Reduction

We describe a result essentially due to Zimmer, which provides some additional information on the Oseledets decomposition at regular points (see Section 3.1). It is given in terms of maximal amenable subgroups of the groups $GL(n, \mathbb{R})$.

Recall that a locally compact topological group G is called *amenable* if any continuous action of G on a compact (nonempty) topological space Z has an invariant probability measure. We are interested here in the case of subgroups of $GL(n, \mathbb{R})$ and the space $Z = \mathbb{R}^n$.

It is known that there exist precisely 2^n conjugacy classes of maximal amenable subgroups of $GL(n, \mathbb{R})$ (see [155]). One can obtain a representative of every such conjugacy class in the following way: fix $r \geq 1$ and let $n = n_1 + \cdots + n_r$ be an ordered partition of n into r positive integers. For every such partition, we construct the group of upper triangular matrices

$$\begin{pmatrix} A_1 & & & \\ 0 & A_2 & & \\ \vdots & \ddots & \ddots & \\ 0 & \cdots & 0 & A_r \end{pmatrix},$$

where each diagonal block A_i has size $n_i \times n_i$ and consists of the scalar multiples of the $n_i \times n_i$ orthogonal matrices; that is, $A_i \in O(n_i, \mathbb{R}) \otimes (\mathbb{R} \setminus \{0\})$. The elements above the diagonal blocks are arbitrary, and those below are zero. Every maximal amenable subgroup of $GL(n, \mathbb{R})$ is conjugate to one of these groups. We choose a representation G_i for every conjugacy class, for $i = 1, \ldots, 2^n$.

The following theorem is a particular case of Corollary 1.8 in [91]. (It is essentially implicit in [239].)

Theorem 3.5.9. *If $A\colon X \to GL(n, \mathbb{R})$ is a measurable cocycle, then there exists an equivalent cocycle $B\colon X \to GL(n, \mathbb{R})$ and a measurable decomposition $X = \bigcup_{i=1}^{2^n} X_i$ such that $B(X_i) \subset G_i$.*

One can apply Theorem 3.5.9 to the Oseledets decomposition to conclude that, in each component of dimension m, the action of the cocycle is equivalent to a subaction of the product action of $(O(n_1, \mathbb{R}) \otimes \mathbb{R}^*) \times \cdots \times (O(n_r, \mathbb{R}) \otimes \mathbb{R}^*)$, where $n_1 + \cdots + n_r = m$ for some $r, n_1, \ldots, n_r \geq 1$.

3.5.5 The Case of Noninvertible Cocycles

Consider a measure preserving transformation $f\colon X \to X$ of a Lebesgue space (X, ν). (The map f need not be invertible.) We assume that ν is a probability measure. Given a measurable function $A\colon X \to GL(n, \mathbb{R})$ and $x \in X$, we define the *one-sided cocycle* $\mathcal{A}: X \times \mathbb{N} \to GL(n, \mathbb{R})$ by

$$\mathcal{A}(x, m) = A(f^{m-1}(x)) \cdots A(f(x))A(x).$$

Note that the cocycle equation (2.1) holds for every $m, k \in \mathbb{N}$. Given $(x, v) \in X \times \mathbb{R}^n$, we define again the *forward upper Lyapunov exponent of* (x, v) *(with respect to the cocycle* $\mathcal{A}$*)* by

$$\chi^+(x, v) = \chi^+(x, v, \mathcal{A}) = \varlimsup_{m \to +\infty} \frac{1}{m} \log\|\mathcal{A}(x, m)v\|.$$

However, since the map f and the matrices $A(x)$ may not be invertible, one may not in general define a backward upper Lyapunov exponent. Therefore, we can discuss only the forward regularity for $\mathcal{A}$. One can establish a version of the Multiplicative Ergodic Theorem in this case. Its statement can be found in the appendix in [110]. The proof of the uniform convergence appeared in [24].

Theorem 3.5.10. *Let f be a measure preserving transformation of a Lebesgue space (X, ν) and $\mathcal{A}$ a measurable cocycle over f such that* $\log^+\|A\| \in L^1(X, \nu)$. *Then the set of forward regular points for $\mathcal{A}$ has full ν-measure and for every forward regular point $x \in X$ and every subspace $F \subset E_i^+(x)$ such that $F \cap E_{i-1}^+(x) = \{0\}$, we have*

$$\lim_{m \to +\infty} \frac{1}{m} \log \inf_v \|\mathcal{A}(x, m)v\| = \lim_{m \to +\infty} \frac{1}{m} \log \sup_v \|\mathcal{A}(x, m)v\| = \chi_i^+(x),$$

with the infimum and supremum taken over $\{v \in F : \|v\| = 1\}$.

When the matrix $A(x)$ is invertible for every $x \in X$ and $\log^+\|A\|, \log^+\|A^{-1}\| \in L^1(X, \nu)$ for some f-invariant Lebesgue measure ν, one can show that for the cocycle induced by $\mathcal{A}$ on the inverse limit of f, the set of *regular* points has full ν-measure.

3.6 More results on Lyapunov–Perron regularity

In the previous section we described the theory of Lyapunov exponents for measurable cocycles over measurable transformations of a Lebesgue space X, that is, the action of the group $\mathbb{Z}$ on X. In this section we describe some similar results for other actions on X, including actions of $\mathbb{R}$ (i.e., flows on X) as well as actions of higher-rank Abelian groups. We also consider cocycles with values in other groups then $GL(n, \mathbb{R})$.

3.6.1 Higher-Rank Abelian Actions

We outline the theory of cocycles over higher-rank Abelian actions. Let (X, ν) be a Lebesgue space, and $p \in \mathbb{N}$.

Definition 3.6.1. A measurable function $\alpha: X \times \mathbb{Z}^d \to X$ is called a *measure preserving* $\mathbb{Z}^d$*-action on* X if the following properties hold:

1. $\alpha(x, 0) = x$ for every $x \in X$.
2. $\alpha(x, m + k) = \alpha(x, m)\alpha(x, k)$ for every $x \in X$ and $m, k \in \mathbb{Z}^d$.
3. The map $x \mapsto \alpha(x, m)$ preserves ν for each $m \in \mathbb{Z}^d$.

Definition 3.6.2. A measurable function $\mathcal{A}\colon X \times \mathbb{Z}^d \to GL(n, \mathbb{R})$ is called a $\mathbb{Z}^d$-*cocycle* over α if for every $x \in X$,

1. $\mathcal{A}(x, 0) = \mathrm{Id}$;
2. for $m, k \in \mathbb{Z}^d$,

$$\mathcal{A}(x, m + k) = \mathcal{A}(\alpha(x, k), m)\mathcal{A}(x, k). \tag{3.53}$$

A $\mathbb{Z}^d$-cocycle over α induces a *linear extension* $F\colon X \times \mathbb{R}^n \times \mathbb{Z}^d \to X \times \mathbb{R}^n \times \mathbb{Z}^d$ by

$$F(x, v, m) = (\alpha(x, m), \mathcal{A}(x, m)v).$$

Given $(x, v, m) \in X \times \mathbb{R}^n \times \mathbb{Z}^d$, we define the *forward Lyapunov exponent of* (x, v) *along* m *(with respect to the cocycle* $\mathcal{A}$*)* by

$$\chi^+(x, v, m) = \overline{\lim_{\tau \to +\infty}} \frac{1}{\tau} \log\|\mathcal{A}(x, \tau m)v\|.$$

Let $\{e_1, \ldots, e_d\}$ denote the standard basis of $\mathbb{R}^d$.

Proposition 3.6.3. *For each $x \in X$, there exist an integer $1 \le p^+(x) \le n$ and distinct linear functionals $\chi_i^+(x)\colon \mathbb{R}^d \to \mathbb{R}$, $i = 1, \ldots, p^+(x)$, such that*

1. *the function p^+ is measurable and f-invariant;*
2. *if*

$$E_i^{\ell+}(x) := \{v \in \mathbb{R}^n : \chi^+(x, v, e_\ell) \le \chi_i^+(x)e_\ell\},$$

then $\chi^+(x, v, e_\ell) = \chi_i^+(x)e_\ell$ for every $v \in E_i^{\ell+}(x) \setminus E_{i-1}^{\ell+}(x)$.

Proof. The functionals $\chi_i^+(x)$ can be constructed by considering successively $\ell = 1, \ldots, p$ and defining $\chi_i^+(x)$ in such a way that $\chi_1^+(x)e_\ell, \ldots, \chi_{p^+(x)}^+(x)e_\ell$ are the values of the Lyapunov exponent associated to the $\mathbb{Z}$-cocycle $(x, \tau) \mapsto \mathcal{A}(x, \tau e_\ell)$ (possibly with repetitions). We leave the details to the reader. □

The spaces $E_i^{\ell+}(x)$ form the *filtration* $\mathcal{V}_x^{\ell+}$ of $\mathbb{R}^n$ associated to $\chi^+(x)$. For each $i = 1, \ldots, p^+(x)$, we set

$$k_i^{\ell+}(x) = \dim E_i^{\ell+}(x) - \dim E_{i-1}^{\ell+}(x).$$

Note that $k_i^{\ell+}(x)$ may be zero. In a similar fashion, for every $(x, v, m) \in X \times \mathbb{R}^n \times \mathbb{Z}^d$, we set

$$\chi^-(x, v, m) = \overline{\lim_{\tau \to -\infty}} \frac{1}{|\tau|} \log\|\mathcal{A}(x, \tau m)v\|,$$

and we call $\chi^-(x, v, m)$ the *backward Lyapunov exponent of* (x, v) *along* m *(with* *respect to the cocycle $\mathcal{A}$).* There exist an integer $1 \le p^-(x) \le n$ and distinct linear functionals $\chi_i^-(x)\colon \mathbb{R}^d \to \mathbb{R}$, $i = 1, \dots, p^-(x)$, such that

1. the function p^- is measurable and f-invariant;
2. if
$$E_i^{\ell -}(x) := \{v \in \mathbb{R}^n : \chi^-(x, v, e_\ell) \le \chi_i^-(x)e_\ell\},$$
then $\chi^-(x, v, te_\ell) = \chi_i^-(x)e_\ell$ for every $v \in E_i^{\ell -}(x) \setminus E_{i-1}^{\ell -}(x)$.

The spaces $E_i^{\ell -}(x)$ form the *filtration* $\mathcal{V}_x^{\ell -}$ of $\mathbb{R}^n$ associated to $\chi^-(x)$. For each $i = 1, \dots, p^-(x)$, we set

$$k_i^{\ell -}(x) = \dim E_i^{\ell -}(x) - \dim E_{i-1}^{\ell -}(x).$$

Note that $k_i^{\ell -}(x)$ may be zero. For every $x \in X$, one can easily show that

$$\chi_i^+(\alpha(x, m)) = \chi_i^+(x) \text{ for every } i = 1, \dots, p^+(x)$$

and that

$$\chi_i^-(\alpha(x, m)) = \chi_i^-(x) \text{ for every } i = 1, \dots, p^-(x).$$

Write $\mathcal{V}^+ = \{\mathcal{V}_x^{\ell +}\}_{x \in X, 1 \le \ell \le p^+(x)}$ and $\mathcal{V}_x^- = \{\mathcal{V}_x^{\ell -}\}_{x \in X, 1 \le \ell \le p^-(x)}$.

Definition 3.6.4. We say that the families of filtrations $\mathcal{V}^+$ and $\mathcal{V}^-$ are *coherent* at the point $x \in X$ if the following properties hold:

1. $p^+(x) = p^-(x) =: p(x)$.
2. There exists a decomposition
$$\mathbb{R}^n = \bigoplus_{j=1}^{p(x)} H_j(x)$$
into subspaces $H_j(x)$ such that $\mathcal{A}(x, m)H_j(x) = H_j(\alpha(x, m))$ for every $m \in \mathbb{Z}^d$, and
(a) if $E_i^{\ell +}(x) \cap H_j(x) \ne \varnothing$ then $H_j(x) \subset E_i^{\ell +}(x)$;
(b) if $E_i^{\ell -}(x) \cap H_j(x) \ne \varnothing$ then $H_j(x) \subset E_i^{\ell -}(x)$.
3. $\chi_i^+(x) = -\chi_i^-(x) =: \chi_i(x)$.
4. For $v \in H_i(x) \setminus \{0\}$ and $m \in \mathbb{Z}^d \setminus \{0\}$,
$$\lim_{\tau \to \pm\infty} \frac{1}{\tau} \log\|\mathcal{A}(x, \tau m)v\| = \chi_i(x)\frac{m}{\|m\|},$$
with uniform convergence on $\{v \in H_i(x) : \|v\| = 1\}$.

Definition 3.6.5. We say that x is a *forward regular point* for $\mathcal{A}$ if for every $m \in \mathbb{Z}^d \setminus \{0\}$ and $\ell = 1, \dots, p$, the following limit exists:

$$\lim_{\tau \to +\infty} \frac{1}{\tau} \log|\det \mathcal{A}(x, \tau m)| = \sum_{i=1}^{p^+(x)} k_i^{\ell +}(x)\chi_i^+(x)\frac{m}{\|m\|}. \tag{3.54}$$

We say that x is a *backward regular point* for $\mathcal{A}$ if for every $m \in \mathbb{Z}^d \setminus \{0\}$ and $\ell = 1, \dots, d$, the following limit exists:

$$\lim_{\tau \to -\infty} \frac{1}{|\tau|} \log|\det \mathcal{A}(x, \tau m)| = \sum_{i=1}^{p^-(x)} k_i^{\ell-}(x)\chi_i^-(x)\frac{m}{\|m\|}. \tag{3.55}$$

We say that x is a *Lyapunov–Perron regular point* or simply a *regular point* if x is simultaneously forward and backward regular, and the filtrations $\mathcal{V}^+$ and $\mathcal{V}^-$ are coherent at x (compare with Definition 3.1.5).

If x is a regular point, then $k_i^{\ell+}(x) = k_i^{\ell-}(x) =: k_i^\ell(x)$ for every $\ell = 1, \dots, d$ and $i = 1, \dots, p(x)$.

Theorem 3.6.6. *Let α be a measure preserving $\mathbb{Z}^d$-action on a Lebesgue space (X, ν) and $\mathcal{A}\colon X \times \mathbb{Z}^d \to GL(n, \mathbb{R})$ a measurable $\mathbb{Z}^d$-cocycle over α such that*

$$\log^+ \|\mathcal{A}(\cdot, m)\| \in L^1(X, \nu) \tag{3.56}$$

for every $m \in \mathbb{Z}^d$. Then the set of regular points for $\mathcal{A}$ has full ν-measure.

Proof. We use induction in ℓ to derive the statement from the Multiplicative Ergodic Theorem 3.4.3 (for $\mathbb{Z}$-cocycles). First, we apply Theorem 3.4.3 to the $\mathbb{Z}$-cocycle

$$X \times \mathbb{Z} \ni (x, m) \mapsto \mathcal{A}(x, me_1), \tag{3.57}$$

over the $\mathbb{Z}$-action

$$X \times \mathbb{Z} \ni (x, m) \mapsto \alpha(x, me_1).$$

We obtain the Oseledets decomposition

$$H_1^1(x) \oplus \cdots \oplus H_{p^1(x)}^1(x) = \mathbb{R}^n$$

on a set of full ν-measure on which (3.54) and (3.55) hold with $\ell = 1$. This decomposition is not only invariant under the $\mathbb{Z}$-cocycle in (3.57) but also under the original cocycle $\mathcal{A}$. Therefore, one can apply Theorem 3.4.3 to each of the $\mathbb{Z}$-cocycles

$$X \times \mathbb{Z} \ni (x, m) \mapsto \mathcal{A}(x, me_2)|H_i^1(x), \quad i = 1, \dots, p^1(x),$$

and obtain the Oseledets decompositions

$$H_{i1}^2(x) \oplus \cdots \oplus H_{ip_i^2(x)}^2(x) = H_i^1(x)$$

on a set of full ν-measure on which (3.54) and (3.55) hold with $\ell = 1, 2$. These decompositions are again invariant under the cocycle $\mathcal{A}$. We finally obtain the decomposition

$$\bigoplus_{i=1}^{p^1(x)} \bigoplus_{j=1}^{p_i^2(x)} H_{ij}^2(x) = \mathbb{R}^n.$$

Repeating this procedure for all the generators of $\mathbb{Z}^d$, we obtain a full ν-measure set on which (3.54) and (3.55) hold with $\ell = 1, \ldots, d$. For every point x in this set, we have a decomposition

$$H_1(x) \oplus \cdots \oplus H_{p(x)}(x) = \mathbb{R}^n,$$

which is invariant under the cocycle $\mathcal{A}$. Furthermore, for every $i = 1, \ldots, p(x)$, $v \in H_i(x) \setminus \{0\}$, and $m \in \mathbb{Z}^d \setminus \{0\}$,

$$\lim_{\tau \to \pm\infty} \frac{1}{\tau} \log \|\mathcal{A}(x, \tau m)v\| = \chi_i^+(x, v) \frac{m}{\|m\|}$$

with uniform convergence on $\{v \in H_i(x) : \|v\| = 1\}$. This concludes the proof of the theorem. □

By (3.53), Condition (3.56) is equivalent to the following condition:

$$\log^+ \|\mathcal{A}(\cdot, m)\| \in L^1(X, \nu)$$

for every m in a subset $S \subset \mathbb{Z}^d$, which generates $\mathbb{Z}^d$ and such that $-m \in S$ whenever $m \in S$.

Given $\varepsilon > 0$ and a regular point $x \in X$, we introduce a family of inner products $\langle \cdot, \cdot \rangle_x$ on $\mathbb{R}^n$ by setting

$$\langle u, v \rangle'_x = \sum_{m \in \mathbb{Z}^d} \langle \mathcal{A}(x, m)u, \mathcal{A}(x, m)v \rangle e^{-2\chi_i(x)m - 2\varepsilon\|m\|}$$

if $u, v \in H_i(x)$, and $\langle u, v \rangle'_x = 0$ if $u \in H_i(x)$ and $v \in H_j(x)$ with $i \neq j$. We call $\langle \cdot, \cdot \rangle'_x$ a *Lyapunov inner product at* x and the corresponding norm $\|\cdot\|'_x$ a *Lyapunov norm at* x. One can show that there exists a tempered function $K_\varepsilon : X \to \mathbb{R}$ such that if $x \in X$ is a regular point and $v \in \mathbb{R}^n$ then

$$n^{-1/2} \|v\|_x \le \|v\|'_x \le K_\varepsilon(x) \|v\|_x.$$

Lyapunov inner products can be used as a convenient technical tool in proving the following generalization to cocycles over $\mathbb{Z}^d$-actions of the Reduction Theorem 3.5.5. Recall that a Lyapunov change of coordinate is a function $C_\varepsilon : X \to GL(n, \mathbb{R})$ satisfying (3.43).

Theorem 3.6.7 (Reduction Theorem for $\mathbb{Z}^d$-actions). *Let α be a measure preserving $\mathbb{Z}^d$-action on a Lebesgue space (X, ν) and $\mathcal{A}$ a measurable $\mathbb{Z}^d$-cocycle over α. Given $\varepsilon > 0$, $m \in \mathbb{Z}^d$, a regular point x for $\mathcal{A}$, and a Lyapunov change of coordinates C_ε, the following properties hold:*

1. *The cocycle $\mathcal{A}_\varepsilon(x, m) = C_\varepsilon(f(x))^{-1} \mathcal{A}(x, m) C_\varepsilon(x)$ has the block form*

$$\mathcal{A}_\varepsilon(x, m) = \begin{pmatrix} \mathcal{A}_\varepsilon^1(x, m) & & \\ & \ddots & \\ & & \mathcal{A}_\varepsilon^{p(x)}(x, m) \end{pmatrix},$$

where each block $\mathcal{A}^i_\varepsilon(x,m) \in GL(\dim H_i(x), \mathbb{R})$*, and the entries are zero above and below the matrices* $\mathcal{A}^i_\varepsilon(x,m)$*.*

2. *Each block* $\mathcal{A}^i_\varepsilon(x,m)$ *satisfies*

$$e^{\chi_i(x)m-\varepsilon\|m\|} \le \|\mathcal{A}^i_\varepsilon(x,m)^{-1}\|^{-1} \le \|\mathcal{A}^i_\varepsilon(x,m)\| \le e^{\chi_i(x)m+\varepsilon\|m\|}.$$

3. *If* A *satisfies the integrability property* (3.56) *then the map* C_ε *is tempered* ν*-almost everywhere, and the spectra of* $\mathcal{A}$ *and* $\mathcal{A}_\varepsilon$ *coincide* ν*-almost everywhere.*

For $x \in X$ and $i = 1, \ldots, p(x)$, we denote by $\ker \chi_i(x) \subset \mathbb{R}^d$ the kernel of the functional $\chi_i(x)$. The hyperplane $\ker \chi_i(x)$ is called a *Lyapunov hyperplane.* (When $p = 2$, it is also called *Lyapunov line.*) The subspaces $[\chi_i(x)]^{-1}(-\infty, 0)$ and $[\chi_i(x)]^{-1}(0, \infty)$ of $\mathbb{R}^d$ are respectively called *negative* and *positive Lyapunov half-spaces*, and any connected component of $\mathbb{R}^d \setminus \bigcup_{i=1}^{p(x)} \ker \chi_i(x)$ is called a *Weyl chamber* of $\mathcal{A}$ at x.

Note that each Weyl chamber W at x is an open convex polyhedral cone in $\mathbb{R}^d$; moreover, for each $i = 1, \ldots, p(x)$, the set $\{\chi_i(x)m : m \in W\}$ is fully contained in $(-\infty, 0)$ or in $(0, +\infty)$.

Example 3.6.8. *Choose matrices* $A_1, \ldots, A_p \in GL(n, \mathbb{R})$ *such that* $A_iA_j = A_jA_i$ *for every* $i, j = 1, \ldots, d$*. We obtain a* $\mathbb{Z}^d$*-cocycle on* $\mathbb{R}^n$ *by setting*

$$\mathcal{A}(x,m) = {A_1}^{m_1} \cdots {A_d}^{m_d},$$

where $m = (m_1, \ldots, m_d)$*. The functions* $\chi_i \colon \mathbb{R}^d \to \mathbb{R}$ *are independent of* x*, and the Lyapunov hyperplanes are given by* $\{m \in \mathbb{R}^d : \chi_i m = 0\}$*.*

3.6.2 The Case of Flows

We briefly discuss counterparts to the results in this chapter for flows. Let (X, ν) be a Lebesgue space.

Definition 3.6.9. The measurable map $\varphi \colon \mathbb{R} \times X \to X$ is called a *measurable flow* on X if

$$\varphi_0 = \mathrm{Id}, \text{ and } \varphi_t \circ \varphi_s = \varphi_{t+s} \text{ for every } t, s \in \mathbb{R}. \tag{3.58}$$

A measurable flow $\varphi \colon \mathbb{R} \times X \to X$ *preserves* a measure ν if for every t, the map $\varphi_t := \varphi(t, \cdot)$ preserves ν.

We note that given a family $\{\varphi_t : t \in \mathbb{R}\}$ of measurable maps $\varphi_t \colon X \to X$ satisfying (3.58), one can define a measurable flow $\varphi \colon \mathbb{R} \times X \to X$ by setting $\varphi(t,x) = \varphi_t(x)$.

Definition 3.6.10. Let φ be a measurable flow on X. A measurable function $\mathcal{A} \colon X \times \mathbb{R} \to GL(n, \mathbb{R})$ is called a *linear multiplicative cocycle over* φ or simply a *cocycle* if for every $x \in X$,

1. $\mathcal{A}(x, 0) = \mathrm{Id}$;
2. $\mathcal{A}(x, t+s) = \mathcal{A}(\varphi_t(x), s)\mathcal{A}(x, t)$ for $t, s \in \mathbb{R}$.

The cocycle $\mathcal{A}$ induces *linear extensions* $F_t \colon X \times \mathbb{R}^n \to X \times \mathbb{R}^n$ by the formula

$$F_t(x, v) = (\varphi_t(x), \mathcal{A}(x, t)v).$$

Given $(x, v) \in X \times \mathbb{R}^n$, we define the *forward Lyapunov exponent of* (x, v) *(with respect to the cocycle* $\mathcal{A}$*)* by

$$\chi^+(x, v) = \chi^+(x, v, \mathcal{A}) = \varlimsup_{t \to +\infty} \frac{1}{t} \log\|\mathcal{A}(x, t)v\|.$$

For every $x \in X$, there exist a positive integer $p^+(x) \le n$, a collection of values

$$\chi_1^+(x) < \chi_2^+(x) < \cdots < \chi_{p^+(x)}^+(x)$$

and linear spaces

$$\{0\} = E_0^+(x) \subsetneqq E_1^+(x) \subsetneqq \cdots \subsetneqq E_{p^+(x)}^+(x) = \mathbb{R}^n, \tag{3.59}$$

such that

1. $E_i^+(x) = \{v \in \mathbb{R}^n : \chi^+(x, v) \le \chi_i^+(x)\}$;
2. if $v \in E_i^+(x) \setminus E_{i-1}^+(x)$, then $\chi^+(x, v) = \chi_i^+(x)$.

We refer to the collection of linear spaces $E_i^+(x)$ as the *filtration* $\mathcal{V}_x^+$ of $\mathbb{R}^n$ associated to χ^+ at x. For each i, the number

$$k_i^+(x) = \dim E_i^+(x) - \dim E_{i-1}^+(x)$$

is called the *multiplicity* of the value $\chi_i^+(x)$.

In a similar way, for every $(x, v) \in X \times \mathbb{R}^n$ we set

$$\chi^-(x, v) = \chi^-(x, v, \mathcal{A}) = \varlimsup_{t \to -\infty} \frac{1}{|t|} \log\|\mathcal{A}(x, t)v\|,$$

and we call $\chi^-(x, v)$ the *backward Lyapunov exponent of* (x, v) *(with respect to the cocycle* $\mathcal{A}$*)*. There exists a positive integer $p^-(x) \le n$, a collection of *values*

$$\chi_1^-(x) > \cdots > \chi_{p^-(x)}^-(x)$$

and the *filtration* $\mathcal{V}_x^-$ of $\mathbb{R}^n$ associated to χ^- at x,

$$\mathbb{R}^n = E_1^-(x) \supsetneqq \cdots \supsetneqq E_{p^-(x)}^-(x) \supsetneqq E_{p^-(x)+1}^-(x) = \{0\}, \tag{3.60}$$

where $E_i^-(x) = \{v \in \mathbb{R}^n : \chi^-(x, v) \le \chi_i^-(x)\}$. The number

$$k_i^-(x) = \dim E_i^-(x) - \dim E_{i+1}^-(x)$$

is called the *multiplicity* of the value $\chi_i^-(x)$.

Write $\mathcal{V}^+ = \{\mathcal{V}_x^+\}_{x \in X}$ and $\mathcal{V}_x^- = \{\mathcal{V}_x^-\}_{x \in X}$.

We say that the families of filtrations $\mathcal{V}^+$ and $\mathcal{V}^-$ are *coherent* at the point $x \in X$ if the following properties hold:

1. $p^+(x) = p^-(x) =: p(x)$.
2. There exists a decomposition

$$\mathbb{R}^n = \bigoplus_{j=1}^{p(x)} H_j(x) \tag{3.61}$$

into subspaces $H_j(x)$ such that $\mathcal{A}(x,t)H_j(x) = H_j(\varphi_t(x))$ for every $t \in \mathbb{R}$ and

$$E_i^+(x) = \bigoplus_{j=1}^{i} H_j(x) \quad \text{and} \quad E_i^-(x) = \bigoplus_{j=i}^{p(x)} H_j(x).$$

3. $\chi_i^+(x) = -\chi_i^-(x) =: \chi_i(x)$.
4. If $v \in H_i(x) \setminus \{0\}$ and $i = 1, \ldots, p(x)$ then

$$\lim_{t\to\pm\infty} \frac{1}{t} \log\|\mathcal{A}(x,t)v\| = \chi_i(x),$$

with uniform convergence on $\{v \in H_i(x) : \|v\| = 1\}$.

The decomposition (3.61) is called the *Oseledets decomposition* at $x \in X$ and the subspaces $H_i(x)$ are called the *Oseledets subspaces*.

We say that x is a *forward regular point* for $\mathcal{A}$ if the following limit exists:

$$\lim_{t\to+\infty} \frac{1}{t} \log|\det \mathcal{A}(x,t)| = \sum_{i=1}^{p^+(x)} \chi_i^+(x) k_i^+(x).$$

We say that x is a *backward regular point* for $\mathcal{A}$ if the following limit exists:

$$\lim_{t\to-\infty} \frac{1}{|t|} \log|\det \mathcal{A}(x,t)| = \sum_{i=1}^{p^-(x)} \chi_i^-(x) k_i^-(x).$$

We say that a point $x \in X$ is *Lyapunov–Perron regular* or simply *regular* for $\mathcal{A}$ if the following conditions hold:

1. x is simultaneously forward and backward regular for $\mathcal{A}$.
2. The filtrations $\mathcal{V}^+$ and $\mathcal{V}^-$ are coherent at x.

One can prove the following result.

Theorem 3.6.11 (Multiplicative Ergodic Theorem for flows). *Let φ be a measure preserving flow of a Lebesgue space (X, ν) such that φ_t is invertible for every $t \in \mathbb{R}$. Let also $\mathcal{A}$ be a measurable cocycle over φ such that*

$$\sup_{-1\le t\le 1} \log^+\|\mathcal{A}(\cdot,t)\| \in L^1(X,\nu). \tag{3.62}$$

Then the set of regular points for $\mathcal{A}$ has full ν-measure.

Given $\varepsilon > 0$ and a regular point $x \in X$, we introduce a family of inner products $\langle\cdot,\cdot\rangle_x$ on $\mathbb{R}^n$ by setting

$$\langle u, v\rangle'_x = \int_{\mathbb{R}} \langle \mathcal{A}(x,t)u, \mathcal{A}(x,t)v\rangle e^{-2\chi_i(x)t-2\varepsilon|t|}\,dt$$

if $u, v \in H_i(x)$, and $\langle u, v\rangle'_x = 0$ if $u \in H_i(x)$ and $v \in H_j(x)$ with $i \neq j$. We call $\langle\cdot,\cdot\rangle'_x$ the *Lyapunov inner product at* x and the corresponding norm $\|\cdot\|'_x$ the *Lyapunov norm at* x. The one-parameter family of weights $e^{-2\chi_i(x)t-2\varepsilon|t|}$ is called the *Pesin tempering kernel*. One can show that for any regular point $x \in X$ and any $v \in \mathbb{R}^n$,

$$n^{-1/2}\|v\|_x \le \|v\|'_x \le K(x)\|v\|_x,$$

where $K\colon X \to \mathbb{R}$ is a *tempered function* on X, that is,

$$\lim_{t\to\pm\infty} \frac{1}{m} \log K(\varphi_t x) = 0.$$

Recall that a Lyapunov change of coordinate is a function $C_\varepsilon\colon X \to GL(n,\mathbb{R})$ satisfying (3.43).

Theorem 3.6.12 (Reduction Theorem for flows). *Let φ be a measure preserving flow of a Lebesgue space (X,ν) such that φ_t is invertible for every $t \in \mathbb{R}$. Let also $\mathcal{A}$ be a measurable cocycle over φ. Given $\varepsilon > 0$, if x is a regular point for $\mathcal{A}$ then for any Lyapunov change of coordinates C_ε, the following properties hold:*

1. *The cocycle $\mathcal{A}_\varepsilon(x,t) = C_\varepsilon(\varphi_t x)^{-1}\mathcal{A}(x,t)C_\varepsilon(x)$ has the block form*

$$\mathcal{A}_\varepsilon(x,t) = \begin{pmatrix} \mathcal{A}^1_\varepsilon(x,t) & & \\ & \ddots & \\ & & \mathcal{A}^{p(x)}_\varepsilon(x,t) \end{pmatrix},$$

where each block $\mathcal{A}^i_\varepsilon(x,t)$ is a $k_i(x) \times k_i(x)$ matrix, and the entries are zero above and below the matrices $\mathcal{A}^i_\varepsilon(x,t)$.

2. *Each block $\mathcal{A}^i_\varepsilon(x)$ satisfies*

$$e^{\chi_i(x)t-\varepsilon|t|} \le \|\mathcal{A}^i_\varepsilon(x,t)^{-1}\|^{-1} \le \|\mathcal{A}^i_\varepsilon(x,t)\| \le e^{\chi_i(x)t+\varepsilon|t|}.$$

3. *If the integrability condition* (3.62) *holds then the map C_ε is tempered ν-almost everywhere, and the spectra of $\mathcal{A}$ and $\mathcal{A}_\varepsilon$ coincide ν-almost everywhere.*

3.6.3 Nonpositively Curved Spaces

In Section 3.1 we presented a geometric interpretation of forward regularity in terms of sequences of points $\{y_n\}$ in the symmetric space $Y = GL(n,\mathbb{R})/O(n,\mathbb{R})$ for which there are a geodesic ray γ and $A \ge 0$ such that the distance from y_n to $\gamma(An)$ grows sublinearly in n. In this section we describe (without proof) a more general geometric situation where the cocycles take on values in a semigroup of semicontractions associated with some nonpositively curved spaces.

Let (Y, ρ) be a complete metric space. The space Y is said to be

1. *convex* if any two points $x, y \in Y$ have a *midpoint*, that is, a point z for which
$$\rho(z, x) = \rho(z, y) = \frac{1}{2}\rho(x, y);$$
2. *uniformly convex* if it is convex and there is a strictly decreasing continuous function g on $[0, 1]$ such that $g(0) = 1$ and for any $x, y, w \in Y$ and any midpoint m_{xy} of x and y,
$$\frac{\rho(m_{xy}, w)}{R} \leq g\Big(\frac{\rho(m_{xy}, w)}{R}\Big),$$
where $R = \max\{\rho(x, w), \rho(y, w)\}$;
3. *nonpositively curved* (in the sense of Busemann) if it is convex and for any $x, y, z \in Y$ and any midpoints m_{xz} of x and z and m_{yz} of y and z,
$$\rho(m_{xz}, m_{yz}) \leq \frac{1}{2}\rho(x, y).$$

If Y is uniformly convex then midpoints are unique.

Examples of nonpositively curved spaces include uniformly convex Banach spaces (e.g., Hilbert spaces or L^p spaces for $1 < p < \infty$), Cartan–Hadamard manifolds (e.g., Euclidean spaces, hyperbolic spaces, or $GL(n, \mathbb{R})/O(n, \mathbb{R})$) and more generally CAT(0) spaces (e.g., Euclidean buildings or $\mathbb{R}$-trees).

A continuous map $\gamma : I \to Y$ (I is an interval) is called a (unit speed minimizing) *geodesic* if for any $s, t \in I$,
$$\rho(\gamma(s), \gamma(t)) = |s - t|.$$

If Y is convex then any two points can be joined by a geodesic and if Y is uniformly convex then this geodesic is unique.

A geodesic $\gamma\colon [0, \infty) \to Y$ is called a *ray* if the limit $\lim_{t\to\infty} \gamma(t)$ does not exist. Two rays γ_1 and γ_2 are called *asymptotic* if $\rho(\gamma_1(t), \gamma_2(t)) \leq$ constant for $t \geq 0$. We denote by $[\gamma]$ the set of all rays asymptotic to γ and by $Y(\infty)$ the ideal boundary of Y, that is, the set of all classes of asymptotic rays.

Let $D \subset Y$ be a nonempty subset. A map $\varphi\colon D \to D$ is called a *semicontraction (or nonexpanding)* if $\rho(\varphi(v), \varphi(z)) \leq d(y, z)$ for all $y, z \in D$. Isometries are semicontractions.

Let us fix a semigroup S of semicontractions and equip it with the Borel σ-algebra associated to the compact-open topology on S. Consider a cocycle $\mathcal{A}$ with values in S over an ergodic transformation f of a measure space (X, μ), and let $A\colon X \to X$ be the generator. Fix $y \in Y$.

Theorem 3.6.13 (Karlsson–Margulis [98]). *Assume that*
$$\int_X \rho(y, A(x)y)\, d\mu(x) < \infty.$$

Then for almost every $x \in X$, the following limit exists:

$$\lim_{m\to\infty} \frac{1}{m} d(y, \mathcal{A}(x,m)y) = a \tag{3.63}$$

and if $a > 0$ then for almost every $x \in X$, there exists a unique geodesic ray $\gamma(\cdot, x)$ in Y starting at y such that

$$\lim_{m\to\infty} \frac{1}{m} d(\gamma(am,x), \mathcal{A}(x,m)y) = 0,$$

and hence $\mathcal{A}(\cdot, m)y$ converges to $[\gamma]$ in $Y \cup Y(\infty)$.

This theorem provides a geometric version of forward regularity, which is more general than the one given in Section 3.1. Reversing the time one can obtain a similar statement for backward regularity. One can wonder if there is a geometric interpretation–in the sense of Theorem 3.6.13–of Lyapunov–Perron regularity, that is, if there always exists a bi-infinite geodesic, which approximates both the forward and backward orbits $m \mapsto \mathcal{A}(x, \pm m)y$. The answer, in general, is negative. For example, consider $Y = \mathbb{R} \times \mathbb{R}$ with the Riemannian metric $(e^{-y} + C)^2 dx^2 + dy^2$. It is known to be a Cartan–Hadamard manifold. Consider the rigid cocycle with the generator A to be the parabolic isometry given by $(x, y) \mapsto (x + 1, y)$. Note that in this case, $a = C$ where a is the constant in (3.63). If $C > 0$ then the forward and backward orbits converge to two different points on the ideal boundary of Y. These two limit points must be fixed by the isometry A. Now assume that they can be connected by a geodesic in Y. Then, since the two end points are fixed by A, the displacement of A is semidecreasing in both directions along this geodesic and hence it is constant. This, however, is impossible as A is parabolic and Y has no parallel bi-infinite geodesics.

We describe an application of Theorem 3.6.13 to Hilbert–Schmidt operators (see [98]). Let $\mathcal{H}$ be a real Hilbert space and $\mathbf{A}$ the algebra of Hilbert–Schmidt operators on $\mathcal{H}$; that is, $a \in \mathbf{A}$ if

$$\|a\|_2^2 = \operatorname{Tr}(aa^*) = \sum_i \|ae_i\|^2 < \infty$$

for some (and hence any) orthonormal basis $\{e_i\}$ of $\mathcal{H}$. It is well known that $\langle a, b\rangle = \operatorname{Tr}(ab^*)$ is an inner product on $\mathbf{A}$ and that $(\mathbf{A}, \langle\cdot,\cdot\rangle)$ is a Hilbert space. Set

$$\operatorname{Sym} = \{a \in \mathbf{A} : a = a^*\}, \qquad \operatorname{Pos} = \exp\{\operatorname{Sym}\} \subset \operatorname{Id} + \operatorname{Sym},$$

where exp is the usual exponential map and Id the identity operator. Pos is an infinite-dimensional Riemannian manifold with the metric

$$\langle v, w\rangle_p = \operatorname{Tr}(p^{-1} v p^{-1} w),$$

where $p \in \operatorname{Pos}$ and $v, w \in \operatorname{Sym} \equiv T_p \operatorname{Pos}$. Let d be the associated distance. One can show that (Pos, d) is a complete metric space satisfying the semiparallelogram law and that the operators $\exp \mathbf{A}$ act on Pos by isometries,

$$p \mapsto [\exp(a)]p = \exp(a)\, p \, \exp(a)^*.$$

Applying Theorem 3.6.13 and (3.1) yields the following result:

Corollary 3.6.14. *Let $\mathcal{A}$ be an integrable cocycle with values in* $\exp \mathbf{A}$. *Then for every x, there exists an operator $\Lambda(x) = \exp(v(x))$, $v(x) \in \mathrm{Sym}$ such that*

$$\frac{1}{m} \| \log([\Lambda^{-n}(x)\mathcal{A}(x, m)]\mathrm{Id}) \|_2 = \frac{1}{m} \left(\sum_i (\log \mu_i(m))^2 \right)^{1/2} \to 0,$$

where $\mu_i(n)$ are the eigenvalues of $[\Lambda^{-n}(x)\mathcal{A}(x, m)]\mathrm{Id} \in \mathrm{Pos}$.

As an immediate consequence of this statement, we have the following result:

Corollary 3.6.15. *Let $\{e_i(x)\}$ be the orthonormal basis of $\mathcal{H}$ consisting of eigenvectors of $\Lambda(x)$ so that $\Lambda(x)e_i(x) = \exp(\lambda_i(x))e_i(x)$. For $z = \sum_i z_i(x)e_i(x) \in \mathcal{H}$, let $\lambda_z(x) = \sup\{\lambda_i(x) : z_i(x) \neq 0\}$. Then*

$$\lim_{m\to\infty} \frac{1}{m} \log \|\mathcal{A}(x, m)^{-1} z\| = -\lambda_z(x).$$

3.7 Notes

The relation between nonuniform hyperbolicity and Lyapunov exponents (see Theorem 3.3.1) was first established in [169] for the case of derivative cocycles (defined in the next chapter). The proof of this result in the general case, presented here, is a slight modification of the arguments in [169].

The name "Multiplicative Ergodic Theorem" was introduced by Oseledets in [162] where he presented the first proof of the theorem. Other proofs showing that the set of forward regular points has full measure (a statement weaker than the Multiplicative Ergodic Theorem 3.4.3) are due to Raghunathan [187] (his approach was also used by Ruelle in [193, 194]), to Mañé [138] (see also [139]) and to Gol'dsheid and Margulis in [76]. The work of Raghunathan also contains an extension to local fields. Millionschikov (see [49] for references) and Sacker, Sell, and Johnson [95, 197–199] discussed related matters.

There are also infinite-dimensional versions of the Multiplicative Ergodic Theorem. Namely, it was extended by Ruelle in [194] to Hilbert spaces, and by Mañé in [138] to compact transformations in Banach spaces (see also [211] for the case of not necessarily compact transformations). The proof due to Gol'dsheid and Margulis also extends to the infinite-dimensional case (see [76]).

The geometric description of forward regularity presented in Section 3.1 is due to Kaĭmanovich [96] who used it to build the theory of Lyapunov exponents for actions of arbitrary noncompact semisimple Lie groups without utilizing linear representations. In particular, he established Theorem 3.6.13 for the action of semisimple Lie groups on symmetric spaces.

In [194], Ruelle obtained a result similar to Corollary 3.6.15 for more general classes of operators.

4

Methods for Estimating Exponents

We saw in Chapter 2 that absence of zero Lyapunov exponents implies nonuniform hyperbolicity. In fact, this seems to be one of the most effective universal ways to establish weak hyperbolic behavior. In this chapter we discuss some methods that allow one to verify that Lyapunov exponents do not vanish.

4.1 Cone and Lyapunov Function Techniques

The *cone* of size $\gamma > 0$ centered around $\mathbb{R}^{n-k}$ in the product space $\mathbb{R}^n = \mathbb{R}^k \times \mathbb{R}^{n-k}$ is the set

$$C_\gamma = \{(v, w) \in \mathbb{R}^k \times \mathbb{R}^{n-k} : \|v\| < \gamma\|w\|\} \cup \{(0, 0)\}.$$

Note that $\{0\} \times \mathbb{R}^{n-k} \subset C_\gamma$ for every γ.

We consider a cocycle $\mathcal{A}$ over an invertible measurable transformation $f\colon X \to X$ preserving a Borel probability measure ν on X, and let $A\colon \mathbb{R}^n \to GL(n, \mathbb{R})$ be its generator. Assume that there exist $\gamma > 0$ and $a > 1$ such that for ν-almost every $x \in \mathbb{R}^n$,

1. $A(x)C_\gamma \subset C_\gamma$,
2. $\|A(x)v\| \geq a\|v\|$ for every $v \in C_\gamma$.

Then the largest Lyapunov exponent can be shown to be positive ν-almost everywhere. Indeed, $n - k$ values of the Lyapunov exponent, counted with their multiplicities, are positive.

Wojtkowski [225] pointed out that Condition 1 alone is in fact sufficient to establish positivity of the values of the Lyapunov exponent. The importance of this observation is that Condition 1 is of pure qualitative character and thus no estimates on the growth of vectors inside the cone are required.

It turns out that Wojtkowski's approach can be described in a more general and more convenient framework elaborated by Burns and Katok in [103]. This approach, in turn, is a further development of that by Lewowicz in [126, 127] and by Markarian in [146] and is based on the notion of infinitesimal Lyapunov function.

We discuss at length this method in several important examples of uniformly hyperbolic and of nonuniformly hyperbolic maps (see Section 6.1).

In later work Wojtkowski himself strengthened his original approach and using results of Potapov on monotone operators of a linear space generated by a quadratic form, obtained estimates of Lyapunov exponents for cocycles with values in the semigroup of matrices preserving the form (see [230]). These results can be used to estimate Lyapunov exponents for Hamiltonian dynamical systems, as well as for the Boltzmann–Sinai gas of hard spheres and the system of falling balls in one dimension (see [230] for more details and references).

In some particular situations (i.e., for cocycles with values in some particular groups), absence of zero Lyapunov exponents can be established using different techniques (see the survey [21] and the references therein).

The cone techniques that we introduce in this chapter will be used throughout the book – in constructing examples of systems with nonzero Lyapunov exponents (see Sections 6.3, 11.4, and 11.5), in studying admissible manifolds (see Section 7.3.2), in proving the Closing Lemma (see Section 15.1), and so on.

4.1.1 Lyapunov Functions

Let $Q\colon \mathbb{R}^n \to \mathbb{R}$ be a continuous function, which is homogeneous of degree 1 (i.e., $Q(\alpha v) = \alpha Q(v)$ for any $v \in \mathbb{R}^n$ and $\alpha \in \mathbb{R}$) and takes on both positive and negative values. We call the subset

$$C^u(Q) := \{0\} \cup Q^{-1}(0, +\infty) \subset \mathbb{R}^n \tag{4.1}$$

the *positive (generalized) cone associated to Q* or simply the *positive cone of Q* (see Fig. 4.1). Similarly,

$$C^s(Q) := \{0\} \cup Q^{-1}(-\infty, 0) \subset \mathbb{R}^n \tag{4.2}$$

is the *negative (generalized) cone associated to Q* or the *negative cone of Q*. The maximal dimension of a linear subspace $L \subset \mathbb{R}^n$ such that $L \subset C^u(Q)$ (respectively, $L \subset C^s(Q)$) is called *positive* (respectively, *negative*) *rank* of Q and is denoted by $r^u(Q)$ (respectively, $r^s(Q)$). We clearly have $r^u(Q) + r^s(Q) \le n$, and since Q takes on both positive and negative values, we have $r^u(Q) \ge 1$ and $r^s(Q) \ge 1$. We call the function Q *complete* if

$$r^u(Q) + r^s(Q) = n. \tag{4.3}$$

For example, consider the function

$$Q(v) = \text{ sign } K(v, v) \cdot |K(v, v)|^{1/2}, \tag{4.4}$$

where K is a nondegenerate indefinite quadratic form. It is easy to see that Q is complete and its positive and negative ranks are equal to the number of positive and negative eigenvalues of the quadratic form K, respectively.

More generally, if λ is a positive real number and K_λ is a real function on $\mathbb{R}^n$, which is homogeneous of degree λ (i.e., $K_\lambda(\alpha v) = \alpha^\lambda K_\lambda(v)$ for any $v \in \mathbb{R}^n$ and

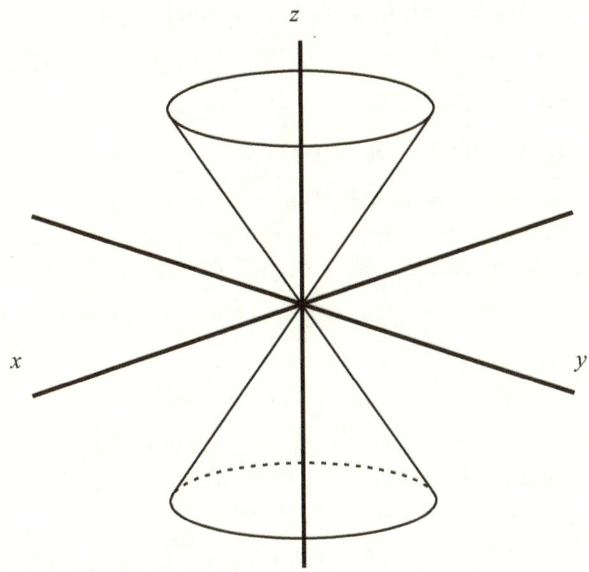

Figure 4.1. The positive cone of the function $Q = \operatorname{sign} K \cdot |K|^{1/2}$ on $\mathbb{R}^3$ where $K(x, y, z) = z^2 - (x^2 + y^2)$.

$\alpha > 0$) and takes on both positive and negative values, we define a homogeneous function Q of degree 1 by

$$Q(v) = \operatorname{sign} K_\lambda(v) \cdot |K_\lambda(v)|^{1/\lambda}.$$

We say that K_λ is *complete* if Q is complete, and we define the positive and negative cones and positive and negative ranks of K_λ as those of Q.

Let $\mathcal{A}\colon X \times \mathbb{Z} \to GL(n, \mathbb{R})$ be a cocycle and $F\colon X \times \mathbb{R}^n \to X \times \mathbb{R}^n$ its linear extension. Recall that

$$F(x, v) = (f(x), A(x)v),$$

where $A(x) = \mathcal{A}(x, 1)$ is the generator of $\mathcal{A}$ (see Chapter 2). Let ν be an f-invariant Borel measure in X.

Definition 4.1.1. A real-valued measurable function Q on $X \times \mathbb{R}^n$ is called a *Lyapunov function for the extension* F or *for the cocycle* $\mathcal{A}$ (with respect to ν) if there exist positive integers r_Q^u and r_Q^s such that for ν-almost every $x \in X$,

1. the function Q_x given by $Q_x(v) = Q(x, v)$ is continuous, homogeneous of degree 1, and takes on both positive and negative values;
2. Q_x is complete, and $r^u(Q_x) = r_Q^u$ and $r^s(Q_x) = r_Q^s$;
3.

$$Q_{f(x)}(A(x)v) \geq Q_x(v) \text{ for all } v \in \mathbb{R}^n. \tag{4.5}$$

We call the numbers r_Q^u and r_Q^s, respectively, the *positive* and *negative ranks* of Q.

When Q is a Lyapunov function, it follows from (4.5) that

$$A(x)C^u(Q_x) \subset C^u(Q_{f(x)}) \quad \text{and} \quad A(f^{-1}(x))^{-1}C^s(Q_x) \subset C^s(Q_{f^{-1}(x)}) \tag{4.6}$$

for ν-almost every $x \in X$.

Definition 4.1.2. A Lyapunov function Q on $X \times \mathbb{R}^n$ is said to be

1. *strict* if the inequality in (4.5) is strict for every $v \neq 0$;
2. *eventually strict* if for ν-almost every $x \in X$, there exists a positive integer $m = m(x)$ such that for every $v \in \mathbb{R}^n \setminus \{0\}$, we have

$$Q_{f^m(x)}(\mathcal{A}(x, m)v) > Q_x(v) \tag{4.7}$$

and

$$Q_{f^{-m}(x)}(\mathcal{A}(x, -m)v) < Q_x(v). \tag{4.8}$$

If a Lyapunov function Q is eventually strict then by (4.5), for ν-almost every $x \in X$, the inequalities (4.7) and (4.8) hold for all $m \geq m(x)$.

If Q is a strict Lyapunov function, it follows from (4.5) that

$$A(x)\overline{C^u(Q_x)} \subsetneqq C^u(Q_{f(x)}) \quad \text{and} \quad A(f^{-1}(x))^{-1}\overline{C^s(Q_x)} \subsetneqq C^s(Q_{f^{-1}(x)}) \tag{4.9}$$

for ν-almost every $x \in X$. Furthermore, if Q is an eventually strict Lyapunov function, it follows from (4.7) and (4.8) that

$$\mathcal{A}(x, m)\overline{C^u(Q_x)} \subsetneqq C^u(Q_{f^m(x)}) \quad \text{and} \quad \mathcal{A}(x, -m)^{-1}\overline{C^s(Q_x)} \subsetneqq C^s(Q_{f^{-m}(x)}) \tag{4.10}$$

for ν-almost every $x \in X$ and every $m \geq m(x)$.

4.1.2 A Criterion for Nonvanishing Lyapunov Exponents

Let $\mathcal{A}\colon X \times \mathbb{Z} \to GL(n, \mathbb{R})$ be a cocycle over an invertible measure preserving transformation $f\colon X \to X$ of a Lebesgue space (X, ν) and $A\colon X \to GL(n, \mathbb{R})$ its generator. We denote by $\chi_1(x) < \cdots < \chi_{p(x)}(x)$ the distinct values of the Lyapunov exponent of $\mathcal{A}$. Assume that

$$\log^+ \|A\|, \log^+ \|A^{-1}\| \in L^1(X, \nu). \tag{4.11}$$

For ν-almost every $x \in X$, we write

$$E^u(x) = \bigoplus_{i:\chi_i(x)>0} H_i(x) \quad \text{and} \quad E^s(x) = \bigoplus_{i:\chi_i(x)<0} H_i(x), \tag{4.12}$$

where $H_i(x)$ are the Oseledets subspaces (see (3.2)).

Theorem 4.1.3. *If the cocycle $\mathcal{A}$ possesses an eventually strict Lyapunov function Q, then*

1. *$\mathcal{A}$ has ν-almost everywhere r^u_Q positive and r^s_Q negative values of the Lyapunov exponent, counted with their multiplicities;*

2. *for ν-almost every $x \in X$, we have*

$$E^u(x) = \bigcap_{m=1}^{\infty} \mathcal{A}(f^{-m}(x), m)\overline{C^u(Q_{f^{-m}(x)})} \subset C^u(Q_x)$$

and

$$E^s(x) = \bigcap_{m=1}^{\infty} \mathcal{A}(f^{m}(x), -m)\overline{C^s(Q_{f^{m}(x)})} \subset C^s(Q_x).$$

Proof. Consider the decomposition of the space (X, ν) into ergodic components $\{(X_\alpha, \nu_\alpha) : \alpha \in Y\}$ where Y is a Lebesgue space. For almost every $\alpha \in Y$ (with respect to the measure on Y induced by the decomposition into ergodic components), we have that $\log^+ \|A\| \in L^1(X_\alpha, \nu_\alpha)$ and that Q is an eventually strict Lyapunov function for the restriction of the cocycle $\mathcal{A}$ to $X_\alpha \times \mathbb{R}^n$. On the other hand, the conclusion of the theorem will hold for f if and only if it holds for $f|X_\alpha$ in almost every ergodic component. Thus, we may assume, without loss of generality, that f is ergodic with respect to ν.

For ν-almost every $x \in X$, we will construct subspaces D^u_x and D^s_x of $\mathbb{R}^n$ of dimension r^u_Q and r^s_Q, respectively, such that for $m \in \mathbb{Z}$,

$$\mathcal{A}(x, m)D^u_x \subset C^u(Q_{f^m(x)}) \quad \text{and} \quad \mathcal{A}(x, m)D^s_x \subset C^s(Q_{f^m(x)}) \tag{4.13}$$

and for every $v \in D^u_x \setminus \{0\}$ and $w \in D^s_x \setminus \{0\}$,

$$\overline{\lim_{m\to+\infty}} \frac{1}{m} \log\|\mathcal{A}(x, -m)v\| < 0 \quad \text{and} \quad \overline{\lim_{m\to+\infty}} \frac{1}{m} \log\|\mathcal{A}(x, m)w\| < 0. \tag{4.14}$$

It follows that $D^u_x = E^u(x)$ and $D^s_x = E^s(x)$ for ν-almost every $x \in X$.

We shall establish the existence of the spaces D^u_x. The argument for the existence of D^s_x is completely similar, with f^{-1} replacing f and the cone $C^s(Q_x)$ playing the role of $C^u(Q_x)$.

Set $C_x = \overline{C^u(Q_x)}$ and for every $m \in \mathbb{Z}$, define

$$C_{m,x} = \mathcal{A}(f^{-m}(x), m)C_{f^{-m}(x)} \subset C^u(Q_x). \tag{4.15}$$

By (4.5), the sequence $\{C_{m,x}\}_{m\geq 1}$ satisfies

$$\cdots \supset C_{-1,x} \supset C_{0,x} \supset C_{1,x} \supset \cdots. \tag{4.16}$$

Each set $C_{m,x}$ contains a subspace of dimension r^u_Q. Using the compactness of the intersection of $C_{m,x}$ with the unit sphere, we conclude that the intersection

$$C_{\infty,x} = \bigcap_{m=1}^{\infty} C_{m,x}$$

also contains a subspace of dimension r^u_Q. By the construction of the set $C_{m,x}$ and (4.5) and (4.7), we see that for ν-almost every $x \in X$, any $v \in C_{\infty,x}$, and any $m \in \mathbb{Z}$, we have

$$\mathcal{A}(x, m)v \in C^u(Q_{f^m(x)}).$$

Let D_x^u be any r_Q^u-dimensional space lying inside the set $C_{\infty,x}$. It follows from (4.15) and (4.16) that (4.13) holds.

By construction, $C_{\infty,x} \subset C^u(Q_x)$ and hence the function Q_x is positive on $C_{\infty,x}$. Since Q_x is a continuous homogeneous function of degree 1,

$$\begin{aligned} a(x) :&= \inf\left\{\frac{Q_x(v)}{\|v\|} : v \in C_{\infty,x} \setminus \{0\}\right\} \\ &= \inf\left\{Q_x(v) : v \in C_{\infty,x},\ \|v\| = 1\right\} > 0 \end{aligned} \tag{4.17}$$

and

$$\begin{aligned} b(x) :&= \sup\left\{\frac{Q_x(v)}{\|v\|} : v \in C_{\infty,x} \setminus \{0\}\right\} \\ &= \sup\left\{Q_x(v) : v \in C_{\infty,x},\ \|v\| = 1\right\} < \infty. \end{aligned} \tag{4.18}$$

For every $x \in X$ and positive integer n, set

$$\kappa_n(x) = \sup\left\{\frac{Q_{f^{-n}(x)}(\mathcal{A}(x,-n)v)}{Q_x(v)} : v \in C_{\infty,x} \setminus \{0\}\right\}$$

and $L(x,n) = \log \kappa_n(x)$. Since $A(x)C_{\infty,x} \subset C_{\infty,f(x)}$, we obtain

$$\kappa_{m+n}(x) \le \kappa_n(x) \cdot \kappa_m(f^{-n}(x)).$$

Therefore, $L(x,n)$ is a subadditive cocycle over f^{-1}.

Inequality (4.5) implies that $\kappa_n(x) \le 1$ for ν-almost every $x \in X$. Since the intersection of $C_{\infty,x}$ with the unit sphere is compact, it follows from (4.7) that for ν-almost every $x \in X$, there exists an integer $n(x) \ge 1$ such that $\kappa_{n(x)}(x) < 1$. Condition (4.11) guarantees that $x \mapsto L(x,1)$ is ν-integrable:

$$\begin{aligned} \left|\log L(x,1)\right| &= \log \inf\left\{\frac{Q_x(v)}{Q_{f^{-1}x}(A(f^{-1}(x))^{-1}v)} : v \in C_{\infty,x} \setminus \{0\}\right\} \\ &= \log \inf\left\{\frac{Q_x(A(f^{-1}(x))w)}{Q_{f^{-1}(x)}(w)} : A(f^{-1}(x))w \in C_{\infty,x} \setminus \{0\}\right\} \\ &\le \log \rho(A \circ f^{-1}) \\ &\le \log\|A \circ f^{-1}\| \in L^1(X,\nu), \end{aligned}$$

where $\rho(B)$ denotes the spectral radius of B. Hence, $\int_X L(x,n)\,d\nu(x) < 0$ for all sufficiently large n. Since f is ergodic, Kingman's Subadditive Ergodic Theorem implies that

$$\begin{aligned} \lim_{m\to+\infty} \frac{L(x,m)}{m} &= \lim_{m\to+\infty} \frac{1}{m}\int_X L(x,m)\,d\nu(x) \\ &= \inf_{m\ge 1} \frac{1}{m}\int_X L(x,m)\,d\nu(x) \\ &\le \frac{1}{n}\int_X L(x,n)\,d\nu(x) < 0 \end{aligned} \tag{4.19}$$

for ν-almost every $x \in X$. By (4.17) and (4.18), any $v \in C_{\infty,x}$ satisfies

$$\begin{aligned}\|\mathcal{A}(x,-m)v\| &\le a(x)^{-1} Q_{f^{-m}(x)}(\mathcal{A}(x,-m)v)\\ &\le a(x)^{-1} Q_x(v)\kappa_m(x)\\ &\le b(x)a(x)^{-1}\kappa_m(x)\|v\|.\end{aligned}$$

Taking logarithms and using (4.19), we obtain

$$\varlimsup_{m\to\infty} \frac{1}{m}\log\|\mathcal{A}(x,-m)v\| \le \inf_{m\ge 1}\frac{1}{m}\int_X L(x,m)\,d\nu(x) < 0,$$

and (4.14) holds. □

4.1.3 Invariant Cone Families

A *(generalized) cone* C in $\mathbb{R}^n$ is a homogeneous set (i.e., $\alpha v \in C$ whenever $v \in C$ and $\alpha \in \mathbb{R}$) such that $C \setminus \{0\}$ is open. In particular, C need not be convex and $\operatorname{int} C$ need not be connected. The *rank* of C is the maximal dimension of a linear subspace $L \subset \mathbb{R}^n$, which is contained in C. We denote it by $r(C)$. The *complementary cone* $\widehat{C}$ to C is defined by

$$\widehat{C} = \left(\mathbb{R}^n \setminus \overline{C}\right) \cup \{0\}.$$

Obviously, the complementary cone to $\widehat{C}$ is C. We have $r(C) + r(\widehat{C}) \le n$ and this inequality may be strict (this is the case, e.g., when $C \ne \mathbb{R}^n$ but $\overline{C} = \mathbb{R}^n$). A pair of complementary cones C and $\widehat{C}$ is called *complete* if $r(C) + r(\widehat{C}) = n$.

Let $\mathcal{A}\colon X \times \mathbb{Z} \to GL(n,\mathbb{R})$ be a cocycle over an invertible transformation $f\colon X \to X$ of a Lebesgue space X and $A\colon X \to GL(n,\mathbb{R})$ its generator. Assume that f preserves a measure ν on X. Consider a measurable family of cones $C = \{C_x : x \in X\}$ in $\mathbb{R}^n$.

Definition 4.1.4. We say that

1. C is *complete* if the pair of complementary cones $(C_x, \widehat{C}_x)$ is complete for ν-almost every $x \in X$;
2. C is $\mathcal{A}$-invariant if for ν-almost every $x \in X$,

$$A(x)C_x \subset C_{f(x)} \quad\text{and}\quad A(f^{-1}(x))^{-1}\widehat{C}_x \subset \widehat{C}_{f^{-1}(x)}. \tag{4.20}$$

Definition 4.1.5. Let C be an $\mathcal{A}$-invariant measurable family of cones. We say that

1. C is *strict* if for ν-almost every $x \in X$,

$$A(x)\overline{C_x} \subset C_{f(x)} \quad\text{and}\quad A(f^{-1}(x))^{-1}\overline{\widehat{C}_x} \subset \widehat{C}_{f^{-1}(x)};$$

2. C is *eventually strict* if for ν-almost every $x \in X$, there exists $m = m(x) \in \mathbb{N}$ such that

$$\mathcal{A}(x,m)\overline{C_x} \subset C_{f^m(x)} \quad\text{and}\quad \mathcal{A}(x,-m)^{-1}\overline{\widehat{C}_x} \subset \widehat{C}_{f^{-m}(x)}. \tag{4.21}$$

The following result is due to Wojtkowski [225]. Its proof can be obtained by modifying arguments in the proof of Theorem 4.1.3 and is omitted.

Theorem 4.1.6. *Let $C = \{C_x : x \in X\}$ be a complete eventually strict $\mathcal{A}$-invariant measurable family of cones. Assume that* (4.11) *holds for some f-invariant measure ν. Then for almost every $x \in X$, we have*

$$E^u(x) = \bigcap_{m=1}^{\infty} \mathcal{A}(f^{-m}(x), m)\overline{C_{f^{-m}(x)}} \subset C_x$$

and

$$E^s(x) = \bigcap_{m=1}^{\infty} \mathcal{A}(f^m(x), -m)\overline{\widehat{C}_{f^m(x)}} \subset \widehat{C}_x,$$

where $E^u(x)$ and $E^s(x)$ are given by (4.12).

Let Q be a Lyapunov function on $X \times \mathbb{R}^n$ for a cocycle $\mathcal{A}$. We consider the family of cones $C = \{C_x : x \in X\}$ in $\mathbb{R}^n$ given by $C_x = C^u(Q_x)$. Conditions 2 and 3 in Definition 4.1.1 imply that C is complete and $\mathcal{A}$-invariant. Note that the complementary cone $\widehat{C}_x$ is not always equal to the cone $C^s(Q_x)$. However, we have $\widehat{C}_x = C^s(Q_x)$ provided that for each v, such that $Q_x(v) = 0$, one can find w arbitrarily close to v such that $Q_x(w) > 0$. Furthermore, if Q is strict (respectively eventually strict) then C is strict (respectively eventually strict).

4.2 Cocycles with Values in the Symplectic Group

Let $\mathcal{A}: X \times \mathbb{Z} \to GL(n, \mathbb{R})$ be a cocycle and Q a homogeneous function of degree 1 on $X \times \mathbb{Z}$. Consider the corresponding families of cones $C^u(Q_x)$ and $C^s(Q_x)$ given by (4.1) and (4.2). If Q is complete (see (4.3)) and (4.6) holds, then Q is a Lyapunov function. Moreover, if (4.9) (respectively (4.10)) holds then Q is strict (respectively eventually strict). However, if condition (4.6) is satisfied only with respect to the family of cones $C^u(Q_x)$ then Q may not be a Lyapunov function. This situation does occur for some interesting classes of cocycles and cones. The most important case for applications involves cocycles with values in the symplectic group $Sp(2m, \mathbb{R})$ for $m \geq 1$ and the so-called symplectic cones that we define later.

We begin with the simple case of $SL(2, \mathbb{R})$ cocycles.

We call a cone in $\mathbb{R}^n$ *connected* if its projection to the projective space $\mathbb{R}P^{n-1}$ is a connected set. A connected cone in $\mathbb{R}^2$ is simply the union of two opposite sectors bounded by two different straight lines intersecting at the origin plus the origin itself. By a linear coordinate change, such a cone can always be reduced to the following standard cone:

$$S = \{(v, w) \in \mathbb{R} : vw > 0\} \cup \{(0, 0)\}.$$

Theorem 4.2.1. *If a cocycle with values in $SL(2, \mathbb{R})$ has an eventually strictly invariant family of connected cones $C = \{C_x : x \in X\}$, then it has an eventually strict Lyapunov function Q such that for ν-almost every $x \in X$, the function Q_x has the form* (4.4) *and its zero set coincides with the boundary of the cone C_x.*

Proof. First assume that $C_x = S$ for ν-almost every $x \in X$. Write

$$A(x) = \begin{pmatrix} a(x) & b(x) \\ c(x) & d(x) \end{pmatrix}.$$

By (4.20), the functions $a, b, c,$ and d are nonnegative. Since $A(x) \in SL(2, \mathbb{R})$, we have $1 = a(x)d(x) - b(x)c(x)$. For each $(u, v) \in S$, put $K(u, v) = uv$. We obtain

$$\begin{aligned} K(A(x)(u, v)) &= a(x)c(x)u^2 + b(x)d(x)v^2 + (a(x)d(x) + b(x)c(x))uv \\ &\geq (a(x)d(x) + b(x)c(x))uv \\ &\geq (a(x)d(x) - b(x)c(x))uv \\ &= K(u, v). \end{aligned} \tag{4.22}$$

Arguing similarly, one can show that all entries of the matrix

$$\mathcal{A}(x, m) = \begin{pmatrix} a(x, m) & b(x, m) \\ c(x, m) & d(x, m) \end{pmatrix}$$

are nonnegative. Moreover, Condition (4.21) implies that there exists $m = m(x) \geq 1$ such that $b(x, m) > 0$ and $c(x, m) > 0$. By (4.22), we conclude that $K(\mathcal{A}(x, m)(u, v)) > K(u, v)$.

In the case of an arbitrary family of connected cones, let us introduce a coordinate change $L : X \to SL(2, \mathbb{R})$ that takes the two lines bounding the cone C_x into the coordinate axes. Then $L(x)C_x = S$. For the cocycle $B : X \to SL(2, \mathbb{R})$ defined by $B(x) = L(f(x))A(x)L(x)^{-1}$, the constant family of cones S is eventually strictly invariant and hence by the previous argument, the function $Q_0(x, (u, v)) = \operatorname{sign}(uv) \cdot |uv|^{1/2}$ is an eventually strict Lyapunov function. Hence for the original cocycle A, the function $Q(x, (u, v)) = Q_0(x, L(x)(u, v))$ has the same properties. □

Let us now proceed with the general symplectic case. We denote by ω the standard symplectic form in $\mathbb{R}^{2m}$,

$$\omega(v, w) = \sum_{i=1}^{m} (v_i w_{m+i} - w_i v_{m+i})$$

and by K the following nondegenerate quadratic form of signature zero:

$$K(v) = \sum_{i=1}^{m} v_i v_{m+i}.$$

The cone

$$S = \{v \in \mathbb{R}^{2m} : K(v) > 0\} \cup \{0\}$$

is called the *standard symplectic cone*. The image of this cone under an invertible linear symplectic map (i.e., an element of $Sp(2m, \mathbb{R})$) is called a *symplectic cone*.

Recall that a subspace L in a $2m$-dimensional symplectic space (H, ω) is called *Lagrangian* if the symplectic form ω vanishes identically on L. Let L_1 and L_2 be two transversal Lagrangian subspaces (i.e., they are complementary of dimension m). Then for any $v \in H$, there is a unique decomposition $v = v_1 + v_2$ with $v_i \in L_i$ for $i = 1, 2$. Let

$$K_{L_1,L_2}(v) = \omega(v_1, v_2) \text{ and } C_{L_1,L_2} = K_{L_1,L_2}^{-1}((0,\infty)) \cup \{0\}.$$

C_{L_1,L_2} is a symplectic cone and K_{L_1,L_2} is the corresponding quadratic form.

It is easy to see (e.g., by a direct calculation for the case of standard cones) that for a given symplectic cone C in a symplectic space, there are exactly two isolated Lagrangian subspaces L_1 and L_2 that belong to the boundary of C and that $C = C_{L_1,L_2}$ or $C = C_{L_2,L_1}$. Thus the cone C canonically determines the form K:

$$K(C) = K_{L_1,L_2} \text{ or } K(C) = K_{L_2,L_1},$$

depending on whether the form K_{L_1,L_2} or the form K_{L_2,L_1} is positive on C. For example, the standard cone S is C_{L_1,L_2}, where

$$L_1 = \{(x, 0) : x \in \mathbb{R}^m\} \text{ and } L_2 = \{(0, x) : x \in \mathbb{R}^m\}.$$

Proposition 4.2.2. *Let H and H' be two $2m$-dimensional spaces, L_1, $L_2 \subset H$ and L_1', $L_2' \subset H'$ pairs of transversal Lagrangian subspaces, and $T\colon H \to H'$ a symplectic linear transformation such that $T\overline{C_{L_1,L_2}} \subset C_{L_1',L_2'}$. Then for all $v \in H \setminus \{0\}$, we have*

$$K_{L_1',L_2'}(Tv) > K_{L_1,L_2}(v).$$

Proof. We first assume that $C_{L_1,L_2} = C_{L_1',L_2'} = S$ is the standard symplectic cone. Write

$$T = \begin{pmatrix} A & B \\ C & D \end{pmatrix}$$

where A, B, C, and D are $m \times m$ matrices. Then $K = K_{L_1,L_2} = K_{L_1',L_2'}$ satisfies

$$\begin{aligned} K(T(u, v)) &= K(Au + Bv, Cu + Dv) \\ &= (Au + Bv)^*(Cu + Dv), \end{aligned}$$

where X^* denotes the transpose of the matrix X. We claim that A is invertible. Otherwise, there would exist $u \in \mathbb{R}^m \setminus \{0\}$ such that $Au = 0$. Therefore since $K(u, 0) = 0$ and $K(T(u, 0)) = K(0, Cu) = 0$, the map T would not take $\overline{S}$ strictly inside S. Using the identity

$$\begin{pmatrix} A & B \\ C & D \end{pmatrix}^* \begin{pmatrix} 0 & -\mathrm{Id} \\ \mathrm{Id} & 0 \end{pmatrix} \begin{pmatrix} A & B \\ C & D \end{pmatrix} = \begin{pmatrix} 0 & -\mathrm{Id} \\ \mathrm{Id} & 0 \end{pmatrix},$$

one can easily show that $A^*D - C^*B = \mathrm{Id}$ and $A^*C = C^*A$. Hence,

$$\begin{aligned} K(T(u, v)) &= (A^{-1}(Au + Bv))^*(A^*(Cu + Dv)) \\ &= (u + A^{-1}Bv)^*(A^*Cu + v + C^*Bv). \end{aligned}$$

Since $A^*Cu + C^*Bv = C^*(Au + Bv)$ and $K(u, v) = u^*v$, we obtain

$$\begin{aligned} K(T(u, v)) =&K(u, v) + v^*(A^{-1}B)^*v \\ &+ (u + A^{-1}Bv)^*C^*A(u + A^{-1}Bv). \end{aligned} \tag{4.23}$$

We now show that the matrices $A^{-1}B$ and C^*A are positive definite. First, observe that if $(u, 0) \in \overline{S}$, then $T(u, 0) = (Au, Cu) \in S$ and $u^*C^*Au > 0$ for all $u \in \mathbb{R}^m \setminus \{0\}$. Hence, C^*A is positive definite. Now, if $A^{-1}B$ is not positive definite, then there exists $v \in \mathbb{R}^m \setminus \{0\}$ such that $(A^{-1}Bv)^*v \leq 0$. Hence, $(-A^{-1}Bv, v) \in \overline{S}$ but

$$T(-A^{-1}Bv, v) = (0, (-CA^{-1}B + D)v) \notin S.$$

This implies that $A^{-1}B$ is also positive definite. Since for every $(u, v) \neq (0, 0)$ we have $v \neq 0$ or $u + A^{-1}Bv \neq 0$, by (4.23), we conclude that

$$K(T(u, v)) > K(u, v)$$

for every $(u, v) \in H \setminus \{0\}$.

In the case of arbitrary cones one can obtain the desired result by reducing it to the previous case using arguments similar to those in the proof of Theorem 4.2.1. □

Proposition 4.2.2 immediately implies the following relation between invariant cone families and Lyapunov functions.

Theorem 4.2.3. *Let $A\colon X \to Sp(2m, \mathbb{R})$ be a cocycle over the measurable transformation $f\colon X \to X$, which preserves a measure ν. If A has an eventually strictly invariant family of symplectic cones $C = \{C_x : x \in X\}$, then it also has an eventually strict Lyapunov function Q, such that for ν-almost every $x \in X$, the function Q_x has the form* (4.4) *with a quadratic form $K = K_x$ of signature zero. Furthermore, the zero set of the function Q_x coincides with the boundary of the cone C_x.*

Combining Theorem 4.2.3 with Theorem 4.1.3 we immediately obtain the following result:

Corollary 4.2.4. *If a cocycle $A\colon X \to Sp(2m, \mathbb{R})$ satisfies* (4.11) *and has an eventually strictly invariant family of symplectic cones, then the linear extension F of f has ν-almost everywhere m positive and m negative values of the Lyapunov exponent.*

4.3 Monotone Operators and Lyapunov Exponents

The cone techniques described above can be formalized and put in a more general setting using the language of Potapov's algebras. We first give a brief introduction to the study of these algebras and then show how to use them to estimate Lyapunov exponents of matrix cocycles.

4.3.1 The Algebra of Potapov

Let V be an n-dimensional real vector space with a pseudo-Euclidean structure, that is, a nondegenerate quadratic form $\mathfrak{J}$ with signature (p, q), where $p + q = n$ and $p, q \in \mathbb{N}$. Consider the cones

$$C^u = \{v \in V | \mathfrak{J}(v) > 0\} \cup \{0\} \quad \text{and} \quad C^s = \{v \in V | \mathfrak{J}(v) < 0\} \cup \{0\},$$

and let $C^0 = \{v \in V | \mathfrak{J}(v) = 0\}$ be their common boundary. A linear subspace is called *positive* (respectively *negative*) if it is contained in C^u (respectively C^s). The maximal dimension of a positive subspace is p and the maximal dimension of a negative subspace is q. Let P be a positive subspace of dimension p (a maximal positive subspace) and let Q be the $\mathfrak{J}$-orthogonal complement of P. Then Q is a negative subspace of dimension q. We can choose coordinates in $P \equiv \mathbb{R}^p$ and $Q \equiv \mathbb{R}^q$ in such a way that $V \equiv \mathbb{R}^p \times \mathbb{R}^q$, and for $v = (v_1, v_2) \in \mathbb{R}^p \times \mathbb{R}^q$,

$$\mathfrak{J}(v) = \langle Jv, v\rangle = \|v_1\|^2 - \|v_2\|^2, \quad \text{where } J = \begin{pmatrix} I_p & 0 \\ 0 & -I_q \end{pmatrix}.$$

Definition 4.3.1. An invertible linear transformation $S \colon V \to V$ is

1. *$\mathfrak{J}$-separated* if $SC^u \subset C^u$;
2. *strictly $\mathfrak{J}$-separated* if $S(C^0 \cup C^u) \subset C^u$;
3. *$\mathfrak{J}$-monotone* if $\mathfrak{J}(Sv) \geq \mathfrak{J}(v)$ for every $v \in V$;
4. *strictly $\mathfrak{J}$-monotone* if $\mathfrak{J}(Sv) > \mathfrak{J}(v)$ for every $v \in V \setminus \{0\}$;
5. *$\mathfrak{J}$-isometric* if $\mathfrak{J}(Sv) = \mathfrak{J}(v)$ for every $v \in V$;
6. *$\mathfrak{J}$-symmetric* if $\langle JS\cdot, \cdot\rangle$ is a symmetric bilinear form on V.

One can introduce similar notions for one-parameter subgroups of linear transformations (elements of the Lie algebra $gl(n)$) by declaring that $G \in gl(n)$ has a given property (among those above) if the linear maps e^{tG} have this property for all $t > 0$. Clearly, a $\mathfrak{J}$-monotone linear transformation is $\mathfrak{J}$-separated, and a strictly $\mathfrak{J}$-monotone linear transformation is strictly $\mathfrak{J}$-separated.

In [180–182], Potapov studied fundamental properties of $\mathfrak{J}$-monotone linear transformations. Here we present some of them:

1. If S is $\mathfrak{J}$-separated then the set of $r > 0$ such that S/r is $\mathfrak{J}$-monotone is a closed interval $[r_-, r_+]$, for some $0 < r_- \leq r_+$ (it can be a point). Moreover, $S = RU$, where U is $\mathfrak{J}$-isometric and $R = (SJS^*J)^{1/2}$ is $\mathfrak{J}$-symmetric and has only positive eigenvalues. We call R the *modulus* of S. The p largest eigenvalues of the modulus R are not smaller than r_+ and the q smallest

eigenvalues are not larger than r_-. We denote the eigenvalues by

$$r_p^+ \geq r_{p-1}^+ \geq \cdots \geq r_1^+ \geq r_1^- \geq \cdots \geq r_{q-1}^- \geq r_q^- > 0,$$

and call them, respectively, *positive* and *negative singular values* of S. (Note that both positive and negative singular values are positive numbers.) We have $r_1^+ = r_+$ and $r_1^- = r_-$.

2. If S is strictly $\mathfrak{J}$-separated then the set of $r > 0$ such that S/r is strictly $\mathfrak{J}$-monotone is an open interval (r_-, r_+), for some $0 < r_- \leq r_+$. Moreover, there is a $\mathfrak{J}$-isometry W such that $W^{-1}RW = \begin{pmatrix} R_+ & 0 \\ 0 & R_- \end{pmatrix}$ with $R_+ \geq r_+ I_p$ and $0 < R_- \leq r_- I_q$.

When S is $\mathfrak{J}$-separated and $L \subset C^u$ is a d-dimensional positive subspace, the set $SL \subset C^u$ is also a positive subspace. Both L and SL are equipped with the scalar product induced by $\mathfrak{J}$ on C^u, thus inducing (unoriented) volume elements that we call $\mathfrak{J}$-*volumes*. Let $\alpha_d(L, S)$ be the coefficient of $\mathfrak{J}$-volume expansion for the linear map S restricted to L. Let also $\sigma_d(S)$ be the infimum of $\alpha_d(L, S)$ over all positive d-dimensional subspaces L.

Proposition 4.3.2. *The following properties hold:*

1. $\sigma_d(S) = r_1^+ r_2^+ \cdots r_d^+$.
2. *If S_1 and S_2 are $\mathfrak{J}$-separated linear transformations, then*

$$r_1^+(S_1S_2) \geq r_1^+(S_1)r_1^+(S_2), \quad r_1^-(S_1S_2) \leq r_1^-(S_1)r_1^-(S_2), \qquad (4.24)$$

and $\sigma_d(S_1S_2) \geq \sigma_d(S_1)\sigma_d(S_2)$.

We now consider the manifold $\mathcal{L}_+$ of all p-dimensional positive subspaces. A $\mathfrak{J}$-separated linear transformation S maps $\mathcal{L}_+$ into itself, and this map is a contraction in the Grassmannian bundle.

Proposition 4.3.3. *If S is a $\mathfrak{J}$-separated linear transformation and $T_1, T_2 \in \mathcal{L}_+$, then*

$$\|ST_1 - ST_2\| \leq \frac{r_1^-}{r_1^+}\|T_1 - T_2\|,$$

and the contraction coefficient is sharp.

One can apply the above general approach to symplectic spaces and symplectic linear transformations (see Section 4.2). In this case we have $p = q$ and $V = \mathbb{R}^p \times \mathbb{R}^p$ with the symplectic form $\omega(v, w) = \langle v_1, w_2 \rangle - \langle v_2, w_1 \rangle$. Consider a special type of $\mathfrak{J}$-form, which we call a Q-form, defined by a pair of transversal Lagrangian subspaces L_1 and L_2. Using the standard splitting $v = v_1 + v_2$, $v_i \in L_i$, $i = 1, 2$, we set $Q(v) = \omega(v_1, v_2)$. The following result was proved by Potapov in [182] and independently by Wojtkowski in [230].

Proposition 4.3.4. *If a symplectic linear transformation is Q-separated (respectively strictly Q-separated) then it is Q-monotone (respectively strictly Q-monotone).*

In the case of Q-separated linear transformations this result if a reformulation of Proposition 4.2.2.

4.3.2 Lyapunov Exponents for $\mathfrak{Z}$-Separated Cocycles

We now translate the properties of $\mathfrak{Z}$-separated operators into properties of Lyapunov exponents of matrix cocycles with values in the $\mathfrak{Z}$-separated matrices.

Let $\mathcal{A}\colon X \times \mathbb{Z} \to GL(n, \mathbb{R})$ be a cocycle over an invertible measure preserving transformation $f\colon X \to X$ of a Lebesgue space (X, ν) and $A\colon X \to GL(n, \mathbb{R})$ its generator. For each regular point $x \in X$, denote by $\chi_1(x) < \cdots < \chi_{p(x)}(x)$ the values of the Lyapunov exponents of $\mathcal{A}$ and let $H_i(x)$ be the spaces in the Oseledets decomposition (3.2). Write $k_i(x) = \dim H_i(x)$. Assume that ν is ergodic and that the integrability condition (4.11) holds.

We say that the cocycle $\mathcal{A}$ is *$\mathfrak{Z}$-separated* if $A(x)$ is $\mathfrak{Z}$-separated for ν-almost every $x \in X$. In this case we denote the positive and negative singular values of $A(x)$ by

$$r_p^+(x) \geq \cdots \geq r_1^+(x) \geq r_1^-(x) \geq \cdots \geq r_q^-(x) > 0.$$

Write $\lambda_\pm = \int_X \log r_1^\pm(x)\, d\nu(x)$.

Proposition 4.3.5. *If $\mathcal{A}$ is $\mathfrak{Z}$-separated, then $\chi^+(x, v) \geq \lambda_+$ and $\chi^-(x, w) \geq -\lambda_-$ for ν-almost every $x \in X$ and every $v \in C^u \setminus \{0\}$ and $w \in C^s \setminus \{0\}$.*

Proof. Since $A(x)/r_1^+(x)$ is $\mathfrak{Z}$-monotone, we have

$$\mathfrak{Z}(A(x)v) \geq (r_1^+(x))^2 \mathfrak{Z}(v)$$

for every v. For $v \in C^u$, we have $\|v\|^2 \geq \mathfrak{Z}(v)$, and hence

$$\|\mathcal{A}(x, m)v\|^2 \geq \mathfrak{Z}(\mathcal{A}(x, m)v) \geq \prod_{i=0}^{m-1} \left(r_1^+(f^i(x))\right)^2 \mathfrak{Z}(v).$$

We claim that all functions $\log r_k^\pm(x)$ are ν-integrable. If $R(x)$ is the modulus of $A(x)$ then

$$(r_p^+(x))^2 = \|R(x)^2\| = \|A(x)JA(x)^*J\| \leq \|A(x)\|^2,$$

$$(r_q^-(x))^{-2} = \|R(x)^{-2}\| = \|J(A(x)^*)^{-1}JA(x)^{-1}\| \leq \|S^{-1}\|^2,$$

which proves the claim due to the integrability assumption. It follows from Birkhoff's Ergodic Theorem that

$$\chi^+(x, v) \geq \int_X \log r_1^+(x)\, d\nu(x) = \lambda_+.$$

To obtain the inequality for $\chi^-(x, v)$, note that each transformation $A(x)^{-1}$ is $(-\mathfrak{Z})$-separated and apply the above argument. □

Proposition 4.3.6. *Assume that $\mathcal{A}$ is $\mathfrak{J}$-separated. If $\lambda_+ > \lambda_-$, then for almost every $x \in X$, there exists $m \in \mathbb{N}$ such that $\mathcal{A}(x, m)$ is strictly $\mathfrak{J}$-separated.*

Proof. Since a $\mathfrak{J}$-separated operator S is strictly $\mathfrak{J}$-separated if and only if $r_1^+(S) > r_1^-(S)$, the desired result follows from (4.24). $\square$

We introduce another labeling of the values of the Lyapunov exponents and the subspaces in the Oseledets decomposition. When $k_1(x) > q$ we set $s_-(x) = 0$, and when $k_{p(x)}(x) > p$ we set $s_+(x) = 0$. Otherwise we define

$$s_-(x) = \max\left\{k : \sum_{i=1}^{k} k_i(x) \le q\right\},$$

$$s_+(x) = \max\left\{k : \sum_{i=p(x)-k+1}^{p(x)} k_i(x) \le p\right\}.$$

When $s_-(x) + s_+(x) = p(x)$, we relabel the subspaces $H_i(x)$ as

$$V_{-s_-(x)}(x) \oplus \cdots \oplus V_{-1}(x) \oplus V_1(x) \oplus \cdots \oplus V_{s_+(x)}(x)$$

and when $s_-(x) + s_+(x) = p(x) - 1$ as

$$V_{-s_-(x)}(x) \oplus \cdots \oplus V_{-1}(x) \oplus V_0(x) \oplus V_1(x) \oplus \cdots \oplus V_{s_+(x)}(x).$$

The subspace $V_0(x)$ may or may not be present. We also denote by $\lambda_i(x)$ the value the Lyapunov exponent $\chi^+(x, \cdot)$ in $V_i(x) \setminus \{0\}$.

Proposition 4.3.7. *Assume that $\mathcal{A}$ is $\mathfrak{J}$-separated.*

1. *If $V_0(x)$ is present, then $\lambda_0(x) = \lambda_+ = \lambda_-$ for ν-almost every $x \in X$.*
2. *If $\lambda_+ > \lambda_-$, then for ν-almost every $x \in X$, the space $V_0(x)$ is not present, $\lambda_1(x) \ge \lambda_+$, $\lambda_{-1}(x) \le \lambda_-$, and*

$$\begin{aligned} V_{-s_-(x)}(x) \oplus \cdots \oplus V_{-1}(x) &\subset C^s, \\ V_1(x) \oplus \cdots \oplus V_{s_+(x)}(x) &\subset C^u. \end{aligned} \tag{4.25}$$

Proof. If the space $V_0(x)$ is present then the dimension of the direct sums $V_{-s_-(x)}(x) \oplus \cdots \oplus V_{-1}(x) \oplus V_0(x)$ and $V_0(x) \oplus V_1(x) \oplus \cdots \oplus V_{s_+(x)}(x)$ exceeds q and p respectively. Hence the first sum must contain positive vectors and the second must contain negative vectors. By Proposition 4.3.5, we conclude that $\lambda_0(x) \ge \lambda_+$ and $-\lambda_0(x) \ge -\lambda_-$ for almost every x. Thus, $\lambda_0(x) = \lambda_+ = \lambda_-$.

Similarly, if $V_0(x)$ is not present then $\lambda_1(x) \ge \lambda_+$ and $\lambda_- \ge \lambda_{-1}(x)$. It remains to establish the inclusions (4.25). First of all, we have

$$V_{-s_-(x)}(x) \oplus \cdots \oplus V_{-1}(x) \subset C^s \cup C^0,$$

$$V_1(x) \oplus \cdots \oplus V_{s_+(x)}(x) \subset C^u \cup C^0.$$

For example, if $V_{-s_-(x)}(x) \oplus \cdots \oplus V_{-1}(x)$ contains a positive vector, then by Proposition 4.3.5, we have $\lambda_{-1}(x) \ge \lambda_+$ contradicting to $\lambda_+ > \lambda_-$. Since

$A(x)H_i(x) = H_i(f(x))$, by Proposition 4.3.6, we obtain the inclusions (4.25) ν-almost everywhere. □

We shall say that the cocycle $\mathcal{A}$ is *eventually strictly $\mathfrak{J}$-separated* if for almost every $x \in X$, there exists $m \in \mathbb{N}$ such that $\mathcal{A}(x, m)$ is strictly $\mathfrak{J}$-separated.

Proposition 4.3.8. *If $\mathcal{A}$ is eventually strictly $\mathfrak{J}$-separated then for ν-almost every $x \in X$, the space $V_0(x)$ is not present, $\lambda_1(x) > \lambda_{-1}(x)$, and the inclusions* (4.25) *hold.*

Proof. By Proposition 4.3.7, the statement holds for strictly $\mathfrak{J}$-separated cocycles, that is, cocycles with $A(x)$ being strictly $\mathfrak{J}$-separated for ν-almost every $x \in X$. The general case can be reduced to this one by considering the subsets

$$X_m = \{x \in X : \mathcal{A}(x, m) \text{ is strictly } \mathfrak{J}\text{-separated}\}.$$

For large $m \in \mathbb{N}$, the set X_m has measure arbitrarily close to 1. Consider the return map to X_m and the corresponding induced cocycle (see (2.8)) that is thus strictly $\mathfrak{J}$-separated. The splitting for the induced cocycle coincides with the original splitting restricted to the subset X_m. By Proposition 3.2.2, the Lyapunov exponents are changed by a factor equal to the average return time (see (2.7)). □

Set $\ell_i(x) = \dim V_i(x)$, and

$$d_0^-(x) = q - \sum_{i=1}^{s_-(x)} \ell_{-i}(x) \quad \text{and} \quad d_0^+(x) = p - \sum_{i=1}^{s_+(x)} \ell_i(x).$$

Clearly, $d_0^-(x) + d_0^+(x) = \dim V_0(x)$.

Theorem 4.3.9. *Assume that $\mathcal{A}$ is $\mathfrak{J}$-separated. For ν-almost every $x \in X$, if $0 \le k \le s_+(x)$ then*

$$d_0^+(x)\lambda_0(x) + \sum_{i=1}^{k} \ell_i(x)\lambda_i(x) \ge \sum_{i=1}^{D(x)} \int_X \log r_i^+ \, d\nu,$$

where $D = d_0^+(x) + \ell_1(x) + \cdots + \ell_k(x)$, and if $0 \le k \le s_-(x)$ then

$$d_0^-(x)\lambda_0(x) + \sum_{i=1}^{k} \ell_{-i}(x)\lambda_{-i}(x) \le \sum_{i=1}^{D(x)} \int_X \log r_i^- \, d\nu,$$

where $D(x) = d_0^-(x) + \ell_{-1}(x) + \cdots + \ell_{-k}(x)$.

Proof. Let

$$L \subset W_k(x) = \bigoplus_{i=-s_-(x)}^{k} V_i(x)$$

be a subspace transversal to $W_{-1}(x)$ and such that $V_0(x) + L = W_k(x)$. We can choose L in the positive cone C^u if its dimension is equal to $D(x)$. Indeed, the family of subspaces of $W_k(x)$ of dimension $D(x)$ (or equivalently of codimension q), which are transversal to $W_{-1}(x)$, is open and dense in the corresponding Grassmannian bundle. Furthermore, the intersection of $W_k(x)$ with any positive subspace of dimension p has dimension at least $D(x)$. Hence there are positive subspaces in $W_k(x)$ of dimension $D(x)$. The family of such subspaces is open in the Grassmannian bundle. Thus there exist positive subspaces of $W_k(x)$ of dimension $D(x)$ that are transversal to $W_{-1}(x)$.

If $L \subset C^u$ then the Euclidean volume on L is not smaller than the $\mathfrak{J}$-volume. By Proposition 4.3.2, we obtain

$$\begin{aligned}\log|\det(\mathcal{A}(x,m)|L)| &\geq \log \alpha_{D(x)}(L, \mathcal{A}(x,m)) + c\\ &= \sum_{i=0}^{m-1} \log \alpha_{D(x)}(L, A(f^i(x))) + c \geq \sum_{i=0}^{m-1} \log \sigma_{D(x)}(A(f^i(x))) + c\\ &= \sum_{i=0}^{m-1} \left[\log r_1^+(f^i(x)) + \cdots + \log r_{D(x)}^+(f^i(x))\right] + c,\end{aligned}$$

where c is the logarithm of the ratio between the $\mathfrak{J}$-volume and the Euclidean volume on L. This readily implies the first inequality. The other inequality can be obtained by observing that each transformation $A(x)^{-1}$ is $(-\mathfrak{J})$-separated. □

We also introduce the values of the Lyapunov exponent counted with their multiplicities,

$$\mu_{-q}(x) \leq \mu_{-q+1}(x) \leq \cdots \leq \mu_{-1}(x) \leq \mu_1(x) \leq \cdots \leq \mu_{p-1}(x) \leq \mu_p(x).$$

Corollary 4.3.10. *Assume that $\mathcal{A}$ is $\mathfrak{J}$-separated. For ν-almost every $x \in X$, $1 \leq k \leq p$, and $1 \leq \ell \leq q$,*

$$\mu_1(x) + \cdots + \mu_k(x) \geq \sum_{i=1}^{k} \int_X \log r_i^+ \, d\nu$$

and

$$\sum_{i=1}^{\ell} \int_X \log r_i^- \, d\nu \geq \mu_{-1}(x) + \cdots + \mu_{-\ell}(x).$$

We now describe what happens in the symplectic case.

Theorem 4.3.11. *If a cocycle $A\colon X \to Sp(2p, \mathbb{R})$ satisfies (4.11), then $\lambda_0(x) = 0$ when $V_0(x)$ is present,*

$$s_+(x) = s_-(x) = s(x), \quad \lambda_{-k}(x) + \lambda_k(x) = 0, \tag{4.26}$$

and the multiplicities of $\lambda_{-k}(x)$ and $\lambda_k(x)$ are equal for $k = 1, \ldots, s(x)$. Moreover, the subspace $W_k(x) = \bigoplus_{i=-s(x)}^{k} V_i(x)$ is the ω-orthogonal complement to $W_{-(k+1)}(x) = \bigoplus_{i=-s(x)}^{-(k+1)} V_i(x)$.

Proof. Let $\mu_1(x) \le \mu_2(x) \le \cdots \le \mu_{2p}(x)$ be the Lyapunov exponents counted with their multiplicities. Note that $\mu_1(x) + \cdots + \mu_{2p}(x) = 0$. We choose subspaces

$$\{0\} = Z_0 \subset Z_1(x) \subset \cdots \subset Z_{2p-1}(x) \subset Z_{2p} = \mathbb{R}^{2p},$$

such that $\dim Z_i = i$ and $\chi^+(x, v) = \mu_i$ for every $v \in Z_i(x) \setminus Z_{i-1}(x)$, $i = 1, \ldots, 2p$. We have

$$\dim Z_i + \dim Z_{2p-i+1} = 2p + 1,$$

and there are vectors $u \in Z_i$ and $v \in Z_{2p-i+1}$ such that $\omega(u, v) \neq 0$. By continuity, there also exist vectors $u \in Z_i \setminus Z_{i-1}$ and $v \in Z_{2p-i+1} \setminus Z_{2p-i}$ such that $\omega(u, v) \neq 0$. We now observe that $\chi^+(x, u) + \chi^+(x, v) \ge 0$. Indeed, for the standard Euclidean norm, we have $|\omega(u, v)| \le \|u\| \cdot \|v\|$, and since the cocycle takes values in $Sp(2p, \mathbb{R})$,

$$\log |\omega(\mathcal{A}(x, m)u, \mathcal{A}(x, m)v)| = \log |\omega(u, v)|.$$

Hence,

$$\frac{1}{m} \log |\omega(u, v)| \le \frac{1}{m} \log \|\mathcal{A}(x, m)u\| + \frac{1}{m} \log \|\mathcal{A}(x, m)v\|,$$

and thus $\mu_i(x) + \mu_{2p-i+1}(x) \ge 0$ for $i = 1, \ldots, p$. Adding these inequalities yields

$$0 = \sum_{i=1}^{p} (\mu_i(x) + \mu_{2p-i+1}(x)) \ge 0,$$

and thus all the inequalities must in fact be identities and (4.26) follows.

To prove the last property, note that $\omega(u, v) = 0$ for any $u \in W_k(x)$ and $v \in W_{-(k+1)}(x)$. Otherwise we could proceed as above to show that $\lambda_k(x) + \lambda_{-(k+1)}(x) \ge 0$. However, this leads to the contradiction

$$0 = \lambda_k(x) + \lambda_{-k}(x) > \lambda_k(x) + \lambda_{-(k+1)}(x) \ge 0.$$

We thus obtain $W_k(x) \subset W$, where W is the ω-orthogonal complement to $W_{-(k+1)}(x)$. Since the dimensions of these subspaces are equal, they must coincide. □

4.3.3 The Lyapunov Spectrum for Conformally Hamiltonian Systems

We describe the Lyapunov spectrum for a special class of dynamical systems called conformally Hamiltonian.

Let $\omega = \sum_{i=1}^{p} dp_i \wedge dq_i$ be the standard linear symplectic form in $\mathbb{R}^n \times \mathbb{R}^n$. We need the following statement:

Proposition 4.3.12. *For any invertible linear mapping S acting on $\mathbb{R}^{2n} = \mathbb{R}^n \times \mathbb{R}^n$, the following are equivalent:*

1. *$\omega(Su, Sv) = \beta\omega(u, v)$ for some scalar β and all $u, v \in \mathbb{R}^{2n}$.*
2. *$\omega(Su, Sv) = 0$ if and only if $\omega(u, v) = 0$; that is, S preserves skew orthogonality of vectors;*
3. *S takes Lagrangian subspaces of $\mathbb{R}^{2n}$ into Lagrangian subspaces.*

We call a linear map from $GL(2n, \mathbb{R})$ *conformally symplectic* if it satisfies one of the properties in Proposition 4.3.12. We denote the group of all conformally symplectic maps by $CSp(2n, \mathbb{R})$.

Consider a cocycle $\mathcal{A}$ with values in $CSp(2n, \mathbb{R})$ over a measure preserving transformation f of a Lebesgue space (X, ν). Let $A\colon X \to GL(n, \mathbb{R})$ be its generator. We assume that f is ergodic and that

$$\int_X \log^+ \|A(x)\| \, d\nu(x) < \infty. \tag{4.27}$$

Define a measurable function $\beta(x)$ such that for all vectors $u, v \in \mathbb{R}^{2n}$,

$$\omega(A(x)u, A(x)v) = \beta(x)\omega(u, v) \tag{4.28}$$

and set

$$b = \int_X \log|\beta(x)| \, d\nu(x). \tag{4.29}$$

Let $\{(\chi_i, k_i) : i = 1, \dots, p\}$ be the Lyapunov spectrum of the measure ν for the cocycle $\mathcal{A}$ (see Section 2.3). We need the following technical lemmas:

Lemma 4.3.13. *We have that $\sum_{i=1}^{p} k_i\chi_i = nb$.*

Proof. Since ω^n is the volume form, it follows from (4.28) that $\det A(x) = \beta(x)^n$ and the result follows from the relation

$$\sum_{i=1}^{p} k_i\chi_i = \int_X \log|\det A(x)| \, d\nu(x).$$

□

Lemma 4.3.14. *For any two vectors $u, v \in \mathbb{R}^{2n}$ that are not skew orthogonal (i.e., $\omega(u, v) \neq 0$) and for ν-almost every x, we have that*

$$\chi(x, u) + \chi(x, v) \geq b.$$

Proof. By (4.28),

$$\omega(\mathcal{A}(x, m)u, \mathcal{A}(x, m)v) = \omega(u, v) \prod_{i=0}^{m-1} \beta(f^i(x)).$$

It follows that

$$\log|\omega(\mathcal{A}(x, m)u, \mathcal{A}(x, m)v)| = \log|\omega(u, v)| + \sum_{i=0}^{m-1} \log|\beta(f^i(x))|.$$

Since $|\omega(u,v)| \le \|u\| \cdot \|v\|$, we find that

$$\frac{1}{m}\log|\omega(\mathcal{A}(x,m)u,\mathcal{A}(x,m)v)| \le \frac{1}{m}\|\mathcal{A}(x,m)u\| + \frac{1}{m}\|\mathcal{A}(x,m)v\|.$$

Combining the above relations and using Birkhoff's Ergodic Theorem, we conclude that for ν-almost every x,

$$b = \int_X \log|\beta(x)|\,d\nu(x) \le \chi(x,u) + \chi(x,v)$$

and the desired result follows. □

For a linear subspace $E \subset \mathbb{R}^{2n}$ denote by $E^{\triangleleft}$ the skew-orthogonal complement of E, that is, the linear subspace in $\mathbb{R}^{2n}$ containing all vectors v for which $\omega(u,v) = 0$ for all $u \in E$. Since ω is nondegenerate,

$$\dim E + \dim E^{\triangleleft} = 2n, \quad (E^{\triangleleft})^{\triangleleft} = E.$$

The proof of the following statement is straightforward and is left to the reader.

Lemma 4.3.15. *Let $E, F \subset \mathbb{R}^{2n}$ be two linear subspaces. If $\omega(u,v) = 0$ for all $u \in E$ and $v \in F$ then*

$$E \subset F^{\triangleleft}, \quad F \subset E^{\triangleleft}, \quad \dim E + \dim F \le 2n$$

Let $\mathcal{A}$ be a measurable cocycle with values in the conformal symplectic group $CSp(2n,\mathbb{R})$ over a measure preserving transformation f of a Lebesgue space (X,ν). We assume that f is ergodic. We describe some results by Liverani and Wojtkowski (see [231]) on relations in the Lyapunov spectrum of ν.

Theorem 4.3.16. *Assume that the cocycle $\mathcal{A}$ satisfies* (4.27). *Then for $i = 1,\dots,p$, we have $\chi_i + \chi_{p-i+1} = b$ and $k_i = k_{p-i+1}$, where b is given by* (4.29). *Moreover, if*

$$\{0\} = E_0^+ \subset E_1^+(x) \subset \cdots \subset E_{p-1}^+(x) \subset E_p^+ = \mathbb{R}^{2n}$$

is the filtration associated with the Lyapunov exponents (and the ergodic measure ν) then the subspace E_{p-i}^+ is the skew-orthogonal complement of E_i^+.

Proof. Let $\chi_1' \le \chi_2' \le \cdots \le \chi_{2n}'$ be the values of the Lyapunov exponent counted with their multiplicities. By Lemma 4.3.13,

$$\chi_1' + \chi_2' + \cdots + \chi_{2n}' = nb.$$

We can choose a filtration of subspaces

$$\{0\} = F_0 \subset F_1(x) \subset \cdots \subset F_{2n-1}(x) \subset F_{2n} = \mathbb{R}^{2n}$$

such that $\dim F_i(x) = i$ and for almost every x and all vector $v \in F_i(x) \setminus F_{i-1}(x)$, the Lyapunov exponent $\chi(x,v) = \chi_i'$ for $i = 1,\dots,2n$. For $i \le n$, we have

$$\dim F_i + \dim F_{2n-i+1} = 2n+1$$

and hence by Lemma 4.3.15, there are $u \in F_i$ and $v \in F_{2n-i+1}$ such that $\omega(u, v) \neq 0$. By continuity, there must also be vectors $\widetilde{u} \in F_i \setminus F_{i-1}$ and $\widetilde{v} \in F_{2n-i+1} \setminus F_{2n-i}$ such that $\omega(\widetilde{u}, \widetilde{v}) \neq 0$. It follows from Lemma 4.3.14 that for $i = 1, \dots, n$,

$$\chi_i' + \chi_{2n-i+1}' \geq b.$$

Summing these inequalities up over i, we obtain that

$$nb = \sum_{i=1}^{n} (\chi_i' + \chi_{2n-i+1}') \geq nb.$$

This shows that all the inequalities are actually equalities. This implies that for $i = 1, \dots, p$, the multiplicities of the values χ_i and χ_{p-i+1} are equal and $\chi_i + \chi_{p-i+1} = b$.

To show that the subspace E_{p-i}^+ is the skew-orthogonal complement of the subspace E_i^+, observe that $\omega(u, v) = 0$ for any $u \in E_i^+$ and $v \in E_{p-i}^+$. Indeed, if this is not the case we could use Lemma 4.3.14 to claim that $\chi_i + \chi_{p-i} \geq b$, which leads to the contradiction

$$b = \chi_i + \chi_{p-i+1} > \chi_i + \chi_{p-i} \geq b.$$

Using the fact that $\omega(u, v) = 0$ for any $u \in E_i^+$ and $v \in E_{p-i}^+$ and applying Lemma 4.3.15, we obtain that $E_{p-i}^+ \subset (E_i^+)^{\triangleleft}$. Since the dimensions of these subspaces are equal, we have $E_{p-i}^+ = (E_i^+)^{\triangleleft}$. □

A smooth manifold M of even dimension is called *conformally symplectic* if it admits a *conformally symplectic structure* given by a nondegenerate differential 2-form Θ satisfying $d\Theta = \gamma \wedge \Theta$, where γ is a closed 1-form. Observe that locally $\gamma = dU$ for some smooth function U and $d(e^{-U}\Theta) = 0$; that is, $e^{-U}\Theta$ defines a bonafide symplectic structure on M.

Given a function $H \colon M \to \mathbb{R}$ called *Hamiltonian*, define a *conformally Hamiltonian vector field* $F = \nabla_\Theta H$ by the usual requirement

$$\Theta(\cdot, F) = dH. \tag{4.30}$$

The flow Φ_t generated by the vector field F is called *conformally Hamiltonian.* The Hamiltonian function H is a first integral since

$$\frac{d}{dt} H = dH(F) = \Theta(F, F) = 0.$$

We restrict the flow Φ_t to a smooth level set of the Hamiltonian, $M^c = \{z \in M : H(z) = c\}$, and assume that dH and F do not vanish on M^c. For $\xi, \eta \in T_z M^c$, we set

$$w(t) = \Theta(d\Phi_t \xi, d\Phi_t \eta).$$

One can show that

$$\frac{d}{dt} w(t) = \gamma(F(\Phi_t(z)))w(t),$$

and hence

$$\Theta(d\Phi_t\xi, d\Phi_t\eta) = \beta\Theta(\xi, \eta) \tag{4.31}$$

with

$$\beta = \exp\left(\int_0^t \gamma(F(\Phi_u(z)))\,du\right).$$

Consider the quotient of the tangent bundle TM^c of M^c by the vector field F, that is, by the one-dimensional subspace spanned by F. We denote the quotient bundle by $\hat{T}M^c$. In view of (4.30), the form Θ factors naturally from TM^c to $\hat{T}M^c$. The factor form defines in each of the quotient tangent spaces $\hat{T}_zM^c$, $z \in M^c$, a linear symplectic form.

The differential of the flow preserves the vector field F; that is, $d\Phi_t(F(z)) = F(\Phi_t(z))$. As a result the derivative can also be factored on the quotient bundle producing the *transversal derivative cocycle*. We denote it by $A_t(z)\colon \hat{T}_zM^c \to T_{\Phi_t(z)}M^c$. It follows from (4.31) that the transversal derivative cocycle is conformally Hamiltonian with respect to Θ (or more precisely, the linear symplectic form it defines in the quotient tangent spaces).

Let us fix an invariant probability measure ν_c on M^c and assume that

$$\int_{M^c} \log^+ \|d\Phi_1\|\,d\nu_c(z) < \infty.$$

Applying Theorem 4.3.16, we obtain the following result that describes some symmetry of the Lyapunov spectrum of the transversal derivative cocycle with respect to ν_c.

Theorem 4.3.17. *Let*

$$\{0\} \subset E_0^+(z) \subset \cdots \subset E_{p-1}^+(z) \subset E_p^+(z) = \hat{T}_zM^c$$

be the filtration of subspaces at z associated with the Lyapunov exponents $\chi_1 < \cdots < \chi_{p-1} < \chi_p$ of the transversal derivative cocycle $A_t(z)$, $z \in M^c$. Then the multiplicities of χ_i and χ_{p-i+1} are equal and

$$\chi_i + \chi_{p-i+1} = a, \quad i = 1, \ldots, p,$$

where $a = \int_{M^c} \gamma(F(z))\,d\nu_c(z)$. Moreover, the subspace $E_{p-i}^+(z)$ is the skew-orthogonal complement of the subspace $E_i^+(z)$.

Note that the invariant measure ν_c can be supported on a single periodic orbit, so that this theorem applies as well to the real parts of the Floquet exponents.

4.4 A Remark on Applications of Cone Techniques

Let $\mathcal{A}$ be a cocycle over an invertible measurable transformation $f\colon X \to X$. We present here some modifications of the above cone technique, which are often used in applications. Assume that there exists a subset $Y \subset X$ with the following properties:

1. For every $x \in Y$, one can construct two cones $C^s(x)$ and $C^u(x)$ such that the angle between them is uniformly away from zero.
2. There is a number $\lambda > 1$ such that

$$\|dfv\| \geq \lambda \|v\| \text{ for } v \in C^u(x),$$

$$\|df^{-1}v\|^{-1} \geq \lambda \|v\| \text{ for } v \in C^s(x).$$

It follows that while the trajectory of a point x stays within Y, every vector v in the unstable cone $C^u(x)$ expands exponentially. Whether the vector v gains positive Lyapunov exponent depends on what happens when the trajectory travels outside Y. Let us look at this problem in more detail. Assume that the trajectory leaves the set Y at time n and then returns to Y at time $m > n$ (so that $f^k(x) \notin Y$ for $n \leq k < m$). To ensure that the vector v has positive Lyapunov exponent, one should be able to control first, the length of the vector $df^m v$ (so that it is not much smaller than the length of the vector $df^n v$) and second, the angle between the vector $df^m v$ and the unstable cone $C^u(f^m(x))$ (so that it is not too big; in particular, that the vector $df^m v$ does not lie in the stable cone $C^u(f^m(x))$).

For example, there are situations where one can guarantee that $df^m v \in C^u(f^m(x))$ and $\|df^m v\| \geq C\|df^n v\|$ where $C > 0$ is a constant. If now ν is an invariant Borel measure for f such that $\nu(Y) \geq 1 - \varepsilon$ with sufficiently small ε then the above properties imply that for almost every $x \in Y$, any vector $v \in C^u(x)$ has positive Lyapunov exponent (due to the fact that the average time the trajectory spends outside Y is very small).

4.5 Notes

One of the most powerful methods to verify that Lyapunov exponents do not vanish was developed by Wojtkowski in [225]. It is a generalization of the initial approach by Sinai [203] and by Alekseev (see [1–3]) to build an invariant family of unstable cones.

Theorem 4.1.3 generalizes results of Wojtkowski, which are described in [225]. Our exposition follows the paper by Katok and Burns [103]. For functions Q as in (4.4), the statement in Theorem 4.1.3 was proved by Markarian in [146, Theorem 1].

Proposition 4.2.2 is a reformulation of Proposition 5.1 in [225].

The approach in Section 4.3 is due to Wojtkowski [230]. It is motivated by the study of geodesic flows on Weyl manifolds in [228, 229]. The symplectic case was studied earlier in [226, 227]. Proposition 4.3.3 extends a result of Bougerol in [33] on the contraction of the space of positive Lagrangian subspaces under the action of monotone symplectic operators.

In [231], Theorem 4.3.17 is used to establish symmetry of the Lyapunov spectrum for Gaussian isokinetic dynamics on Riemannian manifolds and for the Nosé–Hoover Hamiltonian dynamics.

5

The Derivative Cocycle

Every smooth dynamical system as well as nonstationary linear ordinary differential equation generates a special cocycle known as the derivative cocycle. Lyapunov exponents of this cocycle provide an important tool in studying stability of trajectories of dynamical systems and of solutions of ordinary differential equations (see Chapter 7). Throughout this chapter we assume that the phase space of the system is compact. This allows us to simplify some technical arguments. We stress that this requirement can be replaced with a weaker one that the phase space is complete (see Remark 5.4.2 for details).

5.1 Smooth Dynamical Systems and the Derivative Cocycle

Let $f\colon M \to M$ be a C^1 diffeomorphism of a smooth compact Riemannian n-manifold. Given $x \in M$, we identify the tangent spaces $T_{f^m(x)}M$, $m \in \mathbb{Z}$ with $\mathbb{R}^n$. Using this identification we introduce the cocycle $\mathcal{A}_x = \{d_{f^m(x)}f\}_{m\in\mathbb{Z}}$ over the transformation $f\colon X \to X$ where $X = \{f^m(x)\}_{m\in\mathbb{Z}}$. It is called the *derivative cocycle* associated with the diffeomorphism f and the point x.

The "individual" derivative cocycles $\mathcal{A}_x$ depend on the individual trajectories $\{f^m(x)\}_{m\in\mathbb{Z}}$. We now introduce the "global" cocycle associated with f. One can represent M as a finite union $\bigcup_i \Delta_i$ of differentiable copies Δ_i of the n-simplex such that

1. in each Δ_i, one can introduce local coordinates in such a way that $T\Delta_i$ can be identified with $\Delta_i \times \mathbb{R}^n$;
2. all the nonempty intersections $\Delta_i \cap \Delta_j$, for $i \neq j$, are $(n-1)$-dimensional submanifolds.

In each Δ_i, the derivative of f can be interpreted as a linear cocycle. This implies that $df\colon M \to \mathbb{R}^n$ can be interpreted as a measurable linear cocycle $\mathcal{A}$ with $df(x)$ to be the matrix representation of df in local coordinates. We call $\mathcal{A}$ the *derivative cocycle* specified by the diffeomorphism f. It does not depend on the choice of the decomposition $\{\Delta_i\}$. Indeed, if we choose another decomposition $\{\Delta_i'\}$, then the coordinate change in $\Delta_i \cap \Delta'_j$ sending one representation to the other one is effected by maps, which are uniformly bounded together with their derivatives, their inverses, and the inverses of their derivatives. Hence by Proposition 2.1.4, the coordinate change is tempered and the two cocycles corresponding to the two decompositions $\{\Delta_i\}$ and $\{\Delta'_i\}$ are equivalent.

We remark that if ν is an f-invariant Borel probability measure on M then the decomposition $\{\Delta_i\}$ can be chosen such that $\nu(\partial\Delta_i) = 0$ for every i.

5.2 Nonuniformly Hyperbolic Diffeomorphisms

The general notion of hyperbolic cocycles applies to derivative cocycles (see Definitions 2.2.1 and 2.2.3).

Definition 5.2.1. A $C^{1+\alpha}$ diffeomorphism f is said to be *nonuniformly partially hyperbolic in the broad sense* on an f-invariant Borel subset $Y \subset M$ if so is the derivative cocycle generated by f on Y. More precisely, we say that an f-invariant Borel subset $Y \subset M$ is *nonuniformly partially hyperbolic in the broad sense* if there exist: (a) Borel functions $\lambda, \mu, C, K\colon Y \to (0, \infty)$, and $\varepsilon\colon Y \to [0, \varepsilon_0]$ with $\varepsilon_0 > 0$ and (b) subspaces $E^1(x)$ and $E^2(x)$, $x \in Y$, which satisfy the following conditions:

1. The functions λ, μ, and ε are f-invariant; that is,

$$\lambda(f(x)) = \lambda(x), \quad \mu(f(x)) = \mu(x), \quad \varepsilon(f(x)) = \varepsilon(x)$$

   and

$$\lambda(x)e^{\varepsilon(x)} < \mu(x)e^{-\varepsilon(x)}.$$

2. $E^1(x)$ and $E^2(x)$ depend measurably on x and form an invariant splitting of the tangent space; that is,

$$T_xM = E^1(x) \oplus E^2(x),$$

$$dfE^1(x) = E^1(f(x)) \quad \text{and} \quad dfE^2(x) = E^2(f(x)).$$

3. For $v \in E^1(x)$ and $n \geq 0$,

$$\|df^n v\| \leq C(x)\lambda(x)^n\, e^{\varepsilon(x)n}\, \|v\|.$$

4. For $v \in E^2(x)$ and $n \leq 0$,

$$\|df^n v\| \leq C(x)\mu(x)^n\, e^{\varepsilon(x)|n|}\, \|v\|.$$

5. $\sphericalangle(E^1(x), E^2(x)) \geq K(x)$.
6. For $n \in \mathbb{Z}$,

$$C(f^n(x)) \leq C(x)e^{\varepsilon(x)|n|} \quad \text{and} \quad K(f^n(x)) \geq K(x)e^{-\varepsilon(x)|n|}.$$

If $\lambda(x)e^{\varepsilon(x)} < 1$ for $x \in Y$, the subspace $E^1(x)$ is *stable* (*contracting*) and we use the notation $E^s(x)$. If $\mu(x)e^{-\varepsilon(x)} > 1$ for $x \in Y$, the subspace $E^2(x)$ is *unstable* (*expanding*) and we use the notation $E^u(x)$.

Definition 5.2.2. An f-invariant Borel subset $Y \subset M$ is *nonuniformly partially hyperbolic* if there exist

(a) Borel functions $\lambda, \lambda', \mu, \mu' : Y \to (0, \infty)$, and $\varepsilon : Y \to [0, \varepsilon_0]$ with $\varepsilon_0 > 0$ such that for $x \in Y$,

$$\lambda(f(x)) = \lambda(x), \quad \mu(f(x)) = \mu(x), \quad \varepsilon(f(x)) = \varepsilon(x),$$
$$0 < \lambda(x) < \lambda(x)e^{\varepsilon(x)} < \lambda(x)' \le \mu(x)' < \mu'(x)e^{\varepsilon(x)} < \mu(x),$$
$$\lambda(x) < 1 < \mu(x);$$

(b) Borel functions $C, K : \Lambda \to (0, \infty)$;
(c) subspaces $E^c(x)$, $E^s(x)$, and $E^u(x)$, $x \in Y$,

such that

1. $E^c(x)$, $E^s(x)$, and $E^u(x)$ depend measurably on x and form an invariant splitting of the tangent space, that is,
$$T_x M = E^c(x) \oplus E^s(x) \oplus E^u(x),$$
$$df E^c(x) = E^c(f(x)), \quad df E^s(x) = E^s(f(x)),$$
$$df E^u(x) = E^u(f(x));$$
2. for $v \in E^s(x)$ and $n \ge 0$,
$$\|df^n v\| \le C(x)\lambda(x)^n e^{\varepsilon(x)n} \|v\|;$$
3. for $v \in E^u(x)$ and $n \le 0$,
$$\|df^n v\| \le C(x)\mu(x)^n e^{\varepsilon(x)|n|} \|v\|;$$
4. for $v \in E^c(x)$ and $n \ge 0$,
$$C(x)^{-1}(\lambda'(x))^n e^{-\varepsilon(x)n} \|v\| \le \|df^n v\| \le C(x)(\mu'(x))^n e^{\varepsilon(x)n} \|v\|;$$
5.
$$\measuredangle(E^c(x), E^s(x)) \ge K(x), \quad \measuredangle(E^c(x), E^u(x)) \ge K(x),$$
$$\measuredangle(E^s(x), E^u(x)) \ge K(x);$$
6. Condition 6 in Definition 5.2.1 holds.

We say that Y is *nonuniformly (completely) hyperbolic* if it is nonuniformly partially hyperbolic in the broad sense and $\mu(x) > 1$ for every $x \in Y$.

Some results of nonuniform hyperbolicity theory (e.g., construction of stable manifolds and absolute continuity; see Chapters 7 and 8) require that the number ε_0 in Definitions 5.2.1 and 5.2.2 is *sufficiently small.* We will see later (see Theorem 5.4.6) that if f has nonzero Lyapunov exponents then ε_0 can be chosen *arbitrarily small* (but the functions $C(x)$ and $K(x)$ should be chosen appropriately depending on the function $\varepsilon(x)$).

Let Y be an f-invariant Borel subset, which is either nonuniformly partially hyperbolic in the broad sense, nonuniformly partially hyperbolic, or nonuniformly completely hyperbolic. In either case we have the family of nonempty *level sets* $\Lambda_{\lambda\mu\varepsilon j}$ where the positive numbers λ, μ, and ε satisfy (2.15) and $j = \dim E^s(x)$ (i.e, the dimension of the stable subspace is constant on $\Lambda_{\lambda\mu\varepsilon j}$). Furthermore, for each λ, μ, ε, and j we also have the collection of *regular sets* $\Lambda^{\ell}_{\lambda\mu\varepsilon j}$ (see Section 2.2.2). The latter are compact and have properties described in Proposition 2.2.7. We stress that the levels sets $\Lambda_{\lambda\mu\varepsilon j}$ are nonuniformly partially hyperbolic in the broad sense (respectively, nonuniformly partially hyperbolic or nonuniformly completely hyperbolic). In what follows, without loss of generality, we will often restrict f to one of its nonempty level sets and to simplify our notations we will denote this set by Λ and the corresponding regular sets by Λ^{ℓ}.

We adapt the notion of Lyapunov inner product introduced in Section 3.5.1 to the case of derivative cocycles. Let Y be an f-invariant Borel subset, which is nonuniformly partially hyperbolic in the broad sense. Fix a point $x \in Y$ and choose numbers $0 < \lambda' < \mu' < \infty$ such that

$$\lambda(x)e^{\varepsilon(x)} < \lambda' \quad \text{and} \quad \mu' < \mu(x)e^{-\varepsilon(x)}.$$

We define the inner product $\langle\cdot,\cdot\rangle'_x$ by setting

$$\langle v, w\rangle'_x = \sum_{k=0}^{\infty}\langle df^k v, df^k w\rangle_{f^k(x)}\lambda'^{-2k}$$

if $v, w \in E^s(x)$ and

$$\langle v, w\rangle'_x = \sum_{k=0}^{\infty}\langle df^{-k} v, df^{-k} w\rangle_{f^{-k}(x)}\mu'^{2k}$$

if $v, w \in E^u(x)$. Notice that each series converges.

We extend $\langle\cdot,\cdot\rangle'_x$ to all vectors in T_xM by declaring the subspaces $E^s(x)$ and $E^u(x)$ to be mutually orthogonal with respect to $\langle\cdot,\cdot\rangle'_x$; that is, we set

$$\langle v, w\rangle'_x = \langle v^s, w^s\rangle'_x + \langle v^u, w^u\rangle'_x,$$

where $v = v^s + v^u$ and $w = w^s + w^u$ with $v^s, w^s \in E^s(x)$ and $v^u, w^u \in E^u(x)$.

The norm induced by the Lyapunov inner product is denoted by $\|\cdot\|'$ and both depend on the choice of numbers λ' and μ'.

A Lyapunov inner product has several important properties:

1. The angle between the subspaces $E^s(x)$ and $E^u(x)$ in the inner product $\langle\cdot,\cdot\rangle'_x$ is $\pi/2$.

2. $\|A_x\|' \le \lambda'$ and $\|B_x{}^{-1}\|' \le (\mu')^{-1}$.
3. The relation between the Lyapunov inner product and the Riemannian inner product is given by
$$\frac{1}{\sqrt{2}}\|w\|_x \le \|w\|'_x \le D(x)\|w\|_x,$$
where $w \in T_xM$ and
$$D(x) = C(x)K(x)^{-1}[(1 - \lambda e^{\varepsilon}/\lambda')^{-1} + (1 - \mu'/(\mu e^{-\varepsilon}))^{-1}]^{1/2}$$
is a measurable function satisfying
$$D(f^m(x)) \le D(x)e^{2\varepsilon|m|}, \quad m \in \mathbb{Z}. \tag{5.1}$$

Properties 1 and 2 show that the action of the differential df along the trajectory $\{f^n(x)\}$ is *uniformly* partially hyperbolic in the broad sense with respect to any Lyapunov inner product. Note that the function $D(x)$ is bounded from above on any regular set $\Lambda^\ell_{\lambda\mu\varepsilon j}$.

5.3 Hölder Continuity of Invariant Distributions

Let f be a $C^{1+\alpha}$ diffeomorphism of a smooth compact Riemannian manifold M possessing an invariant set Y, which is nonuniformly partially hyperbolic in the broad sense. Let us choose a nonempty level set $\Lambda = \Lambda_{\lambda\mu\varepsilon j}$ and consider the corresponding collection of regular sets $\{\Lambda^\ell = \Lambda^\ell_{\lambda\mu\varepsilon j} : \ell \ge 1\}$ (see the previous section). The numbers λ, μ, ε, and j will be fixed throughout this section. We recall that they satisfy (2.15).

According to Section 2.2.3, the stable subspaces $E^s(x)$ depend continuously on $x \in \Lambda^\ell$. However, since these subspaces depend on the infinite future, they may not be differentiable functions of the point even if f is real analytic. In this section we show that on regular sets the stable subspaces depend Hölder continuously on the point. Observe that the dimension of the stable subspace $E^s(x)$ is constant and hence one can compare the distance in the Grassmannian bundle between $E^s(x)$ and $E^s(y)$ for any $x, y \in \Lambda^\ell$ sufficiently close to each other.

Theorem 5.3.1. *There exist $\varepsilon_0 > 0$ and $\delta_0 > 0$ such that if $\varepsilon \le \varepsilon_0$ then the stable distribution $E^s(x)$ depends Hölder continuously on $x \in \Lambda^\ell$, that is, for any $x, y \in \Lambda^\ell$ with $\rho(x, y) \le \delta_0$,*
$$d(E^s(x), E^s(y)) \le C\rho(x, y)^\beta,$$
where $C > 0$ and $\beta \in (0, 1]$ are constants, ρ is the distance in M, and d is the distance in the Grassmannian bundle of TM generated by the Riemannian metric.

We present the proof of this statement following the arguments by Brin [38]. Indeed, we shall prove a more general statement of which Theorem 5.3.1 is an easy corollary.

A k-dimensional distribution E on a subset Γ of a differentiable manifold M is a family of k-dimensional subspaces $E(x) \subset T_x M$, $x \in \Gamma$. A Riemannian metric on M naturally induces distances in TM and in the space of k-dimensional subspaces in TM. The Hölder continuity of a distribution E can be defined using these distances. By the Whitney Embedding Theorem, every manifold M can be embedded in $\mathbb{R}^N$ with a sufficiently large N. If M is compact, the Riemannian metric on M is equivalent to the distance $\|x - y\|$ induced by the embedding. We assume in Theorem 5.3.2, without loss of generality, that the manifold is embedded in $\mathbb{R}^N$.

For a subspace $A \subset \mathbb{R}^N$ and a vector $v \in \mathbb{R}^N$, set

$$\operatorname{dist}(v, A) = \min_{w \in A} \|v - w\|;$$

that is, $\operatorname{dist}(v, A)$ is the length of the difference between v and its orthogonal projection to A. For subspaces A, B in $\mathbb{R}^N$, define

$$\operatorname{dist}(A, B) = \max\left\{ \max_{v \in A, \|v\|=1} \operatorname{dist}(v, B), \max_{w \in B, \|w\|=1} \operatorname{dist}(w, A) \right\}.$$

A k-dimensional *distribution* E defined on a set $\Gamma \subset \mathbb{R}^N$ is called *Hölder continuous* with *Hölder exponent* $\beta \in (0, 1]$ and *Hölder constant* $L > 0$ if there exists $\varepsilon_0 > 0$ such that

$$\operatorname{dist}(E(x), E(y)) \leq L\|x - y\|^{\beta}$$

for every $x, y \in \Gamma$ with $\|x - y\| \leq \varepsilon_0$.

The subspaces $E_1, E_2 \subset \mathbb{R}^N$ are said to be κ*-transverse* if $\|v_1 - v_2\| \geq \kappa$ for all unit vectors $v_1 \in E_1$ and $v_2 \in E_2$.

Theorem 5.3.2. *Let M be a compact m-dimensional C^2 submanifold of $\mathbb{R}^N$ for some $m < N$ and $f \colon M \to M$ a $C^{1+\alpha}$ map for some $\alpha \in (0, 1)$. Assume that there exist a set $\Gamma \subset M$ and real numbers $0 < \lambda < \mu$, $c > 0$, and $\kappa > 0$ such that for each $x \in \Gamma$, there are κ-transverse subspaces $E^s(x)$, $E^u(x) \subset T_x M$ with the following properties:*

1. $T_x M = E^s(x) \oplus E^u(x)$.
2. $\|df^n v^s\| \leq c\lambda^n \|v^s\|$ *and* $\|df^n v^u\| \geq c^{-1}\mu^n \|v^u\|$ *for every* $v^s \in E^s(x)$, $v^u \in E^u(x)$, *and every positive integer* n.

Then for every $a > \max_{z \in M} \|d_z f\|^{1+\alpha}$, the distribution E^s is Hölder continuous with exponent

$$\beta = \frac{\log \mu - \log \lambda}{\log a - \log \lambda} \alpha.$$

Proof. We begin with two lemmas:

Lemma 5.3.3. *Let A_n and B_n, for $n = 0, 1, \ldots$, be two sequences of real $N \times N$ matrices such that for some $\Delta \in (0, 1)$,*

$$\|A_n - B_n\| \leq \Delta a^n$$

for every positive integer n. Assume that there exist subspaces E_A, $E_B \subset \mathbb{R}^N$ *and numbers* $0 < \lambda < \mu$ *and* $C > 1$ *such that* $\lambda < a$ *and for each* $n \geq 0$,

$$\|A_n v\| \leq C\lambda^n \|v\| \text{ if } v \in E_A; \qquad \|A_n w\| \geq C^{-1}\mu^n \|w\| \text{ if } w \in {E_A}^{\perp};$$
$$\|B_n v\| \leq C\lambda^n \|v\| \text{ if } v \in E_B; \qquad \|B_n w\| \geq C^{-1}\mu^n \|w\| \text{ if } w \in {E_B}^{\perp}.$$

Then

$$\operatorname{dist}(E_A, E_B) \leq 3C^2 \frac{\mu}{\lambda} \Delta^{(\log\mu - \log\lambda)/(\log a - \log\lambda)}.$$

Proof of the lemma. Set

$$Q_A^n = \{v \in \mathbb{R}^N : \|A_n v\| \leq 2C\lambda^n \|v\|\}$$

and

$$Q_B^n = \{v \in \mathbb{R}^N : \|B_n v\| \leq 2C\lambda^n \|v\|\}.$$

For each $v \in \mathbb{R}^N$, write $v = v_1 + v_2$, where $v_1 \in E_A$ and $v_2 \in {E_A}^{\perp}$. If $v \in Q_A^n$, then

$$\|A_n v\| = \|A_n(v_1 + v_2)\| \geq \|A_n v_2\| - \|A_n v_1\| \geq C^{-1}\mu^n \|v_2\| - C\lambda^n \|v_1\|,$$

and hence

$$\|v_2\| \leq C\mu^{-n}(\|A_n v\| + C\lambda^n \|v_1\|) \leq 3C^2 \left(\frac{\lambda}{\mu}\right)^n \|v\|.$$

Therefore,

$$\operatorname{dist}(v, E_A) \leq 3C^2 \left(\frac{\lambda}{\mu}\right)^n \|v\|. \tag{5.2}$$

Set $\gamma = \lambda/a < 1$. There exists a unique nonnegative integer n such that $\gamma^{n+1} < \Delta \leq \gamma^n$. If $w \in E_B$, then

$$\begin{aligned}\|A_n w\| &\leq \|B_n w\| + \|A_n - B_n\| \cdot \|w\| \\ &\leq C\lambda^n \|w\| + \Delta a^n \|w\| \\ &\leq (C\lambda^n + (\gamma a)^n)\|w\| \leq 2C\lambda^n \|w\|.\end{aligned}$$

It follows that $w \in Q_A^n$, and hence $E_B \subset Q_A^n$. By symmetry, $E_A \subset Q_B^n$. By (5.2) and the choice of n, we obtain

$$\operatorname{dist}(E_A, E_B) \leq 3C^2 \left(\frac{\lambda}{\mu}\right)^n \leq 3C^2 \frac{\mu}{\lambda} \Delta^{(\log\mu - \log\lambda)/(\log a - \log\lambda)}.$$

This completes the proof of the lemma. □

Lemma 5.3.4. *Let* $f\colon M \to M$ *be a* $C^{1+\alpha}$ *map of a compact m-dimensional* C^2 *submanifold* $M \subset \mathbb{R}^N$. *Then for every* $a > \max_{z \in M} \|d_z f\|^{1+\alpha}$, *there exists* $D > 1$ *such that for every* $n \in \mathbb{N}$ *and every* $x, y \in M$, *we have*

$$\|d_x f^n - d_y f^n\| \leq D a^n \|x - y\|^\alpha.$$

Proof of the lemma. Let D' be such that

$$\|d_x f - d_y f\| \le D' \|x - y\|^\alpha.$$

Set $b = \max_{z \in M} \|d_z f\| \ge 1$ and observe that for every $x, y \in M$,

$$\|f^n(x) - f^n(y)\| \le b^n \|x - y\|.$$

Fix $a > b$. Then the lemma holds true for $n = 1$ and any $D \ge D'$. For the inductive step, we note that

$$\begin{aligned}\|d_x f^{n+1} - d_y f^{n+1}\| &\le \|d_{f^n(x)} f\| \cdot \|d_x f^n - d_y f^n\| \\ &\quad + \|d_{f^n(x)} f - d_{f^n(y)} f\| \cdot \|d_y f^n\| \\ &\le bDa^n \|x - y\|^\alpha + D'\left(b^n \|x - y\|\right)^\alpha b^n \\ &\le Da^{n+1} \|x - y\|^\alpha \left(\frac{b}{a} + \frac{D'}{D} \frac{(b^{1+\alpha})^n}{a^{n+1}}\right).\end{aligned}$$

If $a > b^{1+\alpha}$, then there exists $D \ge D'$ for which the factor in parentheses is less than 1. □

We proceed with the proof of the theorem:

For $x \in M$, let $(T_x M)^\perp$ denote the orthogonal complement to the tangent plane $T_x M$ in $\mathbb{R}^N$. Since $(TM)^\perp$ is a smooth distribution, it is sufficient to prove the Hölder continuity of $F = E^s \oplus (TM)^\perp$.

Since $E^s(x)$ and $E^u(x)$ are κ-transverse and of complementary dimensions in $T_x M$, there exists $d > 1$ such that $\|df^n w\| \ge d^{-1} \mu^n \|w\|$ for every $x \in \Gamma$ and $w \perp E^s(x)$.

For $x, y \in \Gamma$ and a positive integer n, let A_n and B_n be $N \times N$ matrices such that

$$A_n v = df^n v \text{ if } v \in T_x M \text{ and } A_n w = 0 \text{ if } w \in (T_x M)^\perp,$$

$$B_n v = df^n v \text{ if } v \in T_y M \text{ and } B_n w = 0 \text{ if } w \in (T_y M)^\perp.$$

By Lemma 5.3.4, $\|A_n - B_n\| \le Da^n \|x - y\|^\alpha$. Now the theorem follows from Lemma 5.3.3 with $\Delta = D\|x - y\|^\alpha$, $E_A = F(x)$, $E_B = F(y)$, and $C = \max\{c, d\}$. □

5.4 Lyapunov Exponent and Regularity of the Derivative Cocycle

Let f be a $C^{1+\alpha}$ diffeomorphism of a smooth compact Riemannian manifold M. The Lyapunov exponent χ^+ of the derivative cocycle is given by the following formula (see (2.24)):

$$\chi^+(x, v) = \overline{\lim_{m \to +\infty}} \frac{1}{m} \log\|df^n v\|$$

and is called the *Lyapunov exponent* of the diffeomorphism f at the point x.

We say that a point $x \in M$ is *forward regular*, *backward regular*, or *Lyapunov–Perron regular* (or simply *regular*), respectively, if it is forward regular, backward regular, or regular with respect to the derivative cocycle $\mathcal{A}_x$.

We recall that for any regular point $x \in M$, there exist an integer $p(x) \le n$, numbers

$$\chi_1(x) < \cdots < \chi_{p(x)}(x),$$

and a decomposition

$$T_x M = \bigoplus_{i=1}^{p(x)} H_i(x) \tag{5.3}$$

into subspaces $H_i(x)$ such that $df H_i(x) = H_i(f(x))$, and

$$\lim_{m \to \pm\infty} \frac{1}{m} \log \|df^m v\| = \chi_i(x) \text{ for } v \in H_i(x) \setminus \{0\},$$

with uniform convergence on $\{v \in H_i(x) : \|v\| = 1\}$. The decomposition (5.3) is called the *Oseledets decomposition* at x and the subspaces $H_i(x)$ are called the *Oseledets subspaces* at x. Write $k_i(x) = \dim H_i(x)$.

The following statement is an immediate corollary of Theorem 3.4.3.

Theorem 5.4.1. *Let f be a C^1 diffeomorphism of a smooth compact Riemannian manifold. Then the set of regular points has full measure with respect to any f-invariant Borel probability measure.*

Remark 5.4.2. *The results in this chapter remain true if the requirement that the manifold M is compact is relaxed to M being complete, provided that there exists $C > 0$ such that*

$$\|d_x f\|, \|d_x f^{-1}\| \le C \text{ for every } x \in M. \tag{5.4}$$

(Clearly, this requirement holds when M is compact.) Under Condition (5.4) *the derivative cocycle satisfies* (3.20) *and the Multiplicative Ergodic Theorem 3.4.3 guarantees that the set of regular points (as well as the sets of forward and backward regular points) has full measure with respect to any f-invariant Borel probability measure. It should be noticed that in the case when M is complete the regular sets are closed but are not necessarily compact.*

The set of points that are not Lyapunov–Perron regular is negligible from the measure-theoretic point of view, since it has zero measure with respect to any invariant Borel measure. However, this set may be "large": for example, it may have positive volume (if the system does *not* preserve volume), positive Hausdorff dimension, or positive topological entropy (see Section 6.1.4).

Theorem 5.4.1 provides no effective tool to determine whether a given trajectory is regular (or forward regular or backward regular). We now present a somewhat practical criterion, which guarantees forward and backward regularity of individual trajectories.

Let us first notice that if x is a fixed point or a periodic point for f then the cocycle $\mathcal{A}_x$ is rigid with the generator $A = d_x f$ (if x is a fixed point) or $A = d_x f^p$ (if x is a periodic point of period p).

We now consider the case of an arbitrary point x. Recall that given vectors $v_1, \ldots, v_k$, we denote by $\mathrm{vol}(v_1, \ldots, v_k)$ the volume of the parallelepiped formed by these vectors. The following statement is an immediate consequence of Proposition 3.1.2:

Proposition 5.4.3. *Let f be a diffeomorphism of a smooth compact Riemannian manifold M. The following statements hold:*

1. *If $x \in M$ is such that*

$$\chi^+(x, v_1, \ldots, v_k) = \lim_{m\to+\infty} \frac{1}{m} \log \mathrm{vol}(df^m v_1, \ldots, df^m v_k)$$

(i.e., $\chi^+(x, v_1, \ldots, v_k)$ is exact), for any choice of linearly independent vectors $v_1, \ldots, v_k \in T_xM$ and $k = 1, \ldots, n$, then x is forward regular.

2. *If $x \in M$ is such that*

$$\chi^-(x, v_1, \ldots, v_k) = \lim_{m\to-\infty} \frac{1}{|m|} \log \mathrm{vol}(df^m v_1, \ldots, df^m v_k)$$

(i.e., $\chi^-(x, v_1, \ldots, v_k)$ is exact), for any choice of linearly independent vectors $v_1, \ldots, v_k \in T_xM$ and $k = 1, \ldots, n$, then x is backward regular.

We also formulate a criterion for regularity. It follows immediately from Proposition 3.1.7.

Proposition 5.4.4. *Let f be a diffeomorphism of a smooth compact Riemannian manifold M and $x \in M$. Assume that*

1. *$\chi^+(x, v_1, \ldots, v_k)$ and $\chi^-(x, v_1, \ldots, v_k)$ are exact for any linearly independent vectors $v_1, \ldots, v_k \in T_xM$ and $k = 1, \ldots, n$;*
2. *$p^+(x) = p^-(x) =: p(x)$ and $\chi_i^+(x) = -\chi_i^-(x)$ for $i = 1, \ldots, p(x)$;*
3. *$\bigoplus_{i=1}^{p(x)}(E_i^+(x) \cap E_i^-(x)) = T_xM$ (see (2.25) and (2.27)).*

Then x is regular.

Let ν be an ergodic f-invariant Borel measure. There exist numbers $p = p^\nu$, $\chi_i = \chi_i^\nu$, and $k_i = k_i^\nu$ for $i = 1, \ldots, p$ such that

$$p(x) = p, \quad \chi_i(x) = \chi_i, \quad k_i(x) = k_i$$

for ν-almost every x. The collection of pairs

$$\mathrm{Sp}\,\chi(\nu) = \{(\chi_i, k_i) : 1 \le i \le p\}$$

is called the *Lyapunov spectrum* of the measure ν.

We now consider dynamical systems whose spectrum of the Lyapunov exponent does not vanish on some subset of M. More precisely, let

$$Z = \{x \in M : x \text{ is Lyapunov–Perron regular and there exists}$$
$$1 \leq s(x) < p(x) \text{ with } \chi_{s(x)}(x) < 0 \text{ and } \chi_{s(x)+1}(x) > 0\} \quad (5.5)$$

(compare to (3.19) for general cocycles). This set is f-invariant and is nonuniformly completely hyperbolic.

Definition 5.4.5. We say that f is a *dynamical system with nonzero Lyapunov exponents* if there exists an f-invariant Borel probability measure ν on M such that $\nu(Z) = 1$. The measure ν is called a *hyperbolic measure* for f.

For every $x \in Z$, we set

$$E^s(x) = \bigoplus_{i=1}^{s(x)} H_i(x) \quad \text{and} \quad E^u(x) = \bigoplus_{i=s(x)+1}^{p} H_i(x).$$

Theorem 5.4.6. *The subspaces $E^s(x)$ and $E^u(x)$*

1. *depend Borel measurably on x;*
2. *form a splitting of the tangent space, that is, $T_xM = E^s(x) \oplus E^u(x)$;*
3. *are invariant:*

$$dfE^s(x) = E^s(f(x)) \quad \text{and} \quad dfE^u(x) = E^u(f(x));$$

Furthermore, for every $\varepsilon_0 > 0$, there exist an f-invariant Borel function $\varepsilon\colon \mathcal{R} \to (0, \varepsilon_0]$ and Borel functions $C, K\colon Z \to (0, \infty)$, satisfying

$$\chi_{s(x)}(x) + \varepsilon(x) < \chi_{s(x)+1}(x) - \varepsilon(x)$$

and

4. *the subspace $E^s(x)$ is* stable*: if $v \in E^s(x)$ and $n > 0$, then*

$$\|df^n v\| \leq C(x)e^{(\chi_{s(x)}(x)+\varepsilon(x))n}\|v\|; \quad (5.6)$$

5. *the subspace $E^u(x)$ is* unstable*: if $v \in E^u(x)$ and $n < 0$, then*

$$\|df^n v\| \leq C(x)e^{(\chi_{s(x)+1}(x)-\varepsilon(x))n}\|v\|; \quad (5.7)$$

6.

$$\sphericalangle(E^s(x), E^u(x)) \geq K(x); \quad (5.8)$$

7. *for every $m \in \mathbb{Z}$,*

$$C(f^m(x)) \leq C(x)e^{\varepsilon(x)|m|} \quad \text{and} \quad K(f^m(x)) \geq K(x)e^{-\varepsilon(x)|m|}. \quad (5.9)$$

Proof. The first three statements are immediate consequence of the Multiplicative Ergodic Theorem 3.4.3. The remaining statements follow from Theorem 3.3.1. □

We remark that the requirement (5.9) is crucial and is a manifestation of the regularity property. Roughly speaking, it means that the estimates (5.6)–(5.8) may deteriorate as $|m| \to \infty$ but only with subexponential rate. We stress that the rates of contraction along stable subspaces and expansion along unstable subspaces are exponential and hence are substantially stronger.

An important corollary of Theorem 5.4.6 is that the diffeomorphism f is nonuniformly completely hyperbolic on Z. Moreover, the number ε_0 (and hence the function $\varepsilon(x)$) in Definition 5.2.1 can be chosen arbitrarily small (if the functions $C(x)$ and $K(x)$ are chosen appropriately).

In the case when ν is an f-invariant hyperbolic *ergodic* measure, we have that for every $x \in Z$,

$$s(x) = s = s^{\nu}, \quad \chi_{s(x)}(x) = \chi_s^{\nu}, \quad \chi_{s(x)+1}(x) = \chi_{s+1}^{\nu},$$

and hence the set Z is nonuniformly hyperbolic with

$$\lambda(x) = \lambda = e^{\chi_s^{\nu}}, \quad \mu(x) = \mu = e^{\chi_{s+1}^{\nu}}, \quad \varepsilon(x) = \varepsilon,$$

where ε is any constant satisfying (2.15).

5.5 On the Notion of Dynamical Systems with Nonzero Lyapunov Exponents

In the literature on dynamical systems, there are several (not necessarily equivalent) interpretations of the notion of "systems with nonzero Lyapunov exponents." We discuss some of them. Let f be a $C^{1+\alpha}$ diffeomorphism of a smooth compact Riemannian manifold M. Consider the following properties of the map f:

(P1) There exists a hyperbolic measure ν.

(P2) There exists a nonempty set Z of Lyapunov–Perron regular points with nonzero Lyapunov exponents.

(P3) There is a nonempty set $\widetilde{Z}$, on which f is nonuniformly (completely) hyperbolic, and the number ε_0 in Definition 5.2.2 can be chosen arbitrarily small.

(P4) There is a nonempty set $\hat{Z}$ on which f is nonuniformly (completely) hyperbolic.

Clearly, (P1) implies (P2), and (P3) implies (P4). It follows from Theorem 5.4.6 that (P2) implies (P3). One can show that in the two-dimensional case (P2) is indeed equivalent to (P3). (It follows from (P3) that every point in $\widetilde{Z}$ is Lyapunov–Perron regular.) It is not known whether (P3) is actually stronger than (P4). The notion of dynamical system with nonzero Lyapunov exponents adapted in this book corresponds to the strongest Property (P1) but many results of the book hold if we assume only the weakest Property (P4).

Observe that (P2) does not imply (P1); that is, the presence of a nonempty set Z of Lyapunov–Perron regular points with nonzero Lyapunov exponents does not necessarily imply the existence of a hyperbolic invariant measure supported on Z. To see this let f be a diffeomorphism of a compact surface possessing a linear horseshoe S (see Section 6.1.3). It is easy to see that every point $x \in S$ is Lyapunov–Perron regular. On the other hand, the map $f|S$ is topologically conjugate via a homeomorphism h to the shift map σ on the space Σ_2 of two-sided infinite sequences on two symbols. It is well known that the invariant set $B \subset \Sigma_2$ of points that are not Birkhoff regular is nonempty and does not support any invariant measure for σ. (Recall that a point $\omega \in \Sigma_2$ is Birkhoff regular if for any continuous function $h\colon \Sigma_2 \to \mathbb{R}$ the sequence of Birkhoff's averages

$$h_n = \frac{1}{n}\sum_{k=0}^{n-1} h(\sigma^k(\omega))$$

converges.) The set $Z = h(B)$ is invariant under f, every point x in this set is Lyapunov–Perron regular and for any f-invariant measure ν on S we have that $\nu(Z) = 0$.

On the other hand, consider a diffeomorphism f of a compact surface possessing a nonlinear horseshoe S (see Section 6.1.4). An argument similar to the one in the end of this section shows that the set of points that are not Lyapunov–Perron regular is nonempty. Clearly, this set does not support any invariant measure and every point in this set has two nonzero Lyapunov exponents (but they are not exact).

5.6 Regular Neighborhoods

We recall that for a diffeomorphism $f\colon M \to M$ of a compact smooth n-dimensional Riemannian manifold M, we can consider (by introducing local coordinates) the derivative df as a linear cocycle over f. We shall now use the theory of linear extensions of cocycles (see Chapter 2). Applying the Reduction Theorem 3.5.5, given $\varepsilon > 0$ and a regular point $x \in M$, there exists a linear transformation $C_\varepsilon(x)\colon \mathbb{R}^n \to T_xM$ such that

1. the matrix
$$A_\varepsilon(x) = C_\varepsilon(f(x))^{-1} \circ d_x f \circ C_\varepsilon(x)$$
has the Lyapunov block form (3.44) (see Theorem 3.5.5);
2. $\{C_\varepsilon(f^m(x))\}_{m\in\mathbb{Z}}$ is a tempered sequence of linear transformations.

Our first goal is to construct, for every regular point $x \in M$, a neighborhood $N(x)$ of x such that f acts in $N(x)$ very much like the linear map $A_\varepsilon(x)$ in a neighborhood of the origin.

Denote by $\mathcal{R}$ the set of regular points for f and by $B(0, r)$ the standard Euclidean r-ball in $\mathbb{R}^n$ centered at the origin.

Theorem 5.6.1. *For every $\varepsilon > 0$, the following properties hold:*

1. *There exists a tempered function $q\colon \mathcal{R} \to (0, 1]$ and a collection of embeddings $\Psi_x\colon B(0, q(x)) \to M$ for each $x \in \mathcal{R}$ such that $\Psi_x(0) = x$ and*

$e^{-\varepsilon} < q(f(x))/q(x) < e^{\varepsilon}$; these embeddings satisfy $\Psi_x = \exp_x \circ C_\varepsilon(x)$, where $C_\varepsilon(x)$ is the Lyapunov change of coordinates.

2. *If $f_x := \Psi_{f(x)}^{-1} \circ f \circ \Psi_x : B(0, q(x)) \to \mathbb{R}^n$, then $d_0 f_x$ has the Lyapunov block form* (3.44).
3. *The C^1 distance $d_{C^1}(f_x, d_0 f_x) < \varepsilon$ in $B(0, q(x))$.*
4. *There exist a constant $K > 0$ and a measurable function $A\colon \mathcal{R} \to \mathbb{R}$ such that for every $y, z \in B(0, q(x))$,*

$$K^{-1}\rho(\Psi_x y, \Psi_x z) \leq \|y - z\| \leq A(x)\rho(\Psi_x y, \Psi_x z)$$

with $e^{-\varepsilon} < A(f(x))/A(x) < e^{\varepsilon}$.

Proof. For each $x \in X$, consider the Lyapunov change of coordinates $C_\varepsilon(x)$ from $T_x M$ onto $\mathbb{R}^n$ (see (3.43)). Thus for each $x \in \mathcal{R}$, the matrix

$$A_\varepsilon(x) = C_\varepsilon(f(x))^{-1} \circ d_x f \circ C_\varepsilon(x) \tag{5.10}$$

has the Lyapunov block form.

For each $x \in M$ and $r > 0$, set $T_x M(r) = \{w \in T_x M : \|w\| < r\}$. Choose now $r_0 > 0$ such that for every $x \in M$, the exponential map $\exp_x : T_x M(r_0) \to M$ is an embedding, $\|d_w \exp_x\|$, $\|d_{\exp_x w} \exp_x{}^{-1}\| \leq 2$ for every $w \in T_x M(r_0)$, and $\exp_{f(x)}$ is injective on

$$\exp_{f(x)}^{-1} \circ f \circ \exp_x(T_x M(r_0)).$$

Define the map

$$f_x = C_\varepsilon(f(x))^{-1} \circ \exp_{f(x)}^{-1} \circ f \circ \exp_x \circ C_\varepsilon(x) \tag{5.11}$$

on $C_\varepsilon(x) T_x M(r_0) \subset \mathbb{R}^n$. Observe that the Euclidean ball $\{w \in T_x M : \|w\| < r_0 \|C_\varepsilon(x)^{-1}\|^{-1}\}$ is contained in $C_\varepsilon(x) T_x M(r_0)$ up to identification.

We wish to compare f_x with $A_\varepsilon(x)$. In order to do that we first introduce the maps

$$\begin{aligned} r_x(u) &= f_x(u) - A_\varepsilon(x)u, \\ g_x(w) &= (\exp_{f(x)}^{-1} \circ f \circ \exp_x)w - df w. \end{aligned}$$

Since f is of class $C^{1+\alpha}$ and $d_0 g_x = 0$, there exists a positive constant L, such that $\|d_w g_x\| \leq L\|w\|^\alpha$, and thus

$$\begin{aligned} \|d_u r_x\| &= \|d_u(C_\varepsilon(f(x)) \circ g_x \circ C_\varepsilon(x)^{-1})\| \\ &\leq L\|C_\varepsilon(f(x))\| \cdot \|C_\varepsilon(x)^{-1}\|^{1+\alpha}\|u\|^\alpha. \end{aligned}$$

Hence if $\|w\|$ is sufficiently small, the contribution of the nonlinear part of f_x is negligible. In particular, $\|d_u r_x\| < \varepsilon$ for

$$\|u\| < \delta(x) := \varepsilon^{1/\alpha} L^{-1/\alpha} \|C_\varepsilon(f(x))\|^{-1/\alpha} \|C_\varepsilon(x)^{-1}\|^{-1-1/\alpha}.$$

By the Mean Value Theorem, we also have

$$\|r_x(u)\| < \varepsilon \text{ for } \|u\| \leq \delta(x), \tag{5.12}$$

since $r_x(0) = 0$. It follows from the definition of $\delta(x)$ that

$$\lim_{m\to+\infty} \frac{1}{m} \log \delta(f^m(x)) = - \lim_{m\to+\infty} \frac{1}{m\alpha} \log \left\| C_\varepsilon(f^{m+1}(x)) \right\|$$
$$- \lim_{m\to+\infty} \frac{1}{m(1+1/\alpha)} \log \left\| C_\varepsilon(f^m(x))^{-1} \right\| = 0,$$

since C_ε is tempered. Applying the Tempering Kernel Lemma 3.5.7 to the function δ, we find a measurable function $q\colon X \to \mathbb{R}$ such that $q(x) \geq \delta(x)$ and

$$e^{-\varepsilon} \leq q(f(x))/q(x) \leq e^{\varepsilon}.$$

We define the map $\Psi_x\colon B(0, q(x)) \to M$ by $\Psi_x = \exp_x \circ C_\varepsilon(x)$. It is clear that Ψ_x is an embedding for each $x \in \mathcal{R}$, such that $\Psi_x(0) = x$. This proves Statements 1 and 2. Statement 3 follows from the identity $d_0 f_x = A_\varepsilon(x)$ (which in turn follows immediately from (5.10), (5.11), and the chain rule), and from (5.12).

Now we prove Statement 4. It follows from (3.47) that

$$\|y - z\| = \|\Psi_x^{-1}(\Psi_x y) - \Psi_x^{-1}(\Psi_x z)\| \leq 2\|C_\varepsilon(x)^{-1}\| \rho(\Psi_x y, \Psi_x z),$$

On the other hand, we have

$$\rho(\Psi_x y, \Psi_x z) \leq 2\|C_\varepsilon(x)\| \cdot \|y - z\| \leq 2\|y - z\|.$$

This means that if $K_\varepsilon\colon \mathcal{R} \to \mathbb{R}$ is a function as in the Tempering Kernel Lemma 3.5.7 such that

$$K_\varepsilon(x) \geq \|C_\varepsilon(x)^{-1}\| \quad \text{and} \quad e^{-\varepsilon} \leq K_\varepsilon(f(x))/K_\varepsilon(x) \leq e^{\varepsilon},$$

then in Statement 4 we can take $A(x) = 2K_\varepsilon(f(x))$. □

We note that for each $x \in \mathcal{R}$, there exists a constant $B(x) \geq 1$ such that for every $y, z \in B(0, q(x))$,

$$B(x)^{-1} \rho(\Psi_x y, \Psi_x z) \leq \rho_x'(\exp_x y, \exp_x z) \leq B(x) \rho(\Psi_x y, \Psi_x z),$$

where $\rho_x'(\cdot,\cdot)$ is the distance on $\exp_x B(0, q(x))$ with respect to the Lyapunov norm $\|\cdot\|_x'$. By Luzin's Theorem, given $\delta > 0$, there exists a set of measure at least $1 - \delta$ where $x \mapsto B(x)$ as well as $x \mapsto A(x)$ in Theorem 5.6.1 are bounded.

Definition 5.6.2. For each regular point x, the set

$$R(x) := \Psi_x(B(0, q(x)))$$

is called a *regular neighborhood* of x or a *Lyapunov chart* at x.

We stress that the existence of regular neighborhoods uses the fact that f is of class $C^{1+\alpha}$ in an essential way.

5.7 Cocycles over Smooth Flows

Definition 5.7.1. Let M be a Riemannian manifold. We say that the C^1 map $\varphi\colon \mathbb{R} \times M \to M$ is a C^1 *flow* on M if

1. $\varphi_t := \varphi(t, \cdot)$ is a C^1 diffeomorphism for every $t \in \mathbb{R}$;
2. $\varphi_0 = \mathrm{Id}$, and $\varphi_t \circ \varphi_s = \varphi_{t+s}$ for every $t, s \in \mathbb{R}$.

Let X denote the vector field generated by the flow φ; that is,

$$X(x) = \frac{d}{dt}(\varphi_t(x))|_{t=0}.$$

It satisfies the differential equation

$$\frac{d}{dt}(\varphi_t(x)) = X(\varphi_t(x)).$$

For every $x_0 \in M$, the trajectory $\{x(x_0, t) = \varphi_t(x_0) : t \in \mathbb{R}\}$ represents a solution of the nonlinear differential equation

$$\dot{v} = X(v)$$

on the manifold M. This solution is uniquely determined by the initial condition $x(x_0, 0) = x_0$.

Given a point $x \in M$ and the trajectory $\{\varphi_t(x) : t \in \mathbb{R}\}$ passing through x, we introduce the *variational differential equation*

$$\dot{w}(t) = A(x, t)w(t), \tag{5.13}$$

where

$$A(x, t) = dX(\varphi_t(x)).$$

This is a linear nonautonomous differential equation along the trajectory $\{\varphi_t(x) : t \in \mathbb{R}\}$.

The Lyapunov exponent generated by the cocycle $\mathcal{A}$ is defined by

$$\chi^+(x, v) = \overline{\lim_{t\to+\infty}} \frac{1}{t} \log \|w(t)\|,$$

where $w(t)$ is the solution of (5.13) with initial condition $w(0) = v$. It is called the *Lyapunov exponent of the flow* φ_t. In particular, one can speak of trajectories, which are *forward regular*, *backward regular*, or *(Lyapunov–Perron) regular*.

Note that every periodic trajectory is regular. However, this is not true in general for nonperiodic trajectories. For example, consider a flow on the unit sphere with the North and the South poles to be, respectively, attracting and repelling points, and without other fixed points. If the coefficients of contraction and expansion are different then every trajectory of the flow (except for the North and the South poles) is nonregular.

One can establish a criterion for regularity of individual trajectories.

Proposition 5.7.2. *Let φ be a C^1 flow of a smooth compact Riemannian manifold M. The following statements hold:*

1. *If $x \in M$ is such that*

$$\chi^+(x, v_1, \dots, v_k) = \lim_{t\to+\infty} \frac{1}{t} \log \operatorname{vol}(d\varphi_t v_1, \dots, d\varphi_t v_k)$$

(i.e., $\chi^+(x, v_1, \dots, v_k)$ is exact), for any choice of linearly independent vectors $v_1, \dots, v_k \in T_x M$ and $k = 1, \dots, n$, then x is forward regular.

2. *If $x \in M$ is such that*

$$\chi^-(x, v_1, \dots, v_k) = \lim_{t\to-\infty} \frac{1}{|t|} \log \operatorname{vol}(d\varphi_t v_1, \dots, d\varphi_t v_k)$$

(i.e., $\chi^-(x, v_1, \dots, v_k)$ is exact), for any choice of linearly independent vectors $v_1, \dots, v_k \in T_x M$ and $k = 1, \dots, n$, then x is backward regular.

3. *Assume that*
   (a) *$\chi^+(x, v_1, \dots, v_k)$ and $\chi^-(x, v_1, \dots, v_k)$ are exact for any linearly independent vectors $v_1, \dots, v_k \in T_x M$ and $k = 1, \dots, n$;*
   (b) *$p^+(x) = p^-(x) =: p(x)$ and $\chi_i^+(x) = -\chi_i^-(x)$ for $i = 1, \dots, p(x)$;*
   (c) *$\bigoplus_{i=1}^{p(x)}(E_i^+(x) \cap E_i^-(x)) = T_x M$ (see (3.59) and (3.60)).*
   Then x is regular.

Proof. The proof can be obtained with similar arguments to those in the proofs of Propositions 3.1.2 and 3.1.7. □

It is not an easy task to apply this criterion and check whether a given trajectory is regular. The situation is quite different if we consider regularity of generic trajectories with respect to a Borel measure, which is invariant under the flow φ_t. This means that $\nu(\varphi_t(A)) = \nu(A)$ for any Borel set $A \subset M$ and $t \in \mathbb{R}$. It is easy to see that the derivative cocycle $\mathcal{A}(x, t)$ satisfies

$$\sup_{-1\le t\le 1} \log^+ \|\mathcal{A}(\cdot, t)\| \in L^1(M, \nu).$$

The Multiplicative Ergodic Theorem for flows (see Theorem 3.6.11) implies that almost every trajectory with respect to ν is Lyapunov–Perron regular.

Note that the Lyapunov exponent along the flow direction is zero. We therefore say that an invariant Borel measure ν is *hyperbolic* if all other values of the Lyapunov exponent are nonzero for ν-almost every point, and hence φ_t is a nonuniformly hyperbolic flow. We also assume that ν vanishes on the set of fixed points of φ_t.

5.8 Semicontinuity of Lyapunov Exponents

Let f be a volume preserving C^1 diffeomorphism of a smooth compact n-dimensional Riemannian manifold M. Denote by

$$\chi_1(x, f) \ge \chi_2(x, f) \ge \dots \ge \chi_n(x, f) \tag{5.14}$$

the values of the Lyapunov exponent of f counted with their multiplicities in decreasing order. For each $i = 1, \ldots, n$, set

$$\Lambda_i(f) = \int_M \sum_{j=1}^{i} \chi_i(x, f)\, d\mu(x),$$

where μ is the Riemannian volume. We describe the behavior of the functions $\Lambda_i(f)$ over f in the space $\mathrm{Diff}^1(M, \mu)$ of volume preserving C^1 diffeomorphisms of M endowed with the C^1 topology.

Theorem 5.8.1. *For each $i = 1, \ldots, n$, the function $f \mapsto \Lambda_i(f)$ is upper semicontinuous in $f \in \mathrm{Diff}^1(M, \mu)$.*

Proof. Set

$$a_n(f) = \int_M \log \| \wedge^i (d_x f^n) \| \, d\mu(x)$$

and observe that the sequence a_n is subadditive; that is, $a_{n+m}(f) \le a_n(f) + a_m(f)$. It follows that

$$\lim_{n\to\infty} \frac{a_n(f)}{n} = \inf_{n\ge 1} \frac{a_n(f)}{n}$$

and that

$$\Lambda_i(f) = \inf_{n\ge 1} \frac{1}{n} \int_M \log \| \wedge^i (d_x f^n) \| \, d\mu(x).$$

Since $a_n(f)$ varies continuously with f in the C^1 topology and the infimum of continuous functions is upper semicontinuous, the desired result follows. □

Part II

Examples and Foundations of the Nonlinear Theory

6

Examples of Systems with Hyperbolic Behavior

In this chapter we describe some major examples of nonuniformly hyperbolic diffeomorphisms and flows. There are two principally different situations: (1) nonuniformly hyperbolic horseshoes and (2) systems preserving the Riemannian volume. In the first case, we show how to modify the classical horseshoe map to obtain a diffeomorphism possessing a compact invariant set Λ (of zero volume) such that "almost any" invariant measure on Λ is hyperbolic. In the second case we construct diffeomorphisms on surfaces and flows on three-dimensional manifolds with nonzero Lyapunov exponents with respect to the Riemannian volume.

We stress that in all cases the construction begins with a uniformly hyperbolic system (a horseshoe map, an Anosov diffeomorphism, or an Anosov flow) and goes on to slow down the trajectories of the system near a hyperbolic fixed point or a periodic trajectory. This destroys hyperbolicity and creates orbits with zero Lyapunov exponents. However, for the "majority" of orbits the Lyapunov exponents are nonzero.

Let us point out that absence of zero Lyapunov exponents imposes no obstructions on the topology of the manifold and, as we shall see in Sections 11.4 and 11.5, volume preserving diffeomorphisms and flows with nonzero Lyapunov exponents exist on any manifold (of dimension ≥ 2 in the case of diffeomorphisms and of dimension ≥ 3 in the case of flows).

6.1 Uniformly Hyperbolic Sets

We describe examples of dynamical systems with nonuniformly hyperbolic behavior, which are (large) perturbations of systems with the strongest kind of hyperbolicity.

6.1.1 Hyperbolic Sets for Maps

Let $f: M \to M$ be a C^1 diffeomorphism of a compact Riemannian manifold.

Definition 6.1.1. A compact f-invariant set $\Lambda \subset M$ is said to be *hyperbolic* for f if

1. for every $x \in \Lambda$, there exists a splitting of the tangent space $T_x M = E^s(x) \oplus E^u(x)$, which is invariant under df; that is,
$$df E^s(x) = E^s(f(x)) \quad \text{and} \quad df E^u(x) = E^u(f(x));$$
2. there exist constants $C > 0$ and $\lambda \in (0, 1)$ such that for every $x \in \Lambda$ and $n \geq 1$,
$$\|df^n v\| \leq C\lambda^n \|v\| \text{ for any } v \in E^s(x),$$
$$\|df^{-n} v\| \leq C\lambda^n \|v\| \text{ for any } v \in E^u(x).$$

We then say that f is *hyperbolic* on Λ and that $E^s(x)$ and $E^u(x)$ are respectively the *stable* and *unstable subspaces at* x.

This is a uniform version of the hyperbolicity condition that is much stronger than the nonuniform one given in Definition 5.2.2 – the former corresponds to $\varepsilon = 0$, $C(x) = \text{const}$ and $K(x) = \text{const}$.

It is easy to show that $E^s(x)$ and $E^u(x)$ are uniquely defined. One can show that the number λ does not depend on the choice of equivalent Riemannian metric on M, while the constant $C > 0$ may depend on λ and on the Riemannian metric. Furthermore, we can use a Lyapunov inner product (see Section 5.2) for which $C = 1$. More precisely, given $\lambda' > \lambda$, set

$$\langle v, w\rangle'_x = \begin{cases} \sum_{m=0}^{\infty} \langle df^m v, df^m w\rangle_{f^m(x)} (\lambda')^{-2m}, & v, w \in E^s(x) \\ \sum_{m=0}^{\infty} \langle df^{-m} v, df^{-m} w\rangle_{f^{-m}(x)} (\lambda')^{-2m}, & v, w \in E^u(x) \\ 0, \quad v \in E^s(x),\ w \in E^u(x) \end{cases}$$

and extend it linearly to a metric on Λ. It is easy to show that the new metric is well defined. One can show that

$$\|df^n v\|' \leq (\lambda')^n \|v\|' \text{ for every } v \in E^s(x) \text{ and } n \geq 1$$

and

$$\|df^{-n} v\|' \leq (\lambda')^n \|v\|' \text{ for every } v \in E^u(x) \text{ and } n \geq 1.$$

Repeating arguments in the proof of Proposition 2.2.9, we find that if Λ is a hyperbolic set for a C^1 diffeomorphism f, then

1. the subspaces $E^s(x)$ and $E^u(x)$ vary continuously with $x \in \Lambda$ in the Grassmannian bundle;
2. the subspaces $E^s(x)$ and $E^u(x)$ are uniformly transverse; that is, there exists $c > 0$ such that for any $x \in \Lambda$, the angle between $E^s(x)$ and $E^u(x)$ is at least c.

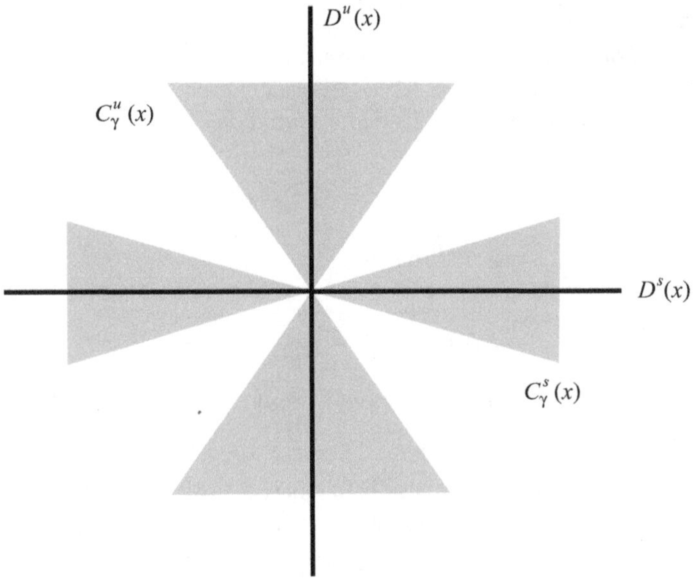

Figure 6.1. The stable and unstable cones at x with respect to the decomposition $T_xM = D^s(x) \oplus D^u(x)$.

If f is a $C^{1+\alpha}$ diffeomorphism then the maps $x \mapsto E^s(x)$ and $x \mapsto E^u(x)$ are in fact Hölder continuous (see Section 5.3).

Let f be a C^1 diffeomorphism of a compact Riemannian manifold and Λ a hyperbolic set for f. If ν is an f-invariant Borel probability measure with $\operatorname{supp}\nu \subset \Lambda$, then the values of the Lyapunov exponent are nonzero ν-almost everywhere. Moreover, for ν-almost every $x \in \Lambda$, we have

$$E^s(x) = \bigoplus_{i:\chi_i(x)<0} H_i(x) \quad \text{and} \quad E^u(x) = \bigoplus_{i:\chi_i(x)>0} H_i(x),$$

where $H_i(x)$ are the Oseledets subspaces (see (5.3)). This follows from the Multiplicative Ergodic Theorem 3.4.3.

While the subspaces $E^s(x)$ and $E^u(x)$ vary continuously with $x \in \Lambda$, the subspaces $H_i(x)$ need not be continuous and are, in general, only measurable. Although for *every* $x \in \Lambda$ the Lyapunov exponents at x are nonzero, in general, there are points, which *are not* Lyapunov–Perron regular (see Section 6.1.4).

If $T_xM = D^s(x) \oplus D^u(x)$ is a splitting of the tangent space at $x \in M$ and $\gamma > 0$, we define the *stable* and *unstable* cones at x with respect to this decomposition of TM (see Fig. 6.1) respectively by

$$C_\gamma^s(x) = \big\{(v, w) \in D^s(x) \oplus D^u(x) : \|w\| < \gamma\|v\|\big\} \cup \{(0, 0)\}$$

and

$$C_\gamma^u(x) = \big\{(v, w) \in D^s(x) \oplus D^u(x) : \|v\| < \gamma\|w\|\big\} \cup \{(0, 0)\}.$$

The following is a characterization of hyperbolic sets in terms of cones.

Theorem 6.1.2. *A compact invariant set $\Lambda \subset M$ of a C^1 diffeomorphism f of a compact manifold is hyperbolic if and only if there exist a Riemannian metric on M, a continuous splitting $T_xM = D^s(x) \oplus D^u(x)$ for each $x \in \Lambda$ (not necessarily df-invariant), a constant $\lambda \in (0,1)$, and a continuous function $\gamma : \Lambda \to \mathbb{R}^+$ such that for $x \in \Lambda$,*

1.
$$df\overline{C^u_{\gamma(x)}(x)} \subset C^u_{\gamma(f(x))}(f(x))$$
and
$$df^{-1}\overline{C^s_{\gamma(x)}(x)} \subset C^s_{\gamma(f^{-1}(x))}(f^{-1}(x));$$
2. *if* $v \in C^u_{\gamma(x)}(x)$, *then* $\|dfv\| \geq \lambda^{-1}\|v\|$, *and if* $v \in C^s_{\gamma(x)}(x)$, *then* $\|df^{-1}v\| \geq \lambda^{-1}\|v\|$.

Proof. If Λ is a hyperbolic set then there exists a sufficiently small $\gamma > 0$ such that the stable and unstable cones, with respect to the decomposition $T_xM = E^s(x) \oplus E^u(x)$, satisfy Conditions 1 and 2. Assume now that Conditions 1 and 2 hold. For each $x \in \Lambda$, set

$$F^s(x) = \bigcap_{m=0}^{\infty} df^{-m}\overline{C^s_{\gamma(f^m(x))}(f^m(x))},$$

$$F^u(x) = \bigcap_{m=0}^{\infty} df^{m}\overline{C^u_{\gamma(f^{-m}(x))}(f^{-m}(x))}.$$

By construction, the families of sets $F^s(x)$ and $F^u(x)$ are df-invariant. Moreover, for any $v \in F^s(x)$ and $m \geq 1$,

$$\|df^{-m}v\| \geq \lambda^{-m}\|v\|$$

and for any $v \in F^u(x)$ and $m \geq 1$,

$$\|df^{m}v\| \geq \lambda^{-m}\|v\|.$$

It remains to show that $F^s(x)$ and $F^u(x)$ are linear subspaces and that $T_xM = F^s(x) \oplus F^u(x)$ for each $x \in \Lambda$. This will immediately imply that Λ is a hyperbolic set for f with stable and unstable subspaces given by $E^s(x) = F^s(x)$ and $E^u(x) = F^u(x)$.

By construction of $F^s(x)$ and $F^u(x)$ and the continuity of the decomposition $T_xM = D^s(x) \oplus D^u(x)$, it follows that $F^s(x)$ and $F^u(x)$ contain subspaces $G^s(x)$ and $G^u(x)$, respectively, of the same dimension as that of $D^s(x)$ and $D^u(x)$. Choose $v \in G^s(x) \cap G^u(x)$. Then for each $m \geq 1$,

$$\|df^{-m}v\| \geq \lambda^{m}\|v\|. \tag{6.1}$$

Since $df^{-m}G^u(x) \subset F^u(f^{-m}(x))$, we have

$$\|v\| = \|df^m\, df^{-m}v\| \geq \lambda^{-m}\|df^{-m}v\|. \tag{6.2}$$

The inequalities (6.1) and (6.2) imply that $v = 0$, and hence $T_x M = G^s(x) \oplus G^u(x)$. It remains to show that $G^s(x) = F^s(x)$ and $G^u(x) = F^u(x)$. Let $v \in G^s(x)$ and $w \in F^u(x)$ be such that $v + w \in F^s(x)$. Observe that for every $m \geq 1$,

$$\begin{aligned}\lambda^{-m}\|w\| &\leq \|df^m w\| = \|df^m(v + w - v)\| \\ &\leq \|df^m(v + w)\| + \|df^m v\| \\ &\leq \lambda^m \|v + w\| + \lambda^m \|v\|.\end{aligned}$$

This implies that

$$\|w\| \leq \lambda^{2m}(\|v + w\| + \|v\|) \to 0 \text{ as } m \to \infty,$$

and hence $w = 0$. This concludes the proof. □

Definition 6.1.3. A C^1 diffeomorphism $f \colon M \to M$ of a compact manifold is called an *Anosov diffeomorphism* if M is a hyperbolic set for f.

It is well known (see [104]) that Anosov diffeomorphisms are *structurally stable*. This means that any sufficiently small C^1 perturbation $g \colon M \to M$ of an Anosov diffeomorphism is still an Anosov diffeomorphism and there exists a Hölder homeomorphism $h \colon M \to M$ such that $g \circ h = h \circ f$. In particular, the set of C^1 Anosov diffeomorphisms is open in the C^1 topology.

6.1.2 Hyperbolic Sets for Flows

Let $\varphi = (\varphi_t)_{t \in \mathbb{R}}$ be a smooth flow on a smooth compact Riemannian n-manifold M. We say that a compact set $\Lambda \subset M$ is *φ-invariant* if $\varphi_t(\Lambda) = \Lambda$ for every $t \in \mathbb{R}$.

Definition 6.1.4. A compact φ-invariant set $\Lambda \subset M$ is called *hyperbolic* for φ if there exist $C > 0$ and $\lambda \in (0, 1)$ such that for every $x \in \Lambda$,

1. there exists a splitting of the tangent space
$$T_x M = E^s(x) \oplus E^u(x) \oplus E^0(x) \tag{6.3}$$
such that for $t \in \mathbb{R}$,
$$d\varphi_t E^s(x) = E^s(\varphi_t(x)) \quad \text{and} \quad d\varphi_t E^u(x) = E^u(\varphi_t(x)); \tag{6.4}$$
2. $X(x) \neq 0$, and $E^0(x)$ is the one-dimensional subspace generated by $X(x)$;
3. for $t > 0$, we have
$$\begin{aligned}\|d\varphi_t v\|_{\varphi_t(x)} &\leq C\lambda^t \|v\|_x \text{ for every } v \in E^s(x), \\ \|d\varphi_{-t} v\|_{\varphi_{-t}(x)} &\leq C\lambda^t \|v\|_x \text{ for every } v \in E^u(x).\end{aligned}$$

We say that φ is hyperbolic on Λ and we call $E^s(x)$ and $E^u(x)$ the *stable* and *unstable subspaces at x*.

It is easy to see that

$$d\varphi_t E^0(x) = E^0(\varphi_t(x)) \text{ for every } x \in \Lambda \text{ and } t \in \mathbb{R}. \tag{6.5}$$

By (6.4) and (6.5), the splitting in (6.3) is invariant under $d\varphi$.

Let Λ be a hyperbolic set for a C^1 flow φ. We introduce an inner product on Λ adapted to φ. Given $\lambda' > \lambda$, set

$$\langle v, w\rangle'_x = \begin{cases} \int_0^\infty \langle d\varphi_t v, d\varphi_t w\rangle_{\varphi_t(x)} (\lambda')^{-2t}\, dt & \text{if } v, w \in E^s(x) \\ \int_0^\infty \langle d\varphi_{-t} v, d\varphi_{-t} w\rangle_{\varphi_{-t}(x)} (\lambda')^{-2t}\, dt & \text{if } v, w \in E^s(x) \end{cases}.$$

We then declare the subspaces $E^s(x)$, $E^u(x)$, and $E^0(x)$ orthogonal at x. One can easily show that the integrals are well defined and that if $v \in E^s(x)$ and $t > 0$ then

$$\|d\varphi_t v\|'_{\varphi_t(x)} \le (\lambda')^t \|v\|'_x,$$

and that if $v \in E^u(x)$ and $t > 0$ then

$$\|d\varphi_{-t} v\|'_{\varphi_{-t}(x)} \le (\lambda')^t \|v\|'_x.$$

Definition 6.1.5. A C^1 flow φ of the compact manifold M is said to be *Anosov* if M is a hyperbolic set for φ.

A simple example of an Anosov flow is a *special flow* over an Anosov diffeomorphism. Let f be an Anosov diffeomorphism of a compact Riemannian manifold M and h a smooth function on M called *roof function*. Consider the direct product $M \times [0, \infty)$ and define the flow φ_t in the following way: starting with a point $(x, 0)$, we set $\varphi_t(x, 0) = (x, t)$ if $t < h(x)$ and $\varphi_t(x, h(x)) = (f(x), 0)$. The phase space for the flow φ_t is diffeomorphic to the direct product $M \times [0, 1]$, where the sections $M \times \{0\}$ and $M \times \{1\}$ are identified by the action of f. Another example of Anosov flows is geodesic flows on compact Riemannian manifolds with negative sectional curvature. Time changes and small C^1 perturbations of Anosov flows are also Anosov flows (see [104] for details).

6.1.3 Linear Horseshoes

Horseshoes were introduced by Smale in [207] and are the most famous examples of hyperbolic sets. The most crucial property of a horseshoe is that the dynamics on it is topologically conjugated to the shift map. We begin with the construction of a linear horseshoe. In the next section we generalize this construction to obtain a nonlinear horseshoe. Although Smale horseshoes are well described in the literature on dynamical systems (see, e.g., [40, 104, 159]), we provide these particular constructions so that in Section 6.2 we can modify them to obtain examples of systems with nonzero Lyapunov exponents.

Let $S = [-1, 1] \times [-1, 1]$ be the unit square. Consider two horizontal strips

$$H_1 = [-1, 1] \times \left[-\frac{3}{4}, -\frac{1}{4}\right] \quad \text{and} \quad H_2 = [-1, 1] \times \left[\frac{1}{4}, \frac{3}{4}\right]$$

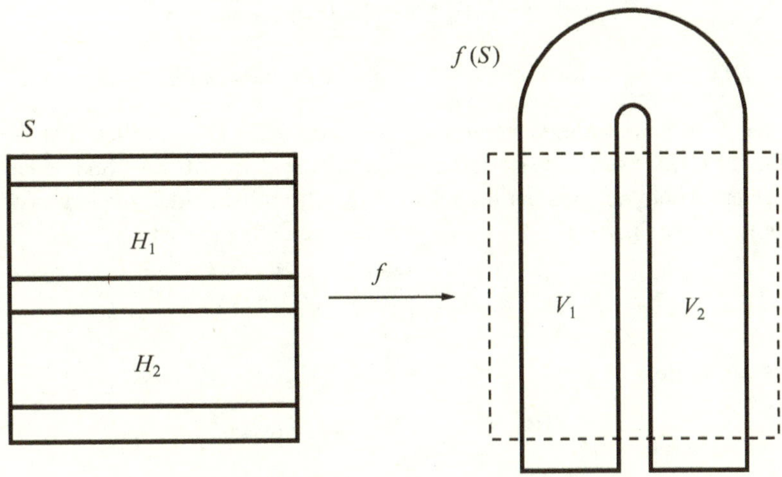

Figure 6.2. Construction of a linear horseshoe.

and two vertical strips

$$V_1 = \left[-\frac{3}{4}, -\frac{1}{4}\right] \times [-1, 1] \quad \text{and} \quad V_2 = \left[\frac{1}{4}, \frac{3}{4}\right] \times [-1, 1]$$

(see Fig. 6.2). Let $f\colon S \to \mathbb{R}^2$ be an injective continuous map, satisfying

1. $f(H_1) = V_1$ and $f(H_2) = V_2$;
2. $f|H_i$ is linear for $i = 1, 2$;
3. $f(S)$ is a horseshoelike set (see Fig. 6.2), with $S \cap f(S) = V_1 \cup V_2$.

By Property 2 we have

$$d(f|H_1) = \begin{pmatrix} 1/4 & 0 \\ 0 & 4 \end{pmatrix} \quad \text{and} \quad d(f|H_2) = \begin{pmatrix} 1/4 & 0 \\ 0 & -4 \end{pmatrix}. \tag{6.6}$$

For each $n \geq 0$, the set $S \cap f^n(S)$ consists of 2^n vertical rectangles of width $1/4^n$, which go from bottom to top of the square S. In a similar way, for each $n \leq 0$, the set $S \cap f^n(S)$ consists of 2^n horizontal rectangles of width $1/4^n$, which go from the left side to the right side of S.

We define the f-invariant set

$$\Lambda = \bigcap_{n=-\infty}^{+\infty} f^n(S)$$

composed of the points whose forward and backward orbits are in S. Notice that Λ is a hyperbolic set for f. The stable and unstable subspaces $E^s(x)$ and $E^u(x)$ at a point $x \in \Lambda$ are the horizontal and vertical lines passing through x. By (6.6), the decomposition $T_x X = E^s(x) \oplus E^u(x)$ is df-invariant and

$$\|dfv\| = \frac{1}{4}\|v\| \text{ for } v \in E^s(x) \quad \text{and} \quad \|dfv\| = 4\|v\| \text{ for } v \in E^u(x).$$

The set Λ is called *Smale horseshoe* or *linear horseshoe*. Notice that Λ is a product of Cantor sets (i.e., perfect nowhere dense compact sets), and hence it is also a Cantor set. We consider the set $\Sigma_2 = \{1, 2\}^{\mathbb{Z}}$ of two-sided sequences of two symbols. For each $\omega = (\cdots i_{-1} i_0 i_1 \cdots) \in \Sigma_2$, we define the set

$$h(\omega) = \bigcap_{n=-\infty}^{+\infty} f^n(H_{i_n}). \tag{6.7}$$

We can write

$$h(\omega) = \lim_{n \to +\infty} (H_{i_{-n} \cdots i_0} \cap V_{i_0 \cdots i_n}), \tag{6.8}$$

where

$$H_{i_{-n} \cdots i_0} = \bigcap_{k=-n}^{0} f^k(H_{i_k}) \quad \text{and} \quad V_{i_0 \cdots i_n} = \bigcap_{k=0}^{n} f^k(H_{i_k}), \tag{6.9}$$

The sets in (6.9) are, respectively, horizontal and vertical rectangles of width $1/4^n$. Therefore, $h(\omega)$ consists of a unique point in Λ, and we obtain a map $h \colon \Sigma_2 \to \Lambda$. One can easily check that h is injective and onto since

$$h(\Sigma_2) = \bigcap_{n=-\infty}^{+\infty} f^n(H_1 \cup H_2) = \bigcap_{n=-\infty}^{+\infty} f^n(X) = \Lambda.$$

We fix $\beta > 1$ and define the metric on Σ_2 by

$$d(\{i_n\}, \{j_n\}) = \sum_{n=-\infty}^{+\infty} \beta^{-|n|} |i_n - j_n|.$$

One can verify that (see [104, Section 2.5])

1. h is a homeomorphism (in fact, both maps h and h^{-1} are Hölder continuous);
2. the shift map $\sigma \colon \Sigma_2 \to \Sigma_2$ (given by $(\sigma\omega)_n = \omega_{n+1}$ for each integer n) is conjugate to $f|\Lambda$; that is, the diagram

$$\begin{array}{ccc} \Sigma_2 & \xrightarrow{\sigma} & \Sigma_2 \\ {\scriptstyle h}\downarrow & & \downarrow{\scriptstyle h} \\ \Lambda & \xrightarrow{f} & \Lambda \end{array}$$

is commutative.

The existence of the topological conjugacy h between σ and $f|\Lambda$ implies that the two maps have the same topological properties. In particular, we obtain the following result:

Proposition 6.1.6.

1. *$f|\Lambda$ is topologically mixing.*
2. *The periodic points of $f|\Lambda$ are dense in Λ.*
3. *$f|\Lambda$ has 2^n periodic points of period n.*
4. *The topological entropy of $f|\Lambda$ is $h(f|\Lambda) = \log 2$.*

6.1.4 Nonlinear Horseshoes

Let $S = [-1, 1] \times [-1, 1]$. Fix $\gamma \in (0, 1)$. Consider a C^1 map $\varphi\colon [-1, 1] \to [-1, 1]$ with $|\varphi'| \le \gamma$. We call the graph $\{(x, \varphi(x)) : x \in [-1, 1]\}$ an *admissible (s, γ)-curve* and the graph $\{(\varphi(x), x) : x \in [-1, 1]\}$ an *admissible(u, γ)-curve.*

Consider two admissible (s, γ)-curves, which are the graphs of functions φ_1 and φ_2 satisfying $\varphi_1 < \varphi_2$. Define an *admissible (s, γ)-rectangle* by

$$\{(x, y) \in X : \varphi_1(x) \le y \le \varphi_2(x)\}.$$

In a similar way, given two admissible (u, γ)-curves, which are the graphs of functions φ_1 and φ_2 satisfying $\varphi_1 < \varphi_2$, define an *admissible (u, γ)-rectangle* by

$$\{(x, y) \in X : \varphi_1(y) \le x \le \varphi_2(y)\}.$$

Given two disjoint admissible (s, γ)-rectangles H_1 and H_2, we consider a C^1 map $f\colon S \to \mathbb{R}^2$ such that

1. the sets $V_1 = f(H_1)$ and $V_2 = f(H_2)$ are admissible (u, γ)-rectangles;
2. if $(x, y) \in H_1 \cup H_2$, then $f(x, y)$ has the form

   $$f(x, y) = (\lambda(x, y)x + g_1(x, y), \mu(x, y)y + g_2(x, y)),$$

   where the *real* functions λ and μ, and the *vector* functions g_1 and g_2 are of class C^1;
3. for $i = 1, 2$ and sufficiently small $\delta = \delta(\lambda_0, \mu_0) > 0$, we have

   $$\lambda_0 := \max|\lambda| < 1 \quad \text{and} \quad \mu_0 := \min|\mu| > 1,$$

   $$\max\|d\lambda\|, \max\|d\mu\| < \delta \quad \text{and} \quad \max|g_i|, \max\|dg_i\| < \delta.$$

We will show that the set

$$\Lambda = \bigcap_{n=-\infty}^{+\infty} f^n(H_1 \cup H_2)$$

is hyperbolic and that $f|\Lambda$ is topologically conjugate to the shift $\sigma : \Sigma_2 \to \Sigma_2$. The main tool is the characterization of hyperbolic sets by invariant families of stable and unstable cones (see Theorem 6.1.2).

Rewrite f in the form

$$f(x, y) = (\lambda_0 x + \tilde{g}_1(x, y), \pm\mu_0 y + \tilde{g}_2(x, y)), \tag{6.10}$$

where

$$\widetilde{g}_1(x,y) = g_1(x,y) + (\lambda(x,y) - \lambda_0)x,$$
$$\widetilde{g}_2(x,y) = g_2(x,y) + (\mu(x,y) \mp \mu_0)y,$$

with a plus or minus sign in (6.10) corresponding to whether $(x,y) \in H_1$ or $(x,y) \in H_2$. Observe that

$$|(\lambda(x,y) - \lambda_0)x|, |(\mu(x,y) \mp \mu_0)y|n \le \sqrt{2}\delta,$$
$$\|d_{(x,y)}[(\lambda(x,y) - \lambda_0)x]\| \le \|d_{(x,y)}\lambda|_{(x,y)}x\| + |\lambda(x,y) - \lambda_0| \le (1+\sqrt{2})\delta,$$
$$\|d_{(x,y)}[(\mu(x,y) \mp \mu_0)y]\| \le \|d_{(x,y)}\mu|_{(x,y)}y\| + ||\mu(x,y) \mp \mu_0| \le (1+\sqrt{2})\delta.$$

This implies that

$$\max|\widetilde{g}_i|, \max\|d\widetilde{g}_i\| < 4\delta \text{ for } i = 1,2 \tag{6.11}$$

For every $(x,y) \in \Lambda$, we identify $T_{(x,y)}S$ with $\mathbb{R}^2$ and consider the continuous decomposition $T_{(x,y)}S = \mathbb{R} \oplus \mathbb{R}$. Set

$$C^u_\gamma(x,y) = \{(v,w) \in \mathbb{R}^2 : |v| < \gamma|w|\} \cup \{(0,0)\}$$

and

$$C^s_\gamma(x,y) = \{(v,w) \in \mathbb{R}^2 : |w| < \gamma|v|\} \cup \{(0,0)\}. \tag{6.12}$$

For every $(v,w) \in \overline{C^u_\gamma(x,y)}$, we have

$$\partial_x f(v,w) = \lambda_0 v + d_{(x,y)}\widetilde{g}_1|_{(x,y)}(v,w)$$

and since $\|(v,w)\| \le |v| + |w| \le (\gamma+1)|w|$, we obtain

$$|\partial_x f(v,w)| \le (\lambda_0\gamma + 4\delta(\gamma+1))|v|. \tag{6.13}$$

In a similar way, we have

$$\partial_y f(v,w) = \mu_0 w + d_{(x,y)}\widetilde{g}_2|_{(x,y)}(v,w),$$

and thus

$$|\partial_y f(v,w)| \ge (\mu_0 - 4\delta(\gamma+1))|w|. \tag{6.14}$$

By (6.13) and (6.14), we conclude that if $\delta > 0$ satisfies

$$\delta < \frac{\gamma(\mu_0 - \lambda_0)}{4(\gamma+1)^2}, \tag{6.15}$$

then $|\partial_x f(v,w)| < \gamma|\partial_y f(v,w)|$. Since $\lambda_0 < 1 < \mu_0$, one can always choose $\delta > 0$ for which the inequality in (6.15) holds. This implies that

$$df\overline{C^u_\gamma(x,y)} \subset C^u_\gamma(f(x,y)).$$

In particular, we conclude that the image under f of any admissible (u,γ)-curve is again an admissible (u,γ)-curve. In addition, for every $(v,w) \in C^u_\gamma(x,y)$, we have

$$\begin{aligned}\|df(v,w)\| &\geq |\partial_y f(v,w)| - |\partial_x f(v,w)| \\ &\geq (\mu_0 - 4\delta(\gamma+1))|w| - (\lambda_0\gamma + 4\delta(\gamma+1))|v| \\ &\geq [\mu_0 - 4\delta(\gamma+1) - \lambda_0\gamma^2 - 4\delta\gamma(\gamma+1)]|w| \\ &= [\mu_0 - \lambda_0\gamma^2 - 4\delta(\gamma+1)]|w| \\ &\geq [\mu_0 - \lambda_0\gamma^2 - 4\delta(\gamma+1)^2](\gamma+1)^{-1}\|(v,w)\|.\end{aligned}$$

It is now straightforward to show that if $\delta > 0$ satisfies (6.15) and $\delta < (\mu_0 - 1)/4$, then

$$[\mu_0 - \lambda_0\gamma^2 - 4\delta(\gamma+1)^2](\gamma+1)^{-1} > 1,$$

provided that $\gamma > 0$ is chosen such that

$$\gamma < \frac{-(1+8\delta) + \sqrt{(1+8\delta)^2 + 4(\lambda_0 + 4\delta)(\mu_0 - 1 - 4\delta)}}{2(\lambda_0 + 4\delta)}.$$

It follows from (6.10) and (6.11) that f is invertible on $H_1 \cup H_2$ and has a differentiable inverse:

$$f^{-1}(x,y) = \left(\frac{x}{\lambda_0} + h_1(x,y), \pm\frac{y}{\mu_0} + h_2(x,y)\right)$$

with a plus or minus sign, respectively, if $(x,y) \in V_1$ or $(x,y) \in V_2$, where

$$\begin{aligned}(h_1(x,y), h_2(x,y)) &= f^{-1}(x,y) \\ &\quad - f^{-1}\left(x + \widetilde{g}_1\left(\frac{x}{\lambda_0}, \pm\frac{y}{\mu_0}\right), y + \widetilde{g}_2\left(\frac{x}{\lambda_0}, \pm\frac{y}{\mu_0}\right)\right).\end{aligned}$$

Since df^{-1} is uniformly continuous, by (6.11), given $\varepsilon > 0$, we obtain

$$\max|h_i|, \max\|dh_i\| < \varepsilon \text{ for } i = 1, 2,$$

provided that $\delta > 0$ is sufficiently small. Arguing similarly, we also obtain that if $\delta, \gamma > 0$ are sufficiently small, then

$$df^{-1}\overline{C^s_\gamma(x,y)} \subset C^s_\gamma(f^{-1}(x,y))$$

for every $(x,y) \in \Lambda$. Moreover, there exists $\widetilde{\lambda}_0 \in (\lambda_0, 1)$ such that

$$\|df^{-1}(v,w)\| \geq \widetilde{\lambda}_0^{-1}\|(v,w)\|$$

for each $(v,w) \in C^s_\gamma(x,y)$ and $(x,y) \in \Lambda$. This implies that the image under f of any admissible (s,γ)-curve is again an admissible (s,γ)-curve. By Theorem 6.1.2, we conclude that Λ is a hyperbolic set for f.

For each $\omega = (\cdots i_{-1}i_0i_1\cdots) \in \Sigma_2$, we define a set $h(\omega) \subset \Lambda$ by (6.7). As in the case of linear horseshoes, one can characterize the set $h(\omega)$ using (6.8) and (6.9) where the sets $H_{i_{-n}\cdots i_0}$ and $V_{i_0\cdots i_n}$ are admissible (s,γ)- and (u,γ)-rectangles. For each $\omega \in \Sigma_2$, the sets

$$H(\omega) = \bigcap_{n=0}^{+\infty} H_{i_{-n}\cdots i_0} \quad \text{and} \quad V(\omega) = \bigcap_{n=0}^{+\infty} V_{i_0\cdots i_n}$$

are admissible (s, γ)- and (u, γ)-rectangles, respectively. We claim that they are, in fact, admissible (s, γ)- and (u, γ)-curves, which intersect transversally at a unique point.

If the graph of a function φ is an admissible (u, γ)-curve, then

$$\begin{aligned} \|f(\varphi(s), s) - f(\varphi(t), t)\| &\geq \mu_0|s - t| - \lambda_0|\varphi(s) - \varphi(t)| \\ &- \sum_{i=1}^{2}|\tilde{g}_i(\varphi(s), s) - \tilde{g}_i(\varphi(t), t)| \geq (\mu_0 - \lambda_0\gamma - 4\delta(\gamma + 1))\,|s - t| \end{aligned}$$

for every $s, t \in [-1, 1]$. Therefore if $\delta, \gamma > 0$ are sufficiently small, then f expands "along" admissible (u, γ)-curves. This implies that $H(\omega)$ is an admissible (s, γ)-curve. In a similar way one can show that f^{-1} expands "along" admissible (s, γ)-curves, and hence the set $V(\omega)$ is an admissible (u, γ)-curve.

Let $H(\omega)$ and $V(\omega)$ be, respectively, the graphs of the functions φ_H and φ_V. Observe that $H(\omega)$ and $V(\omega)$ intersect if and only if there exist s and t such that $(s, \varphi_H(s)) = (\varphi_V(t), t)$, that is, if and only if $t = (\varphi_H \circ \varphi_V)(t)$ for some t. We have

$$|(\varphi_H \circ \varphi_V)(s) - (\varphi_H \circ \varphi_V)(t)| \leq \gamma^2|s - t|.$$

Therefore, $\varphi_H \circ \varphi_V$ is a contraction on $[-1, 1]$ and has a unique fixed point $t = (\varphi_H \circ \varphi_V)(t)$. This shows that $H(\omega)$ and $V(\omega)$ intersect at a unique point $x \in \Lambda$, and hence $h(\omega) = H(\omega) \cap V(\omega)$ defines a map $h\colon \Sigma_2 \to \Lambda$. The map h is a homeomorphism (in fact, both h and h^{-1} are Hölder continuous) and is a topological conjugacy between $f|\Lambda$ and $\sigma : \Sigma_2 \to \Sigma_2$; that is, $f \circ h = h \circ \sigma$. Any set $\Lambda \subset S$ constructed as described in this paragraph is called a *two-dimensional horseshoe*. The following result is a consequence of the above construction:

Theorem 6.1.7.

1. *Any two-dimensional horseshoe is a hyperbolic Cantor set, which is topologically conjugate to the shift map on* Σ_2.
2. *For each map* $f\colon S \to \mathbb{R}^2$ *with a two-dimensional horseshoe, there exists a* C^1 *neighborhood* U *of* f *such that every map in* U *has a two-dimensional horseshoe, which is topologically conjugate to* Λ.

As we noticed above, a nonuniformly hyperbolic set may contain points, which are not Lyapunov–Perron regular. This can even happen when hyperbolicity is uniform. To illustrate this consider an Anosov diffeomorphism f of the torus $\mathbb{T}^2$. Set

$$\lambda^-(x) = \log\|df|E^s(x)\| \quad \text{and} \quad \lambda^+(x) = \log\|df|E^u(x)\|. \tag{6.16}$$

In view of Theorem 5.3.1, $\lambda^-(x)$ and $\lambda^+(x)$ are Hölder continuous functions. For every $x \in \mathbb{T}^2$,

$$\chi(x, v) = \overline{\lim_{n\to\infty}} \frac{1}{n} \sum_{k=0}^{n-1} \lambda^-(f^k(x)), \quad v \in E^s(x)$$

and

$$\chi(x, v) = \overline{\lim_{n\to\infty}} \frac{1}{n} \sum_{k=0}^{n-1} \lambda^+(f^k(x)), \quad v \in E^u(x).$$

Since the stable and unstable subspaces are uniformly transverse, a point x is Lyapunov–Perron regular if and only if the Lyapunov exponent at x is exact (by the Multiplicative Ergodic Theorem 3.4.3). Therefore due to Birkhoff's Ergodic Theorem applied to the functions λ^- and λ^+, almost every point x with respect to any invariant Borel probability measure is Lyapunov–Perron regular. In the case of a toral automorphism A, for which the functions λ^- and λ^+ are constant, we conclude that *every* point $x \in \mathbb{T}^2$ is Lyapunov–Perron regular. Indeed, this statement holds for any two functions λ^- and λ^+, which are cohomologous to constants (i.e., $\lambda^- = a^- + \varphi^- \circ f - \varphi^-$ and $\lambda^+ = a^+ + \varphi^+ \circ f - \varphi^+$ for some constants $a^-, a^+ \in \mathbb{R}$ and some continuous functions φ^- and φ^+ on $\mathbb{T}^2$).

On the other hand, in any arbitrarily small $C^{1+\alpha}$ neighborhood of A, there is a topologically mixing Anosov diffeomorphism f for which at least one of the functions λ^- and λ^+ (or both) is not cohomologous to a constant (see [23]). In this case the set of points that are not Lyapunov–Perron regular is nonempty. Indeed, assuming without loss of generality that the function λ^- is not cohomologous to a constant, there exist two ergodic f-invariant Borel probability measures ν_1 and ν_2 such that

$$a = \int_{\mathbb{T}^2} \lambda^- \, d\nu_1 - \int_{\mathbb{T}^2} \lambda^- \, d\nu_2 \neq 0.$$

By Birkhoff's Ergodic Theorem, for ν_1-almost every $x \in \mathbb{T}^2$,

$$\lim_{n\to\infty} \frac{1}{n} \sum_{j=0}^{n-1} \lambda^-(f^j(x)) = \int_{\mathbb{T}^2} \lambda^- \, d\nu_1$$

and for ν_2-almost every $y \in \mathbb{T}^2$,

$$\lim_{n\to\infty} \frac{1}{n} \sum_{j=0}^{n-1} \lambda^-(f^j(y)) = \int_{\mathbb{T}^2} \lambda^- \, d\nu_2.$$

Let us fix such points x and y, and choose ε, $0 < \varepsilon \le a/4$, and a monotonically increasing sequence of integers n_ℓ, $\ell = 0, 1, 2, \ldots$, such that $n_0 = 0$ and $n_\ell \to \infty$. The Anosov diffeomorphism f is topologically conjugated to a subshift of finite type (Σ_E, σ), where E is a *transitive* $m \times m$ matrix (for some m) and $\Sigma_E \subset \Sigma_m$ consists of those $\omega = (\omega_i)_{i\in\mathbb{Z}}$ for which $e_{\omega_i\omega_{i+1}} = 1$ for all $i \in \mathbb{Z}$ (see, e.g., [104]). Using this symbolic representation of f one can show the existence of a number $s > 0$ and a point $z \in \mathbb{T}^2$ such that the trajectory $\{f^n(z)\}$ follows the trajectory $\{f^n(x)\}$ during periods of time $n_{4k} \le n \le n_{4k+1}$ and follows the trajectory $\{f^n(y)\}$ during periods of time $n_{4k+2} \le n \le n_{4k+3}$ while "jumping" between these trajectories during other periods of time. More precisely, for $k = 0, 1, 2, \dots,$

1. $\rho(f^n(z), f^n(x)) \le \varepsilon$ for $n_{4k} \le n \le n_{4k+1}$;
2. $\rho(f^n(z), f^n(y)) \le \varepsilon$ for $n_{4k+2} \le n \le n_{4k+3}$;
3. $n_{4k+2} - n_{4k+1} \le s$ and $n_{4k+4} - n_{4k+3} \le s$.

Consider the sequence p_k defined by $p_{2k} = n_{4k+1} - n_{4k}$ and $p_{2k+1} = n_{4k+3} - n_{4k+2}$. If this sequence grows sufficiently fast (e.g., if $p_{k+1} = p_k!$) then

$$\left| \frac{1}{n_{4k+1}} \sum_{j=0}^{n_{4k+1}-1} \lambda^-(f^j(z)) - \int_{\mathbb{T}^2} \lambda^- \, d\nu_1 \right| \le \varepsilon,$$

while

$$\left| \frac{1}{n_{4k+3}} \sum_{j=0}^{n_{4k+3}-1} \lambda^-(f^j(z)) - \int_{\mathbb{T}^2} \lambda^- \, d\nu_2 \right| \le \varepsilon.$$

This implies that the point z is not Lyapunov–Perron regular.

In [23], Barreira and Schmeling obtained substantially deeper information about the set of points that are not Lyapunov–Perron regular.

Theorem 6.1.8. *Let f be a $C^{1+\alpha}$ Anosov diffeomorphism of the torus $\mathbb{T}^2$. Assume that at least one of the functions λ^- and λ^+ in* (6.16) *is not cohomologous to a constant. Then the set of points that are not Lyapunov–Perron regular is dense and has Hausdorff dimension 2.*

6.2 Nonuniformly Hyperbolic Perturbations of Horseshoes

We describe several modifications of the constructions in Section 6.1.3. The resulting invariant set Λ is not anymore a hyperbolic set, but the dynamics on it is still conjugate to the shift map and is nonuniformly hyperbolic.

6.2.1 Slow Expansion Near a Fixed Point

For each $\varepsilon \in (0, 1)$, define a map $h_\varepsilon \colon S \to \mathbb{R}^2$ by

$$h_\varepsilon(x, y) = \left(\frac{x}{4}, (1 + \varepsilon(x^2 + y^2))y\right).$$

We have $h_\varepsilon(0, 0) = (0, 0)$, and the preimages of the horizontal sides $y = \pm 1$ of the square S are C^∞ curves $y = \pm\psi_\varepsilon(x)$, with $|x| \le 1$, such that $\max \psi_\varepsilon = \psi_\varepsilon(0) < 1$ and $\max|\psi_\varepsilon'|$ is small if ε is small. Furthermore, vertical lines are invariant under h_ε.

Fix $\varepsilon > 0$ and choose $\sigma, \delta > 0$ such that $\psi_\varepsilon(0) < \sigma - \delta < \sigma + \delta < 1$. Set

$$\begin{aligned} H_1 &= \{(x, y) \in S : -\psi_\varepsilon(x) \le y \le \psi_\varepsilon(x)\}, \\ H_2 &= \{(x, y) \in S : \sigma - \delta \le y \le \sigma + \delta\}, \end{aligned}$$

and note that $H_1 \cap H_2 = \varnothing$. Define the map $f\colon H_1 \cup H_2 \to \mathbb{R}^2$ by

$$f(x, y) = \begin{cases} h_\varepsilon(x, y) & \text{if } (x, y) \in H_1 \\ (x/4 + 3/4, (-y + \sigma)/\delta) & \text{if } (x, y) \in H_2 \end{cases}.$$

The set

$$\Lambda = \bigcap_{n=-\infty}^{+\infty} f^n(H_1 \cup H_2)$$

is invariant under f.

Since $(0, 0) \in H_1$ is a fixed point of f, we have $(0, 0) \in \Lambda$. Moreover, $d_{(0,0)}f = d_{(0,0)}f_\varepsilon = \begin{pmatrix} 1/4 & 0 \\ 0 & 0 \end{pmatrix}$, and hence Λ is not a hyperbolic set. We show that $f|\Lambda$ has nonzero values of the Lyapunov exponent.

Proposition 6.2.1. *For every f-invariant measure ν on Λ, which is not the delta measure at the origin, one value of the Lyapunov exponent is identically equal to $-\log 4$ and the other one is ν-almost everywhere positive.*

Proof. Without loss of generality, we assume that ν is ergodic. Observe that

$$S \setminus \bigcup_{n=-\infty}^{+\infty} f^n(H_2) = \{(0, 0)\}.$$

Hence if $\nu(H_2) = 0$, then $\nu\{(0, 0)\} = 1$. Therefore, $\nu(H_2) > 0$. Since vectors with only vertical component are preserved, the value of the Lyapunov exponent in the vertical direction is given by

$$\begin{aligned} &\int_{H_1} \log|\partial_y[(1 + \varepsilon(x^2 + y^2))y]|\, d\nu(x, y) + \int_{H_2} \log\frac{1}{\delta}\, d\nu \\ &\quad = \int_{H_1} \log(1 + \varepsilon x^2 + 3\varepsilon y^2)\, d\nu(x, y) - \nu(H_2)\log\delta \\ &\quad \ge -\nu(H_2)\log\delta > 0. \end{aligned}$$

We now show that there exists $\gamma > 0$ for which the family of stable cones $C^s_\gamma(x, y)$ (see (6.12)) is preserved by df^{-1}. This is clear for $(x, y) \in f(H_2)$. Let $(x, y) \in f(H_1)$. We have

$$d_{(x',y')}f(v, w) = \left(\frac{v}{4}, w + \varepsilon(x'^2 w + 3y'^2 w + 2x'y'v)\right),$$

where $(x', y') = f^{-1}(x, y)$. If $(v, w) \in C^s_\gamma(x, y)$, then $|w| \le \gamma|v|$ and

$$df^{-1}(v, w) = (4v, w + \varepsilon\widetilde{w})$$

with

$$|\widetilde{w}| = |d_{(x,y)} f^{-1}(0, x'^2 w + 3y'^2 w + 8x'y'v)| \leq 8M(\gamma + 2)|v|$$

and $M = \max \|df^{-1}\|$. Therefore,

$$|w + \varepsilon\widetilde{w}| \leq \frac{\gamma + 8\varepsilon M(\gamma + 2)}{4}|4v| < \gamma|4v|$$

for all sufficiently small ε. This shows that

$$df^{-1}\overline{C_\gamma^s(x, y)} \subset C_\gamma^s(f^{-1}(x, y)),$$

and hence the preimage of any admissible (s, γ)-curve consists of two admissible (s, γ)-curves. This implies that the Lyapunov exponent of a vector $(v, w) \in C_\gamma^s(x, y)$ can be computed by looking only at the horizontal component $\pi_1(df^{-n}(v, w)) = 4^n v$ of a vector (v, w) tangent to an admissible (s, γ)-curve. Therefore the second value of the Lyapunov exponent is equal to $-\log 4$. □

Proposition 6.2.2. *The map $f|\Lambda$ is topologically conjugate to the full two-sided shift on two symbols.*

Proof. We first observe that the intersection $\bigcap_{k=1}^n f^k(H_1 \cup H_2)$ consists of 2^n vertical rectangles of width $1/4^n$. Therefore the set $\bigcap_{n=1}^{+\infty} f^n(H_1 \cup H_2)$ consists of vertical segments $\bigcap_{n=1}^{+\infty} f^n(H_{i_n}) = \{a\} \times [-1, 1]$, where $a = a(i_1 i_2 \cdots) \in (-1, 1)$.

On the other hand, the set $\bigcap_{k=-n+1}^{0} f^k(H_1 \cup H_2)$ consists of 2^n admissible (s, γ)-rectangles. Therefore if $p \in \Lambda$, then the connected component of $\bigcap_{k=-n}^{n} f^k(H_1 \cup H_2)$ containing p is a closed interval of length $d_n(p) > 0$. Since the lower and upper boundaries of $\bigcap_{k=-n+1}^{0} f^k(H_1 \cup H_2)$ are uniformly transversal to the family of vertical lines, the functions $d_n\colon \Lambda \to \mathbb{R}^+$ are continuous on Λ. Moreover, for each $p \in \Lambda$, the sequence $d_n(p)$ is decreasing.

The component of Λ containing p is a closed vertical segment with length $d(p) = \inf_n d_n(p) \geq 0$. To prove that $f|\Lambda$ is topologically conjugate to $\sigma\colon \Sigma_2 \to \Sigma_2$, it remains to show that d is identically zero on Λ_j and then we can define a topological conjugacy between $f|\Lambda$ and σ by (6.7).

Since d is upper semicontinuous, it attains a maximum value $d_{\max} \geq 0$. If $d_{\max} > 0$, let $p \in \Lambda$ be such that $d(p) = d_{\max}$. Since

$$\partial_y[(1 + \varepsilon(x^2 + y^2))y]|_{(x,y)=p} = 1 + \varepsilon(x^2 + 3y^2) > 1$$

for every $p \neq 0$, the image under f of the component of Λ containing p is a closed vertical segment with length $> d_{\max}$. This contradiction implies that $d_{\max} = 0$ and hence that d is identically zero. □

6.2.2 Further Modifications

We describe an example due to Katok and Mendoza [107] where both expansion and contraction slow down at a fixed point. For each $\varepsilon > 0$, we define a map

$g_\varepsilon : S \to \mathbb{R}^2$ by

$$g_\varepsilon(x, y) = ((1 - \varepsilon(x^2 + y^2))x, (1 + \varepsilon(x^2 + y^2))y).$$

As in the previous example, we have $g_\varepsilon(0, 0) = (0, 0)$, and the preimages of the horizontal sides $y = \pm 1$ of S are C^∞ curves $y = \pm\varphi_\varepsilon(x)$ such that $\max \varphi_\varepsilon = \varphi_\varepsilon(0) < 1$ and $\max|\varphi'_\varepsilon|$ is small if ε is small.

Let $\varepsilon > 0$. Choose $\sigma, \delta > 0$ such that $\varphi_\varepsilon(0) < \sigma - \delta < \sigma + \delta < 1$. Set

$$H_1 = \{(x, y) \in S : -\varphi_\varepsilon(x) \le y \le \varphi_\varepsilon(x)\},$$
$$H_2 = \{(x, y) \in S : \sigma - \delta \le y \le \sigma + \delta\},$$

and note that $H_1 \cap H_2 = \varnothing$. Define the map $f : H_1 \cup H_2 \to \mathbb{R}^2$ by

$$f(x, y) = \begin{cases} g_\varepsilon(x, y) & \text{if } (x, y) \in H_1 \\ (x\delta + \sigma, (-y + \sigma)/\delta) & \text{if } (x, y) \in H_2 \end{cases}.$$

The set

$$\bigcap_{n=-\infty}^{+\infty} f^n(H_1 \cup H_2)$$

is f-invariant and contains 0. Since $d_{(0,0)}f = d_{(0,0)}g_\varepsilon = \mathrm{Id}$, it is not hyperbolic. Nevertheless, it is topologically conjugate to the full two-sided shift on two symbols, and the values of the Lyapunov exponent are nonzero almost everywhere with respect to any f-invariant measure, which is different from the delta measure at the origin (see Proposition 6.2.4).

We now modify this example to obtain the slowing down of expansion and contraction near a fixed point, as well as the loss of transversality between expanding and contracting directions near that point.

Consider the linear map

$$T_\delta(x, y) = ((1 - \delta)x + \delta y, -\delta x + (1 + \delta)y).$$

We have $T_\delta C_1^u(x, y) \subset C_1^u(T_\delta(x, y))$, but since $T_\delta(1, 1) = (1, 1)$, the inclusion is not strict. Choose $\delta \in [0, \varepsilon)$ and define the function $f_\delta : S \to \mathbb{R}^2$ in H_1 by

$$f_\delta(x, y) = \begin{cases} (T_\delta \circ g_\varepsilon)(x, y) & \text{if } (x, y) \in H_1 \\ (x\delta + \sigma, (-y + \sigma)/\sigma) & \text{if } (x, y) \in H_2 \end{cases}.$$

Lemma 6.2.3. *If $(x, y) \neq (0, 0)$, then $df_\delta \overline{C_1^u(x, y)} \subset C_1^u(f_\delta(x, y))$.*

Proof. The statement is obvious for $(x, y) \in H_2$. If $(x, y) \in H_1$, then

$$\begin{aligned} d_{(x,y)}f_0(1, 1) &= (1 - 3\varepsilon x^2 - \varepsilon y^2 - 2\varepsilon xy, 2\varepsilon xy + 1 + \varepsilon x^2 + 3\varepsilon y^2) \\ &= (1 - a, 1 + a), \end{aligned}$$

where $a = a(x, y) = \varepsilon(3x^2 + y^2 + 2xy) = \varepsilon[2x^2 + (y + x)^2]$, and

$$\begin{aligned} d_{(x,y)} f_0(-1, 1) &= (-(1 - 3\varepsilon x^2 - \varepsilon y^2) - 2\varepsilon xy, 2\varepsilon xy + 1 + \varepsilon x^2 + 3\varepsilon y^2) \\ &= (-1 + b, 1 + b), \end{aligned}$$

where $b = b(x, y) = \varepsilon(3x^2 + y^2 - 2xy) = \varepsilon[2x^2 + (y - x)^2]$. Notice that $a, b > 0$ for any $(x, y) \neq (0, 0)$. This implies that

$$df_0 \overline{C_1^u(x, y)} \subset C_1^u(f_0(x, y)).$$

For $\delta > 0$ and $(x, y) \in H_1$, we observe that

$$f_\delta(x, y) = (T_\delta \circ f_0)(x, y) = (T_\delta \circ f)(x, y)$$

and

$$\begin{aligned} df_\delta \overline{C_1^u(x, y)} &= T_\delta(df_0 \overline{C_1^u(x, y)}) \\ &\subset T_\delta(C_1^u(f_0(x, y))) \subset C_1^u(f_\delta(x, y)). \end{aligned}$$

This completes the proof of the lemma. □

We define the f_δ-invariant set by

$$\Lambda_\delta = \bigcap_{n=-\infty}^{+\infty} f_\delta{}^n(H_1 \cup H_2).$$

Proposition 6.2.4. *For every $\delta \geq 0$ and every f_δ-invariant measure ν in Λ_δ, which is not the delta measure at the origin, one value of the Lyapunov exponent is ν-almost everywhere positive and the other one is ν-almost everywhere negative.*

Proof. The result clearly holds for every point $(x, y) \in \Lambda_\delta \cap H_2$.

We now show that the second component $f_\delta^{(2)}$ of the map f_δ satisfies $\partial_y f_\delta^{(2)}(x, y) > 1$ for $(x, y) \in \Lambda_\delta \setminus \{(0, 0)\}$. This is clear if $(x, y) \in H_2$. For $(x, y) \in H_1 \setminus \{(0, 0)\}$, we have

$$\begin{aligned} \partial_y f_\delta^{(2)}(x, y) &= (1 + \delta)(1 + 3\varepsilon x^2 + \varepsilon y^2) + 2\,\delta\varepsilon xy \\ &= 1 + \delta[1 + 2\varepsilon x^2 + \varepsilon(y + x)^2] \geq 1 + \delta. \end{aligned}$$

Using Lemma 6.2.3 and following the line of arguments in the proof of Proposition 6.2.1, one can show that the Lyapunov exponent at $(x, y) \in H_1 \setminus \{(0, 0)\}$ of a vector $(0, v) \in T_{(x,y)}X$ can be computed by estimating the vertical components of $df^n(0, v)$ and hence it is positive.

Similar calculations show that for any $(x, y) \in H_1 \setminus \{(0, 0)\}$,

$$\begin{aligned} \partial_x f_\delta^{(1)} &= (1 - \delta)(1 - 3\varepsilon^2 - \varepsilon y^2) + 2\,\delta\varepsilon xy \\ &= 1 - \delta[1 - 2\varepsilon x^2 - \varepsilon(y - x)^2] \leq 1 - \delta. \end{aligned}$$

Arguing as above one can show that the Lyapunov exponent at $(x, y) \in H_1 \setminus \{(0, 0)\}$ of a vector $(v, 0) \in T_{(x,y)}S$ can be computed by estimating the horizontal components of $df^{-n}(v, 0)$ and hence it is negative. □

Proposition 6.2.5. *For all sufficiently small $\delta > 0$, the map $f_\delta|\Lambda_\delta$ is topologically conjugate to the two-sided full shift on two symbols.*

Proof. Note that the map $f_\delta\colon H_1 \cup H_2 \to \mathbb{R}^2$ possesses the following properties:

1. The family of cones $\{C_1^u(x, y) : (x, y) \in \Lambda_\delta\}$ is f_δ-invariant.
2. $\partial_y f_\delta^{(2)} > 1$ and $\partial_x (f_\delta^{-1})^{(1)} > 1$ for every $(x, y) \neq (0, 0)$.
3. For every admissible $(u, 1)$-curve W, the set $f_\delta(W) \cap H_i$ is an admissible $(u, 1)$-curve for $i = 1, 2$.
4. For every admissible $(s, 1)$-curve V, the set $f_\delta^{-1}(V \cap H_i)$ is an admissible $(s, 1)$-curve for $i = 1, 2$.

Property 1 and the first inequality in Property 2 follow from Lemma 6.2.3 and Proposition 6.2.4. The proof of the second inequality is immediate. Finally, Properties 3 and 4 follow from the same properties of f_0 by continuity for all sufficiently small $\delta > 0$.

By Property 3, the set

$$\Lambda_{n,\delta}^{+} = \bigcap_{k=1}^{n} f_\delta^k (H_1 \cup H_2)$$

consists of 2^n admissible $(u, 1)$-rectangles for every $x \in \Lambda_\delta$. We call a connected C^1 curve V a $(s, 1, n)$-*curve* at x if tangent vectors at every point of V lie in their corresponding C_1^u cones, and both x and V are contained in the same connected component of $\Lambda_{n,\delta}^{+}$. We note that any $(s, 1, n)$-curve is the graph of a C^1 function $\psi\colon [a, b] \to [0, 1]$ with $|\psi'| < 1$. Here $[a, b]$ is the projection of V on the x-axis. Put $d^s(V) = b - a$. In a similar way, one defines $(s, 1, \infty)$-curves in the connected components of $\Lambda_{\infty,\delta}^{+}$. Let

$$d_n^s(x) = \sup\{d^s(V) : V \text{ is an admissible } (s, 1, n)\text{-curve at } x\}.$$

Since the boundaries of any component of $\Lambda_{n,\delta}^{+}$ are admissible $(u, 1)$-curves, they are transversal to every $(s, 1, n)$-curve. This implies that d_n^s is continuous on Λ_δ and hence the supremum is attained. Moreover, $d_n^s(x)$ decreases with n, and thus the function $d^s(x) = \inf_n d_n^s(x) \geq 0$ is upper semicontinuous. This implies that d^s attains its supremum, say d. We shall prove that $d = 0$. Assume that $d > 0$ and choose $x \in \Lambda_\delta$ such that $d^s(x) = d$. For each n, there exists a $(s, 1, n)$-curve of length greater than or equal to $d^s(x)$. By the compactness in the Hausdorff topology of the closed sets composed of the $(s, 1, n)$-curves at x, we obtain a $(s, 1, \infty)$-curve V at x with $d^s(V) = d$. Thus, $d = \max_x \max_V d^s(V)$ where V is any $(s, 1, \infty)$-curve at the point $x \in \Lambda_\delta$.

However, the preimage of any $(s, 1, \infty)$-curve at x is a $(s, 1, \infty)$-curve at $f_\delta^{-1}(x)$. By the second inequality in Property 2, we obtain the contradiction $d^s(f_\delta^{-1}(V)) > d^s(V) = d$.

This implies that the connected component of $\Lambda_{n,\delta}^{+}$ containing $x \in \Lambda_\delta$ is itself contained in the cone $C_1^s(x)$.

In a similar way we can define $(u, 1, n)$- and $(u, 1, \infty)$-curves and use the first inequality in Property 2 to show that the connected component

$$\Lambda^-_{n,\delta} = \bigcap_{k=-n+1}^{0} f^k_\delta(H_1 \cup H_2)$$

at $(x_0, y_0) \in \Lambda_\delta$ is itself contained in $C^u_1(x)$. We conclude that the connected component of Λ_δ at (x, y) is itself contained in $C^s_1(x, y) \cap C^u_1(x, y) = \{(x, y)\}$. This implies that the map in (6.7) defines a topological conjugacy between $f_\delta|\Lambda_\delta$ and the two-sided full shift on two symbols. □

The analysis of f_δ can be carried out further to show that for $\delta > 0$ the expanding and contracting directions are not transversal at 0. Namely, for each $x \in \Lambda_\delta$, one can construct stable and unstable distributions (which are similar to those for hyperbolic sets) by

$$L^+_x = \bigcap_{k=0}^{\infty} df^k_\delta \overline{C^u_1(f^{-k}_\delta(x))},$$

$$L^-_x = \bigcap_{k=0}^{\infty} df^{-k}_\delta \overline{C^s_1(f^k_\delta(x))}.$$

In general these sets are cones, but in our example one can prove that L^+_x and L^-_x are one-dimensional subspaces for every $x \in \Lambda_\delta \setminus \{0\}$.

For the map f_0, the sets L^+_x and L^-_x are transverse for every $x \in \Lambda_0 \setminus \{0\}$. At the origin we have $L^+_0 = C^u_1(0)$ and $L^-_0 = C^s_1(0)$ for $\delta = 0$. However, the limits $\lim_{x\to 0} L^+_x$ and $\lim_{x\to 0} L^-_x$ exist and are, respectively, vertical and horizontal subspaces. This means that for f_0 we have a "removable singularity" at the origin.

For the map f_δ with $\delta > 0$ and $x \in \Lambda_\delta$ (including the origin), the sets L^+_x and L^-_x are one-dimensional subspaces. However, $L^+_0 = L^-_0 = \{(v, v) : v \in \mathbb{R}\}$. Thus the origin is a "more substantial singularity."

For every $\delta \geq 0$, the distributions L^+ and L^- are tangential to certain families of invariant curves that are, respectively, connected components of $\Lambda^+_{n,\delta}$ and $\Lambda^-_{n,\delta}$. For $\delta = 0$, these curves are uniformly transverse, while for $\delta > 0$ they intersect tangentially at the origin even though they are transverse everywhere else.

6.3 Diffeomorphisms with Nonzero Lyapunov Exponents on Surfaces

In this section we present examples of area preserving diffeomorphisms of compact surfaces with nonzero Lyapunov exponents, which are not Anosov diffeomorphisms. Indeed, we describe a construction due to Katok [100] showing that any compact surface admits such a diffeomorphism. This is an important first step in proving the remarkable result that every compact manifold of dimension ≥ 2 admits a diffeomorphism with nonzero Lyapunov exponents (see Section 11.4). We begin by describing a construction of a diffeomorphism with nonzero Lyapunov exponents on the two-dimensional torus $\mathbb{T}^2$.

6.3.1 Analytic Nonuniformly Hyperbolic Diffeomorphisms

Let us briefly outline the main idea of the construction. Starting with the hyperbolic toral automorphism A induced by the matrix $\left(\begin{smallmatrix} 2 & 1 \\ 1 & 1 \end{smallmatrix}\right)$, we obtain the desired map by slowing down A near the origin. The new area preserving map G coincides with A outside a small disk around zero but the trajectories of G that enter the disk spend much longer time in it than the corresponding trajectories of A. This produces a substantial drop in the rates of contraction and expansion along trajectories leading to the appearance of orbits with zero Lyapunov exponents (and hence to the destruction of uniform hyperbolicity). However, the Lyapunov exponents along a typical trajectory are nonzero. To see this let us fix a small cone around the unstable direction of A at every point of the torus. Choose a point x outside the disk and a vector v in the cone at x. As long as the trajectory $\{G^k(x)\}_{k=1}^m$ stays outside the disk, the vector $dG^k v = dA^k v$ stays within the cone at $G^k(x)$ and gets expanded with a certain rate. After the trajectory of x enters the disk at time m the rate of expansion drops so that when it exits the disk at time n the vector $dG^n v$ has about the same length as that of the vector $dG^m v$. Furthermore, the vector $dG^n v$ no longer belongs to the cone at $G^n(x)$. However, we will arrange the procedure of slowing down trajectories in the disk in such a way that the vector $dG^n v$ is guaranteed to stay away a certain distance from the stable direction of A at $G^n(x)$; this distance is independent of x. This is crucial–since G preserves area and the disk D is small, the average time a typical trajectory spends outside the disk is big to allow the vector v to gain positive Lyapunov exponent.

We now describe the procedure of slowing down trajectories in detail. It depends upon a real-valued function ψ, which is defined on the unit interval $[0, 1]$ and has the following properties:

1. ψ is a C^∞ function except for the origin.
2. $\psi(0) = 0$ and $\psi(u) = 1$ for $u \geq r_0$ where $0 < r_0 < 1$.
3. $\psi'(u) > 0$ for every $0 < u < r_0$.
4. The following integral converges:

$$\int_0^1 \frac{du}{\psi(u)} < \infty.$$

Consider the disk D_r centered at 0 of radius r. We introduce a coordinate system (s_1, s_2) in D_r obtained from the eigendirections of A such that

$$D_r = \{(s_1, s_2) : {s_1}^2 + {s_2}^2 \leq r^2\}.$$

We observe that A is the time-1 map of the flow generated by the following system of differential equations:

$$\dot{s}_1 = s_1 \log \lambda \quad \text{and} \quad \dot{s}_2 = -s_2 \log \lambda, \tag{6.17}$$

where λ is an eigenvalue of A. Fix sufficiently small numbers $r_0 < r_1$ and consider the time-1 map g generated by the following system of differential equations in

D_{r_1}:

$$\dot{s}_1 = s_1 \psi(s_1{}^2 + s_2{}^2) \log \lambda \quad \text{and} \quad \dot{s}_2 = -s_2 \psi(s_1{}^2 + s_2{}^2) \log \lambda. \tag{6.18}$$

Our choice of function ψ guarantees that $g(D_{r_2}) \subset D_{r_1}$ for some $r_2 < r_0 < r_1$ and that g is of class C^∞ in $D_{r_1} \setminus \{0\}$ and coincides with A in some neighborhood of the boundary ∂D_{r_1}. Therefore the map

$$G(x) = \begin{cases} A(x) & \text{if } x \in \mathbb{T}^2 \setminus D_{r_1}, \\ g(x) & \text{if } x \in D_{r_1}. \end{cases}$$

defines a homeomorphism of the torus $\mathbb{T}^2$, which is a C^∞ diffeomorphism everywhere except for the origin. The map $G(x)$ is a slowdown of the automorphism A at 0.

We denote by W^u and W^s the projections of the eigenlines in $\mathbb{R}^2$ to $\mathbb{T}^2$ corresponding to the eigenvalues λ and λ^{-1}. Set $W = W^u \cup W^s$ and $X = \mathbb{T}^2 \setminus W$. Note that the set W is everywhere dense in $\mathbb{T}^2$.

The map G preserves a probability measure

$$d\nu = \kappa_0^{-1} \kappa \, dm, \tag{6.19}$$

where m is the area and the density κ is a positive C^∞ function and is infinite at 0. It is defined by the formula

$$\kappa(s_1, s_2) = \begin{cases} (\psi(s_1{}^2 + s_2{}^2))^{-1} & \text{if } (s_1, s_2) \in D_{r_1}, \\ 1 & \text{otherwise}, \end{cases} \tag{6.20}$$

where

$$\kappa_0 = \int_{\mathbb{T}^2} \kappa \, dm.$$

To see this one should verify that $\operatorname{div}(\kappa V_\psi) = 0$, where V_ψ is the vector field generated in D_{r_1} by the differential equation (6.18). The calculation is straightforward. It is easy to see that the map G is ergodic and indeed, mixing with respect to ν.

We shall show that at least one Lyapunov exponent at each point in W is zero. In particular, G is not an Anosov diffeomorphism.

Let $x = (0, s_2) \in D_{r_1} \cap W^s$. Consider a vertical vector $v \in T_x\mathbb{T}^2$. We have

$$\begin{aligned} \chi(x, v) &= \overline{\lim_{t\to+\infty}} \frac{\log|\dot{s}_2(t)|}{t} = \overline{\lim_{t\to+\infty}} \frac{\log|s_2(t)|}{t} \\ &= \overline{\lim_{t\to+\infty}} (\log|s_2(t)|)' = \overline{\lim_{t\to+\infty}} (-\psi(s_2(t)^2) \log \lambda), \end{aligned}$$

where $s_2(t)$ is the solution of (6.18) with the initial condition $s_2(0) = s_2$. In view of the choice of the function ψ, we obtain that $\chi(x, v) = 0$. Similarly, $\chi(x, v) = 0$ whenever $x, v \in W^u$.

We shall show now that for almost every $x \in X$, the Lyapunov exponents at x are nonzero. Choose $x \in X \setminus D_{r_1}$ and define the stable and unstable cones in

$T_x\mathbb{T}^2 = \mathbb{R}^2$ by

$$C^s_\gamma(x) = \{(v_1, v_2) \in \mathbb{R}^2 : |v_1| < \gamma|v_2|\} \cup (0,0),$$
$$C^u_\gamma(x) = \{(v_1, v_2) \in \mathbb{R}^2 : |v_2| < \gamma|v_1|\} \cup (0,0),$$

where $v_1 \in W^u$, $v_2 \in W^s$, and $0 < \gamma < 1/4$ (see Section 4.1). Note that for every $x \in D_{r_1} \setminus W$, there exist sequences $m_k = m_k(x) \to \infty$ and $n_k = n_k(x) \to \infty$ such that for every k,

$$G^{m_k}(x), G^{-n_k}(x) \in X \setminus D_{r_1}. \tag{6.21}$$

Therefore,

$$\bigcup_{n\in\mathbb{Z}} G^n(X \setminus D_{r_1}) \supset D_{r_1} \setminus W.$$

Consider the collection of sets

$$X_n = G\left(\bigcup_{j=0}^{n-1} X_j\right) \setminus \bigcup_{j=0}^{n-1} X_j, \quad X_0 = X \setminus D_{r_1}$$

and define for $x \in X_n$,

$$C^u(G(x)) = dGC^u(x).$$

Similarly, consider the collection of sets

$$Y_n = G^{-1}\left(\bigcup_{j=0}^{n-1} Y_j\right) \setminus \bigcup_{j=0}^{n-1} Y_j, \quad Y_0 = X \setminus D_{r_1}$$

and define for $x \in Y_n$,

$$C^s(G^{-1}(x)) = dG^{-1}C^s(x).$$

For every $x \in X$, we have that

$$dG^{-1}(C^s(x)) \subset C^s(G^{-1}(x)) \quad \text{and} \quad dG(C^u(x)) \subset C^u(G(x))$$

and the inclusions are strict on the set $X \setminus D_{r_1}$.

Since $G = A$ outside D_{r_1}, it follows from (6.21) that for every $x \in \mathbb{T}^2 \setminus \{0\}$ the formulas

$$E^s(x) = \bigcap_{j=0}^{\infty} dG^{-j}(C^s(G^j(x))) \quad \text{and} \quad E^u(x) = \bigcap_{j=0}^{\infty} dG^j(C^u(G^{-j}(x)))$$

define the stable and unstable one-dimensional subspaces at x.

Furthermore, the map G is uniformly hyperbolic on $X \setminus D_{r_1}$; more precisely, there is a number $\mu > 1$ such that for every $x \in X \setminus D_{r_1}$,

$$\|dG|E^s(x)\| \le \frac{1}{\mu} \quad \text{and} \quad \|dG^{-1}|E^u(x)\| \le \frac{1}{\mu}.$$

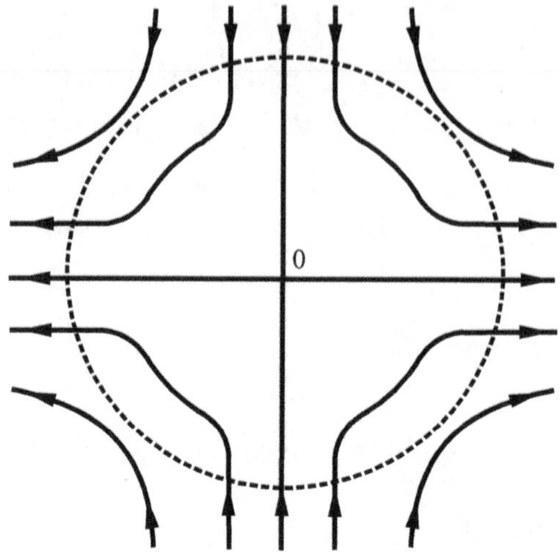

Figure 6.3. Invariant curves for $G_{\mathbb{T}^2}$ near 0.

Indeed, one can show that the stable and unstable subspaces can be extended to $W \setminus \{0\}$ to form two one-dimensional continuous distributions on $\mathbb{T}^2 \setminus \{0\}$. Moreover, for almost every x, we have $\chi(x, v) < 0$ for $v \in E^s(x)$ and $\chi(x, v) > 0$ for $v \in E^u(x)$. Indeed, since G is ergodic, the average time a typical trajectory spends in D_{r_1} is small and the statement readily follows from Theorem 4.1.6.

Our next step is to change the coordinate system in $\mathbb{T}^2$ by a map φ in such a way that the map $G_{\mathbb{T}^2} = \varphi \circ G \circ \varphi^{-1}$ preserves the area. Clearly, it has nonzero Lyapunov exponents almost everywhere and on the set $\varphi(W)$ it has two zero exponents. Furthermore, $G_{\mathbb{T}^2}$ is mixing with respect to area.

We define φ by the following formula:

$$\varphi(s_1, s_2) = \frac{1}{\sqrt{\kappa_0(s_1{}^2 + s_2{}^2)}} \left(\int_0^{s_1{}^2+s_2{}^2} \frac{du}{\psi(u)} \right)^{1/2} (s_1, s_2) \tag{6.22}$$

in D_{r_1} and $\varphi = \mathrm{Id}$ in $\mathbb{T}^2 \setminus D_{r_1}$. It is easy to see that φ is a homeomorphism and is a C^∞ diffeomorphism except at the origin. It also commutes with the involution $I(t_1, t_2) = (1 - t_1, 1 - t_2)$. To check that φ transfers the measure ν into the area, one should verify that $\mathrm{Jac}(\varphi) = d\nu/dm$. In view of (6.20), this is equivalent to the identity

$$\mathrm{Jac}(\varphi)(s_1, s_2) = \frac{1}{\kappa_0 \psi(s_1{}^2 + s_2{}^2)}. \tag{6.23}$$

Write

$$\varphi(s_1, s_2) = \kappa_0{}^{-1/2} f(r)(s_1/r, s_2/r),$$

where $r = \sqrt{s_1{}^2 + s_2{}^2}$. One can obtain (6.23) from the identity $(f(r)^2)' = 2r/\psi(r^2)$. The behavior of the orbits of $G_{\mathbb{T}^2}$ is indicated on Figure 6.3.

We stress that the map $G_{\mathbb{T}^2}$ is a C^∞ diffeomorphism. Indeed, we only need to check that $G_{\mathbb{T}^2}$ is infinitely many times differentiable at the origin. Observe that the system (6.17) is *Hamiltonian with respect to the area m* with the Hamiltonian function $H(s_1, s_2) = s_1 s_2 \log \lambda$. This means (we exploit the fact that the case we consider is two dimensional) that for every continuous vector field V on $\mathbb{T}^2$, we have

$$\Omega(V, V_H) = dH(V),$$

where V_H is the vector field generated by the system (6.17) and Ω is the volume 2-form; that is, $\Omega = ds_1 \wedge ds_2$. The vector field V_ψ, generated by the system (6.18), is obtained from the vector field V_H by a time change, so it is also Hamiltonian with respect to the measure ν with the same Hamiltonian function H (i.e., $\Omega_\psi(V, V_\psi) = dH(V)$ where $\Omega_\psi = \kappa_0^{-1}\kappa\, ds_1 \wedge ds_2$). Near the origin the map $G_{\mathbb{T}^2}$ is the time-1 map generated by the vector field $\varphi_* V_\psi$. Since $\varphi_*\nu = m$, this vector field is Hamiltonian with respect to the area m with the Hamiltonian function

$$H_1(s_1, s_2) = (H \circ \varphi^{-1})(s_1, s_2) = \frac{s_1 s_2 \beta(\sqrt{s_1{}^2 + s_2{}^2}) \log \lambda}{s_1{}^2 + s_2{}^2},$$

where $\beta(u)$ is the function inverse to

$$\gamma(u) = \kappa_0^{-1/2} \sqrt{\int_0^u \frac{d\tau}{\psi(\tau)}}.$$

All the derivatives of the function H_1 can be expressed in terms of the derivatives of the function β, so that if the derivatives of β decrease near zero sufficiently fast then so do the derivatives of the function H_1. The former can be achieved if the derivatives of ψ decrease near zero sufficiently fast.

6.3.2 A Nonuniformly Hyperbolic Diffeomorphism on the Sphere

We now present a construction of a diffeomorphism with nonzero Lyapunov exponents on the two-dimensional sphere. We need a toral automorphism of $\mathbb{T}^2$, which has four fixed points. The automorphism A induced by the matrix $\begin{pmatrix} 5 & 8 \\ 8 & 13 \end{pmatrix}$ is appropriate and the fixed points are $x_1 = (0, 0)$, $x_2 = (1/2, 0)$, $x_3 = (0, 1/2)$, and $x_4 = (1/2, 1/2)$.

For $i = 1, 2, 3, 4$, consider the disk D_r^i centered at x_i of radius r. Repeating the above arguments, we construct a diffeomorphism g_i, which coincides with A outside $D_{r_1}^i$. Therefore the map

$$G_1(x) = \begin{cases} A(x) & \text{if } x \in \mathbb{T}^2 \setminus D_{r_1}^i \\ g_i(x) & \text{if } x \in D_{r_1}^i \end{cases}$$

defines a homeomorphism of the torus $\mathbb{T}^2$, which preserves the measure ν (see (6.19)) and is a C^∞ diffeomorphism. Here $D = \bigcup_{i=1}^4 D_{r_1}^i$. The Lyapunov exponents of G_1 are nonzero almost everywhere with respect to the measure ν.

Using (6.22) in each disk $D^i_{r_2}$, we introduce a coordinate change φ_i, (see (6.22)) such that the map

$$\varphi(x) = \begin{cases} \varphi_i(x) & \text{if } x \in D^i_{r_1} \\ x & \text{otherwise} \end{cases}$$

defines a homeomorphism of $\mathbb{T}^2$, which is a C^∞ diffeomorphism everywhere expect for the points x_i. The map $G_2 = \psi \circ G_1 \circ \psi^{-1}$ is an area preserving C^∞ diffeomorphism.

Using this map we construct a diffeomorphism of the sphere S^2. Consider the map $\zeta : \mathbb{T}^2 \to S^2$ defined by

$$\zeta(s_1, s_2) = \left(\frac{s_1{}^2 - s_2{}^2}{\sqrt{s_1{}^2 + s_2{}^2}}, \frac{2s_1 s_2}{\sqrt{s_1{}^2 + s_2{}^2}} \right). \tag{6.24}$$

This map is a double-branched covering and is regular and C^∞ everywhere except for the points x_i, $i = 1, 2, 3, 4$, where it branches. It commutes with the involution I and preserves the area. Consider the map $G_{S^2} = \zeta \circ G_2 \circ \zeta^{-1}$. One can show that this map is a C^∞ area preserving diffeomorphism, it is mixing, and it has nonzero Lyapunov exponents almost everywhere.

6.3.3 Nonuniformly Hyperbolic Diffeomorphisms on Compact Surfaces

The diffeomorphism G_{S^2} of the sphere S^2 constructed in the previous section can be used as the foundation in building a nonuniformly hyperbolic diffeomorphism on any surface. This is a two-step procedure. First, the sphere can be unfolded into the unit disk and the map G_{S^2} can be carried over to a C^∞ area preserving diffeomorphism g of the disk known as the Katok map. We will describe this procedure as well as some properties of the diffeomorphism g. It is crucial to note that g is the identity on the boundary of the disk and is "sufficiently flat" near the boundary. The second step is to cut the surface in a certain way so that the resulting surface with the boundary is diffeomorphic to the disk. This procedure is well known in topology and indeed, we present its multidimensional version that we will use in Section 11.4 to construct a volume preserving diffeomorphism with nonzero Lyapunov exponents on any manifold. Using the properties of g mentioned above one can show that g can be carried over to become an area preserving C^∞ diffeomorphism with nonzero Lyapunov exponents on the surface.

Set $p_i = \zeta(x_i)$, $i = 1, 2, 3, 4$ (see (6.24)). In a neighborhood of the point p_4, we define the map η by

$$\eta(\tau_1, \tau_2) = \left(\frac{\tau_1\sqrt{1 - \tau_1^2 - \tau_2^2}}{\sqrt{\tau_1^2 + \tau_2^2}}, \frac{\tau_2\sqrt{1 - \tau_1^2 - \tau_2^2}}{\sqrt{\tau_1^2 + \tau_2^2}} \right).$$

One can extend it to a C^∞ diffeomorphism η between $S^2 \setminus \{p_4\}$ and the unit disk $D^2 \subset \mathbb{R}^2$, which preserves the Lebesgue measure. The map

$$g = \eta \circ G_{S^2} \circ \eta^{-1} \tag{6.25}$$

is a diffeomorphism of the disk D^2, which preserves area and has nonzero Lyapunov exponents almost everywhere. It is called the *Katok map*.

We summarize the properties of the diffeomorphism g. They follow immediately from results in Sections 6.3.1 and 6.3.2. Let $\mathcal{Q} = \{q_1, q_2, q_3\} \cup \partial D$ be the *singularity set* for g and let U be a small neighborhood of $\mathcal{Q}$.

Proposition 6.3.1.

1. *The map g is a C^∞ area preserving mixing diffeomorphism with nonzero Lyapunov exponents almost everywhere.*
2. *There exist continuous families of stable cones $C^-(x)$ and unstable cones $C^+(x)$, $x \in D \setminus \mathcal{Q}$, such that*

$$dg^{-1}C^-(x) \subset C^-(g^{-1}(x)) \quad \text{and} \quad dgC^+(x) \subset C^+(g(x))$$

and the inclusions are strict on the closure of the set $D \setminus U$.

3. *The distributions*

$$E^-(x) = \bigcap_j dg^{-j}(C^-(g^j(x))) \quad \text{and} \quad E^+(x) = \bigcap_j dg^j(C^+(g^{-j}(x)))$$

are one-dimensional dg-invariant and continuous on $D \setminus \mathcal{Q}$; the map g is uniformly hyperbolic on the set $D \setminus U$: for $x \in D \setminus U$,

$$\|dg|E^-(x)\| \le \frac{1}{\mu} \quad \text{and} \quad \|dg^{-1}|E^+(x)\| \le \frac{1}{\mu};$$

furthermore, there is an invariant set X of full measure such that for every $x \in X$,

$$E^s(x) = E^-(x) \quad \text{and} \quad E^u(x) = E^+(x).$$

4. *The map g possesses two one-dimensional foliations, W^- and W^+, of the set $D \setminus \mathcal{Q}$ such that*

$$T_xW^-(x) = E^-(x), \quad T_xW^+(x) = E^+(x), \quad x \in D \setminus \mathcal{Q};$$

the sizes of local leaves $V^-(x)$ and $V^+(x)$ are bounded away from zero on the set $D \setminus U$; moreover, for every $x \in X$,

$$W^s(x) = W^-(x) \quad \text{and} \quad W^u(x) = W^+(x).$$

5. *$d^k(g - id)|\partial D = 0$ for any $k \ge 0$; that is, on the boundary of the disk the map g is the identity map and has all its derivatives zero.*
6. *There is a neighborhood $U_1 \supset U$ of ∂D and a vector field V in U_1, which generates an area preserving flow $g^t : U \to D$, $-2 < t < 2$ with $g|U = g^1$.*

The disk D^2 can be embedded into any surface. This is a corollary of a more general statement (see [100]).

Proposition 6.3.2. *Given a p-dimensional compact C^∞ manifold M, there exists a continuous map $h\colon B^p \to M$ on the unit ball in $\mathbb{R}^p$ such that*

1. *the restriction $h|\operatorname{int} B^p$ is a diffeomorphic embedding;*
2. *$h(B^p) = M$;*
3. *the map h moves volume in $\mathbb{R}^p$ into the Riemannian volume in M.*

Note that the map g in (6.25) is identity on the boundary ∂D^2. Moreover, one can choose the function ψ in the construction of the maps $G_{\mathbb{T}^2}$ and G_{S^2} such that g is "sufficiently flat" near the boundary of the disk. More precisely, let $\rho = \{\rho_n\}_n$ be a sequence of nonnegative real-valued continuous functions on B^p, which are strictly positive inside the disk. Let $C^\infty_\rho(B^p)$ be the set of all C^∞ functions $\varphi\colon B^p \to \mathbb{R}$ satisfying the following condition: there exists a sequence of numbers $\varepsilon_n > 0$ such that for every $(x_1, \dots, x_p) \in B^p$ for which $x_1^2 + \cdots + x_p^2 \geq (1 - \varepsilon_n)^2$, we have

$$\left| \frac{\partial^n \varphi(x_1, \dots, x_p)}{\partial x_1^{i_1} \cdots \partial x_p^{i_p}} \right| < \rho_n(x_1, \dots, x_p),$$

where $i_1, \dots, i_p$ are nonnegative integers and $i_1 + \cdots + i_p = n$.

We write each diffeomorphism G of B^p in the form $G(x_1, \dots, x_p) = (G_1(x_1, \dots, x_p), \dots, G_p(x_1, \dots, x_p))$ and we denote by $\operatorname{Diff}^\infty_\rho(B^p)$ the set of diffeomorphisms $G \in \operatorname{Diff}^\infty(B^p)$ such that

$$G_i(x_1, \dots, x_p) - x_i \in C^\infty_\rho(B^p), \quad i = 1, \dots, p.$$

In order to complete our construction we need the following statement:

Proposition 6.3.3 (Katok [100]). *Given a compact C^∞ Riemannian manifold M, there exists a sequence of functions ρ such that for any $G \in \operatorname{Diff}^\infty_\rho(B^p)$ the map k defined by $k(x) = h(G(h^{-1}(x)))$ for $x \in h(\operatorname{int} B^p)$, where h is the map in Proposition 6.3.2, and $k(x) = x$ otherwise, is a C^∞ diffeomorphism of M.*

The function ψ can be chosen so that $g \in \operatorname{Diff}^\infty_\rho(D^2)$ and hence the map f, defined by $f(x) = h(g(h^{-1}(x)))$ for $x \in h(\operatorname{int} D^2)$ and $f(x) = x$ otherwise, has the desired properties: it preserves area and has nonzero Lyapunov exponents.

6.3.4 A Nonuniformly Hyperbolic Diffeomorphism of the Torus

The diffeomorphisms with nonzero Lyapunov exponents constructed in the previous sections are of class C^∞ but are not analytic. In Section 11.4, we show that every compact manifold of dimension ≥ 2 admits a C^∞ volume preserving diffeomorphism with nonzero Lyapunov exponents. It is therefore interesting to know that there are *analytic* volume preserving diffeomorphisms with nonzero Lyapunov exponents, which are not Anosov. The corresponding example is due to Katok and Lewis [105] and is constructed using a version of the well-known blow-up procedure from algebraic geometry.

Setting $X = \{x \in \mathbb{R}^n : \|x\| > 1\}$, consider the map $\varphi\colon \mathbb{R}^n \setminus \{0\} \to X$ given by

$$\varphi(x) = \frac{(\|x\|^n + 1)^{1/n}}{\|x\|} x.$$

It is easy to see that φ has Jacobian 1 with respect to the standard coordinates on $\mathbb{R}^n$.

Let A be a linear hyperbolic transformation of $\mathbb{R}^n$. The diffeomorphism $f = \varphi \circ A \circ \varphi^{-1}$ extends analytically to a neighborhood of the boundary, which depends on A. This follows from the formula

$$\varphi \circ A \circ \varphi^{-1}(x) = \Big(\frac{\|x\|^n - 1}{\|x\|} + \frac{1}{\|Ax\|^n}\Big)^{1/n} Ax.$$

Let (r, θ) be the standard polar coordinates on the set X so that $X = \{(r, \theta) : r > 1, \theta \in S^{n-1}\}$. Introducing new coordinates (s, θ), where $s = r^n - 1$, observe that these coordinates extend analytically across the boundary and have the property that the standard volume form is proportional to $ds \wedge d\theta$. Let B be the quotient of $\bar{X}$ under the identification of antipodal points on the boundary. The map f induces a map F of B, which preserves the volume form $ds \wedge d\theta$, has nonzero Lyapunov exponents, and is analytic.

6.4 Pseudo-Anosov Maps

Pseudo-Anosov maps were singled out by Thurston in connection with the problem of classifying diffeomorphisms of a compact C^∞ surface M up to isotopy (see [69, 213]). According to Thurston's classification, a diffeomorphism of M is isotopic to a homeomorphism f satisfying one of the following properties (see [69, Exposé 9]):

1. f is of finite order and is an isometry with respect to a Riemannian metric of constant curvature on M.
2. f is a "reducible" diffeomorphism, that is, a diffeomorphism leaving invariant a closed curve.
3. f is a *pseudo-Anosov map*.

Pseudo-Anosov maps are surface homeomorphisms that are differentiable except at most at finitely many points called *singularities*. These maps minimize both the number of periodic points (of any given period) and the topological entropy in their isotopy classes. A pseudo-Anosov map is Bernoulli with respect to an absolutely continuous invariant measure with C^∞ density, which is positive except at the singularities (see [69, Exposé 10]).

If M is a torus, then any pseudo-Anosov map is an Anosov diffeomorphism (see [69, Exposé 1]). However, if M has genus greater than 1, a pseudo-Anosov map cannot be conjugated to a diffeomorphism by a conjugating map, which is smooth outside the singularities or even outside a sufficiently small neighborhood of the singularities (see [75, Section 2.4]). Thus in order to find smooth models for the Case 3 with the same dynamical properties as pseudo-Anosov maps, one may have to apply some nontrivial construction, which is global in nature. In [75], Gerber and Katok constructed, for every pseudo-Anosov map f, a C^∞ diffeomorphism g, which is topologically conjugate to f through a homeomorphism isotopic to the identity and which is Bernoulli with respect to a smooth measure (i.e., a measure whose density is C^∞ and positive everywhere).

6.4.1 Definitions and Basic Properties

Let $\{x_1, \dots, x_m\}$ be a finite set of points and ν a Borel measure on M. Write $D_a := \{z \in \mathbb{C} : |z| < a\}$.

Definition 6.4.1. A *measured foliation* $(\mathcal{F}, \nu)$ of M with *singular points* $x_1, \dots, x_m$ is a partition of M for which the following properties hold:

1. There is a collection of C^∞ charts $\varphi_k \colon U_k \to \mathbb{C}$ for $k = 1, \dots, \ell$ and some $\ell \geq m$, with $\bigcup_{k=1}^{\ell} U_k = M$.
2. For each $k = 1, \dots, m$, there is a number $p = p(k) \geq 3$ of elements of $\mathcal{F}$ meeting at x_k such that
   (a) $\varphi_k(x_k) = 0$ and $\varphi_k(U_k) = D_{a_k}$ for some $a_k > 0$;
   (b) if C is an element of $\mathcal{F}$ then components of $C \cap U_k$ are mapped by φ_k to sets of the form
   $$\{z : \operatorname{Im}(z^{p/2}) = \text{constant}\} \cap \varphi_k(U_k);$$
   (c) the measure $\nu|U_k$ is the pullback under φ_k of
   $$|\operatorname{Im}(dz^{p/2})| = |\operatorname{Im}(z^{(p-2)/2}\,dz)|.$$
3. For each $k > m$, we have
   (a) $\varphi_k(U_k) = (0, b_k) \times (0, c_k) \subset \mathbb{R}^2 \equiv \mathbb{C}$ for some $b_k, c_k > 0$.
   (b) If C is an element of $\mathcal{F}$ then components of $C \cap U_k$ are mapped by φ_k to segments of the form
   $$\{z : \operatorname{Im} z = \text{constant}\} \cap \varphi_k(U_k).$$
   (c) The measure $\nu|U_k$ is given by the pullback of $|\operatorname{Im}\, dz|$ under φ_k.

The elements of $\mathcal{F}$ are called *leaves* of the foliation and ν is called a *transverse* measure. We say that x_k is a *$p(k)$-prong singularity* of $\mathcal{F}$ for $k = 1, \dots, m$, and each of the leaves of $\mathcal{F}$ meeting at x_k is called a *prong* of x_k (see Fig. 6.4 for an example). If, in addition, we allow single leaves of $\mathcal{F}$ to terminate in a point (called a *spine*, in which case we set $p = 1$ above), then $(\mathcal{F}, \nu)$ is called a *measured foliation with spines*.

The transverse measure is consistently defined on chart overlaps, because whenever $U_j \cap U_k \neq \varnothing$, the transition functions $\varphi_k \circ \varphi_j{}^{-1}$ are of the form

$$(\varphi_k \circ \varphi_j{}^{-1})(x, y) = (h_{jk}(x, y), c_{jk} \pm y),$$

where h_{jk} is a function and c_{jk} is a constant.

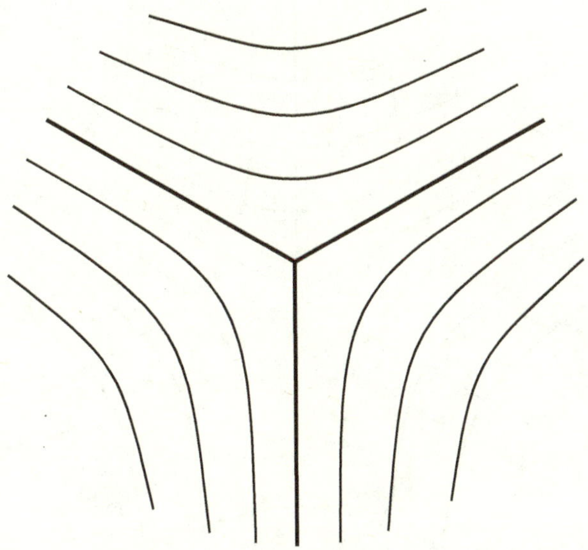

Figure 6.4. A 3-prong singularity.

Definition 6.4.2. A surface homeomorphism f is said to be a *pseudo-Anosov homeomorphism* if it satisfies the following properties:

1. f is differentiable except at a finite number of points $x_1, \ldots, x_m$.
2. There are two measured foliations $(\mathcal{F}^s, \nu^s)$ and $(\mathcal{F}^u, \nu^u)$ with the same singularities $x_1, \ldots, x_m$ and the same number of prongs $p = p(k)$ at each point x_k, $k = 1, \ldots, m$.
3. The leaves of the foliations $\mathcal{F}^s$ and $\mathcal{F}^u$ are transversal at nonsingular points (see Fig. 6.5).
4. There exists a constant $\lambda > 1$ such that
$$f(\mathcal{F}^s, \nu^s) = (\mathcal{F}^s, \nu^s/\lambda) \quad \text{and} \quad f(\mathcal{F}^u, \nu^u) = (\mathcal{F}^u, \lambda\nu^u).$$
5. There are C^∞ charts $\varphi_k \colon U_k \to \mathbb{C}$ for $k = 1, \ldots, \ell$ and some $\ell \geq m$ with $\bigcup_{k=1}^{\ell} U_k = M$.
6. For each $k = 1, \ldots, m$, we have
   (a) $\varphi_k(x_k) = 0$ and $\varphi_k(U_k) = D_{a_k}$ for some $a_k > 0$;
   (b) if C is a leaf of $\mathcal{F}^s$, then components of $C \cap U_k$ are mapped by φ_k to sets of the form
$$\{z : \mathrm{Re}(z^{p/2}) = \text{constant}\} \cap D_{a_k};$$
   (c) if C is a leaf of $\mathcal{F}^u$, then components of $C \cap U_k$ are mapped by φ_k to sets of the form
$$\{z : \mathrm{Im}(z^{p/2}) = \text{constant}\} \cap D_{a_k};$$

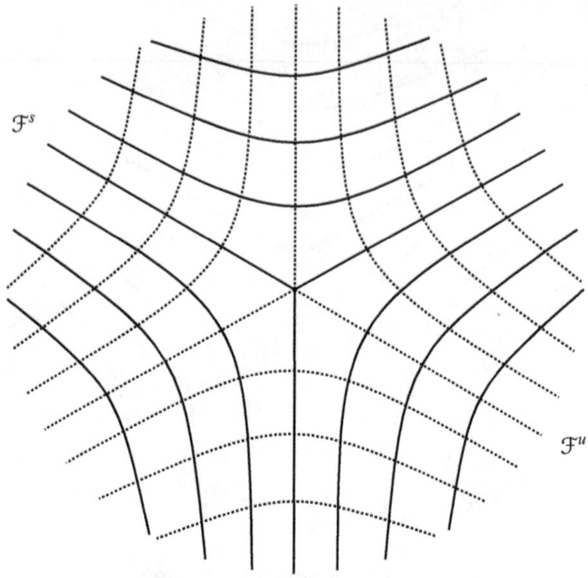

Figure 6.5. Foliations $\mathcal{F}^s$ and $\mathcal{F}^u$ of a pseudo-Anosov map.

(d) the measures $\nu^s|U_k$ and $\nu^u|U_k$ are given by the pullbacks of

$$|\mathrm{Re}(dz^{p/2})| = |\mathrm{Re}(z^{(p-2)/2}\,dz)|$$

and

$$|\mathrm{Im}(dz^{p/2})| = |\mathrm{Im}(z^{(p-2)/2}\,dz)|$$

under φ_k, respectively.

7. For each $k > m$, we have
   (a) $\varphi_k(U_k) = (0, b_k) \times (0, c_k) \subset \mathbb{R}^2 \equiv \mathbb{C}$ for some $b_k, c_k > 0$;
   (b) if C is a leaf of $\mathcal{F}^s$, then components of $C \cap U_k$ are mapped by φ_k to segments

$$\{z : \mathrm{Re}\, z = \text{constant}\} \cap \varphi_k(U_k);$$

   (c) if C is a leaf of $\mathcal{F}^u$, then components of $C \cap U_k$ are mapped by φ_k to segments

$$\{z : \mathrm{Im}\, z = \text{constant}\} \cap \varphi_k(U_k);$$

   (d) the measures $\nu^s|U_k$ and $\nu^u|U_k$ are given by the pullbacks of $|\mathrm{Re}\, dz|$ and $|\mathrm{Im}\, dz|$ under φ_k, respectively.

If, in addition, $(\mathcal{F}^s, \nu^s)$ and $(\mathcal{F}^u, \nu^u)$ are measured foliations with spines (with $p = p(k) = 1$ when there is only one prong at x_k), then f is called a *generalized pseudo-Anosov homeomorphism*.

We call $\mathcal{F}^s$ and $\mathcal{F}^u$ the *stable* and *unstable foliations*, respectively. At each singular point x_k, with $p = p(k)$, the *stable* and *unstable prongs* are, respectively, given by

$$P^s_{kj} = \varphi_k{}^{-1}\left\{\rho\, e^{i\tau} : 0 \le \rho < a_k,\, \tau = \frac{2j+1}{p}\pi\right\},$$
$$P^u_{kj} = \varphi_k{}^{-1}\left\{\rho\, e^{i\tau} : 0 \le \rho < a_k,\, \tau = \frac{2j}{p}\pi\right\},$$

for $j = 0, 1, \ldots, p-1$. We define the *stable* and *unstable sectors* at x_k by

$$S^s_{kj} = \varphi_k{}^{-1}\left\{\rho\, e^{i\tau} : 0 \le \rho < a_k,\, \frac{2j-1}{p}\pi \le \tau \le \frac{2j+1}{p}\pi\right\},$$
$$S^u_{kj} = \varphi_k{}^{-1}\left\{\rho\, e^{i\tau} : 0 \le \rho < a_k,\, \frac{2j}{p}\pi \le \tau \le \frac{2j+2}{p}\pi\right\},$$

respectively, for $j = 0, 1, \ldots, p-1$.

Since f is a homeomorphism, $f(x_k) = x_{\sigma_k}$ for $k = 1, \ldots, m$, where σ is a permutation of $\{1, \ldots, m\}$ such that $p(k) = p(\sigma_k)$ and f maps the stable prongs at x_k into the stable prongs at x_{σ_k} (provided the numbers a_k are chosen such that $a_k/\lambda^{2/p} \le a_{\sigma_k}$).

We describe some properties of the map f and we refer the reader to [69] for the detailed proofs.

Proposition 6.4.3. *The periodic points of f are dense in M.*

Proposition 6.4.4. *The map f possesses an invariant measure ν with the following properties:*

1. *ν has a density, which is smooth everywhere except at the singular points x_k, where it vanishes if $p(k) \ge 3$ and goes to infinity if $p(k) = 1$.*
2. *f is Bernoulli with respect to ν.*

6.4.2 Smooth Models of Pseudo-Anosov Maps

We present a result by Gerber and Katok [75], which gives a smooth realization of pseudo-Anosov maps up to topological equivalence. Our presentation is based on [75], but some parts of the proof have been simplified. The text for this section was provided by M. Gerber.

Theorem 6.4.5. *Let f be a pseudo-Anosov map of M and $\mathcal{O}$ a C^0 neighborhood of f. Then there exists a diffeomorphism $g \in \mathcal{O}$, which is topologically conjugate to f.*

Proof. We split the proof into several steps:

Step I: Construction of g. We apply a local perturbation of f in a neighborhood of each singularity. In each stable sector it consists of a "slowing down" of the orbits as in Section 6.3.1. The coordinate change Φ_{kj} described below brings the invariant measure for the slowed-down map back to the Lebesgue measure. For a generalized pseudo-Anosov map, an additional "blowing-up" procedure as described in Section 2 of [100] is needed to produce a diffeomorphism preserving a smooth measure that is topologically conjugate to the original map. With this additional step the arguments given below will still work, but we will present the proof only for the pseudo-Anosov case.

Henceforth, we assume that $\sigma_k = k$ for $k = 1, \dots, m$, and $f(P^s_{kj}) \subset P^s_{kj}$ for $k = 1, \dots, m$ and $j = 1, \dots, p(k)$. (The arguments in the general case are the same but require more cumbersome notation.) We define $\Phi_{kj} \colon \varphi_k S^s_{kj} \to \{z : \operatorname{Re} z \geq 0\} \cap D_{\tilde{a}_k}$, where $\tilde{a}_k = (2/p)a_k^{p/2}$, by

$$\Phi_{kj}(z) = (2/p)z^{p/2} = w = s_1 + is_2.$$

Note that

$$(\Phi_{kj}^{-1})^* |\operatorname{Re}(z^{(p-2)/2}\, dz)| = |ds_1|, \quad (\Phi_{kj}^{-1})^* |\operatorname{Im}(z^{(p-2)/2}\, dz)| = |ds_2|.$$

Thus in $S^s_{kj} \cap f^{-1} S^s_{kj}$, we have

$$f = \varphi_k^{-1} \Phi_{kj}^{-1} F \Phi_{kj} \varphi_k,$$

where $f(w) = F(s_1, s_2) = (\lambda s_1, s_2/\lambda)$. Since f is expanding linearly in p (≥ 3) directions and contracting in p (≥ 3) directions at x_k, f is not differentiable at x_k.

We may assume that the neighborhoods U_k, $k = 1, \dots, m$, of the singularities in Definition 6.4.2 are chosen to be disjoint and that the closures of the sets U_k do not contain any singularities for $k = m+1, \dots, \ell$. Let r_0 and r_1 be such that $0 < r_1 < r_0 < \min\{a_1, \dots, a_k\}$, and define $\tilde{r}_0$ and $\tilde{r}_1$ by $\tilde{r}_j = (2/p)r_j^{p/2}$ for $j = 1, 2$. We now define the "slowing-down" function Ψ_p corresponding to a p-prong singularity, $p \geq 3$, on the interval $[0, \infty)$ by

1. $\Psi_p(u) = (p/2)^{(2p-4)/p} u^{(p-2)/p}$ for $u \leq \tilde{r}_1^2$;
2. Ψ_p is C^∞ except at 0;
3. $\Psi_p'(u) \geq 0$ for $u > 0$;
4. $\Psi_p(u) = 1$ for $u \geq \tilde{r}_0^2$.

Note that F is the time-1 map of the vector field given by

$$\dot{s}_1 = (\log \lambda) s_1 \quad \text{and} \quad \dot{s}_2 = -(\log \lambda) s_2.$$

Now consider the vector field v_{Ψ_p} given by

$$\dot{s}_1 = (\log \lambda) s_1 \Psi_p(s_1^2 + s_2^2) \quad \text{and} \quad \dot{s}_2 = -(\log \lambda) s_2 \Psi_p(s_1^2 + s_2^2).$$

Let G_p be the time-1 map of v_{Ψ_p}. We assume that r_1 is chosen sufficiently small so that G_p coincides with F in a neighborhood of the boundary of $D_{\tilde{r}_0}$. We also assume that r_0 is chosen sufficiently small so that the neighborhood

$$\mathcal{U}_0 = \bigcup_{k=1}^{m} \varphi_k^{-1} D_{r_0} \tag{6.26}$$

of the set of singularities is disjoint from $\bigcup_{k=m+1}^{\ell} \varphi_k^{-1} D_{a_k}$. We define $g\colon M \to M$ by $g(x) = f(x)$ if $x \notin \mathcal{U}_0$ and by

$$g(x) = \varphi_k^{-1} \Phi_{kj}^{-1} G_{p(k)} \Phi_{kj} \varphi_k(x)$$

if $x \in S^s_{kj} \cap \varphi_k^{-1}(D_{r_0})$, $1 \le k \le m$, $1 \le j \le p(k)$. Then g is a homeomorphism, which is C^∞ everywhere except possibly at the singularities $x_1, \dots, x_m$. In fact, we will see that g is C^∞ everywhere; moreover in a neighborhood of each singularity, g is real analytic.

Step II: Smoothness of g. Fix a singular point x_k and let $p = p(k)$. Consider the coordinate system $z = t_1 + it_2 = \rho\, e^{i\tau}$ on D_{r_0} given by the chart (φ_k, U_k), and the coordinate system $w = s_1 + is_2 = r\, e^{i\theta}$ on $D_{\tilde{r}_0}$, where Φ_{kj}^{-1}, $j = 0, \dots, p-1$, on $D_{\tilde{r}_0} \cap \{(s_1, s_2) : s_1 \ge 0\}$ is defined by

$$z = \Phi_{kj}^{-1}(w) = \left(\frac{p}{2} w\right)^{2/p},$$

or, equivalently,

$$\rho = \left(\frac{p}{2} r\right)^{2/p} \quad \text{and} \quad \tau = \frac{2}{p}\theta + \frac{2j\pi}{p}, \quad \text{where } -\frac{\pi}{2} \le \theta \le \frac{\pi}{2}.$$

In each stable sector $S^s_{kj} \cap D_{r_0}$, g is given, in the z-coordinate system, by the time-1 map of the vector field $(\Phi_{kj}^{-1})_* v_{\Psi_p}$.

The function H defined by

$$H(s_1, s_2) = (\log \lambda) s_1 s_2$$

is Hamiltonian for the vector field v with respect to the volume element $ds_1\, ds_2$. Hence, H is also Hamiltonian for v_{Ψ_p} with respect to $ds_1\, ds_2 / \Psi_p(s_1^2 + s_2^2)$. Then $H_1 = H \circ \Phi_{kj}$ is Hamiltonian for $(\Phi_{kj}^{-1})_* v_{\Psi_p}$ with respect to $\Phi_{kj}^*(ds_1\, ds_2/\Psi_p(s_1^2 + s_2^2))$. We will show that in $D_{\tilde{r}_1}$, $\Phi_{kj}^*(ds_1\, ds_2/\Psi_p(s_1^2 + s_2^2))$ is the Lebesgue measure $\rho\, d\rho\, d\tau = dt_1\, dt_2$ and H_1 is real analytic in the (t_1, t_2)-coordinate system. It then follows that g is real analytic in $\varphi_k^{-1}(D_{r_1})$.

We first compute $\Phi_{kj}^*(ds_1\, ds_2/\Psi_p(s_1^2 + s_2^2))$. We have

$$d\rho = \left(\frac{p}{2}\right)^{(2-p)/p} r^{(2-p)/p}\, dr \quad \text{and} \quad d\tau = \frac{2}{p} d\theta.$$

Thus

$$\left(\frac{p}{2}\right)^{(p-2)/p} r^{(2p-2)/p}\, d\rho = r\, dr \quad \text{and} \quad \frac{p}{2} d\tau = d\theta,$$

and hence

$$\Phi_{kj}^*\left(\frac{ds_1\, ds_2}{\Psi_p(s_1^2 + s_2^2)}\right) = \Phi_{kj}^*\left(\frac{r\, dr\, d\theta}{\Psi_p(r^2)}\right) = \left(\frac{p}{2}\right)^{2/p} r^{2/p}\, d\rho\, d\tau = \rho\, d\rho\, d\tau.$$

Next we compute H_1. In polar coordinates, H is given by

$$H(r\, e^{i\theta}) = (\log \lambda)(r \cos\theta)(r \sin\theta) = \log \lambda \frac{r^2}{2} \sin(2\theta).$$

Thus we have

$$\begin{aligned} H_1(t_1, t_2) = H_1(\rho\, e^{i\tau}) &= \log \lambda \frac{(2\rho^{p/2}/p)^2}{2} \sin[2(\rho\tau/2)] \\ &= \frac{2 \log \lambda}{p^2} \rho^p \sin(p\tau) = \frac{2 \log \lambda}{p^2} \operatorname{Im}(t_1 + it_2)^p, \end{aligned}$$

which is a polynomial in t_1 and t_2, and hence real analytic.

Step III: Cone families and Lyapunov function. In each stable sector S^s_{kj} of a singular point x_k, we use the coordinates (s_1, s_2), with $w = \Phi^s_{kj}(z) = s_1 + is_2$ as above, and in each coordinate chart (φ_k, U_k), $k = m+1, \dots, \ell$, we use the coordinates $z = s_1 + is_2$. In both cases, we have $\nu^s = |ds_1|$ and $\nu^u = |ds_2|$. We define a Riemannian metric $\langle \cdot, \cdot \rangle$ on $M \setminus \{x_1, \dots, x_m\}$ by the pullback of $(ds_1^2 + ds_2^2)/\Psi_p(s_1^2 + s_2^2)$ under $(\Phi^s_{kj} \circ \varphi_k)^*$ in each stable sector and by the pullback of $ds_1^2 + ds_2^2$ under φ_k^* in U_k, for $k = m+1, \dots, \ell$. From the calculation at the end of Step II, we see that $\langle \cdot, \cdot \rangle$ agrees with the Euclidean metric $dt_1^2 + dt_2^2$ in $\mathfrak{U}_0$. Therefore, $\langle \cdot, \cdot \rangle$ extends to a smooth Riemannian metric on M. For $x \in M$, $dg \colon T_x M \to T_{g(x)} M$ is volume preserving in the metric $\langle \cdot, \cdot \rangle$. We define coordinates (ξ_1, ξ_2) in each tangent space $T_x M$ for $x \in M \setminus \{x_1, \dots, x_m\}$ to be the coordinates with respect to

$$(\Phi^s_{kj} \circ \varphi_k)_*^{-1} (\partial/\partial s_i) (\Psi_p(s_1^2 + s_2^2))^{1/2}, \quad i = 1, 2$$

in each stable sector, and with respect to $(\varphi_k)_*^{-1}(\partial/\partial s_i)$, $i = 1, 2$ in U_k, for $k = m+1, \dots, \ell$. In this coordinate system, $\|(\xi_1, \xi_2)\| = \sqrt{\xi_1^2 + \xi_2^2}$, where $\|\cdot\|$ is the norm associated to $\langle \cdot, \cdot \rangle$. For $x \in M \setminus \{x_1, \dots, x_m\}$, let C_x^+ (respectively C_x^-) be the cone in $T_x M$ that is bounded by the lines $\xi_1 = \pm \xi_2$ and contains the tangent line to the $\mathcal{F}^u$ (respectively $\mathcal{F}^s$) leaf. Since the choice of coordinates (ξ_1, ξ_2) is consistent on chart overlaps except possibly for a change in sign in one or both coordinates, the cones C_x^+ and C_x^- are consistently defined.

We also define another coordinate system (η_1, η_2) in $T_x M$ for $x \in M \setminus \{x_1, \dots, x_m\}$ by $\eta_1 = (\xi_1 + \xi_2)/\sqrt{2}$ and $\eta_2 = (\xi_1 - \xi_2)/\sqrt{2}$. We define a Lyapunov function Q_x on $T_x M$ for $x \in M \setminus \{x_1, \dots, x_m\}$ by

$$Q_x(\eta_1, \eta_2) = \operatorname{sign}(\eta_1 \eta_2) \sqrt{|\eta_1 \eta_2|}.$$

Then

$$C_x^+ = \{v \in T_x M : Q_x(v) > 0\}.$$

Note that Q_x is consistently defined on chart overlaps, even though η_1 and η_2 might not be consistently defined. If $v \in T_x M$ is represented as (η_1, η_2), then

$$|Q_x(v)| = \sqrt{|\eta_1 \eta_2|} \le \sqrt{\eta_1^2 + \eta_2^2} = \|v\|. \tag{6.27}$$

The following lemma shows that in the present construction, the cone families are strictly invariant and have intersections that are lines *everywhere*, not only almost everywhere on $x \in M \setminus \{x_1, \dots, x_m\}$.

Lemma 6.4.6. *The cones C_x^+ and C_x^-, $x \in M \setminus \{x_1, \dots, x_m\}$, satisfy the following:*

1. *C_x^+ and C_x^- depend continuously on $x \in M \setminus \{x_1, \dots, x_m\}$.*
2. *C_x^+ (respectively C_x^-) is strictly invariant under dg (respectively dg^{-1}) on $x \in M \setminus \{x_1, \dots, x_m\}$.*
3. *For each $x \in M \setminus \{x_1, \dots, x_m\}$, the intersections*
$$E_x^u := \bigcap_{n \geq 0} dg^n C^+_{g^{-n}(x)} \quad \text{and} \quad E_x^s := \bigcap_{n \leq 0} dg^n C^-_{g^{-n}(x)}$$
are one-dimensional subspaces of $T_x M$; moreover, if $x \in M \setminus \{x_1, \dots, x_m\}$ and x is on an unstable prong, then E_x^u is tangent to the unstable prong (and similarly for x on a stable prong).
4. *E_x^u and E_x^s depend continuously on $x \in M \setminus \{x_1, \dots, x_m\}$.*

Proof of the lemma. Statement 1 follows immediately from the definitions of C_x^+ and C_x^-.

We will prove Statements 2 and 3 for C_x^+. The argument for C_x^- is similar. Let $\mathcal{U}_0$ be as in (6.26). If $x \notin \mathcal{U}_0$, then vectors in the ξ_1 direction expand by a factor of λ and vectors in the ξ_2 direction contract by a factor of $1/\lambda$. Consequently, dgC_x^+ is strictly inside C_{gx}^+ for $x \notin \mathcal{U}_0$. For each x in a stable sector S_{kj}^s, let $(\tilde{\xi}_1, \tilde{\xi}_2)$ be coordinates in $T_x M$ with respect to $(\Phi_{kj}^s \varphi_k)^{-1}(\partial/\partial s_i)$, $i = 1, 2$. As in [100], we consider the linear part of the vector field v_{Ψ_p} to obtain the differential equations

$$d\tilde{\xi}_1/dt = (\log \lambda)(\tilde{\xi}_1(\Psi_p + 2s_1^2 \Psi_p') + 2s_1 s_2 \tilde{\xi}_2 \Psi_p'),$$

$$d\tilde{\xi}_2/dt = -(\log \lambda)(2s_1 s_2 \tilde{\xi}_1 \Psi_p' + \tilde{\xi}_2(\Psi_p + 2s_2^2 \Psi_p')).$$

The functions Ψ_p and Ψ_p' are evaluated at $s_1^2 + s_2^2$ as before. The equation for the tangent $\alpha = \xi_2/\xi_1 = \tilde{\xi}_2/\tilde{\xi}_1$ is

$$d\alpha/dt = -2(\log \lambda)[\alpha(\Psi_p + (s_1^2 + s_2^2)\Psi_p') + s_1 s_2(\alpha^2 + 1)\Psi_p']. \quad (6.28)$$

Substituting $\alpha = 1$ and $\alpha = -1$ in this equation gives

$$d\alpha/dt = -2(\log \lambda)(\Psi_p + (s_1 + s_2)^2 \Psi_p') \leq 0$$

and

$$d\alpha/dt = 2(\log \lambda)(\Psi_p + (s_1 - s_2)^2 \Psi_p') \geq 0.$$

These inequalities are strict everywhere except at the origin. Therefore Statement 2 also holds for $x \in \mathcal{U}_0 \setminus \{x_1, \dots, x_m\}$.

To prove Statement 3, we first consider the case when $x \in M \setminus \{x_1, \dots, x_m\}$ lies on an unstable prong. Let x be a particular point $(s_1(0), 0)$ along the s_1-axis with $s_1(0) > 0$. We consider the flow along the s_1-axis determined by the vector field v_{Ψ_p} for the time interval $-T \leq t \leq 0$, for some large positive T with initial point $(s_1(0), 0)$ at $t = 0$. Substituting $s_2 = 0$ in (6.28), we obtain

$$d\alpha/dt = -2(\log \lambda)(\alpha(\Psi_p + s_1^2 \Psi_p')).$$

Orbits for v_{Ψ_p} spend at least $1/\Psi_p(s_1^2)$ units of time going from $(s_1/\lambda, 0)$ to $(s_1, 0)$. During this time,

$$d\alpha/dt \leq -2(\log\lambda)\alpha\Psi_p(s_1^2/\lambda^2) \text{ for } \alpha > 0,$$

and

$$d\alpha/dt \geq -2(\log\lambda)\alpha\Psi_p(s_1^2/\lambda^2) \text{ for } \alpha < 0.$$

Therefore if $|\alpha| \geq \alpha_0$ during this time, then $|\alpha|$ decreases by at least a constant times α_0.

It follows that if $\alpha(T) = \pm 1$ for sufficiently large positive T, then $\alpha(0)$ is arbitrarily close to 0. Hence Statement 3 holds for $x \in M \setminus \{x_1, \ldots, x_m\}$ on an unstable prong. If there is a positive integer n_0 such that $g^{-n_0}x$ is on an unstable prong, then Statement 3 holds for x replaced by $g^{-n_0}(x)$. But by the invariance of the cones, it follows that the statement also holds for x.

To complete the proof of Statement 3, it remains to consider the case when none of the points $x, g^{-1}(x), g^{-2}(x), \ldots$ lie on an unstable prong. Let $\mathcal{V}_0$ be a neighborhood of $\{x_1, \ldots, x_m\}$ such that $g\mathcal{V}_0 \subset \mathcal{U}_0$ and $g^{-1}\mathcal{V}_0 \subset \mathcal{U}_0$. For each y in $\mathcal{V}_0$ that does not lie on an unstable prong, there is a positive integer L such that $g^{-L}(y) \notin \mathcal{V}_0$. Therefore infinitely many of the points $x, g^{-1}(x), g^{-2}(x), \ldots$ lie in $M \setminus \mathcal{V}_0$. It follows from Statement 2 and Theorem 4.2.1 that $Q_{g(x)}(dg(v)) > Q_x(v)$ for $v \in T_xM$ and $x \in M \setminus \{x_1, \ldots, x_m\}$; that is, Q_x is a strict Lyapunov function. Then

$$B_x := \inf\{Q_{g(x)}(dg(v))/Q_x(v) : v \in \overline{C_x^+} \setminus \{0\}\} > 1$$

for $x \in M \setminus \{x_1, \ldots, x_m\}$. Moreover, since B_x depends continuously on $x \in M \setminus \{x_1, \ldots, x_m\}$, we have

$$B := \inf\{B_x : x \in M \setminus \mathcal{V}_0\} > 1.$$

We also have

$$B_0 := \inf\{Q_x(dg(v)) : v \in \overline{C_{g^{-1}(x)}^+}, \|v\| = 1, x \in M \setminus \mathcal{V}_0\} > 0.$$

Suppose that $0 \leq n_1 < n_2 < \cdots < n_K = N$ are such that $g^{-n_i}(x) \in M \setminus \mathcal{V}_0$ for $i = 1, \ldots, K$. Then for $v \in \overline{C_{g^{-N}(x)}^+}$ with $\|v\| = 1$, we have

$$\|dg^N(v)\| \geq Q_x(dg^N(v)) \geq B^{K-2}B_0,$$

which goes to infinity as K goes to infinity. If we let $v_{1,N}$ and $v_{2,N}$ be unit vectors for $\|\cdot\|$ along the rays $\xi_1 = \pm\xi_2$, $\xi_1 > 0$ in $T_{g^{-N}x}M$, then the area (in $\langle\cdot,\cdot\rangle$) of the span of $dg^N v_{1,N}$ and $dg^N v_{2,N}$ is equal to 1 (since dg preserves area in $\langle\cdot,\cdot\rangle$), but the lengths of the vectors $dg^N v_{1,N}$ and $dg^N v_{2,N}$ go to infinity as K goes to infinity. This proves that the angle between $dg^N v_{1,N}$ and $dg^N v_{2,N}$ goes to zero as K goes to infinity, which proves Statement 3.

Statement 4 follows immediately from Statements 1 and 3. □

For $x \in \{x_1, \dots, x_m\}$, we define E^u_x (respectively E^s_x) $\subset T_xM$ to be the set of rays tangent to the unstable (respectively stable) prongs.

Step IV: Stable curves and equivalence of length measurements. We say that a curve $\gamma(t)$, $a \le t \le b$, contained in M is a *stable curve* if γ is C^1 and

1. $\gamma(t) \in M \setminus \{x_1, \dots, x_m\}$ for $a < t < b$;
2. $\gamma'(t) \in E^s_{\gamma(t)}$ if $\gamma(t) \in M \setminus \{x_1, \dots, x_m\}$;
3. $\gamma'(a) \in E^s_{\gamma(a)}$ if $\gamma(a) \in \{x_1, \dots, x_m\}$;
4. $-\gamma'(b) \in E^s_{\gamma(b)}$ if $\gamma(b) \in \{x_1, \dots, x_m\}$.

Unstable curves are defined analogously, with tangent vectors in the E^u directions. For stable and unstable curves, it is convenient to measure "length" with respect to the Lyapunov function Q_x. In fact, for any C^1 curve $\gamma(t)$, $a \le t \le b$, with $\gamma(t) \in M \setminus \{x_1, \dots, x_m\}$ for $a < t < b$, we define the *Q-length*

$$\mathcal{L}_Q(\gamma) = \int_a^b |Q_{\gamma(t)}(\gamma'(t))|\, dt.$$

Even though Q_x is not defined for $x \in \{x_1, \dots, x_m\}$, the integral exists, because the integrand is continuous and bounded for $a < t < b$. We denote the usual length by $\mathcal{L}(\gamma) = \int_a^b \|\gamma'(t)\|\, dt$. The Q-length, like the usual length, is independent of the parametrization of the curve, because Q_x is homogeneous of degree 1.

We define two length measurements $\mathcal{L}_1$ and $\mathcal{L}_2$ to be *equivalent* for a set of curves $\mathcal{C}$ if for every $\varepsilon > 0$, there exists $\delta > 0$ such that $\mathcal{L}_1(\gamma) < \delta$ for $\gamma \in \mathcal{C}$ implies $\mathcal{L}_2(\gamma) < \varepsilon$ and $\mathcal{L}_2(\gamma) < \delta$ for $\gamma \in \mathcal{C}$ implies $\mathcal{L}_1(\gamma) < \varepsilon$.

Lemma 6.4.7. *If $\mathcal{C}$ is the set of stable curves, then $\mathcal{L}_Q$ and $\mathcal{L}$ are equivalent for $\mathcal{C}$. This is also true if $\mathcal{C}$ is the set of unstable curves.*

Proof of the lemma. We will prove the statement when $\mathcal{C}$ is the set of unstable curves. (The argument for the set of stable curves is similar.)

By (6.27), $|Q_x(v)| \le \|v\|$ for all $x \in M \setminus \{x_1, \dots, x_m\}$ and $v \in T_xM$. Thus, $\mathcal{L}_Q(\gamma) \le \mathcal{L}(\gamma)$ for all C^1 curves $\gamma(t)$, $a \le t \le b$, with $\gamma(t) \in M \setminus \{x_1, \dots, x_m\}$ for $a < t < b$.

Let $\mathcal{U}_0$ be as in (6.26), and let $\mathcal{V}_0 \subset \mathcal{U}_0$ be a neighborhood of the set $\{x_1, \dots, x_m\}$ whose closure is contained in $\mathcal{U}_0$. Due to the strict invariance of the C^+_x cones under dg, there exists $A = A(\mathcal{V}_0)$ such that for $x \in M \setminus \mathcal{V}_0$ and $v \in dgC^+_{g^{-1}(x)}$ (in particular, for $v \in E^u_x$), we have

$$\|v\| \le A|Q_x(v)|.$$

Thus $\mathcal{L}(\gamma) \le A\mathcal{L}_Q(\gamma)$ for any unstable curve γ that lies in $M \setminus \mathcal{V}_0$. It follows that there exists $\tau > 0$ such that any unstable curve of Q-length at most τ that intersects $\mathcal{V}_0$ lies in a single component of $\mathcal{U}_0$.

If an unstable curve γ is contained in a stable sector, then in the (s_1, s_2)-coordinates γ can be parameterized as a graph $\gamma(t) = (t, s_2(t))$, $0 \le a \le t \le b$, where $|s_2'(t)| \le 1$, because $\gamma'(t) \in C^+_{\gamma(t)}$. Let $\widetilde{\gamma}(t)$, $0 \le a \le t \le b$, be the

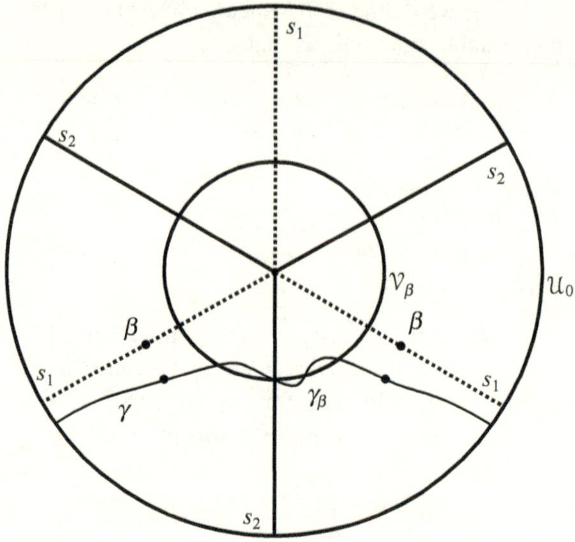

Figure 6.6. To the proof of Lemma 6.4.7.

projection of γ onto the unstable prong; that is, $\widetilde{\gamma}(t) = (t, 0)$. For $\gamma(t) \neq (0, 0)$,

$$\begin{aligned}\|\gamma'(t)\|^2 &= \left\|\left(\frac{\partial}{\partial s_1}\right)_{\gamma(t)}\right\|^2 + \left\|s_2'(t)\left(\frac{\partial}{\partial s_2}\right)_{\gamma(t)}\right\|^2 \\ &\le 2\left\|\left(\frac{\partial}{\partial s_1}\right)_{\gamma(t)}\right\|^2 = \frac{2}{\Psi_p(s_1^2 + s_2^2)},\end{aligned}$$

and for $t \neq 0$,

$$\|\widetilde{\gamma}'(t)\|^2 = \left\|\left(\frac{\partial}{\partial s_1}\right)_{\widetilde{\gamma}(t)}\right\|^2 = \frac{1}{\Psi_p(s_1^2)}.$$

Since $\Psi_p(s_1^2 + s_2^2) \ge \Psi_p(s_1^2)$, it follows that

$$\|\gamma'(t)\| \le \sqrt{2}\|\widetilde{\gamma}'(t)\| = \sqrt{2}(\Psi_p(s_1^2))^{-1/2}.$$

Therefore,

$$\begin{aligned}\mathcal{L}(\gamma) &\le \sqrt{2}\mathcal{L}(\widetilde{\gamma}) = \sqrt{2}\int_a^b (\Psi_p(t^2))^{-1/2}\,dt \\ &= c_p \int_a^b t^{-(p-2)/p}\,dt = c_p(b^{2/p} - a^{2/p}) \le c_p b^{2/p},\end{aligned}$$

for some constant c_p.

Let $\varepsilon > 0$, and choose $\beta > 0$ sufficiently small so that $c_p\beta^{2/p} < \varepsilon/4$ for all $p = p(k)$, $k = 1, \ldots, m$. Note that the cone condition, $\gamma'(t) \in C^+_{\gamma(t)}$, implies that any unstable curve lying in a single component of $\mathcal{U}_0$ can intersect the interiors of at most two stable sectors. We may choose a sufficiently small neighborhood $\mathcal{V}_\beta \subset \mathcal{V}_0$ of $\{x_1, \ldots, x_m\}$ so that any unstable curve γ of Q-length at most τ that intersects $\mathcal{V}_\beta$ contains a curve γ_β such that γ_β lies in the $0 \le s_1 \le \beta$ part of at most two stable sectors and $\gamma \setminus \gamma_\beta$ lies in $M \setminus \mathcal{V}_\beta$ (see Fig. 6.6). As before, there is a constant $A(\mathcal{V}_\beta) > 0$ such that

$$\|v\| \le A(\mathcal{V}_\beta)|Q_x(v)|$$

for $x \in M \setminus \mathcal{V}_\beta$ and $v \in E^u_x$. Let $\delta = \min\{\varepsilon, \beta, \tau, \varepsilon/(2A(\mathcal{V}_\beta))\}$. Let γ be an unstable curve with $\mathcal{L}_Q(\gamma) < \delta$. If γ intersects $\mathcal{V}_\beta$, let γ_β be as described above; otherwise let $\gamma_\beta = \varnothing$. Then

$$\begin{aligned}\mathcal{L}(\gamma) &\le \mathcal{L}(\gamma_\beta) + \mathcal{L}(\gamma \setminus \gamma_\beta)\\ &\le 2c_p\beta^{2/p} + (A(\mathcal{V}_\beta))\mathcal{L}_Q(\gamma \setminus \gamma_\beta) < \varepsilon/2 + \varepsilon/2 = \varepsilon.\end{aligned}$$

This completes the proof of the lemma. □

Step V: Uniform contraction of stable curves.

Lemma 6.4.8 (uniform contraction of stable curves). *For every $\tau > 0$ and $\varepsilon > 0$, there exists a positive integer $N = N(\varepsilon, \tau)$ such that if γ is a stable curve with $\mathcal{L}(\gamma) \le \tau$ and $n \ge N$, then $\mathcal{L}(g^n\gamma) < \varepsilon$.*

Proof of the lemma. Clearly, it suffices to prove this for some $\tau > 0$ and every $\varepsilon > 0$. We will choose τ that is convenient for the proof. By the equivalence of $\mathcal{L}_Q$ and $\mathcal{L}$ for stable curves, we may replace $\mathcal{L}$ by $\mathcal{L}_Q$ in the statement of the lemma. The advantage of working with $\mathcal{L}_Q$ is that $\mathcal{L}_Q(g\gamma) < \mathcal{L}_Q(\gamma)$ for every stable curve γ, by the strict invariance of the C^+_x cones.

Let $\mathcal{U}_0$, $\mathcal{V}_0$, and τ be as in the proof of Lemma 6.4.7. Let $\varepsilon > 0$, and choose β and $\mathcal{V}_\beta$ (depending on ε) as in the proof of Lemma 6.4.7. Let $B = B(\mathcal{V}_\beta) \in (0, 1)$ be such that $Q_{g(x)}(dg(v)) \le BQ_x(v)$ for $v \in E^s_x$ and $x \in M \setminus \mathcal{V}_\beta$.

Let γ be a stable curve with $\mathcal{L}_Q(\gamma) \le \tau$. Let $N = N(\tau, \varepsilon)$ be a sufficiently large positive integer so that $(1 - B)(\varepsilon/2)N > \tau$. For any nonnegative integer k, let $\gamma_k = g^k(\gamma)$. As in the proof of Lemma 6.4.7, if γ_k intersects $\mathcal{V}_\beta$, let $\gamma_{k\beta} \subset \gamma_k$ be such that $\mathcal{L}(\gamma_{k\beta}) < \varepsilon/2$ with $\gamma_k \setminus \gamma_{k\beta}$ lying in $M \setminus \mathcal{V}_\beta$. If γ_k does not intersect $\mathcal{V}_\beta$, let $\gamma_{k\beta} = \varnothing$. We claim that $\mathcal{L}_Q(\gamma_N) < \varepsilon$. For $k = 0, 1, \ldots, N-1$, we have

$$\begin{aligned}\mathcal{L}_Q(\gamma_{k+1}) &= \mathcal{L}_Q(g(\gamma_{k\beta})) + \mathcal{L}_Q(g(\gamma_k \setminus \gamma_{k\beta}))\\ &\le \mathcal{L}_Q(\gamma_{k\beta}) + B\mathcal{L}_Q(\gamma_k \setminus \gamma_{k\beta})\\ &= \mathcal{L}_Q(\gamma_k) - (1 - B)\mathcal{L}_Q(\gamma_k \setminus \gamma_{k\beta}).\end{aligned} \tag{6.29}$$

If $\mathcal{L}_Q(\gamma_k \setminus \gamma_{k\beta}) \ge \varepsilon/2$ for $k = 0, \ldots, N-1$, then (6.29) implies that

$$\mathcal{L}_Q(\gamma_N) \le \tau - (1 - B)(\varepsilon/2)N < 0,$$

which is impossible. Therefore, for some $k \in \{0, \ldots, N-1\}$, we have $\mathcal{L}_Q(\gamma_k \setminus \gamma_{k\beta}) < \varepsilon/2$. For this value of k,

$$\mathcal{L}_Q(\gamma_k) = \mathcal{L}_Q(\gamma_{k\beta}) + \mathcal{L}_Q(\gamma_k \setminus \gamma_{k\beta}) < \varepsilon,$$

and thus $\mathcal{L}_Q(\gamma_n) < \varepsilon$ for all $n \ge k$ and, in particular, for $n \ge N$. □

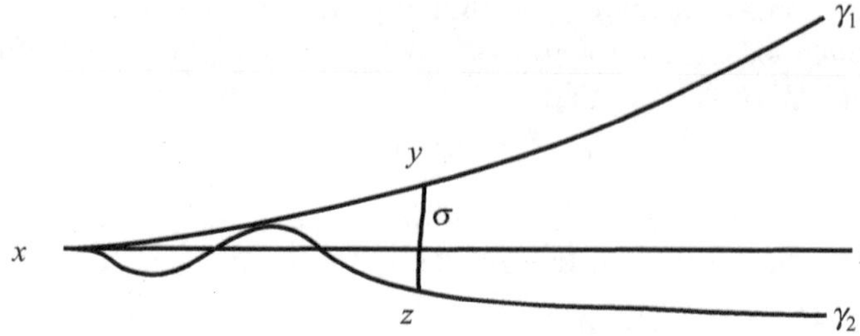

Figure 6.7. To the proof of Lemma 6.4.10.

Lemma 6.4.9 (uniform expansion of unstable curves). *For every* $\varepsilon > 0$ *and* $\tau > 0$, *there exists a positive integer* $N(\varepsilon, \tau)$ *such that for every unstable curve* γ *with* $\mathfrak{L}(\gamma) \geq \varepsilon$ *and every* $n \geq N(\varepsilon, \tau)$, *we have* $\mathfrak{L}(g^n\gamma) > \tau$.

Proof. Let $\varepsilon > 0$ and $\tau > 0$. Note that unstable curves for g are stable curves for g^{-1}. Therefore by Lemma 6.4.8, there exists a positive integer $N(\varepsilon, \tau)$ such that any unstable curve σ for g with $\mathfrak{L}(\sigma) \leq \tau$ satisfies $\mathfrak{L}(g^{-n}\sigma) < \varepsilon$ for $n \geq N(\varepsilon, \tau)$. We apply the contrapositive of this statement to $\sigma = g^n\gamma$ to conclude that $\mathfrak{L}(\gamma) \geq \varepsilon$ for an unstable curve γ implies that $\mathfrak{L}(g^n\gamma) > \tau$ for $n \geq N(\varepsilon, \tau)$. □

Step VI: Uniqueness of stable and unstable curves.

Lemma 6.4.10. *Let* $x \in M$ *and let* L *be one of the rays of* E^s_x *starting at* x. *(There are two such rays if* $x \notin \{x_1, \ldots, x_m\}$ *and* $p(k)$ *such rays if* $x = x_k$, *for some* $k = 1, \ldots, m$*). Let* γ_1 *and* γ_2 *be two stable curves that begin at* x *and are tangent to* L. *Then one of the curves is a part of the other. An analogous result holds for unstable curves.*

Proof of the lemma. Suppose that the lemma were not true. Then by replacing, if necessary, x by another point on $\gamma_1 \cap \gamma_2$, we can keep all the assumptions and also assume that arbitrarily close to x there are points belonging to $\gamma_1 \setminus \gamma_2$ and to $\gamma_2 \setminus \gamma_1$.

Note that the tangents to γ_1 and γ_2 are nearly parallel to L close to x, while the angle between E^u_w and L is bounded away from zero for w close to x and between γ_1 and γ_2. Thus if $y \in \gamma_1 \setminus \gamma_2$ is chosen sufficiently close to x, there is a short unstable curve σ that passes through y and intersects γ_2 at some point z (see Fig. 6.7).

Since $\|d_w g\|$ is bounded for $w \in M$, there exists $J > 0$ such that for any C^1 curve γ,

$$\mathfrak{L}(g\gamma) \leq J\mathfrak{L}(\gamma). \tag{6.30}$$

There exist $\tau > 0$ and $\varepsilon > 0$ such that for any unstable curve with length between τ and $J\tau$, the distance in the Riemannian metric $\langle \cdot, \cdot \rangle$ between the end points is greater than ε. By Lemma 6.4.8 and (6.30), there exists $\delta > 0$ such that $\mathfrak{L}(\gamma) < \delta$ for a stable curve γ implies that $\mathfrak{L}(g^n\gamma) < \varepsilon/2$ for all $n \geq 0$. We take y sufficiently close to x so that the segment $\widetilde{\gamma}_1$ of γ_1 from x to y and the segment $\widetilde{\gamma}_2$ of γ_2 from x to

z satisfy $\mathcal{L}(\widetilde{\gamma}_i) < \delta$, for $i = 1, 2$, and $\mathcal{L}(\sigma) < \tau$. Then, for some positive integer n, $\tau \le \mathcal{L}(g^n\sigma) \le J\tau$ that implies that $d(g^n y, g^n z) > \varepsilon$. But $\mathcal{L}(g^n\widetilde{\gamma}_i) < \varepsilon/2$, $i = 1, 2$ implies that $d(g^n x, g^n y) < \varepsilon/2$ and $d(g^n x, g^n z) < \varepsilon/2$, which contradicts the triangle inequality. □

Step VII: Markov partitions and topological conjugacy between f and g. By the construction in [69], there is a Markov partition $\{R_1, \dots, R_N\}$ for the pseudo-Anosov map f. We will use this Markov partition to prove that f and g are topologically conjugate.

Let $A = (a_{ij})_{1\le i,j\le N}$ be the transition matrix for $\{R_1, \dots, R_N\}$; that is,

$$a_{ij} = \begin{cases} 1 & \text{if } \operatorname{int} f(R_i) \cap \operatorname{int} R_j \ne \varnothing \\ 0 & \text{otherwise} \end{cases}.$$

The union of the boundaries of $R_1, \dots, R_N$ is the union of segments L^u_{kj} and L^s_{kj} of singular leaves extending the unstable and stable prongs, P^u_{kj} and P^s_{kj}, respectively. Let

$$L^u = \bigcup_{k=1}^{m} \bigcup_{j=0}^{p(k)-1} L^u_{kj} \quad \text{and} \quad L^s = \bigcup_{k=1}^{m} \bigcup_{j=0}^{p(k)-1} L^s_{kj}.$$

By Statement 3 of Lemma 6.4.6, the unstable and stable prongs P^u_{kj} and P^s_{kj} are also unstable and stable curves, respectively, for g. If the radius r_0 of the slowing-down region is chosen sufficiently small so that

$$fL^u \cap \varphi_k^{-1}(D_{r_0}) \subset \bigcup_{j=0}^{p(k)-1} P^u_{kj} \quad \text{and} \quad f^{-1}L^s \cap \varphi_k^{-1}(D_{r_0}) \subset \bigcup_{j=0}^{p(k)-1} P^s_{kj}$$

for $k = 1, \dots, m$, then L^u_{kj} and L^s_{kj} are segments of unstable and stable curves for g as well. Moreover, $g(R_i) = f(R_i)$ for $i = 1, \dots, N$ and the boundaries of the sets R_i form the boundaries of a Markov partition for g with the same transition matrix. Each R_i is foliated by the leaves of $\mathfrak{F}^u$ and $\mathfrak{F}^s$ and is also foliated (differently) by unstable and stable curves for g. We let $\partial^u R_i$ (respectively $\partial^s R_i$) denote the two opposite boundaries of R_i, consisting of unstable (respectively stable) curves. A *maximal unstable curve* in R_i is defined to be an unstable curve (for g) that is contained in R_i and goes from one boundary in $\partial^s R_i$ to the other. *Maximal stable curves* are defined similarly. We call a sequence $(\zeta_i)_{i\in I}$ *admissible* if I is a (finite or infinite) set of consecutive integers and $a_{\zeta_i\zeta_{i+1}} = 1$ for all i such that $i, i+1 \in I$. For an admissible sequence $(\zeta_i)_{i\le 0}$, it follows from Lemma 6.4.8 that $\bigcap_{i=-\infty}^{0} g^{-i} R_{\zeta_i}$ is a maximal unstable curve in R_{ζ_0} (see Section 7 of [75] for further details). Similarly, for any admissible sequence $(\zeta_i)_{i\ge 0}$, $\bigcap_{i=0}^{\infty} g^{-i} R_{\zeta_i}$ is a maximal stable curve in R_{ζ_0}. A homeomorphism $h\colon M \to M$ that satisfies $h \circ f = g \circ h$ is defined as follows: if

$$x = \bigcap_{i=-\infty}^{\infty} f^{-i} R_{\zeta_i} \tag{6.31}$$

for an admissible sequence $(\zeta_i)_{i\in\mathbb{Z}}$, then

$$h(x) = \bigcap_{i=-\infty}^{\infty} g^{-i} R_{\zeta_i}. \tag{6.32}$$

If a point x is represented in the form (6.31) for more than one admissible sequence $(\zeta_i)_{i\in\mathbb{Z}}$, then x lies on the boundary of some $f^{-i} R_{\zeta_i}$. In this case $h(x)$ lies on the boundary of the corresponding $g^{-i} R_{\zeta_i}$ and $h(x)$ is represented in the form (6.32) for the same admissible sequences as x is represented in (6.31) (see Section 7 of [75] for details of the proof that h is a homeomorphism). □

6.5 Nonuniformly Hyperbolic Flows

Let φ_t be an Anosov flow on a compact three-dimensional manifold M, which is defined by a vector field X, preserves the volume m, and is ergodic (e.g., φ_t may be the geodesic flow on a compact surface of negative curvature). We will show how to modify the flow φ_t in order to obtain a new flow ψ_t with the following properties:

1. ψ_t is nonuniformly hyperbolic.
2. ψ_t is not uniformly hyperbolic (i.e., it is not an Anosov flow).
3. ψ_t preserves volume and is ergodic.

Our construction is based upon an adaptation to the continuous-time case of the procedure of slowing-down trajectories described in the beginning of Section 6.3.1.

Fix a point $p_0 \in M$. One can introduce a coordinate system (x, y, z) in the ball $B(p_0, d)$ (for some $d > 0$) such that p_0 is the origin (i.e., $p_0 = 0$) and $X = \partial/\partial z$.

For each $\varepsilon > 0$, let $T_\varepsilon = S^1 \times D_\varepsilon \subset B(0, d)$ be the solid torus obtained by rotating the disk

$$D_\varepsilon = \{(x, y, z) \in B(0, d) : x = 0 \text{ and } (y - d/2)^2 + z^2 \le (\varepsilon d)^2\}$$

around the z-axis. Every point on the solid torus can be represented as a pair (θ, y, z) with $\theta \in S^1$ and $(y, z) \in D_\varepsilon$.

For every $0 \le \alpha \le 2\pi$, we consider the cross section of the solid torus $\Pi_\alpha = \{(\theta, y, z) : \theta = \alpha\}$ and construct a new vector field $\widetilde{X}$ on $M \setminus T_\varepsilon$ (see Fig. 6.8).

Lemma 6.5.1. *There exists a smooth vector field $\widetilde{X}$ on $M \setminus T_\varepsilon$ such that the flow $\widetilde{\varphi}_t$ generated by $\widetilde{X}$ satisfies the following properties:*

1. $\widetilde{X}|(M \setminus T_{2\varepsilon}) = X|(M \setminus T_{2\varepsilon})$.
2. *For any $0 \le \alpha, \beta \le 2\pi$, the vector field $\widetilde{X}|\Pi_\beta$ is the image of the vector field $\widetilde{X}|\Pi_\alpha$ under the rotation around the z-axis that moves Π_α onto Π_β.*
3. *For every $0 \le \alpha \le 2\pi$, the unique two fixed points of the flow $\widetilde{\varphi}_t|\Pi_\alpha$ are those in the intersection of Π_α with the hyperplanes $z = \pm\varepsilon d$.*
4. *For every $0 \le \alpha \le 2\pi$ and $(y, z) \in D_{2\varepsilon} \setminus \operatorname{int} D_\varepsilon$, the trajectory of the flow $\widetilde{\varphi}_t|\Pi_\alpha$ passing through the point (y, z) is invariant under the symmetry $(\alpha, y, z) \mapsto (\alpha, y, -z)$.*

5. *The flow* $\widetilde{\varphi}_t|\Pi_\alpha$ *preserves the conditional measure induced by volume on the set* Π_α.

Proof. We shall only describe the construction of $\widetilde{X}$ on the cross section Π_0. The vector field $\widetilde{X}|\Pi_\alpha$ on an arbitrary cross section is obtained from the image of $\widetilde{X}|\Pi_0$ under the rotation around the z-axis that moves Π_0 onto Π_α.

Consider the Hamiltonian $H(y,z) = y(\varepsilon^2 - y^2 - z^2)$. It is straightforward to verify that in the annulus $\varepsilon^2 \leq y^2 + z^2 \leq 4\varepsilon^2$, the corresponding Hamiltonian flow is topologically conjugated to that shown on Figure 6.8. However, the Hamiltonian vector field $(-2yz, 3y^2 + z^2 - \varepsilon^2)$ is not everywhere vertical on the circle $y^2 + z^2 = 4\varepsilon^2$. To correct this, we consider a C^∞ function $p\colon [\varepsilon, \infty) \to [0,1]$ such that $p|[\varepsilon, 3\varepsilon/2] = 1$, $p|[2\varepsilon, \infty) = 0$, and p is strictly decreasing in $(3\varepsilon/2, 2\varepsilon)$. The flow defined by the system of differential equations

$$\begin{cases} y' = -2yzp(\sqrt{y^2+z^2}) \\ z' = (3y^2 + z^2 - \varepsilon^2)p(\sqrt{y^2+z^2}) + 1 - p(\sqrt{y^2+z^2}) \end{cases}$$

has the behavior demonstrated in Figure 6.8. We denote by $\overline{\varphi}_t$ and $\overline{X}$ the corresponding flow and vector field, in coordinates x, y, and z.

We now show that it is possible to effect a time change in the flow $\overline{\varphi}_t$ in the annulus $\varepsilon^2 \leq y^2 + z^2 \leq 4\varepsilon^2$ so that the new flow $\widetilde{\varphi}_t$ preserves volume. We want to determine a C^1 function $\tau\colon (T_{2\varepsilon} \setminus \operatorname{int} T_\varepsilon) \times \mathbb{R} \to \mathbb{R}$ such that $\widetilde{\varphi}_t(x) = \overline{\varphi}_{\tau(x,t)}(x)$. It is easy to see that $\widetilde{X}(x) = h(x)\overline{X}(x)$, where $h(x) = \partial\tau/\partial t(x)$. The flow $\widetilde{\varphi}_t$ preserves volume if and only if

$$\operatorname{div}(h\overline{X}) = \nabla h \cdot \overline{X} + h \operatorname{div} \overline{X} = 0. \tag{6.33}$$

Since the two semicircles defined by $y^2 + z^2 = 4\varepsilon^2$, $y = \pm 2\varepsilon$, are not characteristics, there exists a unique solution h of (6.33) such that $h = 1$ outside $\Pi_0 \cap T_{2\varepsilon}$. This completes the construction of the desired vector field $\widetilde{X}$. □

One can see that the orbits of the flows φ_t and $\widetilde{\varphi}_t$ coincide on $M \setminus T_{2\varepsilon}$, that the flow $\widetilde{\varphi}_t$ preserves volume, and that the fixed points of this flow are only those on the circles $\{(\theta, y, z) : z = -\varepsilon\, d\}$ and $\{(\theta, y, z) : z = \varepsilon\, d\}$.

On the set $T_{2\varepsilon} \setminus \operatorname{int} T_\varepsilon$, we introduce natural coordinates θ_1, θ_2, and r with $0 \leq \theta_1, \theta_2 < 2\pi$, and $\varepsilon d \leq r \leq 2\varepsilon d$ such that the set of fixed points of $\widetilde{\varphi}_t$ is composed of those for which $r = \varepsilon\, d$ and $\theta_1 = 0$ or $\theta_1 = \pi$.

Consider the flow on $T_{2\varepsilon} \setminus \operatorname{int} T_\varepsilon$ defined by

$$(\theta_1, \theta_2, r, t) \mapsto (\theta_1, \theta_2 + [2 - r/(\varepsilon\, d)]^4 t \cos\theta_1, r),$$

and let $\hat{X}$ be the corresponding vector field. Consider now the flow ψ_t on $M \setminus \operatorname{int} T_\varepsilon$ generated by the vector field Y on $M \setminus \operatorname{int} T_\varepsilon$ defined by

$$Y(x) = \begin{cases} X(x), & x \in M \setminus \operatorname{int} T_{2\varepsilon} \\ \widetilde{X}(x) + \hat{X}(x), & x \in \operatorname{int} T_{2\varepsilon} \setminus \operatorname{int} T_\varepsilon \end{cases}.$$

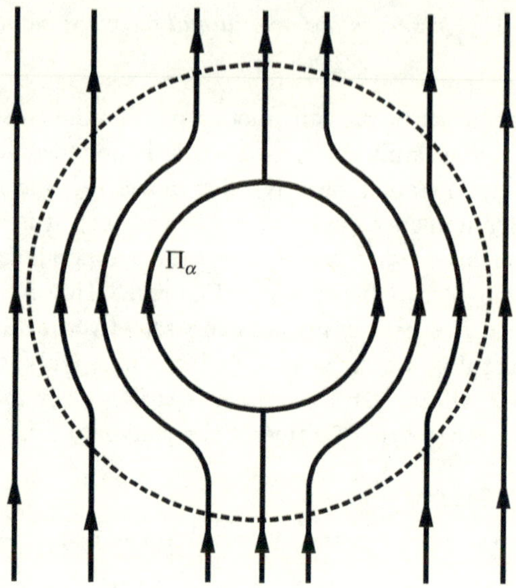

Figure 6.8. A cross section Π_α of the solid torus and the flow $\widetilde{\varphi}_t$.

Proposition 6.5.2. *The following properties hold:*

1. *The flow ψ_t preserves volume and is ergodic.*
2. *The flow ψ_t has no fixed points.*
3. *The flow ψ_t is nonuniformly hyperbolic (but not uniformly hyperbolic). Moreover, for almost every $x \in M \setminus T_{2\varepsilon}$,*

$$\chi(x, v) < 0 \text{ if } v \in E^s(x), \text{ and } \chi(x, v) > 0 \text{ if } v \in E^u(x),$$

where $E^u(x)$ and $E^s(x)$ are respectively stable and unstable subspaces of the flow φ_t at the point x.

Proof. By the construction of ψ_t, it preserves volume. Since the vector fields $\widetilde{X}$ and $\hat{X}$ commute, the flow is ergodic. Therefore the first statement holds. The second statement follows from the construction of the flow ψ_t. In order to prove the third statement, consider the function

$$T(x, t) = \int_0^t I_{T_{2\varepsilon}}(\varphi_\tau x)\, d\tau,$$

where $I_{T_{2\varepsilon}}$ denotes the indicator (the characteristic function) of the set $T_{2\varepsilon}$. By Birkhoff's Ergodic Theorem,

$$\lim_{t \to +\infty} \frac{T(x, t)}{t} = m(T_{2\varepsilon})$$

for almost every $x \in M$.

Fix a point $x \in M \setminus T_{2\varepsilon}$. Consider a moment of time t_1 at which the trajectory $\psi_t(x)$ enters the set $T_{2\varepsilon}$ and the next moment of time t_2 at which this trajectory exits the set $T_{2\varepsilon}$. Given a vector $v \in E^u(x)$ denote by $\widetilde{v}_i$ the orthogonal projection of the vector $d\psi_{t_i} v$ onto the (x, y) plane for $i = 1, 2$. It follows from the construction of the flows $\widetilde{\varphi}_t$ and ψ_t (see Property 4) that $\|\widetilde{v}_1\| \geq \|\widetilde{v}_2\|$. Since the unstable subspaces $E^u(x)$ depend continuously on x, there exists $K \geq 1$ (independent of x, t_1, and t_2) such that

$$\|d\psi_t v\| \geq K \|d\varphi_{t-T(x,t)} v\|.$$

It follows that for almost every $x \in M \setminus T_{2\varepsilon}$ and $v \in E^u(x)$,

$$\chi(x, v) = \overline{\lim_{t\to+\infty}} \frac{1}{t} \log \|d\psi_t v\| \geq (1 - m(T_{2\varepsilon})) \overline{\lim_{t\to+\infty}} \frac{1}{t} \log \|d\varphi_t v\| > 0,$$

provided ε is sufficiently small. Repeating the above argument with respect to the inverse flow ψ_{-t}, one can show that $\chi(x, v) < 0$ for almost every $x \in M \setminus T_{2\varepsilon}$ and $v \in E^s(x)$. □

Set $M_1 = M \setminus T_\varepsilon$ and consider a copy $(\widetilde{M}_1, \widetilde{\psi}_t)$ of the flow (M_1, ψ_t). One can glue the manifolds M_1 and $\widetilde{M}_1$ along their boundaries ∂T_ε and obtain a three-dimensional smooth Riemannian manifold D without boundary. We define a flow F_t on D by

$$F_t x = \begin{cases} \psi_t x, & x \in M_1 \\ \widetilde{\psi}_t x, & x \in \widetilde{M}_1 \end{cases}.$$

It is clear that the flow F_t is nonuniformly hyperbolic and preserves volume.

6.6 Some Other Examples

We shall briefly describe some other examples of dynamical systems with nonzero Lyapunov exponents, without going into details. However, while in the previous sections we considered single dynamical systems, here we deal with one- or two-parameter families of (two-dimensional) maps and we demonstrate a remarkable new phenomenon in the nonuniform hyperbolicity theory: for an appropriate set of parameters of positive Lebesgue measure, the corresponding maps have nonzero Lyapunov exponents. Working with families of systems rather than with a single system changes the setup making it more "practical": in applications the system under consideration often depends on one or more parameters and its behavior may change drastically when the parameters vary. This results in the fact that the set of "good" parameters, for which the corresponding system is nonuniformly hyperbolic, is a Cantor-like set. To determine this set one creates a collection of conditions and then selects parameters so that the corresponding system meets these requirements. Setting up these conditions is a delicate task – on the one hand, they should be sufficiently strong to guarantee that the corresponding system is nonuniformly hyperbolic and on the other hand, they should be sufficiently weak to ensure that the selection procedure does not essentially exhaust all parameters; that is, the remaining Cantor-like set has positive Lebesgue measure.

The selection procedure was first suggested and successfully used by Jakobson in his seminal paper [93] to construct absolutely continuous invariant measures for some families of one-dimensional maps. Since then it has become a powerful tool in studying ergodic properties of one-dimensional maps (see [94]). In the two-dimensional case the above approach has been successfully carried out for very few examples, most remarkably for Henón-like attractors (see below).

1. Maps on the boundary of Anosov diffeomorphisms (see [130, 183]). Given a real number $\alpha \geq -1$, consider the map $f_\alpha \colon \mathbb{T}^2 \to \mathbb{T}^2$ defined by

$$f_\alpha(x, y) = (x + y + h_\alpha(x) \bmod 2\pi, y + h_\alpha(x) \bmod 2\pi),$$

where $h_\alpha(x) = x - (1 + \alpha)\sin x$. Note that the Jacobian of f_α is identically equal to 1 and hence the map preserves area.

For $\alpha = -1$, the map f_α is the well-known linear Anosov automorphism of the torus $\left(\begin{smallmatrix} 2 & 1 \\ 1 & 1 \end{smallmatrix}\right)$, while for α large the map becomes increasingly similar to the classical standard map whose behavior is very hard to describe. Indeed, the change of variables $q = x - y$, $p = y$ reduces f_α to a map in the family of standard maps (see [188, Section 8.5])

$$g_\alpha(q, p) = (q + p \bmod 2\pi, p + h_\alpha(q + p) \bmod 2\pi).$$

Observe that zero is a fixed point for all α; it is hyperbolic for $\alpha < 0$, parabolic for $\alpha = 0$, and elliptic for $\alpha > 0$. Furthermore, for $\alpha \in (-1, 0)$ the map f_α is an Anosov diffeomorphism, while for $\alpha = 0$ it has nonzero Lyapunov exponents almost everywhere with respect to area. To see this, it suffices to notice that the invariant cone family

$$C = \{(u, v) \in \mathbb{R}^2 : uv \geq 0\}$$

is complete and eventually strict, and use Theorem 4.1.6. In particular, the map f_0 lies on the boundary of Anosov maps.

2. The Viana map. We describe an example due to Viana [220] of a smooth (noninvertible) map, which has stably nonzero Lyapunov exponents, that is, any of its sufficiently small perturbations has nonzero Lyapunov exponents almost everywhere. Given real numbers α and $d \geq 4$, consider a C^∞ map $f_\alpha \colon S^1 \times \mathbb{R} \to S^1 \times \mathbb{R}$ defined by

$$f_\alpha(\theta, x) = (d\theta \bmod 1, a_0 + \alpha \sin(2\pi\theta)).$$

We choose $a_0 \in (1, 2)$ in such a way that $x = 0$ is preperiodic for $Q(x) = a_0 - x^2$. It is easy to check that for sufficiently small α, there is an interval $I \subset (-2, 2)$ for which $f_\alpha(S^1 \times I) \subset \operatorname{int}(S^1 \times I)$. It is shown in [219] that for every sufficiently small α and every sufficiently large d, there exists a neighborhood $\mathfrak{U} \subset C^3(S^1 \times I)$ of the map f_α and a constant $c > 0$ such that for every map $f \in \mathfrak{U}$,

$$\varliminf_{n \to \infty} \frac{1}{n} \log \|df^n(\theta, x)v\| \geq c$$

for Lebesgue almost every $(\theta, x) \in S^1 \times I$ and every $v \in T_{(\theta,x)}(S^1 \times I)$. This implies that all the Lyapunov exponents of f are positive almost everywhere with respect to the Lebesgue measure on $S^1 \times I$. Note that this measure is not preserved by f, but it is proved in [4] that f possesses an invariant measure, which is absolutely continuous with respect to the Lebesgue measure.

3. The Hénon map. This map was introduced by Hénon in 1976 (see [82]) as a simplified model for the Poincaré first return time map in the Lorenz system of ordinary differential equations. The Hénon family is given by

$$H_{a,b}(x, y) = (1 - ax^2 + by, x). \tag{6.34}$$

Hénon carried out numerical studies of this family and suggested the presence of a "chaotic" attractor for parameter values near $a = 1.4$ and $b = 0.3$. Observe that for $b = 0$, the family $H_{a,b}$ reduces to the logistic family Q_a. By continuity, given $a \in (0, 2)$, there is a rectangle in the plane, which is mapped by $H_{a,b}$ into itself. It follows that $H_{a,b}$ has an attractor, provided b is sufficiently small. It is called the *Hénon attractor*.

In the seminal paper [25], Benedicks and Carleson, treating $H_{a,b}$ as a small perturbation of Q_a, developed highly sophisticated techniques to describe the dynamics near the attractor. In particular, one can show (see Theorem 13.3.9) that for all sufficiently small b, there is a subset $\Delta_b \subset \mathbb{R}$ of positive Lebesgue measure such that for every $a \in \Delta_b$, the map $H_{a,b}$ possesses an invariant measure with respect to which it has one negative and one positive Lyapunov exponents almost everywhere.

6.7 Notes

The characterization of hyperbolic sets given in Theorem 6.1.2 is due to Alekseev. The horseshoe map was introduced by Smale in [208, p. 759]. The construction of the nonuniformly hyperbolic perturbations of horseshoes described in this chapter is due to Katok and Mendoza [107].

The first example of nonuniformly hyperbolic systems – a volume preserving flow with nonzero Lyapunov exponents (see Section 6.5) – was constructed in [166]. The construction of diffeomorphisms with nonzero Lyapunov exponents on any surface, presented in Section 6.3, is due to Katok.

In [74], Gerber proved the existence of real analytic Bernoulli models of pseudo-Anosov maps as an application of a conditional stability result for the smooth models constructed by Gerber and Katok in [75] (see Section 6.4.2). The proofs rely on the use of Markov partitions. The same results in the analytic case were obtained by Lewowicz and Lima de Sá [128] using a different approach.

7

Stable Manifold Theory

In this chapter we present one of the principal results of the nonuniform hyperbolicity theory – the existence of local stable and unstable manifolds. Roughly speaking, we show that if the differential of the system admits a hyperbolic behavior along a given trajectory then the system itself behaves hyperbolically in a sufficiently small neighborhood of the trajectory. When hyperbolicity is uniform, this is an adaptation of the classical stability theory for ordinary differential equations to dynamical systems with discrete and continuous time. In this case a local stable manifold at a given point x consists of all points in a small neighborhood of x whose trajectories stay within a certain small distance from the trajectory of x. One can then show that trajectories indeed converge to the trajectory of x with an exponential rate, that the local stable manifold is smooth, and that its "size" does not depend on x.

When hyperbolicity is nonuniform, some substantial modifications of the classical theory are needed due to the fact that hyperbolicity conditions may deteriorate along the trajectories (see Condition 6 in Definition 5.2.1). In particular, the size of the local stable manifold may get arbitrarily small.

We also construct global stable and unstable manifolds and we study the "foliations" they form.

7.1 The Stable Manifold Theorem

Let f be a $C^{1+\alpha}$ diffeomorphism of a compact smooth Riemannian manifold M and $Y \subset M$ an f-invariant set, which is nonuniformly partially hyperbolic in the broad sense with $\lambda(x)$, $\mu(x)$, $\varepsilon(x) \in (0, \varepsilon_0]$ to be Borel invariant functions and $T_x M = E^s(x) \oplus E^2(x)$ to be an invariant splitting of the tangent space at $x \in Y$ (see Definition 5.2.1). Recall that a *local smooth submanifold* of M (of dimension $k \leq \dim M$) is the graph of a smooth injective map $\psi \colon B \to M$ where B is the open unit ball in $\mathbb{R}^k$.

Theorem 7.1.1 (Stable Manifold Theorem). *There exist $\varepsilon_0 > 0$, Borel functions $\varepsilon\colon Y \to (0, \varepsilon_0]$ and $T\colon Y \to (0, +\infty)$, and for every $x \in Y$, a local stable manifold $V^s(x)$ such that*

1. $x \in V^s(x)$, $T_x V^s(x) = E^s(x)$, $V^s \in C^{1+\alpha}$;
2. *for* $y \in V^s(x)$ *and* $n \geq 0$,
$$\rho(f^n(x), f^n(y)) \leq T(x)\lambda(x)^n\, e^{\varepsilon(x)n} \rho(x, y), \tag{7.1}$$
where $\lambda(x)$ is the f-invariant Borel function in Definition 5.2.1 and ρ the distance in M induced by the Riemannian metric;
3. *for* $m \in \mathbb{Z}$,
$$T(f^m(x)) \leq T(x)e^{10\varepsilon(x)|m|}. \tag{7.2}$$

In the literature the more expressive notation $V^s_{\mathrm{loc}}(x)$ is sometimes used for the local stable manifold at x. We, however, prefer the shorter notation $V^s(x)$ that we shall consistently and unambiguously use throughout the book.

Inequality (7.2) establishes a crucial property of stable manifolds and is essentially due to Condition 6 in Definition 5.2.1.

Theorem 7.1.1 and the remaining results of Section 7.5 hold if the requirement that M is compact is relaxed to the following: M is complete, there exists $C > 0$ such that (5.4) holds, and the radius of injectivity of the manifold M (i.e., the maximum radius of an open ball in $T_x M$ centered at zero on which the exponential map is a diffeomorphism) is bounded away from zero uniformly.

We obtain a local stable manifold in the form

$$V^s(x) = \exp_x\{(x, \psi^s(x)) : x \in B^s(r)\}, \tag{7.3}$$

where $\psi^s\colon B^s(r) \to E^2(x)$ is a smooth map, which satisfies $\psi^s(0) = 0$ and $d\psi^s(0) = 0$. Here $B^s(r)$ is the ball of radius r in $E^s(x)$ centered at the origin; $r = r(x)$ is called the *size* of the local stable manifold.

Local stable manifolds are not uniquely defined: by varying the size r, one gets different local manifolds at x. Therefore, we shall speak of *a* local stable manifold when we do not specify its size and of *the* local stable manifold when we assume that its size is fixed.

We now describe how we construct the function ψ^s. Fix $x \in M$ and consider the map

$$\widetilde{f}_x = \exp^{-1}_{f(x)} \circ f \circ \exp_x\colon B^s(r) \times B^u(r) \to T_{f(x)}M, \tag{7.4}$$

which is well defined if r is sufficiently small. Here $B^u(r)$ is the ball of radius r in $E^2(x)$ centered at the origin. By Condition (2) in Definition 5.2.1, the map $\widetilde{f}$ can be written in the following form:

$$\widetilde{f}_x(v, w) = (A_x v + g_x(v, w),\ B_x w + h_x(v, w)), \tag{7.5}$$

where $v \in E^s(x)$ and $w \in E^2(x)$. Furthermore,

$$A_x\colon E^s(x) \to E^s(f(x)) \quad \text{and} \quad B_x\colon E^2(x) \to E^2(f(x))$$

are linear maps. In view of Conditions 3 and 4 in Definition 5.2.1, the map A_x is a contraction and the map B_x is an expansion. Since f is of class $C^{1+\alpha}$, we also have

$$\|g_x(v, w)\| \le C_1(\|v\| + \|w\|)^{1+\alpha}, \quad \|h_x(v, w)\| \le C_2(\|v\| + \|w\|)^{1+\alpha}, \tag{7.6}$$

and

$$\begin{aligned} \|dg_x(v_1, w_1) - dg_x(v_2, w_2)\| &\le C_1(\|v_1 - v_2\| + \|w_1 - w_2\|)^\alpha, \\ \|dh_x(v_1, w_1) - dh_x(v_2, w_2)\| &\le C_2(\|v_1 - v_2\| + \|w_1 - w_2\|)^\alpha, \end{aligned} \tag{7.7}$$

where $C_1 > 0$ and $C_2 > 0$ are constants (which may depend on x).

In other words the map $\widetilde{f}_x$ can be viewed as a small perturbation of the linear map $(v, w) \mapsto (A_x v, B_x w)$ by the map $(g_x(v, w), h_x(v, w))$ satisfying Conditions (7.6) and (7.7). Note that the "perturbation map" $\widetilde{f}_x$ (see (7.4) and (7.5)) satisfies Condition (7.6) in a neighborhood U_x of the point x whose size depends on x. Moreover, the size of U_x decays along the trajectory of x with subexponential rate (see (5.1)).

We shall use a Lyapunov inner product $\langle\cdot,\cdot\rangle'$ and the corresponding norm $\|\cdot\|'$ introduced in Section 5.2. Fix $x \in Y$. Consider the trajectory $f^m(x)$ and the family of maps $F_m = \widetilde{f}_{f^m(x)}$. We identify the tangent spaces $T_{f^m(x)}M$ with $\mathbb{R}^p = \mathbb{R}^k \times \mathbb{R}^{p-k}$ (recall that $p = \dim M$ and $1 \le k < p$) via an isomorphism τ_m such that $\tau_m(E^s(x)) = \mathbb{R}^k$ and $\tau_m(E^2(x)) = \mathbb{R}^{p-k}$. The map $\widetilde{F}_m = \tau_{m+1} \circ F_m \circ \tau_m{}^{-1}$ is of the form (compare to the sequence of diffeomorphisms in (7.12)):

$$\widetilde{F}_m(v, w) = (A_m v + g_m(v, w), B_m w + h_m(v, w)), \tag{7.8}$$

where $A_m : \mathbb{R}^k \to \mathbb{R}^k$ and $B_m : \mathbb{R}^{p-k} \to \mathbb{R}^{p-k}$ are linear maps and $g : \mathbb{R}^p \to \mathbb{R}^k$ and $h : \mathbb{R}^p \to \mathbb{R}^{p-k}$ are nonlinear maps defined for each $v \in B^s(r_0) \subset \mathbb{R}^k$ and $w \in B^u(r_0) \subset \mathbb{R}^{p-k}$. (Recall that $B^s(r_0)$ and $B^u(r_0)$ are balls centered at zero of radius r_0.) With respect to the Lyapunov norm, these maps satisfy

$$\|A_m\|' \le \lambda', \quad (\|B_m{}^{-1}\|')^{-1} \ge \mu', \text{ where } 0 < \lambda' < \min\{1, \mu'\} \tag{7.9}$$

(recall that $\lambda(x) < \lambda' < \mu' < \mu(x)$ where $\lambda(x)$ and $\mu(x)$ are the functions in Definition 5.2.1) and

$$g_m(0,0) = 0, \quad dg_m(0,0) = 0, \quad h_m(0,0) = 0, \quad dh_m(0,0) = 0, \tag{7.10}$$

$$\begin{aligned} \|dg_m(v_1, w_1) - dg_m(v_2, w_2)\|' &\le C\gamma^{-m}(\|v_1 - v_2\|' + \|w_1 - w_2\|')^\alpha, \\ \|dh_m(v_1, w_1) - dh_m(v_2, w_2)\|' &\le C\gamma^{-m}(\|v_1 - v_2\|' + \|w_1 - w_2\|')^\alpha, \end{aligned} \tag{7.11}$$

where

$$\lambda'^\alpha < \gamma < 1, \quad 0 < \alpha \le 1, \quad C > 0$$

(see (7.7) and Properties 2 and 3 at the end of Section 7.1).

We conclude that constructing the local stable manifold $V^s(x)$ is now reduced to constructing an invariant family of stable manifolds for the sequence of local diffeomorphisms $\widetilde{F}_m$. This is a particular case of general nonuniformly hyperbolic sequences of diffeomorphisms, which we will introduce in the next section.

7.2 Nonuniformly Hyperbolic Sequences of Diffeomorphisms

Let $f_m \colon U_m \to \mathbb{R}^p$, $m \in \mathbb{Z}$ ($U_m \subset \mathbb{R}^p$ is an open set), be a (two-sided) sequence of C^1 local diffeomorphisms and $\{\langle\cdot,\cdot\rangle_m\}_{m\in\mathbb{Z}}$ a (two-sided) sequence of metrics. We denote by $\{\|\cdot\|_m\}_{m\in\mathbb{Z}}$ the corresponding sequence of norms. Write $\mathcal{F} = \{f_m\}_{m\in\mathbb{Z}}$ and let $0 < \lambda < \mu < \infty$ and $\varepsilon \geq 0$. We assume that $f_m(0) = 0$ and that $\lambda e^{\varepsilon} < \mu e^{-\varepsilon}$.

Definition 7.2.1. We say that the family $\mathcal{F}$ of local diffeomorphisms is *nonuniformly partially hyperbolic in the broad sense* (or admits a *$(\lambda, \mu, \varepsilon)$-splitting*) or is *uniformly partially hyperbolic in the broad sense* (or admits a *(λ, μ)-splitting*) if so is the sequence of matrices $\{d_0 f_m\}_{m\in\mathbb{Z}}$ (see Definition 1.2.1).

A sequence of C^1 local diffeomorphisms admits a (λ, μ)-splitting if and only if it admits a $(\lambda, \mu, 0)$-splitting.

Definition 7.2.2. We say that $\mathcal{F}$ is *nonuniformly completely hyperbolic* if it admits a $(\lambda, \mu, \varepsilon)$-splitting with $\lambda e^{\varepsilon} < 1 < \mu e^{-\varepsilon}$. We say that $\mathcal{F}$ is *uniformly completely hyperbolic* if it admits a $(\lambda, \mu, 0)$-splitting with $\lambda < 1 < \mu$.

Let $\mathcal{F}$ be a sequence of C^1 local diffeomorphisms admitting a $(\lambda, \mu, \varepsilon)$-splitting

$$E_m^1 \oplus E_m^2 = \mathbb{R}^p, \quad df_m E_m^1 = E_{m+1}^1, \quad df_m E_m^2 = E_{m+1}^2$$

for every $m \in \mathbb{Z}$. For $(x, y) \in U_m$, one can write f_m in the form

$$f_m(x, y) = (A_m x + g_m(x, y), B_m y + h_m(x, y)), \tag{7.12}$$

where

1. $A_m = d_0 f_m | E_m^1$ and $B_m = d_0 f_m | E_m^2$ are invertible transformations;
2. $(g_m, h_m) \colon U_m \to \mathbb{R}^p$ is a C^1 map satisfying $(g_m, h_m)(0) = 0$, $d_0(g_m, h_m) = 0$.

Let $A = \{A_m\}_{m\in\mathbb{Z}}$ and $\mathcal{B} = \{B_m\}_{m\in\mathbb{Z}}$. Define new sequences of matrices $\{\mathcal{A}_m\}_{m\in\mathbb{Z}}$ and $\{\mathcal{B}_m\}_{m\in\mathbb{Z}}$ using (1.1). Set

$$\mathcal{F}_m = \begin{cases} f_{m-1} \circ \cdots \circ f_1 \circ f_0 & \text{if } m > 0 \\ \mathrm{Id} & \text{if } m = 0\,, \\ (f_m)^{-1} \circ \cdots \circ (f_{-2})^{-1} \circ (f_{-1})^{-1} & \text{if } m < 0 \end{cases}$$

whenever it is defined. The map $(\mathcal{A}_m, \mathcal{B}_m)$ is a linear approximation of $\mathcal{F}_m$ in a neighborhood of 0. We shall describe in the following sections how the existence of a $(\lambda, \mu, \varepsilon)$-splitting for a sequence of C^1 local diffeomorphisms induces for the nonlinear maps $\mathcal{F}_m$ a behavior similar to those in (7.13) and (7.14) for the linear approximation of $\mathcal{F}_m$.

To simplify our arguments, we shall switch to a Lyapunov inner product (see Section 5.2).

Proposition 7.2.3. *For any numbers $0 < \lambda' < \mu' < \infty$ such that $\lambda e^{\varepsilon} < \lambda'$ and $\mu' < \mu e^{-\varepsilon}$, the following properties hold:*

1. *$\mathcal{F}$ admits a (λ', μ')-uniform splitting with respect to the Lyapunov inner products $\langle\cdot,\cdot\rangle'_m = \langle\cdot,\cdot\rangle'_{\varepsilon,\lambda',\mu',m}$.*
2. *The angle between the subspaces E^1_m and E^2_m in the metric $\langle\cdot,\cdot\rangle'_m$ is $\pi/2$.*

In particular, we have

1. *for $v \in E^1_k$, $k \in \mathbb{Z}$, and $m > 0$,*
$$\|\mathcal{A}_{m+k,k}v\|'_{m+k} \le (\lambda')^m \|v\|'_k; \tag{7.13}$$
2. *for $v \in E^2_k$, $k \in \mathbb{Z}$, and $m < 0$,*
$$\|\mathcal{B}_{m+k,k}v\|'_{m+k} \le (\mu')^m \|v\|'_k. \tag{7.14}$$

In what follows we always assume that $\lambda' < 1$ provided $\lambda e^{\varepsilon} < 1$ and that $\mu' > 1$ provided $\mu e^{-\varepsilon} > 1$.

We shall simply denote by $\|\cdot\|'$ the norm $\|\cdot\|'_m$ in Sections 7.3–7.4.

7.3 The Hadamard–Perron Theorem: Hadamard's Method

There are two well-known methods of building local stable manifolds that originated in works of Hadamard [79] and Perron [164, 165], respectively. Hadamard's approach is more geometrical and can be effected for Lipschitz (not necessarily differentiable) maps, while Perron's approach allows more flexibility.

These methods work well in the case of uniform hyperbolicity and extending them to nonuniformly hyperbolic systems introduces substantial problems. One of them is that the size of local stable manifolds may deteriorate along the trajectory and indeed, may become arbitrarily small. The crucial requirement (2.12) in the definition of nonuniform hyperbolicity provides control of the deterioration: it occurs with at most subexponential rate.

Both the Hadamard and Perron methods allow substantial generalizations to sequences of local diffeomorphisms (instead of iterations of a single diffeomorphism) or maps of Banach spaces (instead of Euclidean spaces), and so on.

Let us emphasize that the construction of stable (unstable) manifolds can be carried out if only one nonuniformly hyperbolic trajectory is present, that is, the nonuniformly partially (or completely) hyperbolic set Λ consists of a single trajectory. In particular, the construction does not involve any invariant measure.

We first describe a construction of local stable manifolds elaborating on Hadamard's approach and we apply it to families of C^1 local diffeomorphisms. Moreover, to simplify our arguments, we only consider the case of nonuniform *complete* hyperbolicity; the proof of the stable manifold theorem in the general case will be given using Perron's method in Section 7.5.

7.3.1 Invariant Cone Families

To a given family of nonuniformly complete hyperbolic C^1 local diffeomorphisms $\mathcal{F} = \{f_m\}_{m\in\mathbb{Z}}$, one can associate two families of cones (see Section 4.1): the *stable cone* of size γ centered around $\mathbb{R}^k$

$$C^s_\gamma = \{(v, w) \in \mathbb{R}^k \times \mathbb{R}^{p-k} : \|w\|' < \gamma\|v\|'\} \cup \{(0, 0)\},$$

and the *unstable cone* of size γ centered around $\mathbb{R}^{p-k}$

$$C^u_\gamma = \{(v, w) \in \mathbb{R}^k \times \mathbb{R}^{p-k} : \|v\|' < \gamma\|w\|'\} \cup \{(0, 0)\},$$

where $\gamma > 0$. In this section we describe properties of these cone families. In what follows we always assume that $\gamma < 1$. Set

$$\sigma_m = \sup\left\{\|d_{(x,y)}(g_m, h_m)\|' : (x, y) \in U_m\right\}, \quad \sigma = \sup\{\sigma_m : m \in \mathbb{Z}\}.$$

Note that σ need not be finite in general.

First we show that stable and unstable cones are invariant under the action of the family $\mathcal{F}$.

Proposition 7.3.1. *Assume that the following condition holds:*

$$\sigma \le \frac{(\mu' - \lambda')\gamma}{(1+\gamma)^2}. \tag{7.15}$$

Then for every $(x, y) \in U_m$, we have

$$d_{(x,y)}f_m C^u_\gamma \subset C^u_\gamma \quad \text{and} \quad d_{(x,y)}(f_m{}^{-1})C^s_\gamma \subset C^s_\gamma. \tag{7.16}$$

Proof. It follows from (7.15) that for every $m \in \mathbb{Z}$,

$$\sigma_m \le \frac{(\mu' - \lambda')\gamma}{(1+\gamma)^2}, \tag{7.17}$$

Note that

$$\begin{aligned}(v', w') :&= d_{(x,y)}f_m(v, w)\\ &= (A_m v + d_{(x,y)}g_m(v, w), B_m w + d_{(x,y)}h_m(v, w)).\end{aligned}$$

Therefore,

$$\|v'\|' \le \lambda'\|v\|' + \sigma_m\|(v, w)\|' \tag{7.18}$$

and

$$\|w'\|' \ge \mu'\|w\|' - \sigma_m\|(v, w)\|'.$$

If $(v, w) \in C^u_\gamma \setminus \{(0, 0)\}$ then $\|v\|' < \gamma\|w\|'$ and

$$\|(v, w)\|' \le \|v\|' + \|w\|' < (1+\gamma)\|w\|'. \tag{7.19}$$

Hence,

$$\|v'\|' < (\lambda'\gamma + \sigma_m(1+\gamma))\|w\|'$$

and

$$\|w'\|' \geq (\mu' - \sigma_m(1+\gamma))\|w\|'. \tag{7.20}$$

In view of (7.17), we obtain that

$$\|v'\|' < \frac{\lambda'\gamma + \sigma_m(1+\gamma)}{\mu' - \sigma_m(1+\gamma)}\|w'\|' \leq \gamma\|w'\|'.$$

Hence, $d_{(x,y)}f_m C^u_\gamma \subset C^u_\gamma$. For the second inclusion in (7.16), we note that

$$\frac{(\mu' - \lambda')(1/\gamma)}{(1+1/\gamma)^2} = \frac{(\mu' - \lambda')\gamma}{(1+\gamma)^2},$$

and hence $d_{(x,y)}(f_m)C^u_{1/\gamma} \subset C^u_{1/\gamma}$. Since $C^s_\gamma = \operatorname{int}(\mathbb{R}^p \setminus C^u_{1/\gamma}) \cup \{(0,0)\}$ and $d_{(x,y)}f_m$ is invertible, we obtain

$$\begin{aligned} d_{(x,y)}f_m C^s_\gamma &= \operatorname{int}(\mathbb{R}^p \setminus d_{(x,y)}f_m C^u_{1/\gamma}) \cup \{(0,0)\} \\ &\supset \operatorname{int}(\mathbb{R}^p \setminus C^u_{1/\gamma}) \cup \{(0,0)\} = C^s_\gamma. \end{aligned}$$

This shows that $d_{(x,y)}({f_m}^{-1})C^s_\gamma \subset C^s_\gamma$. □

The following statement shows that the diffeomorphisms f_m contract vectors in the stable cones and expand vectors in the unstable cones.

Proposition 7.3.2.

1. *If $(x,y) \in U_m$ and $(v,w) \in C^s_\gamma$ then*
$$\|d_{(x,y)}f_m(v,w)\|' \leq (1+\gamma)(\lambda' + \sigma_m)\|(v,w)\|'.$$
2. *If $(x,y) \in U_m$ and $(v,w) \in C^u_\gamma$ then*
$$\|d_{(x,y)}f_m(v,w)\|' \geq (\mu'/(1+\gamma) - \sigma_m)\|(v,w)\|'.$$

Proof. We use the same notation as in the proof of Proposition 7.3.1. If $(v,w) \in C^s_\gamma$ then by (7.18),

$$\|(v',w')\|' \leq (1+\gamma)\|v'\|' \leq (1+\gamma)(\lambda' + \sigma_m)\|(v,w)\|'.$$

On the other hand, if $(v,w) \in C^u_\gamma$ then by (7.20) and (7.19),

$$\|(v',w')\|' \geq \|w'\|' \geq (\mu'/(1+\gamma) - \sigma_m)\|(v,w)\|'.$$

This concludes the proof. □

For each $(x,y) \in U_m$ and $m \in \mathbb{Z}$, set

$$E^s_m(x,y) = \bigcap_{j=0}^{+\infty} d_{(x,y)}(\mathcal{F}_{m+1} \circ {\mathcal{F}_{m+j}}^{-1})C^s_\gamma$$

and

$$E^u_m(x,y) = \bigcap_{j=0}^{+\infty} d_{(x,y)}(\mathcal{F}_m \circ {\mathcal{F}_{m-j}}^{-1})C^u_\gamma.$$

Proposition 7.3.3. *Assume that Condition* (7.15) *holds and, in addition,*

$$\frac{(1+\gamma)(\lambda'+\sigma)}{\mu'/(1+\gamma)-\sigma} < 1. \tag{7.21}$$

Then

1. $E^s_m(0,0) = E^s_m$ *and* $E^u_m(0,0) = E^u_m$;
2. *if* $(x,y) \in U_m$ *then the set* $E^s_m(x,y) \subset C^s_\gamma$ *is a* k*-dimensional subspace and the set* $E^u_m(x,y) \subset C^u_\gamma$ *is a* $(n-k)$*-dimensional subspace; moreover,* $E^s_m(x,y) \oplus E^u_m(x,y) = \mathbb{R}^p$;
3. *if* $(x,y) \in U_m$ *then*
$$df_m E^s_m(x,y) = E^s_{m+1}(x,y) \quad \textit{and} \quad df_m E^u_m(x,y) = E^u_{m+1}(x,y);$$
4. $E^s_m(x,y)$ *and* $E^u_m(x,y)$ *vary continuously with* $(x,y) \in U_m$.

Proof. Note that $\{0\} \times \mathbb{R}^{p-k} \subset C^u_\gamma$. For each $j \geq 0$, consider an orthonormal basis of

$$d_{(x,y)}(\mathcal{F}_m \circ \mathcal{F}_{m-j}{}^{-1})(\{0\} \times \mathbb{R}^{p-k}).$$

Since the unit ball in $\{0\} \times \mathbb{R}^{p-k}$ is compact, one can find a sequence $m_i \to \infty$ such that each of the components of the orthonormal basis of

$$d_{(x,y)}(\mathcal{F}_m \circ \mathcal{F}_{m-m_i}{}^{-1})(\{0\} \times \mathbb{R}^{p-k})$$

converges. Therefore there exists a $(n-k)$-dimensional subspace $F^u_m \subset \mathbb{R}^p$ such that as $i \to \infty$,

$$d_{(x,y)}(\mathcal{F}_m \circ \mathcal{F}_{m-m_i}{}^{-1})(\{0\} \times \mathbb{R}^{p-k}) \to F^u_m.$$

By Proposition 7.3.1, the sequence $\{d_{(x,y)}(\mathcal{F}_m \circ \mathcal{F}_{m-j}{}^{-1})C^u_\gamma\}_{j\geq 0}$ is nonincreasing, and hence $F^u_m \subset E^u_m \subset C^u_\gamma$.

In a similar way one can prove the existence of a k-dimensional subspace $F^s_m \subset \mathbb{R}^p$ such that $F^s_m \subset E^s_m \subset C^s_{1/\gamma}$. Since $C^s_{1/\gamma} \cap C^u_\gamma = \{(0,0)\}$, we conclude that

$$F^u_m \oplus F^s_m = \mathbb{R}^p. \tag{7.22}$$

The inequality in (7.21) implies that

$$\lambda'' := (1+\gamma)(\lambda'+\sigma) < \mu'/(1+\gamma) - \sigma =: \mu''.$$

For $v \in E^u_0$, write $v = v' + v''$ where $v' \in F^u_0$ and $v'' \in F^s_0$. We obtain

$$\begin{aligned}\|v''\|' &\leq (\lambda'')^m \|d_{(x,y)}(\mathcal{F}_m{}^{-1})v''\|' \\ &= (\lambda'')^m \|d_{(x,y)}(\mathcal{F}_m{}^{-1})(v - v')\|' \\ &\leq (\lambda''/\mu'')^m \|v - v'\|' \to 0\end{aligned}$$

as $m \to \infty$. Therefore, $v'' = 0$ and $v \in F^u_0$. This shows that $E^u_0 = F^u_0$ is a $(n-k)$-dimensional subspace. Similarly, given $m \in \mathbb{Z}$, one can show that $E^u_m = F^u_m$ is a

$(n-k)$-dimensional subspace. It is clear that $df_m E^u(x,y) = E^u_{m+1}(x,y)$. Similar arguments apply to show that $E^s_m(x,y) = F^s_m$ is a k-dimensional subspace and that $df_m E^s(x,y) = E^s_{m+1}(x,y)$. By (7.22), we conclude that $E^s_m(x,y) \oplus E^u_m(x,y) = \mathbb{R}^p$.

In order to prove the fourth statement, consider a sequence of points $(x_k, y_k) \in U_0$ such that $(x_k, y_k) \to (x,y) \in U_0$ as $k \to \infty$. Passing to a subsequence $(x'_k, y'_k)_{k\in\mathbb{N}}$, we may assume that the dimensions of $E^s_0(x'_k, y'_k)$ and $E^u_0(x'_k, y'_k)$ do not depend on k. Proceeding as in the proof of Proposition 2.2.9, one can further assume that $E^s_0(x'_k, y'_k)$ and $E^u_0(x'_k, y'_k)$ converge as $k \to \infty$ to some subspaces E^s and E^u (with respect to the topology in the Grassmannian bundle).

For each $m \geq 1$, we have

$$\|d\mathfrak{F}_m | E^s_0(x'_k, y'_k)\|' \leq (\lambda')^m,$$

and hence $\|d_{(x,y)}\mathfrak{F}_m | E^s\|' \leq (\lambda')^m$. Therefore, $E^s \subset E^s_0(x,y)$. In a similar way, one can show that $E^u \subset E^u_0(x,y)$. Since $E^s_0(x_k,y_k) \oplus E^u_0(x_k,y_k) = \mathbb{R}^p$ for every $k \in \mathbb{N}$, we conclude that $E^s = E^s(x,y)$ and $E^u = E^u_0(x,y)$ and that $E^s_0(x_k,y_k) \to E^s_0(x,y)$ and $E^u_0(x_k,y_k) \to E^u_0(x,y)$ as $k \to \infty$. The invariance of the subspaces (see Property 3) guarantees that for each $m \in \mathbb{Z}$, the subspaces $E^s_m(x,y)$ and $E^u_m(x,y)$ vary continuously with $(x,y) \in U_m$. This concludes the proof of the proposition. □

A map $\varphi\colon U \to \mathbb{R}^{p-k}$ with $U \subset \mathbb{R}^k$ is called γ-*Lipschitz* if

$$\|\varphi(x) - \varphi(x')\|' \leq \gamma\|x - x'\|' \quad \text{for every } x, x' \in U.$$

Let $\varphi\colon \mathbb{R}^k \to \mathbb{R}^p$ be a γ-Lipschitz function. Given $x, y \in \mathbb{R}^k$ such that $x \neq y$, set

$$\Delta_{x,y}\varphi = \frac{(y,\varphi(y)) - (x,\varphi(x))}{\|(y,\varphi(y)) - (x,\varphi(x))\|'} \tag{7.23}$$

and

$$t_x\varphi = \{v \in T_{(x,\varphi(x))}\mathbb{R}^p : \Delta_{x,x_\ell}\varphi \to v \text{ for some sequence } x_\ell \to x\}.$$

We define the *tangent set* of $V = \operatorname{graph}(\varphi)$ at $(x,\varphi(x))$ by

$$\tau_{(x,\varphi(x))}V = \{\lambda v : \lambda \in \mathbb{R} \text{ and } v \in t_x\varphi\}.$$

One can easily see that φ is differentiable at x if and only if $\tau_{(x,\varphi(x))}V$ is a k-dimensional subspace.

7.3.2 Admissible Manifolds

One can associate to the cone families described in the previous section, some families of invariant local manifolds, which serve as "prototype" for local stable and unstable manifolds. Let us explain what this means. Consider first a local diffeomorphism $f\colon U \to \mathbb{R}^p$ with a hyperbolic fixed point at the origin. A local stable manifold $V^s(0)$ has two distinctive properties: (1) it is tangent to the stable subspace $E^s(0) = \mathbb{R}^k$ and (2) it is invariant under f; that is, $f(V^s(0)) \subset V^s(0)$ or equivalently, $f^{-1}(V^s(0)) \supset V^s(0)$. However, detecting the local stable manifold is a difficult task and one considers its approximation–a local smooth manifold $V(0)$, which passes through 0 and lies in a small stable cone around 0. More precisely, we require that the angle $\sphericalangle(T_x V(0), \mathbb{R}^k)$ is sufficiently small for all $x \in V(0)$. The local manifold $V(0)$ is call *admissible* and in many situations it can be used instead of the local stable manifold. It is invariant in a weak sense: its image under f is again an admissible manifold. Turning now to the case of nonuniform hyperbolicity, we consider a family $\mathcal{F} = \{f_m\}_{m\in\mathbb{Z}}$ of C^1 local diffeomorphisms and we construct families of admissible manifolds using a variation of the method known in the hyperbolicity theory as *graph transform technique*.

Definition 7.3.4. A set $V \subset \mathbb{R}^p$ is called

1. an *admissible* (s, γ)*-set* if there exists a γ-Lipschitz map $\varphi\colon \mathbb{R}^k \to \mathbb{R}^{p-k}$ such that
$$V = \operatorname{graph}(\varphi) = \{(x, \varphi(x)) : x \in \mathbb{R}^k\}.$$
If, in addition, φ is differentiable then V is called an *admissible* (s, γ)*-manifold.*
2. an *admissible* (u, γ)*-set* if there exists a γ-Lipschitz map $\varphi\colon \mathbb{R}^{p-k} \to \mathbb{R}^k$ such that
$$V = \operatorname{graph}(\varphi) = \{(\varphi(x), x) : x \in \mathbb{R}^{p-k}\}.$$
If, in addition, φ is differentiable then V is called an *admissible* (u, γ)*-manifold.*

We stress that the function φ in the definition of (s, γ)- and (u, γ)-sets is given for all $x \in \mathbb{R}^k$ or $x \in \mathbb{R}^{p-k}$, respectively.

Let $\mathcal{F} = \{f_m\}_{m\in\mathbb{Z}}$ be a sequence of C^1 local diffeomorphisms admitting a $(\lambda, \mu, \varepsilon)$-splitting. We begin with the simpler case when the diffeomorphisms in the sequence are defined on the whole space $\mathbb{R}^p$ and we return to the general case in Section 7.3.4.

Proposition 7.3.5. *Assume that Condition* (7.15) *holds. Then for each $m \in \mathbb{Z}$ if V_m is an admissible (u, γ)-set then so is $f_m(V_m)$.*

Proof. Let $V_m = \operatorname{graph}(\varphi_m)$ and set
$$r_m(x) = B_m x + h_m(\varphi_m(x), x). \tag{7.24}$$
By (7.15), we obtain $\sigma_m < (\mu' - \lambda')/(1 + \gamma)$ and
$$\begin{aligned} \|r_m(x) - r_m(x')\|' &\geq \mu'\|x - x'\|' - \sigma_m(1 + \gamma)\|x - x'\|' \\ &\geq \lambda'\|x - x'\|'. \end{aligned} \tag{7.25}$$
This implies that the function r_m is injective. Therefore, we can define a map $\psi_m\colon \mathbb{R}^{p-k} \to \mathbb{R}^k$ by
$$\psi_m(x) = A_m \varphi_m({r_m}^{-1}(x)) + g_m(\varphi_m({r_m}^{-1}(x)), {r_m}^{-1}(x)). \tag{7.26}$$

We have

$$\begin{aligned} f_m(\varphi_m(x), x) &= (A_m\varphi_m(x) + g_m(\varphi_m(x), x), r_m(x)) \\ &= (\psi_m(r_m(x)), r_m(x)) \end{aligned} \tag{7.27}$$

and so $f_m(V_m) = \text{graph}(\psi_m)$. By (7.25) and (7.15),

$$\begin{aligned} &\|\psi_m(r_m(x)) - \psi_m(r_m(x'))\|' \\ &\leq \lambda'\gamma\|x - x'\|' + \sigma_m(1+\gamma)\|x - x'\|' \\ &\leq \frac{\lambda'\gamma + \sigma_m(1+\gamma)}{\mu' - \sigma_m(1+\gamma)}\|r_m(x) - r_m(x')\|' \leq \gamma\|r_m(x) - r_m(x')\|'. \end{aligned} \tag{7.28}$$

We conclude that $f_m(V_m)$ is an admissible (u, γ)-set. □

Given $\gamma > 0$, let $\Gamma(u, \gamma)$ be the space of sequences $\{V_m\}_{m\in\mathbb{Z}}$ of admissible (u, γ)-sets such that (7.15) holds and $0 \in V_m$ for every $m \in \mathbb{Z}$. We define a metric on $\Gamma(u, \gamma)$ by

$$d_{\Gamma(u,\gamma)}(\{V_{1m}\}_{m\in\mathbb{Z}}, \{V_{2m}\}_{m\in\mathbb{Z}}) = \sup\{d_m(\varphi_{1m}, \varphi_{2m}) : m \in \mathbb{Z}\},$$

where

$$d_m(\varphi_{1m}, \varphi_{2m}) = \sup\left\{\frac{\|\varphi_{1m}(x) - \varphi_{2m}(x)\|'}{\|x\|'} : x \in \mathbb{R}^{p-k} \setminus \{0\}\right\}$$

and $V_{im} = \text{graph}(\varphi_{im})$ for each $m \in \mathbb{Z}$ and $i = 1, 2$. Since $\varphi_{1m}(0) = \varphi_{2m}(0) = 0$, and φ_{1m} and φ_{2m} are γ-Lipschitz, we have $d_m(\varphi_{1m}, \varphi_{2m}) \leq 2\gamma$ and the metric $d_{\Gamma(u,\gamma)}$ is well defined. One can verify that $\Gamma(u, \gamma)$ is a complete metric space.

We define the *graph transform* $G\colon \Gamma(u, \gamma) \to \Gamma(u, \gamma)$ *induced by* $\mathcal{F}$ *on* $\Gamma(u, \gamma)$ by $G(\{V_m\}_{m\in\mathbb{Z}}) = \{f_m(V_m)\}_{m\in\mathbb{Z}}$. Proposition 7.3.5 shows that the map G is well defined.

Proposition 7.3.6. *Assume that Condition* (7.15) *holds. Then the graph transform* G *is a contraction on* $\Gamma(u, \gamma)$.

Proof. Set $V_{im} = \text{graph}(\varphi_{im})$ for each $m \in \mathbb{Z}$ and $i = 1, 2$. In a similar way to that in the proof of Proposition 7.3.5, we define functions

$$\psi_{im}(x) = A_m\varphi_{im}({r_{im}}^{-1}(x)) + g_m(\varphi_{im}({r_{im}}^{-1}(x)), {r_{im}}^{-1}(x)),$$

where

$$r_{im}(x) = B_m x + h_m(\varphi_{im}(x), x).$$

By (7.27), we obtain

$$d_{\Gamma(u,\gamma)}(G(\{V_{1m}\}_{m\in\mathbb{Z}}), G(\{V_{2m}\}_{m\in\mathbb{Z}})) = \sup\{d_m(\psi_{1m}, \psi_{2m}) : m \in \mathbb{Z}\}.$$

We have

$$
\begin{aligned}
&\|\psi_{1m}(r_{1m}(x)) - \psi_{2m}(r_{1m}(x))\|' \\
&\le \|\psi_{1m}(r_{1m}(x)) - \psi_{2m}(r_{2m}(x))\|' + \|\psi_{2m}(r_{2m}(x)) - \psi_{2m}(r_{1m}(x))\|' \\
&\le \lambda'\|\varphi_{1m}(x) - \varphi_{2m}(x)\|' \\
&\quad + \|g_m(\varphi_{1m}(x), x) - g_m(\varphi_{2m}(x), x)\|' + \gamma\|r_{2m}(x) - r_{1m}(x)\|' \\
&\le \lambda'\|\varphi_{1m}(x) - \varphi_{2m}(x)\|' + \sigma_m\|\varphi_{1m}(x) - \varphi_{2m}(x)\|' \\
&\quad + \gamma\|h_m(\varphi_{1m}(x), x) - h_m(\varphi_{2m}(x), x)\|' \\
&\le [\lambda' + \sigma_m(1+\gamma)]\|\varphi_{1m}(x) - \varphi_{2m}(x)\|'.
\end{aligned}
$$

Furthermore, by (7.15), if $x \neq 0$ then

$$
\begin{aligned}
\|r_{1m}(x)\|' &\ge \mu'\|x\|' - \|h_m(\varphi_{1m}(x), x) - h_m(\varphi_{1m}(0), 0)\|' \\
&\ge \mu'\|x\|' - \sigma_m(1+\gamma)\|x\|' > 0.
\end{aligned}
$$

Therefore,

$$
\begin{aligned}
&d_m(\psi_{1m}, \psi_{2m}) \\
&\quad = \sup\left\{\frac{\|\psi_{1m}(r_{1m}(x)) - \psi_{2m}(r_{1m}(x))\|'}{\|r_{1m}(x)\|'} : x \in \mathbb{R}^{p-k} \setminus \{0\}\right\} \\
&\quad \le \frac{\lambda' + \sigma_m(1+\gamma)}{\mu' - \sigma_m(1+\gamma)} d_m(\varphi_{1m}, \varphi_{2m}) \le \beta\, d_m(\varphi_{1m}, \varphi_{2m}),
\end{aligned}
$$

where

$$
\beta := \sup\left\{\frac{\lambda' + \sigma_m(1+\gamma)}{\mu' - \sigma_m(1+\gamma)} : m \in \mathbb{Z}\right\}.
$$

Since $\gamma < 1$, using Condition (7.15), we obtain that $\beta < 1$, and hence

$$
\begin{aligned}
&d_{\Gamma(u,\gamma)}(G(\{V_{1m}\}_{m\in\mathbb{Z}}), G(\{V_{2m}\}_{m\in\mathbb{Z}})) \\
&\qquad \le \beta\, d_{\Gamma(u,\gamma)}(\{V_{1m}\}_{m\in\mathbb{Z}}, \{V_{2m}\}_{m\in\mathbb{Z}}).
\end{aligned}
$$

This concludes the proof of the proposition. □

Theorem 7.3.7. *Assume that Condition* (7.15) *holds. Then there exists a unique family* $\{V_m^u\}_{m\in\mathbb{Z}}$ *of admissible* (u, γ)*-sets such that* $0 \in V_m^u$ *and* $f_m(V_m^u) = V_{m+1}^u$ *for every* $m \in \mathbb{Z}$.

Proof. By Proposition 7.3.6, the contraction mapping principle guarantees that the graph transform G possesses a unique fixed point $\{V_m^u\}_{m\in\mathbb{Z}} \in \Gamma(u, \gamma)$, satisfying the required properties. □

We now briefly describe how to obtain similar results for (s, γ)-manifolds. For every $m \in \mathbb{Z}$ and $(x, y) \in f_m(U_m)$, one can write ${f_m}^{-1}$ in the form

$$
{f_m}^{-1}(x, y) = ({A_m}^{-1}x + \widetilde{g}_m(x, y), {B_m}^{-1}y + \widetilde{h}_m(x, y)),
$$

where $(\widetilde{g}_m, \widetilde{h}_m)\colon f_m(U_m) \to \mathbb{R}^p$ is a C^1 map, satisfying $(\widetilde{g}_m, \widetilde{h}_m)(0) = 0$ and $d_0(\widetilde{g}_m, \widetilde{h}_m) = 0$. Set

$$\tau_m = \sup\left\{\|d_{(x,y)}(\widetilde{g}_m, \widetilde{h}_m)\|' : (x, y) \in f_m(U_m)\right\}, \; \tau = \sup\{\tau_m : m \in \mathbb{Z}\}.$$

Theorem 7.3.8. *Assume that*

$$\tau \le \frac{(\mu' - \lambda')\gamma}{\lambda'\mu'(1+\gamma)^2}. \tag{7.29}$$

Then there exists a unique family $\{V^s_m\}_{m\in\mathbb{Z}}$ of admissible (s, γ)-sets such that $0 \in V^s_m$ and $f_m(V^s_m) = V^s_{m+1}$ for every $m \in \mathbb{Z}$.

Proof. The proof is similar to that of Theorem 7.3.7. □

7.3.3 Existence of (s, γ)- and (u, γ)-Manifolds

The following theorem establishes the existence of (s, γ)-manifolds for sequences of local diffeomorphisms.

Theorem 7.3.9. *Let $\{f_m\}_{m\in\mathbb{Z}}$ be a sequence of C^1 local diffeomorphisms defined on the whole $\mathbb{R}^p$ and admitting a $(\lambda, \mu, \varepsilon)$-splitting for some $0 < \lambda < 1 < \mu$ and $\varepsilon > 0$. Assume that the positive numbers γ, σ, and τ satisfy Conditions* (7.15) *and* (7.29). *Then there exists a unique family $\{V^s_m\}_{m\in\mathbb{Z}}$ of C^1 admissible (s, γ)-manifolds such that*

1. *$0 \in V^s_m$, $f_m(V^s_m) = V^s_{m+1}$, and $T_0 V^s_m = E^s_m$;*
2. *if $(x, y) \in V^s_m$ then*

$$\|f_m(x, y)\|' \le (1+\gamma)(\lambda' + \sigma_m)\|(x, y)\|'; \tag{7.30}$$

3. *assume that*

$$\nu < \mu'/(1+\gamma) - \sigma, \tag{7.31}$$

then for every $(x, y) \in \mathbb{R}^p$, if there exists $C > 0$ such that

$$\|\mathfrak{F}_{m+k} \circ \mathfrak{F}_m{}^{-1}(x, y)\|' \le C\nu^k\|(x, y)\|' \tag{7.32}$$

for every $k \ge 0$, then $(x, y) \in V^s_m$.

Before proving the theorem let us notice that an admissible (s, γ)-manifold is also an admissible (s, γ')-manifold for every $\gamma' > \gamma$. Therefore the uniqueness property in Theorem 7.3.9 implies that the family $\{V^s_m\}_{m\in\mathbb{Z}}$ is independent of γ. This family is called *family of invariant s-manifolds*. It can be characterized as follows.

Proposition 7.3.10. *For each $\gamma \in (0, \sqrt{\mu'/\lambda'} - 1)$ and each sufficiently small $\sigma > 0$,*

1. *if* (7.31) *holds then*

$$V^s_m = \left\{(x, y) \in \mathbb{R}^p : \varlimsup_{k\to+\infty} \frac{1}{k} \log\|\mathfrak{F}_{m+k} \circ \mathfrak{F}_m{}^{-1}(x, y)\|' < \log \nu\right\};$$

2. *if* $(1+\gamma)(\lambda'+\sigma) < 1 < \mu'/(1+\gamma) - \sigma$ *then*

$$\begin{aligned} V_m^s &= \left\{(x,y) \in \mathbb{R}^p : \varlimsup_{k\to+\infty} \frac{1}{k} \log\|\mathfrak{F}_{m+k} \circ \mathfrak{F}_m{}^{-1}(x,y)\|' < 0\right\} \\ &= \left\{(x,y) \in \mathbb{R}^p : \sup_{k\geq 0}\|\mathfrak{F}_{m+k} \circ \mathfrak{F}_m{}^{-1}(x,y)\|' < \infty\right\} \\ &= \left\{(x,y) \in \mathbb{R}^p : \mathfrak{F}_{m+k} \circ \mathfrak{F}_m{}^{-1}(x,y) \to 0 \text{ as } k \to +\infty\right\}. \end{aligned}$$

Proof. Note that $(1+\gamma)(\lambda'+\sigma) < \mu'/(1+\gamma) - \sigma$ for all sufficiently small σ if and only if $\gamma \in (0, \sqrt{\mu'/\lambda'} - 1)$. Statement 1 of the proposition follows immediately from the characterization of V_m^s in Statements 2 and 3 of Theorem 7.3.9. Statement 2 of the proposition can be easily obtained by setting $\nu = 1$ in Statement 1. □

Proof of Theorem 7.3.9. By Theorem 7.3.8 there exists a unique family $\{V_m^s\}_{m\in\mathbb{Z}}$ of admissible (s,γ)-sets such that $0 \in V_m^s$ and $f_m(V_m^s) = V_{m+1}^s$. Choose a sequence of points $x_\ell \to x$ and consider $\Delta_{x,x_\ell}\varphi_m$ given by (7.23). Note that $\Delta_{x,x_\ell}\varphi_m \to v$ as $\ell \to \infty$ if and only if

$$\lim_{\ell\to\infty} \frac{f_m(x_\ell, \varphi_m(x_\ell)) - f_m(x, \varphi_m(x))}{\|f_m(x_\ell, \varphi_m(x_\ell)) - f_m(x, \varphi_m(x))\|'} = \frac{d_{(x,\varphi_m(x))} f_m v}{\|d_{(x,\varphi_m(x))} f_m v\|'}.$$

This implies that

$$(d_{(x,\varphi_m(x))} f_m)\tau_{(x,\varphi_m(x))} V_m^s = \tau_{f_m(x,\varphi_m(x))} f_m(V_m^s).$$

By Proposition 7.3.3, the maximal family with this invariance property is composed of the subspaces $E_m^s(x, \varphi_m(x))$. Hence,

$$\tau_{(x,\varphi_m(x))} V_m^s \subset E_m^s(x, \varphi_m(x)).$$

On the other hand, given $v \in \mathbb{R}^k$, there exists a sequence $t_\ell \to 0$ such that $\Delta_{x,x+t_\ell v}\varphi_m$ converges as $\ell \to +\infty$. This implies that $\tau_{(x,\varphi_m(x))} V_m^s$ projects onto $\mathbb{R}^k$, and hence $\tau_{(x,\varphi_m(x))} V_m^s = E_m^s(x, \varphi_m(x))$. It follows that φ_m is differentiable at x. Furthermore, setting $x = 0$, we conclude that $T_0 V_m^s = E_m^s$.

Since $\{V_m^s\}_{m\in\mathbb{Z}}$ is a fixed point of the graph transform, one can write

$$f_m(x, \varphi_m(x)) = (s_m(x), \varphi_m(s_m(x))),$$

where $V_m^s = \text{graph}(\varphi_m)$ and

$$s_m(x) = A_m x + g_m(x, \varphi_m(x)).$$

Since φ_m is γ-Lipschitz, we obtain

$$\|f_m(x, \varphi_m(x))\|' \leq (1+\gamma)\|s_m(x)\|' \leq (1+\gamma)(\lambda'+\sigma_m)\|(x, \varphi_m(x))\|'.$$

It remains to establish the third property. If $(x,y) \notin V_m^s$ then there exists a (u,γ)-set V_m, containing (x,y) that intersects V_m^s at a unique point (x', y'). It follows from (7.30) that

$$\|\mathfrak{F}_{m+k} \circ \mathfrak{F}_m{}^{-1}(x', y')\|' \leq [(1+\gamma)(\lambda'+\sigma)]^k \|(x', y')\|'.$$

On the other hand, proceeding as in the proof of Proposition 7.3.5, we obtain that

$$\|f_m(x,y) - f_m(x',y')\|' \geq \|r_m(x) - r_m(x')\|' \geq (\mu' - \sigma(1+\gamma))\|x - x'\|'.$$

By Proposition 7.3.5, $f_m(V_m)$ is also a (u, γ)-set. Therefore,

$$\begin{aligned}&\|\mathcal{F}_{m+k} \circ \mathcal{F}_m{}^{-1}(x,y) - \mathcal{F}_{m+k} \circ \mathcal{F}_m{}^{-1}(x',y')\|' \\ &\geq (\mu' - \sigma(1+\gamma))^k \|x - x'\|'/(1+\gamma)^{k-1}.\end{aligned}$$

We conclude that

$$\begin{aligned}\|\mathcal{F}_{m+k} \circ \mathcal{F}_m{}^{-1}(x,y)\|' &\geq \|\mathcal{F}_{m+k} \circ \mathcal{F}_m{}^{-1}(x,y) - \mathcal{F}_{m+k} \circ \mathcal{F}_m{}^{-1}(x',y')\|' \\ &\quad - \|\mathcal{F}_{m+k} \circ \mathcal{F}_m{}^{-1}(x',y')\|' \\ &\geq (\mu' - \sigma(1+\gamma))^k \|x - x'\|'/(1+\gamma)^{k-1} \\ &\quad - [(1+\gamma)(\lambda' + \sigma)]^k \|(x',y')\|'.\end{aligned}$$

In view of (7.31), for sufficiently large k this contradicts to (7.32), and thus $(x, y) \in V^s_m$. □

We can also establish the existence of (u, γ)-manifolds for sequences of local diffeomorphisms.

Theorem 7.3.11. *Let $\{f_m\}_{m\in\mathbb{Z}}$ be a sequence of C^1 local diffeomorphisms defined on the whole $\mathbb{R}^p$ and admitting a $(\lambda, \mu, \varepsilon)$-splitting for some $0 < \lambda < 1 < \mu$ and $\varepsilon > 0$. Assume that the positive numbers γ, σ, and τ satisfy Conditions* (7.15) *and* (7.29). *Then there exists a unique family $\{V^u_m\}_{m\in\mathbb{Z}}$ of C^1 admissible (u, γ)-manifolds such that*

1. *$0 \in V^u_m$, $f_m(V^u_m) = V^u_{m+1}$, and $T_0 V^u_m = E^u_m$;*
2. *if $(x, y) \in V^u_m$ then*

$$\|f_m(x,y)\|' \geq (\mu'/(1+\gamma) - \sigma_m)\|(x,y)\|';$$

3. *if $\nu > (1+\gamma)(\lambda' + \sigma)$ and for some $(x, y) \in \mathbb{R}^p$ there exists $C > 0$ such that*

$$\|\mathcal{F}_{m+k} \circ \mathcal{F}_m{}^{-1}(x,y)\|' \leq C\nu^k \|(x,y)\|'$$

for every $k \leq 0$, then $(x, y) \in V^u_m$.

The family $\{V^u_m\}_{m\in\mathbb{Z}}$ is independent of γ and is called *family of invariant u-manifolds*. It can be characterized as follows.

Proposition 7.3.12. *For each $\gamma \in (0, \sqrt{\mu'/\lambda'} - 1)$ and each sufficiently small $\sigma > 0$,*

1. *if $\nu > (1+\gamma)(\lambda' + \sigma)$ holds then*

$$V^u_m = \left\{(x,y) \in \mathbb{R}^p : \overline{\lim_{k\to-\infty}} \frac{1}{|k|} \log\|\mathcal{F}_{m+k} \circ \mathcal{F}_m{}^{-1}(x,y)\|' < \log\frac{1}{\nu}\right\};$$

2. *if* $(1+\gamma)(\lambda'+\sigma) < 1 < \mu'/(1+\gamma) - \sigma$ *then*

$$\begin{aligned} V_m^u &= \left\{(x,y) \in \mathbb{R}^p : \overline{\lim_{k\to-\infty}} \frac{1}{|k|} \log\|\mathfrak{F}_{m+k} \circ \mathfrak{F}_m{}^{-1}(x,y)\|' < 0\right\} \\ &= \left\{(x,y) \in \mathbb{R}^p : \sup_{k\le 0}\|\mathfrak{F}_{m+k} \circ \mathfrak{F}_m{}^{-1}(x,y)\|' < \infty\right\} \\ &= \left\{(x,y) \in \mathbb{R}^p : \mathfrak{F}_{m+k} \circ \mathfrak{F}_m{}^{-1}(x,y) \to 0 \text{ as } k \to -\infty\right\}. \end{aligned}$$

7.3.4 Invariant Families of Local Manifolds

We now describe local versions of the results in Section 7.3.2 removing the assumption that the diffeomorphisms in the sequence are defined on the whole space $\mathbb{R}^p$. Let $\mathfrak{F}$ be a sequence of C^1 local diffeomorphisms admitting a $(\lambda, \mu, \varepsilon)$-splitting for some $0 < \lambda < 1 < \mu$ and some $\varepsilon > 0$. Set $R_k(\delta) = \{v \in \mathbb{R}^k : \|v\|' \le \delta\}$.

Definition 7.3.13. Let $\gamma, \delta > 0$ and $k, p \in \mathbb{N}$ with $k < p$. A set $V \subset \mathbb{R}^p$ is called

1. an *admissible* (s, γ, δ)*-set near* 0 if there exists a γ-Lipschitz map $\varphi\colon R_k(\delta) \to R_{p-k}(\delta)$ such that $\|\varphi(0)\|' \le \delta/2$ and

$$V = \operatorname{graph}(\varphi) = \{(x, \varphi(x)) : x \in R_k(\delta)\}.$$

2. an *admissible* (u, γ, δ)*-set near* 0 if there exists a γ-Lipschitz map $\varphi\colon R_{p-k}(\delta) \to R_k(\delta)$ such that $\|\varphi(0)\|' \le \delta/2$ and

$$V = \operatorname{graph}(\varphi) = \{(\varphi(x), x) : x \in R_{p-k}(\delta)\}.$$

Proposition 7.3.14. *Let V_m be an admissible (u, γ, δ)-set near 0 for some $\gamma \le 1$ and some $\delta > 0$ satisfying* (7.15). *Assume that $V_m \subset U_m$ and at least one of the following holds:*

1. $0 \in V_m$.
2. $0 \notin V_m$ *and*

$$2\gamma\sigma_m < -(\mu' - \lambda' + \gamma(1+2\lambda') + 2) + \sqrt{(\mu' - \lambda' + \gamma(1+2\lambda') + 2)^2 + 4\gamma\mu'(\mu' - \lambda')}.$$

Then $f_m(V_m)$ is an admissible (u, γ, δ')-set near 0, where $\delta' = [\mu' - \sigma_m(1+\gamma)]\delta$.

Proof. We use the same notation as in the proof of Proposition 7.3.5. By (7.25), we conclude that the function r_m defined by (7.24) on $R_{p-k}(\delta)$ satisfies

$$r_m(R_{p-k}(\delta)) \supset R_{p-k}(\delta').$$

Therefore, we can define a map $\psi_m\colon R_{p-k}(\delta') \to \mathbb{R}^k$ by (7.26). Since $\gamma \le 1$, by (7.15), one can easily show that $\sigma_m \le (\mu' - \lambda')/(3+\gamma)$, and hence

$$\|\psi_m(x)\|' \le \lambda'\delta + 2\sigma_m\delta \le \delta'.$$

This implies that $\psi_m(R_{p-k}(\delta')) \subset R_{n-k}(\delta')$. By (7.27), we conclude that $f_m(V_m) \supset \text{graph}(\psi_m)$.

If $0 \in V_m$ then $\psi_m(0) = 0$ and by (7.28), $f_m(V_m)$ is an admissible (u, γ, δ')-set near 0.

Assume that $0 \notin V_m$. If $x_0 = r_m{}^{-1}(0)$ then

$$\|x_0\|' = \|B_m{}^{-1} h_m(\varphi_m(x_0), x_0)\|' \le (\mu')^{-1}\sigma_m(\|x_0\|' + \delta)$$

and $\|x_0\|' \le \sigma_m\delta/(\mu' - \sigma_m)$. Using Property 2 one can easily show that

$$\begin{aligned}\|\psi_m(0)\|' &\le \lambda'\|\varphi_m(x_0)\|' + \sigma_m(\|\varphi_m(x_0)\|' + \|x_0\|') \\ &\le (\lambda' + \sigma_m)(\|\varphi_m(x_0) - \varphi_m(0)\|' + \delta/2) + \sigma_m\|x_0\|' \\ &\le (\lambda' + \sigma_m)\delta/2 + [(\lambda' + \sigma_m)\gamma + \sigma_m]\sigma_m\delta/(\mu' - \sigma_m) < \delta'/2.\end{aligned}$$

By (7.28), we conclude that $f_m(V_m)$ is an admissible (u, γ, δ')-set near 0. □

We note that since $d_0(g_m, h_m) = 0$, one can always choose the neighborhood U_m of 0 in such a way that (7.15) holds.

For each $\delta > 0$, we denote by $\Gamma(u, \gamma, \delta)$ the space of sequences $\{V_m\}_{m\in\mathbb{Z}}$ of admissible (u, γ, δ)-sets near 0 such that (7.15) holds and $0 \in V_m \subset U_m$ for every $m \in \mathbb{Z}$. One can introduce a metric $d_{\Gamma(u,\gamma,\delta)}$ on $\Gamma(u, \gamma, \delta)$ by

$$d_{\Gamma(u,\gamma,\delta)}(\{V_{1m}\}_{m\in\mathbb{Z}}, \{V_{2m}\}_{m\in\mathbb{Z}}) = \sup\{d_{m\delta}(\varphi_{1m}, \varphi_{2m}) : m \in \mathbb{Z}\},$$

where

$$d_{m\delta}(\varphi_{1m}, \varphi_{2m}) = \sup\left\{\frac{\|\varphi_{1m}(x) - \varphi_{2m}(x)\|'}{\|x\|'} : x \in R_{n-k}(\delta) \setminus \{0\}\right\}$$

and $V_{im} = \text{graph}(\varphi_{im})$ for each $m \in \mathbb{Z}$ and $i = 1, 2$. One can verify that $\Gamma(u, \gamma, \delta)$ is a complete metric space.

Given $\delta > 0$ such that $\mu'/(1+\gamma) - \sigma \ge 1$, we define the *graph transform* $G_\delta \colon \Gamma(u, \gamma, \delta) \to \Gamma(u, \gamma, \delta)$ *induced by* $\mathfrak{F}$ *on* $\Gamma(u, \gamma, \delta)$ by $G_\delta(\{V_m\}_{m\in\mathbb{Z}}) = \{R_n(\delta) \cap f_m(V_m)\}_{m\in\mathbb{Z}}$. Proposition 7.3.14 shows that the map G_δ is well defined.

Proposition 7.3.15. *Assume that $\mu'/(1+\gamma) - \sigma \ge 1$. Then the graph transform G_δ is a contraction in $\Gamma(u, \gamma, \delta)$.*

Proof. We need the following statement:

Lemma 7.3.16. *Let $f \colon U \to \mathbb{R}^p$ be a C^1 map with $0 \in U$ and $f(0) = 0$. Given $\varepsilon > 0$, there exists $\delta > 0$ and a C^1 map $\tilde{f} \colon \mathbb{R}^p \to \mathbb{R}^p$ such that*

1. $\tilde{f} = f$ *on* $B(0, \delta)$*;*
2. $d_{C^1}(\tilde{f}, d_0 f) < \varepsilon$.

Proof of the lemma. Let $\rho \colon \mathbb{R}^p \to [0, 1]$ be a C^∞ map satisfying $\rho|B(0, \delta) = 1$, $\rho|(\mathbb{R}^p \setminus B(0, 2\delta)) = 0$, and $\|d_x\rho\| < \delta$ for each $x \in \mathbb{R}^p$. Set $\tilde{f} = \rho f + (1 - \rho)d_0 f$, where we make the convention that $(\rho f)|(\mathbb{R}^p \setminus U) = 0$. Clearly, $\tilde{f} = f$ on

$B(0, \delta)$. Since $\widetilde{f} - d_0 f = \rho(f - d_0 f)$, we obtain

$$d\widetilde{f} - d_0 f = d\rho(f - d_0 f) + \rho(df - d_0 f)$$

and

$$\begin{aligned} d_{C^1}(\widetilde{f}, d_0 f) &\leq d_{C^0}(f, d_0 f) + d_{C^0}(d\widetilde{f}, d_0 f) \\ &\leq (1 + 1/\delta) d_{C^0}(f, d_0 f) + d_{C^0}(df, d_0 f) \leq C \end{aligned}$$

for some constant $C > 0$ independent of δ. This implies the desired statement. □

The desired result now follows from the lemma and Proposition 7.3.6. □

The unique fixed point of G_δ in $\Gamma(u, \gamma, \delta)$ given by Proposition 7.3.15 is an invariant family of local manifolds $\{V_m\}_{m \in \mathbb{Z}}$.

Theorem 7.3.17. *Assume that $\mu'/(1 + \gamma) - \sigma \geq 1$ and that the numbers γ and σ satisfy* (7.15). *Then the unique family $\{V_m\}_{m \in \mathbb{Z}}$ of admissible (u, γ, δ)-sets near 0 given by Proposition 7.3.15 is composed of C^1 manifolds.*

Proof. The statement can be obtained by repeating arguments in the proof of Theorem 7.3.9. □

7.3.5 Higher Differentiability of Invariant Manifolds

Let $\mathcal{F}$ be a sequence of C^r local diffeomorphisms on the whole $\mathbb{R}^p$, for some $r > 0$, admitting a $(\lambda, \mu, \varepsilon)$-splitting for some $\varepsilon > 0$. In this section we show that the family $\{V_m^u\}_{m \in \mathbb{Z}}$ of C^1 admissible (u, γ)-manifolds in Theorem 7.3.11 is composed of C^r manifolds.

Theorem 7.3.18. *Assume that $\mu\, e^{-\varepsilon} > 1$ and that the numbers γ, σ, and τ satisfy* (7.15) *and* (7.29). *Then the unique family $\{V_m^u\}_{m \in \mathbb{Z}}$ of admissible (u, γ)-manifolds given by Theorem 7.3.11 is composed of C^r manifolds.*

Proof. By Theorem 7.3.11, each V_m^u is a C^1 smooth manifold. We now proceed by induction in r. Assume that the statement of the theorem holds for $r = k$. Define C^{r-1} local diffeomorphisms of $\mathbb{R}^p \times \mathbb{R}^p$ by

$$F_m(x, y, v, w) = (f_m(x, y), d_{(x,y)} f_m(v, w)),$$

where $(v, w) \in E_m^s \oplus E_m^u$ for each $m \in \mathbb{Z}$. Note that $F_m(0) = 0$ and that the family $\{F_m\}_{m \in \mathbb{Z}}$ admits the $(\lambda, \mu, \varepsilon)$-splitting

$$\mathbb{R}^p \times \mathbb{R}^p = (E_m^s \oplus E_m^u) \times (E_m^s \oplus E_m^u).$$

Thus, we can apply the induction hypothesis (i.e., the statement in the theorem for $r = k$) to obtain a unique family $\{V_m^u, E_m(q)_{q \in V_m^u}\}_{m \in \mathbb{Z}}$ of (u, γ)-sets composed of C^k manifolds, where each set $E_m(q)$ is a neighborhood of a linear space, since $(v, w) \mapsto F_m(x, y, v, w)$ is linear. Furthermore, the subspace $E_m(q)$ has the same dimension as E_m^u.

To complete the proof it suffices to show that $E_m(q) = T_q V^u_m$. Indeed, since the map $q \mapsto E_m(q)$ is of class C^k, it follows that V^u_m is of class C^{k+1}. To prove the identity note first that by Theorem 7.3.11, for each $(v, w) \in E_m(q)$,

$$\|d_q f_m(v, w)\|' \geq (\mu'/(1+\gamma) - \sigma_m)\|(v, w)\|'.$$

Write $(v, w) = z_+ + z_-$ with $z_+ \in E^u_m$ and $z_- \in E^s_m$. Since $\mu e^{-\varepsilon} > 1$, we find that

$$d_q(\mathcal{F}_{m+k} \circ \mathcal{F}_m{}^{-1})((v, w) - z_+) \to 0 \quad \text{as} \quad k \to -\infty.$$

On the other hand,

$$\|d_q(\mathcal{F}_{m+k} \circ \mathcal{F}_m{}^{-1})z_-\|' \to \infty \quad \text{as} \quad k \to -\infty,$$

provided that $z_- \neq 0$. Hence, we must have $z_- = 0$, and thus $E_m(q) \subset T_q V^u_m$. Since the subspaces $E_m(q)$ and $T_q V^u_m$ have the same dimension, we conclude that $E_m(q) = T_q V^u_m$. □

7.4 The Graph Transform Property

We describe in the setting of nonuniform hyperbolicity a version of a result that is well known in the uniform hyperbolicity theory (and usually referred to as Inclination Lemma or λ-Lemma). It will be used later to establish the crucial absolute continuity property of local stable manifolds (see Section 8.6). Our approach is essentially an elaboration of the Hadamard method.

Let $\mathcal{F} = \{f_m\}_{m\in\mathbb{Z}}$ be a sequence of C^r local diffeomorphisms for some $r > 0$, admitting a $(\lambda, \mu, \varepsilon)$-splitting for some $\varepsilon > 0$. Choose numbers r_0, b_0, and c_0 and for every $m \geq 0$, set

$$r_m = r_0 e^{-\varepsilon m}, \quad b_m = b_0(\mu')^{-m} e^{\varepsilon m}, \quad c_m = c_0 e^{-\varepsilon m}.$$

Consider the set Ψ of $C^{1+\alpha}$ functions on $\{(m, v) : m \geq 0, v \in B^s(r_m)\}$ such that

$$\psi(m, v) \in E^u(0) \text{ for every } m \geq 0 \text{ and } v \in B^s(r_m)$$

(where $B^s(r_m)$ is the ball in $E^s(0)$ centered at 0 of radius r_m), and satisfying the following conditions:

$$\|\psi(m, 0)\|' \leq b_m, \qquad \max_{v\in B^s(r_m)} \|d\psi(m, v)\|' \leq c_m.$$

Theorem 7.4.1 (Graph Transform Property). *There are positive constants r_0, b_0, and c_0 such that for every function $\psi \in \Psi$, there exists a function $\widetilde{\psi} \in \Psi$ for which*

$$f_m^{-1}(\{(v, \psi(m, v)) : v \in B^s(r_m)\}) \supset \{(v, \widetilde{\psi}(m+1, v)) : v \in B^s(r_{m+1})\}$$

for all $m \geq 0$.

For the proof of this theorem, see the proof of Proposition 7.3.14.

7.5 The Hadamard–Perron Theorem: Perron's Method

We present here a proof of a slightly changed and more general version of the Stable Manifold Theorem using Perron's method. The original idea of Perron has evolved during the last 30 years or so into a powerful method especially due to works of Irwin [92] and so sometimes the method is called Perron–Irwin's method. It allows one to construct local stable manifolds in various and quite general situations (see [50, 53]). It should be stressed that in the papers mentioned the hyperbolicity is uniform or "pointwise" uniform – still a stronger condition than nonuniform hyperbolicity. An adaptation of Perron's argument to the nonuniformly hyperbolic case was obtained in [167]. We closely follow this paper and construct stable (and unstable) invariant manifolds for diffeomorphisms that are nonuniformly hyperbolic in the broad sense.

7.5.1 *An Abstract Version of the Stable Manifold Theorem*

Let f be a $C^{1+\alpha}$ diffeomorphism of a compact smooth Riemannian manifold M and Y is an f-invariant set that is nonuniformly hyperbolic in the broad sense (see Section 2.2). We stress that we allow the case when Y consists of a single trajectory. Consider the family of local diffeomorphisms $\widetilde{F}_m$ given by (7.8). We now state a general version of the Stable Manifold Theorem 7.1.1.

Theorem 7.5.1. *Let κ be any number satisfying*

$$\lambda' < \kappa < \min\{\mu', \gamma^{1/\alpha}\}, \tag{7.33}$$

where the numbers λ', μ', and γ are as in (7.9) and (7.11). Then there exist $D > 0$ and $r_0 > r > 0$, and a map $\psi^s\colon B^s(r) \to \mathbb{R}^{p-k}$ such that

1. *ψ^s is of class $C^{1+\alpha}$, and $\psi^s(0) = 0$ and $d\psi^s(0) = 0$;*
2. *$\|d\psi^s(v_1) - d\psi^s(v_2)\|' \le D(\|v_1 - v_2\|')^\alpha$ for any $v_1, v_2 \in B^s(r)$;*
3. *if $m \ge 0$ and $v \in B^s(r)$ then*

$$\left(\prod_{i=0}^{m-1} \widetilde{F}_i\right)(v, \psi^s(v)) \in B^s(r) \times B^u(r),$$

$$\left\|\left(\prod_{i=0}^{m-1} \widetilde{F}_i\right)(v, \psi^s(v))\right\|' \le D\kappa^m \|(v, \psi^s(v))\|',$$

where $\prod_{i=0}^{m-1} \widetilde{F}_i$ denotes the composition $\widetilde{F}_{m-1} \circ \cdots \circ \widetilde{F}_0$;

4. *given $v \in B^s(r)$ and $w \in B^u(r)$, if there is a number $K > 0$ such that*

$$\left(\prod_{i=0}^{m-1} \widetilde{F}_i\right)(v, w) \in B^s(r) \times B^u(r), \quad \left\|\left(\prod_{i=0}^{m-1} \widetilde{F}_i\right)(v, w)\right\|' \le K\kappa^m$$

for every $m \ge 0$, then $w = \psi^s(v)$;

5. *the numbers D and r depend only on the numbers λ', μ', γ, α, κ, and C.*

Proof. Consider the linear space Γ_κ of sequences of vectors $z = \{z(m) \in \mathbb{R}^p\}_{m\in\mathbb{N}}$, satisfying the following condition:

$$\|z\|_\kappa = \sup_{m\geq 0}(\kappa^{-m}\|z(m)\|') < \infty;$$

Γ_κ is a Banach space with the norm $\|z\|_\kappa$. Given $r > 0$, set

$$W = \{z \in \Gamma_\kappa : z(m) \in B^s(r) \times B^u(r) \text{ for every } m \in \mathbb{N}\}.$$

Since $0 < \kappa < 1$, the set W is open. Consider the map $\Phi_\kappa : B^s(r_0) \times W \to \Gamma_\kappa$ given by

$$\Phi_\kappa(y, z)(0) = \left(y, -\sum_{k=0}^{\infty}\left(\prod_{i=0}^{k} B_i\right)^{-1} h_k(z(k))\right),$$

and for $m > 0$ by

$$\Phi_\kappa(y, z)(m) = -z(m) + \left(\left(\prod_{i=0}^{m-1} A_i\right) y, 0\right)$$
$$+ \left(\sum_{n=0}^{m-1}\left(\prod_{i=n+1}^{m-1} A_i\right) g_n(z(n)), -\sum_{n=0}^{\infty}\left(\prod_{i=0}^{n} B_{i+m}\right)^{-1} h_{n+m}(z(n+m))\right).$$

Using Conditions (7.9)–(7.33), we will show that the map Φ_κ is well defined and is continuously differentiable over y and z. Indeed, by (7.10) and (7.11), for $z \in B^s(r_0) \times B^u(r_0)$ and $n \geq 0$,

$$\begin{aligned} \|g_n(z)\|' &= \|g_n(z) - g_n(0)\|' \leq \|dg_n(\xi)\|' \cdot \|z\|' \\ &= \|dg_n(\xi) - dg_n(0)\|' \cdot \|z\|' \\ &\leq C\gamma^{-n}(\|\xi\|')^\alpha \|z\|' \leq C\gamma^{-n}(\|z\|')^{1+\alpha}, \end{aligned} \tag{7.34}$$

where ξ lies on the interval that joins the points 0 and z. Similarly, we have that

$$\|h_n(z)\|' \leq C\gamma^{-n}(\|z\|')^{1+\alpha}. \tag{7.35}$$

Using (7.9), (7.34), and (7.35), we obtain

$$\|\Phi_\kappa(y, z)\|_\kappa = \sup_{m\geq 0}(\kappa^{-m}\|\Phi_\kappa(y, z)(m)\|') \leq \sup_{m\geq 0}(\kappa^{-m}\|z(m)\|')$$
$$+ \sup_{m\geq 0}\left\{\kappa^{-m}\left[\prod_{i=0}^{m-1}\|A_i\|' \cdot \|y\|' + \sum_{n=0}^{m-1}\left(\prod_{i=n+1}^{m-1}\|A_i\|'\right)\right.\right.$$
$$\times C\gamma^{-n}(\|z(n)\|')^{1+\alpha}$$
$$\left.\left. + \sum_{n=0}^{\infty}\left(\prod_{i=0}^{n}\|B_{i+m}^{-1}\|'\right) C\gamma^{-(m+n)}(\|z(n+m)\|')^{1+\alpha}\right]\right\}$$

$$\begin{aligned}&\le \|z\|_\kappa + \sup_{m\ge 0}(\kappa^{-m}\lambda'^m)\|y\|'\\&\quad+\sup_{m\ge 0}\left(\kappa^{-m}C\|z\|_\kappa{}^{1+\alpha}\left[\sum_{n=0}^{m-1}\lambda'^{m-n-1}\gamma^{-n}\kappa^{(1+\alpha)n}\right.\right.\\&\quad\left.\left.+\sum_{n=0}^{\infty}\mu'^{-(n+1)}\gamma^{-(m+n)}\kappa^{(1+\alpha)(m+n)}\right]\right).\end{aligned}\tag{7.36}$$

In view of (7.33), we have $\sup_{m\ge 0}(\kappa^{-m}\lambda'^m) = 1$. Since the function $x \mapsto xa^x$ attains the maximum $-e^{-1}\log a$ at $x = -1/\log a$, we obtain

$$\begin{aligned}&\sup_{m\ge 0}\left[\kappa^{-m}\lambda'^{m-1}\sum_{n=0}^{m-1}\left(\lambda'^{-1}\gamma^{-1}\kappa^{(1+\alpha)}\right)^n\right]\\&\le\begin{cases}\lambda'^{-1}(\kappa^{-1}\lambda')^m m & \text{if } \lambda'^{-1}\gamma^{-1}\kappa^{(1+\alpha)} \le 1\\ \lambda'^{-1}(\gamma^{-1}\kappa^{\alpha})^m m & \text{if } \lambda'^{-1}\gamma^{-1}\kappa^{(1+\alpha)} \ge 1\end{cases}\\&\le \lambda'^{-1}e^{-1}\left(\log\max\left\{\frac{\kappa^\alpha}{\gamma},\frac{\lambda'}{\kappa}\right\}\right)^{-1} =: M_1.\end{aligned}$$

Furthermore, since $\mu'^{-1}\gamma^{-1}\kappa^{1+\alpha} = (\mu'^{-1}\kappa)(\kappa^\alpha\gamma^{-1}) < 1$, we have

$$\begin{aligned}&\sup_{m\ge 0}\left(\kappa^{-m}\gamma^{-m}\kappa^{(1+\alpha)m}\mu'^{-1}\sum_{n=0}^{\infty}\left(\mu'^{-1}\gamma^{-1}\kappa^{(1+\alpha)}\right)^n\right)\\&=\frac{1}{\mu'-\gamma^{-1}\kappa^{1+\alpha}} =: M_2.\end{aligned}\tag{7.37}$$

Setting $M = M_1 + M_2$, we conclude that

$$\|\Phi_\kappa(y,z)\|_\kappa \le \|z\|_\kappa + \|y\|' + CM\|z\|_\kappa{}^{1+\alpha}.$$

This implies that the map Φ_κ is well defined. Moreover, $\Phi_\kappa(0,0) = (0,0)$. We now show that the map Φ_κ is of class C^1. Indeed, for any $y \in B^s(r_0)$ and $t \in E^s$ such that $y + t \in B^s(r_0)$, any $z \in W$, and $m \ge 0$, we have

$$\Phi_\kappa(y+t,z)(m) - \Phi_\kappa(y,z)(m) = \left(\left(\prod_{i=0}^{m-1}A_i\right)t, 0\right).$$

It follows that

$$\partial_y\Phi_\kappa(y,z)(m) = \left(\prod_{i=0}^{m-1}A_i, 0\right).\tag{7.38}$$

For any $y \in B^s(r_0)$, $z \in W$, and $t \in \Gamma_\kappa$ such that $z + t \in W$, we write

$$\Phi_\kappa(y, z+t) - \Phi_\kappa(y,z) = (\mathcal{A}_\kappa(z) - \mathrm{Id})t + o(z,t),$$

where Id is the identity map,

$$(\mathcal{A}_\kappa(z))t(m) = \left(\sum_{n=0}^{m-1} \left(\prod_{i=n+1}^{m-1} A_i \right) dg_n(z(n))t(n), \right.$$
$$\left. - \sum_{n=0}^{\infty} \left(\prod_{i=0}^{n} B_{i+m} \right)^{-1} dh_{n+m}(z(n+m))t(m+n) \right),$$

and

$$o(z,t)(m)$$
$$= \left(\sum_{n=0}^{m-1} \left(\prod_{i=n+1}^{m-1} A_i \right) o_1(z,t)(n), - \sum_{n=0}^{\infty} \left(\prod_{i=0}^{n} B_{i+m} \right)^{-1} o_2(z,t)(m+n) \right).$$

Here, $o_i(z,t)(m)$, for $i = 1, 2$, are defined by

$$o_1(z,t)(m) = g_m((z+t)(m)) - g_m(z(m)) - dg_m(z(m))t(m),$$
$$o_2(z,t)(m) = h_m((z+t)(m)) - h_m(z(m)) - dh_m(z(m))t(m).$$

If $z_1, z_2 \in W$ and $t \in \Gamma_\kappa$ then

$$\|(\mathcal{A}_\kappa(z_1) - \mathcal{A}_\kappa(z_2))t\|_\kappa$$
$$\le \sup_{m \ge 0} \left\{ \kappa^{-m} \left[\sum_{n=0}^{m-1} \left(\prod_{i=n+1}^{m-1} \|A_i\|' \right) C\gamma^{-n} (\|z_1(n) - z_2(n)\|')^\alpha \|t(n)\|' \right. \right.$$
$$+ \sum_{n=0}^{\infty} \left(\prod_{i=0}^{n} \|B_{i+m}^{-1}\|' \right) C\gamma^{-(n+m)} (\|z_1(n+m) - z_2(n+m)\|')^\alpha$$
$$\left. \left. \times \|t(n+m)\|' \right] \right\}$$
$$\le \sup_{m \ge 0} \left\{ \kappa^{-m} C \left[\sum_{n=0}^{m-1} \lambda'^{m-n-1} \gamma^{-n} \kappa^{(1+\alpha)n} \right. \right.$$
$$\left. \left. + \sum_{n=0}^{\infty} \mu'^{-(n+1)} \gamma^{-(n+m)} \kappa^{(1+\alpha)(n+m)} \right] \right\} \times \|z_1 - z_2\|_\kappa{}^\alpha \|t\|_\kappa.$$

It follows that (see (7.36)–(7.37))

$$\|(\mathcal{A}_\kappa(z_1) - \mathcal{A}_\kappa(z_2))t\|_\kappa \le CM\|z_1 - z_2\|_\kappa{}^\alpha \|t\|_\kappa. \tag{7.39}$$

By (7.11) and the Mean Value Theorem, we find that

$$\|o_i(z,t)(m)\|' \le C\gamma^{-m}(\|t(m)\|')^{1+\alpha}.$$

This implies that

$$\|o(z,t)\|_\kappa \le CM\|t\|_\kappa{}^{1+\alpha}.$$

We conclude that $\partial_z \Phi_\kappa(y, z) = \mathcal{A}_\kappa(z) - \mathrm{Id}$. In particular, $\partial_z \Phi_\kappa(y, 0) = -\mathrm{Id}$. By (7.39), the map $\partial_z \Phi_\kappa(y, z)$ is continuous. Therefore the map Φ_κ satisfies the conditions of the Implicit Function Theorem, and hence there exist a number $r \le r_0$ and a map $\varphi\colon B^s(r) \to W$ of class C^1 with

$$\varphi(0) = 0 \quad \text{and} \quad \Phi_\kappa(y, \varphi(y)) = 0. \tag{7.40}$$

We use a special version of the Implicit Function Theorem, which enables us to obtain an explicit estimate of the number r and thus to show that this number depends only on the numbers λ', μ', γ, α, κ, and C. More precisely, the following statement holds:

Lemma 7.5.2 (Implicit Function Theorem). *Let E_1, E_2, and G be Banach spaces and $g\colon A_1 \times A_2 \to G$ a C^1 map, where $A_i \subset E_i$ is a ball centered at 0 of radius r_i, for $i = 1, 2$. Assume that $g(0, 0) = 0$ and that the partial derivative (over the second coordinate) $D_2 g(0, 0)\colon E_2 \to G$ is a linear homeomorphism. Also assume that Dg is Hölder continuous in $A_1 \times A_2$ with Hölder constant a and Hölder exponent α. Let B be a ball in E_1 centered at zero of radius*

$$\rho_1 = \min\left\{ r_1, r_2, \frac{r_2}{2bc}, \frac{1}{(1 + 2bc)(2ac)^{1/\alpha}} \right\}, \tag{7.41}$$

where

$$b = \sup_{x \in A_1} \|D_1 g(x, 0)\| \quad \textit{and} \quad c = \|(D_2 g(0, 0))^{-1}\|.$$

Then there exists a unique map $u\colon B \to A_2$, satisfying the following properties:

1. *u is of class $C^{1+\alpha}$, $g(x, u(x)) = 0$ for every $x \in B$, and $u(0) = 0$.*
2. *If $x \in B$, then*
$$\left\| \frac{du}{dx}(x) \right\| \le 1 + 2bc.$$
3. *If $x_1, x_2 \in B$, then*
$$\left\| \frac{du}{dx}(x_1) - \frac{du}{dx}(x_2) \right\| \le 8ac(1 + bc)^2 \|x_1 - x_2\|^\alpha.$$

Proof of the lemma. Let $\rho_2 = \max\{\rho_1, 2bc\rho_1\}$. By (7.41), we have that $\rho_1 \le r_1$ and $\rho_2 \le r_2$. Also, let $S_i \subset E_i$ be the open balls centered at zero of radii ρ_i, for $i = 1, 2$. We equip the space $E_1 \times E_2$ with the norm $\|(x, y)\| = \|x\| + \|y\|$ for $(x, y) \in E_1 \times E_2$. By (7.41), for $x \in S_1$ and $y_1, y_2 \in S_2$,

$$\begin{aligned} &\|g(x, y_1) - g(x, y_2) - D_2 g(0, 0)(y_1 - y_2)\| \\ &\le a(\rho_1 + \rho_2)^\alpha \|y_1 - y_2\| \le \frac{1}{2c}\|y_1 - y_2\|. \end{aligned}$$

Define the function $H\colon S_1 \times S_2 \to E_2$ by

$$H(x, y) = y - (D_2 g(0, 0))^{-1} g(x, y).$$

We obtain

$$\begin{aligned}\|H(x,y_1)-&H(x,y_2)\|\\&=\|(D_2g(0,0))^{-1}[D_2g(0,0)(y_1-y_2)\\&\quad-(g(x,y_1)-g(x,y_2))]\|\le\frac{1}{2}\|y_1-y_2\|.\end{aligned}\tag{7.42}$$

Furthermore,

$$\begin{aligned}\|H(x,0)\|&=\|-(D_2g(0,0))^{-1}g(x,0)\|\\&=\|-(D_2g(0,0))^{-1}[g(x,0)-g(0,0)]\|\le cb\rho_1\le\frac{\rho_2}{2}.\end{aligned}\tag{7.43}$$

We need the following statement:

Lemma 7.5.3. *Let E be a Banach space, $S\subset E$ the open ball of radius r centered at 0, and $h\colon S\to E$ a transformation. Assume that there exists $0\le\beta<1$ such that $\|h(y_1)-h(y_2)\|\le\beta\|y_1-y_2\|$ for any y_1, $y_2\in S$, and $\|h(0)\|\le r(1-\beta)$. Then there exists one and only one $y\in S$ such that $h(y)=y$.*

The proof of the lemma can be easily obtained by showing that $h^n(0)$ is a Cauchy sequence converging to a point in S. We leave the details to the reader.

By Lemma 7.5.3 with $r=\rho_2$ and $\beta=1/2$ (see (7.42) and (7.43)), there exists a unique function $u\colon S_1\to S_2$ such that $u(0)=0$ and

$$g(x,u(x))=0$$

for every $x\in S_1$. Write $h_x(y)=H(x,y)$. Then

$$\|(h_x)^n(0)-(h_x)^{n-1}(0)\|\le\frac{\rho_2}{2^n}.$$

This implies that the series

$$\sum_{n=0}^{\infty}[h_x{}^n(0)-h_x{}^{n-1}(0)]=u(x)$$

converges uniformly and defines a continuous function.

We now show that u is differentiable. Choose x, $x+s\in S_1$ and set $t=u(x+s)-u(x)$. For any $\delta>0$, there exists $r>0$ such that

$$\begin{aligned}&\|D_1g(x,u(x))s+D_2g(x,u(x))t\|\\&=\|g(x+s,u(x)+t)-g(x,u(x))-D_1g(x,u(x))s-D_2g(x,u(x))t\|\\&\le\delta\|(s,t)\|\end{aligned}$$

whenever $\|s\|<r$. Set

$$a=2\|(D_2g(x,u(x)))^{-1}D_1g(x,u(x))\|+1.$$

Choosing δ so small that $\delta\|(D_2g(x,u(x)))^{-1}\|\le 1/2$, we find that

$$\|t\|-\frac{a-1}{2}\|s\|\le\|t+(D_2g(x,u(x)))^{-1}D_1g(x,u(x))s\|\le\frac{\|s\|+\|t\|}{2}.$$

This implies that $\|t\| \le a\|s\|$, and thus

$$\|t + (D_2g(x, u(x)))^{-1} D_1g(x, u(x))s\| \le \delta(a + 1)\|(D_2g(x, u(x)))^{-1}\| \cdot \|s\|.$$

We conclude that u is differentiable and its derivative is given by

$$\frac{du}{dx}(x) = -(D_2g(x, u(x)))^{-1} D_1g(x, u(x)). \tag{7.44}$$

Therefore, u is of class C^1. Since Dg is Hölder continuous, it follows from (7.44) that u has a Hölder continuous derivative. One can also see that

$$\begin{aligned}\|D_1g(x, y)\| &\le \|D_1g(x, y) - D_1g(0, 0)\| + \|D_1g(0, 0)\| \\ &\le a(\rho_1 + \rho_2)^\alpha + b \le \frac{1}{2c} + b.\end{aligned} \tag{7.45}$$

Furthermore,

$$\begin{aligned}\|D_2g(x, y)\| &\ge \|D_2g(0, 0)\| - \|D_2g(x, y) - D_2g(0, 0)\| \\ &\ge \frac{1}{c} - a(\rho_1 + \rho_2)^\alpha \ge \frac{1}{2c}.\end{aligned}$$

This implies that $D_2g(x, y)$ is invertible for every $(x, y) \in S_1 \times S_2$ and that

$$\|(D_2g(x, y))^{-1}\| \le 2c. \tag{7.46}$$

It follows from (7.45) and (7.46) that for every $x \in S_1$,

$$\begin{aligned}\left\|\frac{du}{dx}(x)\right\| &= \|-(D_2g(x, u(x)))^{-1} D_1g(x, u(x))\| \\ &\le 2c\left(\frac{1}{2c} + b\right) = 1 + 2bc.\end{aligned} \tag{7.47}$$

In view of (7.44), we obtain that

$$D_1g(x, u(x)) + D_2g(x, u(x))\frac{du}{dx}(x) = 0,$$

and thus

$$\begin{aligned}&D_1g(x_1, u(x_1)) - D_1g(x_2, u(x_2)) \\ &+ [D_2g(x_1, u(x_1)) - D_2g(x_2, u(x_2))]\frac{du}{dx}(x_1) \\ &+ D_2g(x_2, u(x_2))\left(\frac{du}{dx}(x_1) - \frac{du}{dx}(x_2)\right) = 0\end{aligned}$$

for every $x_1, x_2 \in S_1$. Using (7.45) and (7.47), we conclude that

$$\begin{aligned}\left\|\frac{du}{dx}(x_1) - \frac{du}{dx}(x_2)\right\| &\le \|(D_2g(x_2, u(x_2)))^{-1}\| \\ &\quad \times a\|(x_1, u(x_1)) - (x_2, u(x_2))\|^\alpha(2 + 2bc) \\ &\le 4ac(1 + bc)(2 + 2bc)^\alpha\|x_1 - x_2\|^\alpha \\ &\le 8ac(1 + bc)^2\|x_1 - x_2\|^\alpha.\end{aligned}$$

This completes the proof of the lemma. □

The map Φ_κ satisfies the conditions of Lemma 7.5.2 with

$$c = 1, \quad b = 1, \quad a = CM. \tag{7.48}$$

To see this we only need to show that the derivatives $\partial_y \Phi_\kappa$ and $\partial_z \Phi_\kappa$ are Hölder continuous. This holds true for $\partial_y \Phi_\kappa$ in view of (7.38). Furthermore, by (7.39),

$$\begin{aligned}
&\|\partial_z \Phi_\kappa(y_1, z_1) - \partial_z \Phi_\kappa(y_2, z_2)\| \\
&\le \|\partial_z \Phi_\kappa(y_1, z_1) - \partial_z \Phi_\kappa(y_1, z_2)\| + \|\partial_z \Phi_\kappa(y_1, z_2) - \partial_z \Phi_\kappa(y_2, z_2)\| \\
&= 2\|\mathcal{A}_\kappa(z_1) - \mathcal{A}_\kappa(z_2)\| \le CM\|z_1 - z_2\|_\kappa{}^\alpha.
\end{aligned}$$

We now describe some properties of the map φ. Differentiating the second equality in (7.40) with respect to y, we obtain

$$d\varphi(y) = -[\partial_z \Phi_\kappa(y, \varphi(y))]^{-1} \partial_y \Phi_\kappa(y, \varphi(y)).$$

Setting $y = 0$ in this equation yields

$$d\varphi(0)(m) = \left(\prod_{i=0}^{m-1} A_i, 0\right). \tag{7.49}$$

One can write the vector $\varphi(y)(m)$ in the form

$$\varphi(y)(m) = (\varphi_1(y)(m), \varphi_2(y)(m)),$$

where $\varphi_1(y)(m) \in \mathbb{R}^k$ and $\varphi_2(y)(m) \in \mathbb{R}^{p-k}$. It follows from (7.40) that for $m \ge 0$,

$$\varphi_1(y)(m) = \left(\prod_{i=0}^{m-1} A_i\right) y + \sum_{n=0}^{m-1} \left(\prod_{i=n+1}^{m-1} A_i\right) g_n(\varphi(y)(n)), \tag{7.50}$$

and

$$\varphi_2(y)(m) = -\sum_{n=0}^{\infty} \left(\prod_{i=0}^{n} B_{i+m}\right)^{-1} h_{n+m}(\varphi(y)(n+m)). \tag{7.51}$$

This implies that

$$\begin{aligned}
\varphi_1(y)(m+1) &= A_m \varphi_1(y)(m) + g_m(\varphi_1(y)(m), \varphi_2(y)(m)), \\
\varphi_2(y)(m+1) &= B_m \varphi_2(y)(m) + h_m(\varphi_1(y)(m), \varphi_2(y)(m)).
\end{aligned}$$

Indeed, iterating the first equation "forward" one easily obtains (7.50). Rewriting the second equation in the form

$$\varphi_2(y)(m) = B_m^{-1} \varphi_2(y)(m+1) - B_m^{-1} h_m(\varphi_1(y)(m), \varphi_2(y)(m))$$

and iterating it "backward" yields (7.51).

Thus, we obtain that the function $\varphi(y)$ is invariant under the family of maps $\widetilde{F}_m$; that is,

$$\widetilde{F}_m(\varphi(y)(m)) = \varphi(y)(m+1). \tag{7.52}$$

The desired map ψ^s is now defined by

$$\psi^s(v) = \varphi_2(v)(0)$$

for each $v \in B^s(r)$. Note that $\varphi_1(v)(0) = v$. It follows from (7.40), (7.49), and Lemma 7.5.2 that the map ψ^s satisfies Statement 1 of the theorem. It follows from (7.52) that

$$\prod_{i=0}^{m-1} \widetilde{F}_i(v, \psi^s(v)) = \prod_{i=0}^{m-1} \widetilde{F}_i(\varphi_1(v)(0), \varphi_2(v)(0))$$
$$= \prod_{i=0}^{m-1} \widetilde{F}_i(\varphi(v)(0)) = \varphi(v)(m).$$

Applying Lemma 7.5.2 and (7.48), we find that

$$\left\| \prod_{i=0}^{m-1} \widetilde{F}_i(v, \psi^s(v)) \right\|' \leq \kappa^m \|\varphi(v)\|_\kappa = \kappa^m \|\varphi(v) - \varphi(0)\|_\kappa$$
$$\leq \kappa^m \sup_{\xi \in B^s(r)} \|d\varphi(\xi)\|_\kappa \|v\|'$$
$$\leq 3\kappa^m \|v\|' \leq 3\kappa^m \|(v, \psi^s(v))\|'.$$

(We use the fact that for every $v = (v_1, v_2) \in \mathbb{R}^2$, we have $\|v_1\|, \|v_2\| \leq \|v\|$.) This proves Statement 3. Furthermore, for any $u_1, u_2 \in B^s(r)$, we have in view of Lemma 7.5.2 and (7.48) that D depends only on the numbers λ', μ', γ, α, κ, and C and that

$$\|d\psi^s(v_1) - d\psi^s(v_2)\|' = \|d\varphi_2(v_1)(0) - d\varphi_2(v_2)(0)\|'$$
$$\leq \|d\varphi(v_1)(0) - d\varphi(v_2)(0)\|' \leq 32CM(\|v_1 - v_2\|')^\alpha.$$

This establishes Statements 2 and 5. Let $(v, w) \in B^s(r) \times B^u(r_0)$ satisfies the assumptions of Statement 4 of the theorem. Set

$$z(m) = \left(\prod_{i=0}^{m-1} \widetilde{F}_i(v, w) \right).$$

It follows that $z \in \Gamma_\kappa$ (with $\|z\|_\kappa \leq K$) and that $\Phi_k(v, z) = 0$. The uniqueness of the map φ implies that $z = \varphi(v)$, and hence

$$w = \varphi_2(v)(0) = \psi^s(v).$$

This establishes Statement 4. The desired result now follows. □

7.5.2 Smoothness of Local Manifolds

Theorem 7.1.1 fails for C^1 diffeomorphisms (see Pugh's counterexample in Section 7.8). On the other hand, one can obtain a more refined information about the smoothness of the local stable manifold $V^s(x)$ in the case when the diffeomorphism f is of class $C^{r+\alpha}$ with $r \geq 1$ and $0 < \alpha \leq 1$ (i.e., $d^r f$ is Hölder continuous with Hölder exponent α). Namely, in this case $V^s(x)$ is of class $C^{r+\alpha}$; in particular, if f is of class C^r for some $r \geq 2$, then $V^s(x)$ is of class C^{r-1} and its rth derivative is Lipschitz continuous. These results are immediate consequences of the following version of Theorem 7.5.1.

Theorem 7.5.4. *Assume that the conditions of Theorem 7.5.1 hold. In addition, assume that*

1. g_m *and* h_m *are of class* C^r *for some* $r \geq 2$;
2. *there exists* $K > 0$ *such that for* $\ell = 2, \ldots, r$,

$$\sup_{z\in B} \|d^\ell g_m(z)\|' \leq K\gamma^{-m} \quad \text{and} \quad \sup_{z\in B} \|d^\ell h_m(z)\|' \leq K\gamma^{-m},$$

where $B = B^s(r_0) \times B^u(r_0)$ *(see (7.8))*;

3. *for any* $z_1, z_2 \in B$,

$$\|d^r g_m(z_1) - d^r g_m(z_2)\|' \leq K\gamma^{-m}(\|z_1 - z_2\|')^\alpha,$$
$$\|d^r h_m(z_1) - d^r h_m(z_2)\|' \leq K\gamma^{-m}(\|z_1 - z_2\|')^\alpha$$

for some $\alpha \in (0, 1)$.

If $\psi^s(u)$ *is the map constructed in Theorem 7.5.1, then there exists a number* $N > 0$, *which depends only on the numbers* $\lambda', \mu', \gamma, \alpha, \kappa$, *and* K, *such that*

1. ψ^s *is of class* $C^{r+\alpha}$;
2. $\sup_{u\in B^s(r)} \|d^\ell \psi^s(u)\|' \leq N$ *for* $\ell = 1, \ldots, r$.

Proof. We shall show that Φ_κ is of class C^r. A modification of arguments in the proof of Theorem 7.5.1 allows one to prove that

$$d_y^\ell \Phi_\kappa(y, \mathbf{z}) = (0, 0), \quad 2 \leq \ell \leq r.$$

We shall establish that

$$d_z^\ell \Phi_\kappa(y, \mathbf{z})(m) =$$
$$\left(\sum_{n=0}^{m-1} \left(\prod_{i=n+1}^{m-1} A_i\right) d^\ell g_n(\mathbf{z}(n)), -\sum_{n=0}^{\infty} \left(\prod_{i=0}^{n} B_{i+m}^{-1}\right) d^\ell h_{m+n}(\mathbf{z}(m+n))\right).$$

In order to do this it suffices to verify that all partial derivatives of Φ of order r exist and are continuous. We need to check only this for $\partial_{\mathbf{z}}^r \Phi(x, \mathbf{z})$ since all other partial derivatives of order r are zero.

Set

$$(\Psi_\ell(y, \mathbf{z})(\mathbf{v}^1, \ldots, \mathbf{v}^\ell))_m = \left(\sum_{k=0}^{m-1} \mathcal{A}_m \mathcal{A}_{k+1}{}^{-1}\, d_{z_k}^\ell g_k\left(v_k^1, \ldots, v_k^\ell\right),\right.$$
$$\left. -\sum_{k=0}^{\infty} \mathcal{B}_m \mathcal{B}_{k+m+1}{}^{-1}\, d_{z_{k+m}}^\ell h_{k+m}\left(v_k^1, \ldots, v_k^\ell\right)\right),$$

where

$$\mathcal{A}_m = \prod_{i=0}^{m-1} A_i \quad \text{and} \quad \mathcal{B}_m = \prod_{i=0}^{m-1} \mathcal{B}_i,$$

for each $m \geq 0$, $\ell = 2, \ldots, r$, and $\mathbf{v}^1, \ldots, \mathbf{v}^\ell \in \Gamma$.

Lemma 7.5.5. *We have $\partial^\ell_{\mathbf{z}}\Phi(x,\mathbf{z}) = \Psi_\ell(x,\mathbf{z})$ and $\Psi_\ell(x,\mathbf{z})$ is a bounded multilinear operator for every $x \in U^s_0$, $\mathbf{z} \in W$, and $\ell = 2, \dots, r$.*

Proof of the lemma. For each $\ell = 2, \dots, r-1$, we have

$$\begin{aligned}
&\|(d^{\ell-1}_{z_k+v^\ell_k} g_k - d^{\ell-1}_{z_k} g_k)(v^1_k, \dots, v^{\ell-1}_k) - d^\ell_{z_k} g_k(v^1_k, \dots, v^\ell_k)\|' \\
&\qquad\qquad \le \sum_{i=1}^{n} \|d^\ell_{\xi_{ki}} g_{ki} - d^\ell_{z_k} g_{ki}\|' \prod_{j=1}^{\ell} \|v^j_k\|'_k \\
&\qquad\qquad \le n K'_{\ell+1} \gamma^{-\alpha k} \|v^\ell_k\|'_k \prod_{j=1}^{\ell} \|v^j_k\|'_k,
\end{aligned}$$

where ξ_{ki} is a vector in the segment joining z_k and $z_k + v^\ell_k$. In a similar way one can show that

$$\begin{aligned}
&\|(d^{\ell-1}_{z_k+v^\ell_k} h_k - d^{\ell-1}_{z_k} h_k)(v^1_k, \dots, v^{\ell-1}_k) - d^\ell_{z_k} h_k(v^1_k, \dots, v^\ell_k)\|' \\
&\qquad\qquad \le n K'_{\ell+1} \gamma^{-\alpha k} \|v^\ell_k\|'_k \prod_{j=1}^{\ell} \|v^j_k\|'_k.
\end{aligned}$$

For $\ell = 2, \dots, r-1$, we obtain

$$\begin{aligned}
&\|(\Psi_{\ell-1}(y, \mathbf{z} + \mathbf{v}^\ell) - \Psi_{\ell-1}(y,\mathbf{z}))(\mathbf{v}^1, \dots, \mathbf{v}^{\ell-1}) - \Psi_\ell(y,\mathbf{z})(\mathbf{v}^1, \dots, \mathbf{v}^\ell)\|_\Gamma \\
&\quad \le \sup_{m\ge 0} \left\{ \kappa^{-m} \sum_{k=0}^{m-1} \|\mathcal{A}_m \mathcal{A}_{k+1}{}^{-1}\|' n K'_{\ell+1} \gamma^{-\alpha k} \|v^\ell_k\|'_k \prod_{j=1}^{\ell} \|v^j_k\|'_k \right\} \\
&\quad\quad + \sup_{m\ge 0} \left\{ \kappa^{-m} \sum_{k=0}^{\infty} \|\mathcal{B}_m \mathcal{B}_{k+m+1}{}^{-1}\|' n K'_{\ell+1} \gamma^{-\alpha(k+m)} \right. \\
&\qquad\qquad \left. \times \|v^\ell_{k+m}\|'_{k+m} \prod_{j=1}^{\ell} \|v^j_{k+m}\|'_{k+m} \right\} \\
&\quad \le n K'_{\ell+1} \sup_{m\ge 0} \left\{ \kappa^{-m} \lambda'^{m-1} \sum_{k=0}^{m-1} [\lambda'^{-1} \gamma^{-\alpha} \kappa^\ell]^k \right\} \|\mathbf{v}^\ell\|_\Gamma \prod_{j=1}^{\ell} \|\mathbf{v}^j\|_\Gamma \\
&\quad\quad + n K'_{\ell+1} \sup_{m\ge 0} \left\{ \kappa^{-m} \mu'^{-1} \gamma^{-\alpha m} \sum_{k=0}^{\infty} [\mu'^{-1} \gamma^{-\alpha}]^k \kappa^{(m+k)\ell} \right\} \\
&\qquad \times \|\mathbf{v}^\ell\|_\Gamma \prod_{j=1}^{\ell} \|\mathbf{v}^j\|_\Gamma \\
&\quad \le n K'_{\ell+1} M \|\mathbf{v}^\ell\|_\Gamma \prod_{j=1}^{\ell} \|\mathbf{v}^j\|_\Gamma,
\end{aligned}$$

since $\kappa^\ell \le \kappa^{1+\alpha}$ for every $\ell \ge 2$. This implies that $\partial^\ell_{\mathbf{z}}\Phi(y,\mathbf{z})$ exists and that $\partial^\ell_{\mathbf{z}}\Phi(y,\mathbf{z}) = \Psi_\ell(y,\mathbf{z})$ for every $\ell = 2, \dots, r-1$.

Now we show that the derivative of order r exists. We have

$$\begin{aligned}\|(d^{r-1}_{z_k+v^r_k}g_k - d^{r-1}_{z_k}g_k)(v^1_k, \ldots, v^{r-1}_k) - d^r_{z_k}g_k(v^1_k, \ldots, v^r_k)\|' \\ \leq \sum_{i=1}^{n}\|d^r_{\xi_{ki}}g_{ki} - d^r_{z_k}g_{ki}\|'\prod_{j=1}^{r}\|v^j_k\|'_k \\ \leq nK'\gamma^{-\alpha k}\|v^r_k\|'^{\alpha}_k\prod_{j=1}^{r}\|v^j_k\|_k,\end{aligned}$$

where ξ_{ki} is a vector in the segment joining z_k and $z_k + v^r_k$. In a similar way one can show that

$$\begin{aligned}\|(d^{r-1}_{z_k+v^r_k}h_k - d^{r-1}_{z_k}h_k)(v^1_k, \ldots, v^{r-1}_k) - d^r_{z_k}h_k(v^1_k, \ldots, v^r_k)\|' \\ \leq nK'\gamma^{-\alpha k}\|v^r_k\|'^{\alpha}_k\prod_{j=1}^{r}\|v^j_k\|_k.\end{aligned}$$

We obtain

$$\begin{aligned}&\|(\Psi_{r-1}(y, \mathbf{z}+\mathbf{v}^r) - \Psi_{r-1}(y, \mathbf{z}))(\mathbf{v}^1, \ldots, \mathbf{v}^{r-1}) - \Psi_r(y, \mathbf{z})(\mathbf{v}^1, \ldots, \mathbf{v}^r)\|_\Gamma \\ &\leq \sup_{m\geq 0}\left\{\kappa^{-m}\sum_{k=0}^{m-1}\|\mathcal{A}_m\mathcal{A}_{k+1}{}^{-1}\|'nK'\gamma^{-\alpha k}\|v^r_k\|'^{\alpha}_k\prod_{j=1}^{r}\|v^j_k\|'_k\right\} \\ &\quad + \sup_{m\geq 0}\left\{\kappa^{-m}\sum_{k=0}^{\infty}\|\mathcal{B}_m\mathcal{B}_{k+m+1}{}^{-1}\|'nK'\gamma^{-\alpha(k+m)}\right. \\ &\qquad\left.\times\|v^r_{k+m}\|'^{\alpha}_{k+m}\prod_{j=1}^{r}\|v^j_{k+m}\|'_{k+m}\right\} \\ &\leq nK'\sup_{m\geq 0}\left\{\kappa^{-m}\lambda'^{m-1}\sum_{k=0}^{m-1}[\lambda'^{-1}\gamma^{-\alpha}\kappa^{\ell+\alpha}]^k\right\}\|\mathbf{v}^r\|_\Gamma{}^{\alpha}\prod_{j=1}^{r}\|\mathbf{v}^j\|_\Gamma \\ &\quad + nK'\sup_{m\geq 0}\left\{\kappa^{-m}\mu'^{-1}\gamma^{-\alpha m}\sum_{k=0}^{\infty}[\mu'^{-1}\gamma^{-\alpha}]^k\kappa^{(m+k)(\ell+\alpha)}\right\} \\ &\qquad\times\|\mathbf{v}^r\|_\Gamma{}^{\alpha}\prod_{j=1}^{r}\|\mathbf{v}^j\|_\Gamma \\ &\leq nK'M\|\mathbf{v}^r\|_\Gamma{}^{\alpha}\prod_{j=1}^{r}\|\mathbf{v}^j\|_\Gamma,\end{aligned}$$

since $\kappa^{\ell+\alpha} \leq \kappa^{1+\alpha}$ for every $\ell \geq 1$. This implies that $\partial^r_{\mathbf{z}}\Phi(y, \mathbf{z})$ exists and that $\partial^r_{\mathbf{z}}\Phi(y, \mathbf{z}) = \Psi_r(y, \mathbf{z})$.

We now prove that the derivatives computed above are continuous. For each $\ell = 2, \ldots, r$, we have

$$
\begin{aligned}
&\|\Psi_\ell(y, \mathbf{z})(\mathbf{v}^1, \ldots, \mathbf{v}^\ell)\|_\Gamma \\
&\le \sup_{m\ge 0}\left\{\kappa^{-m}\sum_{k=0}^{m-1}\|\mathcal{A}_m\mathcal{A}_{k+1}{}^{-1}\|' K'_\ell \gamma^{-\alpha k}\prod_{j=1}^{\ell}\|v_k^j\|'_k\right\} \\
&\quad + \sup_{m\ge 0}\left\{\kappa^{-m}\sum_{k=0}^{\infty}\|\mathcal{B}_m\mathcal{B}_{k+m+1}{}^{-1}\|' K'_\ell \gamma^{-\alpha(k+m)}\prod_{j=1}^{\ell}\|v_{k+m}^j\|'_{k+m}\right\} \\
&\le K'_\ell \sup_{m\ge 0}\left\{\kappa^{-m}\lambda'^{m-1}\sum_{k=0}^{m-1}[\lambda'^{-1}\gamma^{-\alpha}\kappa^\ell]^k\right\}\prod_{j=1}^{\ell}\|\mathbf{v}^j\|_\Gamma \\
&\quad + K'_\ell \sup_{m\ge 0}\left\{\kappa^{-m}\mu'^{-1}\gamma^{-\alpha m}\sum_{k=0}^{\infty}[\mu'^{-1}\gamma^{-\alpha}]^k\kappa^{(m+k)\ell}\right\}\prod_{j=1}^{\ell}\|\mathbf{v}^j\|_\Gamma \\
&\le K'_\ell M\prod_{j=1}^{\ell}\|\mathbf{v}^j\|_\Gamma,
\end{aligned}
$$

since $\kappa^\ell \le \kappa^{1+\alpha}$ for every $\ell \ge 2$. This shows that the multilinear operator $\Psi_\ell(y, \mathbf{z})$ is bounded and thus the proof of the lemma is complete. □

By Lemma 7.5.5, the map Φ is of class C^r. We have

$$
\begin{aligned}
&\|(\Psi_r(y, \mathbf{z}) - \Psi_r(y', \mathbf{z}'))(\mathbf{v}^1, \ldots, \mathbf{v}^r)\|_\Gamma \\
&\le \sup_{m\ge 0}\left\{\kappa^{-m}\sum_{k=0}^{m-1}\|\mathcal{A}_m\mathcal{A}_{k+1}{}^{-1}\|'\|d^r_{z_k}g_k - d^r_{z'_k}g_k\|'\prod_{j=1}^{\ell}\|v_k^j\|'_k\right\} \\
&\quad + \sup_{m\ge 0}\left\{\kappa^{-m}\sum_{k=0}^{\infty}\|\mathcal{B}_m\mathcal{B}_{k+m+1}{}^{-1}\|'\right. \\
&\qquad\qquad \left.\times\|d^r_{z_{k+m}}h_{k+m} - d^r_{z'_{k+m}}h_{k+m}\|'\prod_{j=1}^{\ell}\|v_{k+m}^j\|'_{k+m}\right\} \\
&\le K'_r \sup_{m\ge 0}\left\{\kappa^{-m}\lambda'^{m-1}\sum_{k=0}^{m-1}[\lambda'^{-1}\gamma^{-\alpha}\kappa^{\ell+\alpha}]^k\right\}\|\mathbf{z}-\mathbf{z}'\|_\Gamma{}^\alpha\prod_{j=1}^{\ell}\|\mathbf{v}^j\|_\Gamma \\
&\quad + K'_r \sup_{m\ge 0}\left\{\kappa^{-m}\mu'^{-1}\gamma^{-\alpha m}\sum_{k=0}^{\infty}[\mu'^{-1}\gamma^{-\alpha}]^k\kappa^{(m+k)(\ell+\alpha)}\right\} \\
&\qquad\qquad \times\|\mathbf{z}-\mathbf{z}'\|_\Gamma{}^\alpha\prod_{j=1}^{\ell}\|\mathbf{v}^j\|_\Gamma \\
&\le K'_r M\|\mathbf{z}-\mathbf{z}'\|_\Gamma{}^\alpha\prod_{j=1}^{\ell}\|\mathbf{v}^j\|_\Gamma,
\end{aligned}
\tag{7.53}
$$

since $\kappa^{\ell+\alpha} \le \kappa^{1+\alpha}$ for every $\ell \ge 1$.

We use the following version of implicit function theorem.

Lemma 7.5.6. *Under the hypothesis of Lemma 7.5.2 assume, in addition, that G is of class C^r and that there is a constant $A > 0$ such that*

$$\|d^r_{(x,y)}G - d^r_{(x',y')}G\| \le A\|(x,y) - (x',y')\|^\alpha$$

for every (x,y), $(x',y') \in B_1(0,r_1) \times B_2(0,r_2)$. Then there exists constants $r_0' \in (0,r_0]$ and L, $L_1, \ldots, L_r > 0$ such that the following statements hold:

1. *ψ is of class C^r on $B(0,r_0')$.*
2. *$\|d^\ell_x\psi\| \le L_\ell$ for every $x \in B_1(0,r_0')$ and $\ell = 1, \ldots, r$.*
3. *For $x_1, x_2 \in B(0,r_0)$,*

$$\|d^r_{x_1}\psi - d^r_{x_2}\psi\| \le L\|x_1 - x_2\|^\alpha.$$

By (7.53), we can apply Lemma 7.5.6 to the map $G = \Phi$. We conclude that

$$\sup\{\|d^\ell_x\varphi\|_0' : x \in B(0,r_0')\} < L_\ell,$$

and for $x_1, x_2 \in B(0,r_0')$,

$$\|d^r_{x_1}\varphi - d^r_{x_2}\varphi\|' \le \|d^r_{x_1}\psi - d^r_{x_2}\psi\| \le L\left(\|x_1 - x_2\|_0'\right)^\alpha,$$

where L, $L_1, \ldots, L_r$ are positive constants. This concludes the proof of the theorem. □

7.6 Local Unstable Manifolds

In the hyperbolic theory there is a symmetry between the objects marked by the index "s" and those marked by the index "u." Namely, when the time direction is reversed, the statements concerning objects with index "s" become the statements about the corresponding objects with index "u." In particular, this allows one to define unstable manifolds.

Let f be a $C^{1+\alpha}$ diffeomorphism of a compact smooth Riemannian manifold M and $Y \subset M$ a nonuniformly partially hyperbolic set for f.

Theorem 7.6.1 (Unstable Manifold Theorem). *There exist an f-invariant Borel function $\varepsilon\colon Y \to (0,\varepsilon_0]$, a Borel function $T\colon Y \to (0,+\infty)$, and for every $x \in Y$, a local unstable manifold $V^u(x)$ such that*

1. *$x \in V^u(x)$, $T_xV^u(x) = E^u(x)$;*
2. *for $y \in V^u(x)$ and $n \le 0$,*

$$\rho(f^n(x), f^n(y)) \le T(x)\mu(x)^n e^{\varepsilon(x)|n|}\rho(x,y),$$

   where $\mu(x)$ is an f-invariant Borel function in Definition 5.2.1;

3. *the function $T(x)$ satisfies (7.2).*

7.7 The Stable Manifold Theorem for Flows

Let φ_t be a smooth flow on a compact smooth Riemannian manifold M and Y a nonuniformly partially hyperbolic set for φ_t; that is, Y is a φ_t-invariant subset, which is nonuniformly partially hyperbolic for the time-1 map φ_1. The following result is an analogue of Theorem 7.1.1 for flows.

Theorem 7.7.1. *There exist (a) f-invariant Borel functions $\varepsilon\colon Y \to (0, \varepsilon_0]$, $\lambda, \mu\colon Y \to (0, \infty)$ satisfying (2.9); (b) a Borel function $T\colon Y \to (0, +\infty)$; and (c) for every $x \in Y$, a local stable manifold $V^s(x)$ such that*

1. $x \in V^s(x)$, $T_x V^s(x) = E^s(x)$, $V^s \in C^{1+\alpha}$;
2. *for $y \in V^s(x)$ and $t > 0$,*

$$\rho(\varphi_t(x), \varphi_t(y)) \le T(x)\lambda(x)^t e^{\varepsilon(x)t} \rho(x, y);$$

3. *for $s \in \mathbb{R}$,*

$$T(\varphi_s(x)) \le T(x) e^{10\varepsilon(x)|s|}.$$

The proof of Theorem 7.7.1 can be obtained applying Theorem 7.1.1 to the diffeomorphism $f = \varphi_1$ (which is partially hyperbolic in the broad sense). We call $V^s(x)$ a *local stable manifold* at x. We also define a local weakly stable manifold at x by

$$V^{s0} = \bigcup_{|t|\le\tau} V^s(\varphi_t(x)),$$

where $\tau > 0$ is a sufficiently small constant independent of x.

By reversing the time one can show that for every $x \in Y$, there exists a *local unstable manifold* $V^u(x)$ at x such that $T_x V^u(x) = E^u(x)$. One can also construct a local weakly unstable manifold $V^{u0}(x)$.

7.8 C^1 Pathological Behavior: Pugh's Example

We shall show that the requirement in Theorem 7.1.1 that the diffeomorphism f is of class $C^{1+\alpha}$ for some $\alpha > 0$ cannot be relaxed. Namely, we describe an example due to Pugh [184] of a nonuniformly hyperbolic C^1 diffeomorphism (which is not $C^{1+\alpha}$ for any $\alpha > 0$) of a four-dimensional manifold with the following property: there exists no manifold tangent to $E^s(x)$ for which (7.1) holds on some open neighborhood of x.

Let S^2 be the sphere, S^2_0 its equator, and S^2_- and S^2_+ the southern and northern hemispheres, respectively. Consider the central projections $\rho_-\colon \mathbb{R}^2 \to S^2_-$ and $\rho_+\colon \mathbb{R}^2 \to S^2_+$ with the South pole at the origin. We have $\rho_+ = i \circ \rho_-$, where i is the antipodal map of S^2. Any map $f\colon \mathbb{R}^2 \to \mathbb{R}^2$ lifts to a map $\rho_- f \cup \rho_+ f\colon S^2 \setminus$

$S_0^2 \to S^2 \setminus S_0^2$ such that the diagram

$$\begin{array}{ccc} S_\pm^2 & \xrightarrow{\rho_\pm f} & S_\pm^2 \\ {\scriptstyle\rho_\pm}\downarrow & & \downarrow{\scriptstyle\rho_\pm} \\ \mathbb{R}^2 & \xrightarrow{f} & \mathbb{R}^2 \end{array}$$

is commutative. The following statement gives sufficient conditions for extending $\rho_- f \cup \rho_+ f$ to a map $\rho_\sharp f$ on S^2.

Proposition 7.8.1 ([184, Lemma 4]). *Let $A = \left(\begin{smallmatrix} a & c \\ 0 & b \end{smallmatrix}\right)$ with $ab \neq 0$, and let $h\colon \mathbb{R} \to \mathbb{R}$ be a C^1 function with compact support. Then the map*

$$f(x, y) = (ax + cy + h(y), by)$$

lifts to a unique continuous map $\rho_\sharp f$ of S^2, which agrees with $\rho_- f \cup \rho_+ f$ on $S^2 \setminus S_0^2$. Moreover, $\rho_\sharp f$ is a C^1 diffeomorphism whose 1-jet at the equator S_0^2 is the same as that of $\rho_\sharp A$.

At the points $x_\pm$, where the x-axis longitude L_x meets S_0^2, this 1-jet is given by $\left(\begin{smallmatrix} 1/a & 0 \\ 0 & b/a \end{smallmatrix}\right)$ with respect to the splitting $T_{x_\pm} S^2 = T_{x_\pm} L_x \oplus T_{x_\pm} S_0^2$.

We proceed with the construction of Pugh's example. Let $g\colon (0, \infty) \to (0, \infty)$ be a smooth function such that

$$g(u) = \frac{u}{\log(1/u)} \qquad \text{for } 0 < u < 1/e,$$

$g'(u) > 1$ for $u \geq 1/e$, and $g'(u)$ is constant for $u \geq 1$. We extend g to $\mathbb{R}$ by setting

$$g_o(u) = \begin{cases} g(u), & u > 0 \\ 0, & u = 0 \\ -g(u), & u < 0 \end{cases}.$$

Then $g_o'(u) = c$ for some constant $c > 1$ provided that $|u| \geq 1$. Choose constants a and b such that $0 < a < ab < 1 < b$. Consider the maps

$$f_\pm(x, y) = (ax \pm g_o(y), by).$$

Since $y \mapsto g_o(y) - cy$ has compact support, the maps $f_\pm$ satisfy the hypotheses of Proposition 7.8.1 and lift to S^2 as $\rho_\sharp(f_\pm)$.

We divide S^2 into two hemispheres $H_\pm$ along the x-axis longitude L_x: $H_\pm$ is the hemisphere containing the quarter sphere $\rho_-\{(x, y) \in \mathbb{R}^2 : \pm y > 0\}$. Define $F_S\colon S^2 \to S^2$ by

$$F_S = \begin{cases} \rho_\sharp f_+(z) & z \in H_+ \\ \rho_\sharp f_-(z) & z \in H_- \end{cases}.$$

One can then show that F_S is a C^1 diffeomorphism.

Let $h\colon M \to M$ be a diffeomorphism of a compact surface M having a hyperbolic invariant set Λ on which h is topologically conjugate to the full 2-shift. If

$T_\Lambda M = E^s \oplus E^u$ is the invariant splitting into stable and unstable subspaces then h can be chosen such that

$$\begin{aligned} \|dh(v)\| &< \|dF_S(u)\| \text{ when } u \in TS^2, v \in E^s, \|u\| = \|v\| = 1, \\ \|dh(v)\| &> \|dF_S(u)\| \text{ when } u \in TS^2, v \in E^u, \|u\| = \|v\| = 1. \end{aligned} \tag{7.54}$$

Let Λ_i, $i = 0, 1$, be the compact sets of points of Λ that correspond to sequences of symbols with i in the initial position. Choose a smooth bump function $\mu\colon M \to [0, \pi/2]$ such that $\Lambda_0 = \mu^{-1}(0) \cap \Lambda$, $\Lambda_1 = \mu^{-1}(\pi/2) \cap \Lambda$, and $\mu^{-1}(\{0, \pi/2\})$ is a neighborhood of Λ. Let now R_θ be the rotation of S^2 by angle θ that fixes the poles. We define a C^1 diffeomorphism $F\colon M \times S^2 \to M \times S^2$ by

$$F(w, z) = (h(w), (R_{\mu(w)} \circ F_S \circ R_{-\mu(w)})(z)).$$

Note that F leaves invariant the foliation $\mathfrak{F}$ by 2-spheres $\{w\} \times S^2$, $w \in M$, and by (7.54), F is normally hyperbolic to $\mathfrak{F}$. This means that for every $w \in M$ and $z \in S^2$, we have the splitting

$$T_{(w,z)}(M \times S^2) = E^{ss}(w, z) \oplus T_{(w,z)}\mathfrak{F} \oplus E^{uu}(w, z),$$

which is invariant under dF. Moreover, dF contracts vectors in $E^{ss}(w, z)$ and expands vectors in $E^{uu}(w, z)$ with uniform rates, which are strictly bigger than the rates of contraction and expansion along $T_{(w,z)}\mathfrak{F}$. The subspaces $E^{ss}(w, z)$ and $E^{uu}(w, z)$ are called the *strongly stable* and *strongly unstable* subspaces at (w, z). They are locally integrable to local strongly stable and local strongly unstable manifolds $V^{ss}(w, z)$ and $V^{uu}(w, z)$.

Proposition 7.8.2. *There is a point $P \in M \times S^2$, which is Lyapunov–Perron regular with nonzero Lyapunov exponents for the diffeomorphism F and for which the stable set*

$$V^s(P) = \left\{x \in M \times S^2 : \lim_{n\to\infty} \frac{1}{n} \log d(F^n(x), F^n(P)) < 0\right\}$$

is not an injectively immersed submanifold tangent to $E^{ss}(P) \oplus T_P\mathfrak{F}$.

Proof. We consider another extension of the function g to $\mathbb{R}$,

$$\tilde{g}(u) = \begin{cases} g(u), & u > 0 \\ 0, & u = 0 \\ g(u), & u < 0 \end{cases}.$$

Then $\tilde{g}$ is of class C^1 on $\mathbb{R}$ and of class C^∞ on $\mathbb{R} \setminus \{0\}$. Consider the C^1 maps f_S and f_T in $\mathbb{R}^2$ defined by

$$f_S(x, y) = (ax + g(y), by) \quad \text{and} \quad f_T(x, y) = (bx, g(x) + ay).$$

One can easily verify that f_S and f_T are C^1 diffeomorphisms of $\mathbb{R}^2$ onto itself and that F_S lifts f_S to S^2 (but not as $\rho_- f_S \cup \rho_+ f_S$; in fact the canonical lift $\rho_\pm f_S$ fails to be C^1 at the equator).

Set $L_k = k(k+1)/2$ and for each $n \in \mathbb{N}$,

$$f_n = \begin{cases} f_S, & n \in \mathcal{S} \\ f_T, & n \in \mathcal{T} \end{cases},$$

where

$$\mathcal{S} = \{n \in \mathbb{N} : L_{k-1} < n \leq L_k \text{ for some odd } k\},$$
$$\mathcal{T} = \{n \in \mathbb{N} : L_{k-1} < n \leq L_k \text{ for some even } k\}.$$

Notice that the sets $\mathcal{S}$ and $\mathcal{T}$ form a partition of $\mathbb{N}$.

Since $h|\Lambda$ is topologically conjugate to the full 2-shift, there exists $p \in \Lambda$ such that $h^n(p) \in \Lambda_0$ whenever $f_n = f_S$ and $h^n(p) \in \Lambda_1$ whenever $f_n = f_T$. Let $P = (p, z_0)$, where z_0 is the south pole of S^2. The F-orbit of P is $\{(h^n(p), z_0)\}$ since z_0 is fixed under both F_S and R_θ. Since F_S is a lift of f_S, using the constancy of μ near Λ, one can show that

$$d_{(h^n(p),z_0)}F = \begin{pmatrix} d_{h^n(p)}h & 0 \\ 0 & d_0 f_n \end{pmatrix}. \tag{7.55}$$

Indeed by the choice of μ and the fact that $f_T = R_{\pi/2} \circ f_S \circ R_{-\pi/2}$, it can be shown that the diagram

$$\begin{array}{ccc} \{h^n(p)\} \times S^2 & \xrightarrow{F} & \{h^n(p)\} \times S^2 \\ \downarrow & & \downarrow \\ S^2 & \xrightarrow{F_n} & S^2 \end{array} \tag{7.56}$$

is commutative, where $F_n = F_S$ when $f_n = f_S$ and $F_n = F_T$ when $f_n = f_T$. It follows from (7.55) that P has one positive Lyapunov exponent corresponding to $dh|E^u$ and three negative Lyapunov exponents: one corresponds to $dh|E^s$ and the other two, being $(1/2)\log(ab)$, correspond to the products of the matrices $d_0 f_n$. Let E^s_P denote the space of vectors with negative Lyapunov exponents. Note that the F-orbit of P is regular since $dF^n|E^s_P$ is diagonal preserving the decomposition $E^s \oplus (x\text{-axis}) \oplus (y\text{-axis})$.

Since F is normally hyperbolic to $\mathcal{F}$, a point $(w, z) \in M \times S^2$ satisfies

$$\lim_{n\to\infty} \frac{1}{n} \log d(F^n(w, z), F^n(P)) < 0$$

if and only if (w, z) lies in the strong stable manifold of some point $(p, z') \in V^s(P) \cap (\{p\} \times S^2)$. This means that

$$V^s(P) = V^{ss}(V^s(P) \cap (\{p\} \times S^2)).$$

By (7.56), the set $V^s(P) \cap (\{p\} \times S^2)$ coincides with

$$\left\{z \in S^2 : \lim_{n\to\infty} \frac{1}{n} \log \|(f_n \circ \cdots \circ f_0)(z)\| < 0\right\}.$$

Examining carefully the explicit formulas for the maps f_n, one can show that if $z = (x, y) \in \mathbb{R}^2$ with $x > 0$ and $y > 0$, then $\|f^n(z)\| \to \infty$ as $n \to \infty$. Thus, $V^s(P)$ is contained in $V^{ss}(\{p\} \times S^2)$ but does not include any neighborhood of P. Hence it cannot be an immersed manifold tangent to E^s_P. □

7.9 Notes

The Stable Manifold Theorem 7.1.1 has a long history that goes back to seminal works of Hadamard [79] and Perron [164]. They developed two somewhat different proofs of the theorem in the simplest case of a hyperbolic fixed point. Since then every version of the Stable Manifold Theorem has been proved using either Hadamard's or Perron's approach.

One of the advantages of Hadamard's approach is that it can be effected for a class of Lipschitz maps, which may not be differentiable. On the other hand, the differentiability of the manifolds must be proved independently and a posteriori.

The Stable Manifold Theorem 7.1.1 for nonuniformly hyperbolic systems was first established by Pesin in [167]. His approach, which we closely follow here, was an elaboration of the classical work of Perron. This approach was extended by Katok and Strelcyn in [110] to smooth maps with singularities. (They include billiard systems and other physical models.) Ruelle [193] provided another proof of the Stable Manifold Theorem 7.1.1, based on the study of the perturbations of the matrix products in the Multiplicative Ergodic Theorem 3.4.3. Fathi, Herman, and Yoccoz [68] provided a detailed exposition of the Stable Manifold Theorem 7.1.1, elaborating on the approaches of Pesin and Ruelle. Another proof of Theorem 7.1.1 was obtained by Pugh and Shub in [185] using graph transform techniques. Proposition 7.8.2 is due to Pugh [184].

One can extend the Stable Manifold Theorem 7.1.1 to infinite-dimensional spaces. Ruelle [194] proved this theorem for Hilbert spaces, closely following his approach in [193], and Mañé [138] considered Banach spaces (under certain compactness assumptions on the dynamics).

In [185], Pugh and Shub showed that, in fact, if f is of class C^p for some $p \geq 2$, then $V^s(x)$ is also of class C^p.

The proof of Theorem 7.3.9 uses Hadamard's method and closely follows the presentation in [104].

8

Basic Properties of Stable and Unstable Manifolds

Let f be a $C^{1+\alpha}$ diffeomorphism of a compact smooth Riemannian manifold possessing a set Y, which is nonuniformly partially hyperbolic in the broad sense (see Section 5.2). We describe some basic properties of local stable manifolds $V^s(x)$ (see Theorem 7.1.1) and we also construct global stable manifolds $W^s(x)$, $x \in Y$. In the case when Y is nonuniformly partially or completely hyperbolic, we introduce local unstable manifolds $V^u(x)$ as well as global unstable manifolds $W^u(x), x \in Y$. We discuss invariant stable and unstable foliations formed by global manifolds. These are "tricky" foliations since their leaves depend only measurably on the point. However, they possess a crucial absolute continuity property. We introduce and study this property in the last few sections of this chapter.

8.1 Characterization and Sizes of Local Stable Manifolds

1. Starting from the set Y, let us choose a nonempty level set $\Lambda = \Lambda_{\lambda\mu\varepsilon j}$ (see (2.21)) where the numbers λ, μ, and ε satisfy (2.15) and j is the dimension of the local stable manifolds that is constant on Λ (see Sections 2.2.2 and 5.2 for details). Consider also the corresponding collection of regular sets $\{\Lambda^\ell = \Lambda^\ell_{\lambda\mu\varepsilon j} : \ell \geq 1\}$. We remind the reader that the set Λ is nonuniformly partially hyperbolic in the broad sense for f and that regular sets are nested compact and exhaust Λ.

Using Statement 4 of Theorem 7.5.1, one can characterize the local stable manifold at a point $x \in \Lambda$ as consisting of all the points y in a sufficiently small neighborhood of x for which there is a constant $C > 0$ such that for every $n > 0$,

$$d(f^n(y), f^n(x)) \leq C\kappa^n\, d(y, x) \tag{8.1}$$

where $0 < \kappa < 1$ is the number in (7.33). As an immediate corollary of (8.1), we obtain that for any $x \in \Lambda^\ell$, $y \in V^s(x)$, and $v \in T_y V^s(x)$, the Lyapunov exponent $\chi(y, v) < 0$.

We present another important property of local stable manifolds, which is an immediate corollary of Statement 2 of Theorem 7.5.1: for every $x \in \Lambda^\ell$, and any points $z_1, z_2 \in V^s(x)$,

$$d(T_{z_1}V^s(x), T_{z_2}V^s(x)) \le C\rho(z_1, z_2)^\alpha, \tag{8.2}$$

where $C > 0$ is a constant and d is the distance in the Grassmannian bundle of TM generated by the Riemannian metric.

2. Consider the nonuniformly hyperbolic set Λ_x that consists of a *single* nonuniformly hyperbolic trajectory $\{f^n(x)\}$. The Stable Manifold Theorem 7.1.1 applies. It characterizes the behavior of trajectories nearby an *individual* nonuniformly hyperbolic trajectory and does not need the presence of any other nonuniformly hyperbolic trajectories. Also, note that one can replace the assumption that the manifold M is compact by the assumption that the diffeomorphism f satisfies (7.4)–(7.7) along a given nonuniformly hyperbolic trajectory.

3. We discuss how the sizes of local stable manifolds vary. Recall that the size of the local stable manifold $V^s(x)$ at a point $x \in \Lambda$ is the number $r = r(x)$ that is determined by Theorem 7.5.1 and such that (7.3) holds.

It follows from Statement 5 of Theorem 7.5.1 and (5.1) that the sizes of the local stable manifolds at a point $x \in \Lambda$ and any point $f^m(x)$, $m \in \mathbb{Z}$, along the trajectory of x are related by

$$r(f^m(x)) \ge K e^{-\varepsilon|m|} r(x), \tag{8.3}$$

where $K > 0$ is a constant.

Using again Statement 5 of Theorem 7.5.1 and (5.1), we find that the sizes are bounded from below on any regular set Λ^ℓ; that is, there exists a number $r_\ell > 0$ that depends only on ℓ such that

$$r(x) \ge r_\ell, \quad x \in \Lambda^\ell. \tag{8.4}$$

Moreover, the number r_ℓ can be chosen such that $r_{\ell+1} \ge r_\ell e^{-\varepsilon}$.

Let ν be an f-invariant Borel measure for which $\nu(\Lambda) = 1$. For any sufficiently large ℓ, the regular set Λ^ℓ has positive measure. Therefore the trajectory of almost every point visits this set infinitely many times. It follows that for typical points x, the function $r(f^m(x))$ is oscillating in m and it exceeds r_ℓ for infinitely many values of m. Nevertheless, for some integers m, the value $r(f^m(x))$ may become as small as it is allowed by (8.3). Let us emphasize that the rate with which the sizes of the local stable manifolds $V^s(f^m(x))$ decrease as $m \to +\infty$ is smaller than the rate with which the trajectories $\{f^m(x)\}$ and $\{f^m(y)\}$, $y \in V^s(x)$, approach each other – while the former is of order $e^{\varepsilon m}$, the latter is of order $\lambda^m e^{\varepsilon m}$.

4. Local stable manifolds depend uniformly continuously on $x \in \Lambda^\ell$ in the C^1 topology. This means that if $x_m \in \Lambda^\ell$ is a sequence of points converging to x then

$$d_{C^1}(V^s(x_m), V^s(x)) \to 0 \tag{8.5}$$

as $m \to \infty$. Recall that given two local smooth submanifolds V_1 and V_2 in M of the same size r (i.e., V_i is of the form (7.3) for some r and some function ψ_i,

$i = 1, 2$), we have

$$d_{C^1}(V_1, V_2) = \sup_{x \in B^s(r)} \|\psi_1(x) - \psi_2(x)\| + \sup_{x \in B^s(r)} \|d_x\psi_1 - d_x\psi_2\|.$$

We briefly explain how to obtain (8.5). Using the continuous dependence of stable subspaces on regular sets and Theorem 7.5.1, we obtain for sufficiently large m that the local stable manifold $V^s(x_m)$ can be viewed as a graph of a function $\psi_m \colon B(r) \to E^2(x)$, where $B(r)$ is a ball in $E^s(x)$ centered at zero of some small radius r. By (8.2), we conclude that the sequence of functions ψ_m is compact in the C^1 topology. Let $\widetilde{\psi}$ be an accumulation point and define $\widetilde{V}$ to be the image under the exponential map at x of the graph of $\widetilde{\psi}$. It is easy to see that given n, inequality (8.1) holds for any $y \in \widetilde{V}$. It follows that $\widetilde{V}$ is a local stable manifold at x.

If f is an Anosov diffeomorphism (i.e., f is uniformly hyperbolic; see Section 6.1.1) then $\Lambda^\ell = M$ for some ℓ, and hence

$$r(x) \geq r_\ell \text{ for every } x \in M. \tag{8.6}$$

5. One can show that local stable manifolds depend Hölder continuously on $x \in \Lambda^\ell$ in the C^1 topology. This means that there exist $c > 0$ and $\beta \in (0, 1]$ such that for $x, y \in \Lambda^\ell$,

$$d_{C^1}(V^s(x), V^s(y)) \leq c\, d(x, y)^\beta.$$

The proof exploits the Hölder continuity property of stable distributions (see Theorem 5.3.1) and the fact that holonomy maps associated with families of local stable manifolds are Hölder continuous (see Theorem 8.4.3 below).

6. Assume that the diffeomorphism f possesses a nonempty set Y, which is nonuniformly partially (completely) hyperbolic (see Section 5.2). Then every nonempty level set $\Lambda = \Lambda_{\lambda\mu\varepsilon j}$ is nonuniformly partially (completely) hyperbolic and for each $x \in \Lambda$, one can construct local unstable manifolds $V^u(x)$. They have similar properties to those of local stable manifolds (with f replaced by the inverse map f^{-1} wherever necessary).

7. We now consider the case when the set Λ is nonuniformly completely hyperbolic. The continuity of local stable and unstable manifolds, $V^s(x)$ and $V^u(x)$, implies that there exists a number $\delta_\ell > 0$ such that for every $x \in \Lambda^\ell$ and $y \in \Lambda^\ell \cap B(x, \delta_\ell)$, the intersection $V^s(x) \cap V^u(y)$ is nonempty and consists of a single point, which depends continuously on x and y. Moreover, there exists a function $\psi = \psi(\ell)$ such that

$$\psi(\ell) \leq \ell\, e^{c\varepsilon} \tag{8.7}$$

for some $c > 0$ and

$$V^s(x) \cap V^u(x) \in \Lambda^{\psi(\ell)}. \tag{8.8}$$

8.2 Global Stable and Unstable Manifolds

Let $f\colon M \to M$ be a $C^{1+\alpha}$ diffeomorphism of a smooth compact Riemannian manifold M. Assume that f possesses a set that is nonuniformly partially hyperbolic in the broad sense. We consider one of the nonempty level sets $\Lambda = \Lambda_{\lambda\mu\varepsilon j}$ and the corresponding collection of regular sets $\{\Lambda^\ell = \Lambda^\ell_{\lambda\mu\varepsilon j} : \ell \geq 1\}$ (see the previous section). Starting with local stable manifolds, we will construct global stable manifolds for f.

Given a point $x \in \Lambda$, consider a local stable manifold $V^s(x)$ at x. If $y \in (V^s(x) \setminus \{x\}) \cap \Lambda$ then the set $V^s(x) \cup V^s(y)$ is a smooth submanifold of M, the result of *gluing* together the manifolds $V^s(x)$ and $V^s(y)$.

In the uniformly hyperbolic case, the process of gluing local manifolds can be continued indefinitely and allows one to construct the global stable manifold at x. The latter is a smooth embedding of an Euclidean space of the appropriate dimension and consists of all the points whose trajectories converge to the trajectory of x (with exponential rate).

In the nonuniformly hyperbolic case, gluing local manifolds may be more difficult. For example, the intersection $(V^s(x) \setminus \{x\}) \cap \Lambda$ may turn out to be empty; even if it is not, the process may end after just finitely many steps. Since the sizes of local manifolds may be arbitrarily small, even if the process of gluing continues indefinitely, it may result in a smooth submanifold of finite size. In fact, one may even expect that $V^s(y) \subset V^s(x)$ for every $y \in V^s(x) \cap \Lambda$ (and hence that the gluing process ends up with $V^s(x)$). Let us stress that the fact that the global manifold has finite size has not been observed yet in any known example of nonuniformly hyperbolic system and although expected, this phenomenon is considered rather pathological then typical.

There is, however, a simpler way of building global stable manifolds, which we exploit here. For every $x \in \Lambda$, we define the *global stable manifold* by

$$W^s(x) = \bigcup_{n=0}^{\infty} f^{-n}(V^s(f^n(x))). \tag{8.9}$$

This is a finite-dimensional immersed smooth submanifold (of class C^r if f is of class $C^{r+\alpha}$) invariant under f.

As we have seen above local stable manifolds are not uniquely defined: one can choose a "smaller" submanifold containing x and lying inside $V^s(x)$ and view it as a "new" local stable manifold at x. However, such variations in the choice of local manifolds do not affect the global stable manifolds in the following sense. Fix $x \in \Lambda$ and consider its trajectory $\{f^n(x)\}_{n\in\mathbb{Z}}$. For each $n \geq 0$, choose a ball $B_n \subset V^s(f^n(x))$ centered at $f^n(x)$ of radius $r_n > 0$ (with respect to the intrinsic metric in $V^s(f^n(x))$). The following result is a corollary of Theorem 7.5.1.

Theorem 8.2.1. *Assume that* $r_{n+1} > r_n\, e^{-\varepsilon n}$. *Then*

$$W^s(x) = \bigcup_{n=0}^{\infty} f^{-n}(B_n).$$

We give another useful characterization of global stable manifolds. Choose a nonempty regular set Λ^ℓ and for $x \in \Lambda^\ell$, let $n_i(x) > 0$ be the successive moments of time for which $f^{n_i(x)}(x) \in \Lambda^\ell$.

Theorem 8.2.2. *If the sequence $\{n_i(x)\}$ is unbounded then*

$$W^s(x) = \bigcup_{n=0}^{\infty} f^{-n_i(x)}(V^s(f^{n_i(x)}(x))).$$

Proof. Let $y \in W^s(x)$. We have for all sufficiently large i that

$$\rho(f^{n_i(x)}(x), f^{n_i(x)}(y)) \leq \frac{1}{2} r_\ell,$$

where r_ℓ is the size of the local stable manifold $V^s(x)$, $x \in \Lambda^\ell$ (see (8.4)). Since $f^{n_i(x)}(x) \in \Lambda^\ell$, it follows that $f^{n_i(x)}(y) \in V^s(f^{n_i(x)}(x))$. This implies the desired result. □

If the map f possesses an invariant Borel measure ν for which $\nu(\Lambda) > 0$, then one can choose a regular set Λ^ℓ such that $\nu(\Lambda^\ell) > 0$. It follows that the sequence $\{n_i(x)\}$ is unbounded for almost every $x \in \Lambda^\ell$.

Assume now that Λ is a nonuniformly partially hyperbolic set for f (see Section 5.2). For $x \in \Lambda$, we define the *global unstable manifold* by

$$W^u(x) = \bigcup_{n=0}^{\infty} f^n(V^u(f^{-n}(x))).$$

This is a finite-dimensional immersed smooth submanifold (of class C^r if f is of class $C^{r+\alpha}$) invariant under f.

Theorem 8.2.3. *If $x \in \Lambda$, then*

1. *for every $y \in W^s(x)$, we have that $\rho(f^n(x), f^n(y)) \to 0$ as $n \to +\infty$ with an exponential rate;*
2. *for every $y \in W^u(x)$, we have that $\rho(f^n(x), f^n(y)) \to 0$ as $n \to -\infty$ with an exponential rate.*

We emphasize that for Anosov diffeomorphisms the global stable and unstable manifolds can be characterized as follows:

$$W^s(x) = \{y \in M : \rho(f^n(x), f^n(y)) \to 0 \text{ as } n \to +\infty\},$$
$$W^u(x) = \{y \in M : \rho(f^n(x), f^n(y)) \to 0 \text{ as } n \to -\infty\}.$$

Indeed, by (8.6) the sizes of local stable manifolds are uniformly bounded from below by a constant r. Therefore if $\rho(f^n(x), f^n(y)) \to 0$ as $n \to +\infty$, the distance between $f^n(x)$ and $f^n(y)$ will be eventually smaller than r. It follows from Theorem 7.5.1 that $f^n(y) \in V^s(f^n(x))$, and hence $y \in W^s(x)$. Similar arguments work for $W^u(x)$.

Let φ_t be a smooth flow on a compact smooth Riemannian manifold M and Y a nonuniformly hyperbolic set for φ_t (see Section 7.7). We again consider one of the nonempty level sets $\Lambda = \Lambda_{\lambda\mu\varepsilon j}$ (see the previous section). For every $x \in \Lambda$, we define the *global stable manifold* at x by

$$W^s(x) = \bigcup_{t>0} \varphi_{-t}(V^s(\varphi_t(x))). \tag{8.10}$$

This is a finite-dimensional immersed smooth submanifold (of class C^r if φ_t is of class $C^{r+\alpha}$). For every $y \in W^s(x)$, we have $\rho(\varphi_t(x), \varphi_t(y)) \to 0$ as $t \to +\infty$ with an exponential rate.

We also define the *global weakly stable manifold* at x by

$$W^{s0}(x) = \bigcup_{t\in\mathbb{R}} W^s(\varphi_t(x)).$$

It follows from (8.10) that

$$W^{s0}(x) = \bigcup_{t\in\mathbb{R}} \varphi_t(W^s(x)).$$

Similarly, we define the *global unstable manifold* at x by

$$W^u(x) = \bigcup_{t>0} \varphi_t(V^u(\varphi_{-t}(x))).$$

This is a finite-dimensional immersed smooth submanifold (of class C^r if φ_t is of class $C^{r+\alpha}$). For every $y \in W^u(x)$, we have $\rho(\varphi_t(x), \varphi_t(y)) \to 0$ as $t \to -\infty$ with an exponential rate.

We also define the *global weakly unstable manifold* at x by

$$W^{u0}(x) = \bigcup_{t\in\mathbb{R}} W^u(\varphi_t(x)).$$

It follows from (8.10) that

$$W^{u0}(x) = \bigcup_{t\in\mathbb{R}} \varphi_t(W^u(x)).$$

8.3 Foliations with Smooth Leaves

A partition W of M is called a *foliation of M with smooth leaves* if there exist $\delta > 0$, $q > 0$, and $k \in \mathbb{N}$ such that for each $x \in M$,

1. the element $W(x)$ of the partition W containing x is a smooth k-dimensional immersed submanifold; it is called the *(global) leaf* of the foliation at x; the connected component of the intersection $W(x) \cap B(x, \delta)$ that contains x is called the *local leaf* at x and is denoted by $V(x)$;
2. there exists a continuous map $\varphi_x : B(x, q) \to C^1(D, M)$ (where $D \subset \mathbb{R}^k$ is the unit ball) such that for every $y \in X \cap B(x, q)$, the manifold $V(y)$ is the image of the map $\varphi_x(y) : D \to M$.

The function $\Phi_x(y, z) = \varphi_x(y)(z)$ is called the *foliation coordinate chart*. This function is continuous and has continuous derivative $\partial\Phi_x/\partial z$.

In this book we deal only with foliations with smooth leaves and simply call them foliations.

We say that the foliation W is *smooth* if the function φ is smooth. Let E be a continuous distribution on M. We call E *integrable* if there exists a foliation W with smooth leaves such that $E(x) = T_x W(x)$ for every $x \in M$. Further, we say that E is *uniquely integrable* if any smooth curve $\sigma : \mathbb{R} \to M$, satisfying $\sigma'(t) \in E(\sigma(t))$ for every $t \in \mathbb{R}$, lies in the leaf $W(\sigma(0))$.

If f is an Anosov diffeomorphism of a compact smooth manifold M then its stable and unstable distributions E^s and E^u are uniquely integrable to foliations W^s and W^u formed by global stable and unstable manifolds, respectively.

We now consider the case when f is a diffeomorphism possessing a nonuniformly partially hyperbolic set, which is nonuniformly partially hyperbolic in the broad sense. Let $\Lambda = \Lambda_{\lambda\mu\varepsilon j}$ be a nonempty level set (see Section 2.2.2).

Theorem 8.3.1. *If $x, y \in \Lambda$, then*

1. $W^s(x) \cap W^s(y) = \varnothing$ *if* $y \notin W^s(x)$;
2. $W^s(x) = W^s(y)$ *if* $y \in W^s(x)$.

This means that global stable manifolds form a partition of Λ, which we denote by W^s. In general this partition may not be measurable nor may it form a foliation (with smooth leaves). Later in the book we will observe situations when there exists a foliation W of M with smooth leaves and such that $W(x) \supset W^s(x)$ for those $x \in \Lambda$ for which $W^s(x)$ is well defined (see Chapters 11 and 12). This property is crucial in studying local ergodicity of f (with respect to a smooth hyperbolic measure on M; see Section 9.2).

In the case when Λ is a nonuniformly partially hyperbolic set for a diffeomorphism f, the global unstable manifolds $W^u(x)$ form a partition W^u of Λ.

8.4 Filtrations of Intermediate Local and Global Manifolds

There are more invariant local and global manifolds associated with the diffeomorphism f. Given a Lyapunov–Perron regular point $x \in M$, consider the values of the Lyapunov exponents at x,

$$\chi_1(x) < \cdots < \chi_{p(x)}(x)$$

and the corresponding Oseledets decomposition at x,

$$T_x M = \bigoplus_{i=1}^{p(x)} H_i(x).$$

Set $s(x) = \max\{i : \chi_i(x) < 0\}$ and $u(x) = \min\{i : \chi_i(x) > 0\}$. Assume that $1 < s(x) < u(x) < p(x)$. For $k = 1, \ldots, s(x)$, define

$$E_k^s(x) = \bigoplus_{i=1}^{k} H_i(x).$$

The Stable Manifold Theorem 7.1.1 applies to the distribution $E_k^s(x)$ and provides a local manifold $V_k^s(x)$, which we call the *kth local stable manifold* at x. Therefore, we obtain the *filtration of local stable manifolds*

$$x \in V_1^s(x) \subset V_2^s(x) \subset \cdots \subset V_{s(x)}^s(x). \tag{8.11}$$

In a similar way, for $k = u(x), \ldots, p(x)$, let us set

$$E_k^u(x) = \bigoplus_{i=k}^{p(x)} H_i(x).$$

The Unstable Manifold Theorem 7.6.1 applies to the distribution $E_k^u(x)$ and provides a local manifold $V_k^u(x)$, which we call the *kth local unstable manifold* at x. We obtain the *filtration of local unstable manifolds*

$$x \in V_{u(x)}^u(x) \subset V_{u(x)+1}^u(x) \subset \cdots \subset V_{p(x)}^u(x).$$

The local manifolds can be characterized as follows: for $k = 1, \ldots, s(x)$,

$$V_k^s(x) = \left\{ y \in B(x, r(x)) : \overline{\lim_{n\to+\infty}} \frac{1}{n} \log \rho(f^n(x), f^n(y)) < \chi_k(x) \right\},$$

and for $k = u(x), \ldots, p(x)$,

$$V_k^u(x) = \left\{ y \in B(x, r(x)) : \overline{\lim_{n\to-\infty}} \frac{1}{|n|} \log \rho(f^n(x), f^n(y)) < -\chi_k(x) \right\}$$

for some measurable function $r(x) > 0$. They are $C^{1+\alpha}$ immersed submanifolds. We also have $V^s(x) = V_{s(x)}^s(x)$ and $V^u(x) = V_{u(x)}^u(x)$.

Given $r \in (0, r(x))$, we denote by $B_k^s(x, r) \subset V_k^s(x)$ the ball centered at x of radius r with respect to the intrinsic metric on $V_k^s(x)$.

By Theorem 5.6.1, there exists a *Lyapunov chart* at x. This means that for any given $\varepsilon > 0$,

1. there exists a local diffeomorphism $\varphi_x : U_x \to \mathbb{R}^n$ with the property that the spaces $\mathbb{H}_i = \varphi_x(\exp_x H_i(x))$ form an orthogonal decomposition of $\mathbb{R}^n$;
2. if $i = 1, \ldots, p(x)$ and $v \in H_i(x)$ then
$$\begin{aligned} e^{\chi_i^+(x)-\varepsilon} \|\varphi_x(\exp_x v)\| &\le \|\varphi_{f(x)}(\exp_{f(x)}(df v))\| \\ &\le e^{\chi_i^+(x)+\varepsilon} \|\varphi_x(\exp_x v)\|; \end{aligned} \tag{8.12}$$
3. there is a constant K and a tempered function $A : \Lambda \to \mathbb{R}$ such that for y, $z \in U_x$,
$$K^{-1}\rho(y, z) \le \|\varphi_x(y) - \varphi_x(z)\| \le A(x)\rho(y, z); \tag{8.13}$$
4. there exists $\bar r(x) \in (0, r(x))$ such that $B_i^s(x, \bar r(x)) \subset V_i^s(x) \cap U_x$ for every $x \in \Lambda$ and $i = 1, \ldots, s(x)$ with $\chi_i(x) < 0$; moreover, for $1 \le i \le s(x)$, the manifolds $\varphi_x(V_i^s(x))$ are graphs of smooth functions $g_i^s : \mathbb{E}_i^s \to \mathbb{E}_{i+1}^s$, where $\mathbb{E}_i^s = \varphi_x(\exp_x E_i^s(x))$; the first derivatives of g_i^s are bounded by $1/3$;

5. there exists $\bar{r}(x) \in (0, r(x))$ such that $B^u_i(x, \bar{r}(x)) \subset V^u_i(x) \cap U_x$ for every $x \in \Lambda$ and $i = u(x), \ldots, p(x)$ with $\chi_i(x) > 0$; moreover, for $u(x) \le i \le p(x)$, the manifolds $\varphi_x(V^u_i(x))$ are graphs of smooth functions $g_i : \mathbb{E}^u_i \to \mathbb{E}^u_{i+1}$, where $\mathbb{E}^u_i = \varphi_x(\exp_x E^u_i(x))$; the first derivatives of g^u_i are bounded by 1/3.

It follows that if $k \le s(x)$ then

$$f(V^s_k(x) \cap U_x) \subset V^s_k(f(x)) \cap U_{f(x)},$$

and if $k \ge u(x)$ then

$$f^{-1}(V^u_k(x) \cap U_x) \subset V^u_k(f^{-1}(x)) \cap U_{f^{-1}(x)}.$$

For each $k = 1, \ldots, s(x)$, we define the *kth global stable manifold* by

$$W^s_k(x) = \bigcup_{n=0}^{\infty} f^{-n}(V^s_k(f^n(x))),$$

and for each $k = u(x), \ldots, p(x)$, we define the *kth global unstable manifold* by

$$W^u_k(x) = \bigcup_{n=0}^{\infty} f^n(V^u_k(f^{-n}(x))).$$

They are finite-dimensional immersed smooth submanifolds (of class C^r if f is of class $C^{r+\alpha}$) invariant under f. They have the following properties that are immediate consequences of the Stable Manifold Theorem 7.5.1 and the Unstable Manifold Theorem 7.6.1.

Theorem 8.4.1. *The following statements hold:*

1. *For $k = 1, \ldots, s(x)$ and $y \in W^s_k(x)$, we have $\rho(f^n(x), f^n(y)) \to 0$ as $n \to +\infty$ with an exponential rate.*
2. *For $k = u(x), \ldots, p(x)$ and $y \in W^u_k(x)$, we have $\rho(f^n(x), f^n(y)) \to 0$ as $n \to -\infty$ with an exponential rate.*

We obtain two *filtrations of global stable and unstable manifolds*

$$x \in W^s_1(x) \subset W^s_2(x) \subset \cdots \subset W^s_{s(x)}(x),$$
$$x \in W^u_{u(x)}(x) \subset W^u_{u(x)+1}(x) \subset \cdots \subset W^u_{p(x)}(x).$$

Note that the kth global manifolds do not depend on the particular choice of local manifolds in the sense of Theorem 8.2.1.

Let $k = k(x)$, $1 \le k(x) \le s(x)$, be an f-invariant Borel function on the set $\mathcal{R} \subset M$ of Lyapunov–Perron regular points. Let also the set Y be given by

$$Y = \{x \in \mathcal{R} : \chi_{k(x)}(x) < 0\} \tag{8.14}$$

(compare to (3.6) for the case of general cocycles). Assuming that this set is not empty, we have that it is nonuniformly partially hyperbolic in the broad sense for f (see Definition 5.2.1) with $\lambda(x) = e^{\chi_{k(x)}(x)}$, $\mu(x) = e^{\chi_{k(x)+1}(x)}$, and an arbitrary

f-invariant Borel function $\varepsilon(x)$ satisfying (2.9) (with an appropriate choice of the functions $C(x)$ and $K(x)$; see Theorem 3.3.1). The kth global manifolds form a "measurable lamination" of Y. More precisely, $W^s_{k(x)}(x) \cap W^s_{k(y)}(y) = \varnothing$ if $y \notin W^s_{k(x)}(x)$ and $W^s_{k(x)}(x) = W^s_{k(y)}(y)$ if $y \in W^s_{k(x)}(x)$ and $k(x) = k(y)$, with similar properties for the global unstable manifolds. These laminations are somewhat "tricky" objects and we will not consider them in the book. Instead, we introduce a technically more useful notion of families of kth local stable (unstable) manifolds.

Let us choose one of the nonempty level sets $\Lambda = \Lambda_{\lambda\mu\varepsilon j}$ and consider the corresponding collection of regular sets $\{\Lambda^\ell = \Lambda^\ell_{\lambda\mu\varepsilon j} : \ell \geq 1\}$ (see Sections 2.2.2 and 5.2; recall that the numbers λ, μ, and ε satisfy (2.15) and $j = \dim V^s_k(x)$). For $x \in \Lambda^\ell$, denote by $n_i(x) > 0$ the successive moments of time for which $f^{n_i(x)}(x) \in \Lambda^\ell$. The following result is an extension of Theorem 8.2.2 to intermediate global manifolds.

Theorem 8.4.2. *If the sequence $\{n_i(x)\}$ is unbounded then*

$$W^s_k(x) = \bigcup_{i=0}^{\infty} f^{-n_i(x)}(V^s_k(f^{n_i(x)}(x))).$$

The assumption of this theorem holds if the map f possesses an invariant Borel measure ν for which $\nu(\Lambda) > 0$. In this case, $\nu(\Lambda^\ell) > 0$ for all sufficiently large ℓ. It follows that the sequence $\{n_i(x)\}$ is unbounded for almost every $x \in \Lambda^\ell$.

Given a nonempty regular set Λ^ℓ and a point $x \in \Lambda^\ell$, consider the ball $B(x, r)$ centered at x of radius r. The collection

$$\mathcal{L}^s_k(x) = \{V^s_k(w) : w \in \Lambda^\ell \cap B(x, r)\}, \tag{8.15}$$

$1 \leq k = k(w) \leq s(w)$, is called a *family of kth local stable manifolds*. In a similar way for $u(w) \leq k = k(w) \leq p(w)$, we obtain a *family $\mathcal{L}^u_k(x)$ of kth local unstable manifolds*. The function $k(w)$ is constant along kth local manifolds (more precisely, for points in $\Lambda \cap V^s_k(w)$ or in $\Lambda \cap V^u_k(w)$) but may vary when w moves from one kth local manifold to another.

Fix $x \in \Lambda^\ell$, a number r, $0 < r \leq r_\ell$, and set

$$Q^\ell_k(x) = \bigcup_{w \in \Lambda^\ell \cap B(x,r)} V^s_k(w), \tag{8.16}$$

where $B(x, r)$ is the ball at x of radius r.

A *local transversal* is an open local submanifold T, which is uniformly transverse to each $V^s_k(w)$ in the family. Given transversals T^1 and T^2, we define the *holonomy map*

$$\pi : Q^\ell_k(x) \cap T^1 \to Q^\ell_k(x) \cap T^2 \tag{8.17}$$

using the relation

$$\pi(y) = T^2 \cap V^s_k(w), \text{ where } y = T^1 \cap V^s_k(w) \text{ and } w \in Q^\ell_k(x) \cap B(x, r).$$

The holonomy map is a homeomorphism onto its image. In fact one can prove a stronger result.

Theorem 8.4.3. *The holonomy map π is Hölder continuous on Λ^ℓ with Hölder constant depending only on ℓ.*

8.5 The Lipschitz Property of Intermediate Foliations

It is shown in [22] that the local manifold $V_k^s(y)$ in (8.11) depends Lipschitz continuously on $y \in V_{k+1}^s(x)$ while it runs over a regular set. In order to state this result, consider the set Y given by (3.6). Note that Y can be decomposed into sets Y_β in which the numbers $k(x)$, $\dim H_i(x)$, and $\chi_i(x)$ are constant for each i. From now on we restrict our consideration to one of the nonempty subset Y_β and set $p(x) = p$, $s(x) = s$, $u(x) = u$, and $\chi_i(x) = \chi_i$ for each i and $x \in Y_\beta$.

Given $\ell > 0$, consider the set $Y'_{\beta\ell}$ of points $x \in Y_\beta$ for which

$$\bar{r}(x) > 1/\ell, \quad A(x) < \ell, \quad \sphericalangle(H_i(x), \bigoplus_{j\neq i} H_j(x)) > \frac{1}{\ell}, \quad i, j = 1\dots, k,$$

where

$$T_x M = \bigoplus_{i=1}^{p} H_i(x)$$

is the Oseledets decomposition. Let $Y_{\beta\ell}$ be the closure of $Y'_{\beta\ell}$. This set is obviously compact. We also have that $Y_{\beta\ell} \subset Y_{\beta(\ell+1)}$ and $Y_\beta \subset \bigcup_{\ell>0} Y_{\beta\ell}$. The functions $\bar{r}(x)$ and $A(x)$ can be extended to $Y_{\beta\ell}$ such that $\bar{r}(x) > 1/\ell$, $A(x) < \ell$, and $\sphericalangle(H_i(x), \bigoplus_{j\neq i} H_j(x)) > 1/\ell$ for $i = 1\dots, p$. For every $x \in Y_{\beta\ell}$, one can construct an invariant ith local stable manifold $V_i^s(x)$.

Let us fix $c > 0$, $\ell > 0$, $x \in Y_{\beta\ell}$, and $y' \in Y_{\beta\ell} \cap B_{i+1}^s(x, c/\ell)$. (Recall that $B_i^s(x, q)$ is the ball in $V_i^s(x)$ centered at x of radius q.) For each $i < s$, consider two local smooth manifolds Σ_x and $\Sigma_{y'}$ in $V_{i+1}^s(x)$, containing x and y', respectively, and transverse to $V_i^s(z)$ for all $z \in Y_{\beta\ell} \cap B_{i+1}^s(x, c/\ell)$. Recall that the holonomy map

$$\pi_i = \pi_i(\Sigma_x, \Sigma_{y'})\colon \Sigma_x \cap Y_{\beta\ell} \cap B_{i+1}^s(x, c/\ell) \to \Sigma_{y'}$$

is defined by $\pi_i(x') = V_i^s(x') \cap \Sigma_{y'}$ with $x' \in \Sigma_x$ and is well defined if c is sufficiently small. (c may depend on ℓ but does not depend on x and y.)

Our goal is to prove the following theorem.

Theorem 8.5.1. *Let f be a $C^{1+\alpha}$ diffeomorphism and Y_β and $Y_{\beta\ell}$ the sets constructed above. There exists $\varepsilon_0 > 0$ such that for every $0 < \varepsilon \le \varepsilon_0$, the following statement holds. For each $\ell > 0$, $i < s$, $x \in Y_{\beta\ell}$, and $y' \in Y_{\beta\ell} \cap B_{i+1}^s(x, c/\ell)$, the holonomy map $\pi_i(\Sigma_x, \Sigma_{y'})$ is Lipschitz continuous with Lipschitz constant depending only on β and ℓ.*

Proof. We briefly outline our approach. For $x \in Y_{\beta\ell}$, let (U_x, φ_x) be a Lyapunov chart at x. For each $z \in M$ and $n \in \mathbb{N}$, we write $z_n = f^n(z)$. Given $x \in Y_\beta$, consider the Lyapunov chart and the map $\Phi_x : B^s_{i+1}(x, \bar{r}(x)) \to \mathbb{E}^s_i \oplus \mathbb{H}_{i+1}$ defined by

$$\Phi_x y = (\pi_1(\varphi_x(y)), \pi_2(\varphi_x(y))),$$

where $\pi_1 : \mathbb{E}^s_{i+1} \to \mathbb{E}^s_i$ and $\pi_2 : \mathbb{E}^s_{i+1} \to \mathbb{H}_{i+1}$ are the orthogonal projections. First using exponential convergence along the $\mathbb{H}_{i+1}$-subspace and the Hölder continuity of the differential, we show that for any two points $y, z \in B^s_{i+1}(x, c/\ell)$, the norm of the vector $\pi_2(\Phi_x y - \Phi_x z)$ (in the strong stable direction $\mathbb{H}_{i+1}$) can be estimated, up to a multiplicative constant, by the norm of the vector $\pi_2 dF^{-n}_{x_n}(0)(\Phi_{x_n} y_n - \Phi_{x_n} z_n)$, obtained by iterating y and z forward and then taking the pullback under the differential of the diffeomorphism. We then use this property to estimate the ratio between the distances $d(x, x')$ and $d(y, y')$ in the transversals Σ_x and $\Sigma_{y'}$, respectively, by the corresponding ratio of the above norms in the $\mathbb{H}_{i+1}$-subspace. The stronger exponential contraction along the $\mathbb{H}_{i+1}$-subspace allows one to neglect the distance between iterates along the V^s_i-foliation when compared to the distance on the $\mathbb{H}_{i+1}$-direction.

We need a series of lemmas. First we observe that the Lipschitz property does not depend on the choice of transversal sections.

Lemma 8.5.2. *For any local smooth manifolds Σ'_x and $\Sigma'_{y'}$ in $V^s_{i+1}(x)$ containing x and y' respectively and transverse to $V^s_i(z)$ for each point $z \in B^s_{i+1}(x, c/\ell) \cap Y_{\beta\ell}$, the holonomy map $\pi_i(\Sigma'_x, \Sigma'_{y'})$ is Lipschitz continuous if and only if the map $\pi_i(\Sigma_x, \Sigma_{y'})$ is Lipschitz continuous (possibly with a different Lipschitz constant depending only on the angles between the tangent spaces to the transversals).*

Proof of the lemma. Since the local manifolds $V^s_i(y)$ vary continuously on the compact set $B^s_{i+1}(x, c/\ell) \cap Y_{\beta\ell}$ and are at least C^1, the angle between any transverse section to $V^s_i(z)$ (in $V^s_{i+1}(x)$) and any local manifold $V^s_i(y)$ is uniformly bounded away from 0. Therefore if $z \in \Sigma_x$ and $z' = \Sigma'_x \cap V^s_i(z)$ then the ratio $d(x, z)/d(x, z')$ is bounded from below and above (uniformly in z) by constants (depending only on those angles). This implies the desired result. □

The following two lemmas are immediate consequences of the properties of the Lyapunov charts (see Section 5.6).

Lemma 8.5.3. *The map Φ_x is a diffeomorphism with all derivatives tempered. Moreover, there is a constant C_1 (depending only on ℓ) such that for every $x \in Y_{\beta\ell}$, $n \geq 0$, and $y, z \in B^s_{i+1}(x, \bar{r}(x))$, we have*

1. $\|\Phi_{x_n} y_n - \Phi_{x_n} z_n\| \leq C_1 e^{(\chi_{i+1}+\varepsilon)n} \|\Phi_x y - \Phi_x z\|$;
2. *if* $\|\pi_1(\Phi_x y - \Phi_x z)\| / \|\pi_2(\Phi_x y - \Phi_x z)\| < 3\ell$ *then for* $j = 1, 2$,

$$\|\pi_j(\Phi_{x_n} y_n - \Phi_{x_n} z_n)\| \leq C_1 e^{(\chi_{i+1}+\varepsilon)n} \|\pi_j(\Phi_x y - \Phi_x z)\|.$$

For $x \in Y_\beta$, let us introduce the map $F_x : \Phi_x(B^s_{i+1}(x, \bar{r}(x))) \to \mathbb{E}^s_i \oplus \mathbb{H}_{i+1}$ given by $F_x(v) = \Phi_{f(x)} f(\Phi_x^{-1}(v))$. We write

$$F_x^n = \begin{cases} F_{x_n} \circ \cdots \circ F_x & n > 0 \\ \mathrm{Id} & n = 0 \, . \\ F_{x_n}^{-1} \circ \cdots \circ F_x^{-1} & n < 0 \end{cases}$$

Lemma 8.5.4. *For every $x \in Y_\beta$, the map F_x is a $C^{1+\alpha}$ diffeomorphism onto its image; the Hölder constant $d(x)$ of the differential $dF_x(0)$ at 0 is a tempered function of x; for every $\ell > 0$, there exists a constant $d_\ell > 0$ such that $d(x) \le d_\ell$ for every $x \in Y_{\beta\ell}$; finally, $dF_x(0)\mathbb{H}_i = \mathbb{H}_i$ for every $x \in Y$ and every i.*

We need the following result:

Lemma 8.5.5. *There is a constant C_2 (depending only on ℓ) such that if $x \in Y_{\beta\ell}$, $y, z \in B^s_{i+1}(x, \bar{r}(x))$, and*

$$\|\pi_1(\Phi_x y - \Phi_x z)\| / \|\pi_2(\Phi_x y - \Phi_x z)\| < 3\ell,$$

then for $n \ge 0$,

$$\begin{aligned} C_2^{-1} \|\pi_2(\Phi_x y - \Phi_x z)\| &\le \|\pi_2 \, dF_{x_n}^{-n}(0)(\Phi_{x_n} y_n - \Phi_{x_n} z_n)\| \\ &\le C_2 \|\pi_2(\Phi_x y - \Phi_x z)\|. \end{aligned} \tag{8.18}$$

Proof of the lemma. We first observe that

$$\begin{aligned} &\|\Phi_{x_n} y_n - \Phi_{x_n} z_n\| \\ &\le \max\{\|\pi_1(\Phi_{x_n} y_n - \Phi_{x_n} z_n)\|, \|\pi_2(\Phi_{x_n} y_n - \Phi_{x_n} z_n)\|\} \\ &\le \max\{C_1 \, e^{(\chi_i + \varepsilon)n} \|\pi_1(\Phi_x y - \Phi_x z)\|, C_1 \, e^{(\chi_{i+1} + \varepsilon)n} \|\pi_2(\Phi_x y - \Phi_x z)\|\} \\ &\le \max\{C_1 \, e^{(\chi_i + \varepsilon)n} 3\ell \|\pi_2(\Phi_x y - \Phi_x z)\|, C_1 \, e^{(\chi_{i+1} + \varepsilon)n} \|\pi_2(\Phi_x y - \Phi_x z)\|\} \\ &\le C_3 \, e^{(\chi_{i+1} + \varepsilon)n} \|\pi_2(\Phi_x y - \Phi_x z)\| \le C_4 \, e^{2\varepsilon n} \|\pi_2(\Phi_{x_n} y_n - \Phi_{x_n} z_n)\|, \end{aligned}$$

where $C_3 > 0$ and $C_4 > 0$ are constants. By Lemma 8.5.4, there is a constant $C_5 > 0$ (depending only on ℓ) such that

$$\begin{aligned} &\|\pi_2(\Phi_{x_{n-1}} y_{n-1} - \Phi_{x_{n-1}} z_{n-1} - dF_{x_n}^{-1}(\Phi_{x_n} y_n)(\Phi_{x_n} y_n - \Phi_{x_n} z_n))\| \\ &\le \|\Phi_{x_{n-1}} y_{n-1} - \Phi_{x_{n-1}} z_{n-1} - dF_{x_n}^{-1}(\Phi_{x_n} y_n)(\Phi_{x_n} y_n - \Phi_{x_n} z_n)\| \\ &\le C_5 \, e^{\varepsilon n} \|\Phi_{x_n} y_n - \Phi_{x_n} z_n\|^{1+\alpha}. \end{aligned}$$

If $\pi_2 \Phi_x y = \pi_2 \Phi_x z$ then the statement follows from the Lyapunov block form of $dF_{x_n}^{-n}(0)$. Assume now that $\pi_2 \Phi_x y \ne \pi_2 \Phi_x z$. By Lemma 8.5.3 and (8.12), there are positive constants C_6 and C_7 (depending only on ℓ) such that, letting

$w_n = \Phi_{x_n} y_n - \Phi_{x_n} z_n,$

$$
\begin{aligned}
&\frac{\|\pi_2\, dF_{x_n}^{-n}(0)(\Phi_{x_n} y_n - \Phi_{x_n} z_n)\|}{\|\pi_2\, dF_{x_{n-1}}^{-(n-1)}(0)(\Phi_{x_{n-1}} y_{n-1} - \Phi_{x_{n-1}} z_{n-1})\|} \\
&\quad \le 1 + \frac{\|\pi_2\, dF_{x_{n-1}}^{-(n-1)}(0)(w_{n-1} - dF_{x_n}^{-1}(0) w_n)\|}{\|\pi_2\, dF_{x_{n-1}}^{-(n-1)}(0) w_{n-1}\|} \\
&\quad \le 1 + \frac{\|dF_{x_{n-1}}^{-(n-1)}(0)[\pi_2(w_{n-1} - dF_{x_n}^{-1}(0) w_n)]\|}{\|dF_{x_{n-1}}^{-(n-1)}(0)[\pi_2 w_{n-1}]\|} \\
&\quad \le 1 + C_6\, e^{\varepsilon n} \frac{\|\pi_2(w_{n-1} - dF_{x_n}^{-1}(0) w_n)\|}{\|\pi_2 w_{n-1}\|} \\
&\quad \le 1 + C_6\, e^{\varepsilon n} \frac{\|\pi_2(w_{n-1} - dF_{x_n}^{-1}(\Phi_{x_n} y_n) w_n)\|}{\|\pi_2 w_{n-1}\|} \\
&\qquad + C_6\, e^{\varepsilon n} \frac{\|\pi_2((dF_{x_n}^{-1}(\Phi_{x_n} y_n) - dF_{x_n}^{-1}(0)) w_n)\|}{\|\pi_2 w_{n-1}\|} \\
&\quad \le 1 + C_6\, e^{\varepsilon n} \frac{c_1\, e^{\varepsilon n} \|w_n\|^{1+\alpha}}{\|\pi_2 w_{n-1}\|} \\
&\qquad + C_6\, e^{\varepsilon n} \frac{\|dF_{x_n}^{-1}(\Phi_{x_n} y_n) - dF_{x_n}^{-1}(0)\| \cdot \|w_n\|}{\|\pi_2 w_{n-1}\|} \\
&\quad \le 1 + C_5 C_6 C_4\, e^{4\varepsilon n} \|w_n\|^{\alpha} + C_6\, e^{\varepsilon n} C_7\, e^{\varepsilon n} \|\Phi_{x_n} y_n\|^{\alpha} \cdot C_4\, e^{2\varepsilon n} \\
&\quad \le 1 + (C_5 + C_7) C_6 C_4 C_1{}^{\alpha}\, e^{[(\chi_{i+1}+\varepsilon)\alpha + 4\varepsilon]n}.
\end{aligned}
$$

Set $C_8 = (C_5 + C_7) C_6 C_3 C_1{}^{\alpha}$. Choosing ε sufficiently small, we may assume that $4\varepsilon + (\chi_{i+1} + \varepsilon)\alpha \le \gamma < 0$. Therefore,

$$
\begin{aligned}
\frac{\|\pi_2\, dF_{x_n}^{-n}(0)(\Phi_{x_n} y_n - \Phi_{x_n} z_n)\|}{\|\pi_2(\Phi_x y - \Phi_x z)\|} &\le \prod_{n=0}^{\infty} \left(1 + C_8\, e^{[4\varepsilon + (\chi_{i+1}+\varepsilon)\alpha] n}\right) \\
&\le \prod_{n=0}^{\infty} (1 + C_8\, e^{\gamma n}) = C_2 < +\infty
\end{aligned}
$$

and the upper estimate in (8.18) follows. The lower estimate can be obtained in a similar way. □

We now proceed with the proof of the theorem. By Lemma 8.5.2, we may specify the choice of the transversals. We say that a transversal section Σ is *good* if $\varphi_x(\Sigma)$ is the graph of a function

$$h^i : \varphi_x(B_{i+1}^s(x, \bar{r}(x))) \cap \mathbb{H}_{i+1} \to \mathbb{E}_{i+1}^s$$

whose first derivative is bounded by $1/3$. If $2\varepsilon < |\chi_i - \chi_{i+1}|$ and $x \in Y_{\beta\ell}$, then by the properties of Lyapunov charts (see Section 5.6), all the images $f^n(\Sigma)$ are good, provided Σ is good. Moreover, there is a constant $C_0 > 0$ such that

$$C_0^{-1} d_{\varphi_x \Sigma}(\varphi_x(y), \varphi_x(z)) \leq \|\pi_2(\varphi_x(y) - \varphi_x(z))\| \leq C_0 \, d_{\varphi_x \Sigma}(\varphi_x(y), \varphi_x(z)) \tag{8.19}$$

for any good transversal section where $d_{\varphi_x \Sigma}$ is the induced distance on $\varphi_x \Sigma$.

Let $x \in Y_{\beta\ell}$. Fix good transversals Σ_x and $\Sigma_{y'}$ with $y' \in B^s_{i+1}(x, c/\ell) \cap Y_{\beta\ell}$. Let $y = \pi_i(x)$ and $y' = \pi_i(x')$. Since the transversals are good, we know that

$$\|\pi_1(\Phi_x y - \Phi_x y')\| / \|\pi_2(\Phi_x y - \Phi_x y')\| < 3\ell.$$

We also have that

$$\|\pi_1(\Phi_x x')\| / \|\pi_2(\Phi_x x')\| < 3\ell.$$

Set $C = 10\ell^2 C_0{}^2 C_2{}^2$. By Lemma 8.5.5 and (8.19), for every $n \geq 0$,

$$\begin{aligned}
\frac{d(y, y')}{d(x, x')} &\leq 10 \frac{d_{\Sigma_y}(y, y')}{d_{\Sigma_x}(x, x')} \\
&\leq 10\ell^2 \frac{d_{\varphi_x \Sigma_y}(\varphi_x(y), \varphi_x(y'))}{d_{\varphi_x \Sigma_x}(0, \varphi_x(x'))} \leq 10\ell^2 C_0{}^2 \frac{\|\pi_2(\Phi_x y - \Phi_x y')\|}{\|\pi_2(\Phi_x x')\|} \\
&\leq C \frac{\|\pi_2 \, dF_{x_n}^{-n}(0)(\Phi_{x_n} y_n - \Phi_{x_n} y_n')\|}{\|\pi_2 \, dF_{x_n}^{-n}(0)\Phi_{x_n} x_n'\|} \\
&\leq C \left(1 + \frac{\|\pi_2 \, dF_{x_n}^{-n}(0)(\Phi_{x_n} x_n' - \Phi_{x_n} y_n' + \Phi_{x_n} y_n)\|}{\|\pi_2 \, dF_{x_n}^{-n}(0)\Phi_{x_n} x_n'\|}\right) \\
&\leq C \left(1 + \frac{\|\pi_2 \, dF_{x_n}^{-n}(0)(\Phi_{x_n} x_n' - \Phi_{x_n} y_n')\| + \|\pi_2 \, dF_{x_n}^{-n}(0)\Phi_{x_n} y_n\|}{\|\pi_2 \, dF_{x_n}^{-n}(0)\Phi_{x_n} x_n'\|}\right) \\
&\leq C \left(1 + \frac{e^{\varepsilon n} \, d(x_n', y_n') + e^{\varepsilon n} \, d(x_n, y_n)}{e^{(\chi_{i+1} - \varepsilon)n}}\right) \\
&\leq C(1 + e^{(3\varepsilon + \chi_i - \chi_{i+1})n}).
\end{aligned}$$

Choosing ε such that $3\varepsilon + \chi_i - \chi_{i+1} < 0$, we conclude that the map π_i is Lipschitz. □

It immediately follows from Theorem 8.5.1 that for every $x \in Y_{\beta\ell}$, $\ell \geq 1$, and every $1 \leq k < p$, the family of local manifolds $V^s_k(y)$, $y \in V^s_{k+1}(x) \cap Y_{\beta\ell}$, is "Lipschitz rectifiable": it is rectifiable up to a Lipschitz local coordinate change.

8.6 The Absolute Continuity Property

Let f be a $C^{1+\alpha}$ diffeomorphism of a compact smooth Riemannian manifold M. We describe one of the most crucial properties of local stable and unstable manifolds that is known as *absolute continuity*.

Consider a foliation with smooth leaves W of M (see Section 8.3). Fix $x \in M$ and let ξ be the partition of the ball $B(x, q)$ by local manifolds $V(y)$, $y \in B(x, q)$.

The absolute continuity is a property of the foliation with respect to volume and does not require the presence of the dynamics (and its invariant measure). This property can be understood in different ways. The one that addresses the above question in the most straightforward way is the following: the foliation W is *absolutely continuous (in the weak sense)* if for almost every $x \in M$, any $q > 0$, any Borel subset $X \subset B(x, q)$ of positive volume, and almost every $y \in X$, the intersection $X \cap V(y)$ has positive volume, with respect to the Riemannian volume $m_{V(y)}$ on $V(y)$ viewed as a smooth submanifold in M–the so-called *leaf volume*. It is not difficult to see that W is absolutely continuous (in the weak sense) if and only if for any Borel subset $X \subset M$ of positive volume, there exists $y \in M$ such that $m_{W(y)}(X \cap W(y)) > 0$.

If the foliation W is smooth then due to the Fubini theorem, the intersection $E \cap V(y)$ has positive measure for almost every $y \in B(x, q)$. If the foliation is only continuous, the absolute continuity property may fail (see the example at the end of this section).

A stronger version of the absolute continuity property imposes some requirements on the Radon–Nikodým derivatives of the conditional measures generated by volume on $V(y)$ viewed as elements of the partition ξ. More precisely, let T be a local transversal through x to the local manifolds $V(y)$, $y \in B(x, r)$. We say that the foliation W is *absolutely continuous (in the strong sense)* if for almost every $x \in M$, any $q > 0$, and any Borel subset $X \subset B(x, q)$ of positive volume,

$$m(X) = \int_T h(y) \int_{V(y)} \chi_X(y, z) g(y, z)\, dm_{V(y)}(z)\, dm_T(y),$$

where χ_X is the characteristic function of the set X, h a Borel function in $y \in T$, g a Borel function in $y \in T$ and $z \in V(y)$, and m_T the leaf volume on T. In other words, the conditional measures on the elements $V(y)$ of partition ξ are absolutely continuous with respect to the leaf volume on $V(y)$ and the factor-measure is absolutely continuous with respect to the leaf volume on the transversal T.

If a foliation is not absolutely continuous (in the weak sense), the Fubini theorem may fail in the strongest possible way–a set of full volume may meet each leaf of the foliation at a single point–the phenomenon often referred to as "Fubini's nightmare" or "Fubini foiled" (see the example at the end of Section 8.6.1). Such pathological foliations seem to be "typical" objects in the theory of stable ergodicity (see Section 11.6).

A celebrated result by Anosov claims that the stable and unstable invariant foliations for Anosov diffeomorphisms are absolutely continuous. We stress that generically (in the space of Anosov diffeomorphisms) these foliations are not smooth and therefore the absolute continuity property is not at all trivial and requires a deep study of the structure of these foliations. (An example of an Anosov diffeomorphism with nonsmooth stable and unstable foliations was constructed in [6]; this phenomenon is generic in a sense; see [80].)

In [7], Anosov and Sinai suggested an approach to establish absolute continuity (in the strong sense), which is based on the study of the holonomy maps associated with the foliation. To explain this, consider a foliation W. Given x, choose two transversals T^1 and T^2 to the family of local manifolds $V(y)$, $y \in B(x,q)$. To every point $z \in T^1$, the holonomy map associates the point $w = V(z) \cap T^2$. This map is a homeomorphism. If it is absolutely continuous (see the definition below) for all points x and transversals T^1 and T^2 then the absolute continuity property (in the strong sense) follows (see Theorem 8.6.8). In what follows, when referring to absolute continuity (without the adjectives "weak" or "strong"), we mean the property of holonomy maps described above.

For nonuniformly hyperbolic diffeomorphisms, the study of absolute continuity is technically much more complicated due to the fact that the global stable and unstable manifolds may not form foliations (they may not even exist for some points in M) and the sizes of local manifolds may vary wildly from point to point. In order to overcome this difficulty one should define and study the holonomy maps associated with local stable (or unstable) manifolds on regular sets.

8.6.1 Absolute Continuity of Holonomy Maps

Let f be a $C^{1+\alpha}$ diffeomorphism of a compact smooth Riemannian manifold M and Y a nonempty set, which is nonuniformly partially hyperbolic in the broad sense for f (see Definition 5.2.1). For every $x \in \Lambda$, we have an invariant splitting $T_xM = E^1(x) \oplus E^2(x)$ and without loss of generality, we assume that $E^1(x)$ is stable. We shall use the notation $E^s(x)$.

Consider a nonempty level set $\Lambda = \Lambda_{\lambda\mu\varepsilon j}$ and the collection of regular sets $\{\Lambda^\ell : \ell \geq 1\}$ (see Sections 2.2.2 and 5.2) where the numbers λ, μ, and ε satisfy (2.15). Throughout this section we assume that these numbers are fixed. We stress that $\dim E^s(x) = j$ is constant. We remind the reader that regular sets are compact, $\Lambda^\ell \subset \Lambda^{\ell+1}$ and for $x \in \Lambda^\ell$,

$$\|df^n v\| \leq C_\ell \lambda^n e^{\varepsilon n} \|v\|, \quad v \in E^s(x),$$
$$\|df^n v\| \geq C_\ell^{-1} \mu^n e^{-\varepsilon n} \|v\|, \quad v \in E^2(x),$$

where $C_\ell > 0$. We also denote by $V^s(x)$ the local stable manifold at x. As we saw in Section 8.1 they depend continuously on $x \in \Lambda^\ell$ and their sizes are bounded away from zero by a number r_ℓ (see (8.4)).

Fix $x \in \Lambda^\ell$, a number r, $0 < r \leq r_\ell$, and consider the family of local stable manifolds

$$\mathcal{L}^s(x) = \{V^s(w) : w \in \Lambda^\ell \cap B(x,r)\} \tag{8.20}$$

(compare to (8.15)), the set

$$Q^\ell(x) = \bigcup_{w \in \Lambda^\ell \cap B(x,r)} V^s(w) \tag{8.21}$$

(compare to (8.16)) and a local transversal T (an open local submanifold T, which is uniformly transverse to each $V^s(w)$ in the family). For sufficiently small r, we can choose T in such a way that the set $\exp_x^{-1} T$ is the graph of a smooth map

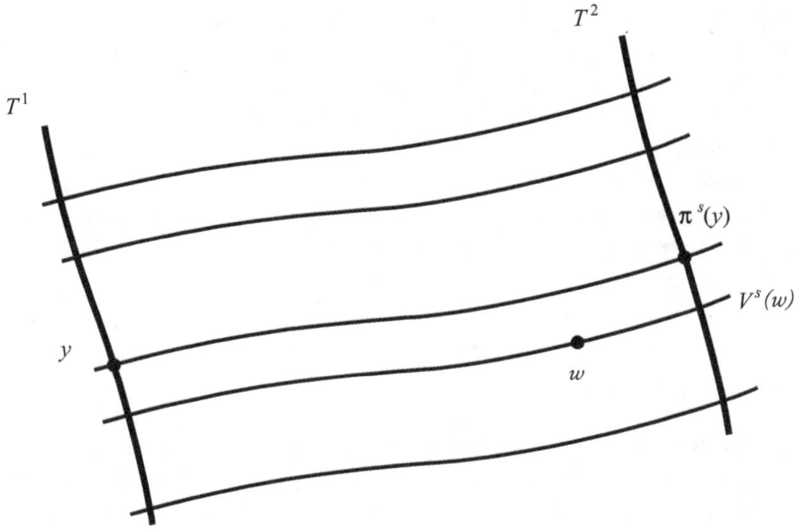

Figure 8.1. A family of stable leaves $V^s(w)$.

$\psi\colon B_2(q) \subset E^2(x) \to E^s(x)$ (where $B_2(q)$ is the ball at 0 of radius $q > 0$) with sufficiently small C^1 norm. We will consider local open submanifolds constructed only in this way and call them *transversals to the family* $\mathcal{L}^s(x)$. We also say that ψ *represents* T.

Let T^1 and T^2 be two transversals to the family $\mathcal{L}^s(x)$. Consider the holonomy map $\pi^s\colon Q^\ell(x) \cap T^1 \to Q^\ell(x) \cap T^2$ (see (8.17) and Fig. 8.1). This map depends on x, ℓ, T^1, and T^2. Set

$$\Delta = \Delta(T^1, T^2) = \|\psi^1 - \psi^2\|_{C^1}, \tag{8.22}$$

where the maps ψ^1 and ψ^2 represent T^1 and T^2 respectively.

We recall that if ν and μ are two measures on a measure space X then ν is said to be *absolutely continuous* with respect to μ if for every $\varepsilon > 0$, there exists $\delta > 0$ such that $\nu(Z) < \varepsilon$ for every measurable set Z for which $\mu(Z) < \delta$. A measurable invertible transformation $G\colon X \to Y$ of measurable spaces (X, ν) and (Y, μ) is said to be *absolutely continuous* if the measure μ is absolutely continuous with respect to the measure $G_*\nu$. In this case one defines the *Jacobian* $\mathrm{Jac}(G)(x)$ of G at a point $x \in X$ (specified by the measures ν and μ) to be the Radon–Nikodým derivative $d\mu/d(G_*\nu)$. If X is a metric space then for μ-almost every $x \in X$, one has

$$\mathrm{Jac}(G)(x) = \lim_{r\to 0} \frac{\mu(G(B(x,r)))}{\nu(B(x,r))}. \tag{8.23}$$

We remind the reader that given a submanifold T in M, we denote by m_T the leaf volume on T. We denote by $\mathrm{Jac}(\pi^s)(y)$ the Jacobian of the holonomy map π^s at the point $y \in Q^\ell(x) \cap T^1$ specified by the measures m_{T^1} and m_{T^2}.

Theorem 8.6.1 (The Absolute Continuity Property; see [169]). *Let f be a $C^{1+\alpha}$ diffeomorphism of a compact smooth manifold M possessing a nonempty set, which is nonuniformly partially hyperbolic in the broad sense. Let also $\Lambda = \Lambda_{\lambda\mu\varepsilon j}$ be a nonempty level set. There exists $\varepsilon_0 >$ such that if $0 < \varepsilon < \varepsilon_0$ then the following statement holds. Fix $\ell \geq 1$, $x \in \Lambda$, and transversals T^1 and T^2 to the family $\mathcal{L}^s(x)$. Assume that $m_{T^1}(Q^\ell \cap T^1) > 0$. Then the holonomy map π^s is absolutely continuous (with respect to the measures m_{T^1} and m_{T^2}) and the Jacobian $\mathrm{Jac}(\pi^s)$ is bounded from above and bounded away from zero. Moreover, there exists a constant $K = K(\ell) > 0$ such that*

$$|\mathrm{Jac}(\pi^s)(y) - 1| \leq K\Delta(T^1, T^2) \tag{8.24}$$

for every $y \in Q^\ell(x) \cap T^1$.

Proof. Before we present the formal argument, let us briefly outline our approach. The original idea goes back to the article of Anosov and Sinai [7] where they established absolute continuity in the case of uniformly hyperbolic systems. Their method needs substantial modifications to take care of some new phenomena associated with nonuniform hyperbolicity.

A very general description of our approach is as follows. First we push the sets $Q^\ell \cap T^i$, $i = 1, 2$, forward by applying the map f^n with sufficiently large n. The set $A_2^n = f^n(Q^\ell \cap T^2)$ lies within exponentially small distance from the set $A_1^n = f^n(Q^\ell \cap T^1)$. Since the contraction along the stable direction is stronger than the (possible) contraction in other directions, we can compare the Lebesgue measures of A_n^1 and A_n^2 and show that their ratio is close to 1. We then obtain the desired result by comparing the Lebesgue measures of the pullbacks of the sets A_n^i under f^{-n}.

Let us take a closer look at this argument. Consider a point $w \in \Lambda^\ell \cap B(x, r)$ and let $y_1 = T^1 \cap V^s(w)$ and $y_2 = \pi^s(y_1) \in T^2 \cap V^s(w)$. Fix a positive number n and observe that the distance between the points $f^n(y_1)$ and $f^n(y_2)$ is of order $\lambda^n e^{\varepsilon n}$. In fact, there are open sets $T_n^i(w) \subset f^n(T^i)$ around $f^n(y_i)$, $i = 1, 2$, each of size of order $\mu^n e^{-\varepsilon n}$, which stay within distance of order $\lambda^n e^{\varepsilon n}$ from each other. The set A_1^n is a Cantor-like set of positive leaf volume in $f^n(T^1))$ with many "holes." Using the Besicovich Covering Lemma, we can find a cover of A_1^n by open sets $T_n^1(w_j)$, $j = 1, \dots, p = p(n)$, which has finite multiplicity depending only on the dimension of T^1. The sets $T_n^2(w_j)$ form a cover of A_2^n. Moreover, we can modify the elements of these covers to obtain "almost" partitions of the sets A_1^n and A_2^n (i.e., the intersection of any two elements has zero leaf volume). This can be done without distorting too much the sizes of the elements of the old covers so that the sizes of the new elements are still of order $\mu^n e^{-\varepsilon n}$. Since $\lambda e^\varepsilon < \mu e^{-\varepsilon}$, we obtain that the ratio of the volumes of the corresponding elements of the partitions is uniformly bounded from above and from below and indeed is sufficiently close to 1 (if the number n is sufficiently large). We then show that the same can be said about the pullbacks under f^{-n} of the corresponding elements of the partitions.

We now proceed with the formal proof and split it into several steps.

Step I. Choose a number γ such that

$$\lambda e^{2\varepsilon} < \gamma < \min\{1, \mu e^{-2\varepsilon}\}.$$

This can be done if ε is sufficiently small. Fix a point $w \in \Lambda^\ell \cap B(x,r)$ and let $y_i = V^s(w) \cap T^i$ for $i = 1, 2$. For $n \geq 0$, $q > 0$, and $i = 1, 2$, we set

$$T_n^i = f^n(T^i), \quad w_n = f^n(w), \quad y_n^i = f^n(y_i), \quad q_n = q\gamma^n. \tag{8.25}$$

Note that $T_0^i = T^i$, $w_0 = w$, $y_0^i = y_i$, and $q_0 = q$. Fix $n > 0$ and consider the point w_n. In view of Theorem 7.4.1 for $i = 1, 2$, there exists an open neighborhood $T_n^i(w, q) \subset T_n^i$ of the point y_n^i such that

$$T_n^i(w, q) = \exp_{w_n}\{(\psi_n^i(v), v) \colon v \in B_2(q_n)\}, \tag{8.26}$$

where the map ψ_n^i represents $T_n^i(w, q)$. If $q = q(n)$ is sufficiently small then there is a number $R = R(\ell, n) > 0$ such that for every $w \in \Lambda^\ell \cap B(x, r)$,

$$B_n^i(w, \frac{1}{2}R) \subset T_n^i(w, q) \subset B_n^i(w, R) \subset T_n^i(w, 10q), \tag{8.27}$$

where $B_n^i(w, R)$ is the ball in T_n^i centered at w of radius R. Moreover, we may assume that for sufficiently large n,

$$\|d\psi_n^i v\| \leq \frac{1}{10}, \quad v \in B_2(10q_n), \quad i = 1, 2. \tag{8.28}$$

It follows that for any $w \in \Lambda^\ell \cap B(x, r)$ and $k = 0, \ldots, n$,

$$f^{-1}(T_k^i(w, q)) \subset T_{k-1}^i(w, q), i = 1, 2. \tag{8.29}$$

We wish to compare the leaf volumes $m_{T_n^1}|T_n^1(w, q)$ and $m_{T_n^2}|T_n^2(w, q)$ for sufficiently large n.

Lemma 8.6.2. *There exists $K_1 > 0$ such that the following holds: for any $n > 0$, there exists $q_0 = q_0(n) > 0$ such that for any $0 < q \leq q_0$, we have*

$$\left|\frac{m_{T_n^2}(T_n^2(w, q))}{m_{T_n^1}(T_n^1(w, q))} - 1\right| \leq K_1\Delta.$$

Proof of the lemma. The result follows from Theorem 7.4.1. □

Step II. We continue with the following Covering Lemma.

Lemma 8.6.3. *For any sufficiently large $n > 0$, there are points $w_j \in \Lambda^\ell \cap B(x, r)$, $j = 1, \ldots, p = p(n)$, and a number $q = q(n) > 0$ such that for $i = 1, 2$, the sets $T_n^i(w_j, q)$ form an open cover of the set $f^n(Q^\ell(x) \cap T^i)$ (see (8.21)) of finite multiplicity, which depends only on the dimension of T^i.*

Proof of the lemma. We recall the Besicovich Covering Lemma. It states that for each $Z \subset \mathbb{R}^k$, if $r \colon Z \to \mathbb{R}^+$ is a bounded function, then the cover $\{B(x, r(x)) :$

$x \in Z\}$ of Z contains a countable *Besicovich subcover* of finite multiplicity depending only on k.

The relations (8.27) and (8.28) allow one to apply the Besicovich Covering Lemma to the cover of the set $Z^1 = f^n(Q^\ell(x) \cap T^1)$ by balls

$$\left\{B^1_m(w, \frac{1}{2}R) : w \in \Lambda^\ell \cap B(x,r)\right\}.$$

Thus, we obtain a Besicovich subcover $\{B^1_m(w_j, R/2) : j = 1, \dots, p\}$, $p = p(n)$ of the set Z^1 of finite multiplicity M (which does not depend on m). The sets $T^1_n(w_j, q)$ also cover Z^1 and the multiplicity M_1 of this cover does not depend on n. Indeed, M_1 does not exceed the multiplicity M_2 of the cover of Z^1 by balls $B^1_n(w_j, R)$. Note that every ball $B^1_n(w_j, R)$ can be covered by not more than CM balls $B^1_n(w, R/2)$, where C depends only on the dimension of T^1. Furthermore, given w, we have $w_j \in B^1_n(w, R/2)$ for at most a number M of points w_j (otherwise at least $M + 1$ balls $B^1_n(w_j, R/2)$ would contain w). Therefore, $w_j \in B^1_n(w, R)$ for at most CM points w_j. This implies that $M_1 \le M_2 \le CM$. Finally, note that the sets $T^2_n(w_j, q)$ form a cover of $Z^2 = f^n(Q^\ell(x) \cap T^2)$ of multiplicity M_2. □

It follows from the definition of the sets $T^i_n(w, q)$ (see (8.26)) that for each sufficiently small $q = q(n) > 0$, there is a number $R = R(\ell, n) > 0$ such that for every $w \in \Lambda^\ell \cap B(x, r)$,

$$B^i_n(w, \frac{1}{2}R) \subset T^i_n(w, q) \subset B^i_n(w, R), \tag{8.30}$$

where $B^i_n(w, R)$ is the ball in T^i_n centered at w of radius R. Therefore, one can choose sets D^i_{jn} for $i = 1, 2$ and $j = 1, \dots, s$ such that

1. $\bigcup_{j=1}^s D^i_{jn} \supset f^n(Q^\ell(x) \cap T^i)$;
2. for each i and j,
$$B^i_n(w_j, \frac{1}{4}R) \subset D^i_{jn} \subset T^i_n(w_j, q); \tag{8.31}$$
3. $m_{T^i_n}(D^i_{jn} \cap D^i_{kn}) = 0$ for every $j \ne k$.

To see this we can apply the Besicovich Covering Lemma to the cover $\{B^i_n(w_j, R), j = 1, \dots, p\}$ to get a collection of points w_k, $k = 1, \dots, s$ such that the sets $B^i_n(w_k, R/4)$ are disjoint and

$$\bigcup_{k=1}^{s} B^i_n(w_k, 4R) \supset \bigcup_{j=1}^{p} B^i_n(w_j, R).$$

One can now "blow up" the sets $B^i_n(w_k, R/4)$, $k = 1, \dots, s$, and use (8.30) to obtain the desired sets D^i_{jn}.

It follows from Lemma 8.6.2 and (8.31) that there exists $K_2 > 0$ such that the following holds: for any $n > 0$, there exists $q_0 = q_0(n)$ such that for any

$0 < q \le q_0$, and any $j = 1, \dots, s$, we have

$$\left| \frac{m_{T_n^2}(D_{jn}^2)}{m_{T_n^1}(D_{jn}^1)} - 1 \right| \le K_2 \Delta. \tag{8.32}$$

Step III. We now compute the measures that are the pullbacks under f^{-n} of the leaf volumes $m_{T_n^i} | D_{jn}^i$, for $i = 1, 2$ and $j = 1, \dots, s$. Note that for every $w \in \Lambda^\ell \cap B(x, r)$, the points y_n^1 and y_n^2 (see (8.25)) lie on the stable manifold $V^s(w_n)$. Choose $z \in f^{-n}(D_{jn}^i)$ (with $i = 1$ or 2). Set $z_n = f^n(z)$ and

$$H^i(z, n) = \operatorname{Jac}(df^{-n} | T_{z_n} D_{jn}^i).$$

(Recall that Jac(G) denotes the Jacobian of the transformation G.) We need the following lemma.

Lemma 8.6.4. *There exist $K_2 > 0$ and $n_1(\ell) > 0$ such that for every $w \in \Lambda^\ell \cap B(x, r)$ and $n \ge n_1(\ell)$, one can find $q = q(n)$ such that*

$$\left| \frac{H^2(y_n^2, n)}{H^1(y_n^1, n)} - 1 \right| \le K_2 \Delta,$$

and if $z \in f^{-n}(T_n^1(w, q))$ then

$$\left| \frac{H^1(z_n, n)}{H^1(y_n^1, n)} - 1 \right| \le K_2 \Delta.$$

Proof of the lemma. For any $0 < k \le n$ and $z \in T_0^1(w_j, q)$, we set

$$V(k, z) = T_{z_k} T_k^1(w_j, q).$$

We transport parallelly the space $V(k, z)$ along the geodesic that connects the points z_k and y_k^1 (this geodesic is uniquely defined since these points are sufficiently close to each other) to obtain a new subspace $\widetilde{V}(k, z) \subset T_{y_k^1} M$. We have

$$\begin{aligned} &\left| \operatorname{Jac}(df^{-1} | V(k, z)) - \operatorname{Jac}(df^{-1} | V(k, y_1)) \right| \\ &\le \left| \operatorname{Jac}(df^{-1} | V(k, z)) - \operatorname{Jac}(df^{-1} | \widetilde{V}(k, z)) \right| \\ &\quad + \left| \operatorname{Jac}(df^{-1} | \widetilde{V}(k, z)) - \operatorname{Jac}(df^{-1} | V(k, y_1)) \right|. \end{aligned}$$

Since $f \in C^{1+\alpha}$ in view of (8.29), we obtain

$$\begin{aligned} &\left| \operatorname{Jac}(df^{-1} | V(k, z)) - \operatorname{Jac}(df^{-1} | V(k, y_1)) \right| \\ &\le C_1 \rho(z_k, y_k^1)^\alpha + C_2 d(V(k, z), V(k, y_1)), \end{aligned}$$

where $C_1 > 0$ and $C_2 > 0$ are constants. (Recall that ρ is the distance in M and d is the distance in the Grassmannian bundle of TM generated by the Riemannian metric.) It follows from Theorem 5.3.1 that

$$d(V(k, z), V(k, y_1)) \le C_3 \rho(z_k, y_k^1)^\beta,$$

where $C_3 > 0$ is a constant. Using (8.1) we find

$$\begin{aligned}&\left|\mathrm{Jac}(df^{-1}|V(k,z)) - \mathrm{Jac}(df^{-1}|V(k,y_1))\right|\\ &\le C_4\left(e^{\varepsilon(n-k)}\kappa^{-(n-k)}\rho(z_n,y_n^1)\right)^{\sigma},\end{aligned}$$

where $\sigma = \min\{\alpha, \beta\}$, $0 < \kappa < 1$, and $C_4 > 0$ is a constant. Note that for any $0 < k \le n$ and $z \in T_0^1(w,q)$, we have

$$C_5^{-1} \le |\mathrm{Jac}(df^{-1}|V(k,z))| \le C_5,$$

where $C_5 > 0$ is a constant. We obtain

$$\begin{aligned}\frac{H^1(z_n,n)}{H^1(y_n^1,n)} &= \prod_{k=1}^{n}\frac{\mathrm{Jac}(df^{-1}|V(k,z))}{\mathrm{Jac}(df^{-1}|V(k,y_1))}\\ &= \exp\sum_{k=1}^{n}\log\frac{\mathrm{Jac}(df^{-1}|V(k,z))}{\mathrm{Jac}(df^{-1}|V(k,y_1))}\\ &\le \exp\sum_{k=1}^{n}\left(\frac{\mathrm{Jac}(df^{-1}|V(k,z))}{\mathrm{Jac}(df^{-1}|V(k,y_1))} - 1\right)\\ &\le \exp\left(\sum_{k=1}^{n} C_4C_5\left(e^{\varepsilon(n-k)}\kappa^{-(n-k)}\rho(z_n,y_n^1)\right)^{\sigma}\right)\\ &\le \exp\left(\frac{C_4C_5q^{\sigma}}{1-(e^{-\varepsilon}\kappa)^{\sigma}}\right).\end{aligned}$$

The last expression can be made arbitrarily close to 1 by choosing q and ε sufficiently small. This proves the second inequality. The first inequality can be proven in a similar fashion. □

Lemma 8.6.4 allows one to compare the measures of the preimages under f^{-n} of D_{jn}^1 and D_{jn}^2, $j = 1, \dots, s$. More precisely, the following statement holds:

Lemma 8.6.5. *There exist $K_3 > 0$ and $n_2(\ell) > 0$ such that for any $n \ge n_2(\ell)$, one can find $q = q(n)$ such that for $j = 1, \dots, s$,*

$$\left|\frac{m_{T^2}(f^{-n}(D_{jn}^2))}{m_{T^1}(f^{-n}(D_{jn}^1))} - 1\right| \le K_3\Delta.$$

Proof of the lemma. For $i = 1, 2$ and $j = 1, \dots, s$, we have

$$\begin{aligned}m_{T^i}(f^{-n}(D_{jn}^i)) &= \int_{T_n^i(w,q)} H^i(z,n)\,dm_{T_n^i}(z)\\ &= H^i(z_n^i,n)m_{T_n^i}(D_{jn}^i),\end{aligned}$$

where $z_n^i \in D_{jn}^i$ are some points. It follows from the assumptions of the lemma, (8.29), (8.32), and Lemma 8.6.4 that for sufficiently large n and small q (assuming

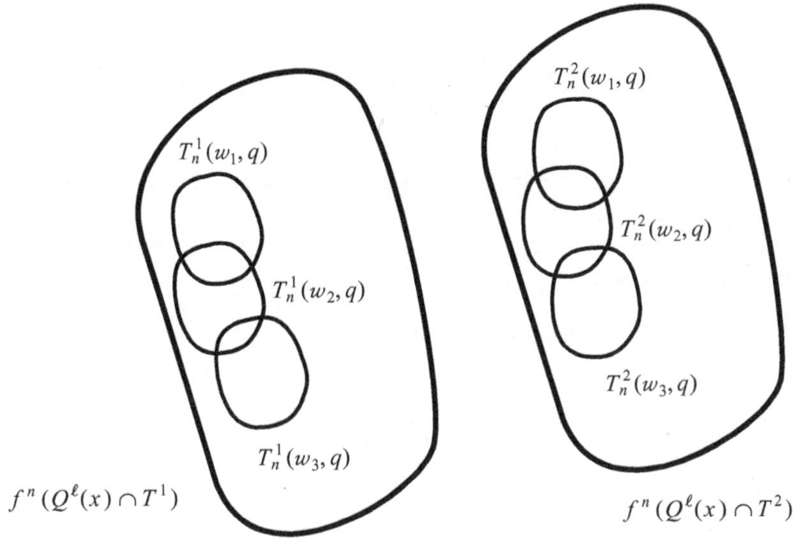

Figure 8.2. Sets $f^n(Q^\ell(x) \cap T^i)$ and their covers by $T_n^i(w_j, q)$.

without loss of generality that $\Delta \le 1$),

$$\begin{aligned}
&\left| \frac{m_{T^2}(f^{-n}(D_{jn}^2))}{m_{T^1}(f^{-n}(D_{jn}^1))} - 1 \right| \\
&\le \left| \frac{H^2(z_n^2, n)}{H^1(z_n^1, n)} - 1 \right| \times \frac{m_{T_n^2}(D_{jn}^2)}{m_{T_n^1}(D_{jn}^1)} + \left| \frac{m_{T_n^2}(D_{jn}^2)}{m_{T_n^1}(D_{jn}^1)} - 1 \right| \\
&\le \left| \frac{H^2(z_n^2, n)}{H^2(y_n^2, n)} \cdot \frac{H^1(y_n^1, n)}{H^1(z_n^1, n)} \cdot \frac{H^2(y_n^2, n)}{H^1(y_n^1, n)} - 1 \right| \times (1 + K_1\Delta) + K_1\Delta \\
&\le ((1 + K_2\Delta)^3 - 1)(1 + K_1\Delta) + K_1\Delta \le K_3\Delta.
\end{aligned}$$

The lemma follows. □

Step IV. Given $n > 0$, choose the points $w_j \in \Lambda^\ell \cap B(x, r)$ and the number $q = q(n)$ as in Lemma 8.6.3. Consider the sets $\hat{T}_n^i = \bigcup_{j=1}^s D_{jn}^i$ and note that $\hat{T}_n^i \supset f^n(Q^\ell(x) \cap T^i)$, $i = 1, 2$ (see Fig. 8.2).

We wish to compare the leaf volumes $m_{T^1}|f^{-n}(\hat{T}_n^1)$ and $m_{T^2}|f^{-n}(\hat{T}_n^2)$. Consider the cover of the set $Q^\ell(x) \cap T^i$ by the sets $\{f^{-n}(D_{jn}^i)\}$. We have

$$\begin{aligned}
\frac{m_{T^2}(f^{-n}(\hat{T}_n^2))}{m_{T^1}(f^{-n}(\hat{T}_n^1))} &\le \frac{\sum_{j=1}^s m_{T^2}(f^{-n}(D_{jn}^2))}{\sum_{j=1}^s m_{T^1}(f^{-n}(D_{jn}^1))} \\
&\le \max\left\{ \frac{m_{T^2}(f^{-n}(D_{jn}^2))}{m_{T^1}(f^{-n}(D_{jn}^1))} : j = 1, \dots, s \right\} \\
&\le 1 + K_3\Delta,
\end{aligned}$$

with a similar bound for the inverse ratio. We conclude that

$$\left|\frac{m_{T^2}(f^{-n}(\hat{T}_n^2))}{m_{T^1}(f^{-n}(\hat{T}_n^1))} - 1\right| \le K_3\Delta. \tag{8.33}$$

Step V. Given $\delta > 0$, denote by U_δ^i the δ-neighborhood of the set $Q^\ell(x) \cap T^i$ for $i = 1, 2$. For any $\delta > 0$ and any sufficiently large $n > 0$, we have $Q^\ell(x) \cap T^i \subset f^{-n}(\hat{T}_n^i) \subset U_\delta^i$ for $i = 1, 2$. Observe that $m_{T^1}(U_\delta^1) > 0$. We need the following lemma from measure theory.

Lemma 8.6.6. *There exists* $\delta_0 > 0$ *such that for every* $0 < \delta \le \delta_0$,

$$\frac{m_{T^1}(Q^\ell(x) \cap T^1)}{m_{T^1}(U_\delta^1)} \ge 1 - \Delta.$$

We have that

$$\frac{m_{T^2}(U_\delta^2)}{m_{T^1}(U_\delta^1)} = \frac{m_{T^2}(U_\delta^2)}{m_{T^2}(f^{-n}(\hat{T}_n^2))} \cdot \frac{m_{T^2}(f^{-n}(\hat{T}_n^2))}{m_{T^1}(f^{-n}(\hat{T}_n^1))} \cdot \frac{m_{T^1}(f^{-n}(\hat{T}_n^1))}{m_{T^1}(U_\delta^1)}.$$

It follows from Lemma 8.6.6 and (8.33) that $m_{T^2}(U_\delta^2) > 0$. In particular, similarly to Lemma 8.6.6 we have that for all sufficiently small $\delta > 0$,

$$\frac{m_{T^2}(Q^\ell(x) \cap T^2)}{m_{T^2}(U_\delta^2)} \ge 1 - \Delta.$$

Applying now (8.33), we find that

$$\left|\frac{m_{T^2}(Q^\ell(x) \cap T^2)}{m_{T^1}(Q^\ell(x) \cap T^1)} - 1\right| \le K_4\Delta,$$

where $K_4 > 0$ is a constant, which does not depend on the size of the transversals T^1 and T^2 but only on their dimension. In particular, for each $y \in Q^\ell(x) \cap T^1$, we have

$$\left|\frac{m_{T^2}(\pi^s(Q^\ell(x) \cap B^1(y,r)))}{m_{T^1}(Q^\ell(x) \cap B^1(y,r))} - 1\right| \le K_4\Delta,$$

where $B^1(y,r) \subset T^1$ is the ball of radius r centered at y. Letting $r \to 0$, we conclude from (8.23) that (8.24) holds. This completes the proof of the theorem. □

Let f be a $C^{1+\alpha}$ diffeomorphism of a compact smooth Riemannian manifold M and Y a nonempty set of nonuniformly partially hyperbolic points for f. Choose a nonempty level set $\Lambda = \Lambda_{\lambda\mu\varepsilon j}$. For every $x \in \Lambda$, we have an invariant splitting $T_xM = E^s(x) \oplus E^c(x) \oplus E^u(x)$, where $E^s(x)$ and $E^u(x)$ are the stable and unstable subspaces at x. Given a regular set Λ^ℓ and a point $x \in \Lambda^\ell$, consider the family of local unstable manifolds

$$\mathcal{L}^u(x) = \{V^u(w) : w \in \Lambda^\ell \cap B(x,r)\}.$$

The following result can be obtained from Theorem 8.6.1 by applying it to the inverse map f^{-1}.

Theorem 8.6.7. *The holonomy map π^u associated with the family of local unstable manifolds is absolutely continuous with respect to the leaf volumes m_{T^1} and m_{T^2} on local transversals T^1 and T^2 and its Jacobian satisfies* (8.24).

8.6.2 Absolute Continuity of Local Stable Manifolds

An important consequence of Theorem 8.6.1 is the fact that local stable manifolds are absolutely continuous (in the strong sense). Roughly speaking, this means that the conditional measure generated by volume on local stable manifolds is absolutely continuous (in fact, equivalent) to the leaf volume (see the beginning of Section 8.6). More precisely, let f be a $C^{1+\alpha}$ diffeomorphism of a compact smooth manifold M and Y a nonempty set, which is nonuniformly partially hyperbolic in the broad sense for f (see Section 5.2). We assume that this set has positive (not necessarily full) volume; that is, $m(Y) > 0$. Let us stress that volume m need not be invariant under f. There is a level set $\Lambda = \Lambda_{\lambda\mu\varepsilon j}$ of positive measure (see (2.21)) and let $\{\Lambda^\ell : \ell \geq 1\}$ be the corresponding collection of regular sets (see (2.2.2)). We choose ℓ so large that this set has positive volume.

Fix $x \in \Lambda^\ell$ and consider the family of local stable manifolds $\mathcal{L}^s(x)$ (see (8.20)). There exists a family of smooth submanifolds $T(w)$ for $w \in B(x, r)$ with the following properties:

1. $T(w)$ is a local transversal to the family $\mathcal{L}^s(x)$.
2. $T(w_1) \cap T(w_2) = \varnothing$ if $w_2 \notin T(w_1)$ and $T(w_1) = T(w_2)$ if $w_2 \in T(w_1)$.
3. $\bigcup_{w \in B(x,r)} T(w) \supset B(x, r)$.
4. $T(w)$ depends smoothly on $w \in B(x, r)$.

It follows that the collection of smooth submanifolds $T(w)$ generates a partition η of the ball $B(x, r)$.

Consider now the measurable partition ξ^s of the set $Q^\ell(x)$ (see (8.21)) into local stable manifolds $V^s(w)$. We use the following notations: $m_{V^s}(w)$ for the leaf volume on the stable manifold $V^s(w)$ at the point w, $m_T(w)$ for the leaf volume on $T(w)$, and $m^s(w)$ for the conditional measure on $V^s(w)$ generated by the partition ξ^s and volume m. The factor-space $Q^\ell(x)/\xi^s$ can be identified with the subset

$$A(w) = \big\{y \in T(w) : \text{there exists } z \in \Lambda^\ell \cap B(x, r) \\ \text{such that } y = T(w) \cap V^s(z)\big\}$$

for every $w \in B(x, r)$.

Theorem 8.6.8 (see [169]). *For m-almost every $x \in \Lambda^\ell$ and $w \in \Lambda^\ell \cap B(x, r)$, the measure $m^s(w)$ is equivalent to the leaf volume $m_{V^s}(w)$ with density to be continuous (hence bounded from above) and strictly positive.*

Proof. For $w \in B(x,r)$, denote by $\{\widetilde{\mu}(w)\}$ the system of conditional measures and by $\hat{\mu}$ the factor-measure generated by the partition η and volume m. One can see that for every $w_0 \in \Lambda^\ell \cap B(x,r)$ and $y \in A(w_0)$, the factor-space $B(x,r)/\eta$ can be identified with the local stable manifold $V^s(y)$. Since the partition T is smooth, by Fubini's theorem, there exists a smooth function $g(w,z)$, which is strictly positive and such that for $w \in \Lambda^\ell \cap B(x,r)$ and $z \in T(w)$,

$$d\widetilde{\mu}(w)(z) = g(w,z)\, dm_T(w)(z).$$

Furthermore, for every $w_0 \in \Lambda^\ell \cap B(x,r)$ and $y \in A(w_0)$, there exists a smooth function $h(y,w)$, which is strictly positive and such that for $w \in V^s(y)$,

$$d\hat{\mu}(w) = h(y,w)\, dm_{V^s}(y)(w).$$

Note that the set $A(w) \subset T(w)$ has positive $\widetilde{\mu}$-measure. Let $B \subset Q^\ell(x)$ be a Borel subset of positive volume. We have

$$\begin{aligned} m(B) &= \int_{B(x,r)/\eta} \int_{T(w)} \chi_B(w,z)\, d\widetilde{\mu}(w)(z)\, d\hat{\mu}(w) \\ &= \int_{B(x,r)/\eta} \int_{A(w)} \chi_B(w,z) g(w,z)\, dm_T(w)(z)\, d\hat{\mu}(w) \\ &= \int_{B(x,r)/\eta} \int_{A(w_0)} \chi_B(w,z) g(w,z)\, \mathrm{Jac}(\pi^s_{w_0 w})(y)\, dm_T(w_0)(y)\, d\hat{\mu}(w), \end{aligned}$$

where $\chi_B(w,z)$ is the characteristic function of the set B at the point $z \in A(w)$, $w_0 \in \Lambda^\ell \cap B(x,r)$ is a point, $\pi^s_{w_0 w}$ is the holonomy map between the transversals $T(w_0)$ and $T(w)$ (generated by the family $\mathcal{L}^s(x)$), $y \in A(w_0)$ is such that $z = \pi^s_{w_0 w}(y)$, and $\mathrm{Jac}(\pi^s_{w_0 w})(y)$ is the Jacobian of the map $\pi^s_{w_0 w}$ at y (see Theorem 8.6.1). Observe that $\pi^s_{w_0 w}(A(w_0)) = A(w)$. Applying Fubini's theorem we conclude that $m(B)$ is equal to

$$\begin{aligned} &\int_{A(w_0)} \int_{B(x,r)/\eta} \chi_B(w,z) g(w,z)\, \mathrm{Jac}(\pi^s_{w_0 w})(y)\, d\hat{\mu}(w)\, dm_T(w_0)(y) \\ &= \int_{A(w_0)} \int_{V^s(y)} \chi_B(w,z) g(w,z) \\ &\qquad \times \mathrm{Jac}(\pi^s_{w_0 w})(y) h(y,w)\, dm_{V^s}(y)(w)\, dm_T(w_0)(y). \end{aligned}$$

This implies that the conditional measure $m^s(y)$ on $V^s(y)$ is equivalent to the leaf volume $m_{V^s}(y)$ with the density function

$$\kappa^s(y,w) = \mathrm{Jac}(\pi^s_{w_0 w})(y) g(w, \pi^s_{w_0 w}(y)) h(y,w) > 0.$$

This implies the desired result. □

As an immediate consequence of Theorem 8.6.8, we obtain the following results. The first ones reflect on the intuitive interpretation of absolute continuity (see the beginning of Section 8.6).

Corollary 8.6.9. *Let $A \subset \Lambda$ be a measurable set of positive volume and for each $x \in A$, let $B_x \subset V^s(x)$ be a measurable set of positive leaf volume. Then the set $B = \bigcup_{x\in A} B_x$ has positive volume.*

Proof. Since $m(A) > 0$, we have that $m(A \cap \Lambda^\ell) > 0$ for some $\ell > 0$. Let $x \in A \cap \Lambda^\ell$ be a density point. Consider the set

$$Q_A^\ell(x) = \bigcup_{w \in B(x,r)\cap\Lambda^\ell\cap A} V^s(w).$$

Applying Fubini's theorem, we obtain that the set

$$\bigcup_{w \in B(x,r)\cap\Lambda^\ell\cap A} B_w \subset Q_A^\ell(x)$$

has positive volume. The desired result follows. □

Corollary 8.6.10. *Let $N \subset M$ be a set of zero volume. Then*

$$m_{V^s}(x)(V^s(x) \cap N) = 0 \textit{ for } m\textit{-almost every point } x \in \Lambda.$$

Proof. Assuming the contrary we find a set A of positive volume such that $m_{V^s}(x)(V^s(x) \cap N) > 0$ for every $x \in A$. Let $x \in A$ be a density point. In view of Corollary 8.6.9, for sufficiently small $r > 0$, the set

$$B = \bigcup_{w \in B(x,r)\cap A} V^s(x) \cap N$$

has positive volume. Clearly, $B \subset N$ and we arrive at a contradiction. □

Consider a set Λ^ℓ of positive volume. Let $x \in \Lambda^\ell$ be a density point and T a local transversal to the family $\mathcal{L}^s(x)$.

Corollary 8.6.11. *Let $N \subset T$ be a set of zero Lebesgue measure in T. We have that $m(\bigcup V^s(w)) = 0$ where the union is taken over all points $w \in \Lambda^\ell \cap B(x,r)$ for which $V^s(w) \cap T \in N$.*

Proof. The claim follows immediately from Theorem 8.6.8. □

The next statement shows that for m-almost every $x \in \Lambda$, the local manifold $V^s(x)$ lies in Λ almost surely.

Corollary 8.6.12. *Assume that $m(\Lambda) = 1$. Then for m-almost every $x \in \Lambda$*

$$m_{V^s}(x)(V^s(x) \setminus \Lambda) = 0.$$

Moreover, the result holds true if we only assume that $m(\Lambda) > 0$ and Λ is open (mod 0).

Proof. Assuming the contrary we find a set A of positive volume such that $m_{V^s}(x)(V^s(x) \setminus \Lambda) > 0$ for every $x \in A$. It follows from Corollary 8.6.9 that the set $B = \bigcup_{x\in A} V^s(x) \setminus \Lambda$ has positive volume. However, any point in this set does not lie in Λ leading to a contradiction. The proof of the second claim is similar and is left to the reader. □

One can strengthen the last result by showing that there is a set $E \subset \Lambda$ of full volume such that $m_{V^s}(x)(V^s(x) \setminus E) = 0$ (see the proof of Lemma 11.4.18).

8.6.3 Foliation That Is Not Absolutely Continuous

We describe an example due to Katok of a foliation that is not absolutely continuous. Consider a hyperbolic automorphism A of the torus $\mathbb{T}^2$ and let $\{f_t : t \in S^1\}$ be a family of diffeomorphisms preserving the area m and satisfying the following conditions:

1. f_t is a small perturbation of A for every $t \in S^1$.
2. f_t depends smoothly on t.
3. The function $h(t) = h_m(f_t)$ is strictly monotone in a small neighborhood of $t = 0$ (where $h_m(f_t)$ is the measure-theoretic entropy of the diffeomorphism f_t; see Chapter 10).

Note that for any family f_t, the entropy is given by

$$h(t) = \int_{\mathbb{T}^2} \log \|df_t | E_t^u(x)\| \, dm(x),$$

where $E_t^u(x)$ denotes the unstable subspace of f_t at the point x (see Chapter 10). Hence, one can modify A in a small neighborhood such that $h(t)$ is strictly monotone.

We introduce the diffeomorphism $F\colon \mathbb{T}^2 \times S^1 \to \mathbb{T}^2 \times S^1$ by $F(x,t) = (f_t(x), t)$. Since f_t is sufficiently close to A, they are conjugate via a Hölder homeomorphism g_t; that is, $f_t = g_t \circ A \circ g_t^{-1}$. Given $x \in \mathbb{T}^2$, consider the set

$$H(x) = \{(g_t(x), t) : t \in S^1\}.$$

It is diffeomorphic to the circle S^1 and the collection of these sets forms an F-invariant foliation H of $\mathbb{T}^2 \times S^1 = \mathbb{T}^3$ with $F(H(x)) = H(A(x))$. Note that $H(x)$ depends Hölder continuously on x. However, the holonomy maps associated with the foliation H are not absolutely continuous. To see this consider the holonomy map

$$\pi_{t_1,t_2}\colon \mathbb{T}^2 \times \{t_1\} \to \mathbb{T}^2 \times \{t_2\}.$$

We have that

$$\pi_{0,t}(x,0) = (g_t(x), t) \text{ and } F(\pi_{0,t}(x,0)) = \pi_{0,t}(A(x), 0).$$

If the map $\pi_{0,t}$ (with t being fixed) were absolutely continuous, the measure $(\pi_{0,t})_* m$ would be absolutely continuous with respect to m. Note that each map f_t is ergodic (it is conjugate to the ergodic map A), and hence m is the only absolutely continuous f_t-invariant probability measure. Thus, $(\pi_{0,t})_* m = m$. In particular, $h(t) = h(0)$. Since the entropy function $h(t)$ is strictly monotone in a small neighborhood of $t = 0$, the map g_t is not absolutely continuous for small t and so is the map $\pi_{0,t}$.

8.6.4 The Jacobian of the Holonomy Map

In this section we strengthen the Absolute Continuity Theorem 8.6.1 and obtain an explicit formula for the Jacobian.

Theorem 8.6.13. *Under the assumptions of Theorem 8.6.1 for every $y \in Q^\ell(x) \cap T^1$,*

1. $$\mathrm{Jac}(\pi^s)(y) = \prod_{k=0}^{\infty} \frac{\mathrm{Jac}(df^{-1}|T_{f^k(\pi^s(y))}f^k(T^2))}{\mathrm{Jac}(df^{-1}|T_{f^k(y)}f^k(T^1))}$$

   (recall that Jac *denotes the Jacobian of the corresponding map); in particular, the infinite product on the right-hand side converges;*
2. *the function* $\mathrm{Jac}(\pi^s)(y)$ *depends Hölder continuously on y.*

Proof. Fix a point $y \in Q^\ell(x) \cap T^1$. Repeating arguments in the proof of Lemma 8.6.4, it is straightforward to verify that the infinite product

$$\prod_{k=0}^{\infty} \frac{\mathrm{Jac}(df^{-1}|T_{f^k(\pi^s(y))}f^k(T^2))}{\mathrm{Jac}(df^{-1}|T_{f^k(y)}f^k(T^1))} =: J$$

converges. Therefore given $\varepsilon > 0$, there exists $m_1 > 0$ such that for every $m \geq m_1$, we have

$$\left| J - \prod_{k=0}^{m-1} \frac{\mathrm{Jac}(df^{-1}|T_{f^k(\pi^s(y))}f^k(T^2))}{\mathrm{Jac}(df^{-1}|T_{f^k(y)}f^k(T^1))} \right| \leq \varepsilon. \tag{8.34}$$

Fix $m > 0$. One can choose small neighborhoods $U_m^1 \subset T^1$ and $U_m^2 \subset T^2$ of the points y and $\pi^s(y)$, respectively, and define the holonomy map $\pi_m^s : f^m(U_m^1) \to f^m(U_m^2)$ by $\pi_m^s = f^m \circ \pi^s \circ f^{-m}$. We have that

$$\mathrm{Jac}(\pi^s)(y) = \prod_{k=0}^{m-1} \frac{\mathrm{Jac}(df^{-1}|T_{f^k(\pi^s(y))}f^k(T^2))}{\mathrm{Jac}(df^{-1}|T_{f^k(y)}f^k(T^1))} \mathrm{Jac}(\pi_m^s)(f^m(y)). \tag{8.35}$$

We now choose $m > m_1$ so large and neighborhoods U_m^1 and U_m^2 so small that $\Delta(f^m(U_m^1), f^m(U_m^2)) \leq \varepsilon$. The first statement (or more precisely the corresponding statement for the holonomy map π_m^s) implies that

$$|\mathrm{Jac}(\pi_m^s)(f^m(y)) - 1| \leq K\varepsilon. \tag{8.36}$$

It follows from (8.34) to (8.36) that

$$|\mathrm{Jac}(\pi^s)(y) - J| \leq \varepsilon + (J + \varepsilon)K\varepsilon.$$

Since ε is arbitrary, this completes the proof of the first statement. The second statement follows from the first one and Theorem 5.3.1. This completes the proof of the theorem. □

We consider two particular cases of Theorem 8.6.13. First, we assume that the set Λ consists of nonuniformly completely hyperbolic points for the diffeomorphism f. For $x \in \Lambda^\ell$, we can choose transversals T^1 and T^2 to the family of local stable manifolds $\mathcal{L}^s(x)$ (see (8.20)) to be local unstable manifolds $V^u(w_1)$ and $V^u(w_2)$ at some points $w_1, w_2 \in \Lambda^\ell \cap B(x, r)$. We have that $\Delta = \Delta(T^1, T^2)$ (see (8.22)) admits the following estimate:

$$\Delta \le C \max_{w \in \Lambda^\ell \cap B(x,r)} d(E^u(x), E^u(w)),$$

where $C > 0$ is a constant.

Theorem 8.6.14. *For every $y \in Q^\ell \cap V^u(w_1)$,*

1. *there exists a constant $K = K(\ell) > 0$ such that*
$$|\mathrm{Jac}(\pi^s)(y) - 1| \le K \max_{w \in \Lambda^\ell \cap B(x,r)} d(E^u(x), E^u(w));$$
2. $$\mathrm{Jac}(\pi^s)(y) = \prod_{k=0}^{\infty} \frac{\mathrm{Jac}(df^{-1}|E^u(f^k(\pi^s(y))))}{\mathrm{Jac}(df^{-1}|E^u(f^k(y)))};$$
3. *the function* $\mathrm{Jac}(\pi^s)(y)$ *depends Hölder continuous on y.*

Our next case is a flow φ_t, which is nonuniformly completely hyperbolic on a set Λ. We can choose transversals T^1 and T^2 to the family of local stable manifolds $\mathcal{L}^s(x)$ to be local weakly unstable manifolds $V^{u0}(w_1)$ and $V^{u0}(w_2)$ at some points $w_1, w_2 \in \Lambda^\ell \cap B(x, r)$. We have that

$$\Delta \le C \max_{w \in \Lambda^\ell \cap B(x,r)} d(E^{u0}(x), E^{u0}(w)),$$

where $C > 0$ is a constant.

Theorem 8.6.15. *For every $y \in Q^\ell \cap V^u(w_1)$,*

1. *there exists a constant $K = K(\ell) > 0$ such that*
$$|\mathrm{Jac}(\pi^s)(y) - 1| \le K \max_{w \in \Lambda^\ell \cap B(x,r)} d(E^{u0}(x), E^{u0}(w));$$
2. $$\mathrm{Jac}(\pi^s)(y) = \prod_{k=0}^{\infty} \frac{\mathrm{Jac}(df^{-1}|E^{u0}(f^k(\pi^s(y))))}{\mathrm{Jac}(df^{-1}|E^{u0}(f^k(y)))};$$
3. *the function* $\mathrm{Jac}(\pi^s)(y)$ *depends Hölder continuous on y.*

8.7 Notes

The construction of global stable and unstable manifolds is due to Pesin (see [167]). He also described properties of local and global stable and unstable manifolds. The Lipschitz property of intermediate foliations was shown by Barreira, Pesin, and Schmeling in [22].

The proof of the absolute continuity property for Anosov diffeomorphisms was given by Anosov [6] and by Anosov and Sinai [7]. The approach in [7] is based on the study of the holonomy maps associated with the foliation. This approach was used in [167] to establish the absolute continuity property for nonuniformly partially hyperbolic diffeomorphisms (in the broad sense). The proof of Theorem 8.6.1 presented above is an elaboration of his proof. In [185], Pugh and Shub established the absolute continuity property using a somewhat different approach that is based on their study of unstable manifolds. A version of the example by Katok (see the end of Section 8.6.1) was presented by Milnor in [152] (see also Section 11.6 and Chapter 6 in [81] where yet another version of Katok's example by Burns and Flaminio is given).

Part III

Ergodic Theory of Smooth and SRB Measures

9

Smooth Measures

In the previous chapters of the book, we developed techniques and obtained a number of principal results on local behavior of trajectories (such as existence of local stable and unstable manifolds and absolute continuity of families of local manifolds), which will allow us to describe ergodic properties of nonuniformly hyperbolic dynamical systems with respect to some "natural" invariant measures. We consider two cases: smooth hyperbolic invariant measures and the physically important Sinai–Ruelle–Bowen (SRB) measures. The former case will be the subject of study in this and in the next three chapters, while the latter case will be discussed in Chapter 13.

Unlike the classical uniform hyperbolicity, the nonuniform hyperbolicity requirements are *weak enough* to allow presence of nonuniformly hyperbolic diffeomorphisms and flows on any smooth manifolds (see Theorems 11.4.1 and 11.5.1). Surprisingly, these requirements are *strong enough* to ensure abundance of ergodic properties: the ultimate result is the Spectral Decomposition Theorem 9.5.11, asserting that ergodic components of the system are of positive measure and the restriction of the system to every such component is Bernoullian (up to the discrete spectrum).

These results constitute the core of the smooth ergodic theory and were established in [168–170]. Our presentation follows these works, with necessary modifications and generalizations to include nonuniformly partially (not necessary completely) hyperbolic systems. This allows us to greatly expand the class of invariant measures under consideration to the so-called s- and u-measures (and, in particular, to SRB measures).

9.1 Ergodic Components

Let f be a $C^{1+\alpha}$ diffeomorphism of a smooth compact Riemannian manifold M without boundary. Throughout this chapter we assume that f preserves an *absolutely continuous measure* ν, that is, a measure, which is absolutely continuous with respect to the Riemannian volume m on M with measurable density bounded from above. More precisely, there are $C > 0$ and a set $X \subset M$ of full volume such that $0 \le d\nu/dm(x) \le C$ for every $x \in X$. An important particular case is when ν is a *smooth measure*; that is, there are $C > 0$ and a set $X \subset M$ of full volume such that $C^{-1} \le d\nu/dm(x) \le C$ for every $x \in X$. Assuming that the measure ν is hyperbolic, our goal is to describe the decomposition into its ergodic components.

In fact, the underlying property of the measure ν we need is that the conditional measures generated by ν on local unstable (or respectively, stable) manifolds are absolutely continuous with respect to the leaf volume on these manifolds. We shall call the measures with this property *u-measures* (respectively, *s-measures*). The precise definitions will be given in Section 9.2. An important example of u-measures is *SRB measures* (see Chapter 13).

Recall that a measurable transformation G of a measurable space X preserving a probability measure ν is *ergodic* if any G-invariant set has measure either 0 or 1. Equivalently, G is ergodic if any G-invariant measurable function is constant almost everywhere. A measurable partition ξ is called the *decomposition into ergodic components* if for almost every element C of ξ, the map $G|C$ is ergodic with respect to the conditional measure ν_C induced by ν.

The following statement is one of the main results of Smooth Ergodic Theory and describes the decomposition of the measure ν into ergodic components. The proof exploits a simple yet deep argument due to Hopf [87] (see the proof of Lemma 9.1.3 below).

Theorem 9.1.1 (see [169]). *Let f be a $C^{1+\alpha}$ diffeomorphism of a smooth compact Riemannian manifold M. Assume that f preserves an absolutely continuous hyperbolic measure ν. Then there exist invariant sets $\Lambda_0, \Lambda_1, \dots$ such that*

1. $\bigcup_{i\ge 0} \Lambda_i = M$, *and* $\Lambda_i \cap \Lambda_j = \varnothing$ *whenever* $i \ne j$;
2. $\nu(\Lambda_0) = 0$, *and* $\nu(\Lambda_i) > 0$ *for each* $i \ge 1$;
3. $f|\Lambda_i$ *is ergodic for each* $i \ge 1$.

Proof. We begin with the following statement:

Lemma 9.1.2. *Given an f-invariant integrable function φ, there exists a set $N \subset M$ of zero measure such that if $y \in M$ and $z, w \in V^s(y) \setminus N$ or $z, w \in V^u(y) \setminus N$, then $\varphi(z) = \varphi(w)$.*

Proof of the lemma. Let ψ be a continuous function. By Birkhoff's Ergodic Theorem, the functions

$$\overline{\psi}(x) = \lim_{n\to\infty} \frac{1}{2n+1} \sum_{k=-n}^{n} \psi(f^k(x)),$$

$$\psi^+(x) = \lim_{n\to\infty} \frac{1}{n} \sum_{k=1}^{n} \psi(f^k(x)), \quad \text{and} \quad \psi^-(x) = \lim_{n\to\infty} \frac{1}{n} \sum_{k=1}^{n} \psi(f^{-k}(x))$$

are defined and $\overline{\psi}(x) = \psi^+(x) = \psi^-(x)$ outside a subset $N \subset M$ of zero measure.

Let $z, w \in V^s(y) \setminus N$. Since $\rho(f^n(z), f^n(w)) \to 0$ as $n \to \infty$ (see Theorem 7.5.1) and ψ is continuous, we obtain

$$\overline{\psi}(z) = \psi^+(z) = \psi^+(w) = \overline{\psi}(w).$$

Similarly, for $z, w \in V^u(y)$,

$$\overline{\psi}(z) = \psi^-(z) = \psi^-(w) = \overline{\psi}(w).$$

Notice that the continuous functions are dense in $L^1(M, \nu)$, and hence the functions of the form $\overline{\psi}$ are dense in the set of f-invariant Borel functions. The lemma follows. □

Since ν is an absolutely continuous hyperbolic measure, f is nonuniformly completely hyperbolic on the set of full measure Z given by (5.5). We consider a collection of level sets $\{\Lambda_{\lambda_i \mu_i \varepsilon_i j}\}$ (see Section 2.2.2; the numbers λ_i, μ_i, and ε_i satisfy (2.15)) that are of positive measure and cover Z. It suffices to prove our theorem for the restriction of f to each of these level sets. Let us fix a level set $\Lambda = \Lambda_{\lambda_i \mu_i \varepsilon_i j}$ and consider the corresponding collection of regular sets $\{\Lambda^\ell = \Lambda^\ell_{\lambda_i \mu_i \varepsilon_i j} : \ell \geq 1\}$. Choose ℓ so large that the set Λ^ℓ has positive measure. Pick a Lebesgue point $x \in \Lambda^\ell$, and consider a family $\mathcal{L}^s(x)$ of local stable manifolds $V^s(y)$, $y \in \Lambda^\ell \cap B(x, r)$ (see (8.20)) where $r = r(\ell) > 0$ is a sufficiently small number and the set $Q^\ell(x)$ is given by (8.21) (see Fig. 9.1).

Lemma 9.1.3. *There exists $r = r(\ell) > 0$ such that the map f is ergodic on the set*

$$Q(x) = \bigcup_{n \in \mathbb{Z}} f^n(Q^\ell(x)). \tag{9.1}$$

Proof of the lemma. Let φ be an f-invariant integrable function and N the set of zero measure constructed in Lemma 9.1.2. Choose $0 < r < \min\{r_\ell, \delta_\ell\}$ (see (8.4)). Consider the family $\mathcal{L}^u(x)$ of local unstable manifolds $V^u(x)$, $y \in \Lambda^\ell \cap B(x, r)$. By Corollary 8.6.12 (applied to this family), there exists $y \in \Lambda^\ell \cap B(x, r)$ such that $m_{V^u}(y)(V^u(y) \cap N) = 0$ (indeed, almost every $y \in \Lambda^\ell \cap B(x, r)$ has this property), where $m_{V^u}(y)$ is the leaf volume on $V^u(y)$. Let

$$R^s = \bigcup V^s(w),$$

where the union is taken over all points $w \in \Lambda^\ell \cap B(x, r_\ell)$ for which $V^s(w) \cap V^u(y) \in N$. By Corollary 8.6.11, we have $\nu(R^s) = 0$.

Let $z_1, z_2 \in Q^\ell(x) \setminus (R^s \cup N)$. There are points $w_i \in \Lambda^\ell \cap B(x, r)$ such that $z_i \in V^s(w_i)$ for $i = 1, 2$. Note that the intersection $V^s(w_i) \cap V^u(y)$ is nonempty and consists of a single point y_i for $i = 1, 2$ (see Fig. 9.1). By Lemma 9.1.2,

$$\varphi(z_1) = \varphi(y_1) = \varphi(y_2) = \varphi(z_2).$$

This completes the proof of the lemma. □

Lemma 9.1.4. *We have $\nu(Q(x)) > 0$ and $Q(x) \subset \Lambda$ (mod 0).*

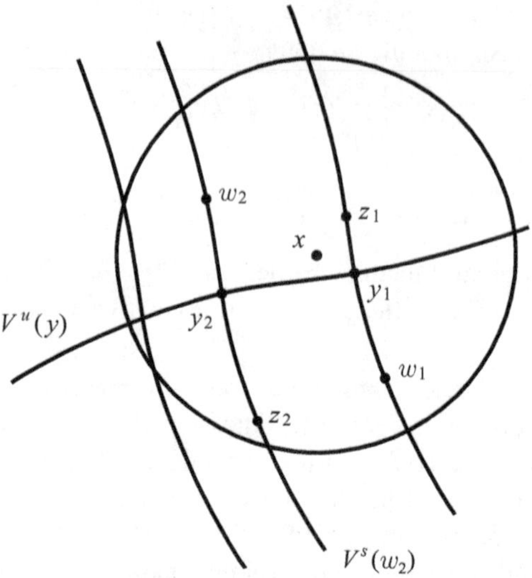

Figure 9.1. Stable manifolds forming the set $Q^\ell(x)$.

Proof of the lemma. Since $Q(x) \supset Q^\ell(x) \supset \Lambda^\ell \cap B(x,r)$, we have

$$\nu(Q(x)) \geq \nu(\Lambda^\ell \cap B(x,r)) > 0.$$

By Lemma 9.1.3, the map $f|Q(x)$ is ergodic. Therefore,

$$Q(x) = Q(x) \cap \Lambda \subset \Lambda \pmod 0$$

and the lemma follows. □

Since almost every point $x \in \Lambda$ is a Lebesgue point of Λ^ℓ for some ℓ, the invariant sets $Q(x)$ (defined by (9.1) for different values of ℓ) cover the set Λ (mod 0). By Lemma 9.1.4, there is at most countable number of such sets. We denote them by $Q_1, Q_2, \dots$. We have $\nu(Q_i) > 0$ for each $i \geq 1$, and the set $\Lambda \setminus \bigcup_{i\geq 1} Q_i$ has zero measure. By Lemma 9.1.3, the map $f|Q_i$ is ergodic for each $i \geq 1$. This yields

$$Q_i \cap Q_j = \varnothing \pmod 0, \tag{9.2}$$

whenever $i \neq j$. For each $n \geq 1$, let us set $\widetilde{\Lambda}_n = Q_n \setminus \bigcup_{i=1}^{n-1} Q_i$. We have $\widetilde{\Lambda}_i \cap \widetilde{\Lambda}_j = \varnothing$ whenever $i \neq j$. It follows from (9.2) that $\nu(Q_i) = \nu(\widetilde{\Lambda}_i) > 0$ for each $i \geq 1$. This completes the proof of the theorem. □

We describe an example of a diffeomorphism with nonzero Lyapunov exponents, which has more than one ergodic component. Consider the diffeomorphism $G_{\mathbb{T}^2}$ of the torus $\mathbb{T}^2$ constructed in Section 6.3. It preserves area and is ergodic. The punched torus $\mathbb{T}^2 \setminus \{0\}$ is C^∞-diffeomorphic to the manifold $\mathbb{T}^2 \setminus \overline{U}$, where U is a small open disk around 0 and $\overline{U}$ denotes its closure. Therefore, we obtain a

C^∞ diffeomorphism $F_{\mathbb{T}^2}$ of the manifold $\mathbb{T}^2 \setminus U$ with $F_{\mathbb{T}^2}|\partial U = \text{Id}$. We have that $F_{\mathbb{T}^2}$ preserves area, has nonzero Lyapunov exponents, and is ergodic.

Let $(\widetilde{M}, \widetilde{F}_{\mathbb{T}^2})$ be a copy of $(M, F_{\mathbb{T}^2})$. By gluing the manifolds M and $\widetilde{M}$ along ∂U, we obtain a smooth compact manifold $\mathcal{M}$ without boundary and a diffeomorphism $\mathcal{F}$ of $\mathcal{M}$, which preserves area and has nonzero Lyapunov exponents almost everywhere. However, the map $\mathcal{F}$ is not ergodic: it has two ergodic components of positive measure, M and $\widetilde{M}$. Similarly, one can obtain a diffeomorphism with nonzero Lyapunov exponents with n ergodic components of positive measure for an arbitrary n.

However, this construction cannot be used to obtain a diffeomorphism with nonzero Lyapunov exponents, which has countably (not finitely) many ergodic components of positive measure. Such an example was constructed in [58] and is presented in Section 11.3. It illustrates that Theorem 9.1.1 cannot be improved. It is therefore an interesting problem to find conditions that guarantee that the number of ergodic components is finite. Note that for each ergodic component Λ_i of positive measure, the values of the Lyapunov exponent are constant for almost every point $x \in \Lambda_i$. It follows that if the number of ergodic components is finite, the values of the Lyapunov exponent are uniformly away from zero for almost every point x on the manifold. There are situations where this requirement is indeed sufficient to guarantee the finiteness of the number of ergodic components (see [5]).

We now consider a smooth flow φ_t on a compact smooth manifold M and we assume that it preserves an absolutely continuous hyperbolic measure ν (see Section 5.7), that is, an invariant measure, which is absolutely continuous with respect to volume and for which all the Lyapunov exponents except for the one along the flow direction are nonzero for almost every point $x \in M$ (in particular, φ_t is a nonuniformly hyperbolic flow). We also assume that ν vanishes on the set of fixed points of φ_t.

The following result describes the partition of ν into its ergodic components and is an analogue of the corresponding result in the discrete-time case. The proof is a simple modification of the proof of Theorem 9.1.1 and is omitted.

Theorem 9.1.5 (see [170]). *Let φ_t be a smooth flow on a compact smooth manifold M. Assume that φ_t preserves an absolutely continuous hyperbolic measure ν. Then there exist invariant sets $\Lambda_0, \Lambda_1, \dots$ such that*

1. $\bigcup_{i\geq 0} \Lambda_i = M$, *and* $\Lambda_i \cap \Lambda_j = \varnothing$ *whenever* $i \neq j$;
2. $\nu(\Lambda_0) = 0$, *and* $\nu(\Lambda_i) > 0$ *for each* $i \geq 1$;
3. $\varphi_t|\Lambda_i$ *is ergodic for each* $i \geq 1$.

Using the flow described in Section 6.5 one can construct an example of a flow with nonzero Lyapunov exponents, which has an arbitrary finite number of ergodic components. This is a simple modification of the construction in the discrete-time case.

9.2 Local Ergodicity

Consider a $C^{1+\alpha}$ diffeomorphism of a compact manifold M preserving a smooth hyperbolic measure ν. In this case f is nonuniformly (completely) hyperbolic on the whole of M (see Section 3.3). For the purpose of this section we allow greater generality when f preserves a smooth measure and is nonuniformly (completely) hyperbolic on some set of positive (not necessarily full) volume, which contains an *open* (mod 0) subset $\mathcal{O}$ – in what follows the term *smooth hyperbolic measure* will be used in this more general sense. This in part is motivated by applications to the study of ergodic properties of geodesic flows (see Theorem 12.2.12). This also allows one to analyze some situations where hyperbolic and nonhyperbolic behavior coexist in an essential way – for example, the complement to $\mathcal{O}$ has positive volume and all Lyapunov exponents on this set are zero (see [90]).

Restricting the measure ν to the set $\mathcal{O}$ produces an absolutely continuous invariant measure for which $\mathcal{O}$ is a nonuniformly completely hyperbolic set for f. It is an open problem whether there exists an absolutely continuous f-invariant hyperbolic measure for which the set of all nonuniformly hyperbolic points contains no open (mod 0) subsets.

As we saw in the previous section, ergodic components of ν are of positive measure and one can wonder whether they are open (mod 0). This problem is often referred to as the *local ergodicity problem.*

The main obstacles for local ergodicity are the following:

1. The stable and unstable distributions are measurable but not necessarily continuous.
2. The global stable (or unstable) leaves may not form a foliation.
3. The unstable leaves may not *expand* under the action of f^n (we remind the reader that they were defined as being exponentially *contracting* under f^{-n}); the same is true for stable leaves with respect to the action of f^{-n}.

There are two different ways to obtain sufficient conditions for local ergodicity. Each of them is based on requirements that eliminate one or more of the above-mentioned obstacles.

We first describe the approach developed in [169]. Roughly speaking, it requires that the stable (or unstable) global leaves form a foliation of a measurable subset of full measure in M. First, we extend the notion of foliations of M with continuous leaves introduced in Section 8.3 to foliations of a measurable subset.

Given a subset $X \subset M$, we call a partition ξ of X a *(δ, q)-foliation of X with smooth leaves* or simply a *(δ, q)-foliation of X* if there exist continuous functions $\delta\colon X \to (0, \infty)$ and $q\colon X \to (0, \infty)$ and an integer $k > 0$ such that for each $x \in X$,

1. there exists a smooth immersed k-dimensional submanifold $W(x)$ containing x for which $\xi(x) = W(x) \cap X$ where $\xi(x)$ is the element of the partition ξ containing x; the manifold $W(x)$ is called the *(global) leaf* of the (δ, q)-foliation at x; the connected component of the intersection

$W(x) \cap B(x, \delta(x))$ that contains x is called the *local leaf* at x and is denoted by $V(x)$;

2. there exists a continuous map $\varphi_x : B(x, q(x)) \to C^1(D, M)$ ($D \subset \mathbb{R}^k$ is the open unit ball) such that for every $y \in X \cap B(x, q(x))$, the manifold $V(y)$ is the image of the map $\varphi_x(y) : D \to M$.

For every $x \in X$ and $y \in B(x, q(x))$, we set

$$U(y) = \varphi(y)(D) \tag{9.3}$$

and we call it the *local leaf* of the (δ, q)-foliation at y. Note that $U(y) = V(y)$ for $y \in X$, and hence $U(y)$ are extensions of local leaves through X.

For a general diffeomorphism preserving a smooth hyperbolic measure, one should not expect the stable (or unstable) leaves to form a (δ, q)-foliation for some functions $\delta(x)$ and $q(x)$. In order to explain why this can happen, consider a local stable manifold $V^s(x)$ passing through a point $x \in \mathcal{O}$. For a typical x, its local stable and unstable manifolds lie almost surely in one ergodic component (see Lemma 9.2.2 below). For sufficiently large ℓ, the set $V^s(x) \cap \Lambda^\ell$ (where Λ^ℓ is a regular set) has positive leaf volume but is, in general, a Cantor-like set. When the local manifold moves backward a given time n, one should expect a sufficiently small neighborhood of the set $V^s(x) \cap \Lambda^\ell$ inside $V^s(x)$ to expand. Other pieces of the local manifold (corresponding to bigger values of ℓ) will also expand but with *smaller* rates. Therefore the global leaf $W^s(x)$ (defined by (8.9)) may bend "uncontrollably" – the phenomenon that has not been observed yet but is thought to be "real" (and even "typical" in some sense). As a result the map $x \mapsto \varphi_x$ in the definition of (δ, q)-foliations may not be, indeed, continuous.

Furthermore, the global manifold $W^s(x)$ may be "bounded"; that is, it may not admit an embedding of an arbitrarily large ball in $\mathbb{R}^k$ (where $k = \dim W^s(x)$). This phenomenon, again, is yet to be observed.

The local continuity of the global unstable leaves often comes up in the following setting. Using some additional information on the system, one can build an invariant foliation whose leaves contain local unstable leaves. This alone may not yet guarantee that global unstable leaves form a foliation. However, one may often find that the local unstable leaves expand in a "controllable" and somewhat uniform way when they are moved forward. As we shall see below, this guarantees the desired properties of unstable leaves. Such a situation occurs, for example, for geodesic flows on compact Riemannian manifolds of nonpositive curvature (see Section 12.2). The above observation applies to global unstable manifolds as well.

1. We now state a formal criterion for local ergodicity.

Theorem 9.2.1. *Let f be a $C^{1+\alpha}$ diffeomorphism of a compact smooth Riemannian manifold, preserving a smooth measure ν, that is, f possesses a nonuniformly hyperbolic set $\mathcal{O}$, which is open* (mod 0). *Let also W be a (δ, q)-foliation of $\mathcal{O}$ with the following properties:*

1. *$W(x) \supset V^s(x)$ for almost every $x \in \mathcal{O}$.*

2. *There exists a number* $\delta_0 > 0$ *and a measurable function* $n(x)$ *on* $\mathcal{O}$ *such that for almost every* $x \in \mathcal{O}$ *and any* $n \geq n(x)$,

$$f^{-n}(V^s(x)) \supset B_U(f^{-n}(x), \delta_0),$$

where $B_U(y, r)$ *is the ball in* $U(y)$ *(with respect to the intrinsic metric; see (9.3)) centered at* y *of radius* r.

Then every ergodic component of f *of positive measure is open* (mod 0).

Proof. Let $Q \subset \mathcal{O}$ be an f-invariant set with $\nu(Q) > 0$ and $f|Q$ is ergodic. We will show that Q is open (mod 0). Clearly, Q is a level set and we consider the corresponding collection of regular sets Λ^ℓ (see Section 2.2.2). These sets are nested and cover Q, and hence we can choose $\ell \geq 1$ so large that $\nu(\Lambda^\ell) > 0$. By Lemma 9.1.3, there exists a density point x of the set Λ^ℓ such that $Q = Q(x)$ (mod 0) (see (9.1)). We need the following result that is a version of Corollary 8.6.12 for hyperbolic smooth measures. It shows that local stable and unstable manifolds lie almost surely in one ergodic component. We leave its proof to the reader.

Lemma 9.2.2. *For almost every* $x \in Q$,

$$m_{V^s}(x)(V^s(x) \setminus Q) = 0 \quad \textit{and} \quad m_{V^u}(x)(V^u(x) \setminus Q) = 0.$$

(Here $m_{V^s}(x)$ *and* $m_{V^u}(x)$ *are the leaf volumes on* $V^s(x)$ *and* $V^u(x)$ *respectively.)*

For a subset $Y \subset V^u(x)$, set

$$R(x, r, Y) = \bigcup_{y \in Y} B_U(y, r).$$

We call the set $R(x, r, Y)$ a *fence* through Y. We also consider the fences (see Fig. 9.2)

$$R(r) = R(x, r, V^u(x)), \quad \widetilde{R}(r) = R(x, r, V^u(x) \cap Q)$$

and

$$R^m(r) = R(x, r, V^u(x) \cap \Lambda^m), \quad m > 0.$$

Clearly,

$$R^m(r) \subset \widetilde{R}(r) \subset R(r).$$

Since W is a (δ, q)-foliation of $\mathcal{O}$, one can find $\delta_0 > 0$ such that $B_U(y, \delta_0) \subset U(y)$ for any $y \in V^u(x)$. By Theorem 7.5.1 and (8.4), there exists $r_m > 0$ such that $B_U(y, r_m) \subset V^s(y)$ for any $y \in V^u(x) \cap \Lambda^m$.

Fix $r \in (0, r_m]$. Given $y \in R^m(r/2)$, we denote by $n_i(y)$ the successive return times of the positive semitrajectory of y to the set $R^m(r/2)$. We also denote by $z_i \in V^u(x) \cap \Lambda^m$ a point for which $f^{n_i(y)}(y) \in B_U(z_i, r/2)$. In view of Poincaré's Recurrence Theorem, one can find a subset $N \subset M$ of zero ν-measure such that the sequence $n_i(y)$ is well defined provided that $y \in R^m(r/2) \setminus N$.

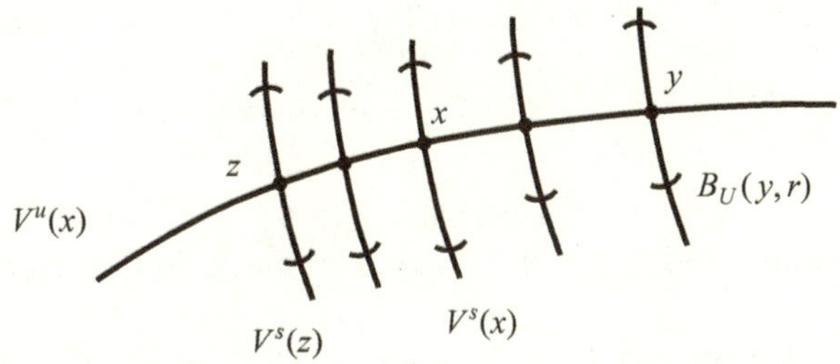

Figure 9.2. The fences of local stable manifolds.

Lemma 9.2.3. *For ν-almost every $y \in R^m(r/2)$, we have*

$$W^s(y) = \bigcup_{i=1}^{\infty} f^{-n_i(y)}(B_U(z_i, r)). \tag{9.4}$$

Proof of the lemma. Let $y \in R^m(r/2) \setminus N$ and $z \in W^s(y)$. By Theorem 8.2.3, we have $\rho(f^{n_i(y)}(y), f^{n_i(y)}(z)) \leq r/2$ for all sufficiently large i. Therefore,

$$f^{n_i(y)}(z) \in B_U(f^{n_i(y)}(y), r/2) \subset B_U(z_i, r)$$

and the lemma follows. □

Denote by $\xi^m(\delta_0)$ the partition of $R^m(\delta_0)$ into the sets $B_U(y, \delta_0)$.

Lemma 9.2.4. *The partition $\xi^m(\delta_0/2)$ is measurable and has the following properties:*

1. *The conditional measure on the element $B_U(y, \delta_0/2)$ of this partition is absolutely continuous with respect to the leaf volume $m_{V^s}(y)$.*
2. *The factor-measure on the factor-space $R^m(\delta_0/2)/\xi^m(\delta_0/2)$ is absolutely continuous with respect to the leaf volume $m_{V^u}(x)|V^u(x) \cap \Lambda^m$.*

Proof of the lemma. Choose $r = \min\{\delta_0/100, r_m\}$. By Lemma 9.2.3 and Corollary 8.6.10, for almost every $x \in \Lambda^m$ and almost every point $w \in V^u(x) \cap \Lambda^m$, one can find a point $y(w) \in B_U(w, r/2) \subset R^m(r/2)$ for which (9.4) holds. Moreover, the point $y(w)$ can be chosen in such a way that the map

$$w \in V^u(x) \cap \Lambda^m \mapsto y(w) \in R^m(r/2)$$

is measurable. For each $n > 0$, set

$$R_n = \bigcup_{w \in V^u(x) \cap \Lambda^m} \bigcup_{n_i(y(w)) \leq n} \left(f^{-n_i(y(w))}(B_U(z_i, r)) \cap R^m(3\delta_0/4)\right).$$

In view of the second assumption of the theorem, we have

$$R^m\left(\frac{3}{4}\delta_0\right) = \bigcup_{n \in \mathbb{N}} R_n.$$

Given $\varepsilon > 0$, there exist $p > 0$ and a set $Y \subset V^u(x) \cap \Lambda^m$ such that

$$m_{V^u}(x)((V^u(x) \cap \Lambda^m) \setminus Y) \le \varepsilon \quad \text{and} \quad R(\delta_0/2, Y) \subset \bigcup_{n \le p} R_n. \tag{9.5}$$

It follows from Theorem 8.6.8 that the partition $\xi^m(\delta_0/2)|R_n$ satisfies Statements 1 and 2 of the lemma for each $n > 0$. Since ε is arbitrary, the desired result follows. □

We proceed with the proof of the theorem. Denote by $\xi(\delta_0)$ and $\widetilde{\xi}(\delta_0)$ the partitions of the sets $R(\delta_0)$ and $\widetilde{R}(\delta_0)$ respectively into the sets $B_U(y, \delta_0)$. Since W is a (δ, q)-foliation of $\mathcal{O}$, the factor-space $R(\delta_0)/\xi(\delta_0)$ can be identified with $V^u(x)$ and the factor-space $\widetilde{R}(\delta_0)/\xi(\delta_0)$ with $V^u(x) \cap Q$.

Letting $m \to \infty$ in Lemma 9.2.4, we conclude that the partition $\widetilde{\xi}(\delta_0/2)$ also satisfies Statements 1 and 2 of the lemma. By Lemma 9.1.3, we have $Q \supset \widetilde{R}(\delta_0/2)$. This implies that $Q = \bigcup_{n\in\mathbb{Z}} f^n(\widetilde{R}(\delta_0/2))$.

Note that the set $R(\delta_0/2)$ is open and that $\widetilde{R}(\delta_0/2) \subset R(\delta_0/2)$. We will show that the set $A = R(\delta_0/2) \setminus \widetilde{R}(\delta_0/2)$ has zero measure. Assuming the contrary, we can find a level set $\Lambda_{\lambda\mu\varepsilon j}$ (see (2.21); the numbers λ, μ, and ε satisfy (2.15)) such that the intersection

$$A_{\lambda\mu\varepsilon j} = \Lambda_{\lambda\mu\varepsilon j} \cap (R(\delta_0/2) \setminus \widetilde{R}(\delta_0/2))$$

has positive measure. Consider the corresponding collection of regular sets $\{\Lambda^m_{\lambda\mu\varepsilon j} : m \ge 1\}$ (see Section 2.2.2). The sets

$$A^m = \Lambda^m_{\lambda\mu\varepsilon j} \cap (R(\delta_0/2) \setminus \widetilde{R}(\delta_0/2))$$

have positive measure for all sufficiently large m. Therefore for ν-almost every $z \in A^m$, we obtain that the leaf volume $m_{V^u}(z)(V^u(z) \cap A^m) > 0$. Consider the set $R(z, \delta_0/2, A^m)$. Clearly,

$$R(z, \delta_0/2, A^m) \subset R(\delta_0/2) \setminus \widetilde{R}(\delta_0/2).$$

Let $B^m = R(z, \delta_0/2, A^m) \cap V^u(x)$. We have $B^m \subset V^u(x) \setminus Q$, and thus $m_{V^u}(x)(B^m) = 0$. Moreover, repeating arguments in the proof of Lemma 9.2.4 (see (9.5)), we find that the holonomy map π^s, which takes $V^u(z) \cap A^m$ onto B^m, is absolutely continuous. Hence, $m_{V^u}(x)(B^m) > 0$. This is a contradiction. □

In the case of one-dimensional (δ, q)-foliations, the second condition of Theorem 9.2.1 holds automatically and hence, can be omitted.

Theorem 9.2.5. *Let W be a one-dimensional (δ, q)-foliation of $\mathcal{O}$, satisfying the following property: $W(x) \supset V^s(x)$ for almost every $x \in \mathcal{O}$. Then every ergodic component of f of positive measure is open* (mod 0).

Proof. Let Q be an ergodic component of positive measure. Clearly, Q is a level set and we consider the corresponding collection of regular sets Λ^ℓ (see Section 2.2.2). Fix $\ell > 1$ sufficiently large. For almost every point $x \in \Lambda^\ell$, the intersection $A(x) = V^s(x) \cap \Lambda^\ell$ has positive Lebesgue measure in $V^s(x)$. For every $y \in A(x)$, let $s(y)$ be the distance between x and y measured along $V^s(x)$. Then there exists a differentiable curve $\gamma : [0, s(y)] \to V^s(x)$ with $\gamma(0) = x$ and $\gamma(s(y)) = y$, satisfying

$$\rho_{W(f^{-n}(x))}(f^{-n}(x), f^{-n}(y)) = \int_0^{s(y)} \|d_{\gamma(t)} f^{-n} \gamma'(t)\|\, dt$$

$$\geq \int_0^{s(y)} \|d_{f^n(\gamma(t))} f^n \gamma'(t)\|^{-1} dt \geq \ell^{-1} e^{-\varepsilon n} \lambda^{-n} \geq \delta_0$$

for sufficiently large n, where δ_0 is a positive constant. Therefore the second condition of Theorem 9.2.1 holds and the desired result follows. □

Theorem 9.2.1 provides a way to establish ergodicity of the map $f|\mathcal{O}$. Recall that a map f is *topologically transitive* if given two nonempty open sets, there exists a positive integer n such that $f^n(A) \cap B \neq \varnothing$.

Theorem 9.2.6. *Assume that the conditions of Theorem 9.2.1 hold and that $f|\mathcal{O}$ is topologically transitive. Then $f|\mathcal{O}$ is ergodic.*

Proof. Let $C, D \subset \mathcal{O}$ (mod 0) be two distinct ergodic components of positive measure. We have that $\nu(f^n(C) \cap D) = 0$ for any integer n. Since the sets C and D are open (mod 0) and f is topologically transitive, we have $f^m(C) \cap D \neq \varnothing$ for some m. Furthermore, the set $f^m(C) \cap D$ is also open (mod 0). Since ν is equivalent to volume, we conclude that $\nu(f^m(C) \cap D) > 0$. This contradiction implies that $f|\mathcal{O}$ is ergodic. □

One can readily extend Theorem 9.2.1 to dynamical systems with continuous time.

Theorem 9.2.7. *Let φ_t be a smooth flow on a compact smooth Riemannian manifold preserving a smooth hyperbolic measure ν; that is, φ_t possesses a nonuniformly hyperbolic set $\mathcal{O}$, which is open* (mod 0). *Let also W be a (δ, q)-foliation of $\mathcal{O}$ with the following properties:*

1. *$W(x) \supset V^s(x)$ for almost every $x \in \mathcal{O}$.*
2. *There exist a number $\delta_0 > 0$ and a measurable function $t(x)$ on $\mathcal{O}$ such that for almost every $x \in \mathcal{O}$ and any $t \geq t(x)$,*

$$\varphi_{-t}(V^s(x)) \supset B_U(\varphi_{-t}(x), \delta_0).$$

Then every ergodic component of the flow $\varphi_t|\mathcal{O}$ of positive measure is open (mod 0). *Furthermore, if φ_1 is topologically transitive then $\varphi_t|\mathcal{O}$ is ergodic.*

2. We illustrate the above statements by presenting a result of Burns and Gerber [44] on ergodicity of flows on three-dimensional manifolds.

Theorem 9.2.8. *Let M be a compact three-dimensional C^3 Riemannian manifold and φ_t a C^2 volume preserving flow on M. Suppose that there is a continuous* *two-dimensional φ_t-invariant distribution $P = \{P_x : x \in M\}$ transverse to the flow and a continuous family of two-dimensional cones $C = \{C_x : x \in M\}$ such that $C_x \subset P_x$ for all $x \in M$. Also, suppose that there is a connected open subset U of M such that*

1. $\bigcup_{t\in\mathbb{R}} \varphi_t(U)$ *has full measure in M;*
2. $d\varphi_t(\overline{C_x}) \subset C_{\varphi_t(x)}$ *whenever $x \in U$, $\varphi_t(x) \in U$, and $t > 0$.*

Then the flow φ^t is ergodic.

Outline of proof. As shown in Chapter 4, Condition 2 implies that for almost every $x \in U$, there is one positive and one negative Lyapunov exponent for the flow φ_t. It then follows from Condition 1 that this is true for almost every $x \in M$. Let us note that the results in Chapter 4 are not formally stated for the continuous-time case, but we can easily pass from the discrete-time case to the flow case by considering the first return map $(\varphi_1)_U$ of the time-1 map φ_1 to U.

Let $W^s(x)$ and $W^u(x)$ be the global stable and unstable leaves at x; they are well defined for almost every $x \in M$. Moreover, since these leaves are one dimensional, repeating the arguments in the proof of Theorem 9.2.5 one can show that they contain arbitrarily long C^1 curves centered at x.

Let E^s_x and E^u_x be the tangent lines to the stable and unstable manifolds at x. We have that $E^u_x \subset C_x$ at almost every $x \in U$. Using the invariance of the complementary cones $\hat{C}_x = (P_x \setminus \overline{C_x}) \cup \{0\}$ under $d\varphi_{-t}$ whenever $x \in U$ and $\varphi_{-t}(x) \in U$, we can construct another continuous family of cones $C^s_x \subset P_x$ for $x \in U$, such that $\overline{C^s_x} \cap \overline{C_x} = \{0\}$ and for almost every $x \in U$, we have $E^s_x \subset C^s_x$. The continuity of the cones C_x and C^s_x then implies that the intersections between stable manifolds and weak unstable manifolds are transversal. Moreover, for every $x \in U$, there are neighborhoods $N_1(x)$ and $N_2(x)$ such that $N_1(x) \subset N_2(x) \subset U$, and any local stable manifold of a point $y \in N_1(x)$ that is sufficiently long (depending on x) and any local weak unstable manifold of a point $z \in N_1(x)$ that is sufficiently large (depending on x and corresponding to a sufficiently long interval in the flow direction and a sufficiently long unstable manifold) intersect in $N_2(x)$. The Hopf argument and the absolute continuity property can then be used to prove that for any continuous function $f\colon M \to \mathbb{R}$, the functions

$$f_+(p) = \lim_{T\to+\infty} \frac{1}{T} \int_0^T f(\varphi_t(p))\, dt$$

and

$$f_-(p) = \lim_{T\to+\infty} \frac{1}{T} \int_0^T f(\varphi_{-t}(p))\, dt$$

are almost every constant on $N_1(x)$. This proves that φ_t is ergodic. □

3. We now describe the approach by Katok and Burns [103] to study local ergodicity. A continuous function $Q\colon TM \to \mathbb{R}$ is called an *infinitesimal eventually strict Lyapunov function* for f over a set $U \subset M$ if

1. for each $x \in U$, the function $Q_x = Q|T_xM$ is homogeneous of degree 1 and takes on both positive and negative values;
2. there exist continuous distributions $D_x^s \subset C^s(x)$ and $D_x^u \subset C^u(x)$ such that $T_xM = D_x^s \oplus D_x^u$ for all $x \in U$, where

$$C^s(x) = Q^{-1}((-\infty, 0)) \cup \{0\} \quad \text{and} \quad C^u(x) = Q^{-1}((0, \infty)) \cup \{0\};$$

3. for every $x \in U$, every number $n \in \mathbb{N}$ for which $f^n(x) \in U$, and every $v \in T_xM$,

$$Q_{f^n(x)}(df^n v) \geq Q_x(v);$$

4. for ν-almost every $x \in U$, there exist $k = k(x)$, $\ell = \ell(x) \in \mathbb{N}$ such that $f^k(x) \in U$, $f^{-\ell}(x) \in U$, and for $v \in T_xM \setminus \{0\}$,

$$Q_{f^k(x)}(df^k v) > Q_x(v) \quad \text{and} \quad Q_{f^{-\ell}(x)}(df^{-\ell} v) < Q_x(v).$$

A function Q is called an *infinitesimal eventually uniform Lyapunov function* for f over a set $U \subset M$ if it satisfies Conditions 1–3 and the following condition: there exists $\varepsilon > 0$ such that for ν-almost every $x \in M$, one can find $k = k(x)$, $\ell = \ell(x) \in \mathbb{N}$ for which $f^k(x) \in U$, $f^{-\ell}(x) \in U$ and if $v \in T_xM \setminus \{0\}$, then

$$Q_{f^k(x)}(df^k v) > Q_x(v) + \varepsilon\|v\|$$

and

$$Q_{f^{-\ell}(x)}(df^{-\ell} v) < Q_x(v) - \varepsilon\|v\|.$$

The following result gives a criterion of local ergodicity in terms of infinitesimal Lyapunov functions.

Theorem 9.2.9. *Let f be a $C^{1+\alpha}$ diffeomorphism of a compact smooth Riemannian manifold M preserving a smooth measure ν. Let also Λ be a nonuniformly (completely) hyperbolic set for f of full measure. The following properties hold:*

1. *If f possesses an infinitesimal eventually strict Lyapunov function Q over an open set $U \subset M$, then almost every ergodic component of f on the set $\bigcup_{n\in\mathbb{Z}} f^n(U)$ is open* (mod 0).
2. *If f possesses an infinitesimal eventually uniform Lyapunov function Q over an open set $U \subset M$, then every connected component of the set $\bigcup_{n\in\mathbb{Z}} f^n(U)$ belongs to one ergodic component of f. In particular, if U is connected then $f|U$ is ergodic.*

Proof. Since the set Λ has full measure, given $\varepsilon > 0$, we can find a level set $\Lambda_{\lambda\mu\varepsilon j}$ and a regular set $\Lambda^\ell = \Lambda^\ell_{\lambda\mu\varepsilon j}$ such that $\nu(M \setminus \Lambda^\ell) < \varepsilon$. Furthermore, for almost every $x \in M$, we have the splitting of the tangent space $T_xM = E^s(x) \oplus E^u(x)$ where $E^s(x)$ and $E^u(x)$ are stable and unstable subspaces at x. Let $r_\ell > 0$ be a lower bound for the sizes of the local stable and unstable manifolds of points in Λ^ℓ (see Section 8.1 and (8.4)). Given $x \in \Lambda^\ell$ and $0 < r < r_\ell$, we consider the local stable manifold at x of size r and we use here the notation $V_r^s(x)$; recall that $V_r^s(x)$

is the image under the exponential map of the ball in $E^s(x)$ centered at the origin of radius r. Similarly, we consider the local unstable manifold at x of size r and we denote it by $V_r^u(x)$. Consider the set

$$A_\varepsilon = \{x \in U \cap \Lambda^\ell : Q_x(v) \geq \varepsilon\|v\| \text{ for all } v \in E^u(x) \\ \text{and } Q_x(w) \leq -\varepsilon\|w\| \text{ for all } w \in E^s(x)\}.$$

Since both Q_x and the norm are homogeneous of degree 1 and vary continuously with x, it is clear that A_ε is compact. Moreover, it follows from Theorem 4.1.3 that $\bigcup_{\varepsilon>0} A_\varepsilon$ has full measure in U.

Lemma 9.2.10. *Let F be a compact subset of U and $\varepsilon > 0$. Then there exist $r = r(F, \varepsilon)$ and $c = c(F, \varepsilon)$ such that*

1. *the manifold $V_r^u(x)$ is defined for every $x \in A_\varepsilon \cap F$;*
2. *if $x \in A_\varepsilon \cap F$, $y \in V_r^u(x)$, $n \geq 0$, $f^n(y) \in F$, and $v \in T_y V_r^u(x)$, then $\|df^n v\| \geq c\|v\|$.*

Proof of the lemma. Note that the manifold $V_r^u(x)$ is defined for all $x \in \Lambda^\ell \supset A_\varepsilon \cap F$, provided that r is sufficiently small. By the definition of A_ε,

$$\inf_{x\in A_\varepsilon\cap F}\ \inf_{v\in E^u(x)\setminus\{0\}} \frac{Q_x(v)}{\|v\|} \geq \varepsilon. \tag{9.6}$$

Since the set $A_\varepsilon \cap F$ is compact, Q_x varies continuously with x, and the local unstable manifolds $V^u(x)$ vary continuously with x in the C^1 topology, (9.6) implies that for $r > 0$ sufficiently small, $V_r^u(x)$ is defined for every $x \in A_\varepsilon \cap F$ and

$$\inf_{x\in A_\varepsilon\cap F}\ \inf_{v\in V_r^u(x)}\ \inf_{v\in T_y V_r^u(x)} \frac{Q_y(v)}{\|v\|} = c_1(F, \varepsilon) > 0. \tag{9.7}$$

On the other hand, the uniform continuity of $x \mapsto Q_x$ on the compact set F implies that there is a constant $c_2(F) > 0$ such that for all $x \in F$ and $v \in T_x M$,

$$Q_x(v) \leq c_2(F)\|v\|. \tag{9.8}$$

Finally, since Q is an infinitesimal Lyapunov function,

$$Q_{f^n(y)}(df^n v) \geq Q_y(v) \tag{9.9}$$

whenever $y \in U$, $n \geq 0$, $f^n(y) \in U$, and $v \in T_y M$. Thus if x, y, and v are as in the statement of the lemma, we see from (9.7) to (9.9) that

$$\|df^n v\| \geq c_2(F)^{-1} Q_{f^n(y)}(df^n v) \\ \geq c_2(F)^{-1} Q_y(v) \geq c_2(F)^{-1} c_1(F, \varepsilon)\|v\|.$$

This gives the desired inequality. □

Let now W be a C^1 submanifold of M and ρ_W the metric induced on W by the Riemannian metric on M. We denote by $r_x(W)$ the injectivity radius of W at a point $x \in W$, that is, the maximum radius of an open ball in $T_x W$ centered at 0 on which the exponential map is defined and is a diffeomorphism.

Lemma 9.2.11. *Let F be a compact subset of U. Then there exists a constant $r = r(\varepsilon) > 0$ such that if $y \in F$ and $f^{-n}(y) \in A_\varepsilon \cap F$ for some $n \in \mathbb{N}$, then $r_y(W^u(x)) \geq r$.*

Proof of the lemma. Choose a compact subset F' of U such that $F \subset \operatorname{int} F'$. Define constants r' and c' applying Lemma 9.2.10 to the set F'. Let $\eta = \inf_{x \in F} \operatorname{dist}(x, \partial F')$. If $y' \in V^u_{r'}(f^{-n}(y))$ and $f^n(y') \in F'$, then the derivative $d_{y'} f^n$ contracts vectors in $T_{y'} V^u(f^{-n}(y))$ by at most the factor c'. Hence, $f^n(V^u_r(f^{-n}(y)))$ contains a ball with radius at least $\min\{\eta, c'r'\}$. □

Lemma 9.2.12. *Let F be a compact subset of U. Then there exists a constant $R = R(F) > 0$ such that $r_x(W^u(x)) \geq R$ for almost every $x \in F \cap \Lambda$.*

Proof of the lemma. Choose a compact subset F' of U such that $F \subset \operatorname{int} F'$. It suffices to show that for almost every $x \in \Lambda \cap F$, there exists $r = r(x) > 0$ such that $r_y(W^u(y)) \geq r$ for all y in a dense subset of $W^u(x) \cap F'$. Then we can take $R = \inf_{x \in F} \operatorname{dist}(x, \partial F')$.

Let E be an ergodic component such that $\nu(E \cap F') > 0$. Choose $\varepsilon > 0$ so that $\nu(E \cap F' \cap A_\varepsilon) > 0$. Since f is ergodic on E, ν-almost every $y \in E$ has the property that there exists $n \in \mathbb{N}$ with $f^{-n}(y) \in F' \cap A_\varepsilon$. By Lemma 9.2.11, there exists $r > 0$ such that $r_y(W^u(y)) \geq r$ for ν-almost every $y \in E \cap F'$. We now see from Corollary 8.6.10 that for ν-almost every $x \in E \cap F'$, the set of points $y \in W^u(y) \cap F'$ with $r_y(W^u(y)) \geq r$ has full $\nu^u(y)$-measure. (Recall that $\nu^u(y)$ is the conditional measure on $V^u(y)$ generated by the measure ν; see Section 8.6.2.) Since ν is a smooth invariant measure, a subset of $W^u(y) \cap F$ with full $\nu^u(y)$-measure is dense in $W^u(x) \cap F'$.

By Theorem 9.1.1, the union of the ergodic components E such that $\nu(E \cap F) > 0$ has full measure in F and the lemma follows. □

We now obtain a certain locally uniform transversality of typical stable and unstable manifolds. Given $\theta > 0$, we define the *θ-interior* of a cone C to be the set $\operatorname{int}_\theta C = \{x \in S \cap C : \operatorname{dist}_S(x, C \cap \partial S) > \varepsilon\}$, where S is the unit sphere.

Lemma 9.2.13. *For almost every $z \in U$, there exist $\theta = \theta(z) > 0$ and a neighborhood N of z such that $T_y W^u(x) \subset \operatorname{int}_\theta C^u(y)$ for almost every $x \in N$ and every $y \in W^u(x) \cap N$.*

Proof of the lemma. Since Q is eventually strict, we may assume that there exists $l > 0$ such that $f^{-l}(x) \in U$ and $df^l C^u(f^{-l}(x)) \subset C^u(x)$. By continuity, we can choose a neighborhood $N \subset U$ of x and $\theta > 0$ such that $f^{-l}(N) \subset U$ and $df^l C^u(f^{-l}(y)) \subset \operatorname{int}_{2\theta} C^u(y)$ for all $y \in N$. For almost every $y \in N$, we have

$$T_{f^{-l}(y)} W^u(f^{-l}(y)) \subset C^u(f^{-l}(x))$$

that implies that

$$T_y W^u(y) \subset df^l C^u(f^{-l}(y)) \subset \operatorname{int}_{2\theta} C^u(y). \tag{9.10}$$

It follows from Corollary 8.6.10 that for almost every $x \in U \cap \Lambda$, Property (9.10) holds for ν-almost every $y \in W^u(x) \cap N$. Since a subset of $W^u(x) \cap N$ of full ν-measure is dense, we conclude that $T_y W^u(x) \subset \operatorname{int}_\theta C^u(y)$ for all $y \in W^u(x) \cap N$. □

When the infinitesimal Lyapunov function Q is eventually uniform, there is a global version of Lemma 9.2.13.

Lemma 9.2.14. *If the Lyapunov function Q is eventually uniform, then there exists $\theta > 0$ such that $T_y W^u(x) \subset \operatorname{int}_\theta C^u(y)$ for almost every $x \in U$ and every $y \in W^u(x) \cap U$.*

Proof of the lemma. For almost every $y \in U$, we have $y \in \Lambda$ and there is a sequence $n_k \to \infty$ such that $f^{-n_k}(y) \in U \cap \Lambda$ and for each k,

$$T_{f^{-n_k}(y)} W^u(f^{-n_k}(y)) \subset C^u(f^{-n_k}(y)).$$

For any such point y, we obtain

$$T_y W^u(y) \subset \bigcap_{k=1}^{\infty} df^{n_k} C^u(f^{-n_k}(y)). \tag{9.11}$$

Since Q is eventually uniform, there exists $\theta > 0$ such that for almost every y the right-hand side of (9.11) lies in $\operatorname{int}_{2\theta} C^u(y)$. Thus, $T_y W^u(y) \subset \operatorname{int}_{2\theta} C^u(y)$ for almost every $y \in U \cap \Lambda$. One can apply essentially the same absolute continuity argument as at the end of the proof of the former lemma to obtain the desired statement. □

Let k and k' be positive integers and $n = k + k'$. Let also C and C' be cones in $\mathbb{R}^n$ of ranks k and k' respectively (see Section 4.1.3). Assume that there is $\varepsilon > 0$ such that

$$\sphericalangle(v, v') \geq \varepsilon \text{ whenever } v \in C \setminus \{0\} \text{ and } v' \in C' \setminus \{0\}.$$

Let W and W' be C^1 manifolds in $\mathbb{R}^n$ of dimensions k and k' respectively. Suppose that

$$T_x W \subset C \text{ and } T_{x'} W' \subset C' \text{ whenever } x \in W \text{ and } x' \in W'.$$

Suppose also that there exist $x_0 \in W$ and $x_0' \in W'$ such that $r_{x_0}(W) \geq 1$ and $r_{x_0'}(W') \geq 1$, where σ denotes the injectivity radius of a submanifold of $\mathbb{R}^n$ defined by the Euclidean distance.

Lemma 9.2.15. *If $\operatorname{dist}(x_0, x_0') < \sin^2(\varepsilon/2)$ then $W \cap W' \neq \varnothing$.*

Proof of the lemma. Take $x \in W$ and $x' \in W$ with $x \neq x'$. Let $u(x, x')$ be the unit vector that points from x to x', $v(x, x')$ the orthogonal projection of $u(x, x')$ onto $T_x W$, and $\theta(x, x') \in [0, \pi/2]$ the angle between $u(x, x')$ and $T_x W$. Similarly, let $u'(x, x')$ be the unit vector that points from x' to x, $v'(x, x')$ the orthogonal projection of $u'(x, x')$ onto $T_{x'} W'$, and $\theta'(x, x') \in [0, \pi/2]$ the angle between $u'(x, x')$ and $T_{x'} W'$. The assumptions imply that

$$\min\{\theta(x, x'), \theta'(x, x')\} \leq \pi/2 - \varepsilon/2.$$

Let $\mathcal{W} = \{(x, x') \in W \times W' : x \neq x'\}$. We define the vector field V on $\mathcal{W}$ by $V(x, x') = (v(x, x'), v'(x, x'))$. We may assume that $(x_0, x_0') \in \mathcal{W}$. Let (x_t, x_t') be the integral curve of V starting at (x_0, x_0'). The curve is defined until $(x_t, x_t') \to \partial(W \times W')$ or $l(t) = \operatorname{dist}(x_t, x_t') \to 0$. Since x_t and x_t' have at most unit speed, the first possibility cannot occur until $t \geq 1$. On the other hand, $l(0) < \sin^2(\varepsilon/2)$ and

$$\frac{dl}{dt} = -\cos^2\theta(x_t, x_t') - \cos^2\theta'(x_t, x_t') \leq -\cos^2(\pi/2 - \varepsilon/2) = -\sin^2(\varepsilon/2).$$

Therefore there exists $\tau \in (0, 1)$ such that $l(t) \to 0$ as $t \nearrow \tau$. It is clear that $\lim_{t \nearrow \tau} x_t$ exists and lies in $W \cap W'$. □

We shall say that an open set $N \subset U$ has the *intersection property* if there exists $r > 0$ such that both $V_r^s(x)$ and $V_r^u(x)$ are defined for almost every $x \in N$ and $V_r^s(y) \cap V_r^u(z) \neq \varnothing$ for $\nu \times \nu$-almost every $(y, z) \in N \times N$. The following is a straightforward consequence of Lemmas 9.2.12–9.2.15.

Lemma 9.2.16.

1. *If f possesses an infinitesimal eventually strict Lyapunov function Q, then almost every $x \in U$ has a neighborhood $N(x)$ with the intersection property.*
2. *If f possesses an infinitesimal eventually uniform Lyapunov function Q, then every $x \in U$ has a neighborhood $N(x)$ with the intersection property.*

The proof of the next statement is an easy application of the Hopf argument (see the proof of Theorem 9.1.1) and is omitted.

Lemma 9.2.17. *Let N be an open subset of U that has the intersection property and let φ be an integrable function that is f-invariant. Then φ is almost everywhere constant on N.*

To conclude the proof of the theorem, observe that by Lemma 9.2.17, each neighborhood $N(x)$ constructed in Lemma 9.2.16 lies (module a set of zero measure) in a single ergodic component of f. This readily implies the desired statements. □

Finally, we outline the approach by Liverani and Wojtkowski [131] to study local ergodicity in the symplectic case.

Let M be a smooth compact symplectic manifold of dimension $2d$ with the symplectic form ω. Let also $f: M \to M$ be a symplectomorphism (i.e., a diffeomorphism of M, which preserves the symplectic structure).

Fix $x \in M$. Recall that a subspace $V \subset T_x M$ is *Lagrangian* if V is a maximal subspace on which ω vanishes. (It has dimension d.) Given two transverse

Lagrangian subspaces V_1 and V_2, define the *sector* $C = C(V_1, V_2)$ between them by

$$C = \{v \in T_x M : \omega(v_1, v_2) \geq 0 \text{ for } v = v_1 + v_2,\ v_i \in V_i,\ i = 1, 2\}.$$

Define further the quadratic form associated with an ordered pair of transverse Lagrangian subspaces V_1 and V_2 by

$$Q(v) = Q(V_1, V_2, v) = \omega(v_1, v_2) \text{ for } v = v_1 + v_2,\ v_i \in V_i,\ i = 1, 2.$$

Using this quadratic form one can write the cone $C(V_1, V_2)$ in the form

$$C(V_1, V_2) = \{v \in T_x M : Q(v) \geq 0\}.$$

Define the interior of the cone by

$$\operatorname{int} C(V_1, V_2) = \{v \in T_x M : Q(v) > 0\}.$$

Assume that two continuous subbundles of transverse Lagrangian subspaces are chosen in an open (not necessarily dense) subset $U \subset M$ and denote them by $\{V_1(x)\}_{x \in U}$ and $\{V_2(x)\}_{x \in U}$ respectively. For $x \in U$, let $C(x) = C(V_1(x), V_2(x))$ and $C'(x) = C(V_2(x), V_1(x))$.

If $x \in U$ and $f^n(x) \in U$, let us set

$$\sigma(d_x f^n) = \inf_{v \in \operatorname{int} C(x)} \sqrt{\frac{Q(V_1(x), V_2(x), df^n v)}{Q(V_1(x), V_2(x), v)}}.$$

Theorem 9.2.18. *Assume that the following conditions hold:*

1. *Monotonicity condition: if $x \in U$ and $f^k(x) \in U$ for $k \geq 0$ then $df^k C(x) \subset C(f^k(x))$.*
2. *Strict monotonicity condition: for almost every point $x \in U$, there exist $n > 0$ and $m < 0$ such that $f^n(x), f^m(x) \in U$ and*

$$\begin{aligned} df^n C(x) &\subset \operatorname{int} C(f^n(x)) \cup \{0\}, \\ df^m C'(x) &\subset \operatorname{int} C'(f^m(x)) \cup \{0\}. \end{aligned} \tag{9.12}$$

Then for any $n \geq 1$ and any point $x \in U$ such that $f^n(x) \in U$ and $\sigma(d_x f^n) > 1$, there is a neighborhood of x, which is contained in one ergodic component of f.

Proof. Take a point $x \in M$. Since strict monotonicity is symmetric both in the past and in the future, we discuss only one of them. We assume that there exists $N \in \mathbb{N}$ such that $f^{-N}(x) \in U$ and $d_x f^{-N}$ is strictly monotone. We consider a neighborhood U_x of x such that f^{-N} is a diffeomorphism of U_x onto a neighborhood of $f^{-N}(x)$, and we assume that both neighborhoods are contained in U. By the Darboux theorem, a symplectic manifold locally can be identified with the standard symplectic space. Hence reducing U_x further if necessary, we can identify it with the set $V(a) = U(a) \times U(a)$, where

$$U(a) = \{(x_1, \ldots, x_d) \in \mathbb{R}^d : |x_i| < a, i = 1, \ldots, d\}.$$

Let $\pi_i : \mathbb{R}^d \times \mathbb{R}^d \to \mathbb{R}^d$, $i = 1, 2$, be the canonical projections.

Given $\rho > 0$, we introduce the cones

$$C_\rho = \left\{(v, w) \in \mathbb{R}^d \times \mathbb{R}^d : \|w\| \leq \rho\|v\|\right\},$$
$$C'_\rho = \left\{(v, w) \in \mathbb{R}^d \times \mathbb{R}^d : \|v\| \leq \rho^{-1}\|w\|\right\}.$$

One can always choose $\rho < 1$ such that for any $z \in V(a)$,

$$C'(z) \subset C'_{\rho^{-1}} \quad \text{and} \quad df^N C(y) \subset C_\rho \text{ for any } y \in f^{-N}(V(a)). \tag{9.13}$$

We say that a point $z \in V(a)$ has k *spaced returns* in a given time interval if there are k moments of time $i_1 < i_2 < \cdots < i_k$ in this interval such that $f^{i_j}(z) \in V(a)$ for $j = 1, \cdots, k$ and $i_{j+1} - i_j \geq N$ for $j = 1, \cdots, k-1$.

Lemma 9.2.19. *If $z \in V(a)$ and $n \geq N$ are such that $z' = f^n(z) \in V(a)$, then*

$$d_z f^n C_{\rho^{-1}} \subset C_\rho \quad \textit{and} \quad d_{z'} f^{-n} C'_\rho \subset C'_{\rho^{-1}}. \tag{9.14}$$

Moreover, for $(v', w') = d_z f^n(v, w)$, we have

$$\|v'\| \geq b\rho^{-k}\|v\| \textit{ whenever } (v, w) \in C_\rho, \tag{9.15}$$

and

$$\|w\| \geq b\rho^{-k}\|w'\| \textit{ whenever } (v', w') \in C'_{\rho^{-1}},$$

where k is the maximal number of spaced returns of z to $V(a)$ from N to n, and $b = \sqrt{1-\rho^4}$.

Proof of the lemma. It follows from the first inclusion in (9.13) that $C_\rho \subset C_{\rho^{-1}} \subset C(x)$ for any $x \in V(a)$. Hence,

$$d_z f^{n-N} C_{\rho^{-1}} \subset C(f^{n-N}(z)),$$

and the first inclusion in (9.14) follows from the second inclusion in (9.13). We also have from (9.13) that for any $x \in V(a)$,

$$d_x f^{-N} C'_\rho \subset C'(f^{-N}(x)).$$

The second inclusion in (9.14) follows by applying $d_{z'} f^{-N}$, followed by $d_{f^{-N}(z')} f^{-n+N}$, and then using again (9.13).

To obtain the inequalities, we use the form Q associated to $C_{\rho^{-1}}$, equal to $\rho^{-1}\|v\|^2 - \rho\|w\|^2$. On every spaced return to $V(a)$, the value of this form on vectors in $C_{\rho^{-1}}$ increases by at least the factor ρ^{-2}. To see this consider two Lagrangian subspaces V_1 and V_2 and define the distance between them by

$$d(V_1, V_2) = \frac{1}{2} \sup_{v \in V_1 \setminus \{0\}} \left|\log Q((\pi_1|V_1)^{-1}(v)) - \log Q((\pi_1|V_2)^{-1}(v))\right|.$$

If L is a strictly monotone linear operator, that is, $LC(V_1, V_2) \subset \operatorname{int} C \cup \{0\}$, then one can show that

$$d(LV_1, LV_2) = \log \frac{\sigma(L)^2 + 1}{\sigma(L)^2 - 1}. \tag{9.16}$$

Applying (9.16) to the form Q above and $L = d_x f^{-N}$, we find that

$$\rho^{-1}\|v\|^2 \geq \rho^{-1}\|v\|^2 - \rho\|w\|^2 \geq (\rho^{-1} - \rho^3)\|v\|^2.$$

This immediately yields (9.15). The remaining inequality can be obtained in a similar manner. □

Let $V^u_\varepsilon(z)$ be the local unstable manifold at a point $z = (z_1, z_2) \in V_1 \times V_2$ of size ε (see Section 8.1)

Lemma 9.2.20. *For $\delta > 0$, almost every $z \in V_1 = V(a - \delta/b)$ has an unstable manifold of size δ.*

Proof of the lemma. Let $A_\varepsilon \subset V_1$ be the set of points that have unstable manifolds of size ε. Clearly, all points in V_1 belong to $\bigcup_{\varepsilon>0} A_\varepsilon$. Consider a set A_ε of positive measure and let k be the smallest positive integer such that $b\rho^{-k}\varepsilon \geq \delta$. Almost every $z \in A_\varepsilon$ has k spaced returns to A_ε in the past. Let z be such a point and let $-N \geq -i_1 > \cdots > -i_k = -n$ be the k times of spaced returns to A_ε. Denote by $\widetilde{V}^u_\varepsilon(f^{-n}(z))$ the connected component of

$$\bigcap_{j=0}^{k} f^{i_j - n}\left[f^{n-i_j} V^u_\varepsilon(f^{-n}(z)) \cap V(a)\right]$$

containing $f^{-n}(z)$. Let δ' be the largest positive number such that $V^u_{\delta'}(z)$ is well defined and is contained in $f^n(\widetilde{V}^u_\varepsilon(f^{-n}(z)))$.

We proceed by contradiction. Assume that $\delta' < \delta$. If follows from the maximality of δ' that the boundary of $V^u_{\delta'}(z)$ must contain a point z' in the boundary of $f^n(\widetilde{V}^u_\varepsilon(f^{-n}(z)))$. The preimage $f^{-n}(z')$ belongs to the boundary of $\widetilde{V}^u_\varepsilon(f^{-n}(z))$. Thus either $f^{-n}(z')$ belongs to the boundary of $V^u_\varepsilon(f^{-n}(z))$ or $f^{-i_j}(z')$ belongs to $\partial V(a)$ for some $0 \leq j \leq k$.

Let $V^u_{\delta'}(z)$ be the graph of $\psi\colon B(\pi_1 z, \delta') \to U(a)$ and let $\gamma_0\colon [0, 1) \to B(\pi_1 z, \delta')$ be the segment connecting $\pi_1 z$ and $\pi_1 z'$. We consider the preimages of the curve

$$\{(\gamma_0(t), \psi(\gamma_0(t))) : 0 \leq t < 1\} \subset V^u_{\delta'}(z)$$

and define $\gamma_j\colon [0, 1) \to U(a)$, $j = 0, \ldots, k$ by the formula

$$\gamma_j(t) = \pi_1(f^{-i_j}(\gamma_0(t), \psi(\gamma_0(t)))).$$

It follows from (9.15) that the length of γ_0 is not smaller than the length of γ_j times $b\rho^{-j}$. If $f^{-n}(z')$ belongs to the boundary of $V^u_\varepsilon(f^{-n}(z))$ then the length of γ_k is at least ε and we come to the contradiction $\delta' \geq b\rho^{-k}\varepsilon \geq \delta$. If $f^{-i_j}(z') \in \partial V(a)$ for some $0 \leq j \leq k$, then γ_j, which connects $\pi_1(f^{-i_j}(z)) \in V_1$ and $\pi_1(f^{-i_j}(z'))$, must have length at least δ/b and we again get the contradiction $\delta' \geq b\rho^{-j}\delta/b \geq \delta$. □

Denote by $\bar{B}(z, r)$ the closed ball of radius r centered at $z = (z_1, z_2)$ and define the rectangle

$$R(z, \delta) = \bar{B}(z_1, \delta/2) \times \bar{B}(z_2, \delta/2) \quad \text{whenever } z = (z_1, z_2).$$

We say that the unstable manifold $V^u_{\delta'}(z')$ of $z' = (z'_1, z'_2)$ of size δ' is *connecting* in the rectangle $R(z, \delta)$ if

$$\bar{B}(z_1, \delta/2) \subset B(z'_1, \delta') \quad \text{and} \quad \pi_2(V^u_{\delta'}(z') \cap R(z, \delta)) \subset B(z_2, \delta/2).$$

For a given rectangle $R(z, \delta)$, we define its *unstable core* as the set of points $z' = (z'_1, z'_2) \in R(z, \delta)$ for which

$$\rho\|z'_1 - z_1\| + \|z'_2 - z_2\| < (1 - \rho)\delta/2.$$

Lemma 9.2.21. *If a manifold $V^u_{\delta'}(z')$ of size $\delta' > \|\pi_1 z' - \pi_2 z\| + \delta/2$ intersects the unstable core of a rectangle $R(z, \delta)$, then it is connecting in the rectangle.*

Proof of the lemma. Let $V^u_{\delta'}(z')$ be the graph of $\psi\colon B(z'_1, \delta') \to U(a)$ and $(x'_1, \psi(x'_1))$ a point in the unstable core of the rectangle. By the condition on δ', we have $\bar{B}(z_1, \delta/2) \subset B(z'_1, \delta')$. Furthermore, if $x_1 \in \bar{B}(z_1, \delta/2)$ then

$$\begin{aligned}
\|\psi(x_1) - z_2\| &\le \|\psi(x_1) - \psi(x'_1)\| + \|\psi(x'_1) - z_2\| \\
&\le \sup\|d\psi\| \cdot \|x_1 - x'_1\| + \|\psi(x'_1) - z_2\| \\
&\le \rho\|x_1 - z_1\| + \rho\|x'_1 - z_1\| + \|\psi(x'_1) - z_2\| \\
&< \rho\delta/2 + (1 - \rho)\delta/2 = \delta/2.
\end{aligned}$$

Thus, $V^u_{\delta'}(z')$ is connecting in the rectangle. □

Given a rectangle R, set

$$V^s(R) = \bigcup\{R \cap V^s_{\delta'}(z') : V^u_{\delta'}(z') \text{ is connecting in } R\}$$

and define $V^u(R)$ analogously.

Lemma 9.2.22. *For any rectangle $R \subset V_1$, if the sets $V^s(R)$ and $V^u(R)$ have positive measure, then $V^s(R) \cup V^u(R)$ belong to one ergodic component of f.*

Proof of the lemma. The statement follows easily applying the Hopf argument (see Theorem 9.1.1). □

We proceed with the proof of the theorem. By Lemma 9.2.21, all points in the stable (respectively unstable) core of a rectangle $R \subset V_1$, which have a stable (respectively unstable) manifold of sufficiently large size, belong to $V^s(R)$ (respectively $V^u(R)$). By Lemma 9.2.20, almost every point in V_1 has both stable and unstable manifolds of size δ. Hence by Lemma 9.2.21, for any rectangle $R \subset V_1$ of size δ, the set $V^s(R)$ contains at least the stable core of R and the set $V^u(R)$ contains at least the unstable core of R. Then the sets $V^s(R)$ and $V^u(R)$ have positive measure, and we can apply Lemma 9.2.22. To complete the proof, we consider a family of rectangles of size δ contained in V_1 whose cores cover a small neighborhood $V_2 \subset V_1$. By Lemma 9.2.22, since the cores form an open cover of the connected set V_2, we conclude that V_2 belongs to one ergodic component. □

It follows from this theorem that if U is connected and every point in it is strictly monotone (i.e., (9.12) holds) then $\bigcup_{k\in\mathbb{Z}} f^k(U)$ belongs to one ergodic component of f. This is a symplectic version of Theorem 9.2.9.

9.3 The s- and u-Measures

Let f be a $C^{1+\alpha}$ diffeomorphism of a compact smooth Riemannian manifold M preserving a Borel probability measure ν. Given a Lyapunov–Perron regular point x, consider the Lyapunov exponents $\chi_1(x) < \cdots < \chi_{p(x)}(x)$ at x. Assume that there exists an f-invariant Borel function $k = k(x)$ such that $1 \le k(x) < p(x)$ and the set Y given by (8.14) has positive measure. This set is nonuniformly partially hyperbolic in the broad sense for f with $\lambda(x) = e^{\chi_{k(x)}(x)}$, $\mu(x) = e^{\chi_{k(x)+1}(x)}$, and an arbitrary f-invariant Borel function $\varepsilon(x)$ satisfying (2.9) (with an appropriate choice of the functions $C(x)$ and $K(x)$; see Theorem 3.3.1). For $x \in Y$, a local stable manifold at x is the kth local manifold $V_k^s(x)$ at x (see Section 8.4).

Consider a nonempty level set $\Lambda_{\lambda\mu\varepsilon j}$ and the collection of regular sets $\{\Lambda^\ell_{\lambda\mu\varepsilon j} : \ell \ge 1\}$ (see Sections 2.2.2 and 5.2) where the numbers λ, μ, and ε satisfy (2.15). We stress that $\dim E_k^s(x) = j$ is constant for $x \in \Lambda_{\lambda\mu\varepsilon j}$. Choose ℓ so large that the corresponding regular set is not empty. For a point $x \in \Lambda^\ell_{\lambda\mu\varepsilon j}$ and a number $0 < r < r_\ell$ (see (8.4)), let the set $Q^\ell(x)$ be given by (8.16) and let $\xi(x)$ be the partition of $Q^\ell(x)$ by the kth local manifolds $V_k^s(w)$, $w \in \Lambda^\ell \cap B(x, r)$.

Definition 9.3.1. The measure ν is called an *(s, k)-measure* for f if for any level set $\Lambda_{\lambda\mu\varepsilon j}$ of positive measure, any regular set $\Lambda^\ell_{\lambda\mu\varepsilon j}$ of positive measure and ν-almost every $x \in \Lambda^\ell_{\lambda\mu\varepsilon j}$, $w \in \Lambda^\ell_{\lambda\mu\varepsilon j} \cap B(x, r)$, the conditional measures $\nu_k^s(w)$ generated by ν on kth local manifolds $V_k^s(w)$ are absolutely continuous with respect to the leaf volume $m_{V_k^s}(w)$ (see Section 8.6.2).

Similarly, one can define the notion of a *(u, k)-measure* for f as an (s, k)-measure for f^{-1}.

Definition 9.3.2. A measure ν is called an *s-measure* for f if it is an (s, k)-measure for f where $k(x)$ is the biggest number of negative Lyapunov exponents at x. The definition of a *u-measure* is similar.

In what follows we assume that the numbers λ, μ, ε, and j are fixed so that $\nu(\Lambda_{\lambda\mu\varepsilon j}) > 0$ and we set $\Lambda = \Lambda_{\lambda\mu\varepsilon j}$. We also choose a number ℓ so large that the regular set $\Lambda^\ell = \Lambda^\ell_{\lambda\mu\varepsilon j}$ has positive measure.

We describe some properties of s- and u-measures. The proofs are easy corollaries of Theorems 8.5.1 and 8.6.8.

Theorem 9.3.3. *The following statements hold:*

1. *If ν is an (s, k)-measure for f then ν is an (s, k')-measure for f whenever $1 \le k' = k'(x) \le k = k(x) < p$; a similar statement holds for (u, k)-measures.*
2. *An s-measure for f is an (s, k)-measure for f where $k = k(x) \le s(x)$ is an f-invariant Borel function.*
3. *An absolutely continuous invariant measure for f is both an s- and a u-measure for f.*

4. *If a hyperbolic measure ν is both an s- and a u-measure for f then ν is an absolutely continuous invariant hyperbolic measure.*

If ν is an (s, k)-measure for f then for any $w \in \Lambda^\ell$,

$$d\nu_k^s(w)(z) = \kappa_k^s(w, z) dm_{V_k^s}(w)(z),$$

where $\kappa_k^s(w, z)$, $z \in V_k^s(w)$, is the density function. We shall assume that it is normalized such that $\kappa_k^s(w, w) = 1$.

Theorem 9.3.4. *The following statements hold:*

1. *The infinite product converges*

$$\widetilde{\kappa}_k^s(w, z) = \prod_{i=0}^{\infty} \frac{\mathrm{Jac}(df | E_k^s(f^i(w)))}{\mathrm{Jac}(df | T_z V_k^s(f^i(z)))} \tag{9.17}$$

(where Jac *is the Jacobian of the corresponding map); the function $\widetilde{\kappa}_k(w, z)$ is Hölder continuous in $w \in \Lambda^\ell$ and is smooth in $z \in V_k^s(w)$.*

2. *If $\kappa_k^s(w, z) > 0$ for every $w \in \Lambda^\ell$ and $z \in V_k^s(w)$ then $\kappa_k^s(w, z) = \widetilde{\kappa}_k^s(w, z)$ for Lebesgue almost every z and there is $C = C(\ell) > 0$ such that*

$$C^{-1} dm_{V_k^s}(w)(z) \le d\nu_k^s(w)(z) \le C\, dm_{V_k^s}(w)(z).$$

Proof. The proof of the first statement is similar to the proof of Theorem 8.6.13 and is omitted. To prove the second statement observe that the density function satisfies the homological equation

$$\kappa_k^s(f(w), f(z)) = \frac{\mathrm{Jac}(df | T_z V_k^s(z))}{\mathrm{Jac}(df | E_k^s(w))} \kappa_k^s(w, z). \tag{9.18}$$

Indeed, since f is a diffeomorphism for $w \in \Lambda$, the conditional measure on $V_k^s(f(w))$ is the image under f of the conditional measure on $V_k^s(w)$. This implies that the function $\mathrm{Jac}(df | T_z V_k^s(z)) \kappa_k^s(w, z)$ coincides with the density function $\kappa_k^s(f(w), f(z))$ up to the normalization. Since $\kappa_k^s(f(w), f(w)) = 1$, the relation (9.18) follows.

Iterating the identities in (9.18) yields

$$\kappa_k^s(w, z) = \kappa_k^s(f^n(w), f^n(z)) \prod_{i=0}^{n-1} \frac{\mathrm{Jac}(df | E_k^s(f^i(w)))}{\mathrm{Jac}(df | T_z V_k^s(f^i(z)))}.$$

Since $\kappa_k^s(w, w) = 1$, all we have to do to complete the proof of the second statement is to show that for ν-almost every $w \in \Lambda$ and $m_{V_k^s}(w)$-almost every $z \in V_k^s(w)$, we have that

$$\kappa_k^s(f^n(w), f^n(z)) \to 1 \tag{9.19}$$

as $n \to \infty$. Fix $\varepsilon > 0$. By Luzin's Theorem, for ν-almost every $w \in \Lambda$, there exists a compact set $X_\varepsilon(w) \subset V_k^s(w)$ such that $m_{V_k^s}(w)(X_\varepsilon(w)) \ge 1 - \varepsilon$ and the function $\kappa_k^s(w, z)$ is continuous in $z \in X_\varepsilon(w)$. Since the measure ν is invariant, applying Poincaré's Recurrence Theorem and using Corollary 8.6.9, we find that there exists a measurable set $Y_\varepsilon \subset \Lambda$ such that

1. $m(Y_\varepsilon) \geq 1 - \varepsilon$;
2. the set $Z_\varepsilon = \bigcup_{w \in Y_\varepsilon} X_\varepsilon(w)$ is measurable and $m(Z_\varepsilon) \geq 1 - 2\varepsilon$;
3. for almost every $w \in Y_\varepsilon$, the frequency of returns of the trajectory of w to Y_ε exceeds $1 - 2\varepsilon$;
4. for almost every $w \in Y_\varepsilon$ and $z \in X_\varepsilon(w)$, the frequency of returns of the trajectory of z to Z_ε exceeds $1 - 2\varepsilon$.

Therefore for almost every $w \in Y_\varepsilon$ and $z \in X_\varepsilon(w)$, there exists a sequence of integers $n_i \to \infty$ such that $f^{n_i}(w) \in Y_\varepsilon$ and $f^{n_i}(z) \in Z_\varepsilon$. It follows that $f^{n_i}(z) \in X_\varepsilon(f^{n_i}(w))$. This implies (9.19).

Finally, observe that for $w \in \Lambda^\ell$, the function $\widetilde{\kappa}_k^s(w, \cdot)$ is bounded from above and bounded away from zero and the second statement follows. □

It is not a priori clear whether the density function of an (s, k)-measure vanishes on a set of positive leaf volume. However, this is not the case due to a deep result of Ledrappier and Strelcyn (see Theorem 13.2.1 below).

Theorem 9.3.5. *For any (s, k)-measure ν, the density function $\kappa_k^s(w, z)$ is strictly positive on a set of full leaf volume and hence is given by* (9.17).

When ν is a smooth hyperbolic measure, this result follows from Theorems 8.6.8, 9.3.3, and 9.3.4 (see also Statement 1 of Theorem 9.3.6 below). It should be noticed that this result (and the formula for the density (9.17)) is not necessary to establish ergodicity, K and Bernoulli properties of absolutely continuous hyperbolic invariant measures (see the rest of this chapter), nor is it used in computing the entropy of the measure (see Chapter 10). However, it is crucial in characterizing measures satisfying the entropy formula and in studying some other properties of ν (see Chapters 13 and 14).

We now consider the case when f is a $C^{1+\alpha}$ diffeomorphism of a compact smooth manifold M preserving an absolutely continuous hyperbolic measure ν. Consider the families $\mathfrak{L}^s$ and $\mathfrak{L}^u$ of local stable and unstable manifolds. For $w \in B(x, r) \cap \Lambda^\ell$, let $m_{V^s}(w)$ be the leaf volume on $V^s(x)$. Let also ξ^s be the partition of $B(x, r)$ by local stable manifolds and $\nu^s(w)$ the conditional measure on $V^s(w)$ generated by ν and the partitions ξ^s. Finally, let $\hat{\nu}^s$ be the factor-measure.

Theorem 9.3.6. *The following statements hold:*

1. *For ν-almost every $w \in \Lambda^\ell \cap B(x, r)$, the measure $\nu^s(w)$ is absolutely continuous with respect to $m_{V^s}(w)$; that is, $d\nu^s(w)(y) = \kappa^s(w, y)\, dm_{V^s}(w)(y)$.*
2. *$\kappa^s(w, y) > 0$ for every $w \in \Lambda^\ell$ and almost every $y \in V^s(w)$ (with respect to the leaf volume) and is given by* (9.17).
3. *For ν-almost every $w \in \Lambda^\ell \cap B(x, r)$, the factor-measure $\hat{\nu}^s$ is absolutely continuous with respect to the leaf volume $m_T(w)$ where $T(w)$ is a local transversal to the family $\mathfrak{L}^s(x)$.*
4. *Similar statements hold for the family of local unstable manifolds.*

In other words any absolutely continuous invariant measure ν is both an s-measure and a u-measure, and hence in view of Theorem 9.3.6 it is also an (s, k)-measure and (u, k)-measure for appropriate functions $k = k(x)$.

9.4 The Leaf-Subordinated Partition and the K-Property

Let $f\colon M \to M$ be a $C^{1+\alpha}$ diffeomorphism of a smooth compact Riemannian manifold M preserving a Borel probability measure ν and $\mathcal{R} \subset M$ the set of Lyapunov–Perron regular points. For $x \in \mathcal{R}$, let $\chi_1(x) < \cdots < \chi_{p(x)}(x)$ be the distinct values of the Lyapunov exponent. Assume that for some function $k = k(x)$, $1 \le k(x) < p(x)$ the set $Y \subset \mathcal{R}$ given by (8.14) has positive measure. Recall that Y is nonuniformly partially hyperbolic in the broad sense for f with $\lambda(x) = e^{\chi_{k(x)}(x)}$, $\mu(x) = e^{\chi_{k(x)+1}(x)}$, and an arbitrary f-invariant Borel function $\varepsilon(x)$ satisfying (2.9). In particular, we have an invariant splitting of the tangent space $T_x M = E_k^s(x) \oplus F'(x)$ and for each $x \in Y$, a kth local manifold $V_k^s(x)$ and the kth global manifold $W_k^s(x)$ (see Section 8.4).

We assume further that there is a level set $\Lambda = \Lambda_{\lambda\mu\varepsilon j}$ of positive measure where the numbers λ, μ, and ε satisfy (2.15) and j is the dimension of local manifolds (therefore constant on Λ). Set

$$\hat{\Lambda} = \bigcup_{x \in \Lambda} W_k^s(x). \tag{9.20}$$

Consider the partition W_k^s of the set $\hat{\Lambda}$ by global manifolds $W_k^s(x)$. In general, this partition may not be measurable. The goal of this section is to construct a subpartition of $\hat{\Lambda}$, which is measurable and satisfies some "nice" properties. In the particular case when ν is an absolutely continuous hyperbolic measure for f (or more generally, a u-measure for f) and $k(x) = s(x)$ is the biggest number of negative Lyapunov exponents, we will obtain a description of the *Pinsker partition* (also called the *π-partition*). This is the maximal partition $\pi(f|\Lambda)$ with zero entropy. Observe that if the Pinsker partition is a trivial partition $\mathcal{N}$ (i.e., the only nontrivial element of the partition is the whole space), then any factor-automorphism over any measurable nontrivial partition has positive entropy (in other words, $f|\Lambda$ is a K-automorphism; see Theorem 9.4.3). An equivalent definition is that there is a measurable partition ξ of Λ such that $f\xi \ge \xi$, $\bigvee_{i=0}^{\infty} f^i\xi = \varepsilon$, and $\bigwedge_{i>-\infty}^{0} f^i\xi = \mathcal{N}$.

We denote the measurable hull of a partition ξ by $\mathcal{H}(\xi)$ and we use the notation ε for the partition by points.

Theorem 9.4.1. *There exists a measurable partition η of $\hat{\Lambda}$ with the following properties:*

1. *For ν-almost every $x \in \Lambda$, the element $C_\eta(x)$ is an open subset of $W_k^s(x)$.*
2. $f\eta \ge \eta$.
3. $\eta^+ = \bigvee_{i=0}^{\infty} f^i\eta = \varepsilon$.
4. $\bigwedge_{i>\infty}^{0} f^i\eta = \mathcal{H}(W_k^s|\hat{\Lambda}) \ge \pi(f|\Lambda)$.

Proof. Given $\ell > 1$, choose a regular set Λ^ℓ of positive measure. For a sufficiently small $0 < r \le r_\ell$ (see (8.4)) and $x \in \Lambda^\ell$, consider the set $Q_k^\ell(x)$ given by (8.16). For each $w \in \Lambda^\ell \cap B(x,r)$, the kth local manifold $V_k^s(w)$ can be viewed as the graph of a function $\psi_k^s(w)\colon B_k^s(0,q) \to F'(w)$, where $B_k^s(0,q) \subset E_k^s(w)$ is the ball centered at zero of radius q. We can choose q so small that the first derivatives of $\psi_k^s(w)$ are bounded by $1/3$ uniformly in $w \in \Lambda^\ell \cap B(x,r_\ell)$. Set

$$Q(x) = \bigcup_{n\ge 0} f^n(Q_k^\ell(x)). \tag{9.21}$$

It suffices to construct the partition η on the set $Q(x)$.

Consider the partition $\tilde{\xi}$ of $Q_k^\ell(x)$ by local manifolds $V^s(w)$, $w \in \Lambda^\ell \cap B(x,r)$. Adding the element $Q(x) \setminus Q_k^\ell(x)$, we obtain a partition of $Q(x)$, which we denote by ξ. Set $\eta = \xi^- = \bigvee_{i\le 0} f^i\xi$. We will show that the partition η has all the desired properties.

1. We need the following result from measure theory:

Lemma 9.4.2. *Let $r_0 > 0$ and let μ be a finite nonnegative Borel measure on $[0,r_0]$. For $0 < a < 1$, the Lebesgue measure of the set*

$$L_a = \left\{ r : 0 \le r \le r_0,\ \sum_{k=0}^{\infty} \mu([r-a^k, r+a^k]) < \infty \right\}$$

is equal to r_0.

Proof of the lemma. Let us set

$$N_{a,k} = \left\{ r : 0 \le r \le r_0,\ \nu([r-a^k, r+a^k]) > \frac{\mu([0,r_0])}{k^2} \right\}.$$

We call an interval of length $2a^k$ centered at a point in $N_{a,k}$ a "bad interval." It is easy to see that $N_{a,k}$ can be covered by bad intervals $J_{i,k}$, $1 \le i \le S(k)$, so that any point meets at most two bad intervals. We have

$$\frac{S(k)\mu([0,r_0])}{k^2} \le \sum_{i=1}^{S(k)} \mu(J_{i,k}) \le 2\mu([0,r_0])$$

and $|N_{a,k}| \le 2S(k)a^k$, where $|W|$ denotes the Lebesgue measure of the set W. We have $|N_{a,k}| \le 4a^k k^2$ and therefore by the Borel–Cantelli Lemma, Lebesgue almost any point in L_a belongs only to a finite number of $N_{a,k}$ and thus the series

$$\sum_{k=0}^{\infty} \nu([r-a^k, r+a^k])$$

converges. □

It follows from the definition of the partition η that for $w \in Q(x)$ and some $n \ge 0$,

$$C_\eta(w) \subset f^n(V_k^s(f^{-n}w)) \subset V_k^s(w).$$

We claim that there exists a function $\beta = \beta(w) \geq 0$ such that if $y \in V_k^s(w)$ and $\rho_{V_k^s}(y, w) \leq \beta(w)$ then $y \in C_\eta(w)$ (where $\rho_{V_k^s}$ is the distance in the local manifold $V_k^s(w)$ induced by the Riemannian metric).

For $w \in \Lambda$, let $\ell(w) = \inf\{\ell' : w \in \Lambda^{\ell'}\}$. Define

$$\beta(w) = \inf_{n\geq 0}\left\{r_\ell, \frac{1}{2\ell(w)}\rho(f^{-n}(w), \partial B(x,r))\lambda^n e^{\varepsilon n}, \frac{r}{\ell(w)}\right\}.$$

Assume that $y \in V_k^s(w)$ and $w \in \Lambda^\ell \cap B(x,r)$ are such that $\rho_{V_k^s}(y, w) \leq \beta(w)$. We shall show that for any $n \geq 0$,

$$C_\xi(f^n(y)) = C_\xi(f^n(w)). \tag{9.22}$$

For any $n \geq 0$, we have

$$\begin{aligned}\rho_{V_k^s}(f^{-n}(y), f^{-n}(w)) &\leq \ell(w)\lambda^n e^{\varepsilon n}\rho_{V_k^s}(y, w)\\ &\leq \frac{1}{2}\rho(f^{-n}(w), \partial B(x,r))\end{aligned} \tag{9.23}$$

and

$$\rho_{V_k^s}(f^{-n}(y), f^{-n}(w)) \leq \ell(w)\lambda^n e^{\varepsilon n}\rho_{V_k^s}(y, w) \leq r. \tag{9.24}$$

We have four cases to consider:

(i) If both $f^{-n}(y)$ and $f^{-n}(w)$ belong to $Q(x)$, then (9.24) implies (9.22).

(ii) If neither $f^{-n}(y)$ nor $f^{-n}(w)$ belong to $Q(x)$, (9.22) follows from the definition of the partition ξ.

(iii)–(iv) If $f^{-n}(y)$ belongs to $Q(x)$ but $f^{-n}(w)$ does not or vice versa, we have

$$\rho(f^{-n}(w), \partial B(x,r)) \leq \rho_{V_k^s}(f^{-n}(y), f^{-n}(w)),$$

which contradicts to (9.23).

Thus only (i) and (ii) occur implying (9.22).

It remains to show that one can choose the number r so small that $\beta > 0$ ν-almost everywhere. We will actually show that this is true for Lebesgue almost every choice of r. For $x \in M$, let μ be the finite nonnegative measure on $[0, r_\ell]$ defined by

$$\mu(A) = \nu(y \in M : \rho(x, y) \in A)$$

By the lemma applied to $a = \lambda^n e^{\varepsilon n}$, we find that $|K| = r_\ell$ where

$$K = \left\{r : 0 \leq r \leq r_\ell, \sum_{k=0}^{\infty}\nu(\{y \in M : |\rho(x,y) - r| < \lambda^n e^{\varepsilon n}\}) < \infty\right\}.$$

As f preserves the measure ν, we also have

$$K = \left\{r : 0 \leq r \leq r_\ell, \sum_{k=0}^{\infty}\nu(\{y \in M : |\rho(x, f^{-k}(y)) - r| < \lambda^n e^{\varepsilon n}\}) < \infty\right\}.$$

From the uniformity of the Riemannian metric on $B(x, r_\ell)$, there is a constant $D > 0$ such that $\rho(w, \partial B(x,r)) \le \tau$ implies $|\rho(x,w) - r| \le D\tau$ for r and τ such that $0 < \tau \le r \le r_\ell$. Thus we have for $r \in K$,

$$\sum_{k=0}^{\infty} \nu\Big(\Big\{y \in M : \rho(f^{-k}(y), \partial B(x,r)) \le \frac{\lambda^n e^{\varepsilon n}}{D}\Big\}\Big) < \infty.$$

By the Borel–Cantelli Lemma, for ν-almost every y there exist only a finite number of k with

$$\rho(f^k(y), \partial B(x,r)) \le \frac{\lambda^n\, e^{\varepsilon n}}{D}.$$

The set

$$T = \Big\{r : \nu\Big(\bigcup_n f^n(\partial B(x,r))\Big) > 0\Big\}$$

is at most countable and for $r \in K \setminus T$, we have that $\beta > 0$ ν-almost everywhere.

2. $f\eta = f\xi \vee \eta \ge \eta$.

3. It is sufficient to show that for almost any two points $z_1, z_2 \in Q_k^\ell(x)$, one can find an integer n such that $z_1 \notin C_{f^n\eta}(z_2)$. By Poincaré's Recurrence Theorem, we obtain that for almost every point $z \in Q_k^\ell(x)$ the diameter of the set $C_{f^n\eta}(z)$ goes to zero as $n \to \infty$. It follows that for almost any two points $z_1, z_2 \in Q_k^\ell(x)$ and all sufficiently large n, the diameters of elements $C_{f^n\eta}(z_1)$ and $C_{f^n\eta}(z_2)$ become smaller than the distance between z_1 and z_2, and the result follows.

4. Set $\widetilde{\eta} = \bigwedge_{i>-\infty}^{0} f^i\eta$. We have that $\widetilde{\eta}|Q_k^\ell(x) \le \xi$. It follows from Theorem 8.2.2 that $W_k^s(z) \subset C_{\widetilde{\eta}}(z)$ for almost every $z \in Q_k^\ell(x)$ and hence for almost every $z \in Q(x)$. This implies that $W_k^s|Q(x) \ge \widetilde{\eta}$ and hence (since the partition $\widetilde{\eta}$ is measurable), $\mathcal{H}(W_k^s|Q(x)) \ge \widetilde{\eta}$. On the other hand, we have that $\eta \ge W_k^s|Q(x)$, and hence $\widetilde{\eta} \ge \mathcal{H}(W_k^s|Q(x))$. The inclusion $\mathcal{H}(W_k^s|Q(x)) \ge \pi(f|Q(x))$ is a corollary of the following property of the Pinsker partition: if a measurable partition η satisfies $f\eta \ge \eta$ and $\eta^+ = \varepsilon$ then $\pi(f) \le \bigwedge_{i>-\infty}^{0} f^i\eta$.

This completes the proof of the theorem. □

We call η a *leaf-subordinated partition* associated with the kth global manifolds.

Recall that a diffeomorphism f is a K-automorphism if its π-partition is trivial. K-automorphisms have an abundance of stochastic properties. In particular, they have positive entropy and are mixing.

As an important manifestation of Theorem 9.4.1, we obtain the K-property of a $C^{1+\alpha}$ diffeomorphism f, which preserves an absolutely continuous hyperbolic measure ν. Assuming that f is ergodic with respect to ν we have that for every point x in a set Λ of full measure the values of the Lyapunov exponents are constant; that is, the distinct values are

$$\chi_1 < \cdots < \chi_s < 0 < \chi_{s+1} < \chi_p.$$

We also have two filtrations of global manifolds at almost every $x \in M$,

$$W_1^s(x) \subset \cdots \subset W_s^s(x) \quad \text{and} \quad W_{s+1}^u(x) \supset \cdots \supset W_p^u(x)$$

with $\dim W_k^s(x) = \text{constant}$ and $\dim W_k^u(x) = \text{constant}$.

Theorem 9.4.3 (see [168, 169]). *Let f be a $C^{1+\alpha}$ diffeomorphism preserving an absolutely continuous hyperbolic measure ν. Assume that f is ergodic with respect to ν. Then*

1. $\mathcal{H}(W_{k_1}^s|\hat{\Lambda}) = \mathcal{H}(W_{k_2}^s|\hat{\Lambda}) \geq \pi(f)$ *for any numbers k_1 and k_2 such that* $1 \leq k_1 < k_2 \leq s$, *and* $\mathcal{H}(W_{k_1}^u|\hat{\Lambda}) = \mathcal{H}(W_{k_2}^u|\hat{\Lambda}) \geq \pi(f)$ *for any numbers k_1 and k_2 such that* $s+1 \leq k_1 < k_2 \leq p$;
2. *the Pinsker partition of f is finite and consists of n_i elements* Λ_i^q, $q = 1, \dots, n_i$, *such that* $f(\Lambda_i^q) = \Lambda_i^{q+1}$, $q = 1, \dots, n_i - 1$ *and* $f(\Lambda_i^n) = \Lambda_i^1$;
3. $f^{n_i}|\Lambda_i^1$ *is a K-automorphism.*

Proof. The first statement follows from Theorem 9.4.1. To prove the second statement fix $x \in \Lambda^\ell$, a number r, $0 < r \leq r_\ell$, and set

$$Q_s^\ell(x) = \bigcup_{w \in \Lambda^\ell \cap B(x,r)} V^s(w) \quad \text{and} \quad Q_u^\ell(x) = \bigcup_{w \in \Lambda^\ell \cap B(x,r)} V^u(w),$$

where $B(x, r)$ is the ball at x of radius r and then

$$P^\ell(x) = Q_s^\ell(x) \cup Q_u^\ell(x).$$

Let also η_s and η_u be the leaf-subordinated partitions associated with global stable and unstable manifolds respectively. (η_u is the leaf-subordinated partition for f^{-1}.) By the construction of these partitions and Statement 4 of Theorem 9.4.1, we have that

$$\pi | Q_s^\ell(x) \leq \eta_s, \quad \pi | Q_u^\ell(x) \leq \eta_u,$$

and hence

$$\pi | P^\ell(x) \leq \eta_s \wedge \eta_u = P^\ell(x).$$

It follows that some element of the partition π contains the set $P^\ell(x)$ of positive measure for some $\ell > 0$ and $x \in \Lambda^\ell$. Since the Pinsker partition is invariant, all elements of this partition have the same positive measure and f cyclicly permutes them. The second statement follows. The last statement is the direct consequence of the second one. □

It follows from the above theorem that if ν is an absolutely continuous hyperbolic ergodic measure for f then $\mathcal{H}(W^u|\hat{\Lambda}) \geq \pi(f)$. We shall see later in Section 10.4 that the entropy of f with respect to the partition η_u is equal to the measure-theoretic entropy of f. By the properties of the Pinsker partition this implies that indeed, $\mathcal{H}(W^u|\hat{\Lambda}) = \pi(f)$. Similarly, $\mathcal{H}(W^s|\hat{\Lambda}) = \pi(f)$.

9.5 The Bernoulli Property

In the ergodic theory there is a hierarchy of ergodic properties of which ergodicity (or the description of ergodic components) is the weakest one. Among the stronger properties are mixing and K-property (including the description of the Pinsker partition). The strongest property that implies all others is the Bernoulli property or the description of Bernoulli components. In connection with smooth ergodic measures the latter is known as the Spectral Decomposition Theorem that we shall discuss below.

We remind the reader of the notion of a Bernoulli automorphism. Let (X, μ) be a Lebesgue space with a probability measure μ. Assume that μ has at most countably many atoms whose union $Y \subset X$ is such that $\mu|(X \setminus Y)$ is metrically isomorphic to the Lebesgue measure on the unit interval $[0, 1]$. One can associate to (X, μ) the two-sided Bernoulli shift $\sigma\colon X^{\mathbb{Z}} \to X^{\mathbb{Z}}$ defined by $(\sigma x)_n = x_{n+1}$, $n \in \mathbb{N}$, which preserves the convolution $\bigoplus_{\mathbb{Z}} \mu$. A *Bernoulli automorphism* (T, ν) is an invertible (mod 0) measure preserving transformation, which is metrically isomorphic to the Bernoulli shift associated to some Lebesgue space (X, μ).

Note that a Bernoulli automorphism is a K-automorphism.

Theorem 9.5.1 (see [169]). *Let f be a $C^{1+\alpha}$ diffeomorphism of a smooth compact Riemannian manifold M preserving an absolutely continuous hyperbolic measure ν. Assume that f is a K-automorphism. Then f is a Bernoulli automorphism.*

Proof. Since f is ergodic, there is a full measure level set Λ and the corresponding collection of regular sets $\{\Lambda^\ell : \ell \geq 1\}$ (see Section 2.2.2). We split the proof into several steps. First we present some characterization and properties of Bernoulli automorphisms, which we will use later. Then we introduce sets of a special kind, which we call rectangles. They have local product structure generated by local stable and unstable manifolds on a given regular set Λ^ℓ. We construct a partition of Λ^ℓ by rectangles and we use this partition to study the evolution of any partition α of the manifold M whose elements have piecewise smooth boundaries. We show that iteration of α under the system intersects each rectangle essentially leafwise, that is, along local stable manifolds. This property is crucial in constructing a Bernoulli partition.

Step I. We provide a characterization of Bernoulli automorphisms, which we will use to establish the Bernoulli property. Let (X, ν) be a Lebesgue space. In what follows we always assume that the measures we consider are normalized. In particular, if $A \subset X$ is a measurable subset then by the measure on A we mean the measure $(1/\nu(A))\nu$.

Let ξ be a finite measurable partition of X. We say that a property (associated with the partition) holds for ε-almost every element of ξ if the union of those elements, for which the property fails, has ν-measure at most ε.

Let $\xi_i = \{A_1^{(i)}, \dots, A_m^{(i)}\}$ and $\eta_i = \{B_1^{(i)}, \dots, B_m^{(i)}\}$, $1 \leq i \leq n$, be two sequences of partitions of Lebesgue spaces (X, ν) and (Y, μ) respectively. We write $\{\xi_i\}_1^n \sim$

$\{\eta_i\}_1^n$ if for any $1 \le k_i \le m$ and $1 \le i \le n$,

$$\nu\left(\bigcap_{i=1}^n A_{k_i}^{(i)}\right) = \mu\left(\bigcap_{i=1}^n B_{k_i}^{(i)}\right).$$

We say that the distance

$$d(\{\xi_i\}_1^n, \{\eta_i\}_1^n) \le \varepsilon$$

if there exist sequences of partitions $\bar{\xi}_i = \{\bar{A}_1^{(i)}, \dots, \bar{A}_m^{(i)}\}$ and $\bar{\eta}_i = \{\bar{B}_1^{(i)}, \dots, \bar{B}_m^{(i)}\}$ of a Lebesgue space (Z, λ) such that

$$\{\xi_i\}_1^n \sim \{\bar{\xi}_i\}_1^n \quad \text{and} \quad \{\eta_i\}_1^n \sim \{\bar{\eta}_i\}_1^n,$$

and

$$\frac{1}{n}\sum_{i=1}^n \sum_{j=1}^m \lambda(\bar{A}_j^{(i)} \Delta \bar{B}_j^{(i)}) \le \varepsilon.$$

A partition ξ of a Lebesgue space (X, ν) is called a *VWB-partition* (abbreviation for very weakly Bernoulli) if for any $\varepsilon > 0$ there exists $N_0 = N_0(\varepsilon)$ such that for any $N' > N \ge N_0$, $n \ge 0$, and ε-almost every element $A \in \bigvee_{k=N}^{N'} f^k\xi$, we have

$$d(\{f^{-i}\xi\}_1^n, \{f^{-i}\xi|A\}_1^n) \le \varepsilon.$$

(Let us stress that the partition $\xi|A$ is considered with respect to the normalized measure.)

Lemma 9.5.2 (see [161]). *Let f be an automorphism of a Lebesgue space (X, ν). Assume that there exists a sequence of VWB-partitions $\xi_1 \le \xi_2 \le \dots$ such that $\xi_n \to \varepsilon$. Then f is a Bernoulli automorphism.*

Given a sequence of partitions $\{\xi_i\}_1^n$ of a Lebesgue space (X, ν), we define the sequence of integer-valued functions $\ell_i(x)$ by the condition $x \in A_{\ell_i(x)}^{(i)}$. We call the sequence of functions $\ell_i(x)$ the *name* of the sequence of partitions $\{\xi_i\}_1^n$.

We say that a transformation $\theta\colon X \to Y$ of Lebesgue spaces (X, ν) and (Y, μ) is *ε-measure preserving* if there exists a set $E \subset X$ such that $\nu(E) \le \varepsilon$ and for every measurable set $A \subset X \setminus E$,

$$\left|\frac{\nu(\theta(A))}{\nu(A)} - 1\right| \le \varepsilon.$$

Define the function $e = e(n)$ on the set of positive integers by $e(0) = 0$ and $e(n) = 1$ for $n > 0$.

Lemma 9.5.3 (see [161]). *Let $\{\xi_i\}_1^n$ and $\{\eta_i\}_1^n$ be two sequences of partitions of Lebesgue spaces (X, ν) and (Y, μ) with name functions $\ell_i(x)$ and $m_i(x)$ respectively. Assume that there exists an ε-measure preserving transformation $\theta\colon X \to Y$, satisfying*

$$\frac{1}{n}\sum_{i=1}^n e(\ell_i(x) - m_i(\theta(x))) \le \varepsilon$$

for all $x \in X$ *with possible exception for a set* $E \subset X$ *with* $\nu(E) \leq \varepsilon$. *Then* $d(\{\xi_i\}_1^n, \{\eta_i\}_1^n) \leq 16\varepsilon$.

We need the following property of K-automorphisms:

Lemma 9.5.4 (see [161]). *Let* f *be a* K*-automorphism of a Lebesgue space* (X, ν), ξ *a finite partition of* X, *and* $B \subset X$ *a measurable subset. Given* $\delta > 0$, *there exists* $N_0 > 0$ *such that for any* $N' > N \geq N_0$ *and* δ*-almost every element* $A \in \bigvee_{k=N}^{N'} f^k\xi$,

$$\left| \frac{\nu(A \cap B)}{\nu(A)} - \nu(B) \right| \leq \delta. \tag{9.25}$$

Step II. Recall that f is nonuniformly hyperbolic on M. Consider a regular set Λ^ℓ of positive measure and let $\psi = \psi(\ell)$ be the function given by (8.7) and (8.8). We construct a partition of the set Λ^ℓ by subsets of a special type: they have local product structure. We call them rectangles.

More precisely, a measurable set $\Pi \subset M$ is called a δ*-rectangle* at a point $w \in \Lambda^\ell$ (or simply a *rectangle*) if

$$w \in \Pi \subset \Lambda^{\psi(\ell)} \cap B(w, \delta) \tag{9.26}$$

and for any $y, z \in \Pi$,

$$V^s(y) \cap V^u(z) \in \Pi. \tag{9.27}$$

(Note that this intersection is nonempty and consists of a single point.)

The following result establishes existence of rectangles at points in Λ^ℓ.

Lemma 9.5.5. *For any sufficiently small* $\delta = \delta(\ell) > 0$, *there is* $r = r(\delta) > 0$ *such that for any* $w \in \Lambda^\ell$, *one can construct a* δ*-rectangle* Π *at* w *satisfying*

$$\Lambda^\ell \cap B(w, r) \subset \Pi. \tag{9.28}$$

Proof of the lemma. For a sufficiently small $\delta > 0$, there is $r > 0$ such that for any $y_1, y_2 \in B(w, r) \cap \Lambda^\ell$,

$$V^s(y_1) \cap V^u(y_2) \in B(w, \delta).$$

Set

$$\Pi = \{y \in M : \text{there are } y_1, y_2 \in B(w, r) \cap \Lambda^\ell \text{ with } y = V^s(y_1) \cap V^u(y_2)\}.$$

We shall show that the set Π is a rectangle. Condition (9.26) follows from the definition of the set Π and (8.8). To prove Condition (9.27) fix points $y, z \in \Pi$. There are points $y_1, y_2, z_1, z_2 \in \Pi$ such that

$$y = V^s(y_1) \cap V^u(y_2) \quad \text{and} \quad z = V^s(z_1) \cap V^u(z_2).$$

If δ is chosen sufficiently small (depending only on ℓ), it follows that

$$V^s(y) \cap V^u(z) = V^s(y_1) \cap V^u(z_2) \in \Pi.$$

Since the requirement (9.28) holds trivially, the lemma follows. □

We wish to construct a partition of the set Λ^ℓ by rectangles.

Lemma 9.5.6. *For any two rectangles* Π *and* Π' *one can find five disjoint sets* $\Pi_1, \dots, \Pi_5$ *such that*

$$\bigcup_{i=1}^{5} \Pi_i = \Pi \cup \Pi'$$

and Π_i *is a rectangle provided* $\Pi_i \cap \Lambda^\ell \neq \varnothing$.

Proof of the lemma. We may assume that $\Pi \cap \Pi' \neq \varnothing$. Set

$$\tilde{\Pi} = \bigcup_{y \in \Pi \cap \Pi'} V^u(y)$$

and then

$$\Pi_1 = \Pi \cap \Pi', \quad \Pi_2 = \Pi \setminus \tilde{\Pi}, \quad \Pi_3 = \Pi' \setminus \tilde{\Pi},$$
$$\Pi_4 = (\tilde{\Pi} \cap \Pi) \setminus (\Pi' \cap \Pi), \quad \Pi_5 = (\tilde{\Pi} \cap \Pi') \setminus (\Pi' \cap \Pi).$$

Clearly, the sets Π_i are disjoint and their union is $\Pi \cap \Pi'$. We shall show that each Π_i is a rectangle. Fix $y, z \in \Pi_i$ and let $w = V^s(y) \cap V^u(z)$. Since $\Pi_i \subset \Pi$ (or $\Pi_i \subset \Pi'$), we have that $w \in \Pi$ (respectively, $w \in \Pi'$). We only consider the cases $i = 1, 2, 4$, as the other cases can be proved in a similar fashion:

(1) $i = 1$; since $y, z \in \Pi'$, we have $w \in \Pi'$, and hence $w \in \Pi \cap \Pi' = \Pi_1$;
(2) $i = 2$; since $z \notin \tilde{\Pi}$, we have $V^u(z) \not\subset \tilde{\Pi}$, and hence $w \in \Pi \setminus \tilde{\Pi} = \Pi_2$; and
(3) $i = 4$; since $z \in \tilde{\Pi}$, we have $w \in V^u(z) \subset \tilde{\Pi}$, and hence either $w \in \Pi_4$ or $w \in \Pi_1$. In the latter case since $y \in \tilde{\Pi}$, one can find a point $y_1 \in \Pi \cap \Pi' = \Pi_1$ such that $y \in V^u(y_1)$. Therefore, $y = V^s(w) \cap V^u(y_1) \in \Pi_1$ (since both $y_1, w \in \Pi_1$). However, this contradicts the fact that $y \notin \Pi_1$.

□

Lemma 9.5.7. *For any sufficiently small* $\delta > 0$, *one can find disjoint rectangles* $\Pi_1, \dots, \Pi_m$ *such that* $\bigcup_{i=1}^m (\Pi_i \cap \Lambda^\ell) = \Lambda^\ell$.

Proof of the lemma. Fix $\delta > 0$ and choose $r > 0$ in accordance with Lemma 9.5.5. Since the set Λ^ℓ is compact, one can find a finite cover of this set by open balls $B(w_i, r)$, $w_i \in \Lambda^\ell$, $i = 1, \dots, m$. Consider δ-rectangles at w_i constructed in Lemma 9.5.5. Their union contains the set Λ^ℓ. Applying Lemma 9.5.6 to each pair of this rectangles one can obtain the desired sets Π_i. □

Step III. Fix $\varepsilon > 0$ and choose a number $\ell > 1$ such that

$$\nu(\Lambda^\ell) \geq 1 - \varepsilon. \tag{9.29}$$

There is a finite measurable partition α of the manifold M whose elements have piecewise smooth boundaries and arbitrarily small diameter. Moreover, it is not difficult to construct a sequence of such partitions $\alpha_1 \leq \alpha_2 \leq \cdots$ such that $\alpha_n \to \varepsilon$. By Lemma 9.5.2, it suffices to show that each partition α_n is a VWB-partition.

Therefore, we shall show that any partition α of M whose elements have piecewise smooth boundaries is a VWB-partition.

We first study sets that intersect a given rectangle in a "nice" way. We say that a measurable set $A \subset M$ intersects a rectangle Π *leafwise* if

$$V^u(w) \cap \Pi \subset A \cap \Pi \text{ for any } w \in A \cap \Pi. \tag{9.30}$$

Lemma 9.5.8. *Given a rectangle Π and $\beta > 0$, one can find $N_1 > 0$ such that for any $N' \geq N \geq N_1$ and β-almost every element $A \in \bigvee_{k=N}^{N'} f^k\alpha$, there exists a subset $E \subset \Pi$, intersecting Π leafwise, for which*

$$\frac{\nu(E)}{\nu(A)} \geq 1 - \beta. \tag{9.31}$$

Proof of the lemma. Given an element D of the partition α and $k > 0$, denote by $B_D^k \subset f^k(D)$ the subset that intersects Π leafwise. Since $\Pi \subset \Lambda^{\psi(\ell)}$, it follows from (7.1), (7.2), (9.30), and the definition of regular sets (see Proposition 2.2.7) that the distance d_k between any point of the set

$$D \cap (f^{-k}(\Pi) \setminus f^{-k}(B_D^k))$$

and the boundary $\partial D'$ does not exceed

$$d_k \leq C_1 a^k, \tag{9.32}$$

where $C_1 = C_1(\ell) > 0$ and $0 < a < 1$ are constants. For any $N_1 > 0$, set

$$G_k = \bigcup_{D \in \alpha} (f^k(D) \setminus B_D^k), \quad G = \bigcup_{k \geq N_1} G_k.$$

Since the boundary of the set D' is piecewise smooth, it follows from (9.32) that

$$\nu(G_k) \leq C_2 a^k,$$

where $C_2 = C_2(\ell) > 0$ is a constant. We now choose the number N_1 so large that

$$\nu(G) \leq \sum_{k=N_1}^{\infty} \nu(G_k) \leq \beta^2.$$

It is easy to see that for any $N' \geq N \geq N_1$ and β-almost every element $A \in \bigvee_{k=N}^{N'} f^k\alpha$, we have that $\nu(A \cap G) \leq \beta\nu(A)$. It suffices to set $E = A \setminus (A \cap G)$. $\square$

Lemma 9.5.9. *For any $\delta > 0$, there exists $0 < \delta_1 < \delta$ with the following property. Let Π be a δ_1-rectangle and $E \subset \Pi$ a subset of positive measure, which intersects Π leafwise. Then one can find a bijective and onto map $\theta\colon E \to \Pi$ for which*

$$|\text{Jac}(\theta)(y) - 1| \leq \delta \tag{9.33}$$

(where $\text{Jac}(\theta)(y)$ is the Jacobian of theta at the point $y \in E$), and for any $k \geq 0$ and $y \in E$,

$$d(F^k(y), F^k(\theta(y))) \leq \delta. \tag{9.34}$$

Proof of the lemma. Choose $w_1 \in \Pi$ and consider the holonomy map

$$\pi^u = \pi^u_{w,w_1} : X \to V^s(w_1),$$

where $X \subset V^s(w)$ is a measurable set (see Section 8.6). By Theorem 8.6.1 (applied to the family of local unstable manifolds), we can choose δ_1 so small that for every $x \in X$, the Jacobian $\mathrm{Jac}(\pi^u)(x)$ satisfies

$$|\mathrm{Jac}(\pi^u)(x) - 1| \le \frac{1}{3}\delta. \tag{9.35}$$

If δ_1 is sufficiently small then in view of (7.1), (7.2), and the definition of regular sets (see Proposition 2.2.7), we also have that for any $y_1, y_2 \in V^s(w_1)$,

$$d(f^k(y_1), f^k(y_2)) \le \delta. \tag{9.36}$$

Let E be a measurable subset of positive measure, which intersects the rectangle Π leafwise. Since the intersection $E \cap V^s(w)$ has positive $m_{V^s}(w)$-measure (where m_{V^s} is the leaf volume), one can construct a bijective and onto map

$$\theta_0 : E \cap V^s(w) \to V^s(w) \cap \Pi.$$

It follows from (9.35) that for any $z \in E \cap V^s(w)$,

$$|\mathrm{Jac}(\theta_0)(z) - 1| \le \delta. \tag{9.37}$$

We have for any $y \in E$,

$$z = V^u(y) \cap V^s(w) \in \Pi \cap V^s(w).$$

Since also $z \in E$, we obtain that $z \in E \cap V^s(w)$. Set for any $y \in E$,

$$\theta(y) = V^u(\theta_0(z)) \cap V^s(y) = \pi^u_{w,y} \circ \theta_0 \circ (\pi^u_{w,y})^{-1}(y).$$

The inequality (9.33) follows from (9.37), while the inequality (9.34) follows from (9.36) and the fact that $\theta(y) \in V^s(y)$ for any $y \in E$. □

Step IV. We now show that the partition α is a VWB-partition.

Lemma 9.5.10. *For any $0 < \varepsilon' < \varepsilon$, one can find $N_2 > 0$ such that for any $N' \ge N \ge N_2$ and ε-almost every element $A \in \bigvee_{k=N}^{N'} f^k\alpha$, there exist a subset $E \subset A$ and a bijective and onto map $\theta : E \to M$ for which*

1. $\nu(E)/\nu(A) \ge 1 - 2\varepsilon$;
2. *the map θ is 13ε-measure preserving;*
3. *for any $k \ge 0$ and $y \in E$,*

$$d(f^k(y), f^k(\theta(y))) \le \varepsilon'. \tag{9.38}$$

Proof of the lemma. Given $\varepsilon' > 0$, choose δ_1 as in Lemma 9.5.9. Let $\eta = \{\Pi_0, \Pi_1, \dots, \Pi_m\}$ be the partition of M where the sets $\Pi_i, i = 1, \dots, m$, are constructed in Lemma 9.5.7 with $\delta = \delta_1$ and

$$\Pi_0 = M \setminus \bigcup_{i=1}^{m} \Pi_i.$$

By (9.29), we have that

$$\nu(\Pi_0) < \varepsilon. \tag{9.39}$$

Setting $\delta = \varepsilon \min\{\nu(\Pi_i) : 1 \le i \le m\}$ in (9.25) choose the number N_0 according to Lemma 9.5.4. Applying then Lemma 9.5.8 to each rectangle Π_i, $i = 1, \dots, m$ with $\delta = \delta_1$ and

$$\beta = \frac{\varepsilon}{m} \min_{1 \le i \le m} \nu(\Pi_i),$$

we obtain the numbers $N_{1i} > 0$. Set $N_2 = \max\{N_1, \max_{1 \le i \le m} N_{1i}\}$. It follows from Lemma 9.5.8 and inequalities (9.31) and (9.39) that for ε-almost every element $A \in \bigvee_{k=N}^{N'} f^k\alpha$, there exists a subset $E \subset A$, which satisfies Statement 1 of the lemma and intersects each rectangle Π_i, $i = 1, \dots, m$, leafwise. In addition, $E \cap \Pi_0 = \varnothing$. Therefore by Lemma 9.5.8 we have

$$\nu((A \setminus E) \cap \Pi_i) \le \beta\nu(A) \le \varepsilon\nu(A)\nu(\Pi_i) \le \varepsilon\nu(E)\nu(\Pi_i)(1 + 3\varepsilon).$$

In view of (9.25) it follows that

$$\left| \frac{\nu(E_i)}{\nu(E)} - \nu(\Pi_i) \right| \le 6\varepsilon\nu(\Pi_i), \tag{9.40}$$

where $E_i = E \cap \Pi_i$. Let $\theta_i : E_i \to \Pi_i$, $i = 1, \dots, m$, be the bijective map constructed in Lemma 9.5.9. Define the map $\theta : E \to M$ by setting $\theta(y) = \theta_i(y)$ for $y \in E_i$. We have for any measurable subset $B \subset E$ that

$$\left| \frac{\nu(\theta_i(B_i))\nu(E_i)}{\nu(\Pi_i)\nu(B_i)} - 1 \right| \le \varepsilon',$$

where $B_i = B \cap \Pi_i$. Therefore using (9.40) we obtain that

$$\left| \frac{\nu(\theta_i(B_i))\nu(E_i)}{\nu(B_i)} - 1 \right| \le \varepsilon' + (1 + \varepsilon')6\varepsilon \le 13\varepsilon.$$

It follows from here and (9.39) that the map θ is 13ε-measure preserving and the second statement of the lemma follows. The third statement is an immediate consequence of Lemma 9.5.9. □

Fix $\varepsilon > 0$, $0 < \varepsilon' < \varepsilon$ and choose the number N_2 as in Lemma 9.5.10. Let $N' > N \ge N_2$ and Y be the exceptional set of measure $\le \varepsilon$ constructed in this lemma. Note that the set Y consists of elements of the partition $\bigvee_{k=N}^{N'} f^k\alpha$. Choose an element A of this partition, which is not in Y, and consider the set E and the map θ constructed in Lemma 9.5.10. Recall that the number ε' can be chosen arbitrarily small and that the elements of the partition α have piecewise smooth boundaries.

It then follows from (9.38) that the sequences of partitions $\{\xi_i\}_1^n = \{f^{-i}\alpha\}_1^n$ and $\{\eta_i\}_1^n = \{f^{-i}\alpha|A\}_1^n$ satisfy the condition of Lemma 9.5.3. Therefore by this lemma,

$$d(\{f^{-i}\alpha\}_1^n, \{f^{-i}\alpha|A\}_1^n) \leq 300\varepsilon.$$

Since ε is arbitrary, the partition α is a VWB-partition. □

Combining Theorems 9.1.1, 9.4.3, and 9.5.1, we obtain the following Spectral Decomposition Theorem.

Theorem 9.5.11. *Let f be a $C^{1+\alpha}$ diffeomorphism of a compact smooth Riemannian manifold M preserving an absolutely continuous hyperbolic measure ν. Then*

1. *M can be decomposed into countably many invariant disjoint subsets Λ_i, $i \geq 0$, such that $\nu(\Lambda_0) = 0$, $\nu(\Lambda_i) > 0$, and $f|\Lambda_i$ is ergodic for $i \geq 1$;*
2. *Λ_i is a disjoint union of sets Λ_i^j, for $j = 1, \dots, n_i$, which are cyclically permuted by f, that is, $f(\Lambda_i^j) = \Lambda_i^{j+1}$ for $j = 1, \dots, n_i - 1$, and $f(\Lambda_i^{n_i}) = \Lambda_i^1$;*
3. *$f^{n_i}|\Lambda_i^1$ is a Bernoulli automorphism.*

The following result is an immediate consequence of Theorem 9.5.11.

Corollary 9.5.12. *Let f be a $C^{1+\alpha}$ diffeomorphism of a compact smooth Riemannian manifold M preserving an absolutely continuous measure ν. Assume that ν is hyperbolic and that f is mixing with respect to ν. Then f is a Bernoulli automorphism.*

9.6 The Continuous-Time Case

Let φ_t be a C^2 flow on a smooth compact Riemannian manifold M preserving an absolutely continuous hyperbolic measure ν (i.e., an absolutely continuous invariant measure with respect to which all Lyapunov exponents except the one in the flow direction are nonzero almost everywhere). By Theorem 9.1.5, M can be decomposed into ergodic components Λ_i, $i = 1, 2, \dots,$ of positive measure. Applying Theorem 9.4.1 to the nonuniformly partially hyperbolic diffeomorphism in the broad sense φ_1, we obtain the following result:

Theorem 9.6.1. *Assume that the flow φ_t preserves an absolutely continuous hyperbolic measure ν and is ergodic. Then there exists a measurable partition η of M for which*

1. *for almost every $x \in M$, the element $C_\eta(x)$ is an open subset of $W^s(x)$;*
2. *$\varphi_1\eta \geq \eta$;*
3. *$\bigvee_{i=0}^{\infty} \varphi_i\eta = \varepsilon$; and*
4. *$\bigwedge_{i>-\infty}^{0} \varphi_i\eta = \mathcal{H}(W^s) \geq \pi(\varphi_1)$.*

The following result establishes the K-property of the flow φ_t. We remind the reader that according to results in [118, 190] a flow φ_t is a *K-flow* if and only if the diffeomorphism φ_1 is a K-automorphism.

Theorem 9.6.2 (see [168, 169]). *Assume that the flow φ_t preserves an absolutely continuous hyperbolic measure ν and has continuous spectrum. Then it is a K-flow.*

Proof. In view of Theorem 9.6.1 we only need to show that the partition W^s by the global stable manifolds is metrically transitive; that is, if a measurable set A is a W^s-set (this means that it consists of the whole global manifolds $W^s(x)$, $x \in A$) of positive measure then $A = M$ (mod 0). Let A be such a set of intermediate measure (i.e., $\nu(A) > 0$ but $\nu(A) \neq 1$).

Lemma 9.6.3. *There exists a Borel W^s-set of positive measure.*

Proof of the lemma. Since the flow φ_t is ergodic, it has only one level set (see (2.21)) and we consider the corresponding collection of regular sets $\{\Lambda^\ell : \ell \geq 1\}$ (see Section 2.2.2). For some $\ell > 0$, we have that $\nu(A \cap \Lambda^\ell) > 0$. By Theorem 9.3.4, there exists a point $x \in A \cap \Lambda^\ell$ such that the set $C = V^u(x) \cap A \cap \Lambda^\ell$ is measurable and $m_{V^u}(x)(C) > 0$ (where m_{V^u} is the leaf volume). There exists a closed subset $D \subset C$ such that $m_{V^u}(x)(D) > 0$. Therefore by the absolute continuity property of the foliation W^s, the set

$$B = \bigcup_{y \in D} W^s(y)$$

has positive measure. Note that in the definition of the global stable manifolds for the flow (see (8.10)), the union can be taken over values of t running over the set of rational numbers $\mathbb{Q}$. We have

$$B = \bigcup_{y \in D} \bigcup_{t \in \mathbb{Q}} \varphi_{-t}(V^s(\varphi_t(y))) = \bigcup_{t \in \mathbb{Q}} \varphi_{-t}\left(\bigcup_{y \in \varphi_t(D)} V^s(y) \right).$$

Since the set $\varphi_t(D)$ is closed and the local stable manifolds $V^s(y)$ depend continuously on the point $y \in \varphi_t(D)$, the set

$$\bigcup_{y \in \varphi_t(D)} V^s(y)$$

is Borel and so is the set B. □

Let $\mathcal{A}$ be the class of Borel W^s-sets. Fix $A \in \mathcal{A}$ and set for $w \in A$,

$$m_\tau(w) = m_\tau(A, w) = \frac{1}{2\tau} \operatorname{mes}\{t \colon |t| \leq \tau, \varphi_t(w) \in A\},$$

where "mes" denotes the Lebesgue measure on $\mathbb{R}$. It is easy to see that the function $m_\tau(w)$ is measurable and is constant on the leaves of the foliation W^s (since this foliation is invariant with respect to the flow). Note that

$$m_\tau(w) = \frac{1}{2\tau}\int_{-\tau}^{\tau} \chi_A(\varphi_t(w))\,dt.$$

Since φ_t is a smooth flow, it follows that

$$m_\tau(w) \to 1 \text{ as } \tau \to 0 \tag{9.41}$$

for almost every $w \in A$. We need the following technical lemma.

Lemma 9.6.4. *There exists a number $\tau > 0$ such that the set*

$$R = \bigcup_{|t|\le\tau} \varphi_t(B)$$

has intermediate measure.

Proof of the lemma. Set $D = M \setminus B$. It follows from (9.41) that for all sufficiently small τ, the sets

$$\widetilde{B} = \Big\{w \in B\colon m_\tau(w) > \frac{9}{10}\Big\} \quad\text{and}\quad \widetilde{D} = \Big\{w \in D\colon m_\tau(w) > \frac{9}{10}\Big\}$$

have positive measure and belong to the class $\mathcal{A}$. It suffices to show that $\widetilde{D} \cap R = \varnothing$. Assuming the contrary one can find a point w and a number s such that

$$|s| \le \tau, \quad w \in \widetilde{B}, \quad w' = \varphi_s(w) \in \widetilde{D}.$$

We may assume that $s > 0$. Then

$$\begin{aligned}&\operatorname{mes}\{t\colon s-\tau \le t \le \tau, \varphi_t(w) \in D\}\\ &\qquad\le \operatorname{mes}\{t\colon -\tau \le t \le \tau, \varphi_t(w) \in D\} \le \frac{1}{10}2\tau = \frac{1}{5}\tau.\end{aligned}$$

On the other hand,

$$\begin{aligned}&\operatorname{mes}\{t\colon -\tau \le t \le \tau - s, \varphi_t(w') \in B\}\\ &\qquad\le \operatorname{mes}\{t\colon -\tau \le t \le \tau, \varphi_t(w') \in B\} \le \frac{1}{10}2\tau = \frac{1}{5}\tau.\end{aligned}$$

Since $B \cup D = M$, every point of the interval

$$\{\varphi_t(w)\colon s-\tau \le t \le \tau\}$$

lies in either B or D, and hence its length does not exceed $(1/5)\tau + (1/5)\tau = (2/5)\tau < \tau$. On the other hand, the length of this interval is $\tau - (s-\tau) = 2\tau - s > \tau$. This contradiction means that $\widetilde{D} \cap R = \varnothing$ and the desired result follows. □

We proceed with the proof of the theorem and introduce the notion of rectangle, which is an extension to flows of the corresponding notion for diffeomorphisms.

Consider a regular set Λ^ℓ. A measurable set $\Pi \subset M$ is called a *δ-rectangle* at a point $w \in \Lambda^\ell$ (or simply a *rectangle*) if

$$w \in \Pi \subset \Lambda^{\psi(\ell)} \cap B(w,\delta)$$

and for any $y, z \in \Pi$,

$$V^{s0}(y) \cap V^u(z) \in \Pi.$$

(Note that this intersection is nonempty and consists of a single point.)

We say that a measurable set $A \subset M$ intersects a rectangle Π *leafwise* if for any $y \in A \cap \Pi$,

$$V^u(y) \cap \Pi \subset A \cap \Pi.$$

The following statements are extensions to flows of similar results for diffeomorphisms (see Lemmas 9.5.5, 9.5.7, and 9.5.8):

Lemma 9.6.5. *For any sufficiently small $\delta = \delta(\ell) > 0$, there is $r = r(\delta) > 0$ such that for any $w \in \Lambda^\ell$ one can construct a δ-rectangle Π at w satisfying*

$$\Lambda^\ell \cap B(w, r) \subset \Pi.$$

Let α be a finite partition of M whose elements have piecewise smooth boundaries.

Lemma 9.6.6. *Given a rectangle Π and $\beta > 0$, one can find $N_1 > 0$ such that for any $N' \geq N \geq N_1$ and β-almost every element $A \in \bigvee_{k=N}^{N'} f^k\alpha$, there exists a subset $E \subset \Pi$, intersecting Π leafwise, for which*

$$\frac{\nu(E)}{\nu(A)} \geq 1 - \beta.$$

Lemma 9.6.7. *For any sufficiently small $\delta > 0$, one can find disjoint rectangles $\Pi_1, \dots, \Pi_m$ such that $\bigcup_{i=1}^m (\Pi_i \cap \Lambda^\ell) = \Lambda^\ell$.*

Fix a number ε such that

$$0 < \varepsilon \leq \frac{1}{4}\nu(R) \tag{9.42}$$

and choose $\ell > 0$ so big that

$$\nu(M \setminus \Lambda^\ell) \leq \varepsilon. \tag{9.43}$$

Consider a cover of Λ^ℓ by sets Π_i, constructed in Lemma 9.6.7, whose diameters are $\leq (1/2)\tau$ where τ is the number from Lemma 9.6.4. It follows from the continuity of the spectrum of the flow φ_t that there exists a number $\Delta > 0$ and a sequence of numbers $t_n \to \infty$ such that for any $j = 1, \dots, m$,

$$\nu(\varphi_{-t_n}(B) \cap \Pi_j) \geq \Delta. \tag{9.44}$$

Fix a rectangle Π_i and consider its partition ξ_i by local weakly stable manifolds $V^{s0}(w)$, $w \in \Pi_i$. Let $\widetilde{\nu}_i$ be the factor-measure in Π_i/ξ_i. (This measure is equivalent to the leaf volume $m_{V^u}(w)$ for some $w \in \Pi_i$.) It follows from (9.43), (9.44), Lemma 9.6.4, and the absolute continuity of local stable manifolds that there exists $\delta > 0$ such that

$$\widetilde{\nu}_i(\{y \in \Lambda^\ell : V^{s0}(y) \subset \varphi_{-t_n}(R) \cap \Pi_i\}) \geq \delta. \tag{9.45}$$

Choose a number γ, $0 < \gamma < (1/4)\min\{\delta, \varepsilon\}$. There exists a measurable partition $\alpha = \{A_1, \dots, A_m\}$ of the manifold M whose elements have piecewise smooth boundaries and an α-set D such that

$$\nu(R \,\triangle\, D) \le \gamma. \tag{9.46}$$

By Lemma 9.6.6, for sufficiently small $\beta > 0$, one can find a number $T > 0$ and for any $t \ge T$, a set F_t such that

$$\nu(F_t) \le \gamma \tag{9.47}$$

and the set $P_t = \varphi_{-t}(D) \setminus F_t$ intersects Π_i leafwise. It follows from Theorem 8.6.13 and (9.45)–(9.47) that for any $t_n \ge T$, one can find $y_n \in \Pi_i$ such that the set

$$Q_n = V^{s0}(y_n) \cap P_{t_n} \cap \Pi_i$$

is measurable and

$$m_{V^{s0}}(y_n)(Q_n) \ge \left(1 - \gamma - \frac{\gamma}{\delta}K\right) \nu^{s0}(y_n)(V^{s0}(y_n) \cap \Pi_i), \tag{9.48}$$

where $m_{V^{s0}}$ is the leaf volume. We also have that

$$\left(\bigcup_{z \in Q_n} V^u(z)\right) \cap \Pi_i \subset P_{t_n} \cap \Pi_i \subset \varphi_{-t}(D) \cap \Pi_i.$$

Therefore using Theorem 8.6.13, (9.46), and (9.48), we obtain that for $t_n \ge T$,

$$\frac{\nu(\varphi_{-t}(C) \cap \Pi_i)}{\nu(\Pi_i)} \ge 1 - (2\gamma + \frac{\gamma}{\delta}K)K.$$

Summing this up over $i = 1, \dots, m$ and applying (9.43), we conclude that for sufficiently large t_n,

$$\nu(\varphi_{-t}(C)) \ge 1 - m\gamma(2 + K\delta^{-1})K - \varepsilon.$$

Since the number γ can be chosen arbitrarily small (independently of m and δ), this inequality contradicts (9.42). Therefore the measurable hull of the partition ξ_i is the trivial partition. This means that the map φ_1 is a K-automorphism. □

We now establish the Bernoulli property of the flow φ_t. Recall that a flow φ_t is said to be Bernoulli if so is the time-1 map φ_1.

Theorem 9.6.8 (see [168, 169]). *Assume that the flow φ_t preserves an absolutely continuous hyperbolic measure ν and is a K-flow. Then it is a Bernoulli flow.*

Proof. Fix $\varepsilon > 0$ and choose $\ell > 1$ such that (9.29) holds. The proof in the continuous-time case is a simple modification of the proof in the discrete-time case and only the proof of Lemma 9.5.9 needs some changes.

Lemma 9.6.9. *For any $\delta > 0$, there exists $0 < \delta_1 < \delta$ with the following property. Let Π be a δ_1-rectangle and $E \subset \Pi$ a subset of positive measure, which intersects*

Π leafwise. Then one can find a bijective and onto map $\theta\colon E \to \Pi$ for which

$$|\mathrm{Jac}(\theta)(y) - 1| \le \delta,$$

(where $\mathrm{Jac}(\theta)(y)$ *is the Jacobian of theta at the point $y \in E$) and for any $t \ge 0$ and $y \in E$,*

$$d(F^t(y), F^t(\theta(y))) \le \delta.$$

Proof of the lemma. Let Π be a δ_1-rectangle at a point w and $w_1 \in \Pi$. Consider the holonomy map

$$\pi^u = \pi^u_{w,w_1} : X \to V^{s0}(w_1),$$

where $X \subset V^{s0}(w)$ is a measurable set (see Section 7.7). Applying Theorem 8.6.1 to the time-1 map φ_1, the family of local unstable manifolds $V^u(z)$, $z \in \Pi$, and the transversals $T^1 = V^{s0}(w)$ and $T^2 = V^{s0}(w_1)$, we obtain that for every $x \in X$ the Jacobian $\mathrm{Jac}(\pi^u)(x)$ satisfies

$$|\mathrm{Jac}(\pi^u)(x) - 1| \le \frac{1}{3}\delta. \tag{9.49}$$

If δ_1 is sufficiently small then for any $y_1, y_2 \in V^s(w)$ and $t > 0$,

$$d(\varphi_t(y_1), \varphi_t(y_2)) \le \delta. \tag{9.50}$$

Therefore, (9.50) follows for any $y_1, y_2 \in V^{s0}(w)$ and $t > 0$.

Let now E be a measurable set of positive measure, which intersects the rectangle Π leafwise. Let also

$$\theta_0 : E \cap V^{s0}(w) \to \Pi \cap V^{s0}(w)$$

be a bijective and onto map preserving the (normalized) measure $\nu^{s0}(w)$. (It is easy to see that such a map exists since the intersection $E \cap V^{s0}(w)$ has positive $\nu^{s0}(w)$-measure.) For $y \in E$,

$$z = V^u(y) \cap V^{s0}(w) \in \Pi \cap V^{s0}(w).$$

Moreover, $z \in E$, and hence $z \in E \subset V^{s0}(w)$. Therefore the formula

$$\theta(y) = V^u(\theta_0(z)) \cap V^{s0}(w) = \pi^u_{w,y} \circ \theta_0 \circ (\pi^u_{w,y})^{-1}(y)$$

defines the map θ properly. The first statement of the lemma follows from (9.49) and the second one from (9.50) and the fact that $\theta(y) \in V^{s0}(y)$ for any $y \in E$. □

The rest of the proof goes as in Step IV of the proof of Theorem 9.5.1. □

The following result is an immediate consequence of Theorems 9.6.8 and 9.6.2.

Corollary 9.6.10. *Let φ_t be a smooth flow on a compact smooth Riemannian manifold M preserving an absolutely continuous measure ν. Assume that ν is hyperbolic and that φ_t is weakly mixing with respect to ν. Then φ_t is a Bernoulli flow.*

9.7 Notes

Theorem 8.6.8 was proved in [169] and is an extension to nonuniformly hyperbolic systems of the corresponding results by Anosov and Sinai [7].

The notions of (s, k)- and (u, k)-measures as well as of s- and u-measures are adaptations to the case of nonuniform hyperbolicity of the corresponding notion (called u-Gibbs measure) for uniformly hyperbolic attractors introduced by Pesin and Sinai [176].

The description of ergodic properties of diffeomorphisms and flows with non-zero Lyapunov exponents with respect to smooth invariant measures was obtained by Pesin in [168–170]. This includes the decomposition into ergodic components (Theorems 9.1.1 and 9.1.5), construction of the Pinsker partition (Theorems 9.4.1 and 9.6.1), the K-property and the Bernoulli property (Theorems 9.4.3, 9.5.1, 9.6.2, and 9.6.8), and the Spectral Decomposition Theorem 9.5.11. Our exposition essentially follows the arguments in the papers [168–170].

The construction of the Pinsker partition presented here combines arguments by Ledrappier and Strelcyn [123] (see the proof of Statement 1 of Theorem 9.4.1) and the adaptation to nonuniformly hyperbolic systems of the corresponding construction for Anosov maps by Sinai [201]. The proof of the K-property for nonuniformly hyperbolic flows (see Theorem 9.5.1) uses some ideas from the proof of the corresponding statement for Anosov flows (see [6]). The proof of the Bernoulli property is a far-reaching generalization of the arguments by Ornstein and Weiss who established this property for geodesic flows on compact manifolds of negative curvature.

(δ, q)-foliations with smooth leaves were introduced by Pesin [169] as a part of his study of local ergodicity (see Theorems 9.2.1, 9.2.5, and 9.2.6). Burns and Gerber [44] used cone techniques to establish local ergodicity for flows in dimension three (see Theorem 9.2.8). This work was a precursor for Katok–Burns' method to local ergodicity (see Theorem 9.2.9). The approach by Liverani and Wojtkowski in Section 9.2 is built upon a method, which was developed by Sinai [202] in his pioneering work on billiard systems. It has been later improved by Sinai and Chernov [205] and by Krámli, Simányi, and Szász [117] who considered semidispersing billiards. We observe that Theorem 9.2.18 is a particular case of a more general result by Liverani and Wojtkowski for smooth dynamical systems with singularities (see [21, Section 19]).

10

Measure-Theoretic Entropy and Lyapunov Exponents

One of the main ideas of smooth ergodic theory is that sufficient instability of trajectories yields rich ergodic properties of the system. The entropy formula is in a sense a "quantitative manifestation" of this idea and is yet another pearl of the theory. It expresses the Kolmogorov–Sinai entropy $h_\nu(f)$ of a diffeomorphism, preserving a smooth measure, in terms of the values of the Lyapunov exponent.

10.1 Entropy of Measurable Transformations

We briefly recall some relevant notions from the entropy theory. Let $(X, \mathcal{B}, \mu)$ be a Lebesgue measure space and ξ a finite or countable measurable partition of X. The *entropy* of the measurable partition ξ (with respect to μ) is given by

$$H_\mu(\xi) = -\sum_{C\in\xi} \mu(C)\log\mu(C),$$

with the convention that $0\log 0 = 0$. Given two measurable partitions ξ and ζ, we also define the *conditional entropy of* ξ *with respect to* ζ by

$$H_\mu(\xi\,|\,\zeta) = -\sum_{C\in\xi}\sum_{D\in\zeta} \mu(C\cap D)\log\frac{\mu(C\cap D)}{\mu(D)}.$$

If ξ, η, and ζ are measurable partitions with $\xi \le \eta$, we have $H_\mu(\xi) \le H_\mu(\eta)$ and $H_\mu(\zeta\,|\,\xi) \ge H_\mu(\zeta\,|\,\eta)$.

Consider a measurable transformation $T\colon X \to X$ and define the *entropy* of T with respect to partition ξ as the limit

$$h_\mu(T,\xi) = \lim_{n\to\infty}\frac{1}{n}H_\mu\left(\bigvee_{k=0}^{n-1} T^{-k}\xi\right), \tag{10.1}$$

which always exists. It can be shown that

$$h_\mu(T,\xi) = \inf_n \frac{1}{n}H_\mu\left(\bigvee_{k=0}^{n-1} T^{-k}\xi\right) = \lim_{n\to\infty} H_\mu\left(\xi\,|\bigvee_{k=1}^{n} T^{-k}\xi\right). \tag{10.2}$$

For the partition $\xi^- = \bigvee_{k=0}^{\infty} T^{-k}\xi$, we have $h_\mu(T,\xi) = H_\mu(\xi|\xi^-)$. We finally define the *entropy* of T with respect to μ by

$$h_\mu(T) = \sup_\xi h_\mu(T,\xi),$$

where the supremum is taken over all measurable partitions ξ with finite entropy. (Indeed, the supremum coincides with the supremum over all finite measurable partitions.) We have $h_\mu(T,\xi) \le H_\mu(\xi)$ (by (10.2)) and $h_\mu(T^m) = mh_\mu(T)$ for every $m \ge 0$.

When T is invertible with measurable inverse, one can replace $T^{-k}\xi$ by $T^k\xi$ in the formulas (10.1) and (10.2). In particular, $h_\mu(T^{-1},\xi) = h_\mu(T,\xi)$ and $h_\mu(T^{-1}) = h_\mu(T)$. Therefore, $h_\mu(T^m) = |m|h_\mu(T)$ for every $m \in \mathbb{Z}$.

10.2 The Margulis–Ruelle Inequality

Let f be a C^1 diffeomorphism of a smooth compact Riemannian manifold and ν an invariant Borel probability measure on M. Consider the Lyapunov spectrum of the exponent $\chi^+(x,\cdot)$ (see Section 5.4)

$$\operatorname{Sp}\chi(x) = \{(\chi_i(x), k_i(x)) : 1 \le i \le p(x)\}.$$

The following result is known as the Margulis–Ruelle Inequality.

Theorem 10.2.1 (Margulis (unpublished), Ruelle [192]). *We have*

$$h_\nu(f) \le \int_M \sum_{i:\chi_i(x)>0} k_i(x)\chi_i(x)\,d\nu(x).$$

Proof. By decomposing ν into its ergodic components, we may assume without loss of generality that ν is ergodic. Then $k_i(x) = k_i$ and $\chi_i(x) = \chi_i$ are constant ν-almost everywhere for each i. Fix $m > 0$. Our strategy to prove this theorem is as follows. First, we construct a special partition ξ of the manifold M and we show that the entropy of f^m with respect to ξ, $h_\nu(f^m,\xi)$, can be estimated from above by the desired expression – the sum of the Lyapunov exponents – up to a small error term. We then conclude that the entropy of f^m, $h_\nu(f^m)$, admits a similar estimate. Dividing by m, using Abramov's formula that connects the entropies of f and f^m and letting m go to infinity, we obtain the desired result.

We now present the formal proof. Since M is compact, there exists $t_m > 0$ such that for every $0 < t \le t_m$, $y \in M$, and $x \in B(y,t)$, we have

$$\frac{1}{2}d_x f^m\left(\exp_x^{-1} B(y,t)\right) \subset \exp_{f^m x}^{-1} f^m(B(y,t)) \subset 2\,d_x f^m\left(\exp_x^{-1} B(y,t)\right), \tag{10.3}$$

where for a set $A \subset T_zM$ and $z \in M$, we write $\alpha A = \{\alpha v : v \in A\}$.

Lemma 10.2.2. *Given $\varepsilon > 0$, there exists a partition ξ of M, satisfying the following conditions:*

1. $\operatorname{diam}\xi \le t_m/10$ *and* $h_\nu(f^m,\xi) \ge h_\nu(f^m) - \varepsilon$.
2. *For every element* $C \in \xi$, *there exist balls* $B(x,r)$ *and* $B(x,r')$, *such that* $r < 2r' \le t_m/20$ *and* $B(x,r') \subset C \subset B(x,r)$.
3. *There exists* $0 < r < t_m/20$ *such that if* $C \in \xi$ *then* $C \subset B(y,r)$ *for some* $y \in M$, *and if* $x \in C$ *then*

$$\frac{1}{2} d_x f^m \left(\exp_x^{-1} B(y,r)\right) \subset \exp_{f^m x}^{-1} f^m(C)$$
$$\subset 2\, d_x f^m \left(\exp_x^{-1} B(y,r)\right).$$

Proof of the lemma. Given $\alpha > 0$, consider a maximal α-separated set Γ, that is, a finite set of points for which $d(x,y) > \alpha$ whenever $x, y \in \Gamma$. For each $x \in \Gamma$, define

$$\mathfrak{D}_\Gamma(x) = \{y \in M : d(y,x) \le d(y,z) \text{ for all } z \in \Gamma \setminus \{x\}\}.$$

Obviously, $B(x,\alpha/2) \subset \mathfrak{D}_\Gamma(x) \subset B(x,\alpha)$. Note that the sets $\mathfrak{D}_\Gamma(x)$ corresponding to different points $x \in \Gamma$ intersect only along their boundaries, that is, at a finite number of submanifolds of codimension greater than zero. Since ν is a Borel measure, if necessary we can move the boundaries slightly so that they have zero measure. Thus, we obtain a partition ξ with $\operatorname{diam}\xi \le \alpha$. Moreover, we can choose a partition ξ such that

$$h_\nu(f^m,\xi) > h_\nu(f^m) - \varepsilon \quad \text{and} \quad \operatorname{diam}\xi < t_m/10.$$

This implies Statements 1 and 2. Statement 3 follows from (10.3). □

We proceed with the proof of the theorem. We have

$$\begin{aligned} h_\nu(f^m,\xi) &= \lim_{k\to\infty} H_\nu(\xi \,|\, f^m\xi \vee \cdots \vee f^{km}\xi) \\ &\le H_\nu(\xi \,|\, f^m\xi) = \sum_{D\in f^m\xi} \nu(D) H(\xi \,|\, D) \\ &\le \sum_{D\in f^m\xi} \nu(D) \log \operatorname{card}\{C \in \xi : C \cap D \ne \varnothing\}, \end{aligned} \tag{10.4}$$

where $H(\xi \,|\, D)$ is the entropy of ξ with respect to the conditional measure on D induced by ν. Our goal is to obtain a uniform exponential estimate for the number of those elements $C \in \xi$, which have nonempty intersection with a given element $D \in f^m\xi$, and then an exponential bound for the number of those sets $D \in f^m\xi$ that contain regular points (see Section 3.1).

Lemma 10.2.3. *There exists a constant* $K_1 > 0$ *such that if* $D \in f^m\xi$, *then*

$$\operatorname{card}\{C \in \xi : D \cap C = \varnothing\} \le K_1 \sup\{\|d_x f\|^{mn} : x \in M\}$$

where $n = \dim M$.

Proof of the lemma. By the Mean Value Theorem,

$$\operatorname{diam}(f^m(C)) \le \sup\{\|d_x f\|^m : x \in M\} \operatorname{diam} C.$$

Thus if $C \cap D \neq \varnothing$ then C is contained in the $4r'$-neighborhood of D. Therefore,

$$\operatorname{diam} C \leq (\sup\{\|d_x f\|^m : x \in M\} + 2)4r'$$

and hence the volume of C is bounded from above by

$$Kr^n \sup\{\|d_x f\|^{mn} : x \in M\},$$

where $K > 0$ is a constant. On the other hand, by Property 2 of the partition ξ (see Lemma 10.2.2), the set C contains a ball $B(x, r')$, and hence the volume of C is at least $K'(r')^n$, where K' is a positive constant. This implies the desired result. □

Fix $\varepsilon > 0$ and let $\mathcal{R}_m$ be the set of regular points (see Section 3.1) $x \in M$, which satisfy the following condition: for $k > m$ and $v \in T_x M$,

$$e^{k(\chi(x,v)-\varepsilon)}\|v\| \leq \|df^k v\| \leq e^{k(\chi(x,v)+\varepsilon)}\|v\|.$$

Lemma 10.2.4. *If $D \in f^m\xi$ has nonempty intersection with $\mathcal{R}_m$ then there exists a constant $K_2 > 0$ such that*

$$\operatorname{card}\{C \in \xi : D \cap C \neq \varnothing\} \leq K_2\, e^{\varepsilon m} \prod_{i:\chi_i>0} e^{m(\chi_i+\varepsilon)k_i}.$$

Proof of the lemma. Let $C' \in \xi$ be such that $C' \cap \mathcal{R}_m \neq \varnothing$ and $f^m(C') = D$. Pick a point $x \in C' \cap f^{-m}(\mathcal{R}_m)$ and let $B = B(x, 2\operatorname{diam} C')$. The set

$$\widetilde{B}_0 = df^m(\exp_x^{-1} B) \subset T_{f^m(x)}M$$

is an ellipsoid and

$$D \subset B_0 = \exp_{f^m(x)}(\widetilde{B}_0).$$

If a set $C \in \xi$ has nonempty intersection with B_0 then it lies in the set

$$B_1 = \{y \in M : d(y, B_0) < \operatorname{diam} \xi\}.$$

Therefore,

$$\operatorname{card}\{C \in \xi : D \cap C \neq \varnothing\} \leq \operatorname{vol}(B_1)(\operatorname{diam} \xi)^{-n},$$

where $\operatorname{vol}(B_1)$ denotes the volume of B_1. Up to a bounded factor, $\operatorname{vol}(B_1)$ is bounded by the product of the lengths of the axes of the ellipsoid $\widetilde{B}_0$. Those of them that correspond to nonpositive exponents are at most subexponentially large. The remaining ones are of size at most $e^{m(\chi_i+\varepsilon)}$, up to a bounded factor, for all sufficiently large m. Thus,

$$\begin{aligned}\operatorname{vol}(B_1) &\leq K\, e^{m\varepsilon}(\operatorname{diam} B)^n \prod_{i:\chi_i>0} e^{m(\chi_i+\varepsilon)k_i}\\ &\leq K\, e^{m\varepsilon}(2\operatorname{diam} \xi)^n \prod_{i:\chi_i>0} e^{m(\chi_i+\varepsilon)k_i},\end{aligned}$$

for some constant $K > 0$. The lemma follows. □

By Lemmas 10.2.3 and 10.2.4 and (10.4), we obtain

$$\begin{aligned} mh_\nu(f) - \varepsilon &= h_\nu(f^m) - \varepsilon \le h_\nu(f^m, \xi) \\ &\le \sum_{D \cap \mathcal{R}_m \neq \varnothing} \nu(D) \left(\log K_2 + \varepsilon m + m \sum_{i:\chi_i > 0} (\chi_i + \varepsilon) k_i \right) \\ &\quad + \sum_{D \cap \mathcal{R}_m = \varnothing} \nu(D)(\log K_1 + nm \log \sup\{\|d_x f\| : x \in M\}) \\ &\le \log K_2 + \varepsilon m + m \sum_{i:\chi_i > 0} (\chi_i + \varepsilon) k_i \\ &\quad + (\log K_1 + nm \log \sup\{\|d_x f\| : x \in M\}) \nu(M \setminus \mathcal{R}_m). \end{aligned}$$

By the Multiplicative Ergodic Theorem 3.4.3, we have

$$\bigcup_{m \ge 0} \mathcal{R}_m = M \pmod 0$$

for all sufficiently small ε. It follows that

$$h_\nu(f) \le \varepsilon + \sum_{i:\chi_i > 0} (\chi_i + \varepsilon) k_i.$$

Letting $\varepsilon \to 0$ we obtain the desired upper bound. □

10.3 The Topological Entropy and Lyapunov Exponents

As an immediate consequence of Ruelle's inequality, we obtain an upper bound for the topological entropy $h(f)$ of a diffeomorphism f. Indeed, by the Variational Principle for the topological entropy (see, e.g., [104, Theorem 4.5.3]), we have

$$h(f) = \sup_\nu h_\nu(f) \le \sup_\nu \int_M \Sigma_+(x)\, d\nu(x), \tag{10.5}$$

where the suprema are taken over all f-invariant Borel probability measures on M and

$$\Sigma_+(x) = \sum_{i:\chi_i(x) > 0} k_i(x) \chi_i(x).$$

In general the Margulis–Ruelle Inequality and the inequality (10.5) can be strict. In fact, as the following example shows, there are C^∞ diffeomorphisms for which $h(f) < \inf_\nu \int_M \Sigma_+ \, d\nu$, and hence $h_\nu(f) < \int_M \Sigma_+ \, d\nu$ for *any* invariant measure ν.

Example 10.3.1 ([101]). *Let f be a diffeomorphism of the two-dimensional sphere S^2 with three repelling fixed points p_1, p_2, and p_3 and one saddle fixed point q (see Fig. 10.1). Suppose that the stable and unstable manifolds of the point q form two loops γ_1 and γ_2 that divide S^2 into three regions A_1, A_2, and A_3. For $i = 1, 2, 3$, we have $p_i \in A_i$ and any point in $A_i \setminus \{p_i\}$ tends, respectively, to γ_1, γ_2, and $\gamma_1 \cup \gamma_2$.*

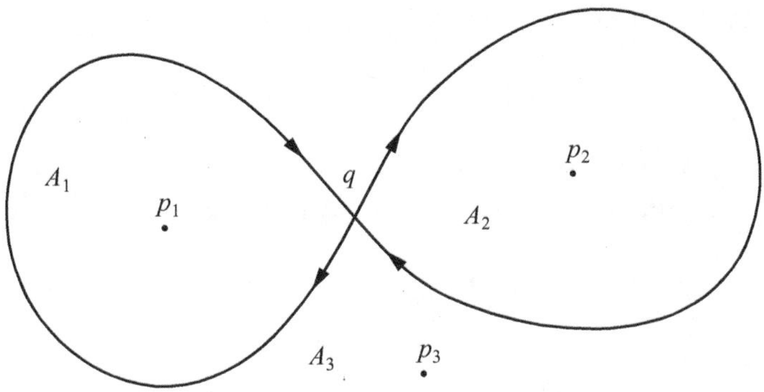

Figure 10.1. A diffeomorphism of the 2-sphere with three repelling points p_1, p_2, and p_3 and one saddle q.

Thus any f-invariant finite *measure ν is supported on the finite set $\{p_1, p_2, p_3, q\}$. Therefore, $h_\nu(f) = 0$, while $\int_M \Sigma_+ \, d\nu > c > 0$ for some c independent of ν. In addition, we have*

$$h(f) = \sup_\nu h_\nu(f) < \inf_\nu \int_M \Sigma_+ \, d\nu,$$

where the supremum and infimum are taken over all f-invariant Borel probability measures on S^2.

The following example assumes some familiarity with the thermodynamic formalism.

Example 10.3.2 (Two-Dimensional Horseshoes). *Let Λ be a locally maximal hyperbolic set of a topologically transitive Axiom A surface diffeomorphism of class C^1 (see Section 6.1.1). The unstable distribution E^u is assumed to be one dimensional. One can show (see [150]) that for every $x \in \Lambda$ the Hausdorff dimension of the set $W^u(x, \Lambda) = W^u(x) \cap \Lambda$ is the unique root s of Bowen's equation*

$$P\big(-s \log\|df|E^u\|\big) = 0.$$

Here P is the topological pressure of the function

$$\varphi_s(x) = -s \log \|df|E^u(x)\|$$

for $f|\Lambda$ (see [104] for the definition). Note that s is independent of x.

Assume that $W^u(x, \Lambda)$ has Hausdorff dimension strictly less than 1. Since $s \mapsto P(-s \log\|df|E^u\|)$ is decreasing, we obtain $P(-\log\|df|E^u\|) < 0$. By the Variational Principle for the topological pressure (see [104]), for every f-invariant measure ν, we have

$$h_\nu(f) - \int_\Lambda \log\|df|E^u(x)\| \, d\nu(x) \le P(-\log \|df|E^u\|) < 0.$$

Using Birkhoff's Ergodic Theorem and the fact that $\dim E^u = 1$, *we conclude that*

$$h_\nu(f) < \int_\Lambda \log \|df|E^u(x)\| \, d\nu(x) = \int_\Lambda \Sigma_+(x) \, d\nu(x).$$

Note that $h_\nu(f^{-1}) = h_\nu(f)$ and the Lyapunov exponents of f^{-1} are those of f taken with opposite sign. Therefore it follows from Theorem 10.2.1 that

$$h_\nu(f) \leq -\int_M \sum_{i:\chi_i(x)<0} \chi_i(x) k_i(x) \, d\nu(x).$$

Set

$$a = \int_M \sum_{i:\chi_i(x)>0} \chi_i(x) k_i(x) \, d\nu(x)$$

and

$$b = -\int_M \sum_{i:\chi_i(x)<0} \chi_i(x) k_i(x) \, d\nu(x).$$

In Example 10.3.1 one can choose the eigenvalues of df at the critical points and the measure ν to guarantee any of the relations: $a < b$, $a = b$, or $a > b$. One can also show that if ν is the Riemannian volume on M then $a = b$.

An important manifestation of the Margulis–Ruelle Inequality is that positivity of measure-theoretic entropy implies the existence of at least one nonzero Lyapunov exponent.

Corollary 10.3.3. *Let* ν *be a measure, which is invariant under a* C^1 *diffeomorphism* f *of a compact manifold. If* $h_\nu(f) > 0$ *then* ν *has at least one positive and one negative Lyapunov exponent.*

Proof. Let $\{\nu_\alpha\}_{\alpha \in A}$ be the ergodic decomposition of ν (see, e.g., [104, Theorem 4.1.12]), so that

$$h_\nu(f) = \int_A h_{\nu_\alpha}(f) \, d\alpha.$$

It follows that $h_{\nu_\alpha}(f) > 0$ for some $\alpha \in A$. Since ν_α is ergodic, Σ_+ is constant ν_α-almost everywhere, and Theorem 10.2.1 implies that $\Sigma_+ > 0$ on a set of full ν_α-measure. (Recall that Σ_+ is nonnegative.) Hence, ν_α has at least one positive Lyapunov exponent.

We can also apply Theorem 10.2.1 to f^{-1} and since $h_{\nu_\alpha}(f^{-1}) = h_{\nu_\alpha}(f) > 0$, we conclude that f^{-1} has at least one positive Lyapunov exponent. Its opposite is a negative Lyapunov exponent for f. □

For surface diffeomorphisms, Corollary 10.3.3 implies that if $h_\nu(f) > 0$ then the Lyapunov exponents of ν are all nonzero; that is, ν is hyperbolic (see Section 8.6 and Chapter 15).

Corollary 10.3.4. *If the topological entropy of a C^1 diffeomorphism f of a compact manifold is positive, then there exists an ergodic f-invariant measure with at least one positive and one negative Lyapunov exponent.*

Proof. It is sufficient to notice that by the Variational Principle for the topological entropy, there exists an f-invariant Borel probability measure ν with $h_\nu(f) > 0$. □

For surface diffeomorphisms, Corollary 10.3.4 means that any diffeomorphism with positive topological entropy possesses an ergodic invariant measure whose Lyapunov exponents are all nonzero.

We present another corollary of the Margulis–Ruelle Inequality.

Corollary 10.3.5. *We have*

$$\begin{aligned} h(f) &\le \dim M \times \inf_{m\ge 1} \frac{1}{m} \log^+ \sup_{x\in M} \|d_x f^m\| \\ &= \dim M \times \lim_{m\to\infty} \frac{1}{m} \log^+ \sup_{x\in M} \|d_x f^m\|. \end{aligned}$$

Proof. If $n, m \in \mathbb{Z}$ and $x \in M$ then $\|d_x f^{n+m}\| \le \|d_{f^m(x)} f^n\| \cdot \|d_x f^m\|$. Therefore,

$$\log \sup_{x\in M} \|d_x f^{n+m}\| \le \log \sup_{x\in M} \|d_x f^n\| + \log \sup_{x\in M} \|d_x f^m\|.$$

This implies that

$$\begin{aligned} \chi^+(x, v) &= \overline{\lim_{m\to+\infty}} \frac{1}{m} \log \|d_x f^m v\| \\ &\le \lim_{m\to+\infty} \frac{1}{m} \log \sup_{x\in M} \|d_x f^m\| \\ &= \inf_{m\ge 1} \frac{1}{m} \log \sup_{x\in M} \|d_x f^m\|, \end{aligned}$$

and hence

$$\begin{aligned} \Sigma_+(x) &\le \dim M \times \lim_{m\to+\infty} \frac{1}{m} \log^+ \sup_{x\in M} \|d_x f^m\| \\ &= \dim M \times \inf_{m\ge 1} \frac{1}{m} \log^+ \sup_{x\in M} \|d_x f^m\|. \end{aligned}$$

Now the desired property follows from the Variational Principle for the topological entropy and Theorem 10.2.1. □

10.4 The Entropy Formula

Let $f\colon M \to M$ be a $C^{1+\alpha}$ diffeomorphism for some $\alpha > 0$ and ν an f-invariant measure that is absolutely continuous with respect to volume. The main result of this section is the *entropy formula* (often called *Pesin's entropy formula*) that expresses the entropy of f with respect to ν via its Lyapunov exponents.

In Section 10.5 we present another proof of the entropy formula due to Mañé that does not explicitly use invariant manifold theory or absolute continuity.

Theorem 10.4.1. *The following formula for the entropy of f holds:*

$$h_\nu(f) = \int_M \sum_{i:\chi_i(x)>0} k_i(x)\chi_i(x)\, d\nu(x).$$

In view of the Margulis–Ruelle Inequality, we need to establish only the lower bound

$$h_\nu(f) \geq \int_M \sum_{i:\chi_i(x)>0} k_i(x)\chi_i(x)\, d\nu(x),$$

or equivalently (by replacing f by f^{-1} and using Theorem 3.4.3),

$$h_\nu(f) \geq -\int_M \sum_{i:\chi_i(x)<0} k_i(x)\chi_i(x)\, d\nu(x).$$

We obtain this inequality as a simple corollary of a more general result about u-measures (in view of Theorem 8.6.8, ν is a u-measure for f), which we now state. In fact, we consider the even more general case when ν is a (u, m)-measure for f with $m = m(x)$ and $1 \leq m < p$. This means that the set Y given by (8.14) (with $k(x) = m(x)$) has positive measure and is nonuniformly partially hyperbolic in the broad sense for f with $\lambda(x) = e^{\chi_{m(x)}(x)}$, $\mu(x) = e^{\chi_{m(x)+1}(x)}$, and an arbitrary f-invariant Borel function $\varepsilon(x)$ satisfying (2.9) (see Section 9.3). Further, there is a level set $\Lambda = \Lambda_{\lambda\mu\varepsilon j}$ of positive measure where the numbers λ, μ, and ε satisfy (2.15) and j is the dimension of mth local manifolds V_m (i.e., therefore constant on Λ). We denote by η the leaf-subordinated partition of the set $\hat{\Lambda}$ (see (9.20)) associated with global manifolds W_m (see Theorem 9.4.1). Finally, let $\{\Lambda^\ell = \Lambda^\ell_{\lambda\mu\varepsilon j} : \ell \geq 1\}$ be the corresponding collection of regular sets.

Theorem 10.4.2. *The entropy of f with respect to the (u, m)-measure ν and the partition η admits the following estimate from below:*

$$h_\nu(f, \eta) \geq -\int_\Lambda \sum_{i=1}^m k_i(x)\chi_i(x)\, d\nu(x).$$

Proof. For sufficiently small $r = r(\ell)$ and $x \in \Lambda^\ell$, consider the sets $Q^\ell(x)$ and $Q(x)$ defined by (9.21). Let the measure $\widetilde{\nu}$ be the restriction of ν to $Q(x)$; that is, $\widetilde{\nu} = (\nu(Q(x)))^{-1}\nu|Q(x)$. It suffices to show that

$$h(f|Q(x), \eta) \geq -\int_{Q(x)} \sum_{i=1}^m k_i(z)\chi_i(z)\, d\widetilde{\nu}(z). \tag{10.6}$$

Consider the function

$$g(z) = \prod_{i=1}^m \exp(\chi_i(z))^{k_i(z)}.$$

Fix $\varepsilon > 0$, and let

$$R(x, p) = \{z \in Q(x) : p\varepsilon < g(z) \leq (p+1)\varepsilon\}.$$

It suffices to show the inequality (10.6) for the restriction $\bar{f} = f|R(x, p)$ and the measure $\bar{\nu} = (\nu(R(x, p)))^{-1}\nu|R(x, p)$.

Set $J_n(z) = \text{Jac}(df^n|T_z W_m(z))$. It follows from the Multiplicative Ergodic Theorem 3.4.3 that there exists a positive Borel function $T(z, \varepsilon)$, $z \in R(x, p)$, such that for $n > 0$,

$$J_n(z) \leq T(z, \varepsilon)g(z)\exp(\varepsilon n).$$

Set for $t \geq 0$,

$$R(x, p, t) = \{z \in R(x, p) : T(z, \varepsilon) \leq t\}.$$

We have that for any $\alpha > 0$ and all sufficiently large t,

$$\bar{\nu}(R(x, p, t)) \geq 1 - \alpha. \tag{10.7}$$

Since ν is a (u, m)-measure, for every $z \in R(x, p, t)$ and $n \geq 0$, we have that

$$\nu^m(z)(f^n(C_\eta(z))) \leq c(z)J_n(z),$$

where $c(z) > 0$ is a measurable function and ν^m the conditional measure generated by ν on the mth local manifolds. In view of Luzin's Theorem and (10.7), there exist $C_1 = C_1(t) > 0$ and a set $R(x, p, t, C_1) \subset R(x, p, t)$ such that

$$\bar{\nu}(R(x, p, t, C_1)) \geq 1 - 2\alpha. \tag{10.8}$$

and for any $z \in R(x, p, t, C_1)$ and $n > 0$,

$$\nu^m(z)(f^n(C_\eta(z))) \leq C_1 J_n(z). \tag{10.9}$$

Denote by $B_\eta(z, r)$ the ball in $C_\eta(z)$ centered at z of radius r.

Lemma 10.4.3. *There exists $q = q(t)$ and a subset $A^t \subset R(x, p, t, C_1)$ such that $\bar{\nu}(A^t) \geq 1 - 3\alpha$ and for any $z \in A^t$ the element $C_\eta(z)$ contains the ball $B_\eta(z, q)$.*

Proof of the lemma. For sufficiently small $q > 0$, consider the set

$$D_q = \{z \in R(x, p, t, C_1) : \text{there is } y \in B_\eta(z, q) \text{ for which } y \notin C_\eta(z)\}.$$

For $z \in D_q$, there exists $n > 0$ and a point $y \in B_\eta(z, q)$ such that $y \in C_{f^{-n}\eta}(z)$. It follows that one can find a point $w \in B_\eta(z, q)$ for which

$$f^n(w) \in Y = \bigcup_{z \in B(x,r) \cap R(x,p,t,C_1) \cap Q^\ell(x)} \partial V_m(z).$$

Therefore the point $f^n(z)$ lies in the neighborhood of the set Y of radius $C_2\lambda^n q$, where $C_2 > 0$ and $0 < \lambda < 1$ are constants. This implies that

$$\bar{\nu}(D_q) \leq \sum_{n=0}^{\infty} C_2\lambda^n q = C_3 q,$$

where $C_3 = C_3(t) > 0$ is a constant. To complete the proof of the lemma, we set $q(t) = 3\alpha C_3^{-1}$ and $A^t = R(x, p, t, C_1) \setminus D_q$. □

We proceed with the proof of the theorem. Denote by $\nu_\eta(z)$ the conditional measure on the element $C_\eta(z)$ of the partition η generated by the measure ν. By (10.8), there exist $C_4 > 0 = C_4(t) > 0$ and a set $\widetilde{A}^t$, such that

$$\bar{\nu}(\widetilde{A}^t) \geq 1 - 4\alpha. \tag{10.10}$$

and for every $z \in \widetilde{A}^t$, we have that

$$C_4^{-1} \leq \frac{d\nu_\eta(z)}{d\,\mathrm{vol}(z)} \leq C_4 \tag{10.11}$$

(where vol is the leaf volume on the local manifold V_m). For any $n > 0$, we obtain that

$$h_{\bar{\nu}}(f) = \frac{1}{n} h_{\bar{\nu}}(f^n) \geq \frac{1}{n} H(f^n \eta | \eta).$$

We use here the fact that $\eta = \eta^- = \bigvee_{i \leq 0} f^i \eta$ (see Theorem 9.4.1). It follows from (10.7), (10.9), and (10.11) that for every $x \in \widetilde{A}^t$ and $n > 0$,

$$\begin{aligned} H(\bar{f}^n \eta | C_\eta(z)) &= -\int_{C_\eta(x)} \frac{\nu_\eta(C_\eta(x) \cap C_{f^n \eta}(z))}{\nu_\eta(C_\eta(y))}\, d\nu_\eta(y) \\ &\geq -\log[C_4{}^2 C_1 t((p+1)\varepsilon)^n e^{\varepsilon n}\, \mathrm{vol}(B_\eta(z, q(t)))^{-1}] = I_n. \end{aligned} \tag{10.12}$$

We have that $\mathrm{vol}(B_\eta(z, q(t))) \geq C_5 q^m(t)$, where $C_5 > 0$ is a constant. It follows that

$$\begin{aligned} I_n &\geq -\log(C_4{}^2 C_1 t)(C_5 q^j(t))^{-1} - n(\log((p+1)\varepsilon) + \varepsilon) \\ &\geq C_6 - n(\log g(z) + \varepsilon), \end{aligned} \tag{10.13}$$

where j is the dimension of the local manifolds V_m. Therefore integrating inequality (10.12) over the elements $C_\eta(x)$ and taking (10.13) and (10.10) into account, we conclude that

$$\begin{aligned} \frac{1}{n} H(\bar{f}^n \eta | \eta) &\geq \frac{1}{n} I_n \bar{\nu}(A^t) \geq \frac{1}{n} \int_{R(x,p)} I_n\, d\bar{\nu}(1 - 4\alpha) \\ &\geq \int_{R(x,p)} \sum_{i=1}^{m} k_i(z) \chi_i(z)\, d\bar{\nu}(z) - \gamma, \end{aligned}$$

where γ can be made arbitrarily small if ε and α are chosen sufficiently small and n sufficiently large. The desired result follows. □

We present another version of the entropy formula. Consider the set Y of Lyapunov–Perron regular points for which there exists a splitting of the tangent bundle $T_x M = E(x) \oplus E^u(x)$ such that $\chi(x, v) \leq 0$ for $v \in E(x)$ and $\chi(x, v) > 0$ for $x \in E^u(x)$. Without loss of generality, we may assume that $\nu(Y) = 1$.

Theorem 10.4.4. *Let f be a $C^{1+\alpha}$ diffeomorphism of a compact smooth Riemannian manifold M and ν an f-invariant measure that is absolutely continuous with respect to volume. Then the entropy of f with respect to ν can be computed by the formula*

$$h_\nu(f) = \int_M \log|\mathrm{Jac}(df|E^u(x))|\, d\nu(x),$$

where Jac *is the Jacobian of the corresponding map.*

Proof. By Proposition 3.1.6, for every Lyapunov–Perron regular point x, we have that

$$\lim_{n\to\infty} \frac{1}{n} \log|\mathrm{Jac}(df^n|E^u(x))| = \sum_{i:\chi_i(x)>0} k_i(x)\chi_i(x).$$

Since

$$\lim_{n\to\infty} \frac{1}{n} \log|\mathrm{Jac}(df^n|E^u(x))| = \lim_{n\to\infty} \frac{1}{n} \sum_{k=0}^{n-1} \log|\mathrm{Jac}(df|E^u(f^k(x)))|,$$

the desired result follows from the entropy formula and Birkhoff's Ergodic Theorem. □

10.5 Mañé's Proof of the Entropy Formula

We present another proof of the entropy formula due to Mañé [137].

Theorem 10.5.1. *Let M be a compact Riemannian manifold and $f\colon M \to M$ a $C^{1+\alpha}$ diffeomorphism for some $\alpha > 0$. If ν is an f-invariant smooth measure then*

$$h_\nu(f) = \int_M \Sigma_+ \, d\nu,$$

where $\Sigma_+ = \sum_{i:\chi_i(x)>0} k_i(x)\chi_i(x)$.

Proof. In view of Theorem 10.2.1 we only need to show that

$$h_\nu(f) \geq \int_M \Sigma_+ \, d\nu.$$

Mañé's proof is based on a general lower estimate of the measure-theoretic entropy that avoids the use of partitions. Let $g\colon M \to M$ be a measurable transformation and $\rho\colon M \to (0, 1)$ a Borel function. Given $x \in M$ and $n \geq 0$, define

$$S_n(g, \rho, x) = \{y \in M : d(g^k(x), g^k(y)) \leq \rho(g^k(x)) \text{ for } 0 \leq k \leq n\}.$$

For a measure ν on M, define

$$h_\nu(g, \rho, x) = \overline{\lim_{n\to\infty}} -\frac{1}{n} \log \nu(S_n(g, \rho, x)).$$

Proposition 10.5.2. *If ν is a g-invariant probability measure on M and $\mu \gg \nu$ is another measure on M (not necessarily g-invariant) then*

$$h_\nu(g) \geq \int_M h_\mu(g, \rho, x)\, d\nu(x)$$

for every function $\rho\colon M \to (0, 1)$ *such that* $\log \rho \in L^1(M, \nu)$.

Proof of the proposition. We need the following two lemmas:

Lemma 10.5.3. *If $x_n \in (0, 1)$ for each $n \geq 1$ and $\sum_{n=1}^\infty n\, x_n < +\infty$, then $\sum_{n=1}^\infty x_n \log(1/x_n) < +\infty$.*

Proof of the lemma. Let S be the set of integers $n \geq 1$ such that $\log(1/x_n) < n$. Then

$$\begin{aligned} \sum_{n=1}^\infty x_n \log \frac{1}{x_n} &= \sum_{n\in S} x_n \log \frac{1}{x_n} + \sum_{n\notin S} x_n \log \frac{1}{x_n} \\ &\leq \sum_{n\in S} n x_n + \sum_{n\notin S} x_n \log \frac{1}{x_n} \\ &\leq \sum_{n=1}^\infty n x_n + \sum_{n\notin S} x_n \log \frac{1}{x_n}. \end{aligned} \tag{10.14}$$

Notice that $e^{-n} \geq x_n$ for every $n \notin S$. We have that $\sqrt{t} \log(1/t) \leq 2/e$ for all $0 < t < 1$. Hence,

$$\begin{aligned} \sum_{n\notin S} x_n \log \frac{1}{x_n} &= \sum_{n\notin S} \sqrt{x_n} \left(\sqrt{x_n} \log \frac{1}{x_n} \right) \\ &\leq \frac{2}{e} \sum_{n\notin S} \sqrt{x_n} \leq \frac{2}{e} \sum_{n=1}^\infty e^{-n/2} < +\infty. \end{aligned} \tag{10.15}$$

By (10.14) and (10.15) we obtain the desired statement. □

If ξ is a partition, we denote by $\xi(x)$ the element of ξ containing x.

Lemma 10.5.4. *If ν is a probability measure on M and $\rho\colon M \to (0, 1)$ is such that $\log \rho \in L^1(M, \nu)$, then there exists a partition ξ of M with entropy $H_\nu(\xi) < +\infty$ such that* $\operatorname{diam} \xi(x) \leq \rho(x)$ *for ν-almost every x.*

Proof of the lemma. There exist $C > 0$ and $r_0 > 0$ such that for $0 < r < r_0$ one can find a partition ξ_r of M whose elements have diameter at most r and the number of elements $N(r)$ satisfies

$$N(r) \leq C r^{-\dim M}.$$

For each $n \geq 0$, consider the disjoint sets $U_n = \{x \in M : e^{-(n+1)} < \rho(x) \leq e^{-n}\}$. We have

$$\sum_{n=1}^{m} n\nu(U_n) \leq -\sum_{n=1}^{m} \int_{U_n} \log \rho \, d\nu \leq -\int_M \log \rho \, d\nu, \tag{10.16}$$

and since $\log \rho$ is integrable, we obtain

$$\sum_{n=1}^{\infty} n\nu(U_n) < +\infty. \tag{10.17}$$

Let ξ be the partition of M whose elements are the sets $Q \cap U_n$, where $Q \in \xi_{r_n}$ for $r_n = e^{-(n+1)}$ and $n \geq 1$. Then

$$H_\nu(\xi) = -\sum_{n=0}^{\infty} \sum_{P \in \xi : P \subset U_n} \nu(P) \log \nu(P). \tag{10.18}$$

Consider the function

$$\varphi(x) = \begin{cases} x \log x & \text{if } x > 0 \\ 0 & \text{if } x = 0 \end{cases}.$$

Since $\varphi''(x) = 1/x > 0$ for every $x > 0$, the function φ is strictly convex; that is, $\varphi(\sum_{i=1}^m \alpha_i x_i) \leq \sum_{i=1}^m \alpha_i \varphi(x_i)$ for each $\alpha_i, x_i \geq 0$ such that $\sum_{i=1}^m \alpha_i = 1$, with equality if and only if all the numbers x_i corresponding to nonzero α_i are equal. If $\alpha_i = 1/m$ for every i, we conclude that

$$-\sum_{i=1}^{m} x_i \log x_i \leq \left(\sum_{i=1}^{m} x_i\right)\left(\log m - \log \sum_{i=1}^{m} x_i\right).$$

Thus,

$$\begin{aligned} -\sum_{P \in \xi : P \subset U_n} \nu(P) \log \nu(P) &\leq \nu(U_n)(\log N(r_n) - \log \nu(U_n)) \\ &\leq \nu(U_n)(\log C - \dim M \log r_n - \log \nu(U_n)). \end{aligned}$$

By (10.16), (10.18), and the identity $-\log r_n = n + 1$, we obtain

$$\begin{aligned} H_\nu(\xi) &\leq \sum_{n=0}^{\infty} \nu(U_n)(\log C + (n+1) \dim M - \log \nu(U_n)) \\ &\leq \log C + \dim M \left(1 - \int_M \log \rho \, d\nu\right) - \sum_{n=0}^{\infty} \nu(U_n) \log \nu(U_n). \end{aligned}$$

It follows from (10.17) and Lemma 10.5.3 that the series in the last expression converges. Hence, $H_\nu(\xi) < +\infty$. Moreover, if $x \in U_n$, then $\xi(x)$ is contained in some element of ξ_{r_n}. This shows that

$$\operatorname{diam} \xi(x) \leq r_n = e^{-(n+1)} < \rho(x)$$

and the statement of the lemma follows. □

To prove the proposition take the partition ξ given by Lemma 10.5.4. Denote by $\xi_n(x)$ the element of $\xi_n = \bigvee_{j=0}^{n} g^{-j}\xi$ that contains x. By Shannon–McMillan–Breiman's Theorem, we obtain

$$h_\nu(g) \geq h_\nu(g, \xi) = \int_M \lim_{n\to\infty} -\frac{1}{n} \log \nu(\xi_n(x))\, d\nu(x),$$

where the limit exists ν-almost everywhere and is in $L^1(X, \nu)$. The desired result will follow if we show that

$$\lim_{n\to\infty} -\frac{1}{n} \log \nu(\xi_n(x)) \geq h_\mu(g, \rho, x) \text{ for } \nu\text{-almost every } x.$$

To this end observe that by the Radon–Nikodým Theorem, there exists a μ-integrable function $k\colon M \to [0, +\infty]$ such that

$$\int_A k\, d\mu = \nu(A)$$

for every measurable set A.

Let $\mathcal{B}$ be a σ-subalgebra of the σ-algebra where ν and μ are defined. Recall that the *conditional expectation* of k with respect to $\mathcal{B}$ is a $\mathcal{B}$-measurable function $k_{\mathcal{B}} \in L^1(X, \mu)$ such that

$$\int_A k_{\mathcal{B}}\, d\mu = \int_A k\, d\mu$$

for every $A \in \mathcal{B}$. We obtain (see, e.g., [139, Section 0.V]),

$$\lim_{n\to\infty} \frac{\nu(\xi_n(x))}{\mu(\xi_n(x))} = \lim_{n\to\infty} \frac{1}{\mu(\xi_n(x))} \int_{\xi_n(x)} k\, d\mu = k_{\mathcal{B}}(x) < +\infty$$

for μ-almost every $x \in M$. By Shannon–McMillan–Breiman's Theorem,

$$\begin{aligned}\lim_{n\to\infty} \frac{1}{n} \log \nu(\xi_n(x)) &= \lim_{n\to\infty} \left(\frac{1}{n} \log \mu(\xi_n(x)) + \frac{1}{n} \log \frac{\nu(\xi_n(x))}{\mu(\xi_n(x))} \right) \\ &= \lim_{n\to\infty} \frac{1}{n} \log \mu(\xi_n(x))\end{aligned}$$

for μ-almost every $x \in M$. Moreover, since $\operatorname{diam} \xi(x) \leq \rho(x)$, we have $\xi_n(x) \subset S_n(g, \rho, x)$. Thus,

$$\lim_{n\to\infty} -\frac{1}{n} \log \mu(\xi_n(x)) \geq h_\mu(g, \rho, x),$$

and

$$\lim_{n\to\infty} -\frac{1}{n} \log \nu(\xi_n(x)) \geq h_\mu(g, \rho, x)$$

for μ-almost all $x \in M$. Since $\nu \ll \mu$, the last inequality also holds for ν-almost all $x \in M$. □

In what follows we will avoid the use of coordinate charts, treating M as if it were $\mathbb{R}^n$. It is easy to see that all arguments can be formalized in a straightforward manner. Furthermore, we put the norm $\|(x_1, \ldots, x_n)\| = |x_1| + \cdots + |x_n|$ in $\mathbb{R}^n$ and denote by λ the volume in $\mathbb{R}^n$.

Let E be a normed space and $E = E_1 \oplus E_2$ a splitting. Set $\gamma(E_1, E_2) = \max\{\|\pi_1\|, \|\pi_2\|\}$, where $\pi_i \colon E \to E_i$ for $i = 1, 2$ are the associated projections. We say that a subset $G \subset E$ is a (E_1, E_2)-*graph* if there exist an open set $U \subset E_2$ and a C^1 map $\psi \colon U \to E_1$ such that $G = \{(\psi(x), x) : x \in U\}$; the number

$$\sup\{\|\psi(x) - \psi(y)\|/\|x - y\| : x, y \in U\}$$

is called the *dispersion* of G.

Proposition 10.5.5. *For every sufficiently small $\varepsilon > 0$, there exists a function $\rho \colon M \to (0, 1)$ with $\log \rho \in L^1(M, \lambda)$, an integer $N > 0$, and a set K with $\nu(M \setminus K) < \varepsilon$ such that for every $x \in K$,*

$$h_\lambda(f^N, \rho, x) \geq N(\Sigma_+(x) - \varepsilon).$$

Proof of the proposition. We start with the following lemma:

Lemma 10.5.6. *If $\eta > \beta > 1$, $\alpha > 0$, and $c > 0$, then for every $\delta > 0$ satisfying*

$$\delta < \min\left\{\frac{\eta}{\alpha(1+c)}, \frac{c(\eta - \beta)}{\alpha(1+c)^2}\right\} \tag{10.19}$$

the following property holds. Let E be a finite-dimensional normed space, $E = E_1 \oplus E_2$ a splitting of E such that $\gamma(E_1, E_2) \leq \alpha$, and F an embedding of the ball $B(0, r)$ into another normed space E' such that

1. *d_0F is an isomorphism and $\gamma\big((d_0F)E_1, (d_0F)E_2\big) \leq \alpha$;*
2. *$\|d_0F - d_xF\| \leq \delta$ for all $x \in B(0, r)$;*
3. *$\|d_0Fv\| \geq \eta\|v\|$ for all $v \in E_2$;*
4. *$\|d_0F|E_1\| \leq \beta$.*

Then for every (E_1, E_2)-graph G with dispersion at most c, contained in the ball $B(0, r)$, the image $F(G)$ is a $\big((d_0F)E_1, (d_0F)E_2\big)$-graph with dispersion at most c.

Proof of the lemma. The proof is based on the standard graph transform technique; (see Sections 7.1 and 7.4). Let $G = \{(\psi(x), x) : x \in U\}$ be an (E_1, E_2)-graph with dispersion at most c. By Conditions 1 and 2, one can write F in the form

$$F(x, y) = \big(Lx + p(x, y), Ty + q(x, y)\big),$$

where $L = d_0F|E_1$, $T = d_0F|E_2$, and the partial derivatives of p and q have norm at most $\alpha\delta$. Then

$$F(G) = \big\{\big(L\psi(u) + p(\psi(u), u), Tu + q(\psi(u), u)\big) : u \in U\big\}.$$

To study this set, we define the function $\Phi \colon U \to (d_0F)E_2$ by

$$\Phi(u) = Tu + q(\psi(u), u).$$

If $u, v \in U$, then

$$\begin{aligned}\|\Phi(u)-\Phi(v)\| &\geq \|T(u-v)\| - \|q(\psi(u),u)-q(\psi(v),v)\| \\ &\geq \eta\|u-v\| - \alpha\delta(\|\psi(u)-\psi(v)\|+\|u-v\|) \\ &\geq (\eta-\alpha\delta(c+1))\|u-v\|.\end{aligned}$$

Therefore if $0 < \delta < \eta\alpha^{-1}(c+1)^{-1}$, then Φ is a homeomorphism onto $\Phi(U)$ and its inverse is Lipschitz with Lipschitz constant $(\eta - \alpha\delta(1+c))^{-1}$. Now define $\Psi\colon \Phi(U) \to (d_0F)E_1$ by

$$\Psi(u) = L\psi(\Phi^{-1}u) + p(\psi(\Phi^{-1}u), \Phi^{-1}u).$$

Clearly, $F(G) = \{(\Psi(u), u) : u \in \Phi(U)\}$. Observe that

$$(\Psi \circ \Phi)(u) = L\psi(u) + p(\psi(u), u).$$

Therefore,

$$\begin{aligned}&\|(\Psi\circ\Phi)(u) - (\Psi\circ\Phi)(v)\| \\ &\quad\leq \|L\|\cdot\|\psi(u)-\psi(v)\| + \|p(\psi(u),u)-p(\psi(v),v)\| \\ &\quad\leq \beta c\|u-v\| + \alpha\delta(\|\psi(u)-\psi(v)\|+\|u-v\|) \\ &\quad\leq (\beta c + \alpha\delta(c+1))\|u-v\|.\end{aligned}$$

Since $\Psi = (\Psi\circ\Phi)\circ\Phi^{-1}$, using (10.19) we conclude that the dispersion of Ψ is at most

$$\frac{\beta c + \alpha\delta(c+1)}{\eta - \alpha\delta(c+1)} < c.$$

This completes the proof of the lemma. □

For every regular point x of f, we define

$$F^s(x) = \bigoplus_{\chi_i(x)\leq 0} E_i(x) \quad\text{and}\quad F^u(x) = \bigoplus_{\chi_i(x)>0} E_i(x).$$

For each nonnegative integer j, let $\mathcal{R}^j$ be the set of regular points (see Section 3.1) x of f such that $\dim E^u(x) = j$. By the Multiplicative Ergodic Theorem 3.4.3 and Egorov's Theorem, for each j such that $\nu(\mathcal{R}^j) > 0$ there exist $K_j \subset \mathcal{R}^j$ with $\nu(K_j) > \nu(\mathcal{R}^j) - \varepsilon 2^{-j}$, real numbers $1 < \beta < \eta$, and an integer $N > 0$ such that the map $g = f^N$ has the following properties for every $x \in K_j$ and $n \geq 0$:

(P1) $\|dg^n v\| \geq \eta^n\|v\|$ for every $v \in E^u(x)$.
(P2) $\|dg^n | E^s(x)\| \leq \beta^n$.
(P3) $\log|\det(dg^n | E^u(x))| \geq nN(\Sigma_+(x) - \varepsilon)$.
(P4) The maps $x \mapsto E^u(x)$ and $x \mapsto E^s(x)$ are continuous on K_j.

Lemma 10.5.7. *For every $c > 0$, there exists $\xi > 0$ such that if $x \in K_j$ and $g^m(x) \in K_j$ for some $m \geq 0$, then every $(E^s(x), E^u(x))$-graph with dispersion at most c, which is contained in the ball $B(x, \xi^m)$, is taken by g^m into an $(E^s(g^m(x)), E^u(g^m(x)))$-graph with dispersion at most c.*

Proof of the lemma. Let $\alpha = \sup\{\gamma(E^s(x), E^u(x)) : x \in K_j\}$. We first apply Lemma 10.5.6 to this α and to η and β satisfying Properties (P1) and (P2). We claim that there exist $C > 0$ and $0 < t \leq 1$ such that

$$\|d_x g^m - d_y g^m\| \leq C^m \|x - y\|^t \tag{10.20}$$

for all $x, y \in M$ and $m \geq 1$. To prove this claim choose $C_0 > 0$ satisfying $\|d_x g - d_y g\| \leq C_0 \|x - y\|^t$ for all $x, y \in M$, where $t = \alpha^N$. This is possible because $x \mapsto d_x f$, and hence also $x \mapsto d_x g$ are Hölder continuous. Set $A = \sup\{\|d_x g\| : x \in M\}$. Choose $C > C_0$ such that for all $m \geq 0$,

$$C \geq A + C_0 \left(\frac{A^{t+1}}{C}\right)^m. \tag{10.21}$$

Then (10.20) holds for $m = 1$. Assume that (10.20) holds for $m = n$. Then using (10.21) we obtain that

$$\begin{aligned}
\|d_x g^{n+1} - d_y g^{n+1}\| &\leq \|d_{g^n x} g(d_x g^n - d_y g^n)\| + \|(d_{g^n x} g - d_{g^n y} g) d_y g^n\| \\
&\leq AC^n \|x - y\|^t + A^n C_0 \|g^n x - g^n y\|^t \\
&\leq (AC^n + A^n C_0 A^{nt}) \|x - y\|^t \\
&\leq C^{n+1} \|x - y\|^t.
\end{aligned}$$

Fix $\delta > 0$ and choose $0 < \xi < 1$ such that $(C\xi^t)^m < \delta$ for all $m \geq 1$. Then

$$\|d_x g^m - d_y g^m\| \leq \delta$$

for every $x \in M$, $y \in B(x, \xi^m)$, and $m \geq 1$. Assume that $\delta > 0$ satisfies (10.19); then it also satisfies (10.19) with η and β replaced by η^n and β^n. Therefore, we can apply Lemma 10.5.6 to obtain the desired result. □

For each $x \in \mathcal{R}^j$, we denote by $\ell(x)$ the least positive integer such that $g^{\ell(x)}(x) \in K_j$. We define a function $\ell \colon M \to \mathbb{N}$ by putting $\ell(x) = 0$ for $x \notin K_j$. The function ℓ is defined ν-almost everywhere and is integrable: if $W_k = \{x \in K_j : \ell(x) = k\}$ then we have

$$\bigcup_{n \geq 0} g^n(K_j) = \bigcup_{k=1}^{\infty} \bigcup_{i=0}^{k-1} g^i(W_k) \quad (\text{mod } 0),$$

where the sets $g^i(W_k)$ are disjoint, and

$$\int_M \ell \, d\nu = \sum_{k=1}^{\infty} k\nu(W_k) = \sum_{k=1}^{\infty} \sum_{i=0}^{k-1} \nu(g^i(W_k)) = \nu\left(\bigcup_{n \geq 0} g^n(K_j)\right) \leq 1.$$

Now choose a sufficiently small $c > 0$ such that there exists $a > 0$ with the following property:

> if $x \in K_j$, $y \in M$, and $d(x, y) < a$, then every subspace $E \subset T_y M$, which is a $(E^s(x), E^u(x))$-graph with dispersion at most c, satisfies
>
> $$\big|\log|\det(dg|E)| - \log|\det(dg|E^u(x))|\big| \leq \varepsilon. \tag{10.22}$$

Choose a number $\xi = \xi(c) > 0$ as in the statement of Lemma 10.5.7 and define a function $\rho\colon M \to (0,1)$ by

$$\rho(x) = \min\left\{a, \frac{k_1}{k_2}\xi^{\ell(x)}\right\}.$$

Since ℓ is integrable, $\log \rho$ is also integrable.

Notice that $\mathcal{R}^j$ is f-invariant. By Birkhoff's Ergodic Theorem, the function

$$\tau(x) = \lim_{n\to\infty} \frac{1}{n} \operatorname{card}\{0 \le k \le n : g^k x \in \mathcal{R}^j \setminus K_j\}$$

is defined for ν-almost all $x \in \mathcal{R}^j$, and

$$\begin{aligned}\varepsilon \ge \nu(\mathcal{R}^j \setminus K_j) &= \int_{\mathcal{R}^j\setminus K_j} \tau \, d\nu \\ &\ge \int_{\{x\in\mathcal{R}^j:\tau(x)>\sqrt{\varepsilon}\}} \tau \, d\nu > \sqrt{\varepsilon}\,\nu(\{x \in \mathcal{R}^j : \tau(x) > \sqrt{\varepsilon}\}).\end{aligned}$$

Therefore,

$$\nu(\{x \in \mathcal{R}^j : \tau(x) \le \sqrt{\varepsilon}\}) \ge 1 - \sqrt{\varepsilon}.$$

Then there exists a compact set $K_j' \subset K_j$ with $\nu(K_j') > \nu(K_j) - \varepsilon 2^{-j}$ and an integer $N_0 > 0$, such that if $n > N_0$, then

$$\operatorname{card}\{0 \le k \le n : g^k(x) \in \mathcal{R}^j \setminus K_j\} \le 2n\sqrt{\varepsilon} \tag{10.23}$$

for all $x \in K_j'$. Observe that there exists $B > 0$, satisfying

$$\lambda(S_n(g,\rho,x)) \le B \int_{E^s(x)} \lambda((y + E^u(x)) \cap S_n(g,\rho,x))\, d\lambda(y) \tag{10.24}$$

for every $n \ge 0$, where the symbol λ is also used for the Lebesgue measure on the subspaces $E^s(x)$ and $y + E^u(x)$.

For $x \in K_j$, write

$$D(x,r) = \{x + y_1 + y_2 : y_1 \in E^s(x),\, y_2 \in E^u(x),\, \|y_1\| \le r,\, \|y_2\| \le r\}$$

and let $k_2 > k_1 > 0$ and $r_1 > 0$ be such that

$$B(x, k_1 r) \subset D(x,r) \subset B(x, k_2 r)$$

for all $x \in K_j$ and $0 < r \le r_1$. For $y \in E^s(x)$, we denote by $\Lambda_n(y)$ the set of points $w \in y + E^u(x)$ such that

$$g^k(w) \in D(g^k(x), \rho(g^k(x))/k_1)$$

for all $0 \le k \le n$. This implies that

$$\Lambda_n(y) \supset (y + E^u(x)) \cap S_n(g,\rho,x). \tag{10.25}$$

Lemma 10.5.8. *If $g^n(x) \in K_j$, $y \in E^s(x)$, and $\Lambda_n(y) \ne \varnothing$, then $g^n(\Lambda_n(y))$ is a $(E^s(g^n(x)), E^u(g^n(x)))$-graph with dispersion at most c.*

Proof of the lemma. The property clearly holds for $n = 0$.

Suppose it holds for some $m > 0$. Then, if $g^n(x) \in K_j$, $g^k(x) \notin K_j$ for $m < k < n$, and $\Lambda_n(y) \neq \varnothing$, we have

$$g^n(\Lambda_n(y)) \subset g^{n-m}(g^m(\Lambda_m(y))) \cap D(g^n(x), \rho(g^n(x))/k_1).$$

But $g^m(\Lambda_m(y))$ is a $(E^s(g^m(x)), E^u(g^m(x)))$-graph with dispersion at most c, and

$$\begin{aligned} g^m(\Lambda_m(y)) &\subset D(g^m(x), \rho(g^m(x))/k_1) \\ &\subset B(g^m(x), k_2\rho(g^m(x))/k_1) \\ &\subset B(g^m(x), \xi^{N(g^m(x))}). \end{aligned}$$

Since $N(g^m(x)) = n - m$, it follows from Lemma 10.5.7 that $g^n(\Lambda_m(y)) = g^{n-m}(g^m(\Lambda_m(y)))$ is an $(E^s(g^n(x)), E^u(g^n(x)))$-graph with dispersion at most c, and the same holds for $g^n(\Lambda_n(y)) \subset g^n(\Lambda_m(y))$. □

Choose now $\kappa > 0$ such that every $(E^s(w), E^u(w))$-graph G with dispersion at most c, contained in $D(w, \rho(w))$ for each $w \in K_j$, has volume $\lambda(G) < \kappa$. It follows from Lemma 10.5.8 that

$$\kappa > \lambda(g^n(\Lambda_n(y))) = \int_{\Lambda_n(y)} |\det(dg^n|T_z\Lambda_n(y))|\, d\lambda(z),$$

for every $y \in E^s(x)$, $x \in g^{-n}(K_j)$, and $n \geq 0$. Let $S_n = \{0 \leq k < n : g^k(x) \in K_j\}$. Set

$$C = \sup\{\log|\det(df|E)| : p \in M, E \subset T_pM\}.$$

If $n \geq N_0$, it follows from (10.22) and (10.23) that for every $x \in K'_j$,

$$\begin{aligned} &\log|\det(dg^n|T_z\Lambda_n(y))| \\ &\quad= \sum_{k=0}^{n-1} \log|\det(dg|T_{g^k(z)}(g^k(\Lambda_n(y))))| \\ &\quad\geq \sum_{k\in S_n} \log|\det(dg|T_{g^k(z)}(g^k(\Lambda_n(y))))| - NC(n - \operatorname{card} S_n) \\ &\quad\geq \sum_{k\in S_n} \log|\det(dg|E^u(g^k(x)))| - \varepsilon n - NC(n - \operatorname{card} S_n) \\ &\quad\geq \sum_{k=0}^{n-1} \log|\det(dg|E^u(g^k(x)))| - \varepsilon n - 2NC(n - \operatorname{card} S_n) \\ &\quad\geq nN(\Sigma_+(x) - \varepsilon) - \varepsilon n - 2NC(n - \operatorname{card} S_n) \\ &\quad\geq nN(\Sigma_+(x) - \varepsilon) - \varepsilon n - 4NCn\sqrt{\varepsilon}. \end{aligned}$$

By (10.24), we obtain for every $x \in K'_j$,

$$\kappa > \lambda(\Lambda_n(y)) \exp[nN(\Sigma_+(x) - \varepsilon')],$$

where $\varepsilon' = \varepsilon + \varepsilon/N + 4C\sqrt{\varepsilon}$. It follows from (10.24) and (10.25) that

$$\lambda(S_n(g,\rho,x)) \le B \int_{E^s(x)} \lambda(\Lambda_n(y))\, d\lambda(y) \le B\kappa \exp[-nN(\Sigma_+(x) - \varepsilon')].$$

Therefore for every $x \in K'_j$,

$$h_\lambda(g,\rho,x) = \overline{\lim_{n\to\infty}} -\frac{1}{n} \log \lambda(S_n(g,\rho,x)) \ge N(\Sigma_+(x) - \varepsilon').$$

Set $K = \bigcup_j K'_j$. We have

$$\nu(M \setminus K) = \sum_{j=0}^{\infty} \nu(\mathcal{R}^j \setminus K'_j) \le 2\varepsilon.$$

This completes the proof of the proposition. □

There exists a subset $K' \subset K$ and a number $A > 0$ such that $\nu(K \setminus K') < \varepsilon$ and $\Sigma_+(x) \le A$ for every $x \in K'$. It follows from Propositions 10.5.2 and 10.5.5 that

$$\begin{aligned} Nh_\nu(f) = h_\nu(g) &\ge \int_M h_\lambda(g,\rho,x)\, d\nu(x) \\ &\ge \int_{K'} N(\Sigma_+(x) - \varepsilon)\, d\nu(x) \\ &= N\int_M \Sigma_+\, d\nu - N\int_{M\setminus K'} \Sigma_+\, d\nu - N\varepsilon\nu(K') \\ &\ge N\int_M \Sigma_+\, d\nu - 2\varepsilon NA - N\varepsilon. \end{aligned}$$

Thus,

$$h_\nu(f) \ge \int_M \Sigma_+\, d\nu - 2\varepsilon A - \varepsilon.$$

Since ε is arbitrary, we obtain the desired result. □

10.6 Notes

In [192], Ruelle established the upper bound for the measure-theoretic entropy of a C^1 map preserving an arbitrary Borel probability measure, in terms of the Lyapunov exponents. Earlier the same upper bound was found by Ruelle [191] for axiom A diffeomorphisms and the time-1 map of Axiom A flows. Margulis (unpublished) established the upper bound for volume preserving diffeomorphisms.

The entropy formula (see Theorem 10.4.1) was established by Pesin in [169]. A somewhat different proof can be found in [168]. The arguments presented here follow essentially the approach in [168]. Tahzibi [209] extended the entropy formula to an open and dense subset in the space of area preserving C^1 diffeomorphisms of a compact surface.

Let us point out that the positivity of topological entropy can sometimes be determined using pure topological information. For example, theorems of Manning [142], Misiurewicz and Przytycki [153, 154], and Yomdin [234, 235] relate the topological entropy to the action of the diffeomorphism on the homology groups (see also [99]).

11

Stable Ergodicity and Lyapunov Exponents. More Examples of Systems with Nonzero Exponents

In Chapter 6 we constructed some examples of systems with nonzero Lyapunov exponents by slowing down trajectories of Anosov systems. Replacing Anosov systems with uniformly partially hyperbolic systems allows greater flexibility and leads to new and interesting examples.

11.1 Uniform Partial Hyperbolicity and Stable Ergodicity

A C^2 diffeomorphism f preserving a smooth measure ν is said to be *stably ergodic* if there exists a C^1-open neighborhood $\mathcal{U}$ of f in $\mathrm{Diff}^1(M, \nu)$ such that any C^2 diffeomorphism $g \in \mathcal{U}$ is ergodic with respect to ν. Volume preserving Anosov diffeomorphisms are stably ergodic. We describe a class of partially hyperbolic diffeomorphisms that are stably ergodic.

A diffeomorphism f is said to be *(uniformly) partially hyperbolic* if there exist (a) numbers λ, λ', μ, and μ' such that $0 < \lambda < 1 < \mu$ and $\lambda < \lambda' \leq \mu' < \mu$; (b) numbers $C > 0$ and $K > 0$; and (c) subspaces $E^c(x)$, $E^s(x)$, and $E^u(x)$, $x \in M$, such that

1.
$$df E^c(x) = E^c(f(x)), \quad df E^s(x) = E^s(f(x)),$$
$$df E^u(x) = E^u(f(x))$$
and
$$T_x M = E^s(x) \oplus E^c(x) \oplus E^u(x);$$
2. for $v \in E^s(x)$ and $n \geq 0$,
$$\|df^n v\| \leq C\lambda^n \|v\|;$$
3. for $v \in E^u(x)$ and $n \leq 0$,
$$\|df^n v\| \leq C\mu^n \|v\|;$$

4. for $v \in E^c(x)$ and $n \geq 0$,

$$C^{-1}(\lambda')^n \|v\| \leq \|df^n v\| \leq C(\mu')^n \|v\|;$$

5. $\sphericalangle(E^c(x), E^s(x)) \geq K$, $\sphericalangle(E^c(x), E^u(x)) \geq K$, $\sphericalangle(E^s(x), E^u(x)) \geq K$.

In other words, f is uniformly partially hyperbolic if it is nonuniformly partially hyperbolic on M in the sense of Definition 5.2.2 with the functions $\lambda(x)$, $\lambda'(x)$, $\mu(x)$, and $\mu'(x)$ to be constant, the function $C(x)$ bounded from above, and the function $K(x)$ bounded from below.

We refer the reader to [174] for some basic information and examples of partially hyperbolic systems.

The subspaces $E^c(x)$, $E^s(x)$, and $E^u(x)$ are called, respectively, *central, stable,* and *unstable*. They depend continuously (and, in fact, Hölder continuously) on $x \in M$.

The stable and unstable distributions are uniquely integrable and generate two f-invariant foliations of M with smooth leaves, W^s and W^u (see Section 8.3). We denote by $V^s(x)$ and $V^u(x)$ the corresponding local stable and unstable manifolds; they depend continuously on $x \in M$; in particular, their sizes are uniformly bounded away from zero. Note that the central distribution may not be integrable (see [174]).

We point out that partially hyperbolic C^r diffeomorphisms, $r \geq 1$, form an open set in the space of $\text{Diff}^r(M)$.

In studying ergodic properties of diffeomorphisms, which are uniformly partially hyperbolic, an important role is played by the accessibility property that we now introduce.

We call two points $p, q \in M$ *accessible* if there are points $p = z_0, z_1, \dots, z_{\ell-1}$, $z_\ell = q$ such that $z_i \in W^u(z_{i-1})$ or $z_i \in W^s(z_{i-1})$ for $i = 1, \dots, \ell$. The collection of points $z_0, z_1, \dots, z_\ell$ is called a *us-path* connecting p and q, and is denoted by $[p, q] = [z_0, z_1, \dots, z_\ell]$. Accessibility is an equivalence relation. The diffeomorphism f is said to have the *accessibility property* if the partition into accessibility classes is trivial (i.e., any two points $p, q \in M$ are accessible) and to have the *essential accessibility property* if the partition into accessibility classes is ergodic (i.e., a measurable union of equivalence classes must have zero or full measure). We also say that f is *stably accessible* if there exists a C^1-open neighborhood $\mathcal{U}$ of f in $\text{Diff}^1(M)$ (or $\text{Diff}^1(M, \nu)$) such that any diffeomorphism $g \in \mathcal{U}$ is accessible. Finally, we call f *center bunched* if $\max\{\lambda, \mu\} < \lambda'/\mu'$.

Theorem 11.1.1 (Burns–Wilkinson [48]). *Let f be a C^2 diffeomorphism of a compact smooth manifold preserving a smooth measure ν. Assume that f is uniformly partially hyperbolic, essentially accessible, and center bunched. Then f is ergodic.*

This theorem is a generalization of an earlier result by Pugh and Shub [186] who required, in addition, the central distribution E^c to be integrable (as well as the distributions $E^s \oplus E^c$ and $E^u \oplus E^c$).

It is easy to see that center bunching is an open property in the space of C^1 diffeomorphisms. This implies the following result.

Theorem 11.1.2. *Under the conditions of Theorem 11.1.1 assume that f is stably accessible. Then f is stably ergodic.*

The assumption that f is center bunched is technical and can be dropped if $\dim E^c = 1$ (see [48, 189]).

We present some results on accessibility and stable accessibility. The first one shows that stable accessibility is C^1 dense in the space of partially hyperbolic diffeomorphisms of M, volume preserving or not.

Theorem 11.1.3 (Dolgopyat–Wilkinson [60]). *Let $f \in \mathrm{Diff}^r(M, \nu)$ (or $f \in \mathrm{Diff}^r(M)$) be partially hyperbolic. Then for every neighborhood $\mathfrak{U} \subset \mathrm{Diff}^1(M, \nu)$ (respectively, $\mathfrak{U} \subset \mathrm{Diff}^1(M)$) of f, there exists a C^r diffeomorphism $g \in \mathfrak{U}$, which is stably accessible.*

In the case $\dim E^c = 1$, the density statement can be ensured in the space $\mathrm{Diff}^r(M, \nu)$ (respectively, $\mathrm{Diff}^r(M)$) for any $r \geq 1$.

We consider some examples of partially hyperbolic systems.

1. Let M and N be compact smooth Riemannian manifolds and $f\colon M \to M$ a C^r Anosov diffeomorphism, $r \geq 1$. The diffeomorphism

$$F = f \times \mathrm{Id}_N\colon M \times N \to M \times N$$

is partially hyperbolic. Any small perturbation $G \in \mathrm{Diff}^r(M \times N)$ of F is partially hyperbolic; the central distribution E^c_G is integrable and the corresponding central foliation W^c_G has compact smooth leaves, which are diffeomorphic to N (see [86]). If f preserves a smooth measure ν on M then the map F preserves the measure $\nu \times m$, where m is the Riemannian volume in N.

In the particular case $N = S^1$ one has the following result.

Theorem 11.1.4 (Niţică–Török [160]). *There is a neighborhood $\mathfrak{U}$ of the map F in $\mathrm{Diff}^2(M \times S^1)$ or $\mathrm{Diff}^2(M \times S^1, \nu \times m)$ such that stable accessibility is open and dense in $\mathfrak{U}$.*

2. Let T^t be the special flow over a C^r Anosov diffeomorphism $f\colon M \to M$ with a roof function $H(x)$. The flow is an Anosov flow on the space that is diffeomorphic to the product $M \times [0, 1]$, where the manifolds $M \times \{0\}$ and $M \times \{1\}$ are identified by the action of f.

Theorem 11.1.5 (Bowen [34], Brin [35,36], Talitskaya [210]). *There exist $\varepsilon > 0$ and an open and dense set of C^r functions $\widetilde{H}\colon M \to \mathbb{R}^+$ with $|\widetilde{H}(x)| \leq 1$ such that the special flow T^t with the roof function*

$$H(x) = H_0 + \varepsilon \widetilde{H}(x)$$

($H_0 > 0$ is a constant) is stably accessible.

3. Let G be a compact Lie group and $f: M \to M$ a C^r Anosov diffeomorphism. Each C^r function $\varphi: M \to G$ defines a skew product transformation $f_\varphi: M \times G \to M \times G$, or *G-extension of f*, defined by the formula

$$F_\varphi(x, y) = (f(x), \varphi(x)y).$$

Left transformations are isometries of G in bi-invariant metric, and therefore, F_φ is partially hyperbolic. If f preserves a smooth probability measure ν then f_φ preserves the smooth probability measure $\nu \times \nu_G$, where ν_G is the (normalized) Haar measure on G.

Theorem 11.1.6 (Burns–Wilkinson [47]). *For every neighborhood $\mathfrak{U} \subset C^r(M, G)$ of the function φ, there exists a function $\psi \in \mathfrak{U}$ such that the diffeomorphism F_ψ is stably accessible.*

The proofs of Theorems 11.1.3–11.1.6 are based on the *quadrilateral argument* first introduced by Brin (see [36]). We describe this argument in Section 11.3 (see the proof of Proposition 11.3.2).

11.2 Partially Hyperbolic Systems with Nonzero Lyapunov Exponents

1. A somewhat different approach to the study of ergodic properties of uniformly partially hyperbolic systems preserving a smooth measure is based on taking into account the Lyapunov exponents along the central direction E^c. We consider the case when these central exponents are all negative on a subset of positive measure. Similarly, one can assume that central exponents are all positive on a subset of positive measure.

We first consider some examples of diffeomorphisms with negative (positive) central exponents. We begin with an example where the central direction is one dimensional. Let A be a linear hyperbolic automorphism of the two-dimensional torus $\mathbb{T}^2$. Consider the map $F = A \times \mathrm{Id}$ of the three-dimensional torus $\mathbb{T}^3 = \mathbb{T}^2 \times S^1$, where Id is the identity map. This map preserves volume and is uniformly partially hyperbolic. It has zero Lyapunov exponent in the central direction. Any sufficiently small C^1 perturbation G of F is also uniformly partially hyperbolic with one-dimensional central direction. The following result due to Shub and Wilkinson (see [200]) shows that this perturbation can be made in such a way to ensure negative (positive) Lyapunov exponents in the central direction.

Observe that if a diffeomorphism has positive Lyapunov exponents on a set of positive measure then the inverse diffeomorphism has negative Lyapunov exponents on this set.

Theorem 11.2.1. *For any $k \geq 2$ and $\delta > 0$, there exists a volume preserving C^∞ perturbation G of F such that G is δ-close to F in the C^k-topology and has negative (positive) central Lyapunov exponents on a set of positive measure.*

Proof. We use the approach in [58] that in turn is built on the argument in [196]. Fix a point $x \in \mathbb{T}^2$ and consider the disk D centered at x of a sufficiently small radius $r_0 > 0$. Let $\Omega = D \times (-\varepsilon_0, \varepsilon_0)$ be an open domain in $\mathbb{T}^3$. We introduce a coordinate system (x_1, x_2, y) in Ω where x_1 and x_2 run over stable and unstable lines of A respectively, $y \in (-\varepsilon_0, \varepsilon_0)$ and $x = (0, 0, 0)$.

Choose two nonnegative C^∞ functions $\rho = \rho(r)$, $0 < r < r_0$, and $\psi = \psi(y)$, $y \in (-\varepsilon_0, \varepsilon_0)$, which satisfy

1. $\rho(r) > 0$ if $0.2r_0 \leq r \leq 0.8r_0$ and $\rho(r) = 0$ if $0 < r \leq 0.1r_0$ or $0.9r_0 \leq r \leq r_0$;
2. $\psi(y) = \psi_0$ if $y \in (-0.8\varepsilon_0, 0.8\varepsilon_0)$, where $\psi_0 > 0$ is a constant; $\psi(y) = 0$ if $|y| \geq 0.9\varepsilon_0$;
3. $\|\rho\|_{C^k} \leq \delta$, $\|\psi\|_{C^k} \leq \delta$.

Define the map h_τ on Ω by

$$\begin{aligned} &h_\tau(x_1, x_2, y) \\ &\quad = (x_1, x_2\cos(2\pi\tau\alpha) - y\sin(2\pi\tau\alpha), x_2\sin(2\pi\tau\alpha) + y\cos(2\pi\tau\alpha)) \end{aligned} \tag{11.1}$$

with $\alpha(x_1, x_2, y) = \rho(x_1^2)\psi(x_2^2 + y^2)$ and set $h_\tau = \mathrm{Id}$ on $M \setminus \Omega$. For every sufficiently small τ, the map h_τ is a C^∞ volume preserving diffeomorphism of M.

We shall show that for any sufficiently small τ, the perturbation $G_\tau = F \circ h_\tau$ has nonzero Lyapunov exponents on a set of positive measure.

Denote by $E^s_\tau(w)$, $E^u_\tau(w)$, and $E^c_\tau(w)$ the stable, unstable, and central subspaces at a point $w \in \mathbb{T}^3$ for the map G_τ. Set $\kappa_\tau(w) = \|dG_\tau|E^u_\tau(w)\|$. The subspace $E^u_\tau(w)$ depends continuously on τ (for a fixed w) and hence so does κ_τ. Furthermore, $G_0 = F$, and hence $\kappa_0(w) = \eta$ where $\eta > 1$ is the eigenvalue of the matrix A. However, by Proposition 11.3.6, for all sufficiently small $\tau > 0$,

$$\int_{\mathbb{T}^3} \log \kappa_\tau(w)\, dw < \log \eta.$$

Denote by $\chi^s_\tau(w)$, $\chi^u_\tau(w)$, and $\chi^c_\tau(w)$ the Lyapunov exponents of G_τ at the point $w \in \mathbb{T}^3$ in the stable, unstable, and central directions respectively. (Since these directions are one dimensional, the Lyapunov exponents do not depend on the vector.) For $\tau > 0$, we have

$$\begin{aligned} \int_{\mathbb{T}^3} \chi^u_\tau(w)\, dw &= \int_{\mathbb{T}^3} \lim_{n\to\infty} \frac{1}{n} \log \prod_{i=0}^{n-1} \kappa_\tau(G^i_\tau(w))\, dw \\ &= \int_{\mathbb{T}^3} \log \kappa_\tau(w)\, dw < \log \eta. \end{aligned}$$

In view of (11.1), $E^s_\tau(w) = E^s_0(w) = E^s_F(w)$ for every τ and we conclude that $\chi^s_\tau(w) = -\log \eta$ for every $w \in M$. Since G_τ is volume preserving,

$$\int_{\mathbb{T}^3} \left(\chi^s_{i\tau}(w) + \chi^u_{i\tau}(w) + \chi^c_{i\tau}(w)\right) dw = 0.$$

It follows that

$$\int_{\mathbb{T}^3} \chi_\tau^c(w) \neq 0.$$

This completes the proof of the theorem. □

One can modify the construction in Theorem 11.2.1 to ensure that the perturbation G has the accessibility property and indeed, is stably accessible (see Proposition 11.3.2). In this case by Theorem 11.1.1 the map G is ergodic and hence has negative (positive) central Lyapunov exponents almost everywhere. Moreover, since G is center bunched, by Theorem 11.1.2, it is stably ergodic. In view of Theorem 11.2.4, any sufficiently small perturbation of G has nonzero Lyapunov exponents almost everywhere (see Section 11.7 for more details and more general results).

2. Our next example is a partially hyperbolic diffeomorphism with two-dimensional central distribution and negative central exponents (for a similar result in higher dimensions see [57]). Consider the four-dimensional torus $\mathbb{T}^4 = \mathbb{T}^2 \times \mathbb{T}^2$. Fix a point $z_0 = (x_0, y_0) \in \mathbb{T}^4$ with $x_0 \in T^2$ and $y_0 \in T^2$, and consider a small ball $B(z_0, r) \in \mathbb{T}^4$. Define the skew product map F by

$$F(z) = F(x, y) = (A(x), R_{\alpha(x)}(y)), \quad z = (x, y),$$

where $R_{\alpha(x)}$ is the translation by the vector $\alpha(x)$ and $\alpha\colon \mathbb{T}^4 \to \mathbb{R}$ a nonnegative C^∞ function, which is equal to zero outside $B(z_0, r)$ and is strictly positive otherwise.

By Theorem 11.1.6, one can choose the function $\alpha(x)$ to guarantee that the map F is accessible and indeed, stably accessible. By Theorem 11.1.1, F is ergodic. Moreover, since it is center bunched, by Theorem 11.1.2, it is stably ergodic.

Theorem 11.2.2. *There exists an arbitrarily small C^1 perturbation G of F such that G preserves a smooth measure on $\mathbb{T}^4$ and has negative central Lyapunov exponents almost everywhere.*

Proof. We elaborate on the approach in [59]. Since any sufficiently small perturbation G of F is ergodic, the central Lyapunov exponents of G are constant:

$$\chi_1^c(z, G) = \chi_1^c(G) \quad \text{and} \quad \chi_2^c(z, G) = \chi_2^c(G)$$

for almost every $z \in \mathbb{T}^4$. We assume that $\chi_1^c(G) \geq \chi_2^c(G)$.

We first construct a perturbation $\tilde{G}$ of F such that

$$\chi_1^c(\tilde{G}) + \chi_2^c(\tilde{G}) = \int \log \det(d\tilde{G}|E_{\tilde{G}}^c)(z)\, dz = -a < 0. \tag{11.2}$$

Choose a smooth coordinate system $\xi = \{\xi_1, \xi_2, \xi_3, \xi_4\}$ in a small neighborhood of a point z_0 such that

1. $dz = d\xi$;
2. $E_R^c(z_0) = \{\partial/\partial\xi_2, \partial/\partial\xi_3\}$, $E_R^s(z_0) = \partial/\partial\xi_1$, $E_R^u(z_0) = \partial/\partial\xi_4$.

Let $\psi(t)$ be a C^∞ function with compact support. Fix $0 < \gamma < 1$, set $\tau = \|\xi\|^2/\gamma^2$, and define the map $\varphi(\xi) = \varphi(\xi_1, \xi_2, \xi_3, \xi_4)$ by

$$(\xi_1 \cos(\varepsilon\psi(\tau)) + \xi_2 \sin(\varepsilon\psi(\tau)), -\xi_1 \sin(\varepsilon\psi(\tau)) + \xi_2 \cos(\varepsilon\psi(\tau)), \xi_3, \xi_4). \tag{11.3}$$

The proof of (11.2) is similar to the proof of Lemma 11.4.11 (with sufficiently small γ).

We further perturb the map $\widetilde{G}$ to obtain the desired map G. The proof of the following result is similar to the proof of Proposition 11.4.22.

Lemma 11.2.3. *For any positive ε_1 and ε_2, there exists a perturbation G of $\widetilde{G}$ such that $d_{C^1}(G, \widetilde{G}) \le \varepsilon_1$ and*

$$|\chi_1^c(G) - \chi_2^c(G)| \le \varepsilon_2.$$

If ε_1 and ε_2 are sufficiently small then $\chi_1^c(G) < 0$ and $\chi_2^c(G) < 0$ and the desired result follows. □

In view of Theorem 11.2.4, any sufficiently small perturbation of G has nonzero Lyapunov exponents almost everywhere.

3. The following result by Burns, Dolgopyat, and Pesin [42] shows that partially hyperbolic diffeomorphisms with the accessibility property and negative (positive) central exponents are stably ergodic.

Theorem 11.2.4. *Let f be a C^2 diffeomorphism of a compact smooth Riemannian manifold preserving a smooth measure ν. Assume that f is partially hyperbolic. Assume also that f is accessible and has negative central exponents at every point x of a set A of positive measure. Then*

1. *A has full measure; in particular, f has nonzero Lyapunov exponents almost everywhere;*
2. *f is stably ergodic;*
3. *every C^2 diffeomorphism g, which is sufficiently close to f in the C^1 topology, has nonzero Lyapunov exponents almost everywhere.*

Proof. We adopt the following notations:

1. $V^s(x)$ and $V^u(x)$ are for the local stable and unstable manifolds tangent to $E^s(x)$ and $E^u(x)$ respectively; since the stable and unstable subspaces depend continuously on $x \in M$, the local stable and unstable manifolds depend continuously on x and their sizes are bounded away from zero uniformly in x; furthermore, V^s and V^u are local leaves of the global stable and unstable foliations W^s and W^u.
2. $V^{sc}(x)$ is for the local stable manifold tangent to $E^{sc}(x)$; since the Lyapunov exponents in the center direction are negative, $V^{sc}(x)$ is defined for almost every $x \in A$; its size depends measurably on x.

3. $m^s(x)$, $m^u(x)$, and $m^{sc}(x)$ are for the leaf volumes on $V^s(x)$, $V^u(x)$, and $V^{sc}(x)$ respectively.

We split the proof into several steps:

Step I. Since Lyapunov exponents are invariant functions, without loss of generality we may assume that the set A is invariant under f. We first show that every ergodic component of $f|A$ is open (mod 0) and so is the set A. For almost every $x \in A$, the manifold $V^{cs}(x)$ is transverse to the unstable foliation W^u. This and the uniform size of local unstable manifolds $V^u(z)$ ensure that the set

$$P(x) = \bigcup_{y \in V^{sc}(x)} V^u(y)$$

is a neighborhood of x. Therefore the set

$$Q(x) = \bigcup_{n \in \mathbb{Z}} f^n(P(x))$$

is open in M and by Lemma 9.1.3, f is ergodic on $Q(x)$.

Step II. We first establish stable ergodicity of f under the additional assumption

$$\int_M \log \|df|E_f^c(x)\| \, d\nu(x) < 0 \tag{11.4}$$

(see Step V below for the general case).

Lemma 11.2.5. *There are a neighborhood $\mathcal{U}$ of f in* $\mathrm{Diff}^1(M, \nu)$ *and a constant $\alpha > 0$ with the following property. For any $g \in \mathcal{U}$, there is a subset $A_g \subset M$ with $\nu(A_g) > 0$ such that for every $x \in A_g$,*

$$\overline{\lim_{n\to\infty}} \frac{1}{n} \sum_{j=0}^{n-1} \log \|dg|E_g^c(g^j(x))\| \le -\alpha. \tag{11.5}$$

Proof of the lemma. Choose $\alpha > 0$ such that

$$\int_M \log \|df|E_f^c(x)\| \, d\nu(x) < -\alpha. \tag{11.6}$$

Since the central bundle E_g^c depends continuously on the diffeomorphism g in the C^1 topology, there is a neighborhood $\mathcal{U} \subset \mathrm{Diff}^1(M, \nu)$ of f such that for any $g \in \mathcal{U}$,

$$\int_M \log \|dg|E_g^c(x)\| \, d\nu(x) < -\alpha.$$

For $g \in \mathcal{U}$, let A_g be set of points x where the forward Birkhoff's average of $\log \|dg|E_g^c(\cdot)\|$ is defined and is less than $-\alpha$. It follows from Birkhoff's Ergodic Theorem that $\nu(A_g) > 0$. If $x \in A_g$, we have

$$\lim_{n\to\infty} \frac{1}{n} \sum_{j=0}^{n-1} \log \|dg|E_g^c(g^j(x))\| \le -\alpha.$$

This completes the proof. $\square$

Since

$$\frac{1}{n}\log\|dg^n|E^c_g(x)\| \leq \frac{1}{n}\sum_{j=0}^{n-1}\log\|dg|E^c_g(g^j(x))\|,$$

we see that g has negative central Lyapunov exponents for any $x \in A_g$. It follows that the set A_g is open (mod 0) and $g|A_g$ has at most countably many ergodic components, which are open (mod 0).

Step III. We introduce the important notion of σ-hyperbolic times. First we need the following technical lemma:

Lemma 11.2.6. *Let $A \geq c_2 \geq c_1 > 0$ be some numbers and $\theta = (c_2 - c_1)/(A - c_1)$. Given any real numbers $a_1, \dots, a_N \leq A$ with*

$$\sum_{j=1}^{N} a_j \geq c_2 N,$$

there exist $\ell > \theta N$ and $1 < n_1 < \cdots < n_\ell \leq N$ such that for every $0 \leq n < n_i$,

$$\sum_{j=n+1}^{N-i} a_j \geq c_1(n_i - n). \tag{11.7}$$

Proof of the lemma. For every $1 \leq n \leq N$, set

$$S(n) = \sum_{j=1}^{n}(a_j - c_1)$$

and also $S(0) = 0$. Define the numbers $1 < n_1 < \cdots < n_\ell \leq N$ to be the maximal sequence such that $S(n_i) \geq S(n)$ for every $0 \leq n < n_i$ and $i = 1, \dots, \ell$. Note that ℓ cannot be zero since $S(N) > S(0)$ and (11.7) follows. We shall show that $\ell > \theta N$. Observe that $S(n_i - 1) < S(n_{i-1})$, and hence $S(n_i) < S(n_{i-1}) + (A - c_1)$ for every $1 < i \leq \ell$. Moreover, $S(n_1) \leq (A - c_1)$ and $S(n_i) \geq S(N) \geq N(c_2 - c_1)$. This implies that

$$N(c_2 - c_1) \leq S(n_i) = \sum_{i=2}^{\ell}(S(n_i) - S(n_{i-1})) + S(n_1) < \ell(A - c_1)$$

and the desired result follows. □

Set $\sigma = \exp(-\alpha/3)$, where α is defined in (11.6). We call the number n a *σ-hyperbolic time for g at x* if

$$\|dg^j|E^c_g(g^{-n}(x))\| \leq \sigma^j, \tag{11.8}$$

for $0 \leq j \leq n$. The following result shows that if $g \in \mathcal{U}$ and $x \in A_g$, then there are infinitely many σ-hyperbolic times for g at x.

Lemma 11.2.7. *There is $\theta > 0$, which depends only on the number σ and the map f, such that for any $x \in A_g$ and any sufficiently large $N \geq 1$, there exist σ-hyperbolic times $1 \leq n_1 < \cdots < n_\ell \leq N$ for x with $\ell \geq \theta N$.*

Proof of the lemma. It follows from (11.5) that for all sufficiently large $N \geq 1$,

$$\sum_{j=0}^{N-1} \log \|dg|E_g^c(g^j(x))\| \leq -\frac{\alpha}{3}.$$

Now it suffices to set $c_1 = \alpha/3$, $c_2 = 2c_1$, $A = \sup_{x \in M} \left|\log \|dg|E_g^c(x)\|\right|$, and $a_j = -\log \|dg|E_g^c(g^{-j}(x))\|$ and to apply Lemma 11.2.6. □

Denote by $B^{sc}(y, r)$ the ball in $V^{sc}(y)$ centered at y of radius r. It follows from (11.8) that we can choose $r_0 > 0$ and a neighborhood $\mathcal{V} \subset \mathcal{U}$ of f (see Lemma 11.2.5) such that for every C^2 diffeomorphism $g \in \mathcal{V}$ and any σ-hyperbolic time n for g at x,

$$d_n = \operatorname{diam}(g^j(B^{sc}(g^{-n}(x), r_0))) \leq \sigma^{j/2} \quad \text{for } 0 \leq j \leq n.$$

Lemma 11.2.8. *There are a neighborhood $\mathcal{V}$ of f in* $\text{Diff}^1(M, \nu)$ *and a number $r_0 > 0$ such that if $g \in \mathcal{V}$ is a C^2 diffeomorphism and $x \in A_g$, then there is $n \geq 0$ such that the size of the global stable manifold $W^{cs}(g^{-n}(x))$ is at least r_0.*

Proof of the lemma. Since $\sigma < 1$, we can make d_n as small as we wish by choosing n to be a sufficiently large σ-hyperbolic time for g at x. In particular, we can ensure that $g^n(B^{sc}(g^{-n}(x), r_0))$ lies in the local stable manifold $V^{sc}(x)$. Then $B^{sc}(g^{-n}(x), r_0)$ is contained in the global stable manifold $W^{sc}(g^{-n}(x))$, as claimed. □

We now repeat the above argument for $g|A_g$. Instead of using $V^{sc}(x)$ to construct a neighborhood of x, we use $B^{sc}(g^{-n}(x), r_0)$ to construct a neighborhood of $g^{-n}(x)$, where n is a σ-hyperbolic time for g at x. Note that n can be chosen arbitrarily large due to Lemma 11.2.7. This gives us the following statement:

Lemma 11.2.9. *There is a positive number $\tilde{r}_0 = \tilde{r}_0(f)$ such that any ergodic component of $g|A_g$ contains a ball of radius $\tilde{r}_0$.*

Step IV. It remains to show that the set A_g has full measure and $g|A_g$ is ergodic.

Given $\varepsilon > 0$, we say that a diffeomorphism g is *ε-accessible* if for every open ball B of radius ε, the union of accessibility classes passing through B is M. An equivalent requirement is that the accessibility class of any point should enter every open ball of radius ε.

Lemma 11.2.10. *Assume that f has the accessibility property. Then the following properties hold for every $\varepsilon > 0$:*

1. *There exist $\ell > 0$ and $R > 0$ such that for any $p, q \in M$, one can find a us-path that starts at p, ends within distance $\varepsilon/2$ of q, and has at most ℓ legs, each of them with length at most R.*

2. *There exists a neighborhood* $\mathcal{U}$ *of* f *in the space* $\text{Diff}^2(M)$ *such that every* $g \in \mathcal{U}$ *is* ε*-accessible.*

Proof of the lemma. Let $q_1, \dots, q_N$ be an $\varepsilon/4$-net in M. For each $p \in M$ and each q_k, choose a us-path from p to q_k; let $\ell(p, k)$ be the number of legs and $R(p, k)$ the length of the longest leg in this path. Set

$$R(p) = \max_k R(p, k) \quad \text{and} \quad \ell(p) = \max_k \ell(p, k).$$

By continuity of the foliations W^u and W^s, every point $p \in M$ has a neighborhood $U(p)$ such that for each k, any point in $U(p)$ is joined to some point in $B(q_k, \varepsilon/4)$ by a us-path, which has at most $\ell(p)$ legs, each of length at most $2R(p)$. The sets $\{U(p)\}$ form an open cover of M. Let $\{U(p_1), \dots, U(p_m)\}$ be a finite subcover. Then $R = \max_i 2R(p_i)$ and $\ell = \max_i \ell(p_i)$ satisfy Statement 1.

The second statement follows from the first one and the continuous dependence of the leaves of W^u and W^s on g. □

Given $\varepsilon > 0$, we say that an orbit $\mathcal{O}(x) = \{f^n(x) : n \in \mathbb{Z}\}$ is *ε-dense* if the points of the orbit form an ε-net.

Lemma 11.2.11. *If* g *is* ε*-accessible, then almost every orbit is* ε*-dense.*

Proof of the lemma. It suffices to show that if B is an open ball of radius ε, then the orbit of almost every point enters B. To this end, let us call a point *good* if it has a neighborhood in which the orbit of almost every point enters B. We now wish to show that an arbitrary point p is good. Since g is ε-accessible, there is a us-path $[z_0, \dots, z_k]$ with $z_0 \in B$ and $z_k = p$. We shall show by induction on j that each point z_j is good. This is obvious for $j = 0$.

Now suppose that z_j is good. Then z_j has a neighborhood N such that $\mathcal{O}(x) \cap B \neq \varnothing$ for almost every $x \in N$. Let S be the subset of N, consisting of points with this property that are also both forward and backward recurrent. It follows from Poincaré's Recurrence Theorem that S has full measure in N. If $x \in S$, any point $y \in W^s(x) \cup W^u(x)$ has the property that $\mathcal{O}(y) \cap B \neq \varnothing$. The absolute continuity of the foliations W^s and W^u means that the set

$$\bigcup_{x \in S} W^s(x) \cup W^u(x)$$

has full measure in the set

$$\bigcup_{x \in N} W^s(x) \cup W^u(x).$$

The latter is a neighborhood of z_{j+1}. Hence, z_{j+1} is good. □

It follows from Lemmas 11.2.9 to 11.2.11 that the map f satisfying Condition (11.4) is stably ergodic.

Step V. We proceed with the proof of the theorem and establish Condition (11.4). If f is essentially accessible, the accessibility class of almost every point is dense in M. With minor modifications, the proof of Lemma 11.2.11 shows that this property implies that almost every point has a dense orbit. It follows that the set A has full measure and f is ergodic. Hence there exists $\beta > 0$ such that for almost every $x \in M$,

$$\lim_{n\to+\infty} \frac{1}{n} \log \|df^n | E^c_f(x)\| \leq -\beta.$$

Integrating over M we obtain

$$\lim_{n\to\infty} \frac{1}{n} \int_M \log \|df^n | E^c_f(x)\| \, d\nu(x) \leq -\beta.$$

In particular, there exists $n_0 > 0$ such that

$$\int_M \log \|df^{n_0} | E^c_f(x)\| \, d\nu(x) < 0.$$

Hence, f^{n_0} satisfies the Condition (11.4) and therefore is stably ergodic. It follows that the map f itself is stably ergodic. □

11.3 Hyperbolic Diffeomorphisms with Countably Many Ergodic Components

We present the example by Dolgopyat, Hu, and Pesin [58] of a diffeomorphism with nonzero Lyapunov exponents with respect to a smooth invariant measure, which has countably many ergodic components.

Theorem 11.3.1. *There exists a C^∞ volume preserving diffeomorphism f of the three-dimensional torus $\mathbb{T}^3$ such that f has nonzero Lyapunov exponents almost everywhere and has countably many ergodic components that are open* (mod 0).

Let $A\colon \mathbb{T}^2 \to \mathbb{T}^2$ be a linear hyperbolic automorphism. Passing if necessary to a power of A we may assume that A has at least two fixed points p and p'. Consider the map $F = A \times \mathrm{Id}$ of the three-dimensional torus $\mathbb{T}^3 = \mathbb{T}^2 \times S^1$. We will perturb F to obtain the desired map f.

Consider a countable collection of intervals $\{I_n\}_{n=1}^\infty$ on the circle S^1, where

$$I_{2n} = [(n+2)^{-1}, (n+1)^{-1}] \quad \text{and} \quad I_{2n-1} = [1-(n+1)^{-1}, 1-(n+2)^{-1}].$$

Clearly, $\bigcup_{n=1}^\infty I_n = (0,1)$ and the sets $\mathrm{int}(I_n)$ are pairwise disjoint.

Set $I = [0,1]$. By Proposition 11.3.2, for each n one can construct a C^∞ volume preserving ergodic diffeomorphism $f_n\colon \mathbb{T}^2 \times I \to \mathbb{T}^2 \times I$, which satisfies

1. $\|F - f_n\|_{C^n} \leq e^{-n^2}$;
2. $d^m f_n | \mathbb{T}^2 \times \{z\} = d^m F | \mathbb{T}^2 \times \{z\}$ for $z = 0, 1$ and all $0 \leq m < \infty$;
3. f_n has nonzero Lyapunov exponents almost everywhere.

Let $L_n\colon I_n \to I$ be the affine map and $\pi_n = (\mathrm{Id}, L_n)\colon \mathbb{T}^2 \times I_n \to \mathbb{T}^2 \times I$. We define the map f by setting $f | \mathbb{T}^2 \times I_n = \pi_n^{-1} \circ f_n \circ \pi_n$ for all n and $f | \mathbb{T}^2 \times \{0\} =$

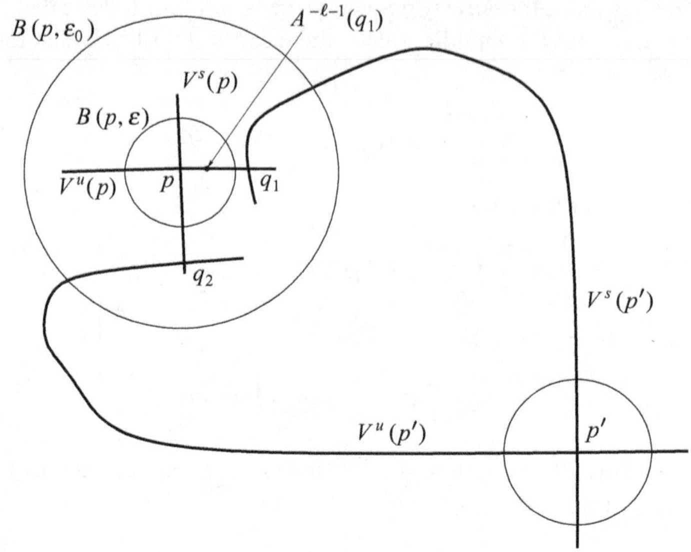

Figure 11.1. To the proof of Proposition 11.3.2.

$F|\mathbb{T}^2 \times \{0\}$. Note that for every $n > 0$ and $0 \leq m \leq n$, we have

$$\begin{aligned}&\|d^m F|\mathbb{T}^2 \times I_n - \pi_n^{-1} \circ d^m f_n \circ \pi_n\|_{C^n}\\ &\leq \|\pi_n^{-1} \circ (d^m F - d^m f_n) \circ \pi_n\|_{C^n} \leq e^{-n^2} \cdot (n+1)^n \to 0\end{aligned}$$

as $n \to \infty$. It follows that f is C^∞ on M and has the required properties.

Proposition 11.3.2. *For any $k \geq 2$ and $\delta > 0$, there exists a map g of the three-dimensional manifold $M = \mathbb{T}^2 \times I$ such that*

1. *g is a C^∞ volume preserving diffeomorphism of M;*
2. *$\|F - g\|_{C^k} \leq \delta$;*
3. *$d^m g|\mathbb{T}^2 \times \{z\} = d^m F|\mathbb{T}^2 \times \{z\}$ for $z = 0, 1$ and all $0 \leq m < \infty$;*
4. *g is ergodic and has nonzero Lyapunov exponents almost everywhere.*

Proof. We achieve this result by arranging two perturbations of the map F to ensure nonzero *average* central exponent $\int_M \chi^c(x)\,dx \neq 0$ (see Proposition 11.3.6) and the essential accessibility property (see Proposition 11.3.3).

Let $\eta > 1$ and η^{-1} be the eigenvalues of A. Choose a small number $\varepsilon_0 > 0$ such that $d(p, p') \geq 3\varepsilon_0$, and consider the local stable and unstable one-dimensional manifolds $V^s(p)$, $V^u(p)$, $V^s(p')$, and $V^u(p')$ of "size" ε_0.

Let us choose the smallest positive number n_1 such that the intersection $A^{-n_1}(V^s(p')) \cap V^u(p) \cap B(p, \varepsilon_0)$ consists of a single point, which we denote by q_1. Similarly, we choose the smallest positive number n_2 such that the intersection $A^{n_2}(V^u(p')) \cap V^s(p) \cap B(p, \varepsilon_0)$ consists of a single point, which we denote by q_2 (see Fig. 11.1).

Given a sufficiently small $\varepsilon \in (0, \varepsilon_0)$, $\varepsilon \leq (1/2)\min\{d(p, q_1), d(p, q_2)\}$, there is $\ell \geq 2$ such that (see Fig. 11.1)

$$A^{-\ell}(q_1) \notin B(p, \varepsilon) \quad \text{and} \quad A^{-\ell-1}(q_1) \in B(p, \varepsilon). \tag{11.9}$$

We now choose $\varepsilon' \in (0, \varepsilon)$ such that $A^{-\ell-1}(q_1) \in B(p, \varepsilon')$.

Finally, we assume ε to be so small that for some $q \in \mathbb{T}^2$,

$$B(p, \varepsilon) \cap (A^{-n_1}(V^s(p')) \cup A^{n_2}(V^u(p'))) = \varnothing,$$

and for $i = 1, \ldots, N$,

$$A^i(B(q, \varepsilon)) \cap B(q, \varepsilon) = \varnothing \quad \text{and} \quad A^i(B(q, \varepsilon)) \cap B(p, \varepsilon) = \varnothing,$$

where the number $N > 0$ will be determined later and $\varepsilon = \varepsilon(N)$.

Set $\Omega_1 = B(p, \varepsilon_0) \times I$ and $\Omega_2 = B^{uc}(\bar{q}, \varepsilon_0) \times B^s(\bar{q}, \varepsilon_0)$, where $\bar{q} = (q, 1/2)$, and the sets $B^{uc}(\bar{q}, \varepsilon_0) \subset V^u(q) \times I$ and $B^s(\bar{q}, \varepsilon_0) \subset V^s(q)$ are balls of radius ε_0 about $\bar{q}$.

After this preliminary consideration, we describe the construction of the map g. Consider the coordinate system in Ω_1 originated at $(p, 0) \in M$ with x-, y-, and z-axes to be unstable, stable, and central directions respectively for the map F. If a point $w = (x, y, z) \in \Omega_1$ and $F(w) \in \Omega_1$ then $F(w) = (\eta x, \eta^{-1} y, z)$.

Choose a C^∞ function $\xi : I \to \mathbb{R}^+$, satisfying

1. $\xi(z) > 0$ on $(0, 1)$;
2. $d^i\xi(0) = d^i\xi(1) = 0$ for $i = 0, 1, \ldots, k$;
3. $\|\xi\|_{C^k} \leq \delta$.
   We also choose two C^∞ functions $\varphi = \varphi(x)$ and $\psi = \psi(y)$, which are defined on the interval $(-\varepsilon_0, \varepsilon_0)$ and satisfy:
4. $\varphi(x) = \varphi_0$ if $x \in (-\varepsilon', \varepsilon')$ and $\psi(y) = \psi_0$ if $y \in (-\varepsilon', \varepsilon')$, where φ_0 and ψ_0 are positive constants;
5. $\varphi(x) = 0$ if $|x| \geq \varepsilon$; $\psi(y) \geq 0$ for any y and $\psi(y) = 0$ if $|y| \geq \varepsilon$;
6. $\|\varphi\|_{C^k} \leq \delta$, $\|\psi\|_{C^k} \leq \delta$;
7. $\int_0^{\pm\varepsilon} \varphi(s)\, ds = 0$.
   We now define the vector field X on Ω_1 by
   $$X(x, y, z) = \left(-\psi(y)\xi'(z) \int_0^x \varphi(s) ds, 0, \psi(y)\xi(z)\varphi(x)\right).$$
   It is straightforward to check that X is a divergence-free vector field supported on $(-\varepsilon, \varepsilon) \times (-\varepsilon, \varepsilon) \times I$.
   We define the map h_t on Ω_1 to be the time-t map of the flow generated by X and we set $h_t = \mathrm{Id}$ on the complement of Ω_1. We have that h_t is a C^∞ diffeomorphism of M, which preserves volume and the y-coordinate (the stable direction for the map F).
   Consider now the coordinate system in Ω_2 originated at $(q, 1/2)$ with x-, y-, and z-axes to be unstable, stable, and central directions respectively. We then switch to the cylindrical coordinate system (r, θ, y), where $x = r\cos\theta$, $y = y$, and $z = r\sin\theta$.

Consider a C^∞ function $\rho\colon(-\varepsilon_0,\varepsilon_0)\to\mathbb{R}^+$, satisfying

8. $\rho(r)>0$ if $0.2\varepsilon'\le r\le 0.9\varepsilon$ and $\rho(r)=0$ if $r\le 0.1\varepsilon'$ or $r\ge\varepsilon$;
9. $\|\rho\|_{C^k}\le\delta$.

For a sufficiently small τ, define the map $\widetilde{h}_\tau$ on Ω_2 by (compare to (11.1))

$$\widetilde{h}_\tau(r,\theta,y)=(r,\theta+\tau\psi(y)\rho(r),y) \tag{11.10}$$

and set $\widetilde{h}_\tau=\mathrm{Id}$ on $M\setminus\Omega_2$. The map $\widetilde{h}_\tau$ is a C^∞ volume preserving diffeomorphism of M.

Let us set $g=g_{t\tau}=h_t\circ F\circ\widetilde{h}_\tau$. For all sufficiently small $t>0$ and τ, the map $g_{t\tau}$ is C^k-close to F and hence is a partially hyperbolic C^∞ diffeomorphism of M. Clearly, it is center bunched. It also preserves volume in M. We shall show below (see Proposition 11.3.3) that g has the essential accessibility property. Therefore by Theorem 11.1.1, g is ergodic. It remains to explain that g has nonzero Lyapunov exponents almost everywhere.

Denote by $E^s_{t\tau}(w)$, $E^u_{t\tau}(w)$, and $E^c_{t\tau}(w)$ the stable, unstable, and central subspaces at a point $w\in M$ for the map $g_{t\tau}$. Set $\kappa_{t\tau}(w)=dg_{t\tau}|E^u_{t\tau}(w)$, $w\in M$. By Proposition 11.3.6, for all sufficiently small $\tau>0$,

$$\int_M\log\kappa_{0\tau}(w)\,dw<\log\eta.$$

The subspace $E^u_{t\tau}(w)$ depends continuously on t and τ (for a fixed w) and hence, so does $\kappa_{t\tau}$. It follows that for all sufficiently small $\tau>0$, there is $t>0$ such that

$$\int_M\log\kappa_{t\tau}(w)\,dw<\log\eta.$$

Denote by $\chi^s_{t\tau}(w)$, $\chi^u_{t\tau}(w)$, and $\chi^c_{t\tau}(w)$ the Lyapunov exponents of $g_{t\tau}$ at the point $w\in M$ in the stable, unstable, and central directions respectively. (Since these directions are one dimensional, the Lyapunov exponents do not depend on the vector.) By ergodicity of $g_{t\tau}$ and Birkhoff's Ergodic Theorem, we have that for almost every $w\in M$,

$$\chi^u_{t\tau}(w)=\lim_{n\to\infty}\frac{1}{n}\log\prod_{i=0}^{n-1}\kappa_{t\tau}(g^i_{t\tau}(w))=\int_M\log\kappa_{t\tau}(w)\,dw<\log\eta.$$

Since $E^s_{t\tau}(w)=E^s_{00}(w)=E^s_F(w)$ for every t and τ, we conclude that $\chi^s_{t\tau}(w)=-\log\eta$ for almost every $w\in M$. Since $g_{t\tau}$ is volume preserving,

$$\chi^s_{t\tau}(w)+\chi^u_{t\tau}(w)+\chi^c_{t\tau}(w)=0$$

for almost every $w\in M$. It follows that $\chi^c_{t\tau}(w)\ne 0$ for almost every $w\in M$, and hence $g_{t\tau}$ has nonzero Lyapunov exponents almost everywhere. This completes the proof of the proposition. □

We shall establish the essential accessibility property.

Proposition 11.3.3. *For every sufficiently small $t > 0$ and $\tau \geq 0$, the map $g_{t\tau}$ has the essential accessibility property.*

Proof. Given a point $w \in M$, denote by $\mathcal{A}(w)$ the accessibility class of w, that is, the set of points $q \in M$ such that w and q are accessible. Set $I_p = \{p\} \times (0, 1)$.

Lemma 11.3.4. *For every $z \in (0, 1)$,*

$$\mathcal{A}(p, z) \supset I_p. \tag{11.11}$$

Proof of the lemma. We use the coordinate system (x, y, z) in Ω_1 described above. Since the map h_t preserves the center leaf I_p, we have that for $z \in (0, 1)$,

$$h_t(0, 0, z) = (h_t^{(1)}(0, 0, z), h_t^{(2)}(0, 0, z), h_t^{(3)}(0, 0, z)) = (0, 0, h_t^{(3)}(0, 0, z)).$$

It suffices to show that for every $z \in (0, 1)$,

$$\mathcal{A}(p, z) \supset \{(p, a) : a \in [(h_t^{-\ell})^{(3)}(p, z), z]\}, \tag{11.12}$$

where ℓ is chosen by (11.9). In fact, since accessibility is a transitive relation and $h_t^{-n}(p, z) \to (p, 0)$ for any $z \in (0, 1)$, Condition (11.12) implies that $\mathcal{A}(p, z) \supset \{(p, a) : a \in (0, z]\}$. Since this holds true for all $z \in (0, 1)$ and accessibility is a reflexive relation, we obtain (11.11).

Now we proceed with the proof of (11.12).

Denote by $V_{t\tau}^s(p)$ and $V_{t\tau}^u(p)$ the local stable and unstable manifolds at p for the map $g_{t\tau}$. Let $q_1 \in V_{t\tau}^u(p)$ and $q_2 \in V_{t\tau}^s(p)$ be two points constructed above (see Fig. 11.1). The intersection $V_{t\tau}^s(q_1) \cap V_{t\tau}^u(q_2)$ is nonempty and consists of a single point q_3. We will show that for any $z_0 \in (0, 1)$, there exist $z_i \in (0, 1)$, $i = 1, 2, 3, 4$, such that

$$(q_1, z_1) \in V_{t\tau}^u((p, z_0)), \quad (q_3, z_3) \in V_{t\tau}^s((q_1, z_1)),$$
$$(q_2, z_2) \in V_{t\tau}^u((q_3, z_3)), \quad (p, z_4) \in V_{t\tau}^s((q_2, z_2))$$

and

$$z_4 \leq (h_t^{-\ell})^{(3)}(p, z_0) \tag{11.13}$$

(see Fig.11.2). This will imply that $(p, z_4) \in \mathcal{A}(p, z_0)$. By continuity, we conclude that

$$\{(p, a) : a \in [z_4, z_0]\} \subset \mathcal{A}(p, z_0)$$

and (11.12) will follow.

Since $g_{t\tau}$ preserves the xz-plane, we have $V_{t\tau}^{uc}((p, z_0)) = V_F^{uc}((p, z_0))$. Hence, there is a unique $z_1 \in (0, 1)$ such that $(q_1, z_1) \in V_{t\tau}^u((p, z_0))$. Notice that for $n \leq \ell$,

$$g_{t\tau}^{-n}(p, z_0) = (p, h_t^{-n}((p, z_0))) \quad \text{and} \quad g_{t\tau}^{-n}(q_1, z_1) = (A^{-n}q_1, z_1).$$

This is true because the points $A^{-n}q_1$, $n = 0, 1, \ldots, \ell$, lie outside the ε-neighborhood of I_p, where the perturbation map $h_t = \mathrm{Id}$. Similarly, since the

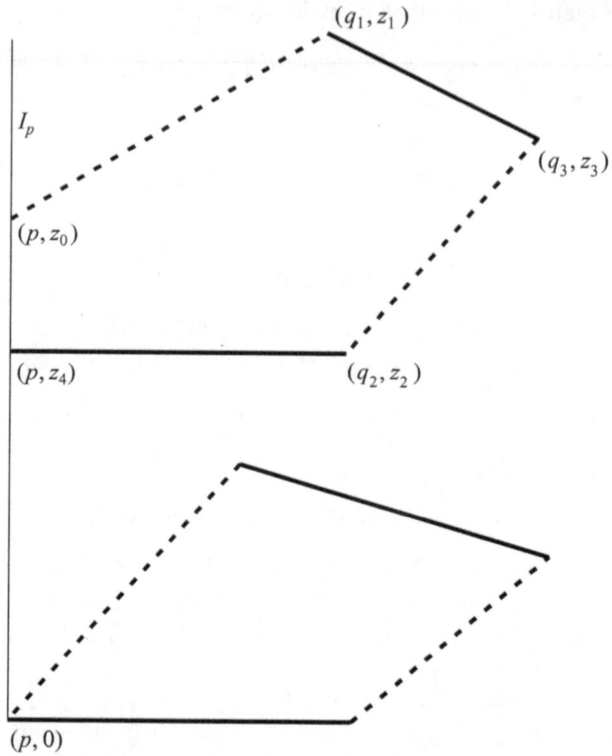

Figure 11.2. To the proof of Lemma 11.3.4: bold lines are stable manifolds and dotted lines are unstable manifolds.

points $A^{-n}q_1, n > \ell$, lie inside the ε'-neighborhood of I_p and the third component of h_t depends only on the z-coordinate, we have

$$g_{t\tau}^{-n}(q_1, z_1) = (A^{-n}q_1, h_t^{-n+\ell}z_1).$$

Since

$$d(g_{t\tau}^{-n}((p, z_0)), g_{t\tau}^{-n}((q_1, z_1))) \to 0$$

as $n \to \infty$, we have

$$d(h_t^{-n}((p, z_0)), h_t^{-n+\ell}((p, z_1))) \to 0$$

as $n \to \infty$. It follows that $z_1 = (h_t^{-\ell})^{(3)}((p, z_0))$.

By the construction of the map h_t (i.e., $h_t = \mathrm{Id}$ outside Ω_1), the sets $A^{-n_1}V_{t\tau}^s(p')$ and $A^{n_2}V_{t\tau}^u(p')$ are pieces of horizontal lines, and hence $z_2 = z_3 = z_1$. Since the third component of h_t does not decrease when a point moves from (q_2, z_2) to (p, z_4) along $V_{t\tau}^s(p)$, we conclude that $z_4 \leq z_3 = z_1 = (h_t^{-\ell})^3(p, z_0)$ and thus (11.13) holds. □

The essential accessibility property follows from Lemma 11.3.4 and the following statement:

Lemma 11.3.5. *Assume that any two points in I_p are accessible. Then the map $g_{t\tau}$ satisfies the essential accessibility property.*

Proof of the lemma. It is easy to see that for any two points $x, y \in M$, which do not lie on the boundary of M, one can find points $x', y' \in I_p$ such that the points x and x' are accessible and so are the points y and y'. By Lemma 11.3.4, the points x' and y' are accessible. Since accessibility is a transitive relation, the result follows. □

This completes the proof of the proposition. □

We now show that for all sufficiently small τ, the map $g_{0\tau}$ has nonzero average Lyapunov exponent in the central direction. Since this map is ergodic, this implies that $g_{0\tau}$ has nonzero Lyapunov exponents almost everywhere.

Proposition 11.3.6. *For any sufficiently small $\tau > 0$,*

$$\int_M \log \kappa_{0\tau}(w) dw < \log \eta.$$

Proof. For any $w \in M$, we introduce the coordinate system in $T_w M$ associated with the splitting $E^u_F(w) \oplus E^s_F(w) \oplus E^c_F(w)$. Given $\tau \geq 0$ and $w \in M$, there exists a unique number $\alpha_\tau(w)$ such that the vector $v_\tau(w) = (1, 0, \alpha_\tau(w))^\perp$ lies in $E^u_{0\tau}(w)$ (where $\perp$ denotes the transpose). Since the map $\widetilde{h}_\tau$ preserves the y-coordinate, by the definition of the function $\alpha_\tau(w)$, one can write the vector $dg_{0\tau}(w)v_\tau(w)$ in the form

$$dg_{0\tau}(w)v_\tau(w) = (\bar{\kappa}_\tau(w), 0, \bar{\kappa}_\tau(w)\alpha_\tau(g_{t0}(w)))^\perp \tag{11.14}$$

for some $\bar{\kappa}_\tau(w) > 1$. Taking into account that the expanding rate of $dg_{0\tau}(w)$ along its unstable direction is $\kappa_{0\tau}(w)$, we obtain that

$$\kappa_{0\tau}(w) = \bar{\kappa}_\tau(w) \frac{\sqrt{1 + \alpha_\tau(g_{0\tau}(w))^2}}{\sqrt{1 + \alpha_\tau(w)^2}}.$$

Since $E^u_{0\tau}(w)$ is close to $E^u_{00}(w)$, the function $\alpha_\tau(w)$ is uniformly bounded. Using the fact that the map $g_{0\tau}$ preserves volume, we find that

$$L_\tau = \int_M \log \kappa_{0\tau}(w)\, dw = \int_M \log \bar{\kappa}_\tau(w)\, dw. \tag{11.15}$$

Consider the map $\widetilde{h}_\tau$. Since it preserves the y-coordinate using (11.10), we can write that

$$\widetilde{h}_\tau(x, y, z) = (r \cos \sigma, y, r \sin \sigma),$$

where $\sigma = \sigma(\tau, r, \theta, y) = \theta + \tau\psi(y)\rho(r)$. Therefore the differential

$$d\widetilde{h}_\tau \colon E^u_F(w) \oplus E^c_F(w) \to E^u_F(g_{0\tau}(w)) \oplus E^c_F(g_{0\tau}(w))$$

can be written in the matrix form

$$d\widetilde{h}_\tau(w) = \begin{pmatrix} A(\tau, w) & B(\tau, w) \\ C(\tau, w) & D(\tau, w) \end{pmatrix}$$
$$= \begin{pmatrix} r_x \cos\sigma - r\sigma_x \sin\sigma & r_y \cos\sigma - r\sigma_y \sin\sigma \\ r_x \sin\sigma + r\sigma_x \cos\sigma & r_y \sin\sigma + r\sigma_y \cos\sigma \end{pmatrix},$$

where

$$r_x = \frac{\partial r}{\partial x} = \frac{x}{r} = \cos\theta, \quad r_z = \frac{\partial r}{\partial z} = \frac{y}{r} = \sin\theta,$$

$$\sigma_x = \frac{\partial \sigma}{\partial x} = \frac{-z}{r^2} + \frac{z}{r}\tau\widetilde{\rho}_r(y, r) = \frac{\sin\theta}{r} + \tau\widetilde{\rho}_r(y, r)\cos\theta,$$

$$\sigma_z = \frac{\partial \sigma}{\partial z} = \frac{x}{r^2} + \frac{x}{r}\tau\widetilde{\rho}_r(y, r) = \frac{\cos\theta}{r} + \tau\widetilde{\rho}_r(y, r)\sin\theta,$$

and $\widetilde{\rho}(y, r) = \psi(y)\rho(r)$. It is easy to check that

$$\begin{aligned} A = A(\tau, w) &= 1 - \tau r\widetilde{\rho}_r \sin\theta\cos\theta - \frac{\tau^2\widetilde{\rho}^2}{2} - \tau^2 r\widetilde{\rho}\widetilde{\rho}_r \cos^2\theta + O(\tau^3), \\ B = B(\tau, w) &= -\tau\widetilde{\rho} - \tau r\widetilde{\rho}_r \sin^2\theta - \tau^2 r\widetilde{\rho}\widetilde{\rho}_r \sin\theta\cos\theta + O(\tau^3), \\ C = C(\tau, w) &= \tau\widetilde{\rho} + \tau r\widetilde{\rho}_r \cos^2\theta - \tau^2 r\widetilde{\rho}\widetilde{\rho}_r \sin\theta\cos\theta + O(\tau^3), \\ D = D(\tau, w) &= 1 + \tau r\widetilde{\rho}_r \sin\theta\cos\theta - \frac{\tau^2\widetilde{\rho}^2}{2} - \tau^2 r\widetilde{\rho}\widetilde{\rho}_r \sin^2\theta + O(\tau^3). \end{aligned} \tag{11.16}$$

Lemma 11.3.7.

$$L_\tau = \log\eta - \int_M \log(D(\tau, w) - \eta B(\tau, w)\alpha_\tau(g_{0\tau}(w)))dw.$$

Proof of the lemma. Since $g_{0\tau} = h_0 \circ F \circ \widetilde{h}_\tau = F \circ \widetilde{h}_\tau$, we have that

$$\mathcal{D}_\tau(w) = dg_{0\tau}(w)|E^u_{0\tau}(w) \oplus E^c_{0\tau}(w) = \begin{pmatrix} \eta A(\tau, w) & \eta B(\tau, w) \\ C(\tau, w) & D(\tau, w) \end{pmatrix}.$$

By (11.14),

$$\begin{aligned} \mathcal{D}_\tau(w)\begin{pmatrix} 1 \\ \alpha_\tau(w) \end{pmatrix} &= \begin{pmatrix} \eta A(\tau, w) + \eta B(\tau, w)\alpha_\tau(w) \\ C(\tau, w) + D(\tau, w)\alpha_\tau(w) \end{pmatrix} \\ &= \begin{pmatrix} \kappa_\tau(w) \\ \kappa_\tau(w)\alpha_\tau(g_{0\tau}(w)) \end{pmatrix}. \end{aligned} \tag{11.17}$$

Since $\widetilde{h}_\tau$ is volume preserving, $AD - BC = 1$ and therefore,

$$A + B\alpha = \frac{1}{D} + \frac{B}{D}(C + D\alpha).$$

Comparing the components in (11.17), we obtain

$$\begin{aligned}\kappa_\tau(w) &= \eta(A(\tau, w) + B(\tau, w)\alpha_\tau(w)) \\ &= \eta\left(\frac{1}{D(\tau, w)} + \frac{B(\tau, w)}{D(\tau, w)}(C(\tau, w) + D(\tau, w)\alpha_\tau(w))\right) \\ &= \eta\left(\frac{1}{D(\tau, w)} + \frac{B(\tau, w)}{D(\tau, w)}(\kappa_\tau(w)\alpha_\tau(g_{0\tau}(w)))\right).\end{aligned}$$

Solving for $\kappa_\tau(w)$, we get

$$\kappa_\tau(w) = \frac{\eta}{D(\tau, w) - \eta B(\tau, w)\alpha_\tau(g_{0\tau}(w))}.$$

The desired result follows from (11.15). □

Lemma 11.3.8.

$$\left.\frac{\partial\alpha}{\partial\tau}(g_{0\tau}(w))\right|_{\tau=0} = \sum_{n=0}^{\infty} \frac{C'_\tau(0, g_{00}^{-n}(w))}{\eta^{n+1}}.$$

Proof of the lemma. Define

$$R(\tau, w, \alpha) = \frac{C(\tau, w) + D(\tau, w)\alpha}{\eta(A(\tau, w) + B(\tau, w)\alpha)}.$$

It follows from (11.14) that

$$\alpha_\tau(g_{0\tau}(w)) = R(\tau, w, \alpha_\tau(w)). \tag{11.18}$$

By (11.16),

$$\begin{aligned}\left.\frac{\partial R}{\partial\tau}\right|_{\tau=0} &= \left.\frac{(C'_\tau + D'_\tau\alpha)(A + B\alpha) + (C + D\alpha)(A'_\tau + B'_\tau\alpha)}{\eta(A + B\alpha)^2}\right|_{\tau=0} \\ &= \frac{C'_\tau(0, w)}{\eta}.\end{aligned}$$

Since $AD - BC = 1$,

$$\left.\frac{\partial R}{\partial\alpha}\right|_{\tau=0} = \left.\frac{AD - BC}{\eta(A + B\alpha)^2}\right|_{\tau=0} = \frac{1}{\eta}.$$

It follows from (11.18) that

$$\left.\frac{\partial\alpha}{\partial\tau}(g_{0\tau}(w))\right|_{\tau=0} = \frac{C'_\tau(0, w)}{\eta} + \frac{1}{\eta}\cdot\left.\frac{\partial\alpha}{\partial\tau}(w)\right|_{\tau=0}.$$

Since this inequality holds for any w, replacing w with $g_{0\tau}^{-1}(w)$ we obtain

$$\left.\frac{\partial\alpha}{\partial\tau}(w)\right|_{\tau=0} = \frac{C'_\tau(0, g_{0\tau}^{-1}(w))}{\eta} + \frac{1}{\eta}\cdot\left.\frac{\partial\alpha}{\partial\tau}(g_{0\tau}^{-1}(w))\right|_{\tau=0}.$$

The result follows by induction. □

Lemma 11.3.9.

$$\frac{dL_\tau}{d\tau}\Big|_{\tau=0} = 0, \quad \frac{d^2 L_\tau}{d\tau^2}|_{\tau=0} > 0. \tag{11.19}$$

Proof of the lemma. In order to simplify notations, we set

$$D'_\tau = \frac{\partial D}{\partial \tau}, \quad B'_\tau = \frac{\partial B}{\partial \tau}, \quad C'_\tau = \frac{\partial C}{\partial \tau}, \quad D''_{\tau\tau} = \frac{\partial^2 D}{\partial \tau^2}, \text{ and } B''_{\tau\tau} = \frac{\partial^2 B}{\partial \tau^2}.$$

Since the function $\alpha_\tau(w)$ is differentiable over τ (see [45]) by Lemma 11.3.7, we find

$$\frac{dL_\tau}{d\tau} = -\int_M \frac{D'_\tau - \eta B'_\tau \alpha(g_{0\tau}(w)) - \eta B(\tau, w)(\partial\alpha_\tau(w)/\partial\tau)(g_{0\tau}(w))}{D(\tau, w) - \eta B(\tau, w)\alpha_\tau(w)(g_{0\tau}(w))} dw$$

and therefore,

$$\begin{aligned}\frac{d^2 L_\tau}{d\tau^2} = &\int_M \left(\frac{D'_\tau - \eta B'_\tau \alpha(g_{0\tau}(w)) - \eta B(\tau, w)(\partial\alpha_\tau(w)/\partial\tau)(g_{0\tau}(w))}{D(\tau, w) - \eta B(\tau, w)\alpha_s(g_{0\tau}(w))}\right)^2 dw \\ &- \int_M \frac{E(\tau, w)}{D(\tau, w) - \eta B(\tau, w)\alpha_\tau(g_{0\tau}(w))} dw,\end{aligned}$$

where

$$\begin{aligned}E(\tau, w) =& D''_{\tau\tau} - \eta B''_{\tau\tau}\alpha(g_{0\tau}(w)) \\ &- \eta B(\tau, w)\frac{\partial^2\alpha_\tau(w)}{\partial\tau^2}(g_{0\tau}(w)) - 2\eta B'_\tau \frac{\partial\alpha_\tau(w)}{\partial\tau}(g_{0\tau}(w)).\end{aligned}$$

Note that for all $w \notin \Omega_2$,

$$A(\tau, w) = D(\tau, w) = 1 \quad \text{and} \quad C(\tau, w) = B(\tau, w) = 0$$

and for all $w \in M$,

$$A(0, w) = D(0, w) = 1, \quad C(0, w) = B(0, w) = 0, \quad \alpha_0(w) = 0.$$

It follows that

$$\frac{dL_\tau}{d\tau}\Big|_{\tau=0} = \int_{\Omega_2} D'_\tau \, dw, \tag{11.20}$$

and also that

$$\frac{d^2 L_\tau}{d\tau^2}\Big|_{\tau=0} = \int_{\Omega_2} \left[(D'_\tau)^2 - D''_{\tau\tau} + 2\eta B'_\tau \frac{\partial\alpha_\tau(w)}{\partial\tau}(g_{0\tau}(w))\right]_{\tau=0} dw. \tag{11.21}$$

By (11.16), we obtain that $D'_\tau(0, w) = r\widetilde{\rho}_r(r)\sin\theta\cos\theta$, and hence

$$\int_{\Omega_2} D'_\tau \, dw = 0.$$

Therefore (11.20) implies the equality in (11.19).

We now proceed with the proof of the inequality in (11.19). Applying Lemma 11.3.8, we obtain that

$$\frac{\partial \alpha}{\partial \tau}(g_{0\tau}(w))\Big|_{\tau=0} = \frac{C'_\tau(0, w)}{\eta} + \sum_{n=1}^{\infty} \frac{C'_\tau(0, g_{00}^{-n}(w))}{\eta^{n+1}}.$$

It follows that

$$\begin{aligned} 2\eta B'_\tau(0, w)\frac{\partial \alpha}{\partial \tau}(g_{0\tau}(w))\Big|_{\tau=0} &= 2B'_\tau(0, w)C'_\tau(0, w) \\ &\quad + 2B'_\tau(0, w)\sum_{n=1}^{\infty} \frac{C'_\tau(0, g_{00}^{-n}(w))}{\eta^n}. \end{aligned}$$

First, we evaluate the term

$$\mathcal{F}(w) = D'_\tau(0, w)^2 - D''_{\tau\tau}(0, w) + 2B'_\tau(0, w)C'_\tau(0, w).$$

Using (11.16), we find that

$$\begin{aligned} \mathcal{F}(w) =&(r\widetilde{\rho}_r \sin\theta \cos\theta)^2 + (\widetilde{\rho}^2 + 2r\widetilde{\rho}\widetilde{\rho}_r \sin^2\theta) \\ &- 2(\widetilde{\rho} + r\widetilde{\rho}_r \sin^2\theta)(\widetilde{\rho} + r\widetilde{\rho}_r \cos^2\theta) \\ =& -\widetilde{\rho}^2 - (r\widetilde{\rho}_r \sin\theta\cos\theta)^2 - 2r\widetilde{\rho}\widetilde{\rho}_r \cos^2\theta. \end{aligned} \tag{11.22}$$

Recall that $\Omega_2 = B^{uc}(\bar{q}, \varepsilon_0) \times B^s(\bar{q}, \varepsilon_0)$ and $\widetilde{\rho}(r) = 0$ if $r \geq \varepsilon$. We have

$$\int_{\Omega_2} 2r\widetilde{\rho}\widetilde{\rho}_r \cos^2\theta \, dw = \int_{-\varepsilon_0}^{\varepsilon_0} dy \int_0^{2\pi} 2\cos^2\theta \, d\theta \int_0^{\varepsilon} r^2\widetilde{\rho}\widetilde{\rho}_r \, dr. \tag{11.23}$$

Since $0 = \widetilde{\rho}(0) = \widetilde{\rho}(\varepsilon)$ (by the definition of the function ρ), we find that

$$\int_0^{\varepsilon} r^2\widetilde{\rho}\widetilde{\rho}_r \, dr = \frac{1}{2}r^2\widetilde{\rho}^2\Big|_0^{\varepsilon} - \int_0^{\varepsilon} r\widetilde{\rho}^2 \, dr = -\int_0^{\varepsilon} r\widetilde{\rho}^2 \, dr.$$

We also have that

$$\int_0^{2\pi} 2\cos^2\theta \, d\theta = \int_0^{2\pi} d\theta. \tag{11.24}$$

It follows from (11.23) and (11.24) that

$$-\int_{\Omega_2} 2r\widetilde{\rho}\widetilde{\rho}_r \cos^2\theta \, dw = \int_{\Omega_2} r\widetilde{\rho}^2 \, dw \leq \varepsilon \int_{\Omega_2} \widetilde{\rho}^2 \, dw. \tag{11.25}$$

Arguing similarly, one can show that

$$-\int_{\Omega_2} (r\widetilde{\rho}_r \sin\theta\cos\theta)^2 \, dw = -\frac{1}{8}\int_{\Omega_2} (r\widetilde{\rho}_r)^2 \, dw \tag{11.26}$$

Thus we conclude using (11.22), (11.25), and (11.26) that

$$\int_{\Omega_2} \mathcal{F}(w) \, dw \leq -(1 - \varepsilon)\int_{\Omega_2} \widetilde{\rho}_r^2 \, dw - \frac{1}{8}\int_{\Omega_2} (r\widetilde{\rho}_r)^2 \, dw < 0. \tag{11.27}$$

We now evaluate the remaining term

$$\mathcal{G}(w) = \sum_{n=1}^{\infty} \frac{1}{\eta^i} \int_{\Omega_2} 2B'_\tau(0, w)C'_\tau(0, g_{00}^{-n}(w))\, dw.$$

Since the map $g_{00} = F$ preserves volume, we obtain that

$$\begin{aligned}
&\int_{\Omega_2} 2B'_\tau(0, w)C'_\tau(0, g_{00}^{-n}(w))\, dw \\
&\leq \int_{\Omega_2} B'_\tau(0, w)^2 dw + \int_{\Omega_2} C'_\tau(0, g_{00}^{-n}(w))^2 dw \\
&= \int_{\Omega_2} B'_\tau(0, w)^2 dw + \int_{\Omega_2} C'_\tau(0, w)^2\, dw
\end{aligned}$$

Applying (11.16), we find that

$$\begin{aligned}
&\int_{\Omega_2} B'_\tau(0, w)^2\, dw + \int_{\Omega_2} C'_\tau(0, w)^2\, dw \\
&= \int_{\Omega_2} (\widetilde{\rho} + r\widetilde{\rho}_r \sin^2\theta)^2\, dw + \int_{\Omega_2} (\widetilde{\rho} + r\widetilde{\rho}_r \cos^2\theta)^2\, dw \\
&\leq 4\left(\int_{\Omega_2} \widetilde{\rho}^2\, dw + \int_{\Omega_2} r^2\widetilde{\rho}_r^2\, dw\right).
\end{aligned}$$

It follows that for sufficiently large $N > 0$ (which does not depend on ε),

$$\begin{aligned}
&\sum_{i=N}^{\infty} \frac{1}{\eta^i} \int_{\Omega_2} 2B'_\tau(0, w)C'_\tau(0, g_{00}^{-i}(w))\, dw \\
&\leq \frac{1}{10}\left(\int_{\Omega_2} \widetilde{\rho}^2\, dw + \int_{\Omega_2} r^2\widetilde{\rho}_r^2\, dw\right).
\end{aligned} \tag{11.28}$$

Note that if $g_{00}^{-n}(\Omega_2) \cap \Omega_2 = \varnothing$, then $B'_\tau(0, w)C'_\tau(0, g_{00}^{-n}(w)) = 0$ for all w. Hence,

$$\int_{\Omega_2} 2B'_\tau(0, w)C'_\tau(0, g_{00}^{-n}(w))\, dw = 0.$$

We may choose the point q and a small ε such that

$$g_{00}^{-n}(\Omega_2) \cap \Omega_2 = F^{-n}(\Omega_2) \cap \Omega_2 = \varnothing$$

for all $n = 1, 2, \ldots, N$. It follows from (11.21), (11.27), and (11.28) that

$$\begin{aligned}
\left.\frac{d^2 L_\tau}{d\tau^2}\right|_{\tau=0} &= \int_{\Omega_2} \mathcal{F}(w)\, dw + \int_{\Omega_2} \mathcal{G}(w)\, dw \\
&\leq -\left(\frac{9}{10} - \varepsilon\right) \int_{\Omega_2} \widetilde{\rho}^2\, dw - \frac{1}{40} \int_{\Omega_2} r^2\widetilde{\rho}_r^2\, dw < 0.
\end{aligned}$$

The desired result follows. □

To conclude the proof of the proposition it remains to notice that by Lemma 11.3.9, we have $L_\tau < \log\eta$ for all sufficiently small τ. □

11.4 Existence of Hyperbolic Diffeomorphisms on Compact Manifolds

Existence of an Anosov diffeomorphism f on a compact Riemannian manifold M imposes strong conditions on the topology of the manifold. For example, M should admit two foliations with smooth leaves (invariant under f). Anosov diffeomorphisms are known to exist only on multidimensional tori or more generally on factors of nilpotent Lie groups. On the contrary nonuniform hyperbolicity imposes no restrictions on the topology of M. In Chapter 6 we described some examples of volume preserving diffeomorphisms and in fact, proved that any surface admits such a diffeomorphism. In this section we present the construction by Dolgopyat and Pesin [59] showing that such diffeomorphisms exist on any manifold of dimension greater than 2.

Theorem 11.4.1 (Dolgopyat–Pesin [59]). *Given a compact smooth Riemannian manifold $M \neq S^1$, there exists a C^∞ volume preserving Bernoulli diffeomorphism f of M, which has nonzero Lyapunov exponents at almost every point.*

Proof of Theorem 11.4.1: the case $\dim M \geq 5$.

We split the proof into several steps:

Step I. Let g be the C^∞ diffeomorphism of the unit disk D constructed in Section 6.3.3 (see (6.25)) and U a sufficiently small neighborhood of the *singularity* set $\Omega = \{q_1, q_2, q_3\} \cup \partial D$ where $q_i = \eta(p_i)$, $i = 1, 2, 3$ (see Section 6.3.3 for the definition of the map η).

Proposition 11.4.2. *There is $\gamma_0 > 0$ such that for every $\gamma_0 > \gamma > 0$, one can find a point $x_0 \in U$ such that*

$$g^j(B(x_0, \gamma)) \cap B(x_0, \gamma) = \varnothing, \quad -N < j < N, \quad j \neq 0,$$

$$g^j(B(x_0, \gamma)) \cap U = \varnothing, \quad -N < j < N,$$

where $N = N(\gamma) = -(\log \gamma / \log \mu) - C$ and $C > 0$ is a constant.

Proof. Note that the statement holds true for the hyperbolic automorphism A and the desired result follows from results in Sections 6.3.1 and 6.3.2. □

Step II. Given a positive integer $n \geq 5$, set $k = \lfloor (n-3)/2 \rfloor$ and consider the $(n-3) \times (n-3)$ block diagonal matrix $A = (A_i)$, where $A_i = \left(\begin{smallmatrix} 2 & 1 \\ 1 & 1 \end{smallmatrix}\right)$ for $i < k$ and

$$A_k = \begin{cases} \begin{pmatrix} 2 & 1 \\ 1 & 1 \end{pmatrix} & \text{if } n \text{ is odd} \\ \begin{pmatrix} 2 & 1 & 1 \\ 1 & 1 & 1 \\ 0 & 1 & 2 \end{pmatrix} & \text{if } n \text{ is even} \end{cases}.$$

It is easy to see that $\det A = 1$ and that A generates a volume preserving hyperbolic automorphism of the torus $\mathbb{T}^{n-3}$. Let T^t be the special flow over A with the roof function

$$H(x) = H_0 + \varepsilon \widetilde{H}(x),$$

where H_0 is a constant and the function $\widetilde{H}(x)$ is such that $|\widetilde{H}(x)| \le 1$. The flow T^t is an Anosov flow on the phase space $\mathcal{Y}^{n-2}$ that is diffeomorphic to the product $\mathbb{T}^{n-3} \times [0, 1]$, where the tori $\mathbb{T}^{n-3} \times 0$ and $\mathbb{T}^{n-3} \times \{1\}$ are identified by the action of A.

One can choose the function $\widetilde{H}(x)$ such that the flow T^t has the stable accessibility property (see Theorem 11.1.5). Consider the following skew product map R of the manifold $\mathcal{M} = D \times \mathcal{Y}^{n-2}$

$$R(z) = R(x, y) = (g(x), T^{\alpha(x)}(y)), \quad z = (x, y), \tag{11.29}$$

where $\alpha\colon D \to \mathbb{R}$ is a nonnegative C^∞ function, which is equal to zero in the neighborhood U of the singularity set $\mathcal{Q}$ and is strictly positive otherwise.

We define the *singularity* set for the map R by $\mathcal{S} = \mathcal{Q} \times \mathcal{Y}^{n-2}$. We also set $\mathcal{N} = (D \setminus U) \times \mathcal{Y}^{n-2}$ and $Z = X \times \mathcal{Y}^{n-2}$ (see Proposition 6.3.1).

For every $z = (x, y) \in D \times \mathcal{Y}^{n-2}$, set

$$C_R^-(z) = C_g^-(x) \times C_{T^t}^s(y) \quad \text{and} \quad C_R^+(z) = C_g^+(x) \times C_{T^t}^u(y),$$

where $C_{T^t}^s(y)$ and $C_{T^t}^u(y)$ are the stable and unstable cones at the point y for the flow T^t and $C_g^-(x)$ and $C_g^+(x)$ are the stable and unstable cones at the point x for the map g (see Section 6.3.3).

Proposition 11.4.3. *The following statements hold:*

1. *We have*

$$dR^{-1}C_R^-(z) \subset C_R^-(R^{-1}(z)) \quad \textit{and} \quad dRC_R^+(z) \subset C_R^+(R(z)) \tag{11.30}$$

and inclusions are strict on the closure of the set $\mathcal{N}$*; moreover, there exists* $\mu > 1$ *such that for all* $z \in \mathcal{N}$*,*

$$\begin{aligned} \|dR(v)\| &\ge \mu\|v\|, \ v \in C_R^+(z), \\ \|dR(v)\| &\le \frac{1}{\mu}\|v\|, \ v \in C_R^-(z). \end{aligned} \tag{11.31}$$

2. *For every* $z \in Z$*,*

$$E_R^s(z) = \bigcap_j dR^{-j}(C_R^-(R^j(z))) \quad \textit{and} \quad E_R^u(z) = \bigcap_j dR^j(C_R^+(R^{-j}(z)))$$

are dR*-invariant stable and unstable continuous distributions and*

$$T_z\mathcal{M} = E_R^s(z) \oplus E_R^c(z) \oplus E_R^u(z),$$

where $E_R^c(z)$ *is the one-dimensional central (flow) direction.*

3. *For every* $z \in \mathcal{N} \cap Z$*,*

$$\|dR|E_R^s(z)\| \le \frac{1}{\mu} \quad \textit{and} \quad \|dR^{-1}|E_R^u(z)\| \le \frac{1}{\mu}.$$

4. *For every* $z = (x, y) \in Z$,

$$\pi_1 E^s_R(z) = E^s_g(x), \quad \pi_1 E^u_R(z) = E^u_g(x),$$
$$\pi_2 E^s_R(z) = E^s_{T^t}(y), \quad \pi_2 E^u_R(z) = E^u_{T^t}(y),$$

where $\pi_1 : T_z\mathcal{M} \to T_x D$ *and* $\pi_2 : T_z\mathcal{M} \to T_y\mathcal{Y}^{n-2}$ *are the natural projections.*

5.

$$m(\{x \in \mathcal{M} : R^n(x) \in U \text{ for all } n \in \mathbb{Z}\}) = 0.$$

Proof. Statements 1–3 follow from Proposition 6.3.1. Statements 4 and 5 are obvious. □

Proposition 11.4.4. *The distributions* $E^s_R(z)$ *and* $E^u_R(z)$ *generate two foliations,* W^s_R *and* W^u_R, *of* $\mathcal{M} \setminus S$*; the sizes of local leaves* $V^s_R(z)$ *and* $V^u_R(z)$ *are bounded away from zero on the set* $\mathcal{N} \setminus (\mathcal{M} \setminus S)$.

Proof. Let $z = (x, y) \in Z$. Set

$$W^s_R(z) = \bigcup_{\hat{x} \in W^s_g(x)} (\hat{x}, W^s_{T^t}(T^{t_s(\hat{x})}(y))),$$
$$W^u_R(z) = \bigcup_{\hat{x} \in W^u_g(x)} (\hat{x}, W^u_{T^t}(T^{t_u(\hat{x})}(y))),$$

where

$$\begin{aligned} t_s(\hat{x}) &= \sum_{n=0}^{\infty} [\alpha(g^n(\hat{x})) - \alpha(g^n(x))], \\ t_u(\hat{x}) &= \sum_{n=0}^{\infty} [\alpha(g^{-n}(\hat{x})) - \alpha(g^n(x))]. \end{aligned} \tag{11.32}$$

Note that each series in (11.32) converges for every $x \in Z$. Indeed, the point $(\psi \circ \zeta \circ \eta)^{-1}(x)$ does not lie on a separatrix of any of the fixed points x_1, x_2, x_3, and x_4 (see Sections 6.3.1 and 6.3.2 for notations). Therefore the series converge exponentially fast. The desired properties of the foliations W^s_R and W^u_R follow from Propositions 6.3.1 and 11.4.3. □

Proposition 11.4.5. *There exists a smooth embedding of the manifold* $\mathcal{Y}^{n-2}$ *into* $\mathbb{R}^n$.

Proof. Since $\mathcal{Y}^{n-2}$ is topologically equivalent to $\mathbb{T}^{n-3} \times [0, 1]$ with the top and the bottom identified by the hyperbolic automorphism A, it suffices to show that the latter can be embedded in $\mathbb{R}^n$. We will construct a deformation $\mathbb{T}^{n-3}_t$ of the torus $\mathbb{T}^{n-3}$, which transfers the torus into its image under A. This deformation is a composition of k deformations, each dealing only with the corresponding block of A and with the corresponding two- or three-dimensional torus.

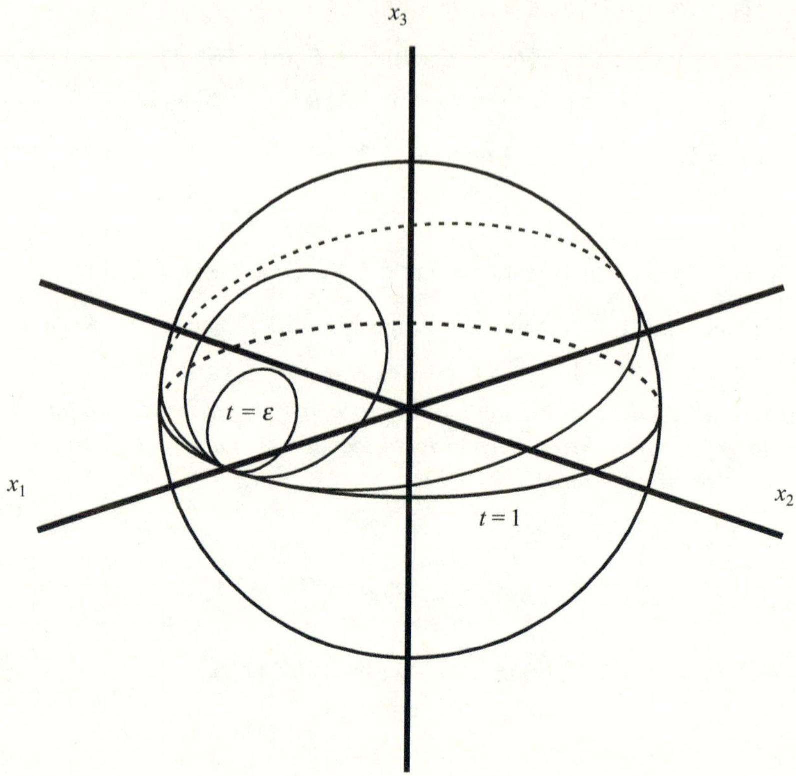

Figure 11.3. The construction of the deformation $\mathbb{T}_t^2$.

Lemma 11.4.6. *There exists a deformation $\mathbb{T}_t^2$ of the torus $\mathbb{T}^2$ in $\mathbb{R}^4$, which transfers $\mathbb{T}_0^2$ into its image under $\left(\begin{smallmatrix}1&1\\0&1\end{smallmatrix}\right)$.*

Proof of the lemma. Let

$$B^3(R) = \{(x_1, x_2, x_3) : x_1^2 + x_2^2 + x_3^2 < R\}$$

be a ball in $\mathbb{R}^3$. The product $S^1 \times B^3(R)$ can be embedded into $\mathbb{R}^4$ in a natural way. We describe a deformation D_t, $0 \le t \le 1$ in $S^1 \times B^3(R)$, which transforms the strip $D_0 = S^1 \times \{(x_1, 0, 0) \colon 0 < x_1 < 1\}$ into the strip

$$D_1 = \{(\varphi, r\cos\varphi, r\sin\varphi, 0) : 0 < r < 1, 0 \le \varphi \le 2\pi\}.$$

For each t, let $D_t = \{I_\varphi(t) : \varphi \in S^1\}$, where $I_\varphi(0) = \{\varphi\} \times \{(x_1, 0, 0) \colon 0 < x_1 < 1\}$ and for small t, $I_\varphi(t)$ describes a small circle near $I_\varphi(0)$ as φ varies from 0 to 2π. For larger t the circles are larger, and the last one corresponding to $t = 1$ coincides with the equator in the (x_1, x_2)-plane (see Fig. 11.3). Observe that $I_\varphi(t) \subset \{\varphi\} \times B^3(R)$ for every t, and that D_t has codimension 2, is orientable, and has trivial normal bundle. It follows that there exist two nonzero vector fields $v_1(t)$ and $v_2(t)$ normal to D_t. Let $\bar{v}(t, \varphi)$ be the restriction of $v_1(t)$ to the middle line of D_t. The pair $I_\varphi(t)$ and $\bar{v}(t, \varphi)$ determines an oriented circle in $\varphi \times B^3(R)$. We obtain a 2-torus for each t, the last being the image of the first one under the matrix $\left(\begin{smallmatrix} 1 & 1 \\ 0 & 1 \end{smallmatrix}\right)$. This implies the desired result. □

A similar statement holds for the matrix $\left(\begin{smallmatrix} 1 & 0 \\ 1 & 1 \end{smallmatrix}\right)$ and hence there exists a deformation $\mathbb{T}^2$ in $\mathbb{R}^4$, which transforms $\mathbb{T}_0^2$ into its image under the action of

$$\begin{pmatrix} 2 & 1 \\ 1 & 1 \end{pmatrix} = \begin{pmatrix} 1 & 1 \\ 0 & 1 \end{pmatrix} \begin{pmatrix} 1 & 0 \\ 1 & 1 \end{pmatrix}.$$

Consider now the manifold $K = \{t \times \mathbb{T}_t^2 : 0 \leq t \leq 1\}$ (lying in $\mathbb{R}^5 = \mathbb{R} \times \mathbb{R}^4$). It has codimension 2 and its normal bundle is trivial. The manifold $K \times S^1 = \{\{t\} \times \mathbb{T}_t^2 \times S^1 : 0 \leq t \leq 1\}$ is also embedded into $\mathbb{R} \times \mathbb{R}^4$. Any torus $\mathbb{T}^m$ can be embedded into $\mathbb{R}^{m+1}$. Hence there is an embedding of $K \times \mathbb{T}^{n-5} = (K \times S^1) \times \mathbb{T}^{n-6}$ into $\mathbb{R}^5 \times \mathbb{R}^{n-5} = \mathbb{R}^n$. (This requires $n \geq 6$ but for $n = 5$, we have seen that K is embedded into $\mathbb{R}^5$.) Thus there exists a deformation $\mathbb{T}_{t,i}^{n-3}$, $i - 1 \leq t \leq i$, of $\mathbb{T}^{n-3}$ into $\mathbb{R}^{n-1}$, which transforms the torus $\mathbb{T}^{n-3}$ into its image under the action of the matrix

$$\begin{pmatrix} E_{2i-2} & 0 & \cdots & 0 \\ 0 & 2 & 1 & \vdots \\ \vdots & 1 & 1 & 0 \\ 0 & \cdots & 0 & E_{n-3-2i} \end{pmatrix},$$

where E_m is the unit matrix of order m. (The last step may be slightly different.) The composition $\mathbb{T}_t^{n-3}$ of these deformations corresponds to the action of A. If we assume now that t belongs to a circle S^1 of length k and $\mathbb{T}_t^{n-3}$ belongs to $t \times B^{n-1}(R)$ (where $B^{n-1}(R)$ is a ball in $\mathbb{R}^n$ of a large radius R) then we have an embedding of the phase space $\mathfrak{Y}^{n-2}$ into $S^1 \times B^{n-1}(R)$, which in turn is embedded into $\mathbb{R}^n$. This completes the proof of the proposition. □

Proposition 11.4.7. *Given a compact smooth Riemannian manifold M of dimension $n \geq 5$, there exists a C^∞ volume preserving diffeomorphism h of M such that*

1. *for almost every $z \in M$, there exists a decomposition*

$$T_z M = E_h^s(z) \oplus E_h^c(z) \oplus E_h^u(z)$$

into dh-invariant stable, neutral, and unstable subspaces with $\dim E_h^c(z) = 1$ *and the Lyapunov exponents at the point z of a vector $v \in T_z M$*

$$\chi(z, v) \begin{cases} < 0 & \text{if } v \in E_h^s(z) \\ = 0 & \text{if } v \in E_h^c(z) \, ; \\ > 0 & \text{if } v \in E_h^u(z) \end{cases}$$

2. *h satisfies the essential accessibility property and is a Bernoulli diffeomorphism.*

Proof. Using Proposition 11.4.5 one can construct a smooth embedding $\chi_1 : D \times \mathcal{Y}^{n-2} \to \mathcal{B}^n$ (where $\mathcal{B}^n$ is the unit ball in $\mathbb{R}^n$), which is a diffeomorphism except for the boundary $\partial D \times \mathcal{Y}^{n-2}$. Then using results in Section 6.3.3 one can find a smooth embedding $\chi_2 : \mathcal{B}^n \to M$, which is a diffeomorphism except for the boundary $\partial\mathcal{B}^n$. Since the map R is identity on the boundary $\partial D \times \mathcal{Y}^{n-2}$, the map $h = (\chi_1 \circ \chi_2) \circ R \circ (\chi_1 \circ \chi_2)^{-1}$ has all the properties stated in Proposition 11.4.7. □

Step III. We construct a small perturbation P of the map R, which satisfies all the desired properties.

Fix a number $\gamma > 0$ and a point $y_0 \in \mathcal{Y}^{n-2}$, and set $\Delta = B(x_0, \gamma) \times B(y_0, \gamma)$ (where the point x_0 is chosen in Proposition 11.4.2).

Proposition 11.4.8. *Given $\varepsilon > 0$, there is a C^∞ volume preserving diffeomorphism $P : \mathcal{M} \to \mathcal{M}$ such that*

1. $d_{C^1}(P, R) \le \varepsilon$ *where the map R is defined by* (11.29)*; moreover,* $P|(\mathcal{M} \setminus \Delta) = R|(\mathcal{M} \setminus \Delta)$*;*
2. *for almost every $z \in \mathcal{M}$, there exists a decomposition*

$$T_z\mathcal{M} = E^s_P(z) \oplus E^c_P(z) \oplus E^u_P(z)$$

into dP-invariant subspaces such that $\dim E^c_P(z) = 1$ *and the Lyapunov exponent at the point z of a vector $v \in T_z\mathcal{M}$ is*

$$\chi(z, v) \begin{cases} < 0 & \text{if } v \in E^s_P(z) \\ > 0 & \text{if } v \in E^u_P(z) \end{cases};$$

3. *the Lyapunov exponent $\chi^c_P(z)$ in the central direction satisfies*

$$\int_{\mathcal{M}} \chi^c_P(z)\, dz < 0.$$

Proof. Let $\varphi_x : \mathcal{Y}^{n-2} \to \mathcal{Y}^{n-2}$, $x \in \mathcal{M}$, be a family of volume preserving C^∞ diffeomorphisms satisfying

$$d_{C^1}(\varphi_x, \mathrm{Id}) \le \varepsilon, \quad \varphi_x(y) = y \quad \text{for } (x, y) \in \mathcal{M} \setminus \Delta. \tag{11.33}$$

A particular choice of such a family will be specified below (see Lemma 11.4.11). Set

$$\varphi(x, y) = (x, \varphi_x(y)), \quad P = \varphi \circ R. \tag{11.34}$$

It is easy to see that the map P is C^∞, volume preserving, and

$$P|(\mathcal{M} \setminus \Delta) = R|(\mathcal{M} \setminus \Delta), \quad d_{C^1}(P, R) \le \varepsilon. \tag{11.35}$$

It follows from Proposition 11.4.3 and the first relation in (11.35) that for every $z \in \mathcal{M} \setminus \mathcal{S}$,

$$P^{-1}(C^-_R(z)) \subset C^-_R(P^{-1}(z)) \quad \text{and} \quad P(C^+_R(z)) \subset C^+_R(P(z)),$$

and inclusions are strict on the set $\mathcal{M} \setminus \mathcal{S}$. Therefore the formulas

$$E_P^s(z) = \bigcap_j dP^{-j}(C_R^-(P^j(z))) \quad \text{and} \quad E_P^u(z) = \bigcap_j dP^j(C_R^+(P^{-j}(z)))$$

define subspaces at every point $z \in Z$. Clearly, these subspaces are dP-invariant. Moreover, since the first coordinate of the point $P(x, y)$ depends only on x (see (11.34)), we obtain that

$$\pi_1 E_P^s(z) = E_g^s(x) \quad \text{and} \quad \pi_1 E_P^u(z) = E_g^u(x), \tag{11.36}$$

where $z = (x, y)$. (Recall that $\pi_1 : T_z\mathcal{M} \to T_x D$ is the natural projection.)

We stress that the property (11.36) holds only due to the special form of the perturbation (see (11.34)). This property is crucial for our further study (see Proposition 11.4.13).

Lemma 11.4.9.

1. *There is a number $C > 0$ such that for every sufficiently small $\gamma > 0$ and $z = (x, y) \in Z$ with $x \in B(x_0, \gamma)$, we have that*

$$\begin{aligned} \sphericalangle(E_P^u(z), dP E_R^u(P^{-1}(z))) &\le C\gamma^{\log\mu/\log\lambda}, \\ \sphericalangle(E_P^s(z), E_R^s(z)) &\le C\gamma^{\log\mu/\log\lambda}. \end{aligned} \tag{11.37}$$

2. *There is a number $\nu > 1$ such that for every $z \in \mathcal{N} \cap Y$,*

$$\|dP|E_P^s(z)\| \le \frac{1}{\nu} \quad \textit{and} \quad \|dP^{-1}|E_P^u(z)\| \le \frac{1}{\nu}. \tag{11.38}$$

Proof of the lemma. The second statement follows immediately from the first one and Statement 3 of Proposition 11.4.3. We will prove the first inequality in (11.37); the proof of the second one is similar. Consider the point

$$z^* = (x^*, y^*) = R^{-(N-1)}(P^{-1}(z)),$$

where $N = N(\gamma)$ is defined in Proposition 11.4.2. By (11.35),

$$d(E_P^u(z^*), E_R^u(z^*)) \le \delta,$$

where d is the distance in the Grassmannian bundle and $\delta = \delta(\varepsilon) > 0$ is sufficiently small. Since

$$P^j(z^*) = R^j(z^*) \text{ for } 0 \le j \le N-1, \tag{11.39}$$

we obtain using Statement 3 of Proposition 11.4.3 that

$$d(dR^{N-1}E_P^u(z^*), dR^{N-1}E_R^u(z^*)) \le \frac{\delta}{\mu^{N-1}}.$$

Using again (11.39) we rewrite the last inequality as

$$d(E_P^u(P^{-1}(z)), E_R^u(P^{-1}(z))) \le \frac{\delta}{\mu^{N-1}} \le \delta\mu\gamma^{\log\mu/\log\lambda}.$$

Applying dP we obtain the desired result. □

Set $\Lambda^+(z, R) = \sum_i \chi_i(z)$, where $\chi_i(z)$ are the values of the Lyapunov exponent of the map $dR|E_R^+(z)$. Similarly, we define $\Lambda^-(z, R)$, $\Lambda^+(z, P)$, and $\Lambda^-(z, P)$. Since the maps R and P preserve volume, we have for every $z \in \mathcal{M} \setminus \mathcal{S}$,

$$\Lambda^+(z, R) + \Lambda^-(z, R) + \chi_R^c(z) = \Lambda^+(z, R) + \Lambda^-(z, R) = 0,$$
$$\Lambda^+(z, P) + \Lambda^-(z, P) + \chi_P^c(z) = 0.$$

It follows that

$$\begin{aligned}\int_{\mathcal{M}} \chi_P^c(z)\, dz = \int_{\mathcal{M}} \Lambda^+(z, R)\, dz - \int_{\mathcal{M}} \Lambda^+(z, P)\, dz \\ + \int_{\mathcal{M}} \Lambda^-(z, R)\, dz - \int_{\mathcal{M}} \Lambda^-(z, P)\, dz.\end{aligned} \tag{11.40}$$

Lemma 11.4.10. *We have*

$$\begin{aligned}\int_{\mathcal{M}} \Lambda^+(z, P)\, dz - \int_{\mathcal{M}} \Lambda^+(z, R)\, dz \\ = \int_{\Delta} \left[\log \det(\Phi^u)(z) + O(\gamma^{\log \mu / \log \lambda})\right] dz,\end{aligned}$$

$$\begin{aligned}\int_{\mathcal{M}} \Lambda^-(z, P)\, dz - \int_{\mathcal{M}} \Lambda^-(z, R)\, dz \\ = -\int_{\Delta} \left[\log \det(\Phi^{-1})^s(z) + O(\gamma^{\log \mu / \log \lambda})\right] dz,\end{aligned}$$

where

$$\Phi^u(z) = d\varphi|E_R^u(z) \quad \text{and} \quad (\Phi^{-1})^s(z) = d\varphi|E_R^s(z). \tag{11.41}$$

Proof of the lemma. We will establish the first relation. The proof of the second one is similar. Consider the induced maps $\widetilde{R}$ and $\widetilde{P}$ generated by the maps R and P respectively on the set Δ. These maps are well defined for almost every $z \in \Delta$. Let $\widetilde{\Delta}$ be the set of such points. By Kac's formula,

$$\int_{\mathcal{M}} \Lambda^+(z, R)\, dz = \int_{\widetilde{\Delta}} \Lambda^+(z, \widetilde{R})\, dz,$$
$$\int_{\mathcal{M}} \Lambda^+(z, P)\, dz = \int_{\widetilde{\Delta}} \Lambda^+(z, \widetilde{P})\, dz.$$

It follows that

$$\int_{\mathcal{M}} \left[\Lambda^+(z, P) - \Lambda^+(z, R)\right] dz = \int_{\widetilde{\Delta}} \left[\Lambda^+(z, \widetilde{P}) - \Lambda^+(z, \widetilde{R})\right] dz.$$

Fix $z = (x, y) \in \widetilde{\Delta}$. Every vector $v \in E_P^u(z)$ can be written in the form $v = v_R + w$ with $v_R \in E_R^u(z)$ and $w \in E_R^s(z) \oplus E_R^c(z)$. Denote by $L = L(z)$ the first return time of the point z to $\widetilde{\Delta}$ under the map R. By (11.34), we have that the first return time of Z to $\widetilde{\Delta}$ under the map P is also L. Moreover,

$$\begin{aligned}\|dP^L v\| &= \|d\varphi\, dR^L(v_R + w)\| \\ &= \|dR^L v_R\| \cdot \left\| d\varphi \left(\frac{dR^L v_R}{\|dR^L v_R\|} \right) \right\| (1 + O(\mu^{-L})) \\ &= (1 + O(\mu^{-L}))\|dR^L v_R\| \cdot \left\| \Phi^u \frac{dR^L v_R}{\|dR^L v_R\|} \right\|.\end{aligned}$$

Notice that

$$\begin{aligned}\int_{\widetilde{\Delta}} \Lambda^+(z, \widetilde{P})\, dz &= \int_{\widetilde{\Delta}} \log \det(d\widetilde{P} | E^u_P(z))\, dz, \\ \int_{\widetilde{\Delta}} \Lambda^+(z, \widetilde{R})\, dz &= \int_{\widetilde{\Delta}} \log \det(d\widetilde{R} | E^u_R(z))\, dz.\end{aligned}$$

It follows from Lemma 11.4.9 that

$$\begin{aligned}&\int_{\widetilde{\Delta}} (\Lambda^+(z, \widetilde{P}) - \Lambda^+(z, \widetilde{R}))\, dz \\ &\qquad = \int_{\widetilde{\Delta}} \log \frac{\det(dP^L | E^u_P(z))}{\det(dR^L | E^u_R(z))}\, dz \\ &\qquad = \int_{\widetilde{\Delta}} \left(\log \det \Phi^u(R^L(z)) + O(\mu^{-L})\right) dz \\ &\qquad = \int_{\widetilde{\Delta}} \left(\log \det \Phi^u(z) + O(\gamma^{\log \mu / \log \lambda})\right) dz,\end{aligned}$$

since $L \geq N$, where $N = N(\gamma)$ is defined in Proposition 11.4.2. The desired result now follows: □

For $z = (x, y) \in Z$, we set

$$\widetilde{\Phi}^u(z) = \frac{\partial \varphi_x}{\partial y} | (E^u_R(z) \cap T_z \mathcal{Y}^{n-2}) \quad \text{and} \quad (\widetilde{\Phi}^{-1})^s(z) = \frac{\partial \varphi_x}{\partial y} | (E^s_R(z) \cap T_z \mathcal{Y}^{n-2}).$$

It follows from the definition of the map φ (see (11.34)) that

$$\det \Phi^u(z) = \det \widetilde{\Phi}^u(z) \quad \text{and} \quad \det \Phi^s(z) = \det \widetilde{\Phi}^s(z).$$

Therefore using (11.40) and Lemma 11.4.10, we obtain that

$$\begin{aligned}&\int_{\mathcal{M}} \chi^c_{\widetilde{P}}(z)\, dz \\ &= \int_{\widetilde{\Delta}} \left(\left[\log \det \widetilde{\Phi}^u(z) - \log \det (\widetilde{\Phi}^{-1})^s(z)\right] + O(\gamma^{\log \mu / \log \lambda})\right) dz.\end{aligned} \tag{11.42}$$

Lemma 11.4.11. *There is a family of diffeomorphisms $\varphi_x : \mathcal{Y}^{n-2} \to \mathcal{Y}^{n-2}$ satisfying* (11.33) *and such that*

$$\begin{aligned}&\int_{\widetilde{\Delta}} \left[- \log \det \widetilde{\Phi}^u(z) + \log \det (\widetilde{\Phi}^{-1})^s(z)\right] dz \\ &\qquad \leq -C\varepsilon^2 \gamma^n + O(\varepsilon^3)\gamma^n + O(\gamma^{n + (\log \mu / \log \lambda)}),\end{aligned}$$

where $C > 0$ is a constant.

Proof of the lemma. Choose a coordinate system

$$\{x, y\} = \{x_1, x_2, y_1, y_2, \dots, y_{n-2}\}$$

in Δ such that $dz = dx\,dy$, $E^c_{T'}(y_0) = \partial/\partial y_1$, and

$$E^s_{T'}(y_0) = \left(\frac{\partial}{\partial y_2}, \dots, \frac{\partial}{\partial y_k}\right) \quad \text{and} \quad E^u_{T'}(y_0) = \left(\frac{\partial}{\partial y_{k+1}}, \dots, \frac{\partial}{\partial y_{i_{n-2}}}\right)$$

for some k, $2 \le k < n-2$. Let $\psi(t)$ be a C^∞ function with compact support. Set $\tau = (1/\gamma^2)(\|x\|^2 + \|y_1\|^2 + \cdots + \|y_{n-2}\|^2)$ and define

$$\begin{aligned}\varphi_x^{-1}(y) =& (x, y_1 \cos(\varepsilon\psi(\tau)) + y_2 \sin(\varepsilon\psi(\tau)),\\ & - y_1 \sin(\varepsilon\psi(\tau)) + y_2 \cos(\varepsilon\psi(\tau)), y_3, \dots, y_{n-2}).\end{aligned}$$

Since the distributions $E^u_R(z)$ and $E^s_R(z)$ are continuous (see Statement 2 of Proposition 6.3.1) by (11.41), we find that

$$\int_{\tilde{\Delta}} \log\det \tilde{\Phi}^u(z)\,dz = o(1)m(\Delta) = o(1)O(\gamma^n)$$

and

$$\begin{aligned}\int_{\tilde{\Delta}} &\log\det(\tilde{\Phi}^{-1})^s(z)\,dz = \int_{\tilde{\Delta}} \log\det(d\varphi_x^{-1}|E^s_R)(z)\,dz\\ &= \int_{\tilde{\Delta}} \log\det\left(d\varphi_x^{-1}|\left(\frac{\partial}{\partial y_2}, \dots, \frac{\partial}{\partial y_k}\right)\right)(x, y)\,dx\,dy + o(1)m(\Delta)\\ &= \int_{\tilde{\Delta}} \log\det\left(d\varphi_x^{-1}|\left(\frac{\partial}{\partial y_2}, \dots, \frac{\partial}{\partial y_k}\right)\right)(x, y)\,dx\,dy + o(1)O(\gamma^n).\end{aligned}$$

It is easy to see that

$$\begin{aligned}\det&\left(d\varphi_x^{-1}|\left(\frac{\partial}{\partial y_2}, \dots, \frac{\partial}{\partial y_k}\right)\right)(x, y)\\ &= -\frac{2y_1y_2}{\gamma^2}\varepsilon\psi'(\tau)\cos(\varepsilon\psi(\tau)) + \cos(\varepsilon\psi(\tau)) - \frac{2y_2^2}{\gamma^2}\varepsilon\psi'(\tau)\cos(\varepsilon\psi(\tau)).\end{aligned}$$

It follows that

$$\begin{aligned}\log\det&\left(d\varphi_x^{-1}|\left(\frac{\partial}{\partial y_2}, \dots, \frac{\partial}{\partial y_k}\right)\right)(x, y)\\ &= -\frac{2y_1y_2}{\gamma^2}\varepsilon\psi'(\tau) - \frac{2y_1^2y_2^2}{\gamma^4}\varepsilon^2(\psi'(\tau))^2\\ &\quad - \frac{1}{2}\varepsilon^2(\psi(\tau))^2 - \frac{2y_2^2}{\gamma^2}\varepsilon^2\psi(\tau)\psi'(\tau) + O(\varepsilon^3).\end{aligned}$$

Making the coordinate change $\eta = y/\gamma$, we compute that

$$\begin{aligned}
&\int_{\tilde{\Delta}} \log\det\left(d\varphi_x^{-1}|\left(\frac{\partial}{\partial y_2},\dots,\frac{\partial}{\partial y_k}\right)\right)(x,y)\,dx\,dy \\
&= \gamma^{n-2}\int_{B(x_0,\gamma)} dx \int_{\mathbb{R}^{n-2}} [-2\eta_1\eta_2\varepsilon\psi(\tau)']\,d\eta \\
&+ \gamma^{n-2}\int_{B(x_0,\gamma)} dx \int_{\mathbb{R}^{n-2}} [-2\eta_1^2\eta_2^2\varepsilon^2(\psi(\tau)')^2]\,d\eta \\
&+ \gamma^{n-2}\int_{B(x_0,\gamma)} dx \int_{\mathbb{R}^{n-2}} [-\frac{1}{2}\varepsilon^2(\psi(\tau))^2 - 2\varepsilon^2\psi(\tau)\psi(\tau)'\eta_2^2]\,d\eta \\
&+ O(\varepsilon^3)\gamma^{n-2}.
\end{aligned} \tag{11.43}$$

Since the function ψ has compact support, the first integral in (11.43) is zero. Integrating by parts we obtain that

$$\int_{\mathbb{R}^{n-2}} \varepsilon^2\psi(\tau)\psi(\tau)'\eta_2^2\,d\eta = -\frac{1}{4}\int_{\mathbb{R}^{n-2}} \varepsilon^2(\psi(\tau))^2\,d\eta.$$

Hence the third integral in (11.43) is also zero. The second integral is a strictly negative number of order $O(\varepsilon^2\gamma^{n-2})$. The desired result follows. □

Using Lemma 11.4.11 and (11.42), we obtain that

$$\int_{\mathcal{M}} \chi_{\tilde{P}}^c(z)\,dz = -C\varepsilon^2\gamma^{n-2} + O(\varepsilon^3)\gamma^{n-2} + o(1)O(\gamma^n) + O\left(\gamma^{\log\mu/\log\lambda+n}\right).$$

In order to complete the proof of the proposition, we choose the number γ so small that $\gamma^2 \le \varepsilon^3$. □

Step IV. We establish some additional crucial properties of the diffeomorphism P given by (11.34). We call a perturbation P of the map R *gentle* if $P = R$ on $U \times \mathcal{Y}^{n-2}$. Any such map P that is sufficiently close to R satisfies (11.30) and (11.31). In what follows we assume that P has these properties. Set

$$E_P^+(z) = \bigcap_j dP^j(C_R^+(P^{-j}(z))) \quad\text{and}\quad E_P^-(z) = \bigcap_j dP^{-j}(C_R^-(P^j(z))). \tag{11.44}$$

Proposition 11.4.12. *The following statements hold:*

1. *$E_P^+(z)$ and $E_P^-(z)$ are dP-invariant distributions that depend continuously over $z \in \mathcal{M}\setminus\mathcal{S}$.*
2. *The distributions $E_P^-(z)$ and $E_P^+(z)$ are integrable and the corresponding global leaves $W_P^-(z)$ and $W_P^+(z)$ form foliations of the set $\mathcal{M}\setminus\mathcal{S}$.*
3. *For every $z \in Z$, we have*

$$E_P^s(z) = E_P^-(z), \quad E_P^u(z) = E_P^+(z),$$
$$W_P^s(z) = W_P^-(z), \quad W_P^u(z) = W_P^+(z);$$

moreover, the sizes of local leaves $V_P^-(z)$ and $V_P^+(z)$ are uniformly bounded away from zero on the set $\mathcal{N}$.

4. *The distributions and the foliations depend continuously on P.*

Proof. Consider the set

$$\mathcal{M}^+ = \{z \in \mathcal{M} \setminus \mathcal{S} : P^n(z) \to \mathcal{S} \text{ as } n \to +\infty\}.$$

Note that

(a) for every $z \in \mathcal{M} \setminus \mathcal{M}^+$, there exists a sequence of numbers $n_k \to +\infty$ such that $P^{n_k}(z) \in \mathcal{N}$; and
(b) for every $z \in \mathcal{M}^+$, there exists a number $n_0 = n_0(z)$ such that for every $n \geq n_0$ if we write $P_n(z) = (x_n, y_n)$ then $x_n = g^{n-n_0}(x_{n_0})$.

It follows that $E_P^-(z)$ is a dP-invariant distribution. We shall show that it is continuous. Fix $z \in \mathcal{M} \setminus \mathcal{S}$ and $\varepsilon > 0$. Let z_m be a sequence of points, which converges to z. There exists $n > 0$ such that $dP^{-n}(C_R^-(P^n(z)))$ is contained in a cone around $E_P^-(z)$ of angle ε. By (*a*), (*b*), and the continuity of the cone family C_R^-, one can find $M > 0$ such that for every $m \geq M$, the angle of the cone $dP^{-n}(C_R^-(P^n(z_m)))$ does not exceed 2ε. Since $E_P^-(z_m) \subset dP^{-n}(C_R^-(P^n(z_m)))$, we conclude that the distance between $E_P^-(z_m)$ and $E_P^-(z)$ in the Grassmannian bundle does not exceed 3ε.

We shall show that the distribution $E_P^-(z)$ is integrable. Fix $z \in \mathcal{M} \setminus \mathcal{M}^+$. We say that a local smooth submanifold V^- passing through a point z is *u-admissible* if $T_w V^- \subset C_R^-(w)$ for every $w \in V^-$. We have for $z \in \mathcal{M}^+$,

$$W_P^-(z) = \bigcup_{n_i \geq 0} P^{-n_k}(V^-(P^{n_k}(z))) = W_P^s(z).$$

For $z \in \mathcal{M}^+$ the existence of the manifold $W_P^-(z)$ follows from Property (a) and Proposition 6.3.1. The desired properties of the foliation W_P^- follow from continuity of the distribution $E_P^-(z)$, Lemma 11.4.9 (see (11.38)), and Proposition 6.3.1. Using similar arguments one can establish the desired properties of other distributions in (11.44) and the corresponding foliations. □

It is easy to see that the perturbation P given by (11.34) is gentle and hence Proposition 11.4.12 applies. Furthermore, due to the special form of the perturbation we will obtain an additional crucial information.

For every $z = (x, y) \in \mathcal{M} \setminus \mathcal{S}$, we define *traces* of stable and unstable global leaves for the maps R and P on the fiber $(\mathcal{Y}^{n-2})_x$ by

$$\widetilde{W}_R^s(y) = W_R^s(z) \cap (\mathcal{Y}^{n-2})_x, \quad \widetilde{W}_P^-(y) = W_P^-(z) \cap (\mathcal{Y}^{n-2})_x$$
$$\widetilde{W}_R^u(y) = W_R^u(z) \cap (\mathcal{Y}^{n-2})_x, \quad \widetilde{W}_P^+(y) = W_P^+(z) \cap (\mathcal{Y}^{n-2})_x.$$

Proposition 11.4.13. *The following properties hold:*

1. *For every $z \in \mathcal{M} \setminus \mathcal{S}$, the collections of manifolds $\widetilde{W}_R^s(y)$, $\widetilde{W}_R^u(y)$, $\widetilde{W}_P^-(y)$, and $\widetilde{W}_P^+(y)$ form four foliations of $(\mathcal{Y}^{n-2})_x$; for $x \in \mathcal{N}$, the sizes of local*

leaves $\widetilde{V}^s_R(y)$, $\widetilde{V}^u_R(y)$, $\widetilde{V}^-_P(y)$, *and* $\widetilde{V}^+_P(y)$ *are uniformly bounded away from zero.*

2. *Given* $\delta > 0$, *there exists* $\varepsilon > 0$ *such that if* $d_{C^1}(P, R) \le \varepsilon$ *then for every* $z = (x, y) \in \mathcal{N}$,

$$\rho(\widetilde{V}^s_R(y), \widetilde{V}^-_P(y)) \le \delta \quad \text{and} \quad \rho(\widetilde{V}^u_R(y), \widetilde{V}^+_P(y)) \le \delta.$$

Proof. The result follows from Propositions 11.4.3, 11.4.4, and 11.4.12, and Lemma 11.4.9. □

Choose a point $z_0 \in \mathcal{N}$. We now establish the absolute continuity property of local manifolds $V^+_P(z), z \in B(z_0, r) \cap Z$ for sufficiently small number $r > 0$. Since the manifolds depend continuously on $z \in \mathcal{N} \cap Z$, there is a local submanifold W passing through z_0 and transversal to $V^+_P(z)$. Set

$$A = \bigcup_{z \in B(z_0, r) \cap Z} V^+_P(z).$$

Denote by ξ the partition of A by $V^+_P(z)$, $z \in B(z_0, r) \cap Z$. Note that the factor-space A/ξ can be identified with $W \cap A$. Finally, we denote by m^+_z and m_W the leaf volume on $V^+_P(z)$ and respectively on W. Since the set Z has full volume for almost every point $z_0 \in Z$, we have that $m_W(W \cap A) = 1$. The following result is an immediate corollary of Theorem 8.6.1.

Proposition 11.4.14. *The foliation* W^+_P *of the set* $\mathcal{N} \cap Z$ *is absolutely continuous: for almost every point* $z \in \mathcal{N} \cap Z$,

1. *the conditional measure on the element* $V^+_P(z)$ *of this partition is equivalent to the measure* m^+_z;
2. *the factor-measure on the factor-space* A/ξ *is equivalent to the measure* m_W.

A similar statement holds for the foliation W^-_P *of* $\mathcal{N} \cap Z$.

We now establish the essential accessibility property of the map P.

Proposition 11.4.15. *If the perturbation* P *is sufficiently close to* R *then any two points* $p, q \in Z \cap \mathcal{N}$ *are accessible.*

Proof. Let $p = (p_1, p_2)$ and $q = (q_1, q_2)$ with $p_1, p_2 \in D$ and $q_1, q_2 \in \mathcal{Y}^{n-2}$. One can connect points p_1 and q_1 by a path $[x_0, \dots, x_\ell]_g$ such that $x_0 = p_1, x_\ell = q_1$, and each point $x_i \in X$. Without loss of generality we may assume that $x_1 \in V^-_g(x_0)$. The local stable manifold $V^-_P(p)$ intersects the fiber $(\mathcal{Y}^{n-2})_{x_0}$ at a single point $y_1 \in Z$. Proceeding by induction we construct points $y_2, \dots, y_\ell$, such that each point $z_i = (x_i, y_i) \in Z$, $i = 0, 1, \dots, y_\ell$, and the path $[z_0, z_1, \dots, z_\ell]_P$ connects the points p and z_ℓ. Note also that $y_\ell \in (\mathcal{Y}^{n-2})_{q_1}$. Fix a number $r > 0$ and consider the interval $[y^-, y^+]$ on the trajectory $T^t(q_2)$ centered at q_2 of radius r. Since the flow T^t has the accessibility property for every $s \in [y^-, y^+]$, one can find a path $[y_\ell, s]_{T^t}$. Moreover, paths corresponding to different s are homotopic to each other.

By Propositions 11.4.4 and 11.4.13 and Statement 4 of Proposition 11.4.3, one can find a family of homotopic paths $[z_\ell, (q_1, s)]_P$ such that s runs an interval on the trajectory $T^t(q_2)$. For sufficiently small ε, this interval contains a subinterval centered at q_2 of length $r - \delta > 0$. The desired result follows. □

We now show that the map P is topologically transitive; indeed, we prove a stronger statement.

Proposition 11.4.16. *For almost every point $z \in \mathcal{N}$, the trajectory $\{P^n(z)\}$ is dense in $\mathcal{N}$ (i.e., $\overline{\{P^n(z)\}} \supset \mathcal{N}$).*

Proof. Consider a maximal set $E_0 \subset \mathcal{N}$ of points z for which

(c) z is topologically recurrent; that is, for any $r > 0$ there exits $n \in \mathbb{Z}$ such that $P^n(z) \in B(z, r)$;
(d) for any $w \in E_0$, the points z and w are accessible.

Lemma 11.4.17. $m(E_0) = 1$.

Proof of the lemma. Since the set of topologically recurrent points has full measure, the desired result follows from Proposition 11.4.15. □

Lemma 11.4.18. *There exists a set E such that $m(E) = 1$, E satisfies Properties (c) and (d), and the following condition holds: for every $z \in E$ the sets $V_P^\alpha(z) \cap E$, $\alpha \in \{-, +\}$, have full measure with respect to the leaf volume on $V_P^\alpha(z)$.*

Proof of the lemma. Given a set $F \subset M$, let

$$F^* = \{z \in F\colon \text{ the sets } F \cap V_P^\alpha(z), \alpha \in \{+, -\} \text{ have full measure}\}$$

(with respect to the leaf volume on $V_P^\alpha(z)$). Starting with the set E_0 define inductively $E_n = E_{n-1}^*$. It follows from the absolute continuity of stable and unstable foliations (see Proposition 11.4.14) that $m(E_n) = 1$. Let $E = \bigcap_{n=0}^\infty E_n$. Then $m(E) = 1$ and Properties (c) and (d) are satisfied since $E \subset E_0$. Also, if $z \in E$ then $z \in E_n$ for each n. Therefore the sets $V_P^\alpha(z) \cap E_n, \alpha \in \{+, -\}$, have full measure and hence so do the sets $V_P^\alpha(z) \cap E$. □

Choose any two points $z, w \in E$ and let $[z_0, \dots, z_\ell]$ be a path connecting them.

Lemma 11.4.19. *Given $\delta > 0$, there are points $z'_j \in E$, $j = 0, \dots, \ell$, such that $z'_0 = z$ and $d(z_j, z'_j) \le \delta$ for $j = 1, \dots, \ell$.*

Proof of the lemma. Without loss of generality we may assume that $z_1 \in V_P^+(z_0)$. If $z_1 \in E$, we set $z'_1 = z_1$. Otherwise, fix $0 < \delta_1 \le \delta$ and let $z'_1 \in E$ be a point such that $z'_1 \in V_P^+(z_0)$ and $d(z_1, z'_1) \le \delta_1$ (such a point exists for every δ_1 in view of Lemma 11.4.18). If δ_1 is sufficiently small, for any $0 < \delta_2 \le \delta_1$ one can find a point $z'_2 \in E$ such that $z'_2 \in V_P^-(z'_1)$ and $d(z_2, z'_2) \le \delta_2$. Since the length of the path ℓ is uniformly bounded over z and w, it remains to use induction to complete the proof. □

We proceed with the proof of the proposition. Choose $z, w \in E$ and let $z'_j \in E$, $j = 0, \dots, \ell$, be points constructed in Lemma 11.4.19. Fix $\delta > 0$ and numbers $0 < \delta_1 < \dots < \delta_\ell \le \delta$. There is $m_1 > 0$ such that $d(P^n(z_0), P^n(z'_1)) \le (1/2)\delta_1$ for every $n \ge m_1$. By Property (*c*), there is $n_1 \ge m_1$ for which $d(P^{n_1}(z_1), z_1) \le (1/2)\delta_1$. It follows that $d(P^{n_1}(z_0), z'_1) \le \delta_1$.

There is $m_2 > 0$ such that for every $n \ge m_2$, $d(P^{-n}(z'_1), P^{-n}(z'_2)) \le (1/3)\delta_2$. By Property (*c*), there is $n_2 \ge m_2$ for which $d(P^{-n_2}(z'_2), z'_2) \le (1/3)\delta_2$. It follows that $d(P^{-n_2}(z'_1), z'_2) \le (2/3)\delta_2$. Note that if δ_1 is chosen sufficiently small (depending only on n_2) and n_1 is chosen accordingly then $d(P^{n_1-n_2}(z_0), z'_2) \le \delta_2$. Proceeding by induction we find numbers n_i, $i = 1, \dots, \ell$, such that

$$d(P^{n_1-n_2+\dots\pm n_\ell}(z_0), z'_\ell) \le \delta_\ell.$$

This implies that for almost every point $z \in \mathcal{N} \cap E$, the orbit $\{P^n(z)\}$ is everywhere dense. The desired result for almost every point $z \in \mathcal{M}$ follows from Statement 1 of Proposition 11.4.8 and Statement 5 of Proposition 11.4.3. □

Step V. We now finish the proof of the theorem in the case $\dim M \ge 5$.

Consider the set Λ of points for which $\chi^c(z) < 0$ and hence all values of the Lyapunov exponent at z are nonzero. By Theorem 9.1.1, ergodic components of $P|\Lambda$ have positive measure. Let Q be such a component. In view of Statement 5 of Proposition 11.4.3, the set $Q \cap \mathcal{N}$ has positive measure. Let z_0 be a Lebesgue point of the set $Q \cap \mathcal{N}$. Fix $r > 0$ and consider the set A defined by Properties (*c*) and (*d*). Using Proposition 11.4.14 and applying the standard Hopf argument (see the proof of Theorem 9.1.1), one can show that $Q \supset A$ for sufficiently small r. This implies that Q is open (mod 0) and so is the set Λ. Applying Proposition 11.4.16 we conclude that $P|\Lambda$ is ergodic. Note that the same arguments can be used to show that the map P^n is ergodic for all n. Hence, P is a Bernoulli diffeomorphism. It also follows from Proposition 11.4.16 that $m(\Lambda) = 1$.

Set $f = (\chi_1 \circ \chi_2) \circ P \circ (\chi_1 \circ \chi_2)^{-1}$ where the maps χ_1 and χ_2 are constructed in Proposition 11.4.7. It follows that the map f satisfies all the desired properties.

Proof of Theorem 11.4.1: the case $\dim M = 3$.

We proceed with the case $\dim M = 3$. Consider the manifold $\mathcal{M} = D \times S^1$ and the skew product map R

$$R(z) = R(x, y) = (g(x), R_{\alpha(x)}(y)), \quad z = (x, y), \tag{11.45}$$

where the diffeomorphism g is constructed in Section 6.3.3, $R_{\alpha(x)}$ the rotation by the angle $\alpha(x)$, and $\alpha : D \to \mathbb{R}$ a nonnegative C^∞ function, which is equal to zero on the set U (see Proposition 6.3.1) and is strictly positive otherwise.

We define the *singularity* set for the map R by $\mathcal{S} = \mathcal{Q} \times S^1$, where $\mathcal{Q}$ is the singularity set of the map g, and we also set $\mathcal{N} = (D \setminus U) \times S^1$ and $Z = X \times S^1$ (see Proposition 6.3.1).

As before we have two cone families $C_R^+(z)$ and $C_R^-(z)$ that satisfy (11.30) and (11.31).

Although the map R is not stably accessible, it has this property on the set $\mathcal{N}$.

Proposition 11.4.20. *The function $\alpha(x)$ can be chosen such that the following statement holds. For any $p, q \in \mathcal{N}$ and any pair of foliations with smooth leaves W^+ and W^- of $\mathcal{M}$, which are sufficiently close to W_R^+ and W_R^- respectively, there is a path $[p, q] = [z_0 z_1 \dots z_\ell]$ such that $z_{j+1} \in W^\alpha(z_j)$ for $\alpha \in \{+, -\}$.*

Proof. Consider the Anosov automorphism of $\mathbb{T}^2$ given by the matrix $\left(\begin{smallmatrix} 5 & 8 \\ 8 & 13 \end{smallmatrix}\right)$. By Theorem 11.1.6, a generic skew product over A has the required property. Now the statement follows from the arguments in Section 6.3. □

Choose the function $\alpha(x)$ such that the map R satisfies Proposition 11.4.20. Then any gentle perturbation of R satisfies the proposition as well. Repeating the proof of Proposition 11.4.16 we obtain the following result.

Corollary 11.4.21. *No gentle perturbation P of R that is sufficiently close to R has any open invariant sets.*

We consider a gentle perturbation P of R in the form $P = \varphi \circ R$. We wish to choose φ such that

$$\int_{\mathcal{M}} \log \det(dP|E_P^c)(z)\, dz = -\rho < 0. \tag{11.46}$$

Indeed, consider a coordinate system $\xi = \{\xi_1, \xi_2, \xi_3\}$ in a small neighborhood of a point $z_0 \in \mathcal{M}$ such that $dz = d\xi$ and

$$E_R^c(z_0) = \frac{\partial}{\partial \xi_1}, \quad E_R^s(z_0) = \frac{\partial}{\partial \xi_2}, \quad \text{and} \quad E_R^u(z_0) = \frac{\partial}{\partial \xi_3}.$$

Let $\psi(t)$ be a C^∞ function with compact support. Set $\tau = \|\xi\|^2/\gamma^2$ and define $\varphi^{-1}(\xi) = \varphi^{-1}(\xi_1, \xi_2, \xi_3)$ by

$$(\xi_1 \cos(\varepsilon\psi(\tau)) + \xi_2 \sin(\varepsilon\psi(\tau)), -\xi_1 \sin(\varepsilon\psi(\tau)) + \xi_2 \cos(\varepsilon\psi(\tau)), \xi_3). \tag{11.47}$$

The proof of (11.46) is similar to the proof of Lemma 11.4.11 (with γ chosen such that $\gamma \le \varepsilon^3$). The remaining part of the proof repeats the arguments in the case $\dim M \ge 5$ (see Propositions 11.4.12 and 11.4.14–11.4.16).

Proof of Theorem 11.4.1: the case $\dim M = 4$

Consider the manifold $\mathcal{M} = D \times \mathbb{T}^2$ and the skew product map R defined by (11.45), where $R_{\alpha(x)}$ is the translation by the vector $\alpha(x)$, and $\alpha\colon D \to \mathbb{R}^2$ a nonnegative C^∞ function, which is equal to zero on the set U and is strictly positive otherwise.

We define the *singularity* set for the map R by $\mathcal{S} = \Omega \times \mathbb{T}^2$, where Ω is the singularity set of the map g, and we set $\mathcal{N} = (D \setminus U) \times \mathbb{T}^2$ and $Z = X \times \mathbb{T}^2$. We have two cone families $C_R^+(z)$ and $C_R^-(z)$ that satisfy (11.30) and (11.31).

One can choose the function $\alpha(x)$ such that the map R is stably accessible (in the sense of Proposition 11.4.20; see the proof of this proposition). We conclude that any gentle perturbation P of R, which is sufficiently close to R, has no open invariant sets.

We consider a gentle perturbation P of R in the form $P = \varphi \circ R$ and we choose the map φ in the form (11.3) to ensure that (11.46) holds. We further perturb the map P to $\bar{P}$ to obtain a set of positive measure on which $\bar{P}$ has three negative Lyapunov exponents.

Proposition 11.4.22. *Suppose that the support of the map φ is sufficiently small. Then for all positive ε_1 and ε_2, there exists a gentle perturbation $\bar{P}$ of P such that $d_{C^1}(P, \bar{P}) \le \varepsilon_1$ and*

$$\int_{\mathcal{M}} \left[\chi_1^c(z, \bar{P}) - \chi_2^c(z, \bar{P})\right] dz \le \varepsilon_2,$$

where $\chi_1^c(z, \bar{P}) \ge \chi_2^c(z, \bar{P})$ are the Lyapunov exponents of $\bar{P}$ along the subspace $E_{\bar{P}}^c(z)$.

If ε_1 and ε_2 are sufficiently small then $\chi_1^c(z, \bar{P}) < 0$ and $\chi_2^c(z, \bar{P}) < 0$ on a set of positive measure. Indeed, there exist $\varepsilon_1 > 0$ and $C > 0$ such that for any perturbation $\bar{P}$ of P with $d_{C^1}(\bar{P}, P) \le \varepsilon_1$, we have

$$\int (\chi_1^c(z, \bar{P}) + \chi_2^c(z, \bar{P}))\, dz \le -\frac{\rho}{2}$$

and

$$|\chi_1^c(z, \bar{P}) \pm \chi_2^c(z, \bar{P})| \le C.$$

Hence,

$$\chi_1^c(z, \bar{P}) + \chi_2^c(z, \bar{P}) < -\frac{\rho}{4}$$

on a set of measure at least $\rho/(4C)$ and

$$\chi_1^c(z, \bar{P}) - \chi_2^c(z, \bar{P}) > \frac{\rho}{8}$$

on a set of measure at most $8\varepsilon_2/C$. To complete the proof one now proceeds as in the case $\dim M \ge 5$.

We now prove Proposition 11.4.22. In what follows we make the following convention: if at some step we use a statement of the type: "for any positive ε_{ℓ_1}, ..., ε_{ℓ_p}, there exist positive $\varepsilon_{k_1}, \dots, \varepsilon_{k_q}$ such that ..." then each time thereafter we assume that ε_{k_j} $(j = 1, \dots, q)$ are functions of ε_{ℓ_i} $(i = 1, \dots, p)$ satisfying the condition above.

Consider the set $B = \{z \in \mathcal{M} \setminus \mathcal{S} : \chi_1^c(z, P) \ne \chi_2^c(z, P)\}$. If $m(B) = 0$, the desired result follows. (It suffices to choose $\bar{P} = P$.) From now on we assume that $m(B) > 0$. Let $E_1^c(z)$ and $E_2^c(z)$ be the one-dimensional Lyapunov directions corresponding to $\chi_1^c(z, P)$ and $\chi_2^c(z, P)$ (i.e., they are generated by a normal basis in T_zM; see Definition 1.3.6). They are defined for almost every $z \in B$ (i.e., for all Lyapunov–Perron regular points).

Lemma 11.4.23. *For every $\varepsilon_3 > 0$, there is a measurable function $n_0 : \mathcal{M} \setminus \mathcal{S} \to \mathcal{N}$ such that for any $z \in \mathcal{M} \setminus \mathcal{S}$ and two one-dimensional subspaces $E', E'' \in E^c_P(z)$, one can find maps*

$$L_j(z, E', E'') : E^c_P(P^{j-1}(z)) \to E^c_P(P^j(z)), \quad 1 \le j \le n_0(z),$$

satisfying

1. $L_j(z, E', E'') = R_{\beta_j(z,E',E'')}(dP|E^c_P(z))$ *where R_β denotes the rotation by angle β and $\beta_j = \beta_j(z, E', E'')$ is such that*

$$\|\beta_j\| \le \varepsilon_3, \quad \beta_j = 0 \text{ on } U; \tag{11.48}$$

2. *if*

$$\hat{L}(z, E', E'') = L_{n_0(z)}(z, E', E'') \circ \cdots \circ L_1(z, E', E'')$$

then $\hat{L}(z, E', E'')E' = dP^{n_0(z)}E''$.

Proof of the lemma. Let A be the set of points $z \in \mathcal{M} \setminus \mathcal{S}$ for which the statements of Lemma 11.4.23 hold. It is easy to see that A is invariant. Since the number $n_0(z)$ does not depend on the choice of subspaces E' and E'' by continuity of dP, we find that the set A is open. In view of Corollary 11.4.21 (which is applicable to our case as well) if A is nonempty, it coincides with $\mathcal{M} \setminus \mathcal{S}$. We shall show that $A \neq \varnothing$.

Let $x \in D \setminus \mathcal{Q}$ be a periodic point of the map g of period r whose trajectory does not intersect supp φ (such a point always exists if supp φ is sufficiently small). We have that $P^r\mathbb{T}^2(x) = \mathbb{T}^2(x)$ where $\mathbb{T}^2(x)$ is a fiber over x. Moreover, $P^r|\mathbb{T}^2(x)$ is a translation. Therefore the desired result holds for any $z \in \mathbb{T}^2(x)$. □

Given positive ε_3, ε_4, and N, define the set $B_1(\varepsilon_3, \varepsilon_4, N)$ of points $z \in \mathcal{M}$ such that $n_0(z, \varepsilon_3) \le N$,

$$\left|\frac{1}{n}\log\|dP^n|E^c_\ell(z, P)\| - \chi^c_\ell(z, P)\right| \le \varepsilon_4, \ell = 1, 2,$$

$$\sphericalangle(E^c_1(P^n(z), P), E^c_2(P^n(z), P)) \ge e^{-\varepsilon_4|n|} \text{ for any } |n| \ge N.$$

Lemma 11.4.24. *For any positive ε_3, ε_4, and ε_5, one can find $N_1 > 0$ such that $m(B \setminus B_1(\varepsilon_3, \varepsilon_4, N)) \le \varepsilon_5$ for any $N \ge N_1$.*

Proof of the lemma. The result follows from Birkhoff's Ergodic Theorem and the Multiplicative Ergodic Theorem 3.4.3. □

Fix $z \in B_1(\varepsilon_3, \varepsilon_4, N)$. Since $\chi^c_1(z, P) \ge \chi^c_2(z, P)$, we obtain from the definition of the set $B_1(\varepsilon_3, \varepsilon_4, N)$ that for every point z in this set, $v \in E^c_2(z, P)$, $\|v\| = 1$, and $|n| \ge N$,

$$\left|\frac{1}{n}\log\|dP^n v\| - \chi^c_2(z, P)\right| \le \varepsilon_4 \tag{11.49}$$

and for $v \in E_1^c(z, P)$, $\|v\| = 1$ such that $\sphericalangle(v, E_2^c(z, P)) \geq e^{-\varepsilon_4}$, and $|n| \geq N$,

$$\left|\frac{1}{n} \log \|dP^n v\| - \chi_1^c(z, P)\right| \leq 2\varepsilon_4. \tag{11.50}$$

Lemma 11.4.25. *For any positive $\varepsilon_3, \varepsilon_4, \varepsilon_6, \varepsilon_7$, and N_2, there exist positive N_3 and ε_5 such that*

1. *for any $\varepsilon_8 > 0$ and $N \geq N_3$, one can find a set $\Omega = \Omega(N)$ for which $P^j(\Omega) \cap \Omega = \varnothing$, $|j| \leq N$ and, if $\bar{\Omega} = \bigcup_{j=0}^N P^j(\Omega)$, then $m(D \setminus \bar{\Omega}) \leq \varepsilon_8$;*
2. *if $B_2(\varepsilon_3, \varepsilon_4, \varepsilon_6, N, M)$ is the set of points $z \in \bar{\Omega}$ such that $z = P^{j_0}(y)$ for some $y \in \Omega$, $|j_0| \leq N$, and*

   $$\text{card}\{j : |(N-j)/j - 1| \leq \varepsilon_6 \text{ and } f^j(y) \in B_1(\varepsilon_3, \varepsilon_4, N)\} \leq M$$

   then

   $$m(B_2(\varepsilon_3, \varepsilon_4, \varepsilon_6, N, N_2)) \leq \varepsilon_7.$$

Proof of the lemma. The first statement is just the Rokhlin–Halmos Lemma (see, e.g., [178, Lemma 4.7]). Note that the measure of each set $R^j(\Omega)$ is of order $1/N$ and that the number

$$\text{card}\left\{j : \left|\frac{N-j}{j} - 1\right| \leq \varepsilon_6\right\}$$

is of order $\varepsilon_6 N$. The second statement follows. □

The set $\Omega(N)$ is called a *tower of height N*.

Lemma 11.4.26. *For any positive ε_3, ε_7, and ε_9, there exist positive ε_4 and ε_6 such that the following statement holds. Fix $z \in B_1(\varepsilon_3, \varepsilon_4, N_1)$, positive n_1 and n_2 satisfying*

$$\left|\frac{n_2}{n_1} - 1\right| \leq \varepsilon_6, \quad n = n_1 + n_2 \geq N_3,$$

and maps $L_j(z) = L_j(z, E_1^c(z, P), E_2^c(z, P))$, $j = 1, \dots, k \leq \varepsilon_6 N_3$, satisfying (11.48) and such that $\hat{L}(z) = L_k(z) \circ \cdots \circ L_1(z)$ takes $E_1^c(z, P)$ into $E_2^c(P^k(z), P)$. Then

$$\begin{aligned}
\exp&\left[n\left(\frac{\chi_1^c(z, P) + \chi_2^c(z, P)}{2} - \varepsilon_9\right)\right] \\
&\leq \left\|\left(dP^{n-k} \circ \hat{L}(z) \circ dP^{n_1} | E_P^c(P^{-n_1}(z))\right)\right\| \\
&\leq \exp\left[n\left(\frac{\chi_1^c(z, P) + \chi_2^c(z, P)}{2} + \varepsilon_9\right)\right].
\end{aligned}$$

Proof of the lemma. Set

$$\mathcal{P} = dP^{n-k} \circ \hat{L}(z) \circ dP^{n_1} | E_P^c(z).$$

Let $e_1 \in E_1^c(z, P)$ and $e_2 \in E_2^c(z, P)$ be a normalized basis in $E_P^c(z)$. Then by (11.49) and (11.50),

$$\frac{1}{n} \log \|\mathcal{P} e_\ell\| = \chi_\ell^c(z, P) n_1 + \chi_{3-\ell}^c(z, P) n_2 + O(\varepsilon_4 n)$$

for $\ell = 1, 2$. Let $\Pi(z) : E^c(z) \to E^c(z)$ be a linear map satisfying $\det \Pi(z) = 1$ and the vectors $\Pi(z)e_1$ and $\Pi(z)e_2$ are orthogonal. Then

$$\begin{aligned}\log\| \exp\left(n \frac{\chi_1^c(z, P) + \chi_2^c(z, P)}{2}\right) \mathcal{P}\| &= \log \|\Pi^{-1}(P^n(z))\| \\ &+ \log \|\Pi(P^n(z)) \circ \exp\left(n \frac{\chi_1^c(z, P) + \chi_2^c(z, P)}{2}\right) \mathcal{P} \circ \Pi^{-1}(P^{n_1}(z))\| \\ &+ \log \|\Pi(P^{-n_1}(z))\|\end{aligned}$$

and each term is of order $O((\varepsilon_6 + \varepsilon_4)n)$. The desired result follows. □

Lemma 11.4.27. *For any positive ε_{10}, ε_{11}, ε_{12}, and ε_{13}, there exist positive ε_3, ε_7, ε_9, and N_2 such that the following holds. Let $\Omega_1 = \Omega \setminus B_2(\varepsilon_3, \varepsilon_4, \varepsilon_6, N_2, N_3)$ where $\Omega = \Omega(N_3)$ is a tower of height N_3 and*

$$\Omega_2 = \{f^j(z) : z \in \Omega_1, \text{ and } j \text{ is the smallest number for which } \left|\frac{N_3 - j}{j} - 1\right| \le \varepsilon_6 \text{ and } f^j(z) \in B_1(\varepsilon_3, \varepsilon_4, N_2)\}.$$

Let also $k = \varepsilon_6 N_3$. Then

1. *there exists an open set Ω_3 satisfying $m(\Omega_3 \Delta \Omega_2) \le \varepsilon_{10}$ and a map $\hat{P} = P \circ \hat{\varphi}$ such that*
$$\operatorname{supp} \hat{\varphi} = \left(\bigcup_{j=0}^{k-1} \widetilde{P}^j(\Omega_3)\right) \setminus (U \times \mathbb{T}^2);$$
2. $d_{C^1}(\hat{\varphi}, \mathrm{Id}) \le \varepsilon_1$;
3. *there exists $\Omega_4 \subset \Omega_2$ such that $m(\Omega_2 \setminus \Omega_4) \le \varepsilon_{11}$ and for all $z \in \Omega_4$,*
$$\|(d\hat{P}^n | E_{\hat{P}}^c)(z) - \hat{L}(z)\| \le \varepsilon_{12} \quad \text{for some} \quad n \le k, \tag{11.51}$$
where the map $\hat{L}(z) : E_P^c(z) \to E_P^c(P^n(z))$ takes $E_1^c(z, P)$ to $E_2^c(P^n(z), z)$ (see Lemma 11.4.26);
4. *for any $z \in \Omega$,*
$$d(E_P^c(z), E_{\hat{P}}^c(z)) \le \varepsilon_{13}.$$

Proof of the lemma. Consider a finite atlas $\Phi = \{\Phi_1, \dots, \Phi_n\}$ such that in each chart Φ_i we can introduce a coordinate system $\{\xi_1, \xi_2, \xi_3, \xi_4\}$ satisfying $dz = d\xi_1\, d\xi_2\, d\xi_3\, d\xi_4$. Approximate Ω_2 by the finite union of balls $\bigcup_j B(z_j, r_j)$, with $r_j \le \rho$ where ρ is sufficiently small. By coordinate rotation we may assume that $E_P^c(z_j) = \langle \partial/\partial\xi_1, \partial/\partial\xi_2 \rangle|_{z_j}$. We can apply Lemma 11.4.26 to each $z \in \Omega_2$ and construct the maps $L_1(z), \dots, L_{N_1}(z)$ such that $\hat{L}(z) = L_{N_1}(z) \circ \cdots \circ L_1(z)$ takes

$E_1^c(z, P)$ to $E_2^c(P^n(z), P)$. By slightly shrinking the set Ω_2, if necessary, we may assume that the maps $L_i(z)$ are continuous on Ω_2. Recall that each map $L_\ell(w)$ is a twist of the form

$$L_\ell(w) = R_{\beta_\ell(w)}(dP|E_P^c(w)).$$

We define $\hat{\varphi}$ on each $B(z_j, r_j)$ to be

$$\hat{\varphi}(\xi_1, \xi_2, \xi_3, \xi_4) = (R_{\psi(\|\xi\|/r_j)\beta_1(z_j)}(\eta_1, \eta_2), \xi_3, \xi_4),$$

where $\{\eta_1, \eta_2, \eta_3, \eta_4\} = \exp_{z_j}^{-1}(\xi_1, \xi_2, \xi_3, \xi_4)$ and the function $\psi(x)$ is supported on $[0, 1]$ and

$$\psi(x) = 1, \quad x \in \left[0, \frac{1}{2}\right]. \tag{11.52}$$

Continuing by induction for each $\ell \leq N_1$, we can approximate the sets $P^\ell(B(z_j, r_j))$ by balls and define $\hat{\varphi}$ on each ball to be an appropriate twist generated by the maps $L_\ell(z)$. This construction allows us to define $\hat{\varphi}$ in such a way that (11.51) holds for $n = N_1$ on a set Δ_1 for which $m(\Delta_1) > c(N_1)m(\Omega_2)$. Here $c(N_1)$ is a constant, which can be made arbitrarily close to $(1/16)^{N_1}$ if the approximation by balls is chosen appropriately; we exploit here the fact that in view of (11.52),

$$\frac{m(B(z, r/2))}{m(B(z, r))} = \frac{1}{16}.$$

Consider a point $z \in \Omega_2 \setminus \Delta_1$. Let $\bar{N}_1(z) > N_1$ be the first moment when the trajectory $\{P^j(z)\}$ visits the set B_1. Define $\hat{\varphi}$ along the orbit $\{f^{j+N}(z)\}$ with $\bar{N}_1(z) \leq j \leq \bar{N}_1(z) + N_1$ to be appropriate twists such that the map $dP^{\bar{N}_1(z)-N_1} \circ d\bar{P}^{N_1}$ takes $E_1^c(z, P)$ to $dP^{\bar{N}_1(z)} \circ d\bar{P}^{N_1} E_2^c(z, P)$. Thus, we obtain a set Δ_2 for which $m(\Delta_2) > m(\Omega_2 \setminus \Delta_1) \geq c$ and $n = N_1 + \bar{N}_1(z)$ on Δ_2. Repeating this procedure (N_2/N_1) times, we obtain the required map $\hat{\varphi}$. All properties of the map $\hat{P}$ can now be verified by the arguments similar to those in Lemma 11.4.11. □

It remains to show that ε_{10}, ε_{11}, ε_{12}, and ε_{13} can be chosen such that

$$\left|\frac{1}{N_3} \log \|\hat{P}^{N_3}(z)|E_{\hat{P}}^c(z)\| - \frac{1}{2}\int_{\mathcal{M}} \log \det(d\hat{P}(z)|E_{\hat{P}}^c(z))\, dz\right| \leq \varepsilon_2.$$

This again is similar to the proof of Lemma 11.4.11 and we leave the details to the reader.

11.5 Existence of Hyperbolic Flows on Compact Manifolds

We present the result by Hu, Pesin, and Talitskaya [89] establishing a continuous-time version of Theorem 11.4.1.

Theorem 11.5.1. *Given a compact smooth Riemannian manifold M with $\dim M \geq 3$, there exists a C^∞ volume preserving flow φ_t, which has nonzero Lyapunov exponents almost everywhere (except for the exponent along the flow direction) and for which φ_t is a Bernoulli diffeomorphism for every $t \neq 0$.*

Proof. We split the proof into several steps:

Step I. We show that the C^∞ area preserving Bernoulli diffeomorphism g of the two-dimensional unit disk D constructed in Section 6.3.3 is diffeotopic to the identity map.

The map g has nonzero Lyapunov exponents almost everywhere, on the boundary of the disk g is the identity map, and it has all its derivatives zero (see Proposition 6.3.1). The *singularity set* for g is $\mathcal{Q} = \{q_1, q_2, q_3\} \cup \partial D$.

Proposition 11.5.2. *There exists a map* $G\colon D \times [0,1] \to D$ *such that*

1. $G(x,t)$ *is* C^∞ *in* (x,t)*;*
2. $G(\cdot,0) = \mathrm{Id}$ *and* $G(\cdot,1) = g$*;*
3. $G(x,t) = g^t(x)$ *for any* $x \in U$ *and* $t \in [0,1]$*, where* U *is a neighborhood of the boundary* ∂D *and* $g^t(x)$ *is the flow defined in Statement 6 of Proposition 6.3.1;*
4. *for any* $t \in [0,1]$*, the map* $G(\cdot,t)\colon D \to D$ *is an area preserving diffeomorphism;*
5. $d^k G(x,1) = d^k G(g(x),0)$ *for any* $k \geq 0$.

Proof of the proposition. In a neighborhood U of the boundary ∂D, the map g is the time-1 map of the flow generated by the vector field V. We extend V to a smooth vector field $\widehat{V}$ on the whole D, and let $\widehat{g}^t$ be the flow generated by $\widehat{V}$. Note that $g|U = \widehat{g}^1|U$. We need the following result of Smale (see [206, Theorem B]).

Lemma 11.5.3. *Let* $\mathcal{A}$ *be the space of* C^∞ *diffeomorphisms of the unit square, which are equal to the identity in some neighborhood of the boundary. Endow* $\mathcal{A}$ *with the* C^r *topology,* $1 < r \leq \infty$*. Then* $\mathcal{A}$ *is contractible to a point.*

The statement also holds if the unit square is replaced by the unit disk. Applying the lemma to the diffeomorphism $g \circ \widehat{g}^{-1}$, which is equal to the identity on U, we obtain a homotopy $\widetilde{G}\colon D \times [0,1] \to D$ such that $\widetilde{G}(\cdot,0) = \mathrm{Id}|D$ and $\widetilde{G}(\cdot,1) = g \circ \widehat{g}^{-1}$. Moreover, $\widetilde{G}$ is C^∞ in (x,t), that is, $\widetilde{G}$ is a diffeotopy in $\mathcal{A}$ (see [206, Theorem 4]). Therefore for each $t \in [0,1]$, there is a neighborhood U_t of ∂D such that $\widetilde{G}(\cdot,t)|U_t = \mathrm{Id}|U_t$. One can see that the set

$$U = \mathrm{int} \bigcap_{t\in[0,1]} U_t$$

is not empty and is a neighborhood of ∂D. Denote $\widetilde{g}^t = \widetilde{G}(\cdot,t)$. It follows that the map $G(\cdot,t) = \widetilde{g}^t \circ \widehat{g}^t$ satisfies Statements 1–3 of the proposition. To prove Statements 4 and 5, we need the following result by Moser [157] (see also [104, Theorem 5.1.27] and [89]).

Lemma 11.5.4. *Let* $\{\Omega_0^t\}$ *and* $\{\Omega_1^t\}$ *be two families of volume forms on* D *that are* C^∞ *in* (x,t)*. Assume that* $\Omega_0^t|U = \Omega_1^t|U$ *for any* t *and* $\Omega_0^t = \Omega_1^t$ *for* $t \in [0,\varepsilon) \cup (1-\varepsilon,1]$*. Then there exists a map* $\bar{G}\colon D \times [0,1] \to D$ *with the following properties:*

1. $\bar{G}(x,t)$ *is* C^∞ *in* (x,t).
2. $\bar{G}(\cdot,0) = \bar{G}(\cdot,1) = \mathrm{Id}$.
3. *For any* $t \in [0,1]$, *the map* $\bar{G}(\cdot,t)\colon D \to D$ *is a diffeomorphism with* $\bar{G}(\cdot,t)^*\Omega_1^t = \Omega_0^t$.
4. $\bar{G}(x,t) = x$ *for any* $t \in [0,1]$ *and* x *in some neighborhood* $U' \subset U$ *of* ∂D.

Consider the map $\bar{G}$ as in the lemma with $\Omega_0^t = dx_1 \wedge dx_2$ and $\Omega_1^t = d\tilde{g}^{t*} d\hat{g}^{t*}\Omega_0^t$ on D for any $t \in [0,1]$. Let $\bar{g}^t = \bar{G}(\cdot,t)$. Then the map $G\colon D \times [0,1] \to D$ given by $G(x,t) = \bar{g}^t \circ \hat{g}^t \circ \tilde{g}^t$ satisfies Statements 1–4 of the proposition. If necessary one can change the map $\widetilde{G}(\cdot,t)$ in a small neighborhood of the sets $D \times 0$ and $D \times 1$ so that it will also satisfy Statement 5. For example, one can choose $\widetilde{G}$ in such a way that

$$\widetilde{G}\big(\widehat{G}(x,t),t\big) = \widetilde{G}\big(\widehat{G}(x,0),0\big),\ t \in [0,\varepsilon),$$
$$\widetilde{G}\big(\widehat{G}(x,t),t\big) = \widetilde{G}\big(\widehat{G}(x,1),1\big),\ t \in (1-\varepsilon,1].$$

This completes the proof of the proposition. □

Step II. We proceed with the proof of the theorem and we first consider the case $\dim M \geq 5$. Define the map

$$R = g \times A\colon D \times \mathbb{T}^{n-3} \to D \times \mathbb{T}^{n-3},$$

where A is the automorphism of the torus $\mathbb{T}^{n-3}$ constructed in the proof of Theorem 11.4.1. Consider the suspension flow over R with the roof function $H = 1$ and the suspension manifold $K = D \times \mathbb{T}^{n-3} \times [0,1]/\sim$, where $\sim$ is the identification $(x,y,1) = (g(x), Ay, 0)$. Denote by Z the vector field of the suspension flow and by φ_Z^t the suspension flow itself. It preserves volume in K. In the "natural" coordinate system $(x_1, x_2, y_1, \dots, y_{n-3}, t) = (x,y,t)$ with $x = (x_1,x_2) \in D$, $y = (y_1,\dots,y_{n-3}) \in \mathbb{T}^{n-3}$ and $t \in [0,1]$, we have $Z = (0,0,1)$.

Let $h^t\colon L \to L$ be the suspension flow over the automorphism A with the roof function $H = 1$. The suspension manifold L is diffeomorphic to $\mathbb{T}^{n-3} \times [0,1]/\sim$, where $\sim$ is the identification $(y,1) = (Ay,0)$. The flow h^t preserves volume on L. Set $N = D \times L$ and consider the map $F\colon K \to N$ given by

$$F(x,y,t) = (G(x,t), y, t),$$

where $G\colon D \times [0,1] \to D$ is the diffeotopy constructed in Proposition 11.5.2. We have

$$F(x,y,1) = (g(x), y, 1) = (g(x), A(y), 0) = F(g(x), A(y), 0).$$

Therefore the map F is well defined. It is easy to see that F preserves volume, is one to one, and is continuous. Hence it is a homeomorphism. Formal differentiation yields

$$dF(x,y,t) = \begin{pmatrix} G_x(x,t) & 0 & G_t(x,t) \\ 0 & 1 & 0 \\ 0 & 0 & 1 \end{pmatrix}.$$

It follows from Statement 5 of Proposition 11.5.2 that for every $k \geq 1$,

$$d^k F(x, y, 1) = d^k F(g(x), Ay, 0),$$

and hence F is a C^∞ diffeomorphism.

Let us choose a C^∞ function $\alpha \colon D \to [0, 1]$ such that

(A1) α and all its partial derivatives of any order are equal to zero on ∂D;
(A2) $\alpha(x) > 0$ for $x \in \operatorname{int} D$ and $\alpha(x) = 1$ for $x \in D \setminus U$;
(A3) $\alpha(x)^{-1} V(x) \to 0$ as $x \to \partial D$, where V is the vector field defined in Statement 6 of Proposition 6.3.1.

Define the vector field X on N by

$$X(G(x, t), y, t) = \left(\frac{\partial G}{\partial t}(x, t), 0, \alpha(G(x, t)) \right), \quad (x, y, t) \in K. \tag{11.53}$$

Note that X is well defined regarding the identification

$$(x, y, 1) = (g(x), Ay, 0)$$

in K. In fact, by Lemma 11.5.4,

$$X(G(x, 1), y, 1) = X(g(x), 0, \alpha(g(x))) = X(G(g(x), 0), y, 0).$$

Note also that by Statement 3 of Proposition 11.5.2, we have $\partial G/\partial t(x, t) = V(G(x, t))$ for $x \in U$. Therefore Equality (11.53) implies that for $(x, y, t) \in K$ with $x \in U$,

$$X(G(x, t), y, t) = (V(x), 0, \alpha(x)).$$

It follows that X vanishes on the boundary $\partial D \times L$ along with all its derivatives. Let $\varphi_t = \varphi_X^t$ be the flow on N generated by the vector field X. Clearly, it is of class C^∞. We show that this flow is volume preserving, has nonzero Lyapunov exponents almost everywhere (except for the exponent along the flow direction), and that φ_X^t is a Bernoulli diffeomorphism for every $t \neq 0$.

Step III. First we observe that the flow φ_X^t preserves volume. Consider the vector field $Y = dFZ$ on N and let φ_Y^t be the corresponding flow. We have

$$Y(G(x, t), y, t) = \left(\frac{\partial G}{\partial t}(x, t), 0, 1 \right), \quad (x, y, t) \in K. \tag{11.54}$$

The vector field Y is divergence free since it is the image of the divergence-free vector field Z under the volume preserving map F. The desired result follows from (11.53) and (11.54).

Step IV. We show that all Lyapunov exponents of the flow φ_X^t, except for the exponent along the flow direction, are nonzero almost everywhere. We begin with a construction of a map, which is similar to the map g: it has the same topological and hyperbolic properties and it is better adapted to the flow but does not preserve area.

Consider the map $g^*\colon D \to D$ such that $g^* = g$ on $D \setminus U$ and g^* is the time-1 map of the flow $(g^*)^t$ generated by the vector field $V^* = \alpha^{-1}V$ on U_1 (see Statement 6 of Proposition 6.3.1). By (A1)–(A3), the map g^* is well defined and is a diffeomorphism. Furthermore, it preserves a measure μ^*, which is absolutely continuous with respect to the area with positive density $\rho^*(x)$; the latter is unbounded as x approaches ∂D.

We now proceed as before replacing g by g^*. Namely, define $G^*\colon D \times [0,1] \to D$ by $G^*(x,t) = G(x,t)$ if $x \in D \setminus U$, and $G^*(x,t) = (g^*)^t(x)$ otherwise. Let $\varphi^t_{Z^*}$ be the suspension flow over $g^* \times A$ with the roof function $H = 1$, the suspension manifold $K^* = D \times \mathbb{T}^{n-3} \times [0,1]/\sim$, (where $\sim$ is the identification $(x,y,1) = (g^*(x), A(y), 0)$), and the vector field Z^* of the suspension flow. Define the map $F^*\colon K^* \to N$ by

$$F^*(x,y,t) = (G^*(x,t), y, t).$$

Define the vector field $\widetilde{Z}$ on K^* by

$$\widetilde{Z}(x,y,t) = (dF^*)^{-1}X(F^*(x,y,t)).$$

It is easy to see that $\varphi^t_X = F^* \circ \varphi^t_{\widetilde{Z}} \circ (F^*)^{-1}$ and the vector fields Z^* and $\widetilde{Z}$ have the same orbits. In other words, there is a function $\beta\colon K^* \to \mathbb{R}^+$ such that for every $(x,y,t) \in K^*$,

(B1) $\widetilde{Z}(x,y,t) = \beta(x,y,t)Z^*(x,y,t)$;
(B2) $\beta(x,y,t) = 0$ if $x \in \partial D$;
(B3) $\beta(x,y,t) = 1$ if $x \notin U$.

By construction, the flow $\varphi^t_{Z^*}$ has nonzero Lyapunov exponents almost everywhere. Denote by $E^u_{Z^*}(x,y,t)$ and $E^{u0}_{Z^*}(x,y,t)$ respectively the unstable and weakly unstable invariant subspaces at the point $(x,y,t) \in K^*$. Observe that the subspaces $E^{u0}_{Z^*}(x,y,t)$ are also invariant under the flow $\varphi^t_{\widetilde{Z}}$. Choose a point (x_0,y_0,t_0) and a vector $v \in E^u_{Z^*}(x_0,y_0,t_0)$. Note that for almost every (x_0,y_0,t_0) (with respect to the Riemannian volume), the proportion of time that the trajectory $\{\varphi^t_{\widetilde{Z}}(x_0,y_0,t_0)\}$ spends in the set $\{(x,y,t)\colon x \notin U\}$ is strictly positive. It follows that the Lyapunov exponent $\chi(v)$ at (x_0,y_0,t_0) with respect to the flow $\varphi^t_{\widetilde{Z}}$ is positive.

Step V. We establish the essential accessibility property of the time-t map of the flow φ^t_X for every fixed $t \neq 0$.

Almost every point $(x_0,y_0,t_0) \in N$ has global stable, unstable, weakly stable, and weakly unstable manifolds

$$W^s_X(x_0,y_0,t_0), \qquad W^u_X(x_0,y_0,t_0), \qquad W^{s0}_X(x_0,y_0,t_0), \quad \text{and} \quad W^{u0}_X(x_0,y_0,t_0)$$

with respect to the flow φ^t_X. Similarly, almost every point $(x_0,y_0,t_0) \in K^*$ has global stable, unstable, weakly stable, and weakly unstable manifolds

$$W^s_{\widetilde{Z}}(x_0,y_0,t_0), \qquad W^u_{\widetilde{Z}}(x_0,y_0,t_0), \qquad W^{s0}_{\widetilde{Z}}(x_0,y_0,t_0), \quad \text{and} \quad W^{u0}_{\widetilde{Z}}(x_0,y_0,t_0)$$

with respect to the flow $\varphi^t_{\widetilde{Z}}$. By Proposition 6.3.1, these foliations can be extended to foliations, which are continuous everywhere except for the singularity set $Q \times$

$\mathbb{T}^{n-3} \times [0,1]/\sim$. We use the same notations for these foliations. Observe that

$$\pi_x(W^{s0}_{\mathcal{Z}}(x_0, y_0, t_0)) = W^s_{g^*}(x_0), \quad \pi_x(W^{u0}_{\mathcal{Z}}(x_0, y_0, t_0)) = W^u_{g^*}(x_0) \qquad (11.55)$$

$$\pi_y(W^{s0}_{\mathcal{Z}}(x_0, y_0, t_0)) = W^s_A(y_0), \quad \pi_y(W^{u0}_{\mathcal{Z}}(x_0, y_0, t_0)) = W^u_A(y_0), \qquad (11.56)$$

where $\pi_x : K^* \to D$ and $\pi_y : K^* \to \mathbb{T}^{n-3}$ are the natural projections.

It suffices to establish the essential accessibility property of the time-t map of the flow $\varphi^t_{\mathcal{Z}}$. Note that the map g^* has the essential accessibility property: indeed, any two points outside the singularity set $\mathcal{Q}$ are accessible. Note also that the automorphism A is accessible. Denote

$$\mathcal{Q}^* = \{(x, y, t) \in K^* : x \in \mathcal{Q}, y \in \mathbb{T}^{n-3}, t \in [0,1]\}.$$

Fix $t \neq 0$ and consider the time-t map of the flow $\varphi^t_{\mathcal{Z}}$. The relations (11.55) and (11.56) imply that for any two points (x, y) and $(x', y') \in D \times \mathbb{T}^{n-3}$ and any $t \in [0,1]$, the point (x, y, t) is accessible to a point (x', y', t'). In particular, any point $(x, y, t) \in K^* \setminus \mathcal{Q}^*$ is accessible to a point in $\Pi_{p,q} = \{(p, q, t') \in K^* : t' \in [0,1]\}$ for some $p \in D$ and $q \in \mathbb{T}^{n-3}$. It remains to show that any two points in $\Pi_{p,q}$ are accessible.

Let $q \in \mathbb{T}^{n-3}$ be a periodic point of the automorphism A, and p and p' be two periodic points of g^* such that the orbit of p and the orbit of p' except for p' itself (under g^*) lie outside the neighborhood U. Consider local stable and unstable manifolds (curves) at p and p', $V^s_{g^*}(p)$, $V^u_{g^*}(p)$, $V^s_{g^*}(p')$, and $V^u_{g^*}(p')$. One can choose points p and p' such that the local unstable curve from p to $V^u_{g^*}(p) \cap V^s_{g^*}(p')$ and the local stable and unstable curves from p to $V^s_{g^*}(p) \cap V^u_{g^*}(p')$ also lie outside the neighborhood U.

For a point $(p, q, t) \in \Pi_{p,q}$, let $\gamma^u_p \subset W^u_{\mathcal{Z}}(p, q, t)$ and $\gamma^s_p \subset W^s_{\mathcal{Z}}(p, q, t)$ be the curves such that $\pi_x(\gamma^u_p) \subset V^u_{g^*}(p)$, $\pi_y(\gamma^u_p) = q$ and $\pi_x(\gamma^s_p) \subset V^s_{g^*}(p)$, $\pi_y(\gamma^s_p) = q$. Since γ^u and γ^s lie outside U and the point (p, q) is periodic, the t-coordinate of a point remains unchanged when this point moves along γ^u_p from (p, q, t) toward the point

$$\gamma^u_p \cap \pi_x^{-1}(V^u_{g^*}(p) \cap V^s_{g^*}(p')) \cap \pi_y^{-1}(q),$$

so does the t-coordinate of a point moving along γ^s_p from (p, q, t) toward the point

$$\gamma^s_p \cap \pi_x^{-1}(V^u_{g^*}(p') \cap V^s_{g^*}(p)) \cap \pi_y^{-1}(q).$$

On the other hand, for a point $(p', q, \tau) \in \Pi_{p',q}$, if we choose $\gamma^u_{p'}$ and $\gamma^s_{p'}$ in a similar way, then the third coordinate τ increases when a point moves along $\gamma^u_{p'}$ from (p, q, τ) toward the point

$$\gamma^u_p \cap \pi_x^{-1}(V^u_{g^*}(p') \cap V^s_{g^*}(p)) \cap \pi_y^{-1}(q)$$

and the third coordinate τ decreases when a point moves along $\gamma^s_{p'}$ from (p, q, τ) toward the point

$$\gamma^u_p \cap \pi_x^{-1}(V^u_{g^*}(p) \cap V^s_{g^*}(p')) \cap \pi_y^{-1}(q).$$

(We assume that $\alpha(p') < 1$). We can now use the argument in the proof of Lemma 11.3.4 to obtain that the point (p, q, t) is accessible to some point $(p, q, t') \in \Pi_{p,q}$ with $t' < t$, and then any two points in $\Pi_{p,q}$ are accessible. This shows that for every $t \neq 0$, the time-t map of the flow φ_X^t has the essentially accessibility property.

Step VI. We show that the flow φ_X^t is Bernoulli. We have established the following properties of the time-t map of the flow:

1. It is nonuniformly partially hyperbolic.
2. The stable and unstable foliations are continuous and absolutely continuous; the time-t map is essentially accessible with respect to these foliations.
3. The central foliation is smooth. (This foliation is formed by the trajectories of the flow.)

Recall that the diffeomorphism $g: D \to D$ is ergodic and has nonzero Lyapunov exponents. Let $\Lambda^\ell \subset D$ be a regular set for g. For $x \in D$ and outside the singularity set $\mathfrak{Q}$, let $\gamma^s(x, r)$ and $\gamma^u(x, r)$ be arcs in $V_g^s(x)$ and $V_g^u(x)$ centered at x of length r.

Choose two points x and x' outside the singularity set, which are sufficiently close to each other. Consider the holonomy map $\theta: V_g^s(x) \to W_g^s(x')$ generated by the family of local leaves $V_g^u(y)$, $y \in V_g^s(x)$. Since the foliation W_g^s is absolutely continuous, the map θ moves the conditional measure (length) m_x^s on $V_g^s(x)$ to a measure on $V_g^s(x')$, which is absolutely continuous with respect to $m_{x'}^s$. The Jacobian $\mathrm{Jac}(\theta)(y)$ is not bounded. It is, however, bounded if we allow y to run over the set $V_g^s(x) \cap \Lambda^\ell$. More precisely, for every $x \in \Lambda^\ell$, there is $\bar{J} = \bar{J}(\ell)$ such that

$$m_{x'}^s(\theta(V_g^s(x) \cap \Lambda^\ell)) \leq \bar{J} m_x^s(V_g^s(x) \cap \Lambda^\ell). \tag{11.57}$$

To see this without loss of generality, we may assume that for sufficiently small r,

$$m_{x'}^s(\gamma^s(x', 3\bar{J}r) \cap \Lambda^\ell) \geq 0.9 m_{x'}^s(\gamma^s(x', 3\bar{J}r)). \tag{11.58}$$

In particular, for sufficiently small r,

$$\theta(\gamma^s(x, r)) \subset \gamma^s(x', 3\bar{J}r). \tag{11.59}$$

Indeed, if (11.59) does not hold then $m_{x'}^s(\gamma') \geq 3\bar{J}r$, where $\gamma' = \theta(\gamma)$ is the longer component of $\theta(\gamma^s(x, r)) \setminus \{0\}$ and $\gamma = \theta^{-1}(\gamma') \subset \gamma^s(x, r)$. By (11.58), we have

$$m_{x'}^s(\gamma^s(x', 3\bar{J}r) \cap \gamma' \cap \Lambda^\ell) \geq 0.8 m_{x'}^s(\gamma^s(x', 3\bar{J}r) \cap \gamma').$$

It follows that

$$\begin{aligned} m_{x'}^s(\theta(\gamma^s(x, r)) \cap \Lambda^\ell) &\geq m_{x'}^s(\gamma^s(x', 3\bar{J}r) \cap \gamma' \cap \Lambda^\ell) \\ &\geq 0.8 m_{x'}^s(\gamma^s(x', 3\bar{J}r) \cap \gamma') \\ &\geq 0.8 \cdot 0.5\, m_{x'}^s(\gamma^s(x', 3\bar{J}r)) \\ &\geq 0.4 \cdot 3\bar{J} m_{x'}^s(\gamma^s(x', 3\bar{J}r)) > \bar{J} m_x^s(\gamma^s(x, r)), \end{aligned}$$

contradicting to (11.57).

Choose a point z outside the singularity set and a point $z' \in V_X^u(z)$. (Recall that $V_X^u(z)$ is a local unstable manifold at z for the flow φ_X^t.) Consider the holonomy map $\pi: V_X^{s0}(z) \to W_X^{s0}(z')$ generated by the family of local unstable manifolds for φ_X^t (where $V_X^{s0}(z)$ is a local weakly stable manifold at z and $W_X^{s0}(z')$ is the global weakly stable manifold at z' for the flow φ_X^t). Consider a small ball $B^{s0}(z,r) \subset V_X^{s0}(z)$ centered at z. By (11.59), the size of $\pi(B^{s0}(z,r))$ in the stable direction $E_{g*}^s(z)$ of the disk is bounded by $C_1 r$ for some $C_1 > 3\bar{J}$ independent of r. Since the stable foliation W_A^s of the torus is smooth, the size of $\pi(B^{s0}(z,r))$ in the stable direction $E_A^s(z)$ is bounded by $C_2 r$ for some $C_2 > 0$. Also, the size of $\pi(B^{s0}(z,r))$ in the central direction is bounded by $C_3 r$ for some $C_3 > 0$. We conclude that $\pi(B^{s0}(z,r)) \subset B^{s0}(z', C_4 r)$, where $C_4 > 0$ is a constant independent of r. Applying the above arguments to the inverse map π^{-1}, we obtain that $\pi(B^{s0}(z,r)) \supset B^{s0}(z', C_5 r)$ where $C_5 > 0$ is a constant independent of r. Since the time-t map is essentially accessible, we can now apply a version of the Hopf argument presented in [87] to show that this map is ergodic. This implies that the flow φ_X^t is weakly mixing and hence by Corollary 9.6.10, it is Bernoulli.

Step VIII. In the case $\dim M = 3$, the proof is essentially the same. Consider the suspension flow over g with the roof function $H = 1$. The suspension manifold $K = D \times [0,1]/\sim$ (where $\sim$ is the identification $(x,1) = (g(x),0)$) is diffeomorphic to

$$N = D \times S^1 = \{(x,t) : x \in M, t \in [0,1]\}/\sim$$

(where $\sim$ is the identification $(x,0) = (x,1)$). Define the vector field X on N by

$$X(G(x,t),t) = \left(\frac{\partial G}{\partial t}(x,t), \alpha(G(x,t))\right),$$

where $\alpha(x)$ is a C^∞ function on D satisfying Conditions (A1)–(A3). The vector field X is divergence free and the flow φ_X^t has all the desired properties.

Step IX. In the case $\dim M = 4$, we start with the Bernoulli map with nonzero Lyapunov exponents on a 3-manifold constructed in Theorem 11.4.1. More precisely, define

$$S = g \times \mathrm{Id}\colon D \times S^1 \to D \times S^1.$$

Let $T(x,y) = (g(x), T_{\gamma(x)}y)\colon D \times S^1 \to D \times S^1$, where $T_{\gamma(x)}$ is the rotation by $\gamma(x)$. Here $\gamma\colon D \to \mathbb{R}$ is a nonnegative C^∞ function, which is equal to zero in a small neighborhood of the singularity set $\widetilde{\mathcal{Q}} = \{q_1, q_2, q_3, \partial D\} \times S^1$ and is positive elsewhere.

The function γ can be chosen so that the map T is robustly accessible (i.e., any C^1 perturbation R of T is accessible provided R coincides with T in a small neighborhood of the singularity set $\widetilde{\mathcal{Q}}$). Moreover, there is a perturbation R of T, which has nonzero Lyapunov exponents and is of the form $R = \varphi \circ T$. Here the map φ differs from the identity map in a small neighborhood of a point $z_0 \in D \times S^1$, which lies outside a small neighborhood $\overline{U}$ of the set $\widetilde{\mathcal{Q}}$. To describe the perturbation φ, consider the coordinate system $\xi = \{\xi_1, \xi_2, \xi_3\}$ in an open disk $B(z_0, \varepsilon)$ of sufficiently small radius ε centered at z_0 such that

1. $dm = d\xi$;
2. $E^c_T(z_0) = \partial/\partial\xi_1$, $E^s_T(z_0) = \partial/\partial\xi_2$, and $E^u_T(z_0) = \partial/\partial\xi_3$.

Let $\psi(t)$ be a C^∞ function with support in $(-\varepsilon, \varepsilon)$. Set $\tau = \|\xi\|^2/\gamma^2$ and define $\varphi^{-1}(\xi) = \varphi^{-1}(\xi_1, \xi_2)$ by

$$(\xi_1 \cos(\varepsilon\psi(\tau)) + \xi_2 \sin(\varepsilon\psi(\tau)), -\xi_1 \sin(\varepsilon\psi(\tau)) + \xi_2 \cos(\varepsilon\psi(\tau)), \xi_3).$$

Choose a function $\widehat{\psi}(x, t)\colon \mathbb{R} \times [0, 1] \to \mathbb{R}$ such that

1. $\widehat{\psi}(x, 0) = 0$ and $\widehat{\psi}(x, 1) = \psi(x)$ for any $x \in \mathbb{R}$;
2. $\widehat{\psi}$ is C^∞ in x and t, and $\widehat{\psi}'_t(x, t) = 0$ for $t = 0, 1$ and $x \in \mathbb{R}$;
3. $\|\widehat{\psi}(x, t)\|_r < \delta$ for all $t \in [0, 1]$.

Define the map $\Psi\colon D \times S^1 \times [0, 1] \to D \times S^1$ by the formula $\Psi(x, y, t) = \varphi^t(x, y)$, where $\varphi^t(\xi) = \varphi^t(\xi_1, \xi_2)$ is defined by

$$(\xi_1 \cos(\varepsilon\widehat{\psi}(\tau, t)) + \xi_2 \sin(\varepsilon\widehat{\psi}(\tau, t)), -\xi_1 \sin(\varepsilon\widehat{\psi}(\tau, t)) + \xi_2 \cos(\varepsilon\widehat{\psi}(\tau, t)), \xi_3).$$

Note that Ψ has properties similar to those of the map G in Proposition 11.5.2. Namely,

1. Ψ is C^∞ in (x, y, t);
2. $\Psi(\cdot, 0) = \mathrm{Id}$ and $\Psi(\cdot, 1) = \varphi$;
3. for any $t \in [0, 1]$, the map $\Psi(\cdot, t)\colon D \times S^1 \to D \times S^1$ is area preserving;
4. $\Psi_t(x, y, 1) = \Psi_t(\varphi(x, y), 0)$;
5. $\Psi(x, y, t) = \mathrm{Id}$ for $x \in \overline{U}$.

Consider the manifolds

$$H = \{(x, y, t) : x \in D, y \in S^1, t \in [0, 1]\}/\sim_1,$$

where $\sim_1$ is the identification

$$\sim_1\colon (x, y, 1) = (T(x, y), 0) = (g(x), T_{\gamma(x)}(y), 0),$$

and

$$K = \{(x, y, t) : x \in D, y \in S^1, t \in [0, 1]\}/\sim_2,$$

where $\sim_2$ is the identification

$$\sim_2\colon (x, y, 1) = (\varphi \circ T(x, y), 0) = (\varphi(g(x), T_{\gamma(x)}(y)), 0).$$

Let $\widetilde{G}\colon H \to K$ be the diffeomorphism given by

$$\widetilde{G}(x, y, t) = (\Psi^{-1}(x, y, t), t),$$

where $\Psi^{-1}(x, y, t)$ for each t is the inverse of Ψ. Note that T itself is diffeotopic to S. Let also $\bar{G}$ be the diffeomorphism $K \to K'$ where K' is the suspension manifold of the suspension flow over S. The manifold K' is diffeomorphic to $N = D \times$

$S^1 \times S^1$. If $\widehat{G}\colon K' \to N$ then $F = \widetilde{G} \circ \bar{G} \circ \widehat{G}\colon H \to N$ is a diffeomorphism. Define the vector field X on N by the formula

$$X = \left(\frac{\partial G}{\partial t}(x, t), 0, \alpha(x) \right),$$

where $\alpha(x)$ satisfies Conditions (A1)–(A3). Repeating the above arguments, one can show that X is divergence free and the flow generated by X has all the desired properties. □

11.6 Foliations That Are Not Absolutely Continuous

Let A be a linear hyperbolic automorphism of the two-dimensional torus $\mathbb{T}^2$. Consider the map $F = A \times \mathrm{Id}$ of the three-dimensional torus $\mathbb{T}^3 = \mathbb{T}^2 \times S^1$, where A is a linear hyperbolic automorphism of $\mathbb{T}^2$. Any sufficiently small C^1 perturbation G of F is uniformly partially hyperbolic with one-dimensional central distribution. The latter is integrable and its integral curves form a continuous foliation W^c of M with compact leaves. (They are diffeomorphic to the unit circle; this foliation can be shown to be indeed Hölder continuous; see [200].) If the perturbation G has nonzero Lyapunov exponents in the central direction, the central foliation is not absolutely continuous (compare to the example at the end of Section 8.6.1). More precisely, for almost every $x \in M$ the conditional measure (generated by the Riemannian volume) on the leaf $W^c(x)$ of the central foliation passing through x is atomic.

We describe a far more general version of this result. Let (X, ν) be a probability space and $f\colon X \to X$ an invertible transformation, which preserves the measure ν and is ergodic with respect to ν. Let also M be a smooth compact Riemannian manifold and $\varphi\colon X \to \mathrm{Diff}^{1+\alpha}(M)$ a map. Consider the skew product transformation $F\colon X \times M \to X \times M$ given by

$$F(x, y) = (f(x), \varphi_x(y))$$

and assume that it is Borel measurable. We also assume that F possesses an invariant ergodic measure μ on $X \times M$ such that $\pi_*\mu = \nu$ where $\pi : X \times M \to X$ is the projection.

Fix $x \in X$ and define the maps $\varphi_x^{(k)}$, $k \in \mathbb{Z}$, on M by

$$\varphi_x^{(k+1)} = \varphi_{f^k(x)} \circ \varphi_x^{(k)},$$

where $\varphi_x^{(0)}$ is the identity map. Since the tangent bundle to M is measurably trivial, the derivative map of φ along the M direction gives a cocycle

$$\mathcal{A}\colon X \times M \times \mathbb{Z} \to GL(n, \mathbb{R}), \quad n = \dim M,$$

where $\mathcal{A}(x, y, k) = d_y\varphi_x^{(k)}$. If $\log^+ \|d\varphi\| \in L^1(X \times M, \mu)$ then the Multiplicative Ergodic Theorem 3.4.3 and ergodicity of μ imply that the Lyapunov exponents of this cocycle are constant for μ-almost every (x, y),

$$\chi_1 < \cdots < \chi_\ell, \quad 1 \le \ell \le n.$$

Theorem 11.6.1. *Assume that for some $\gamma > 0$ the function φ satisfies*

$$\log^+ \|d\varphi\|_\gamma \in L^1(X, \nu),$$

where $\|\cdot\|_\gamma$ is the γ-Hölder norm. Assume also that $\chi_\ell < 0$. Then there exists a set $S \subset X \times M$ of full measure and an integer $k \ge 1$ such that

$$\mathrm{card}(S \cap \{x\} \times M) = k$$

for every $(x, y) \in S$.

Proof. We need the following adaptation of the regular neighborhood construction in Section 5.6 to the cocycle $\mathcal{A}$. Applying the Reduction Theorem 3.5.5 we obtain the following description of regular neighborhoods at typical points in $X \times M$.

Lemma 11.6.2. *There exists a set $\Lambda_0 \subset X \times M$ of full measure such that for any $\varepsilon > 0$ the following statements hold:*

1. *There exist a measurable function $r\colon \Lambda_0 \to (0, 1]$ and a collection of embeddings $\Psi_{x,y}\colon B(0, r(x, y)) \to M$ such that $\Psi_{x,y}(0) = y$ and*

$$\exp(-\varepsilon) < \frac{r(F(x, y))}{r(x, y)} < \exp(\varepsilon).$$

2. *If*

$$\varphi_{(x,y)} = \Psi^{-1}_{F(x,y)} \circ \varphi_x \circ \Psi_{(x,y)}\colon B(0, r(x, y)) \to \mathbb{R}^n$$

then

$$\exp(\chi_1 - \varepsilon) \le \|d_0\varphi^{-1}_{(x,y)}\|^{-1}, \|d_0\varphi_{(x,y)}\| \le \exp(\chi_\ell + \varepsilon).$$

3. *The C^1 distance $d_{C^1}(\varphi_{(x,y)}, d_0\varphi_{(x,y)}) < \varepsilon$ in $B(0, r(x, y))$.*
4. *There exist $K > 0$ and a measurable function $A\colon \Lambda_0 \to \mathbb{R}$ such that for any $z, w \in B(0, r(x, y))$,*

$$\begin{aligned} K^{-1} d(\Psi_{(x,y)}(z), \Psi_{(x,y)}(w)) &\le \|z - w\| \\ &\le A(x) d(\Psi_{(x,y)}(z), \Psi_{(x,y)}(w)) \end{aligned}$$

with

$$e^{-\varepsilon} < \frac{A(F(x, y))}{A(x, y)} < e^{\varepsilon}.$$

Decomposing μ into a system of conditional measures

$$d\mu(x, y) = d\mu_x(y) d\nu(x)$$

and using invariance of μ with respect to F, we obtain for ν-almost every $x \in X$,

$$(\varphi_x)_* \mu_x = \mu_{f(x)}.$$

The following result is an immediate consequence of Lemma 11.6.2.

Lemma 11.6.3. *There exists a set $\Lambda \subset \Lambda_0$ and real numbers $R > 0$, $C > 0$, and $0 < \alpha < 1$ such that*

1. $\mu(\Lambda) > 1/2$ *and if* $(x, y) \in \Lambda$ *then* $\mu_x(\Lambda_x) > 1/2$ *where* $\Lambda_x = \{y \in M : (x, y) \in \Lambda\}$;
2. *if* $(x, y) \in \Lambda$ *and* $z \in M$ *are such that the distance in* M, $d_M(y, z) \le R$, *then for all* $m \ge 0$,

$$d_M(\varphi_x^{(m)}(y), \varphi_x^{(m)}(z)) \le C\alpha^m \, d_M(y, z).$$

To prove the theorem it suffices to show that there is a set $B \subset X$ of positive ν-measure such that for any $x \in B$, the measure μ_x has an atom. Indeed, for $x \in X$, set $d(x) = \sup_{y \in M} \mu_x(y)$. Clearly, the function $d(x)$ is measurable, f-invariant, and positive on B. Since f is ergodic, we have $d(x) = d > 0$ for almost every $x \in X$. Let $S = \{(x, y) \in X \times M : \mu_x(y) \ge d\}$. Note that S is F-invariant, has measure at least d, and hence has measure 1. The desired result follows.

Let Λ be the set constructed in Lemma 11.6.3 and $B = \pi(\Lambda)$. Clearly, $\nu(B) > 0$. We shall show that for any $x \in B$, the measure μ_x has an atom.

Let $\mathfrak{U}$ be a cover of M by N closed balls of radius $R/10$. For $x \in X$, set

$$m(x) = \inf \sum \operatorname{diam} U_j,$$

where the infimum is taken over all collections of closed balls $U_1, \ldots, U_k$ in M such that $k \le N$ and $\mu_x(\bigcup_{j=1}^k U_j) \ge 1/2$. We also set $m = \operatorname{ess\,sup}_{x \in B} m(x)$. We will show that $m = 0$. Otherwise, there is $p > 0$ such that

$$C \Delta N \alpha^p < \frac{m}{2}, \tag{11.60}$$

where Δ is the diameter of M. For $x \in B$, let $U_1(x), \ldots, U_{k(x)}(x)$ be those balls in the cover $\mathfrak{U}$ that meet Λ_x. Since these balls cover Λ_x and $\mu_x(\Lambda_x) > 1/2$, we have that

$$\mu_x\left(\bigcup_{j=1}^{k(x)} U_j(x)\right) \ge \frac{1}{2}. \tag{11.61}$$

Taking into account that $(\varphi_x^{(i)})_* \mu_x = \mu_{f^i(x)}$, for all i we obtain

$$\mu_{f^i(x)}\left(\bigcup_{j=1}^{k(x)} \varphi_x^{(i)} U_j(x)\right) \ge \frac{1}{2}.$$

Since the balls $U_j(x)$ meet Λ_x and have diameter less than $R/10$ by Lemma 11.6.3, we obtain that

$$\operatorname{diam}(\varphi_x^{(i)} U_j(x)) \le C \Delta \alpha^i. \tag{11.62}$$

Let $\tau(x)$ be the first return time of the point $x \in B$ to B under the map f^p and $B_i = \{x \in B : \tau(x) = i\}$. We have that $B = \bigcup_{i=1}^{\infty} B_i$ and since f is invertible and its inverse f^{-1} preserves ν, we also have that

$$B' = \bigcup_{i=1}^{\infty} f^{pi}(B_i) = B \pmod 0.$$

If $z \in B'$ then $z = f^{pi}(x)$ for some $x \in B_i$ and some $i \geq 1$. It follows from the definition of $m(z)$ and inequalities (11.60)–(11.62) that

$$m(z) \leq \sum_{j=1}^{k(x)} \operatorname{diam}(\varphi_x^{(pi)} U_j(x)) \leq Ck(x)\Delta\alpha^{pi} \leq CN\Delta\alpha^p \leq \frac{m}{2}.$$

This implies that

$$m = \operatorname{ess\,sup}_{x\in B} m(x) = \operatorname{ess\,sup}_{z\in B'} m(z) \leq \frac{m}{2},$$

contradicting the assumption that $m > 0$.

We conclude that $m = 0$, and hence $m(x) = 0$ for ν-almost every $x \in B$. It implies that for such an x there is a sequence of closed balls $U^1(x)$, $U^2(x)$, ... with

$$\lim_{i\to\infty} \operatorname{diam} U^i(x) = 0$$

and $\mu_x(U^i(x)) \geq 1/(2N)$ for all i. Take $z_i \in U^i(x)$. Any accumulation point of the sequence $\{z_i\}$ is an atom for μ_x. □

Corollary 11.6.4. *Let μ be an ergodic measure for a $C^{1+\alpha}$ diffeomorphism of a compact Riemannian manifold M. If μ has all of its exponents negative then μ is concentrated on the orbit of a periodic sink.*

Proof. The result follows from Theorem 11.6.1 by taking $X = \{x\}$ and ν to be the point mass. □

Let f be a $C^{1+\alpha}$ diffeomorphism of a compact smooth Riemannian manifold M preserving a Borel probability measure μ. Assume that W is a foliation of M with smooth leaves, which is invariant under f. We say that f is *W-dissipative* if $S_W(x) \neq 0$ for μ-almost every $x \in M$, where $S_W(x)$ denotes the sum of the Lyapunov exponents of f at the point x along the subspace $T_x W(x)$.

For $x \in M$, we denote by $\operatorname{vol}(W(x))$ the leaf volume of the leaf $W(x)$ through x. We say that the foliation W has *finite-volume leaves almost everywhere* if the set of those $x \in M$, for which $\operatorname{vol}(W(x)) < \infty$, has full volume. An example of a foliation whose leaves have finite volume almost everywhere is a foliation with smooth compact leaves. If W is such a foliation then the function $M \ni x \mapsto \operatorname{vol}(W(x))$ is well defined (finite) but may not be bounded (see [67]).

Theorem 11.6.5. *Let f be a $C^{1+\alpha}$ diffeomorphism of a compact smooth Riemannian manifold M preserving a smooth measure μ. Let also W be an f-invariant foliation of M with smooth leaves. Assume that W has finite-volume leaves almost* *everywhere. If f is W-dissipative then the foliation W is not absolutely continuous in the weak sense (see Section 8.6).*

Proof. Let $A^- \subset M$ be the set of points for which $S_W(x) < 0$ and $A^+ \subset M$ the set of points for which $S_W(x) > 0$. They both are f-invariant and either $m(A^-) > 0$ or $m(A^+) > 0$ or both. (Here we use the fact that the invariant measure μ is smooth and hence equivalent to volume.) Without loss of generality we may assume that $m(A^+) > 0$. Then for a sufficiently small $\lambda > 0$, sufficiently large integer ℓ, and every small $\varepsilon > 0$, there exists a Borel set $A^+_{\lambda,\ell,\varepsilon} \subset A^+$ of positive μ-measure such that for every $x \in A^+_{\lambda,\ell,\varepsilon}$ and $n \geq 0$,

$$|\text{Jac}(df^n|T_xW(x))| \geq \ell^{-1} e^{\lambda n} e^{-\varepsilon n}. \tag{11.63}$$

Given $V > 0$, consider the set

$$Y_V = \{y \in M : \text{vol}(W(y)) \leq V\}.$$

Observe that the set $A^+_{\lambda,\ell,\varepsilon}$ has positive volume and let $x \in A^+_{\lambda,\ell,\varepsilon}$ be a density point of m. One can choose $V > 0$ such that the set

$$R = A^+_{\lambda,\ell,\varepsilon} \cap B(x,r) \cap Y_V$$

has positive volume.

Assume on the contrary that the foliation W is absolutely continuous. Then for almost every $y \in R$, the set $R_y = R \cap W(y)$ has positive volume in $W(y)$. Observe that $\mu(R) > 0$ (where μ is an invariant smooth measure for f). Therefore the trajectory of almost every point $y \in R$ returns to R infinitely often. Let y be such a point and $\{n_k\}$ the sequence of successive returns to R. We may also assume that $m_{W(y)}(R_y) > 0$.

Since $f^{n_k}(y) \in R \subset Y_V$, we have that for every $k > 0$,

$$m_{W(f^{n_k}(y))}(f^{n_k}(R_y)) \leq W(f^{n_k}(y)) \leq V.$$

On the other hand, by (11.63),

$$\begin{aligned}\text{vol}(W(f^{n_k}(y))) &\geq m_{W(f^{n_k}(y))}(f^{n_k}(R_y)) \\ &= \int_{R_y} \text{Jac}(df^{n_k}|T_zW(z))\,dm(z) \\ &\geq \ell^{-1} e^{(\lambda-\varepsilon)n_k} m(R_y) > V\end{aligned}$$

if n_k is sufficiently large. This yields a contradiction and completes the proof of the theorem. □

11.7 An Open Set of Diffeomorphisms with Nonzero Lyapunov Exponents on Tori

It is shown in [31] that any partially hyperbolic diffeomorphism f_0 with one-dimensional central direction preserving a smooth measure μ can be slightly perturbed such that the new map f is partially hyperbolic, preserves μ, and has negative (or positive) central exponents (hence the results of Section 11.2 apply to f). This result was first obtained by Shub and Wilkinson [200] in the particular case when f_0 is the direct product of a hyperbolic automorphism of the 2-torus and the identity map of the circle. The perturbation that removes zero exponents can be arranged in the form (11.47). The proof in the general case is a modification of the argument in [200] (see also [17] and [54]).

One can use this observation to obtain an open set of non-Anosov diffeomorphisms with nonzero Lyapunov exponents on multidimensional tori. Consider a diffeomorphism $f_0 = A \times \mathrm{Id}$ of the torus $\mathbb{T}^p = \mathbb{T}^{p-1} \times S^1$, $p \geq 3$, where A is a linear hyperbolic automorphism of $\mathbb{T}^{p-1}$. It is partially hyperbolic and preserves volume. Let f be a C^2 diffeomorphism, which is a small perturbation of f_0 in the C^1 topology, preserving volume and having negative central exponents. One can arrange the perturbation f to have the accessibility property. Then any volume preserving C^2 diffeomorphism g, which is sufficiently close to f, is ergodic and has nonzero Lyapunov exponents almost everywhere.

Note that g is partially hyperbolic and the central distribution E^c is one dimensional. By a result in [86] this distribution is integrable and the leaves W^c of the corresponding foliation are smooth closed curves, which are diffeomorphic to circles. The foliation W^c is continuous (indeed, it is Hölder continuous) but is not absolutely continuous (see [200]). Moreover, there exists a set E of full measure and an integer $k > 1$ such that E intersects almost every leaf $W^c(x)$ at exactly k points (see [196]; the example in Section 8.6.1 is of this type).

11.8 Notes

Viana [220] gave the first robust non-Axiom A examples of attractors with nonzero central exponents. The central foliation in the example by Shub and Wilkinson [200] described in Section 11.2 is not absolutely continuous—the phenomenon known as "Fubini's nightmare." The result in [200] was extended by Ruelle [195] who showed that for an open set of one-parameter families of (not necessarily volume preserving) maps F_ε through F_0, each map F_ε possesses a u-measure with negative central exponent.

Our proof of Theorem 11.2.4 follows the argument in [42]. The notion of σ-hyperbolic times was introduced in [4]. Our proofs of Lemmas 11.2.6 and 11.2.7 follow [4, 5].

Theorem 11.3.1 provides the only known example of a volume preserving diffeomorphism with nonzero Lyapunov exponents and countably many ergodic components. Our proof closely follows [58].

Theorem 11.4.1 provides an affirmative solution to the problem posed in [39]. For surface diffeomorphisms Theorem 11.4.1 was proved by Katok [100]. In [37], for any compact smooth Riemannian manifold M of dimension at least 5, Brin constructed a C^∞ Bernoulli diffeomorphism, which preserves volume and has all but one Lyapunov exponent nonzero. Our proof of Theorem 11.4.1 is a simple modification of arguments in [59]. The argument in the proof of

Proposition 11.4.22 follows some ideas of Mañé [140] (see also the work of Bochi and Viana [28]).

Our proof of Theorem 11.5.1 is a slight improvement of arguments in [89].

Theorem 11.6.1 was proved by Ruelle and Wilkinson in [196]. A somewhat similar result was obtained by Katok (unpublished). Our proof follows closely [196]. The proof of Theorem 11.6.5 presented here is an adaptation of the argument in [85].

12

Geodesic Flows

For a long time geodesic flows have played an important stimulating role in the development of hyperbolic theory. In the beginning of the twentieth century, Hadamard and Morse, while studying the statistics of geodesics on surfaces of negative curvature, pointed out that the local instability of trajectories gives rise to some global properties of dynamical systems such as ergodicity and topological transitivity. The further study of geodesic flows and some objects closely related to them (e.g., frame flows) later inspired the introduction of different classes of hyperbolic dynamical systems (e.g., Anosov systems, uniformly partially hyperbolic systems, and nonuniformly hyperbolic systems preserving volume). On the other hand, geodesic flows always were a touchstone for applying new advanced methods of the general theory of dynamical systems. This, in particular, has led to some new interesting results in differential and Riemannian geometry. In this chapter we will present some of these results. While describing ergodic properties of geodesic flows in the spirit of the book, we consider the Liouville measure that is invariant under the flow (we allow other invariant measures when discussing an upper bound for the measure-theoretic entropy), and we refer the reader to the excellent survey [115] where measures of maximal entropy are studied.

12.1 Hyperbolicity of Geodesic Flows

Throughout the chapter we assume that M is a compact smooth p-dimensional connected Riemannian manifold without boundary with a Riemannian metric of class C^3.

The *geodesic flow* g_t acts on the tangent bundle TM by the formula

$$g_t(v) = \dot{\gamma}_v(t),$$

where $\dot{\gamma}_v(t)$ is the unit tangent vector to the geodesic $\gamma_v(t)$ defined by the vector v (i.e., $\dot{\gamma}_v(0) = v$; this geodesic is uniquely defined) (see Fig. 12.1). The geodesic

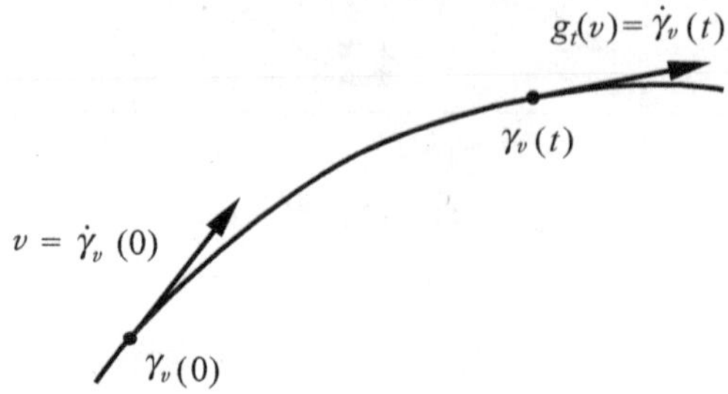

Figure 12.1. Geodesic flow.

flow generates a vector field V on TM given by

$$V(v) = \frac{d(g_t(v))}{dt}\bigg|_{t=0}.$$

Since M is compact, the flow g_t is well defined for all $t \in \mathbb{R}$ and is a smooth flow.

We recall some necessary notions from Riemannian geometry (see [61, 63]). We endow the second tangent bundle $T(TM)$ with a special Riemannian metric. Let $\pi : TM \to M$ be the natural projection (i.e., $\pi(x, v) = x$ for each $x \in M$ and each $v \in T_xM$) and $K : T(TM) \to TM$ the linear (connection) operator defined by $K\xi = (\nabla Z)(t)|_{t=0}$, where $Z(t)$ is any curve in TM such that $Z(0) = d\pi\xi$, $(d/dt)Z(t)|_{t=0} = \xi$, and ∇ is the covariant derivative. (One can show that $K\xi$ does not depend on the choice of the curve $Z(t)$.) The canonical (Sasaki) metric in $T(TM)$ is given by

$$\langle \xi, \eta \rangle_v = \langle d\pi\xi, dv\pi\eta \rangle_{\pi(v)} + \langle K\xi, K\eta \rangle_{\pi(v)}.$$

The set $SM \subset TM$ of the unit vectors is invariant under the geodesic flow and is a compact manifold of dimension $2p - 1$. In what follows we consider the geodesic flow restricted to SM.

The study of hyperbolic properties of the geodesic flow is based upon the description of solutions of the variational equation for the flow. This equation along a given trajectory $g_t(v)$ of the flow is the Jacobi equation along the geodesic $\gamma_v(t)$:

$$Y''(t) + R_{XY}X(t) = 0. \tag{12.1}$$

Here $Y(t)$ is a vector field along $\gamma_v(t)$, $X(t) = \dot{\gamma}(t)$, and R_{XY} is the curvature operator. More precisely, the relation between the variational equation and the Jacobi equation (12.1) can be described as follows. Fix $v \in SM$ and $\xi \in T_vSM$. Let $Y_\xi(t)$ be the unique solution of (12.1) satisfying the initial conditions $Y_\xi(0) = d\pi\xi$ and $Y'_\xi(0) = K\xi$. One can show that the map $\xi \mapsto Y_\xi(t)$ is an isomorphism for which $d\pi\, dg_t\xi = Y_\xi(t)$ and $K\, dg_t\xi = Y'_\xi(t)$. This map establishes the identification between solutions of the variational equation and solutions of the Jacobi equation (12.1).

Recall that the Fermi coordinates $\{e_i(t)\}, i = 1, \ldots, p$, along the geodesic $\gamma_v(t)$ are obtained by the time-t parallel translation along $\gamma_v(t)$ of an orthonormal basis $\{e_i(0)\}$ in $T_{\gamma_v(0)}M$ where $e_1(t) = \dot{\gamma}(t)$. Using these coordinates one can rewrite (12.1) in the matrix form

$$\frac{d^2}{dt^2}A(t) + K(t)A(t) = 0, \tag{12.2}$$

where $A(t) = (a_{ij}(t))$ and $K(t) = (k_{ij}(t))$ are matrix functions and the entries of $K(t)$ are $k_{ij}(t) = K_{\gamma_v(t)}(e_i(t), e_j(t))$.

Two points $x = \gamma(t_1)$ and $y = \gamma(t_2)$ on the geodesic γ are called *conjugate* if there exists a nonidentically zero Jacobi field Y along γ such that $Y(t_1) = Y(t_2) = 0$. Two points $x = \gamma(t_1)$ and $y = \gamma(t_2)$ are called *focal* if there exists a Jacobi field Y along γ such that $Y(t_1) = 0$, $Y'(t_1) \neq 0$, and $(d/dt)\|Y(t)\|^2|_{t=t_2} = 0$.

We say that the manifold M has

1. *no conjugate points* if on each geodesic no two points are conjugate;
2. *no focal points* if on each geodesic no two points are focal;
3. *nonpositive curvature* if for any $x \in M$ and any two vectors $v_1, v_2 \in T_xM$, the sectional curvature $K_x(v_1, v_2)$ satisfies

$$K_x(v_1, v_2) \leq 0. \tag{12.3}$$

If the manifold has no focal points then it has no conjugate points, and if it has nonpositive curvature then it has no focal points. Strictly speaking, the property of having no conjugate or focal points, or of having nonpositive curvature, is a property of the Riemannian metric. However, since the metric is fixed, with some abuse of language, we shall refer to these properties as properties of the manifold M itself.

From now on we consider only manifolds without conjugate points. The boundary-value problem for (12.2) has a unique solution; that is, for any numbers s_1 and s_2 and any matrices A_1 and A_2, there exists a unique solution $A(t)$ of (12.2), satisfying $A(s_1) = A_1$ and $A(s_2) = A_2$.

Given $s \in \mathbb{R}$, let $A_s(t)$ be the unique solution of (12.2) satisfying the boundary conditions: $A_s(0) = \mathrm{Id}$ (where Id is the identity matrix) and $A_s(s) = 0$. Then there exists a limit (see [63])

$$\lim_{s\to+\infty} \frac{d}{dt}A_s(t)\Big|_{t=0} = A^+. \tag{12.4}$$

We define the *positive limit solution* $A^+(t)$ of (12.2) as the solution that satisfies the initial conditions:

$$A^+(0) = \mathrm{Id} \quad \text{and} \quad \frac{d}{dt}A^+(t)\Big|_{t=0} = A^+.$$

This solution is nondegenerate (i.e., $\det A^+(t) \neq 0$ for every $t \in \mathbb{R}$) and $A^+(t) = \lim_{s\to+\infty} A_s(t)$. Similarly, letting $s \to -\infty$ one can define the *negative limit solution* $A^-(t)$ of (12.2).

For $v \in SM$ and w orthogonal to v, we call the Jacobi field $Y_w^+(t) = A^+(t)w$ a *positive limit Jacobi field*. Similarly, one defines a *negative limit Jacobi field*, $Y_w^-(t) = A^-(t)w$.

For every $v \in SM$, set

$$E^+(v) = \{\xi \in T_vSM : \langle \xi, V(v)\rangle = 0 \text{ and } Y_\xi(t) = A^+(t)d\pi\xi\}, \quad (12.5)$$
$$E^-(v) = \{\xi \in T_vSM : \langle \xi, V(v)\rangle = 0 \text{ and } Y_\xi(t) = A^-(t)d\pi\xi\}, \quad (12.6)$$

where V is the vector field generated by the geodesic flow.

Proposition 12.1.1 (Eberlein [61, 63]). *The subspaces $E^+(v)$ and $E^-(v)$ have the following properties:*

1. *$E^+(v)$ and $E^-(v)$ are linear subspaces of T_vSM, each of dimension $p-1$.*
2. *$d\pi E^+(v) = d\pi E^-(v) = \{w \in T_{\pi(v)}M : w$ is orthogonal to $v\}$.*
3. *The subspaces $E^+(v)$ and $E^-(v)$ are invariant under the differential dg_t; that is,*

$$dg_t E^+(v) = E^+(g_t(v)) \quad \text{and} \quad dg_t E^-(v) = E^-(g_t(v)).$$

4. *If $\tau : SM \to SM$ is the involution defined by $\tau v = -v$, then*

$$E^-(-v) = d\tau E^+(v) \quad \text{and} \quad E^+(-v) = d\tau E^-(v).$$

5. *If $K_x(v_1, v_2) \geq -a^2$ for some $a > 0$ and all $x \in M$ then $\|K\xi\| \leq a\|d\pi\xi\|$ for every $\xi \in E^-(v)$ or $\xi \in E^+(v)$.*
6. *If $\xi \in E^-(v)$ or $\xi \in E^+(v)$, then $Y_\xi(t) \neq 0$ for every $t \in \mathbb{R}$.*
7. *$\xi \in E^+(v)$ (respectively, $\xi \in E^-(v)$) if and only if*

$$\langle \xi, V(v)\rangle = 0 \quad \text{and} \quad \|d\pi\, dg_t\xi\| \leq c$$

for every $t > 0$ (respectively, $t < 0$) and some $c > 0$.

8. *If M has no focal points then for $\xi \in E^+(v)$ (respectively, $\xi \in E^-(v)$), the function $t \mapsto \|Y_\xi(t)\|$ is nonincreasing (respectively, nondecreasing).*

In view of Properties 5 and 8, we have $\xi \in E^+(v)$ (respectively, $\xi \in E^-(v)$) if and only if $\langle \xi, V(v)\rangle = 0$ and $\|dg_t\xi\| \leq c$ for $t > 0$ (respectively, $t < 0$), for some constant $c > 0$. This observation and Property 3 justify to call $E^+(v)$ and $E^-(v)$ the *stable* and *unstable subspaces*.

In general, the subspaces $E^+(v)$ and $E^-(v)$ do not span the whole second tangent space T_vSM. If they do span T_vSM for every $v \in SM$, then the geodesic flow is Anosov (see [63,64]). This is the case when the curvature is strictly negative (but there are examples of Anosov geodesic flows on manifolds whose curvature is positive in some parts of the manifold). For a general manifold without conjugate points, one can only expect that the geodesic flow is nonuniformly hyperbolic. To see this consider the sets

$$\Delta^{\pm} = \left\{ v \in SM : \overline{\lim_{t\to\pm\infty}} \frac{1}{t} \int_0^t K_{\pi(g_s(v))}(g_s(v), Y_w^{\pm}(s))\, ds < 0 \right.$$
$$\left. \text{for every } w \in SM \text{ orthogonal to } v \right\}.$$

$$\Delta = \Delta^+ \cap \Delta^-. \tag{12.7}$$

Here $Y_w^{\pm}(s) = A^{\pm}(s)w = Y_\xi(s)$, where $\xi \in E^{\pm}(v)$ is the only vector for which $d\pi\xi = w$. It is easy to see that the sets Δ^+, Δ^-, and Δ are Borel measurable and invariant under the flow g_t. The following result shows that the Lyapunov exponents are nonzero on Δ:

Theorem 12.1.2 ([169,170]). *Let M be a Riemannian manifold without conjugate points. Then $\chi(v,\xi) < 0$ for any $v \in \Delta^+$ and $\xi \in E^+(v)$ or for any $v \in \Delta^-$ and $\xi \in E^-(v)$.*

Proof. Let $\psi : \mathbb{R}^- \to \mathbb{R}$ be a continuous function. Set

$$\overline{\psi} = \overline{\lim_{t\to+\infty}} \frac{1}{t} \int_0^t \psi(\tau)\, d\tau, \quad \underline{\psi} = \underline{\lim_{t\to+\infty}} \frac{1}{t} \int_0^t \psi(\tau)\, d\tau,$$
$$\widetilde{\psi} = \underline{\lim_{t\to\infty}} \frac{1}{t} \int_0^t \psi(s)^2\, ds.$$

We need the following lemma:

Lemma 12.1.3. *Assume that $c = \sup_{t\geq 0} |\psi(t)| < \infty$. Then*

1. *if $\psi(t) \leq 0$ for all $t \geq 0$ and $\widetilde{\psi} > 0$ then $\overline{\psi} < 0$;*
2. *if $\psi(t) \geq 0$ for all $t \geq 0$ and $\widetilde{\psi} > 0$ then $\underline{\psi} > 0$.*

Proof of the lemma. Assume that $\psi(t) \leq 0$. Then $\overline{\psi} \leq 0$. On the other hand, if $c > 0$ then

$$-\frac{\overline{\psi}}{c} = \underline{\left|\frac{\psi}{c}\right|} \geq \widetilde{\left(\frac{\psi}{c}\right)} = \frac{\widetilde{\psi}}{c^2} > 0.$$

This implies that $\overline{\psi} < 0$ and completes the proof of the first statement. The proof of the second statement is similar. □

We proceed with the proof of the theorem. Fix $v \in \Delta^+$ and $\xi \in E^+(v)$, and consider the function $\varphi(t) = (1/2)\|Y_\xi(t)\|^2$. Using (12.2) we obtain

$$\frac{d^2}{dt^2}\varphi(t) = \frac{1}{2}\frac{d^2}{dt^2}\langle Y_\xi, Y_\xi\rangle = -K(t)\varphi(t) + \|Y_\xi'(t)\|^2,$$

where $K(t) = K_{\gamma_v(t)}(Y_\xi(t), w(t))$ and $w(t) = \dot{\gamma}_v(t)$. It follows from Proposition 12.1.1 that $\varphi(t) \neq 0$ for all $t \in \mathbb{R}$. Set

$$z(t) = (\varphi(t))^{-1} \frac{d}{dt}\varphi(t).$$

It is easy to check that the function $z(t)$ satisfies the Ricatti equation

$$\frac{d}{dt}z(t) + z(t)^2 - (\varphi(t))^{-1}\|Y_\xi'(t)\|^2 + K(t) = 0. \tag{12.8}$$

By Proposition 12.1.1,

$$\begin{aligned}\left|\frac{d}{dt}\varphi(t)\right| &= \frac{1}{2}\left|\frac{d}{dt}\langle Y_\xi, Y_\xi\rangle\right| = |\langle Y_\xi, Y_\xi'\rangle| \\ &= |\langle d\pi\, dg_t\xi, K\, dg_t\xi\rangle| \leq a\|d\pi\, dg_t\xi\|^2 = 2a\varphi(t).\end{aligned}$$

It follows that $\sup_{t\geq 0}|z(t)| \leq 2a$. Integrating the Ricatti equation (12.8) on the interval $[0, t]$, we obtain that

$$z(t) - z(0) + \int_0^t z(s)^2\, ds = \int_0^t (\varphi(s))^{-1}\|Y_\xi'(s)\|^2\, ds - \int_0^t K(s)\, ds.$$

It follows that for $v \in \Delta^+$,

$$\begin{aligned}\underline{\lim}_{t\to\infty}\frac{1}{t}\int_0^t z(s)^2\, ds \geq \underline{\lim}_{t\to\infty}\frac{1}{t}\int_0^t (\varphi(s))^{-1}\|Y_\xi'(s)\|^2\, ds \\ - \overline{\lim}_{t\to\infty}\frac{1}{t}\int_0^t K(s)\, ds > 0.\end{aligned}$$

Therefore in view of Lemma 12.1.3, we conclude that

$$\overline{\lim}_{t\to\infty}\frac{1}{t}\int_0^t z(s)\, ds < 0.$$

On the other hand, using Proposition 12.1.1 we find that

$$\begin{aligned}\chi(v, \xi) &= \overline{\lim}_{t\to\infty}\frac{1}{t}\log\|dg_t\xi\| = \overline{\lim}_{t\to\infty}\frac{1}{t}\log\|d\pi\, dg_t\xi\| \\ &= \overline{\lim}_{t\to\infty}\frac{1}{t}\log\|Y_\xi(t)\| = \frac{1}{2}\overline{\lim}_{t\to\infty}\frac{1}{t}\int_0^t z(s)\, ds.\end{aligned}$$

This completes the proof of the first statement of the theorem. The second statement can be proved in a similar way. □

The geodesic flow preserves the Liouville measure μ on the tangent bundle that is induced by the Riemannian metric. We denote by m the Lebesgue measure on M. It follows from Theorem 12.1.2 that if the set Δ has positive Liouville measure then almost every trajectory of the geodesic flow g_t originated in Δ (in particular, every Lyapunov–Perron regular trajectory originated in Δ) is nonuniformly hyperbolic. It is therefore crucial to find conditions, which would guarantee that Δ has positive Liouville measure.

We begin with the two-dimensional case.

Theorem 12.1.4. *Let M be a smooth compact surface. The following statements hold:*

1. *Assume that M has no conjugate points and*
$$\int_M K(x)\,dm(x) < 0. \tag{12.9}$$
Then $\mu(\Delta) > 0$.
2. *Assume that M has nonpositive curvature and genus ≥ 1. Then $\mu(\Delta) > 0$.*

Proof. Choose two orthogonal vectors $v, w \in SM$. It is easy to see that
$$\overline{\lim_{t\to\pm\infty}}\,\frac{1}{t}\int_0^t K_{\pi(g_s(v))}(v,w)\,ds = \overline{\lim_{t\to\pm\infty}}\,\frac{1}{t}\int_0^t K(\pi(g_s(v)))\,ds.$$
By Birkhoff's Ergodic Theorem applied to the function $v \to K(\pi(v))$, we obtain that for μ-almost every $v \in SM$, there exists the limit
$$\lim_{t\to+\infty}\frac{1}{t}\int_0^t K(\pi(g_s v))\,ds = \lim_{t\to-\infty}\frac{1}{t}\int_0^t K(\pi(g_s v))\,ds = \Phi(v)$$
and that
$$\int_{SM}\Phi(v)\,d\mu(v) = \int_{SM}K(\pi(v))\,d\mu(v) = \int_M K(x)\,dm(x).$$
The first statement follows from (12.9). The second statement follows from the first one and the Gauss–Bonnet Formula: the Euler characteristic of M is $\frac{1}{2\pi}\int_M K(x)\,dm(x) = 2 - 2g$, where g is the genus of M. □

In the multidimensional case we have the following criterion for positivity of the Liouville measure of the set Δ. Define the continuous function $\kappa : SM \to (-\infty, 0]$ by
$$\kappa(v) = \max\{K_{\pi(v)}(v,w) : w \in S_{\pi(v)}M \text{ is orthogonal to } v\}.$$

Theorem 12.1.5. *The following statements hold:*

1. *Let M be a smooth compact Riemannian manifold without conjugate points. Assume that*
$$\int_{SM}\kappa(v)\,d\mu(v) < 0. \tag{12.10}$$
Then $\mu(\Delta) > 0$.
2. *Let M be a smooth compact Riemannian manifold of nonpositive curvature. Assume that there exist $x \in M$ and a vector $v \in S_x M$ such that*
$$K_x(v,w) < 0 \tag{12.11}$$
for any $w \in S_x M$, which is orthogonal to v. Then $\mu(\Delta) > 0$.
3. *Let M be a smooth compact Riemannian manifold of nonpositive curvature. Assume that there is a geodesic γ such that there is no nonzero parallel Jacobi field along γ orthogonal to $\dot\gamma$. Then $\mu(\Delta) > 0$ (see [15]).*

Proof. By Birkhoff's Ergodic Theorem applied to the function $v \to K(v)$, for μ-almost every $v \in SM$ there exists the limit

$$\lim_{t\to+\infty} \frac{1}{t}\int_0^t \kappa(g_s(v))\,ds = \lim_{t\to-\infty} \frac{1}{t}\int_0^t \kappa(g_s(v))\,ds = \Psi(v),$$

and

$$\int_{SM} \Psi(v)\,d\mu(v) = \int_{SM} \kappa(v)\,d\mu(v) < 0.$$

Therefore, $\Psi(v) < 0$ for v in some set of positive μ-measure, and hence

$$\overline{\lim_{t\to\pm\infty}} \frac{1}{t}\int_0^t K_{\pi(g_s(v))}(v, w)\,ds \le \lim_{t\to\pm\infty} \frac{1}{t}\int_0^t \kappa(g_s(v))\,ds = \Psi(v) < 0$$

for v in some set of positive μ-measure and any $w \in SM$, which is orthogonal to v. This implies the first statement of the theorem.

To prove the second statement observe that (12.11) holds in a small neighborhood of v. Since the curvature is nonpositive, (12.11) implies (12.10).

To prove the third statement we will show that there exists $v \in SM$ such that for every negative limit Jacobi field $Y(t)$ along $\gamma_v(t)$, one can find a number $t = t(Y) \ge 0$ for which

$$K(\dot\gamma_v(t), Y(t)) < 0. \tag{12.12}$$

Indeed, if this is not true then for any $v \in SM$, there is a negative limit Jacobi field $Y(v)$ along γ_v such that for all $t \ge 0$,

$$K(\dot\gamma_v(t), Y(v)(t)) = 0.$$

Set $v_n = g_{-n}(v)$, $n \ge 0$. Renormalizing $Y(v_n)$, if necessary, we can assume that $\|Y(v_n)(n)\| = 1$. The vectors $Y(v_n)(n)$ are orthogonal to v, and a subsequence of $\{Y(v_n)(n)\}$ converges. The negative limit Jacobi field Y along γ_v determined by the limit satisfies

$$K(\dot\gamma_v(t), Y(t)) = 0$$

for all $t \in \mathbb{R}$. It follows that Y is parallel. Indeed, since the curvature is nonpositive, $\langle R(X, \dot\gamma_v)\dot\gamma_v, X\rangle \le 0$ for all X. Hence, $\langle R(X, \dot\gamma_v)\dot\gamma_v, X\rangle = 0$ implies that $R(X, \dot\gamma_v)\dot\gamma_v = 0$. Therefore, $\nabla^2 Y = \nabla^2 Y + R(X, \dot\gamma_v)\dot\gamma_v = 0$ and thus ∇Y is parallel. Let $\{X_i\}$ be a basis of parallel vector fields along γ_v. Then $\nabla Y = \sum_i \alpha_i X_i$ and hence $Y = \sum_i(\alpha_i t + \beta_i)X_i$. Since Y is negative limit Jacobi field, $\|Y(t)\|$ is bounded and hence $\alpha_i = 0$ for all i and Y is parallel. Conversely, if Y is a parallel Jacobi field along a geodesic γ, then $K(\dot\gamma(t), Y(t)) = 0$ for all $t \in \mathbb{R}$.

We conclude that for the geodesic γ in the hypothesis of Statement 3 the relation (12.12) holds. Since Y depends continuously on $Y(0)$, by compactness there exists $T > 0$ and $b < 0$ such that

$$\int_0^T K(\dot\gamma(t), Y(t))\,dt < b$$

for all negative limit Jacobi fields Y along γ. Since the limit of a sequence of negative limit Jacobi fields is an asymptotic Jacobi field, there is an open neighborhood U of $\dot\gamma(0)$ in SM such that for every $v \in U$,

$$\int_0^T K(g_t(v), Y(t))\, dt < b$$

for all negative limit Jacobi fields Y along γ. By Birkhoff's Ergodic Theorem, for almost every $v \in U$, the trajectory $g_t(v)$ will return to U regularly; that is,

$$\varliminf_{t\to\infty} \frac{1}{t}\int_0^t \chi_U(g_s(v))\, ds > 0,$$

where χ_U is the indicator of U. Every such vector v is contained in Δ. □

We remark that Condition (12.11) is a multidimensional version of Condition (12.9).

We conclude this section with the following statement. Let

$$\Lambda^+ = \{v \in SM : \chi(v, \xi) < 0 \text{ for any } \xi \in E^+(v)\},$$
$$\Lambda^- = \{v \in SM : \chi(v, \xi) > 0 \text{ for any } \xi \in E^-(v)\},$$

$$\Lambda = \Lambda^+ \cap \Lambda^-.$$

Theorem 12.1.6. *Let M be a compact smooth Riemannian manifold without focal points. Assume that the set Λ has positive volume. Then Λ is open* (mod 0).

Proof. Given an open set $U \subset SM$ and $w \in SM$, we say that the orbit $g_t(w)$ has positive frequency of return to U if $\varliminf_{t\to\infty}(|T(t)|/t) > 0$, where

$$T(t) = \{\tau : 0 \le \tau \le t \text{ and } g_\tau(w) \in U\}$$

and $|T(t)|$ is the Lebesgue measure of the set $T(t)$.

Since the set Λ has positive volume for almost every $v \in \Lambda$, there exists $t > 0$ such that

$$\|dg_t | E^+(v)\| \le \frac{1}{2} \quad \text{and} \quad \|dg_{-t} | E^-(v)\| \le \frac{1}{2}.$$

Since the subspaces $E^+(v)$ and $E^-(v)$ vary continuously with $v \in SM$, it follows that there is a neighborhood $U \subset SM$ of v such that for every $w \in U$,

$$\|dg_t | E^+(w)\| \le \frac{3}{4} \quad \text{and} \quad \|dg_{-t} | E^-(w)\| \le \frac{3}{4}. \tag{12.13}$$

Let $\mathcal{U}$ be the set of points that have positive frequency of returns to U. By Poincaré's Recurrence Theorem, $\mathcal{U}$ is not empty. It is easy to see that in fact, it is open (mod 0). For $w \in \mathcal{U}$, let t_0 be a moment of time when the orbit $g_t(w)$ "exits" the closure $\overline{U}$. Since the manifold M has no focal points, by Statement 9 of Proposition 12.1.1, the function $t \mapsto \|dg_t | E^+(w)\|$ is nonincreasing, and hence

$\|dg_t|E^+(w)\| \le \|dg_{t_0}|E^+(w)\|$ for all $t \ge t_0$. It follows from (12.13) that for every $\xi \in E^+(w)$,

$$\chi(w,\xi) = \overline{\lim_{t\to\infty}} \frac{1}{t} \log \|dg_t\xi\| \le \lim_{t\to\infty} \frac{|T(t)|}{t} \log \frac{3}{4} < 0.$$

Similar arguments show that for every $w \in \mathcal{U}$ and $\eta \in E^-(w)$, the Lyapunov exponent satisfies $\chi(w,\eta) > 0$. We conclude that for almost every $v \in \Lambda$, almost every w in a small neighborhood around v lies in Λ. The desired result now follows. □

12.2 Ergodic Properties of Geodesic Flows

Consider the geodesic flow g_t on a compact smooth Riemannian p-dimensional manifold M without conjugate points. Let the set Λ be given by (12.7). We assume that it has positive Liouville measure. By Theorem 12.1.2, the geodesic flow is nonuniformly hyperbolic on Λ and hence the results of Section 9.1 apply and show that the ergodic components of $g_t|\Lambda$ are of positive Liouville measure (see Theorem 9.1.1). We shall show that under some mild geometric assumptions the geodesic flow on Λ is indeed ergodic. To achieve this we first prove that every ergodic component of positive measure is open (mod 0) and then use a result by Eberlein on topological transitivity of geodesic flows.

To establish local ergodicity of $g_t|\Lambda$, we construct two invariant foliations of SM, W^+ and W^-, such that $W^s(x) = W^+(x)$ and $W^u(x) = W^-(x)$ for almost every $x \in \Lambda$.

Denote by H the universal Riemannian cover of M, that is, a simply connected complete Riemannian manifold for which $M = H/\Gamma$ where Γ is a discrete subgroup of the group of isometries of H, isomorphic to $\pi_1(M)$. According to the Hadamard–Cartan Theorem, any two points $x, y \in H$ are joined by a single geodesic, which we denote by γ_{xy}. For any $x \in H$, the exponential map $\exp_x : \mathbb{R}^p \to H$ is a diffeomorphism. Hence the map

$$\varphi_x(y) = \exp_x\left(\frac{y}{1-\|y\|}\right) \tag{12.14}$$

is a homeomorphism of the open unit ball D onto H.

Two geodesics $\gamma_1(t)$ and $\gamma_2(t)$ in H are asymptotic if

$$\sup_{t>0} \rho(\gamma_1(t), \gamma_2(t)) < \infty,$$

where ρ is the distance in H induced by the Riemannian metric. The asymptoticity is an equivalence relation, and the equivalence class $\gamma(\infty)$ corresponding to a geodesic γ is a point at infinity. The set of these classes is denoted by $H(\infty)$ and is the ideal boundary of H.

Fix a point $x \in H$ and a vector $v \in SH$. Consider a sequence of vectors $v_n \in SH$, $v_n \to v$, a sequence of points $x_n \in H$, $x_n \to x$, and a sequence of numbers $t_n \to \infty$. Denote by γ_n the geodesic joining the points x_n and $\gamma_{v_n}(t_n)$. Since the sequence of vectors $\dot{\gamma}_n(0)$ is compact, the sequence of geodesics has a limit geodesic. Following [170] we say that the manifold M satisfies the *Asymptoticity Axiom* if for any choice of $x_n, x \in H, v_n, v \in SH, x_n \to x, v_n \to v$, and $t_n \to \infty$, any limit geodesic of the sequence of geodesics γ_n is asymptotic to the geodesic γ. Of course, this is a property of the Riemannian metric on M.

If the manifold M satisfies the Asymptoticity Axiom then the sequence γ_n, indeed, converges to γ. (Otherwise, there would exist two geodesic rays passing through x and asymptotic to each other.) Moreover, given a geodesic γ and a point $x \in H$, there exists a geodesic σ passing through x and asymptotic to γ. (The vector $v = \dot{\sigma}(0)$ is the limit of vectors $v_n = \dot{\sigma}_n(0)$, where σ_n is the geodesic connecting x with the point $\gamma(t_n)$ for some sequence $t_n \to \infty$.) This asymptotic geodesic is uniquely defined. (Indeed, for any two geodesic rays $\gamma_1(t)$ and $\gamma_2(t)$ starting from a point $x = \gamma_1(0) = \gamma_2(0)$, we have that $\rho(\gamma_1(t), \gamma_2(t)) \to \infty$ as $t \to \infty$; see [63].) One can show that if the manifold M has no focal points (in particular, if it has nonpositive curvature) then it satisfies the Asymptoticity Axiom (see [170]).

Using (12.14) one can extend the topology of the space H to $\overline{H} = H \cup H(\infty)$ so that $\overline{H}$ becomes a compact space. The map φ_x can be extended to a homeomorphism (still denoted by φ_x) of the closed ball $\overline{D} = D \cup S^{p-1}$ onto $\overline{H}$ by the equality

$$\varphi_x(y) = \gamma_y(+\infty), \qquad y \in S^{p-1}.$$

In particular, φ_x maps S^{p-1} homeomorphically onto $H(\infty)$.

Following [62] we say that the manifold M (more precisely, the Riemannian metric in M) satisfies *Axiom* 1 if for any two distinct points x and y on the ideal boundary, there is a geodesic, which joins them. This geodesic is uniquely defined if the Riemannian metric is of strictly negative curvature (i.e., if inequality (12.3) is strict). Otherwise, there may exist a pair of distinct points $x, y \in H(\infty)$, which can be joined by more than one geodesic. If the manifold has no focal points, any two geodesics on the universal cover, which are asymptotic for both $t > 0$ and $t < 0$ (i.e., they join two distinct points on the ideal boundary), bound a flat strip – a statement known as the *flat strip theorem* (see Proposition 12.2.6).

The fundamental group $\pi_1(M)$ of the manifold M acts on the universal cover H by isometries. This action can be extended to the ideal boundary $H(\infty)$. Namely, if $p = \gamma_v(+\infty) \in H(\infty)$ and $\zeta \in \pi_1(M)$, then $\zeta(p)$ is the equivalence class of geodesics that are asymptotic to the geodesic $\zeta(\gamma_v(t))$.

Following [170] we now describe the invariant foliations for the geodesic flow. We consider the distributions E^+ and E^- introduced in (12.5) and (12.6).

Proposition 12.2.1. *Assume that the manifold M has no conjugate points and satisfies the Asymptoticity Axiom and Axiom* 1. *Then the distributions E^+ and E^- are integrable. Their integral manifolds form continuous foliations W^+ and W^- of SM with smooth leaves. These foliations are invariant under the geodesic flow.*

Proof. We establish integrability of the distribution E^+. The proof of the integrability of the distribution E^- is similar. Consider the sphere $S^{p-1}(x,t)$, $x \in H$, $t > 0$. Let $y \in S^{p-1}(x,t)$, $w \in T_t S^{p-1}(x,t)$, $\|w\| = 1$ and let $\delta_t(s)$, $\varepsilon \le s \le \varepsilon$, be a geodesic segment on the sphere such that $\delta_t(0) = y$ and $\dot\delta_t(0) = w$. Consider the variation

$$r_t(u,s) = \gamma_{\delta_t(s)x}(u), \quad \gamma_{\delta_t(0)x} = \delta_t(s).$$

The vector field

$$Y_t(u) = \frac{\partial}{\partial s} r_t(u,s)|_{s=0} \tag{12.15}$$

has the following properties (see [63]): it is a Jacobi field along the geodesic $\gamma_{yx}(u)$ satisfying

$$Y_t(0) = w, \quad Y_t(t) = 0 \tag{12.16}$$

and

$$\langle Y_t(u), \dot\gamma_{yx}(u)\rangle = 0, \quad 0 \le u \le t. \tag{12.17}$$

It is known (see [63]) that a Jacobi field satisfying (12.16) and (12.17) admits the following estimate for $t > 0$,

$$\|Y_t'(s)\| \le a \coth(as)\|Y_t(s)\|. \tag{12.18}$$

Let $n_t(s)$ be a vector at the point $\delta_t(s)$ normal to $S^{p-1}(x,t)$. Since $\langle \dot\delta_t(s), n_t(s)\rangle = 0$ for any $s \in [-\varepsilon, \varepsilon]$, one has

$$\langle \dot\delta_t(s), n_t(s)\rangle = -\langle \dot\delta_t(s), \dot n_t(s)\rangle, \quad -\varepsilon \le t \le \varepsilon.$$

Since the parameter u is the length of the geodesic, we have

$$\frac{\partial}{\partial u} r_t(u,s)|_{u=0} = \gamma_{\delta_t(s)x}(0) = n_t(s).$$

It follows that

$$Y'(0) = \frac{\partial}{\partial u}\frac{\partial}{\partial s} r_t(u,s)|_{s=0,u=0} = \frac{\partial}{\partial s}\frac{\partial}{\partial u} r_t(u,s)|_{u=0,s=0} = \frac{\partial}{\partial s} n_t(s)|_{s=0}.$$

Hence the curvature $K(t,w)$ of the curve $\delta_t(s)$ for $s = 0$ admits the estimate

$$|K(t,w)| = |\langle \dot\delta_t'(s), n_t(s)\rangle|_{s=0}| \le \|w\|\|Y_t'(0)\| = \|Y_t'(0)\|.$$

By (12.18), we have

$$|K(t,w)| \le a \coth(at). \tag{12.19}$$

Fix $v \in SM$ with $x = \pi(v)$ and a number $\varepsilon > 0$. Consider the geodesic element $\Pi(v,\varepsilon)$ that is orthogonal to v, that is, the set of geodesic segments of length 2ε whose initial vector is orthogonal to v. (This is a submanifold of H of class C^2.) In a sufficiently small neighborhood (corresponding to small ε), the set $\exp^{-1}(S^{p-1}(\gamma_v(t),t))$ is a graph of a function $\varphi_t(u)$, $u \in \exp^{-1}(\Pi(v,\varepsilon))$, of class

C^2, such that $\varphi_t(0) = 0$ and $(d/du)\varphi_t(0) = 0$. It follows from (12.19) that for sufficiently small ε the curvature of the manifold $\exp^{-1}(\Pi(v, \varepsilon))$ is bounded from above uniformly in $v \in SM$ and $t \geq 1$. (We also use here the fact that M is a smooth compact manifold of class C^3.) Hence for any $t \geq 1$ and $u \in \exp^{-1}(\Pi(v, \varepsilon))$,

$$\left\| \frac{d^2}{du^2} \varphi_t(u) \right\| \leq Ca, \tag{12.20}$$

where $C > 0$ is a constant independent of v. We conclude that the family $\{\varphi_t(u)\}$ is relatively compact in the C^1 topology. Let φ be a limit function of this family.

Lemma 12.2.2. *The following statements hold:*

1. $\varphi \in C^1$.
2. *For* $y \in \text{graph}(\varphi)$, *the geodesic passing through the point* $\exp_x y$, *orthogonal to the submanifold* $L(x) = \exp_x \text{graph}(\varphi)$, *is asymptotic to the geodesic* $\gamma_v(t)$.
3. *If* $\delta_t(s)$, $-\varepsilon \leq s \leq \varepsilon$, $\delta_t(0) = x$, *is a geodesic segment on* $L(x)$ *and if* $Z(s)$ *is the framing of* $\delta_t(s)$ *by vectors normal to* $L(x)$, *then the variation*

$$r(t, s) = \exp(tZ(s)), \quad t \geq 0, \quad -\varepsilon \leq s \leq \varepsilon,$$

defines by (12.15) *a Jacobi field* $Y(t) = Y_\xi(t)$ *for some* $\xi \in E^+(v)$ *with* $d\pi\xi = \delta_t(0)$.

Proof of the lemma. The first statement follows from the construction of the function φ.

To prove the second statement observe that the function φ is the limit in the C^1 topology of the sequence of functions $\{\varphi_{t_n}\}$ for some sequence of numbers $t_n \to \infty$. For $y \in \text{graph}(\varphi)$, there is a sequence of points $y_n \in \text{graph}(\varphi_{t_n})$. Set $z_n = \exp_x y_n$ and $z = \exp_x y$. Since the geodesic $\gamma_n = \gamma_{z_n \gamma_v(t_n)}$ is orthogonal to the sphere $S^{p-1}(\gamma_v(t_n), t_n)$ centered at z_n, the sequence of geodesics $\{\gamma_n\}$ converges to a geodesic, which is orthogonal to $L(x)$ at z. Since M satisfies the Asymptoticity Axiom, this geodesic is asymptotic to $\gamma_v(t)$.

We shall prove the third statement. Choose a vector $w \in T_x M$, orthogonal to v, and let $\delta_t(s)$, $-\varepsilon \leq s \leq \varepsilon$, be a geodesic segment on the sphere $S^{p-1}(\gamma_v(t), t)$ such that $\delta_t(0) = x$ and $\dot{\delta}_t(0) = w$. Taking an orthonormal system of parallel vector fields along the geodesic $\gamma_v(t)$, we write the Jacobi field (12.15) in the form $Y_t(u) = A_t(u)w$ where $A_t(u)$ is a solution of the matrix equation (12.2) satisfying the boundary conditions $A_t(0) = \text{Id}$ and $A_t(t) = 0$. By (12.4), the limit

$$\lim_{t_n \to \infty} Y_{t_n}(u) = \lim_{t_n \to \infty} A_{t_n}(u)w = A_t w = Y_\xi(u)$$

exists. The desired result follows. □

For $y \in L(x)$ denote by $\gamma(y, t)$ the geodesic through y, orthogonal to $L(x)$, and such that $\gamma(y, 0) = y$. Since M satisfies the Asymptoticity Axiom, by Statement 2 of Lemma 12.2.2, the geodesics $\gamma(y_1, t)$ and $\gamma(y_2, t)$ do not intersect if $y_1 \neq y_2$.

Hence the set

$$U(x) = \bigcup_{y \in L(x)} \bigcup_{-\varepsilon < t < \varepsilon} \gamma(y, t)$$

is an open neighborhood of x. To each point $z \in U(x)$, we assign the vector $v(z)$ that is the unit tangent vector to the geodesic $\gamma(y, t)$ passing through z. Let $\hat{\varphi}(u)$ be any other limit function to the family $\{\varphi_t(u)\}$ and let $\hat{L}(x) = \text{graph}(\hat{\varphi})$. By Lemma 12.2.2, for $x \in \hat{L}(x)$ and $z \in \hat{L}(x)$ the geodesic γ passing though z and orthogonal to $\hat{L}(x)$ is asymptotic to $\gamma_v(t)$. Hence, $\gamma = \gamma(y, \cdot)$ where y is the point of intersection of γ and $\hat{L}(x)$. In other words, the manifold $\hat{L}(x)$ is orthogonal to the vector field $v(z)$. Introducing a local coordinate system in $U(x)$ and using the classical uniqueness theorem in the theory of differential equations, we conclude that $\varphi(u) = \hat{\varphi}(u)$, $u \in \exp_x^{-1}(\Pi(v, \varepsilon))$. In particular, $\varphi_t \to \varphi$ as $t \to \infty$.

Denote by $\widetilde{W}^+(v)$ the framing of $L(x)$ by orthonormal vectors directed as v. It follows from Lemma 12.2.2 that $\widetilde{W}^+(v)$ is a $(p-1)$-dimensional submanifold of SH of class C^1 having the following characteristic property:

> *there is a neighborhood $U(v)$ of v such that for $w \in \widetilde{W}^+(v) \cap U(v)$, the geodesic $\gamma_w(t)$ is orthogonal to the submanifold $\pi(\widetilde{W}^+(v))$ and is asymptotic to the geodesic $\gamma_v(t)$.*

It follows from Statement 3 of Lemma 12.2.2 that

$$v \in \widetilde{W}^+(v), \quad T_v \widetilde{W}^+(v) = E^+(v).$$

Lemma 12.2.3. *The submanifold $\widetilde{W}^+(v)$ depends continuously in the C^1 topology on $v \in SH$.*

Proof of the lemma. Let $v_n, v \in SH$, $v_n \to v$ as $n \to \infty$. Set $\gamma_n(t) = \gamma_{v_n}(t)$, $x_n = \pi(v_n)$, $x = \pi(v)$, and $\gamma(t) = \gamma_v(t)$, and consider the sphere $S^{p-1}(x_n, t)$. For sufficiently small $\varepsilon > 0$ and any $t \geq 1$, the surface $L_{n,t} = \exp_{x_n}^{-1}(S^{p-1}(x_n, t))$ can be represented as the graph of a function of class C^2, denoted by $\psi(n, t, u)$, $u \in \exp_{x_n}^{-1}(\Pi(v_n, \varepsilon))$, and the surface $L_n = \exp_{x_n}^{-1}(\pi(\widetilde{W}^+(v_n)))$ as the graph of a function of class C^1, denoted by $\psi(n, u)$, $u \in \exp_{x_n}^{-1}(\Pi(v_n, \varepsilon))$. Furthermore, we have that $\psi(n, t) \to \psi(n)$ as $t \to \infty$ in the C^1 topology for any $n > 0$. By (12.20), for sufficiently small $\varepsilon > 0$ the surfaces $L_{n,t}$ and L_n can be represented as graphs of functions denoted by $\bar{\psi}(n, t, u)$ and $\bar{\psi}(n, u)$ for $u \in \exp^{-1}(\Pi(v, \varepsilon))$. Moreover, the family of functions $\{\bar{\psi}(n, t), \bar{\psi}(n)\}$ is compact in the C^1 topology. Let $\bar{\psi}$ be the limit function of the family of functions $\{\bar{\psi}(n)\}$. Choose sequences of numbers $n_k \to \infty$ and $t_k \to \infty$ such that $\bar{\psi}(n_k, t_k) \to \bar{\psi}$ and set

$$L = \{\exp_x(u, \bar{\psi}(u)) : u \in \exp_x^{-1}(\Pi(v, \varepsilon))\}.$$

Fix $y \in L$ and let $y_k \in L_{n_k, t_k}$ be a sequence of points converging to y as $k \to \infty$. By the Asymptoticity Axiom, the geodesics $\bar{\gamma}_k$, joining the points y_k and $\gamma_{n_k}(t_k)$, converge to some geodesic $\bar{\gamma}$ asymptotic to γ. Moreover, $\bar{\gamma}$ is orthogonal to the submanifold L (because the geodesics $\bar{\gamma}_k$ are orthogonal to L_{n_k, t_k}). The desired result follows from the characteristic property of $\widetilde{W}^+(v)$. □

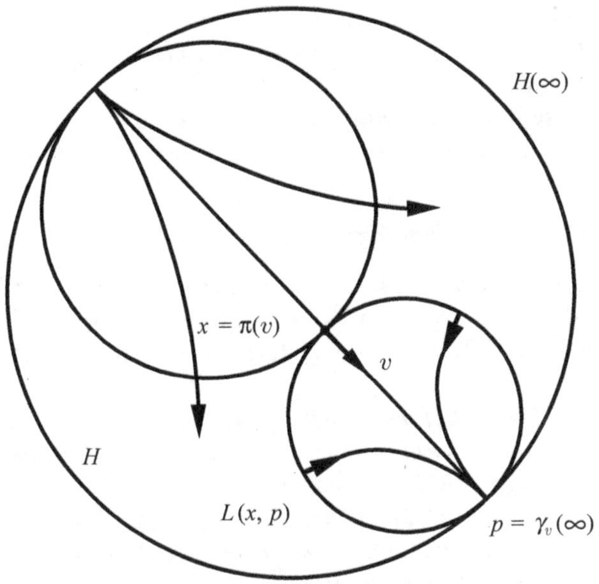

Figure 12.2. Horocycles.

As a corollary of the characteristic property of $\widetilde{W}^+(v)$, we have that for any $v \in SH$ and $w \in \widetilde{W}^+(v)$,

$$\widetilde{W}^+(w) \cap U(v) \subset \widetilde{W}^+(v).$$

This allows one to "glue together" the submanifolds $\widetilde{W}^+(v)$ passing through different $v \in SH$. More precisely, we call v, $w \in SH$ *equivalent* if there are $v = v_1, v_2, \dots, v_m = w$ such that $v_i \in \widetilde{W}^+(v_{i-1})$, $i = 2, \dots, m$. The set of equivalence classes form the required continuous foliation with smooth leaves. □

We call W^+ and W^- the *horospherical foliations* of SM and we denote their lifts from SM to SH by $\widetilde{W}^+$ and $\widetilde{W}^-$ respectively. We summarize the properties of these foliations in the following proposition. Throughout the remaining part of the chapter, we always assume that the manifold M satisfies the Asymptoticity Axiom and Axiom 1.

Proposition 12.2.4. *The following statements hold:*

1. $W^+(-v) = W^-(v)$ *and* $W^-(-v) = W^+(v)$.
2. *There is* $\delta > 0$ *such that for any* $v \in SM$,
   (a) *for* $w \in W^+(v) \cap B(v, \delta)$, *the geodesic* $\gamma_w(t)$ *is orthogonal to the submanifold* $\pi(W^+(v) \cap B(v, \delta))$ *and is asymptotic to the geodesic* $\gamma_v(t)$;
   (b) *for* $w \in W^-(v) \cap B(v, \delta)$, *the geodesic* $\gamma_w(t)$ *is orthogonal to the submanifold* $\pi(W^-(v) \cap B(v, \delta))$ *and is asymptotic to the geodesic* $\gamma_{-v}(t)$.
3. $\widetilde{W}(v)$ *is a connected* $(p-1)$*-dimensional closed submanifold of* SH.
4. $d\zeta \widetilde{W}^+(v) = \widetilde{W}^+(d\zeta v)$ *and* $d\zeta \widetilde{W}^-(v) = \widetilde{W}^-(d\zeta v)$ *for every isometry* $\zeta \in \pi_1(M)$.

Given $x \in H$ and $p \in H(\infty)$, set $L(x, p) = \pi(\widetilde{W}^+(v))$ where $x = \pi(v)$ and $p = \gamma_v(\infty)$. The set $L(x, p)$ is called the *horosphere* centered at p and passing through x. (In the case of surfaces, $L(x, p)$ is called the *horocycle*; however, we shall use the term "horosphere" in the two-dimensional case as well; see Fig. 12.2). We describe some properties of horospheres in the following proposition.

Proposition 12.2.5. *The following statements hold:*

1. *For any $x \in H$ and $p \in H(\infty)$, there exists a unique horosphere $L(x, p)$ centered at p, which passes through x; it is a limit in the C^1 topology of spheres $S^p(\gamma(t), t)$ as $t \to +\infty$ where γ is the unique geodesic joining x and p.*
2. *The leaf $W^+(v)$ is the framing of the horosphere $L(x, p)$ ($x = \pi(v)$ and $p = \gamma_v(+\infty)$) by orthonormal vectors that have the same direction as the vector v (i.e., they are "inside" the horosphere). The leaf $W^-(v)$ is the framing of the horosphere $L(x, p)$ ($x = \pi(v)$ and $p = \gamma_v(-\infty) = \gamma_{-v}(+\infty)$) by orthonormal vectors that have the same direction as the vector $-v$ (i.e., they are "outside" the horosphere).*
3. *$\zeta(L(x, p)) = L(\zeta(x), \zeta(p))$ for every isometry $\zeta \in \pi_1(M)$.*
4. *For every $v, w \in SH$, for which $\gamma_v(+\infty) = \gamma_w(+\infty) = p$, the geodesic $\gamma_w(t)$ intersects the horosphere $L(\pi(v), p)$ at some point.*

We call $v \in SH$ a *nonuniqueness* vector if there exists a vector $w \in SH$ such that

$$\gamma_v(+\infty) = \gamma_w(+\infty) \quad \text{and} \quad \gamma_v(-\infty) = \gamma_w(-\infty).$$

Other vectors $v \in SH$ are called uniqueness vectors.

We denote by $\rho_{W^+(v)}$ the distance in $W^+(v)$ induced by the Riemannian metric.

Proposition 12.2.6. *Assume that the manifold M has no focal points.*

1. *If v is a nonuniqueness vector then there is a vector $w \in SH$, orthogonal to v, and a geodesic segment $\delta(s)$, $0 \le s \le a$, in SH such that*
   (a) *$\delta(0) = \pi(v)$, $\dot{\delta}(0) = w$, $\delta(s) \in L(\pi(v), \gamma_v(+\infty))$ for $s \in [0, a]$;*
   (b) *the union of the geodesics passing through points of $\delta(s)$, orthogonal to $L(\pi(v), \gamma_v(+\infty))$, is the image under a global geodesic isometric embedding of the strip*

$$\{(s, t) \in \mathbb{R} : 0 \le s \le a, -\infty < t < \infty\}.$$

2. *If $v \in SH$ is a uniqueness vector then for any $w \in W^+(v)$, the function*

$$\psi(t) = \rho_{W^+(g_t(v))}(g_t(v), g_t(w))$$

is monotonically increasing and tends to $+\infty$ as $t \to -\infty$.

The following statement shows that global stable and unstable manifolds are contained in the leaves of the horospherical foliations through vectors $v \in \Lambda$. It also expresses expanding properties of local stable and unstable manifolds (see Theorem 9.2.6).

Theorem 12.2.7. *Assume that the manifold M has no focal points and that the set Λ has positive Liouville measure. Then*

1. $W^+(v) \supset V^s(v)$ *and* $W^-(v) \supset V^u(v)$ *for almost every* $v \in \Lambda$;
2. *for any sufficiently small* $\delta > 0$, *there exists a measurable function* $t(v) = t(v, \delta)$ *on* Λ *such that for almost every* $v \in \Lambda$ *and any* $t \geq t(v)$,

$$g_{-t}(V^s(v)) \supset B^+(g_{-t}(v), \delta_0) \quad \text{and} \quad g_t(V^u(v)) \supset B^-(g_t(v), \delta_0),$$

where $B^+(w, r)$ *and* $B^-(w, r)$ *are balls in* $W^+(v)$ *and* $W^-(v)$, *respectively, centered at* v *of radius* r.

Proof. For μ-almost every $v \in \Lambda$, consider local stable and unstable manifolds $V^s(v)$ and $V^u(v)$ and denote by $\widetilde{V}^s(\widetilde{v})$ and $\widetilde{V}^u(\widetilde{v})$ their lifts to SH (with $\widetilde{v}$ being a lift of v). Given $\widetilde{w} \in \widetilde{V}^s(\widetilde{v})$, we have

$$\rho(\pi g_t(\widetilde{v}), \pi g_t(\widetilde{w})) \to 0 \text{ as } t \to \infty. \tag{12.21}$$

It follows that the geodesics $\gamma_{\widetilde{v}}(t)$ and $\gamma_{\widetilde{w}}(t)$ are asymptotic, and hence $\gamma_{\widetilde{v}}(+\infty) = \gamma_{\widetilde{w}}(+\infty)$. We wish to show that $\widetilde{w} \in \widetilde{W}^-(\widetilde{v})$. Assuming the contrary, consider the horosphere $L(\pi\widetilde{v}, \gamma_{\widetilde{v}}(+\infty))$. Let $\widetilde{z}$ be the point of intersection of the geodesic $\gamma_{\widetilde{w}}(t)$ and this horosphere (such a point exists by Proposition 12.2.5). We have

$$\rho(\pi g_t(\widetilde{v}), \pi g_t(\widetilde{w})) \geq \rho(\pi\widetilde{w}, \widetilde{z}) > 0,$$

which contradicts to (12.21). It follows that $\widetilde{V}^s(\widetilde{v}) \subset \widetilde{W}^+(\widetilde{v})$ for every $v \in \Lambda$ and every lift $\widetilde{v}$ of v. Arguing similarly, one can show that $\widetilde{V}^u(\widetilde{v}) \subset \widetilde{W}^-(\widetilde{v})$ for every $v \in \Lambda$ and every lift $\widetilde{v}$ of v. The first statement follows.

The second statement is an immediate corollary of the following lemma. Let Λ^ℓ be a regular set of positive volume.

Lemma 12.2.8. *For any sufficiently small* $\delta > 0$ *and* $\alpha > 0$, *one can find a number* $T = T(\delta, \alpha)$ *with the following property: for any* $v \in \Lambda^\ell$, *any* $w_1, w_2 \in V^s(v)$ *satisfying* $\rho_{W^s(v)}(w_1, w_2) = \alpha$, *and any* $t \geq T$,

$$\rho_{W^s(g_{-t}(v))}(g_{-t}(w_1), g_{-t}(w_2)) \geq \delta$$

(where $\rho_{W^s(z)}$ *is the distance in* $W^s(z)$ *induced by the Riemannian metric).*

Proof of the lemma. Assuming the contrary, one can find $\delta > 0, \alpha > 0$, sequences of vectors $v_n \in \Lambda^\ell$, $w_n^1, w_n^2 \in V^s(v_n)$, and a sequence of numbers $t_n \to \infty$ such that $\rho_{W^s(v_n)}(w_n^1, w_n^2) = \alpha$ and

$$\rho_{W^s(g_{-t_n}(v_n))}(g_{-t_n}(w_n^1), g_{-t_n}(w_n^2)) < \delta. \tag{12.22}$$

Since the set Λ^ℓ is compact, passing to a subsequence if necessary we may assume that the sequence $\{v_n\}$ converges to a vector $v \in \Lambda^\ell$. Let $\widetilde{v}_n$ and $\widetilde{v}$ be lifts of v_n and v respectively. We can choose a sequence of isometries $\zeta_n \in \pi_1(M)$ such that

$d\zeta_n\widetilde{v}_n \to \widetilde{v}$. Let $\widetilde{w}_n^1$ and $\widetilde{w}_n^2$ be the lifts of w_n^1 and w_n^2 such that $\widetilde{w}_n^1, \widetilde{w}_n^2 \in \widetilde{V}^s(\widetilde{v}_n)$. Passing to a subsequence if necessary we may assume that $d\zeta_n\widetilde{w}_n^1 \to \widetilde{w}^1$ and $d\zeta_n\widetilde{w}_n^2 \to \widetilde{w}^2$, where $\widetilde{w}^1, \widetilde{w}^2 \in \widetilde{V}^s(\widetilde{v})$ and $\rho_{W^s(\widetilde{v})}(\widetilde{w}^1, \widetilde{w}^2) = \alpha$.

Note that for every $\widetilde{w} \in \widetilde{V}^s(\widetilde{v})$ and every $\xi \in E^+(\widetilde{v})$, the Lyapunov exponent satisfies $\chi(\widetilde{v}, \xi) < 0$ and hence by Statement 1 of Proposition 12.2.6, every vector $\widetilde{w} \in \widetilde{V}^s(\widetilde{v})$ is a uniqueness vector. Therefore, by Statement 2 of this proposition, there exists $R(\delta, \alpha) > 0$ such that

$$\rho_{\widetilde{W}^s(g_{-R(\delta,\alpha)}(\widetilde{v}))}(g_{-R(\delta,\alpha)}(\widetilde{w}^1), g_{-R(\delta,\alpha)}(\widetilde{w}^2)) \geq 2\delta.$$

Hence for sufficiently large n,

$$a = \rho_{\widetilde{W}^s(g_{-R(\delta,\alpha)}(\widetilde{v}_n))}(g_{-R(\delta,\alpha)}(\widetilde{w}_n^1), g_{-R(\delta,\alpha)}(\widetilde{w}_n^2)) \geq \frac{3}{2}\delta.$$

Using the monotonicity of $\rho_{\widetilde{W}^s(g_{-t}(\widetilde{v}_n))}(g_{-t}(\widetilde{w}_n^1), g_{-t}(\widetilde{w}_n^2))$ at uniqueness vectors (see Statement 2 of Proposition 12.2.6), we obtain that for sufficiently large $t_n \geq R(\delta, \alpha)$,

$$\rho_{\widetilde{W}^s(g_{-t_n}(\widetilde{v}_n))}(g_{-t_n}(\widetilde{w}_n^1), g_{-t_n}(\widetilde{w}_n^2)) \geq a > \frac{3}{2}\delta.$$

However, this contradicts to (12.22). □

The proposition is proven. □

Following Eberlein [62] we say that the manifold M satisfies the *Uniform Visibility Axiom* if for any $\varepsilon > 0$ there exists $R = R(\varepsilon) > 0$ such that from each point $x \in H$, any geodesic segment γ with $d(x, \gamma) \geq R$ is visible at an angle less than ε.

Proposition 12.2.9 (Eberlein [62]). *The following statements hold:*

1. *If a manifold satisfies the Uniform Visibility Axiom with respect to a Riemannian metric then it satisfies this axiom with respect to any other Riemannian metric with no conjugate points.*
2. *A manifold of nonpositive curvature satisfies the Uniform Visibility Axiom if its universal cover does not admit an isometric geodesic embedding of* $\mathbb{R}^2$.
3. *A compact two-dimensional manifold M of genus ≥ 1 satisfies the Uniform Visibility Axiom.*
4. *If M satisfies the Uniform Visibility Axiom then it also satisfies the Asymptoticity Axiom and Axiom* 1 *(see [170]).*

The following statement by Eberlein [62] establishes topological transitivity of geodesic flows.

Theorem 12.2.10. *Assume that the compact manifold M satisfies the Uniform Visibility Axiom. Then the geodesic flow g_t is topologically transitive.*

Proof. Two points $x, y \in H(\infty)$ are said to be *dual* if for any open sets U and V, containing x and y respectively, there exists an element $\xi \in \pi_1(M)$ such that

$\xi(\bar{H} \setminus U) \subset V$. Clearly, $\xi(\bar{H} \setminus U) \subset V$ if and only if $\xi^{-1}(\bar{H} \setminus V) \subset U$. Since the manifold M is compact and satisfies the Uniform Visibility Axiom, one can show that any two points $x, y \in H(\infty)$ are dual (see [62]).

Lemma 12.2.11. *If the points $x, y \in H(\infty)$ are dual then there exists a sequence $\xi_n \in \pi_1(M)$ such that $\xi_n^{-1}(p) \to x$ and $\xi_n(p) \to y$ as $n \to \infty$ for any $p \in H$.*

Proof of the lemma. Let $\{U_n\}$ and $\{V_n\}$ be local bases for the cone topology at x and y respectively. For each $n > 0$, there is $\xi_n \in \pi_1(M)$ such that $\xi_n(\bar{H} \setminus U_n) \subset V_n$ and $\xi_n^{-1}(\bar{H} \setminus V_n) \subset U_n$. For any $p \in H$ and any sufficiently large n, we have $p \in (\bar{H} \setminus U_n) \cap (\bar{H} \setminus V_n)$, and hence $\xi_n(p) \in V_n$ and $\xi_n^{-1}(p) \in U_n$. The lemma follows. □

Given $v, w \in SM$, let $\widetilde{v}$ and $\widetilde{w}$ be their lifts to SH. Denote by $\zeta : H \to M$ the covering map. Then $d\zeta$ maps SH onto SM.

Since the points $x = \gamma_{\widetilde{v}}(+\infty)$ and $y = \gamma_{\widetilde{w}}(-\infty)$ are dual, by the lemma, there exists a sequence $\xi_n \in \pi_1(M)$ such that $\xi_n^{-1}(p) \to x$ and $\xi_n(p) \to y$ as $n \to \infty$ for any $p \in H$. Let $p = \pi\widetilde{v}$ and $q = \pi\widetilde{w}$ where $\pi : SM \to M$ is the projection, and set $t_n = d(\xi_n(p), q)$ and $\widetilde{v}_n = \dot{\gamma}_{\xi_n(p),q}(0)$. (Recall that $\gamma_{\xi_n(p),q}(t)$ is the geodesic connecting the points $\xi_n(p)$ and q.) Then $g_{t_n}(\widetilde{v}_n) = -\dot{\gamma}_{q,\xi_n(p)}(0)$. Since $\xi_n(p) \to y$, it follows that $g_{t_n}(\widetilde{v}_n) \to \widetilde{w}$. Observing that $\xi_n^{-1}(q) \to x$, we obtain that $d\xi_n^{-1}\widetilde{v}_n = \dot{\gamma}_{p,\xi_n^{-1}(q)}(0) \to \widetilde{v}$. It follows that

$$d\zeta\widetilde{v}_n = d\zeta\ d\xi_n^{-1}\widetilde{v}_n \to d\zeta\widetilde{v} = v$$

and that

$$g_{t_n}(d\zeta\widetilde{v}_n) = d\zeta g_{t_n}(\widetilde{v}_n) \to d\zeta\widetilde{w} = w.$$

This means that for any open neighborhoods U and V of v and w respectively, the set $g_t(U)$ meets V for arbitrarily large positive values of t, implying topological transitivity of the geodesic flow g_t. □

We are now ready to prove the main result describing ergodic properties of geodesic flows.

Theorem 12.2.12. *Let M be a compact smooth Riemannian manifold without focal points satisfying the Uniform Visibility Axiom. Assume that the set Λ has positive measure. Then Λ is open* (mod 0), *everywhere dense, and the geodesic flow $g_t|\Lambda$ is nonuniformly hyperbolic, ergodic, and indeed, is Bernoulli.*

Proof. By Theorem 12.1.6, Λ is open. In view of the topological transitivity of the geodesic flow (see Theorem 12.2.10), Λ is everywhere dense. It follows from Theorems 12.2.7 and 9.2.6 that the flow $g_t|\Lambda$ is ergodic. The Bernoulli property follows from Theorem 9.6.2 and from a result by Arnold (see [6, Section 23] and [10]) implying that if the manifold is not homeomorphic to the 2-torus then the geodesic flow has continuous spectrum. □

As a corollary of the above theorem, we have the following result:

Theorem 12.2.13.

1. *Let M be a compact Riemannian manifold of nonpositive curvature satisfying Condition 12.11. Then the set Λ is open* (mod 0), *everywhere dense, and the geodesic flow $g_t|\Lambda$ is nonuniformly hyperbolic, ergodic, and indeed, is Bernoulli.*
2. *Let M be a compact surface of nonpositive curvature and genus ≥ 1. Then the set Λ is open* (mod 0), *everywhere dense, and the geodesic flow $g_t|\Lambda$ is nonuniformly hyperbolic, ergodic, and indeed, is Bernoulli.*

Proof. The first statement follows from Theorems 12.1.5 and 12.2.12. (Observe that since M has nonpositive curvature and satisfies Condition 12.11, it also satisfies the Uniform Visibility Axiom; see Proposition 12.2.9.) The second statement follows from Theorems 12.1.4 and 12.2.12. (Observe that since the genus of M is ≥ 1, the Uniform Visibility Axiom holds; see Proposition 12.2.9.) □

Further results on ergodic properties of geodesic flows on manifolds of nonpositive curvature were obtained by Knieper (see [115]) who established existence and uniqueness of the measure of maximal entropy for the geodesic flow on any compact Riemannian manifold of nonpositive curvature.

12.3 Entropy of Geodesic Flows

For $v \in SM$ let $v^\perp$ be the set of vectors $w \in SM$ that are orthogonal to v. Consider the linear map $S_v : v^\perp \to v^\perp$ defined by the equality $S_v w = K_\xi(w)$, where $\xi(w)$ is the vector in $E^+(v)$ such that $d\pi\xi(w) = w$. In the case of Riemannian metrics of class C^4 without conjugate points satisfying Asymptoticity Axiom, S_v is a linear self-adjoint operator of the second quadratic form for the horosphere $L(\pi(v), \gamma_v(+\infty))$ at the point $\pi(v)$ (which is a submanifold in H of class C^2).

Denote by $\{e_i(v)\}$, $i = 1, \ldots, p-1$, the orthonormal basis in $v^\perp$ consisting of eigenvectors of S_v. Let $K_i(v)$ be the corresponding eigenvalues. The numbers $K_i(v)$ are called the *principal curvatures* and the directions determined by the vectors $e_i(v)$ the *directions of principal curvatures*.

Theorem 12.3.1.

1. *The measure-theoretic entropy of the geodesic flow with respect to the Liouville measure μ is*
$$h_\mu(g_1) = -\int_{SM} \sum_{i=1}^{p-1} K_i(v)\, d\mu(v) = -\int_{SM} \operatorname{Tr} S_v \, d\mu(v),$$
where $\operatorname{Tr} S_v$ *is the trace of the operator S_v.*
2. *Let ν be a g_t-invariant probability measure. Then the measure-theoretic entropy of the geodesic flow with respect to ν is*
$$h_\nu(g_1) \leq -\int_M \operatorname{Tr} S_v \, d\nu(v).$$
The equality holds if and only if ν is equivalent to μ.

Proof. Our approach is based on the argument in [171]. For $v \in SM$ choose a basis $\{\xi_i(v), i = 1, \dots, p-1\}$ in $E^+(v)$ and consider the parallelogram $\Pi(v)$ formed by the vectors $\xi_i(v)$. For any Lyapunov–Perron regular point $v \in SM$ if the basis $\{\xi_i(v)\}$ is normal then

$$\chi^-(\Pi(v)) = \lim_{t\to\infty} \frac{1}{t} \log \mathrm{vol}(dg_s(\Pi(v))) = \sum_{i=1}^{s(v)} k_i(v)\chi_i^-(v), \tag{12.23}$$

where $\mathrm{vol}(dg_t(\Pi(v)))$ is the volume of the parallelepiped $dg_t(\Pi(v))$ and $\chi_i(v)$, $i = 1, \dots, s(v)$, are the distinct values of the Lyapunov exponent at v with multiplicities $k_i(v)$.

For $v \in SM$ and $\xi \in E^+(v)$ by Statement 5 of Proposition 12.1.1, we have that for $t \geq 1$,

$$\|dg_t\xi\| \leq (1+a)\|d\pi \circ dg_t\xi\|.$$

This implies that

$$\chi(v,\xi) = \lim_{t\to\infty} \frac{1}{t} \log \|d\pi \circ dg_t\xi\|,$$

and hence

$$\lim_{t\to\infty} \frac{1}{t} \log \mathrm{vol}(dg_t(\Pi(v))) = \lim_{t\to\infty} \frac{1}{t} \log \mathrm{vol}(d\pi \circ dg_t(\Pi(v))). \tag{12.24}$$

For any $s \neq 0$, let

$$\lambda_s(v) = \frac{\mathrm{vol}(d\pi \circ dg_s(\Pi(v)))}{\mathrm{vol}(d\pi\, \Pi(v))}$$

be the coefficient of the volume growth in $v^\perp$ (i.e., the Jacobian of the map $d\pi \circ dg_s|E^+(v)$). By the entropy formula (see Section 10.4) for every $s \neq 0$,

$$h_\mu(g_s) = -\int_{SM} \sum_{i=1}^{s(v)} k_i(v)\chi_i^-(v)\, d\mu(v) = -\int_{SM} \chi^-(\Pi(v))\, d\mu(v).$$

Hence by Birkhoff's Ergodic Theorem, (12.23), and (12.24) for $s \neq 0$,

$$\begin{aligned} h_\mu(g_s) &= -\int_{SM} \lim_{n\to\infty} \frac{1}{sn} \log \mathrm{vol}(dg_{sn}(\Pi(v)))\, d\mu(v) \\ &= -\int_{SM} \lim_{n\to\infty} \frac{1}{sn} \log \mathrm{vol}(d\pi \circ dg_{sn}(\Pi(v)))\, d\mu(v) \\ &= -\int_{SM} \lim_{n\to\infty} \frac{1}{sn} \sum_{i=1}^{n-1} \log \lambda_s(g_{sn}(v))\, d\mu(v) \\ &= -\int_{SM} \log \lambda_s(v)\, d\mu(v). \end{aligned} \tag{12.25}$$

Since $h_\mu(g_s) = |s|h_\mu(g_1)$, we have by (12.23) and (12.25) that for $s > 0$,

$$\begin{aligned} h_\mu(g_1) &= \frac{1}{s} h_\mu(g_s) = \lim_{s\to 0} \frac{1}{s} h_\mu(g_s) \\ &= -\int_{SM} \lim_{s\to 0} \frac{1}{s} \log \lambda_s(v)\, d\mu(v). \end{aligned} \tag{12.26}$$

We shall compute the expression under the integral. For $v \in SM$, choose the basis $\{\xi_i(v), i = 1, \ldots, p-1\}$ in $E^+(v)$ comprised of the vectors $\xi_i(v)$ for which $d\pi\xi_i(v) = e_i(v)$ are the directions of principle curvatures. For any $i, j = 1, \ldots, p-1$ and $s > 0$,

$$\begin{aligned} \langle Y_{\xi_i}(s), Y_{\xi_j}(s)\rangle &= \langle Y_{\xi_i}(0), Y_{\xi_j}(0)\rangle + \int_0^s \langle Y_{\xi_i}(u), Y_{\xi_j}(u)\rangle'\, du \\ &= \langle Y_{\xi_i}(0), Y_{\xi_j}(0)\rangle \\ &\quad + \int_0^s [\langle Y'_{\xi_i}(u), Y_{\xi_j}(u)\rangle + \langle Y_{\xi_i}(u), Y'_{\xi_j}(u)\rangle]\, du. \end{aligned} \tag{12.27}$$

We also have that

$$\langle Y_{\xi_i}(0), Y_{\xi_j}(0)\rangle = \langle e_i(v), e_j(v)\rangle = \delta_{ij}. \tag{12.28}$$

Let $A(s) = (a_{ij}(s))$ be the matrix with entries

$$a_{ij}(s) = \frac{1}{s} \int_0^s [\langle Y'_{\xi_i}(u), Y_{\xi_j}(u)\rangle + \langle Y_{\xi_i}(u), Y'_{\xi_j}(u)\rangle]\, du.$$

For $s = 0$, we have

$$\begin{aligned} a_{ij}(0) &= \langle Y'_{\xi_i}(0), Y_{\xi_j}(0)\rangle + \langle Y_{\xi_i}(0), Y'_{\xi_j}(0)\rangle \\ &= \langle S_v e_i(v), e_j(v)\rangle + \langle e_i(v), S_v e_j(v)\rangle \\ &= (K_i(v) + K_j(v))\delta_{ij}. \end{aligned} \tag{12.29}$$

Using (12.27) and (12.28), we obtain that for $s \neq 0$,

$$\begin{aligned} V(d\pi \circ dg_s(\Pi(v))) &= \sqrt{\det |\langle d\pi \circ dg_s\xi_i, d\pi \circ dg_s\xi_j\rangle|} \\ &= \sqrt{\det |\langle Y_{\xi_i}(s), Y_{\xi_j}(s)\rangle|} \\ &= \sqrt{\det[\mathrm{Id} + sA(s)]} \\ &= 1 + \frac{1}{2} s \operatorname{Tr} A(s) + O(s^2). \end{aligned}$$

It follows that

$$\lim_{s\to 0} \frac{1}{s} \log \lambda_s(v) = \frac{1}{2} \operatorname{Tr} A(0). \tag{12.30}$$

Taking into account the relation (12.29), we conclude that the ith diagonal element of the matrix $A(0)$ is $2K_i(v)$ and the first statement of the theorem follows from (12.26).

To prove the second statement note that by the Margulis–Ruelle Inequality (see Section 10.2),

$$h_\nu(g_s) \le -\sum_{i=1}^{s(v)} k_i(v)\chi_i^-(v)\,d\mu(v) = -\int_{SM} \chi^-(\Pi(v))\,d\mu(v).$$

By (12.25) and (12.26), we obtain that

$$h_\nu(g_1) \le -\int_{SM} \lim_{s\to 0} \frac{1}{s}\lambda_s(v)\,d\mu(v).$$

Statement 2 of the theorem now follows from (12.29), (12.30), and Theorem 13.2.1. □

Using Theorem 12.3.1, one can obtain an interesting upper bound for the measure-theoretic entropy of the geodesic flow with respect to an invariant probability measure ν on a compact smooth Riemannian manifold M without conjugate points (see [73]):

$$h_\nu(g_1) \le (n-1)\left(-\int_M \mathrm{Ric}(\theta)\,d\nu\right)^{1/2},$$

where $\mathrm{Ric}(\theta)$ is the Ricci tensor. Furthermore, if the equality holds then the sectional curvatures at points of the support of ν are constant.

12.4 Topological Properties of Geodesic Flows

Consider the geodesic flow g_t on a compact Riemannian manifold M without conjugate points. If M satisfies the Uniform Visibility Axiom then g_t is topologically transitive (see Theorem 12.2.10). Furthermore, as we saw the flow is nonuniformly hyperbolic on the set Λ, provided this set has positive volume. Therefore results of Chapter 15 apply and give some information on topological properties of $g_t|\Lambda$, including the growth rate of the number of periodic orbits (see Corollary 15.4.4). Much stronger results can be obtained for geodesic flows on manifolds without conjugate points.

We begin by considering the topological entropy $h(g_t)$ of the geodesic flow and relating it to the exponential growth rate of the volume of balls.

Let M be a compact smooth Riemannian manifold. We denote by $\mathrm{vol}(B(x,r))$ the Riemannian volume of the ball $B(x,r)$, in the universal cover $\widetilde{M}$ of M, centered at x of radius r. Define

$$\lambda(M) = \lim_{r\to\infty} \frac{\log \mathrm{vol}(B(x,r))}{r}.$$

Manning [143] proved that this limit exists and is independent of x. Moreover, he showed that

$$\lambda(M) \le h(g_t). \tag{12.31}$$

He also proved that $\lambda(M) = h(g_t)$, provided the manifold has nonpositive curvature. This result can be extended to manifolds without conjugate points.

Theorem 12.4.1 (Freire–Mañé [73]). *For the geodesic flow g_t on a compact smooth Riemannian manifold M without conjugate points*

$$\lambda(M) = h(g_t). \tag{12.32}$$

We will derive this statement from a general result by Mañé [141] on the topological entropy of the geodesic flow on a smooth compact connected manifold M via the exponential growth rate of the number $n_T(x, y)$ of geodesics of length $\leq T$ joining the points x and y. The function $n_T(x, y)$ is finite and locally constant on an open full-measure subset of $M \times M$.

Theorem 12.4.2 (Mañé [139]). *For the geodesic flow g_t on M,*

$$h(g_t) = \lim_{T\to\infty} \frac{1}{T} \log \int_{M\times M} n_T(x, y)\, dx\, dy.$$

We assume now that the manifold M has no conjugate points. Then the exponential map $\exp_x \colon T_x M \to M$ is a covering map admitting a fundamental domain of diameter $\leq \operatorname{diam} M = c$. Theorem 12.4.2 implies then that for any $x_1, y_1, x_2, y_2 \in M$,

$$n_T(x_1, y_1) \leq n_{T+2c}(x_2, y_2).$$

It follows that

$$h(g_t) = \lim_{T\to\infty} \frac{1}{T} \log n_T(x, y).$$

To prove (12.32) observe that by (12.31), it suffices to show that $\lambda(M) \leq h(g_t)$. If $p \colon \widetilde{M} \to M$ is a covering map and $x \in \widetilde{M}$, then the number $n_T(p(x), p(x))$ is just the number of points in the set

$$G_T = \{z \in M : p(z) = p(x),\ d(z, x) \leq T\}.$$

Observe that there exists $r_0 > 0$ such that any two distinct points z' and z'' with $p(z') = p(z'')$ satisfy $d(z', z'') \geq r_0$. Clearly,

$$B(x, T + r_0) \supset \bigcup_{z\in G_T} B(z, r_0),$$

and the sets $B(z, r_0/2)$, $z \in G_T$ are disjoint. Hence if

$$k = \min\left\{\operatorname{vol}(B(a, \frac{r_0}{2})) : a \in \widetilde{M}\right\},$$

then

$$\operatorname{vol}(B(x, T + r_0)) \geq k \operatorname{card} G_T = k\, n_T(p(x), p(x)).$$

Therefore,

$$\begin{aligned}\lambda(M) &= \lim_{T\to\infty} \frac{1}{T+r_0} \log \operatorname{vol}(B(x, T+r_0))\\ &\geq \lim_{T\to\infty} \frac{1}{T+r_0} \log k\, n_T(p(x), p(x)) = h(g_t),\end{aligned}$$

thus completing the proof of Theorem 12.4.1.

We now consider the case when the manifold has nonpositive curvature. Given a tangent vector $v \in SM$, rank(v) is the dimension of the space of parallel Jacobi fields along the geodesic γ_v. The minimum of rank(v) over all $v \in SM$ is called the *rank* of M, rank(M). If rank(v) = rank(M), the geodesic γ_v and the corresponding vector v are called *regular*. It is easy to see that $1 \leq \operatorname{rank}(M) \leq \dim M$.

The following result describes the fundamental rank rigidity for nonpositively curved manifolds. It was obtained independently by Ballmann, Brin, and Eberlein [16] and by Burns and Spatzier [46].

Theorem 12.4.3. *Let M be a compact smooth Riemannian manifold of nonpositive curvature with irreducible universal cover H. Then the manifold either has rank 1 or H is a symmetric space of higher rank.*

In other words, the universal cover of a nonpositively curved manifold can be represented as a product of Euclidean, symmetric, and rank 1 spaces.

We shall present (without proofs) somewhat "ultimate" asymptotic estimates for the volume growth of balls and the growth of the number of homotopically different periodic orbits for the geodesic flow. Set $h = h(g_t)$.

Theorem 12.4.4 (Gunesch [77, 78]). *Let M be a compact smooth Riemannian manifold of nonpositive curvature and rank 1. Then the following statements hold:*

1. *Given $x \in H$, we have*

$$\operatorname{vol}(B(x, r)) \sim c(x)e^{hr}.$$

2. *If $P(t)$ is the number of homotopy classes of periodic orbits of length at most t, then*

$$P(t) \sim \frac{1}{ht}e^{ht}.$$

Observe that unlike in the case of negatively curved manifolds, for nonpositively curved manifolds there may be uncountably many periodic geodesics homotopic to a given one but they all have the same length.

12.5 The Teichmüller Geodesic Flow

In this section we give a concise description of some recent results on the dynamics of Teichmüller geodesic flows and in particular, we describe their Lyapunov spectrum. We refer the reader to the lecture notes by Zorich [243] and to the surveys by

Masur [148] and Forni [72] where various aspects of the Teichmüller flow theory and its relations to geometry of flat surfaces are discussed in great detail.

1. *Translation surfaces.* A *translation surface* is a pair (M, V) where M is a compact smooth surface endowed with a flat metric (i.e., of zero curvature) with finitely many singularities of conical type (see below) and V a parallel unit vector filed. Any translation surface can be constructed from a planar polygon Π, which has an even number $2n \geq 4$ of sides and along with each side v_j (considered as a vector) the polygon contains a side $v'_j = -v_j$. Identifying each side v_j with the corresponding side v'_j by translation one gets a translation surface M, where the singularities correspond to the points obtained by identification of the vertices of the polygon, and $V = (0, 1)$ is the parallel vertical vector field. Conversely, given a translation surface M, one can unwrap it into a polygon making cuts along some geodesic segments, which connect singularities.

One can also describe translation surfaces using "complex-analytic" language. Recall that a *Riemann surface* M is a smooth surface endowed with an atlas whose coordinate changes are holomorphic maps of the plane. We shall assume that M is compact. An *Abelian differential* ω is a holomorphic complex 1-form on a Riemann surface. The pair (M, ω) determines a translation surface. Indeed, ω has finitely many zeros that are singularities of the flat metric and in a neighborhood of any nonsingular point, one can introduce a coordinate system w such that $\omega_w = dw$. This introduces a *translation atlas* on M making the surfaces to be flat outside singularities that are of conical type. The vector field V is the pullback of the vertical field under the local charts.

Conversely, given a translation surface (M, V) and any nonsingular point $q \in M$, one can identify a neighborhood of q with an open subset of the complex plane $\mathbb{C}$ via an isometry, which sends the vector $V(q)$ to the vertical vector $(0, 1)$. These isometries coincide in the intersection of their domains and determine a Riemann surface atlas on the complement of the singularities. They also transport the canonical Abelian differential dz from $\mathbb{C}$ to the surface.

Given a holomorphic 1-form ω and its zero p, one can introduce a local coordinate system w in a neighborhood of p such that the form can be represented as $\omega(w) = w^d\, dw$ where d is called the *degree* (or *multiplicity*) of the zero. If the form ω has a zero of degree d at a conical singularity then the cone angle is $2\pi(d + 1)$. Moreover, we have that

$$\sum d_j = 2\mathbf{g} - 2,$$

where $\mathbf{g}$ is the genus of the Riemann surface, d_j is the degree of ω at the jth zero p_j, and the sum is taken over all zeros. (This is a version of the Gauss–Bonnet formula for translation surfaces.)

2. *Stratification of the moduli space of Abelian differentials.* Two Riemann surfaces M and N are called *conformally equivalent* if there exists a conformal homeomorphism between them, that is, a homeomorphism $f\colon M \to N$ whose representations in local coordinate charts are holomorphic maps of the plane. Given any $\mathbf{g} \geq 1$, this determines the space $\mathcal{M}_{\mathbf{g}}$ of complex structures as the space of Riemann surfaces of genus $\mathbf{g}$ up to conformal equivalence. It has complex dimension $3\mathbf{g} - 3$. Further, we denote by $\mathcal{A}_{\mathbf{g}}$ the moduli space of Abelian differentials on Riemann surfaces of genus $\mathbf{g}$ modulo conformal equivalence. This is a noncompact complex algebraic orbifold of complex dimension $4\mathbf{g} - 3$. It is naturally stratified by the subsets $\mathcal{A}_{\mathbf{g}}(d_1, \ldots, d_m)$ of Abelian differentials whose zeros have degrees $d_1, \ldots, d_m$. Here $m \geq 0$ and as we saw above $d_1 + \cdots + d_m = 2\mathbf{g} - 2$. Each *stratum* $\mathcal{A}_{\mathbf{g}}(d_1, \ldots, d_m)$ is a noncompact complex algebraic suborbifold of the moduli space $\mathcal{A}_{\mathbf{g}}$ of complex dimension $2\mathbf{g} + m - 1$. It always contains "cusps," that is, regions corresponding to translation surfaces with short saddle connections or short closed geodesics.

Each stratum admits a natural holomorphic affine structure, that is, an atlas whose coordinate changes are holomorphic affine maps. This affine structure generates a natural volume μ on the stratum. Consider now a real hypersurface (the "unit sphere"):

$$\mathcal{A}^1_{\mathbf{g}}(d_1, \ldots, d_m) \subset \mathcal{A}_{\mathbf{g}}(d_1, \ldots, d_m)$$

defined by the equation area$(M) = 1$, where M is the corresponding surface. It is also given by the equation $A(\omega) = 1$, where the function $A\colon \mathcal{A}_{\mathbf{g}}(d_1, \ldots, d_m) \to \mathbb{R}$ is defined by

$$A(\omega) = \frac{1}{2i} \int_M \omega \wedge \overline{\omega}$$

and measures the area of the surface M in the flat metric associated to ω.

The volume element $d\mu$ induces a volume element on $\mathcal{A}^1_{\mathbf{g}}(d_1, \ldots, d_m)$, which we denote by $d\mu_1$. Masur [147] and Veech [216] (see also [218]) showed that every stratum has finite μ_1-volume.

Let us turn back to the geometric interpretation of the translation surface as a planar polygon Π with sides $v_1, \ldots, v_n$ and $v'_1, \ldots, v'_n$, and conical singularities $P_1, \ldots, P_m$. In this "geometric" language one can identify the stratum $\mathcal{A}_{\mathbf{g}}(d_1, \ldots, d_m)$ with the family of translation surfaces, which share the same types $2\pi(d_1 + 1), \ldots, 2\pi(d_m + 1)$ of cone angles. Observe that the topological characteristics of the translation surface (e.g., genus $\mathbf{g}$, number m, and types of conical singularities) do not change under a small deformation of the vectors v_j.

3. *The Teichmüller geodesic flow.* Let M be a translation surface of genus $\mathbf{g}$ and Π a polygonal pattern obtained by unwrapping the surface along some geodesic cuts. A linear transformation $g \in GL^+(2, \mathbb{R})$ of the plane acts on Π changing its shape. However, the sides of the new polygon $g(\Pi)$ are again arranged into pairs, where the sides in each pair are parallel and have equal length (but different from the initial one). Identifying the sides in each pair by a parallel translation we obtain a new translation surface $g(M)$. It is easy to check that this surface does not depend on the way in which M was unwrapped to a polygonal pattern Π and that all its topological characteristics (including genus, number m, and types of conical singularities) are the same as those of the initial surface M. Thus we obtain a continuous action of the group $GL^+(2,\mathbb{R})$ on the moduli space $\mathcal{A}_\mathbf{g}$ of Abelian differentials.

The subgroup $SL(2,\mathbb{R})$ of area preserving linear transformations acts continuously on $\mathcal{A}_\mathbf{g}$ and leaves invariant every stratum $\mathcal{A}_\mathbf{g}(d_1,\ldots,d_m)$ and its "unit sphere" $\mathcal{A}^1_\mathbf{g}(d_1,\ldots,d_m)$. The action of its one-parameter diagonal subgroup

$$\begin{pmatrix} e^t & 0 \\ 0 & e^{-t} \end{pmatrix} \subset SL(2,\mathbb{R})$$

generates a flow on $\mathcal{A}^1_\mathbf{g}(d_1,\ldots,d_m)$, which is called the *Teichmüller flow*. The following result of Masur [147] and Veech [216] (see also [218]) establishes the ergodicity of this flow.

Theorem 12.5.1. *The actions of the group $SL(2,\mathbb{R})$ and its diagonal subgroup preserve the measure μ_1. Both actions are ergodic with respect to this measure on each connected component of every stratum $\mathcal{A}^1_\mathbf{g}(d_1,\ldots,d_m)$.*

One can view the Teichmüller flow as the geodesic flow. To see this we recall that a *quadratic differential* $\varphi\, dz^2$ on a Riemann surface assigns to each point a complex quadratic form on the corresponding tangent space, depending meromorphically on the point. A quadratic differential is said to be holomorphic if $\varphi(z)$ is a holomorphic function relative to any local coordinate. Now each element of the cotangent space $\mathcal{Q}_\mathbf{g}$ to the Riemann moduli space $\mathcal{M}_\mathbf{g}$ can be identified with a pair, which consists of a class of equivalent complex structures and a holomorphic quadratic differential. The space $\mathcal{Q}_\mathbf{g}$ has complex dimension $6\mathbf{g}-6$ and is the space where the Teichmüller flow acts. A holomorphic 1-form $\omega(z)dz$ canonically defines a quadratic differential $\omega^2(z)(dz)^2$, allowing us to consider $\mathcal{A}_\mathbf{g}$ as an invariant subspace in $\mathcal{Q}_\mathbf{g}$.

Further, we introduce a metric on the space $\mathcal{M}_\mathbf{g}$. Let M_1 and M_2 be two compact Riemann surfaces of the same genus $\mathbf{g}$. This is to say that we have a compact topological surface of genus $\mathbf{g}$ endowed with two different Riemann structures. Consider a diffeomorphism $f\colon M_1 \to M_2$. When f is conformal, it sends infinitesimal circles at a point $z \in M_1$ into infinitesimal circles at any nonsingular point $f(z) \in M_2$. A general f sends infinitesimal circles at $z \in M_1$ into infinitesimal ellipses at $f(z) \in M_2$. The *coefficient of quasi-conformality* (or the *dilatation*) *of f at z* is the ratio $K_z(f) = a/b$, where a and b are semiaxes of the ellipse. More precisely,

$$K_z(f) = \frac{|\partial_z f| + |\partial_{\bar{z}} f|}{|\partial_z f| - |\partial_{\bar{z}} f|}(z).$$

Finally, the *coefficient of quasi-conformality of f* is $K(f) = \sup K_z(f)$ where the supremum is taken over all nonsingular points $z \in M_1$. The map f is said to be *quasi-conformal* if $K(f) < \infty$.

The celebrated theorem of Teichmüller (see [243]) claims that given any two Riemann surfaces M_1 and M_2, there exists a quasi-conformal map $f_{M_1,M_2}\colon M_1 \to M_2$, which minimizes the coefficient of quasi-conformality. For this map,

$K_z(f_{M_1,M_2}) = K(f_{M_1,M_2})$ for every nonsingular point $z \in M_1$. Now the *Teichmüller metric* is defined as

$$\text{dist}(M_1, M_2) = \frac{1}{2} \log K(f_{M_1,M_2}).$$

As can be immediately seen from the definition, this metric measures the amount of quasi-conformal distortion between two different complex structures. In the case $\mathbf{g} = 1$, the Teichmüller metric coincides with the Poincaré metric on the modular surface. (Given a torus whose shortest geodesic has length 1, take the closed geodesic of length $r > 1$ that is next to the shortest one in the length spectrum and measure the angle φ where $0 \leq \varphi \leq \pi$ between these two geodesics; the pair determines a unique point on the modular surface.) In particular, the Teichmüller metric is Riemannian with constant negative curvature so that the Teichmüller flow is an Anosov flow. In the higher genus case the Teichmüller metric is only Finsler (i.e., the norm in each tangent space does not come from an Euclidean inner product).

4. *Asymptotic cycles and the Lyapunov spectrum for the Teichmüller geodesic flow.* One can recover substantial information on the geometry of the translation surface (M, V) and the dynamics of the *directional (vertical) flow* generated by V by studying the trajectory of the Teichmüller geodesic flow originated at M. The point is to use measure-theoretic properties of the Teichmüller geodesic flow (e.g., ergodicity; see Theorem 12.5.1) to gain some information on translation surfaces for almost every surface with respect to the measure μ. In particular, some techniques involving the Teichmüller geodesic flows can be used to study ergodicity and minimality of the directional flow.

Recall that a flow φ_t on a compact topological space M is *minimal* if every trajectory of the flow is dense in M. Given a surface M with a flat metric, there are directions for which the corresponding directional flow is minimal. In the case when M is a torus, this automatically implies that the flow is ergodic with respect to area. However, for surfaces of higher genus, the directional flow may be minimal but not ergodic as the following theorem of Masur (see [148]) shows.

Theorem 12.5.2. *Given a translation surface (M, V), if the vertical flow is not ergodic with respect to area then the trajectory $g_t(M)$ of the Teichmüller flow emitted from $(M, V) \in \mathcal{A}_g$ is divergent; that is, its projection to $\mathcal{M}_g$ eventually leaves any given compact subset in the moduli space $\mathcal{M}_g$ and never comes back.*

We now state a remarkable result of Kerchkoff, Masur, and Smillie [111] on ergodicity of the directional flow.

Theorem 12.5.3. *Given a compact surface M with a flat metric, the directional flow in almost any direction V is ergodic with respect to area.*

As an immediate corollary of this result we describe how a typical geodesic winds around a surface M with a flat metric. Let γ be a geodesic segment. Consider the geodesic through a point $x_0 \in \gamma$ that is orthogonal to γ. From time to time this geodesic intersects γ and let $x_1, x_2, \ldots$ be the successive points of intersection. We can close up the corresponding pieces of the geodesic by joining the end points x_0, x_j with a path going along γ and obtain a sequence of cycles $c_1, c_2, \ldots$. Viewing these cocycles as elements of the first homology group $H_1(M, \mathbb{R}) \simeq \mathbb{R}^{2\mathbf{g}}$ and endowing this group with the Euclidean metric, we obtain the following result.

Corollary 12.5.4. *Let M be a compact surface with a flat metric and of area 1. For almost any direction V on M, the limit*

$$\lim_{n\to\infty} \frac{1}{n} c_n = \frac{1}{|\gamma|} c$$

exists and the asymptotic cycle c does not depend on the starting point $x_0 \in M$.

The following result provides a crucial information on distribution of asymptotic cycles via the Lyapunov spectrum of the Teichmüller geodesic flow restricted to a stratum $\mathcal{A}_{\mathbf{g}}(d_1, \ldots, d_m)$ with respect to the measure μ_1. In particular, it shows that the Lyapunov exponents are simple, that is, have multiplicity 1. Let M be a compact surface with a flat metric and genus $\mathbf{g}$ and ω the holomorphic 1-form representing M.

Theorem 12.5.5. *For almost every translation surface M in any stratum $\mathcal{A}_{\mathbf{g}}(d_1, \ldots, d_m)$ (with respect to the measure μ_1), there exists a filtration of subspaces*

$$\mathcal{V}_1 \subset \mathcal{V}_2 \subset \cdots \mathcal{V}_{\mathbf{g}} \subset H_1(M, \mathbb{R}) \tag{12.33}$$

such that for any starting point x_0 in a "typical" geodesic segment γ and the corresponding sequence of cycles,

1. *the following limit exists*

$$|\gamma| \lim_{n\to\infty} \frac{1}{n} c_n = c,$$

   where the nonzero asymptotic cycle $c \in H_1(M, \mathbb{R})$ is dual to the cohomology class of $\mathrm{Re}[\omega]$ *and the one-dimensional subspace $\mathcal{V}_1$ is spanned by c;*

2. *for any $j = 1, \ldots, \mathbf{g} - 1$, the following limit exists*

$$\overline{\lim_{n\to\infty}} \frac{\log|\mathrm{dist}(c_n, \mathcal{V}_j)|}{\log n} = \nu_{j+1}$$

   and

$$|\mathrm{dist}(c_n, \mathcal{V}_{\mathbf{g}})| \le C,$$

   where $C > 0$ is a constant depending only on M and on the choice of the Euclidean structure in the homology space.

Moreover, $\dim \mathcal{V}_j = j$ *and the numbers*

$$2 > 1 + \nu_2 > \cdots > 1 + \nu_{\mathbf{g}}$$

are the largest values of the Lyapunov exponent of the measure μ_1 *for the Teichmüller geodesic flow on the corresponding connected component of the stratum* $\mathcal{A}_{\mathbf{g}}^1(d_1, \ldots, d_m)$*. In particular, they do not depend on the individual surface* M *in the connected component for* μ_1*-almost every* M *and* (12.33) *is the filtration associated with the Lyapunov exponent.*

We remark on the choice of the segment γ. The theorem is valid if the trajectories emitted in the vertical direction from the end points of γ hit the conical singularities before their first return to γ. If we allow any continuous family of geodesic segments γ_s of variable length then the theorem stays valid for a set of γ_s of full measure.

It follows from Theorem 12.5.5 that the complete set of values of the Lyapunov exponent of the measure μ_1 for the Teichmüller geodesic flow on the corresponding connected component of the stratum $\mathcal{A}_{\mathbf{g}}^1(d_1, \ldots, d_m)$; that is, the Lyapunov spectrum of μ_1 is

$$\begin{aligned} 2 &> 1 + \nu_2 > 1 + \nu_3 > \cdots > 1 + \nu_{\mathbf{g}} > 1 = \cdots = 1 \\ &> 1 - \nu_{\mathbf{g}} > \cdots > 1 - \nu_2 > 0 \\ &> -(1 - \nu_2) > \cdots > -(1 - \nu_{\mathbf{g}}) > -1 = \cdots = -1 \\ &> -(1 + \nu_{\mathbf{g}}) > -(1 + \nu_{\mathbf{g}-1}) > \cdots > -(1 + \nu_2) > -2. \end{aligned}$$

There are $m - 1$ exponents equal to 1 and respectively to -1 that are related to m singularities. (Note that the real dimension of the stratum $\mathcal{A}_{\mathbf{g}}^1(d_1, \ldots, d_m)$ is $4\mathbf{g} + 2m - 3$; note also that the symmetric form of the spectrum is a corollary of Theorem 4.3.16, where one should set $b = 0$; see (4.29).) We conclude that *the Teichmüller geodesic flow is nonuniformly hyperbolic* and, in particular, that its entropy is

$$h_{\mu_1}(g_t) = 2\mathbf{g} + m - 1$$

(this can easily be deduced from the entropy formula; see Theorem 10.4.1). The above description of the Lyapunov spectrum of the measure μ_1 (except for the statement on its simplicity) was obtained by Zorich in [243] and Theorem 12.5.5 was proved in [242]. Veech [217] showed that $\nu_2 < 1$, thus providing for four strict inequalities in the spectrum. Forni [72] proved that $\nu_{\mathbf{g}} > 0$ implying, in particular, that the measure μ_1 is hyperbolic and that $\dim \mathcal{V}_{\mathbf{g}} = \mathbf{g}$. The statement that the Lyapunov spectrum is simple had been known as the Kontsevich–Zorich conjecture (see [116]) until it was recently proved by Avila and Viana in [12].

12.6 Notes

The main results in Section 12.1 (Theorems 12.1.2, 12.1.4, and Statements 1 and 2 of Theorem 12.1.5) were proved by Pesin in [169] and [170]. We follow the presentation in these papers. Statement 3 of Theorem 12.1.5 was proved by Ballmann and Brin in [15]. The construction of horospherical foliations (Propositions 12.2.1 and 12.2.4–12.2.6) is an adaptation of the arguments in [169] and [170]. Other methods of constructing these foliations using the Busemann function can be found in [14]. The proofs of Theorems 12.2.7 and 12.2.12 presented here are an elaboration of arguments in [170]. The proof given there contains a mistake and it is still an open problem whether the set Λ (or rather the set Δ) has full Liouville measure. Brin and Burago proved this under the additional assumption that the set of negative curvature in M has finitely many connected components. The same result was obtained by Rodríguez Hertz who used different methods. None of these results is published.

Theorem 12.2.12 was proved by Pesin [171] and independently by Freire and Mañé [73]. This theorem is analogous to the result of [205] for dispersing billiards.

Some estimates from below for the entropy of the geodesic flow were obtained in [163].

Theorem 12.4.4 extends results of Knieper on multiplicative asymptotic bounds for the growth of volume of spheres (and hence also volume of balls) and the number of periodic orbits (see [115] and references therein). It also extends results by Margulis [144], who considered manifolds of negative curvature, to the case of nonpositive curvature. The proof of Theorem 12.4.4 uses the mixing property obtained by Babillot [13] for the measure of maximal entropy of a geodesic flow on a compact smooth Riemannian manifold of nonpositive curvature and rank 1.

13

SRB Measures

In the previous chapters we considered dynamical systems, which are nonuniformly hyperbolic with respect to smooth invariant measures. We now study hyperbolic invariant measures, which are not smooth. This includes, in particular, dissipative systems for which the supports of such measures are attracting invariant sets. A general hyperbolic measure may not have "nice" ergodic properties: its ergodic components may be of zero measure and it may have zero measure-theoretic entropy. There is, however, an important class of hyperbolic measures that appear naturally in applications and have an abundance of ergodic properties. They are called SRB measures (after Sinai, Ruelle, and Bowen).

13.1 Definition and Ergodic Properties of SRB Measures

Let f be a $C^{1+\alpha}$ diffeomorphism of a compact smooth Riemannian manifold M and ν a hyperbolic invariant measure for f. Denote by Z the set of Lyapunov–Perron regular points with nonzero Lyapunov exponents. We have that $\nu(Z) = 1$. The set Z is nonuniformly (completely) hyperbolic for f, and we can choose a level set $\Lambda = \Lambda_{\lambda\mu\varepsilon j}$ of positive measure and consider the corresponding collection of regular sets $\{\Lambda^\ell = \Lambda^\ell_{\lambda\mu\varepsilon j} : \ell \geq 1\}$ (see Sections 2.2.2 and 5.2, (5.5), and Theorem 5.4.6). Fix a regular set Λ^ℓ of positive measure, a point $x \in \Lambda^\ell$, and a number $0 < r < r_\ell$ (see (8.4)). Set

$$Q^\ell(x) = \bigcup_{w \in \Lambda^\ell \cap B(x,r)} V^u(w),$$

where $V^u(w)$ is a local unstable manifold at w, and denote by $\xi(x)$ the partition of $Q^\ell(x)$ by these manifolds.

Definition 13.1.1. A hyperbolic measure ν is called an *SRB measure* if for any level set Λ of positive measure, any regular set Λ^ℓ of positive measure and almost any $x \in \Lambda^\ell$, $w \in \Lambda^\ell \cap B(x,r)$, the conditional measure $\nu^u(w)$, generated by ν

on $V^u(w)$, is absolutely continuous with respect to the leaf volume $m_{V^u}(w)$ on $V^u(w)$.

Clearly, an invariant Borel measure ν is an SRB measure if and only if it is a hyperbolic u-measure for f (see Section 9.3).

For almost every $x \in \Lambda^\ell$ and $w \in \Lambda^\ell \cap B(x, r)$, there is a measurable density function $\kappa^u(w, z)$, $z \in V^u(w)$, such that

$$d\nu^u(w)(z) = \kappa^u(w, z)\, dm_{V^u}(w)(z).$$

For almost every $w \in \Lambda^\ell$, the function $\kappa^u(w, \cdot)$ is bounded from above almost everywhere with respect to the leaf volume $m_{V^u}(w)$ on $V^u(w)$. We assume that the density function is normalized such that $\kappa^u(w, w) = 1$. The following result gives a description of the density function $\kappa^u(w, z)$.

Theorem 13.1.2. *For any $w \in \Lambda^\ell \cap B(x, r)$ and $z \in V^u(w)$, we have that $\kappa^u(w, z) > 0$ and*

$$\kappa^u(w, z) = \prod_{i=1}^{\infty} \frac{\operatorname{Jac}(df^{-1}|E^u(f^{-i}(z)))}{\operatorname{Jac}(df^{-1}|E^u(f^{-i}(w)))}. \tag{13.1}$$

Moreover, the function $\kappa^u(w, z)$ is Hölder continuous in $w \in \Lambda^\ell$ and $z \in V^u(w)$.

Proof. Apply Theorem 9.3.6 to the inverse map f^{-1} and the family of local unstable manifolds. □

An important manifestation of Theorem 13.1.2 is that the density function $\kappa^u(w, z)$ is Hölder continuous and strictly positive on $w \in \Lambda^\ell \cap B(x, r)$ and $z \in V^u(w)$.

We describe some properties of SRB measures that are corollaries of the absolute continuity property (see Theorems 8.6.1 and 8.6.8 and Corollaries 8.6.9–8.6.11).

Proposition 13.1.3. *Let $A \subset M$ be a measurable set of positive ν-measure and for each $x \in A$, let $B_x \subset V^u(x)$ be a measurable set of positive leaf volume. Then the set $B = \bigcup_{x \in A} B_x$ has positive volume.*

Proposition 13.1.4. *Given a set $N \subset M$ of zero ν-measure, we have that $\nu^u(V^u(x) \cap N) = 0$ for ν-almost every $x \in M$.*

Proposition 13.1.5. *For almost every $x \in M$ and every set N of zero volume in $V^u(x)$, we have that $\nu(\bigcup V^s(w)) = 0$ where the union is taken over all points $w \in \Lambda^\ell \cap B(x, r)$ (for sufficiently small r) for which $V^s(w) \cap V^u(x) \in N$.*

SRB measures have ergodic properties similar to those of smooth measures. The proofs of the corresponding results use Theorem 13.1.2 and Propositions 13.1.4 and 13.1.5, and are simple modifications of arguments in the case of smooth measures, since those proofs use absolute continuity of only local unstable manifolds.

We summarize the results on ergodic properties and describe the necessary modifications in the proofs.

In the following theorems we assume that f is a $C^{1+\alpha}$ diffeomorphism of a compact smooth Riemannian manifold M and ν an invariant SRB measure for f.

Theorem 13.1.6. *There exist invariant sets $\Lambda_0, \Lambda_1, \ldots$ such that*

1. $\bigcup_{i\geq 0}\Lambda_i = M$, *and* $\Lambda_i \cap \Lambda_j = \varnothing$ *whenever* $i \neq j$;
2. $\nu(\Lambda_0) = 0$, *and* $\nu(\Lambda_i) > 0$ *for each* $i \geq 1$;
3. $f|\Lambda_i$ *is ergodic for each* $i \geq 1$.

Proof. See the proof of Theorem 9.1.1. Note that in the proof of this theorem only absolute continuity of local unstable manifolds is used and that families of local unstable manifold possess this property with respect to any u-measure. □

Theorem 13.1.7. *There exists a measurable partition η of M with the following properties:*

1. *For almost every $x \in M$, the element $C_\eta(x)$ is an open* (mod 0) *subset of* $W^u(x)$.
2. $f^{-1}\eta \geq \eta$.
3. $\eta^+ = \bigvee_{i>-\infty}^{0} f^i\eta = \varepsilon$.
4. $\bigwedge_{i=0}^{\infty} f^i\eta = \mathcal{H}(W^u) = \pi(f)$.
5. *The entropy of f is*

$$h_\nu(f) = h_\nu(f,\eta) = -\int_\Lambda \sum_{i=1}^{j} k_i(x)\chi_i(x)\,d\nu(x).$$

6. *For each $i = 1, 2, \ldots$, the Pinsker partition of $f|\Lambda_i$ is finite and consists of n_i elements Λ_i^k, $k = 1, \ldots, n_i$, such that $f(\Lambda_i^k) = \Lambda_i^{k+1}$, $k = 1, \ldots, n_i - 1$, and $f(\Lambda_i^{n_i}) = \Lambda_i^1$.*
7. $f^{n_i}|\Lambda_i^k$ *is a Bernoulli automorphism.*
8. *If the map f is mixing then it is a Bernoulli automorphism.*

Proof. Statements 1–3 follow from Theorem 9.4.1. We also have that $\bigwedge_{i=0}^{\infty} f^i\eta = \mathcal{H}(W^u) \geq \pi(f)$. Statement 5 follows from the Margulis–Ruelle Inequality (see Theorem 10.2.1) and Theorem 10.4.2. This implies that indeed $\bigwedge_{i=0}^{\infty} f^i\eta = \mathcal{H}(W^u) = \pi(f)$ and Statement 4 follows. For Statement 6 see the proof of Theorem 9.4.3. It also shows that $f^{n_i}|\Lambda_i^k$ is a K-automorphism. Statement 7 can now be proven using arguments in the proof of Theorem 9.5.1. The last statement is a corollary of Statement 7. □

We describe a result on local ergodicity for SRB measures in the following situation. Let f be a $C^{1+\alpha}$ diffeomorphism of a compact smooth manifold M, possessing a partially hyperbolic attractor Λ. This means that Λ is a compact invariant uniformly partially hyperbolic subset, and there exists a neighborhood U of Λ such that $f(U) \subset U$ and $\Lambda = \bigcap_{n\geq 0} f^n(U)$. For every $x \in \Lambda$, there is a decomposition of the tangent space $T_xM = E^s(x) \oplus E^c(x) \oplus E^u(x)$ into the stable

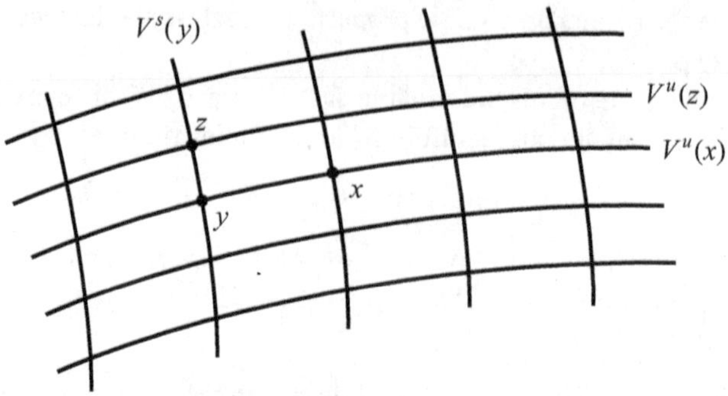

Figure 13.1. Construction of the set C.

$E^s(x)$, central $E^c(x)$, and unstable $E^u(x)$ subspaces. They depend continuously (in fact, Hölder continuously) in $x \in \Lambda$. The size of the local unstable manifold $V^u(x)$ is uniform in x and $V^u(x) \subset \Lambda$ for every $x \in \Lambda$ (hence for the global unstable manifold $W^u(x)$ through x we have $W^u(x) \subset \Lambda$).

Theorem 13.1.8. *Let ν be a u-measure for f on Λ. Assume that f has negative central exponents at every point x of an invariaut set A of positive measure (see Section 11.2). Then A is open* (mod 0).

Proof. By Theorem 13.1.6, every ergodic component of $f|A$ has positive measure and it suffices to show that each such component is open (mod 0) in Λ. Let φ be an integrable f-invariant function on Λ and N the set of zero measure constructed in Lemma 9.1.2. For almost every $x \in A$ and any sufficiently large $\ell > 0$, the set $V^u(x) \cap \Lambda^\ell$ has positive leaf volume while the set $V^u(x) \cap \Lambda^\ell \cap N$ has zero leaf volume. Let us choose such a point x and consider the set $B = \bigcup V^s(y)$ where the union is taken over all $y \in (V^u(x) \cap \Lambda^\ell) \setminus N$. We may also assume that $x \notin N$. Fix a point $y \in (V^u(x) \cap \Lambda^\ell) \setminus N$ and observe that the set

$$C = \bigcup_{z \in B^s(y,\varepsilon)\cap\Lambda} V^u(z)$$

is open in Λ (see Fig. 13.1; here $B^s(y, \varepsilon)$ is the ball in $V^s(y)$ centered at y of radius ε). By Propositions 13.1.4 and 13.1.5, for almost every $w \in C$ the intersection $V^u(w) \cap B$ has positive leaf volume and hence contains a point, which does not lie in N. For any such point w, choose $p \in B \setminus N$ and let $y \in (V^u(x) \cap \Lambda^\ell) \setminus N$ be such that $p \in V^s(y)$. By Lemma 9.1.2, we have that

$$\varphi(w) = \varphi(p) = \varphi(y) = \varphi(x).$$

This implies that φ is constant (mod 0) on C and hence the open set $\bigcup_{n\in\mathbb{Z}} f^n(C)$ is an ergodic component of f. The desired result follows. ☐

13.2 A Characterization of SRB Measures

The following result provides a characterization of SRB measures in terms of the entropy formula. Let f be a $C^{1+\alpha}$ diffeomorphism of a compact smooth manifold M.

Theorem 13.2.1. *A hyperbolic measure ν invariant under a $C^{1+\alpha}$ diffeomorphism f is an SRB measure if and only if it satisfies the entropy formula.*

Proof. In view of Theorem 10.4.2 all we have to show is that a measure ν satisfying the entropy formula is a u-measure for f. Clearly, we may assume that ν has at least one positive Lyapunov exponent. Let η be a leaf-subordinated partition associated with global stable manifolds. We have that $h_\nu(f) = H_\nu(\eta | f\eta)$. Indeed, if ν is ergodic, the result follows from Lemma 14.1.19. The general case can easily be reduced to the ergodic one and we leave the details to the reader.

For almost every $w \in M$, consider the function $\kappa^u(w, z)$ defined by (13.1), $w \in \Lambda^\ell$, and $z \in V^u(w)$. By Theorem 9.3.4, this function is Hölder continuous and so is the function $z \mapsto \log \kappa^u(w, z)$ for $z \in \eta(w)$. In particular, the function $z \mapsto \kappa^u(w, z)$ is uniformly bounded away from 0 and $+\infty$ on $\eta(w)$. Set

$$\rho(w, z) = \kappa^u(w, z) \Big/ \int_{\eta(w)} \kappa^u(w, \tau)\, dm_{V^u}(w)(\tau).$$

We define a new measure μ on M such that (1) the conditional measures $\mu^u(w)$ generated by μ on $V^u(w)$ are equivalent to the leaf volume $m_{V^u}(w)$ with the density function $\rho(w, z)$, that is,

$$d\mu^u(w) = \rho(w, z)\, dm_{V^u}(w)$$

and (2) μ coincides with ν on the σ-algebra $\mathcal{B}_\eta$ of measurable sets that are unions of elements of η.

Lemma 13.2.2.

$$\int_M -\log \mu^u(w)((f^{-1}\eta)(w))\, d\nu(w) = \int_M \log|\mathrm{Jac}(df | E^u(w))|\, d\nu(w).$$

Proof of the lemma. Set

$$q(w) = \mu^u(w)((f^{-1}\eta)(w)) \quad \text{and} \quad L(w) = \int_{\eta(w)} \kappa^u(w, \tau)\, dm_{V^u}(w)(\tau).$$

Then

$$q(w) = \frac{\int_{(f^{-1}\eta)(w)} \kappa^u(w, \tau)\, dm_{V^u}(w)(\tau)}{\int_{\eta(w)} \kappa^u(w, \tau)\, dm_{V^u}(w)(\tau)} = \frac{L(f(w))}{L(w)} \cdot \frac{1}{|\mathrm{Jac}(df | E^u(w))|}.$$

Since L is a positive finite-valued measurable function with

$$\int_M \log^+ \frac{L(f(w))}{L(w)}\, d\nu(w) \le \int_M \log^+ |\mathrm{Jac}(df | E^u(w))|\, d\nu(w) < \infty,$$

it follows that $\log q$ is integrable with

$$\int_M \log q \, d\nu = -\int_M \log|\mathrm{Jac}(df|E^u(w))| \, d\nu(w),$$

as desired. □

We will show that the entropy formula yields that $\mu = \nu$ on the whole σ-algebra of Borel sets in M. This implies that μ is a u-measure in fact, with strictly positive densities.

Lemma 13.2.3. *$\mu = \nu$ on $\mathcal{B}_{f^{-1}\eta}$.*

Proof of the lemma. For ν-almost every z, the partition $(f^{-1}\eta)|\eta(z)$ is countable. For $w \in \eta(z)$, consider the function

$$\frac{d\mu}{d\nu}|_{f^{-1}\eta}(w) = \frac{\mu^u(w)((f^{-1}\eta)(w))}{\nu^u(w)((f^{-1}\eta)(w))}.$$

For ν-almost every $w \in \eta(z)$, we have that $\nu^u(w)((f^{-1}\eta)(w)) \neq 0$ and hence this function is well defined. By the convexity of the logarithm,

$$\int_M \log \frac{d\mu}{d\nu}|_{f^{-1}\eta} \, d\nu \leq \log \int_M \frac{d\mu}{d\nu}|_{f^{-1}\eta} \, d\nu \leq 0. \tag{13.2}$$

The last inequality is obtained as follows: write $\eta(z) = \bigcup_{j \in J} A_j$ with $A_j \in f^{-1}\eta$ for each j. Since $\mu^u(w) = \mu^u(z)$ and $\nu^u(w) = \nu^u(z)$ for $w \in \eta(z)$, we obtain

$$\begin{aligned}
\int_{\eta(z)} \frac{d\mu}{d\nu}|_{f^{-1}\eta}(w) \, d\nu(z)(w) &= \int_{\eta(z)} \frac{\mu^u(w)((f^{-1}\eta)(w))}{\nu^u(w)((f^{-1}\eta)(w))} \, d\nu^u(z)(w) \\
&= \int_{\eta(z)} \frac{\mu^u(z)((f^{-1}\eta)(w))}{\nu^u(z)((f^{-1}\eta)(w))} \, d\nu^u(z)(w) \\
&= \sum_{j:\nu^u(z)(A_j)\neq 0} \frac{\mu^u(z)(A_j)}{\nu^u(z)(A_j)} \nu^u(z)(A_j) \\
&\leq \mu^u(z)(\eta(z)) \\
&= 1.
\end{aligned}$$

On the other hand, it follows from Lemma 13.2.2 and the entropy formula that

$$\begin{aligned}
-\int_M \log \mu^u(z)((f^{-1}\eta)(z)) \, d\nu(z) &= \int_M \log|\mathrm{Jac}(df|E^u(z))| \, d\nu(z) \\
&= H_\nu(f^{-1}\eta|\eta) \\
&= -\int_M \log \nu^u(z)((f^{-1}\eta)(z)) \, d\nu(z),
\end{aligned}$$

and thus

$$\int_M \log \frac{d\mu}{d\nu}|_{f^{-1}\eta} \, d\nu = 0.$$

It follows from (13.2) that $(d\mu/d\nu)|_{f^{-1}\eta}(w) = 1$ for ν-almost every w. Thus, $\mu = \nu$ on the σ-algebra $\mathcal{B}_{f^{-1}\eta}$. □

By induction, we have that $\mu = \nu$ on $\mathcal{B}_{f^{-n}\eta}$ for each $n \in \mathbb{N}$, and hence by Theorem 13.1.7, $\mu = \nu$ on the Borel σ-algebra. This completes the proof of the theorem when ν is ergodic.

We reduce the general case to the ergodic one. Let $\mathcal{J}$ be the sub-σ-algebra of the Borel σ-algebra composed of all invariant subsets and let ζ be a measurable partition such that $\mathcal{B}_\zeta = \mathcal{J}$ (mod 0), and the partition η refines ζ. Consider the family of conditional measures $\bar{\mu}(z)$ generated by μ on $\zeta(z)$. Then there is an invariant set N_1 of full ν-measure, which is composed of regular points (see Section 3.1) such that $\bar{\mu}(z)$ is an ergodic invariant measure for every $z \in N_1$.

Assume that the entropy formula holds. By the Margulis–Ruelle Inequality (see Theorem 10.2.1),

$$h_{\bar{\mu}(z)}(f) \leq \int_M \sum_{i:\chi_i(x)>0} k_i(x)\chi_i(x)\, d\bar{\mu}(z)(x)$$

for every $z \in N_1$. Since $h_\nu(f) = \int_M h_{\bar{\mu}(z)}(f)\, d\nu(z)$, there is an invariant set $N_2 \subset N_1$ of full ν-measure such that for every $z \in N_2$,

$$h_{\bar{\mu}(z)}(f) = \int_M \sum_{i:\chi_i(x)>0} k_i(x)\chi_i(x)\, d\bar{\mu}(z)(x).$$

By Theorem 13.1.7 and since η refines ζ, there is an invariant set $N_3 \subset N_2$ of full ν-measure such that for every $z \in N_3$,

1. $\eta(x) \subset W^u(x)$ and contains a neighborhood of x in $W^u(x)$ for $\bar{\mu}(z)$-almost every x;
2. the family of conditional measures $\mu_\eta(x)$ generated by μ on $\eta(x)$ is a family of conditional measures associated to η in the space $(\eta(z) \cap N_3, \bar{\mu}(z))$.

For $z \in N_3$ and for the transformation f on the measure space $(M, \bar{\mu}(z))$, we can apply the above argument in the case of ergodic measures to conclude that $\mu_\eta(x)$ is equivalent to the leaf volume $m_{V^u}(x)$ on $W^u(x)$ for $\bar{\mu}(z)$-almost every x; that is, $\bar{\mu}(z)(A) = 1$, where

$$A = \{x : \mu_\eta(x) \text{ is equivalent to } m_{V^u}(x)\}.$$

Therefore, $\nu(A) = 1$ and ν is an SRB measure. □

13.3 Constructions of SRB Measures

Any topologically transitive Anosov diffeomorphism f preserves a hyperbolic ergodic SRB measure. To see this consider the evolution of the Riemannian volume m under the system, that is, the sequence of measures

$$\nu_n = \frac{1}{n}\sum_{k=0}^{n-1} f_*^k m \tag{13.3}$$

where the measure f_*m is defined by $f_*m(E) = m(f(E))$ for any Borel set E. One can show that the sequence ν_n converges and that the limit measure ν^+ is an SRB measure for f. (Indeed, ν^+ is the limit of $f_*{}^n m$.) Similarly, let us consider the sequence of measures

$$\nu_n = \frac{1}{n} \sum_{k=0}^{n-1} f_*^{-k} m.$$

It converges to a measure ν^-, which is in general another SRB measure for f^{-1}. Applying Livshitz's theorem (see, e.g., [104, Theorem 19.2.1]) one can show that $\nu^+ = \nu^-$ if and only if ν^+ is a smooth hyperbolic measure for f.

Consider the set $B(\nu) = \bigcup W^s(x)$ where the union is taken over ν-almost every point x. By the absolute continuity property of stable manifolds, the set $B(\nu)$ has positive Lebesgue measure. Furthermore, for every $x \in B(\nu)$ and any continuous function φ on M, we have

$$\lim_{n\to\infty} \frac{1}{n} \sum_{k=0}^{n-1} \varphi(f^k(x)) = \int_M \varphi \, d\nu. \tag{13.4}$$

This property determines the role that SRB measures play in applications. (This is also why SRB measures are often called *physical measures*.) The set $B(\nu)$ is called the *basin* of ν. Observe that for any ergodic hyperbolic measure ν (not necessarily an SRB measure), the relation (13.4) holds for any $x \in B(\nu)$. However, in general the set $B(\nu)$ has zero volume.

The existence of SRB measures for diffeomorphisms other than Anosov involves the study of the sequence (13.3) and is a difficult problem. Only a few examples are available (see [88] for details and references).

We describe some results omitting proofs:

1. We begin with the case when Λ is a partially hyperbolic attractor for a $C^{1+\alpha}$ diffeomorphism f. Clearly, any limit measure ν of the sequence of measures ν_n is concentrated on Λ.

Theorem 13.3.1 (Pesin–Sinai [176]). *Any limit measure ν of the sequence ν_n is a u-measure for f; that is, for almost every $x \in \Lambda$, the conditional measure $\nu^u(x)$ generated by ν on the global strongly unstable manifold $W^u(x)$ is equivalent to the leaf volume $m_{V^u}(x)$ on $W^u(x)$.*

Bonatti, Díaz, and Viana [30] extended this theorem to limit measures of the evolution that starts from any measure, which is absolutely continuous with respect to the Riemannian volume.

We describe another way for constructing u-measures. Fix a point $x \in \Lambda$ and consider a local strongly unstable leaf $V^u(x)$ through x. One can view the leaf volume $m_{V^u}(x)$ on $V^u(x)$ as a measure on the whole set Λ. Consider the sequence of measures on Λ

$$\mu_n = \frac{1}{n}\sum_{k=0}^{n-1} f_*^k m_{V^u} m_{V^u}(x). \tag{13.5}$$

Clearly, any limit measure μ of the sequence μ_n is concentrated on Λ. It is shown in [176] that any such measure is a u-measure. Indeed, one can prove that this remains true if we replace the local unstable manifold $V^u(x)$ with any local manifold passing through x and is sufficiently close to $V^u(x)$ in the C^1 topology.

Moreover, for any ergodic u-measure μ and almost every $x \in \Lambda$, the sequence of measures (13.5) converges to μ. Therefore the class of all the limit measures for sequences of type (13.5) coincides with the class of all u-measures, while the class of limit measures for sequences of type (13.3) may be smaller.

The idea of constructing invariant measure by iterating unstable leaves goes back to pioneering work of Margulis (see the English translation of his dissertation in [145]) and Sinai [204].

Given an invariant measure ν on Λ, consider its basin $B(\nu)$.

Theorem 13.3.2 (Dolgopyat [55]). *Any measure with basin of positive volume is a u-measure.*

While Theorem 13.3.1 guarantees that any partially hyperbolic attractor has a u-measure, measures with basin of positive volume need not exist (just consider the map that is the product of a diffeomorphism with a hyperbolic attractor and the identity map). However, the following partial converse to Theorem 13.3.2 holds true.

Theorem 13.3.3 (Dolgopyat [55]). *If there is a unique u-measure for f in Λ then its basin has full volume in the topological basin of Λ.*

In the case of a hyperbolic attractor, topological transitivity of $f|\Lambda$ guarantees that there is a unique u-measure for f on Λ. In contrast, in the partially hyperbolic situation, even topological mixing is not enough to guarantee that there is a unique u-measure. The situation is, however, different if we require that f possesses an invariant measure on Λ with negative central exponents.

Theorem 13.3.4 (Bonatti–Viana [32], Burns–Dolgopyat–Pesin–Pollicott [43]). *Assume that for every $x \in \Lambda$ the global strongly unstable manifold $W^u(x)$ is dense in Λ_f. Also, assume that there exists a u-measure ν for f with negative central exponents on a subset $A \subset \Lambda$ of positive measure. Then ν is the only u-measure for f with negative central exponents at ν-almost every $x \in \Lambda$. The map f is ergodic and, indeed, Bernoulli with respect to ν, and the basin $B(\nu)$ has full volume in the topological basin of Λ. Finally, ν is the only SRB measure for f on Λ.*

It is well known that any C^1 diffeomorphism g, which is sufficiently close to f in the C^1 topology, possesses a partially hyperbolic attractor Λ_g, which lies in a small neighborhood of Λ_f.

Theorem 13.3.5 (Burns–Dolgopyat–Pesin–Pollicott [43]). *Let f be a $C^{1+\alpha}$ diffeomorphism possessing a partially hyperbolic attractor Λ_f. Assume that there exists a u-measure $\nu = \nu_f$ for f with negative central exponents on a subset $A \subset \Lambda_f$ of positive measure. Also, assume that for every $x \in \Lambda_f$ the global strongly unstable manifold $W^u(x)$ is dense in Λ_f. Then any $C^{1+\alpha}$ diffeomorphism g that is sufficiently close to f in the C^1-topology has negative central exponents almost everywhere with respect to a u-measure ν_g. This measure is the unique u-measure for g with respect to which g is ergodic, and the basin $B(\nu_g)$ has full volume in the topological basin of Λ_g. Finally, ν_g is the only SRB measure for g on Λ_g.*

Examples of partially hyperbolic systems whose strongly unstable foliation is minimal (i.e., every global strongly unstable leaf is dense) are constructed in [29]. Examples of partially hyperbolic systems satisfying the conditions of Theorem 13.3.5 are discussed in [43].

Ruelle [195] showed that for an open set of one-parameter families of (not necessarily volume preserving) maps F_ε through F_0, each map F_ε possesses a u-measure with negative central exponent.

Dolgopyat [57] studied skew products $F = f \times \mathrm{Id}$ where f is an Anosov diffeomorphism and Id is the identity map of any manifold. He showed that there is an open set of one-parameter families of skew products near F where the central exponents are negative with respect to any u-measure.

Dolgopyat [56] also considered one-parameter families f_ε where f_0 is the time-1 map of the geodesic flow on the unit tangent bundle over a negatively curved surface. It is shown that in the volume preserving case, generically, either f_ε or f_ε^{-1} has negative central exponent for small ε and that there is an open set of nonconservative families where the central exponent is negative for any u-measure.

Given diffeomorphisms f and g, let $F_n = f^n \circ g$. It is shown in [56] that if f is either an S^1 extension of an Anosov diffeomorphism or a time-1 map of an Anosov flow and g is close to Id then for typical g and any sufficiently large n, either F_n or F_n^{-1} has negative central exponent with respect to any u-measure.

We now consider partially hyperbolic attractors whose Lyapunov exponents in the central direction are all positive with respect to some invariant measure. Since the stable and unstable directions play different roles in a dissipative system, this case cannot be reduced to the case in which the central exponents are negative by simply replacing f by f^{-1}, as was done in the conservative situation (see Section 11.2). It turns out that a straightforward analogue of the above results is not true. To see this consider the direct product F of two volume preserving Anosov diffeomorphisms f_1 and f_2. Let E_i^s and E_i^u, $i = 1, 2$, be the corresponding stable and unstable subspaces. Assume that the expansion rates along E_1^u are stronger than those along E_2^u. Set

$$E^- = E_1^s \oplus E_2^s, \quad E^c = E_2^u, \quad E^+ = E_1^u.$$

Then $TM = E^- \oplus E^c \oplus E^+$ is the partially hyperbolic splitting for F. Consider an F-invariant measure $\nu = \nu_1 \times \nu_2$, where ν_1 is the volume and ν_2 is a nontrivial mixture of any two ergodic measures of full support. It is easy to see that ν is a u-measure for F with positive central exponents whose ergodic components are not open (mod 0).

Nevertheless, Alves, Bonatti, and Viana [5] obtained an analogue of Theorem 13.3.4 under the stronger assumption that there is a set of positive volume in a neighborhood of the attractor with positive central exponents.

Theorem 13.3.6 (Alves–Bonatti–Viana [5]). *Assume that f is nonuniformly expanding along the center unstable direction; that is,*

$$\overline{\lim_{n\to\infty}} \frac{1}{n} \sum_{j=1}^{n} \log \|df^{-1}|E^{cu}_{f^j(x)}\| < 0 \tag{13.6}$$

for all x in a set of positive Lebesgue measure $A \subset M$. Then f has an ergodic SRB measure. Moreover, if the limit in (13.6) *is bounded away from zero then A is contained* (mod 0) *in the union of the basins of finitely many SRB measures.*

2. The following general statement links convergence of the sequence of measures (13.3) to the existence of SRB measures.

Theorem 13.3.7 (Tsujii [214]). *Let f be a $C^{1+\alpha}$ diffeomorphism of a compact smooth Riemannian manifold M and $A \subset M$ a set of positive volume such that for every $x \in A$, the sequence of measures*

$$\frac{1}{n} \sum_{i=0}^{n-1} \delta_{f^k(x)}$$

converges weakly to an ergodic hyperbolic measure ν_x. If the Lyapunov exponents at x coincide with those of ν_x, then ν_x is an SRB measure for Lebesgue almost every $x \in A$.

3. In [238], L.-S. Young suggested an axiomatic approach for constructing SRB measures. It is built upon her work on tower constructions for nonuniformly hyperbolic systems and presents the system as a Markov extension (see the appendix in [21]). This approach is a basis for establishing existence of SRB measures for Hénon-type attractors, as well as existence of absolutely continuous invariant measures for some piecewise hyperbolic maps and logistic maps.

Let f be a $C^{1+\alpha}$ diffeomorphism of a compact smooth Riemannian manifold M.

An embedded disk $\gamma \subset M$ is called an *unstable disk* if for any $x, y \in \gamma$, the distance $\rho(f^{-n}(x), f^{-n}(y)) \to 0$ exponentially fast as $n \to \infty$; it is called a *stable disk* if for any $x, y \in \gamma$, the distance $\rho(f^n(x), f^n(y)) \to 0$ exponentially fast as $n \to \infty$.

We say that a set Λ has a *hyperbolic product structure* if there exist a continuous family of unstable disks $\Gamma^u = \{\gamma^u\}$ and a continuous family of stable disks $\Gamma^s = \{\gamma^s\}$ such that

1. $\dim \gamma^u + \dim \gamma^s = \dim M$;
2. the γ^u-disks are transversal to the γ^s-disks with the angles between them bounded away from zero;
3. each γ^u-disk meets each γ^s-disk at exactly one point;
4. $\Lambda = (\bigcup \gamma^u) \cap (\bigcup \gamma^s)$.

We impose some conditions on the map f (see Conditions (P1)–(P5) below), which guarantee the existence of an SRB measure for f. Roughly speaking, they mean that there exists a set Λ with a hyperbolic product structure and a return map f^R from Λ to itself such that f is a Markov extension over f^R (see the appendix in [21]). More precisely, we assume the following:

(P1) There exists $\Lambda \subset M$ with a hyperbolic product structure and such that $m_{\gamma^u}(\gamma^u \cap \Lambda) > 0$ for every $\gamma^u \in \Gamma^u$, where m_{γ^u} is the leaf volume on γ^u.

(P2) There are pairwise disjoint subsets $\Lambda_1, \Lambda_2, \ldots \subset \Lambda$ with the following properties:
   - (a) Each Λ_i has a hyperbolic product structure and its defining families can be chosen to be Γ^u and $\Gamma^s_i \subset \Gamma^s$; we call Λ_i an *s-subset*; similarly, one defines *u-subsets*.
   - (b) On each γ^u-disk, $m_{\gamma^u}((\Lambda \setminus \cup \Lambda_i) \cap \gamma^u) = 0$.
   - (c) There exists $R_i \geq 0$ such that $f^{R_i}(\Lambda_i)$ is a u-subset of Λ; moreover, for all $x \in \Lambda_i$ we require that $f^{R_i}(\gamma^s(x)) \subset \gamma^s(f^{R_i}(x))$ and $f^{R_i}(\gamma^u(x)) \supset \gamma^u(f^{R_i}(x))$.
   - (d) For each n, there are at most finitely many i's with $R_i = n$.
   - (e) $\min_i R_i \geq R_0$ for some $R_0 > 0$ depending only on f.

Condition (P2) means that the set Λ has the structure of a "horseshoe," however, with infinitely many branches returning at variable times.

In order to state the remaining Conditions (P3)–(P5), we assume that there is a function $s_0(x, y)$ – a *separation time* of the points x and y – which satisfies that

1. $s_0(x, y) \geq 0$ and it depends only on the γ^s-disks containing the two points;
2. the maximum number of orbits starting from Λ that are pairwise separated before time n is finite for each n;
3. for $x, y \in \Lambda_i$, $s_0(x, y) \geq R_i + s_0(f^{R_i}(x), f^{R_i}(y))$; in particular, $s_0(x, y) \geq R_i$;
4. for $x \in \Lambda_i$, $y \in \Lambda_j$, $i \neq j$ with $R_i = R_j$, we have $s_0(x, y) < R_i - 1$.

Conditions 3 and 4 describe the relations between $s_0(x, y)$ and returns to Λ, namely, that points in the same Λ_i must not separate before they return, while points in distinct Λ_i's must first separate if they are to return simultaneously.

Let f^u be the restriction of f to γ^u. We assume that there exist $C > 0$ and $\alpha < 1$ such that for all $x, y \in \Lambda$, the following conditions hold:

(P3) *Contraction along γ^s-disks*: $\rho(f^n(x), f^n(y)) \le C\alpha^n$ for all $n \ge 0$ and $y \in \gamma^s(x)$.

(P4) *Backward contraction and distortion along γ^u*: for $y \in \gamma^u(x)$ and $0 \le k \le n < s_0(x, y)$, we have

(a) $\rho(f^n(x), f^n(y)) \le C\alpha^{s_0(x,y)-n}$;

(b)

$$\log \prod_{i=k}^{n} \frac{\det df^u(f^i(x))}{\det df^u(f^i(y))} \le C\alpha^{s_0(x,y)-n}.$$

(P5) *Convergence of $d(f^i|\gamma^u)$ and absolute continuity of Γ^s*:

(a) for $y \in \Gamma^s(x)$ and $n \ge 0$,

$$\log \prod_{i=k}^{\infty} \frac{\det df^u(f^i(x))}{\det df^u(f^i(y))} \le C\alpha^n;$$

(b) for $\gamma, \gamma' \in \Gamma^u$, define $\Theta\colon \gamma \cap \Lambda \to \gamma' \cap \Lambda$ by $\Theta(x) = \gamma^s(x) \cap \gamma'$; then Θ is absolutely continuous and

$$\frac{d(\Theta_*^{-1} m_{\gamma'})}{dm_\gamma}(x) = \prod_{i=0}^{\infty} \frac{\det df^u(f^i(x))}{\det df^u(f^i(\Theta(x)))}.$$

In [238], L.-S. Young showed that a map f satisfying Conditions (P1)–(P5) admits a Markov extension (see the appendix in [21]). As an important corollary one has the following result.

Theorem 13.3.8 (L.-S. Young [238]). *Assume that for some $\gamma \in \Gamma^u$,*

$$\int_{\gamma \cap \Lambda} R \, dm_\gamma < \infty.$$

Then f admits an SRB measure.

One of the celebrated manifestations of Theorem 13.3.8 is a construction of SRB measures for the Hénon family $H_{a,b}$ of two-dimensional maps (see (6.34)). For $a \in (0, 2)$, there is a rectangle in the plane, which is mapped by $H_{a,b}$ into itself. It follows that $H_{a,b}$ has an attractor provided b is sufficiently small—the Hénon attractor. Building on the analysis of the map $H_{a,b}$ in [25], Benedicks and L.-S. Young [26] established existence of SRB measures for the Hénon attractors.

Theorem 13.3.9. *There exist $\varepsilon > 0$ and $b_0 > 0$ such that for every $0 < b \le b_0$, one can find a set $\Delta_b \in (2 - \varepsilon, 2)$ of positive Lebesgue measure with the property that for each $a \in \Delta_b$, the map $H_{a,b}$ admits a unique SRB measure $\nu_{a,b}$.*

In [222, 223], Wang and L.-S. Young introduced and studied more general two-parameter families of maps to which the above results extend. In another direction, Mora and Viana [156] modified Benedicks and Carleson's approach in a way that allowed them to treat Hénon-like maps using some techniques from the general bifurcation theory such as homoclinic tangencies. Later Viana [219] extended results from [156] to higher dimensions (see [133] for a more detailed account of these results and further references).

13.4 Notes

The term "SRB measure" was coined by Eckmann and Ruelle in [65]. Theorems 13.1.6 and 13.1.7 were proved by Ledrappier [120] and the entropy formula (Theorem 5) by Ledrappier and Strelcyn [123]. Theorem 13.1.8 was implicit in work of Bonatti and Viana [32]. A different proof was given by Burns, Dolgopyat, Pesin, and Pollicott [43]. We follow the approach in this paper. The fact that a measure invariant under a C^2 diffeomorphism satisfying the entropy formula is a u-measure (see Theorem 13.2.1) is due to Ledrappier [120] in the case of hyperbolic measures. The general case was proved by Ledrappier and L.-S. Young [124].

Part IV

General Hyperbolic Measures

14

Hyperbolic Measures: Entropy and Dimension

In this and the next chapters we discuss some ergodic and topological properties of diffeomorphisms with respect to general hyperbolic measures. Although such measures do not have such strong ergodic properties as smooth and SRB measures do, one can obtain a number of deep results revealing a highly nontrivial structure of these measures. Among them are the Ledrappier–Young formula for the entropy of a diffeomorphism and the result by Barreira, Pesin, and Schmeling establishing an "almost" local direct product structure of general hyperbolic measures. We shall describe these two results in this chapter.

14.1 Pointwise Dimensions and the Ledrappier–Young Entropy Formula

Let f be a $C^{1+\alpha}$ diffeomorphism of a smooth compact Riemannian manifold M preserving a hyperbolic measure ν. Denote by Z the set of Lyapunov–Perron regular points with nonzero Lyapunov exponents. Given $x \in Z$, consider the values of the Lyapunov exponents at x,

$$\chi_1(x) < \cdots < \chi_{s(x)}(x) < 0 < \chi_{s(x)+1}(x) < \chi_{p(x)}(x)$$

and for $i = s(x)+1, \ldots, p(x)$, the corresponding filtrations of local unstable manifolds

$$x \in V^u_{s(x)+1}(x) \subset \cdots \subset V^u_{p(x)}(x)$$

and global unstable manifolds

$$x \in W^u_{s(x)+1}(x) \subset \cdots \subset W^u_{p(x)}(x).$$

We choose an f-invariant Borel function $i = i(x)$ such that $s(x) \le i(x) < p(x)$. We can view the set Z as nonuniformly partially hyperbolic in the broad sense for f with $\lambda(x) = e^{\chi_{i(x)}(x)}$, $\mu(x) = e^{\chi_{i(x)+1}(x)}$, and an arbitrary f-invariant Borel function $\varepsilon(x)$ satisfying (2.9) (with an appropriate choice of the functions $C(x)$

and $K(x)$; see Theorem 3.3.1). We can choose a level set $\Lambda = \Lambda_{\lambda\mu\varepsilon j}$ of positive measure and then a regular set $\Lambda^\ell = \Lambda^\ell_{\lambda\mu\varepsilon j}$ of positive measure.

14.1.1 Local Entropies

Let η_i be the leaf-subordinated partition associated with ith global unstable manifolds (see Section 9.4). Denote by $\nu_i^u(w)$ the conditional measure on $W_i^u(w)$ generated by the partition η_i and the measure ν. For each $x \in \Lambda^\ell$ and $\varepsilon > 0$, set

$$\underline{h}_i(x, \varepsilon) = \varliminf_{n\to\infty} -\frac{1}{n} \log \nu_i^u(x)(D_i(x, n, \varepsilon)),$$

$$\overline{h}_i(x, \varepsilon) = \varlimsup_{n\to\infty} -\frac{1}{n} \log \nu_i^u(x)(D_i(x, n, \varepsilon)),$$

where

$$D_i(x, n, \varepsilon) = \left\{y \in W_i^u(x) : \rho_i(f^k(x), f^k(y)) < \varepsilon \text{ for } 0 \le k < n\right\}$$

and $\rho_i = \rho_{W_i^u}$ is the induced distance on $W_i^u(x)$. We also write

$$\underline{h}_i(x) = \lim_{\varepsilon\to 0} \varliminf_{n\to\infty} -\frac{1}{n} \log \nu_i^u(x)(D_i(x, n, \varepsilon)),$$

$$\overline{h}_i(x) = \lim_{\varepsilon\to 0} \varlimsup_{n\to\infty} -\frac{1}{n} \log \nu_i^u(x)(D_i(x, n, \varepsilon)).$$

Clearly, for every x,

$$\underline{h}_i(f(x)) = \underline{h}_i(x) \quad \text{and} \quad \overline{h}_i(f(x)) = \overline{h}_i(x).$$

Theorem 14.1.1.

1. *For ν-almost every $x \in M$, we have*

$$h_i(x) := \underline{h}_i(x) = \overline{h}_i(x),$$

and

$$\int_M h_i(x)\, d\nu(x) = H_\nu(\eta_i | f\eta_i).$$

2. *If ν is ergodic, then for ν-almost every $x \in M$,*

$$h_i(x) = H_\nu(\eta_i | f\eta_i).$$

We call the number $h_i(x)$ the *local entropy "along" the leaf* $W_i^u(x)$.

Proof of Theorem 14.1.1. Given $\delta > 0$, set $A_\delta = \{x : B_i^u(x, \delta) \subset \eta_i(x)\}$. (Recall that $B_i^u(x, \delta)$ denotes the ball in $V_i^u(x)$ centered at x of radius δ with respect to the induced distance on $V_i^u(x)$.) Clearly, $\nu(A_\delta)$ approaches 1 as $\delta \to 0$. Given $\varepsilon > 0$, choose $\delta > 0$ sufficiently small such that

$$\int_{A_\delta} -\log \nu_i^u(x)((f^{-1}\eta_i)(x))\, d\nu(x) \ge H_\nu(f^{-1}\eta_i | \eta_i) - \varepsilon.$$

Define

$$U_i(x, n, \delta) = \bigcup_{0 \le j \le n:\, f^j(x) \in A_\delta} (f^{-j}\eta_i)(x).$$

Then $D_i(x, n, \delta) \subset U_i(x, n, \delta)$, and

$$-\log \nu_i^u(x)(U_i(x, n, \delta)) \ge \sum_{j=0}^{n} (\chi_{A_\delta} g)(f^j(x)),$$

where $g(x) = -\log \nu_i^u(x)((f^{-1}\eta_i)(x))$. Hence for ν-almost every x,

$$\underline{h}_i(x) \ge \underline{h}_i(x, \delta) \ge \varliminf_{n\to\infty} -\frac{1}{n} \log \nu_i^u(x)(U_i(x, n, \delta)),$$

and

$$\int_M \underline{h}_i(x)\, d\nu(x) \ge \int_{A_\delta} g\, d\nu \ge H_\nu(f^{-1}\eta_i | \eta_i) - \varepsilon = H_\nu(\eta_i | f\eta_i) - \varepsilon.$$

Since ε is arbitrarily small, this implies that

$$\int_M \underline{h}_i(x)\, d\nu(x) \ge H_\nu(\eta_i | f\eta_i). \tag{14.1}$$

To prove the inverse inequality we start with an auxiliary result. Given a partition α of M and $m < n$, denote by $\alpha_m^n(x)$ the element of the partition $\bigvee_{j=m}^{n-1} f^{-j}\alpha$ that contains x.

Lemma 14.1.2. *If $\mathcal{P}$ is a partition of M with $H_\nu(\mathcal{P}) < \infty$, then for ν-almost every x,*

$$\int_M \lim_{n\to\infty} -\frac{1}{n} \log \nu_i^u(x)((\mathcal{P} \vee \eta_i)_0^n(x))\, d\nu(x) = H_\nu(\eta_i | f\eta_i).$$

Proof of the lemma. We define the information function of a partition $\mathcal{P}$ given a partition ξ by

$$I(\mathcal{P}|\xi)(x) = -\log[\nu_\xi(x)(\mathcal{P}(x))],$$

where $\nu_\xi(x)$ is a family of conditional measures associated with ξ. Then

$$\frac{1}{n} I((\mathcal{P} \vee \eta_i)_0^n | \eta_i)(x) = \frac{1}{n} I(\mathcal{P}|\eta_i)(x) + \frac{1}{n} \sum_{k=1}^{n-1} I(\mathcal{P} \vee \eta_i | f\eta_i \vee \mathcal{P}_{-k}^0)(f^k(x)).$$

Since the function $\sup_n I(\mathcal{P} \vee \eta_i | f\eta_i \vee \mathcal{P}_{-n}^0)$ is integrable, one can show that

$$\frac{1}{n} \sum_{k=1}^{n-1} I(\mathcal{P} \vee \eta_i | f\eta_i \vee \mathcal{P}_{-k}^0)(f^k(x))$$

converges ν-almost everywhere and converges in L^1. Therefore for ν-almost every x,

$$\int_M \lim_{n\to\infty} \frac{1}{n} I((\mathcal{P} \vee \eta_i)_0^n | \eta_i)(x)\, d\nu(x) = \lim_{n\to\infty} \frac{1}{n} H_\nu((\mathcal{P} \vee \eta_i)_0^n | \eta_i).$$

The right-hand side can be written as

$$\lim_{n\to\infty}\frac{1}{n}H_\nu((\eta_i)_0^n|\eta_i)+\lim_{n\to\infty}\frac{1}{n}H_\nu(\mathcal{P}_0^n|f^{-n+1}\eta_i).$$

The first summand is equal to $H_\nu(\eta_i|f\eta_i)$ while the second is zero since the sequence of partitions $f^{-n+1}\eta_i$ is generating. This yields the desired result. □

For $x \in \Lambda^\ell$ consider the Lyapunov chart $R(x)$ at x (see Section 5.6) and denote by $W_i'(x)$ the subset of $B(0, q(x))$ such that its Ψ_x-image is the component of $W_i'(x) \cap R(x)$ containing x (see Theorem 5.6.1 for the definition of the map Ψ). Write $f_x = \Psi_{f(x)}^{-1} \circ f \circ \Psi_x$. Observe that $\Psi_x(W_i'(x))$ is an ith local manifold $V_i^u(x)$ at x.

Let $w \in \Lambda^\ell$ be a density point of the measure ν and $Q^\ell = Q_i^\ell(w)$ is given by (8.16). We now construct a partition $\mathcal{P}$ such that the element $\mathcal{P}_0^n(x)$ can be approximated by the set $D_i(x, n, \varepsilon)$. For each $x \in Q^\ell$, we let

$$n_+(x) = \inf\{n \in \mathbb{N} : f^n(x) \in Q^\ell\},$$
$$n_-(x) = \inf\{n \in \mathbb{N} : f^{-n}(x) \in Q^\ell\}.$$

Given $\delta > 0$, define the function $\rho\colon M \to \mathbb{R}$ by

$$\rho(x) = \frac{\delta}{K\ell^2}e^{-\max\{\chi_p^+(x)+3\varepsilon,-\chi_1^+(x)+3\varepsilon\}\cdot\max\{n_+(x),n_-(x)\}}$$

for $x \in Q^\ell$, and $\rho(x) = 1$ otherwise, where ε and K are as in (8.12)–(8.13). Since

$$\int_M \rho\, d\nu \le \sum_{\ell\ge 1}\frac{\delta}{K}\int_{\Lambda^{\ell+1}\setminus\Lambda^\ell}\frac{1}{\ell^2}\, d\nu \le \frac{\delta}{K}\sum_{\ell\ge 1}\frac{1}{\ell^2} < \infty,$$

we can apply Lemma 10.5.4 to obtain a partition $\mathcal{P}$ of M with entropy $H_\nu(\mathcal{P}) < +\infty$ such that $\mathcal{P}(x) \subset B(x, \rho(x))$ for ν-almost every x. We also define the function $\rho_+\colon M \to \mathbb{R}$ by

$$\rho_+(x) = \frac{\delta}{K\ell^2}e^{-\max\{\chi_p^+(x)+3\varepsilon,-\chi_1^+(x)+3\varepsilon\}n_+(x)} \quad \text{for } x \in Q^\ell$$

and $\rho_+(x) = 1$ otherwise.

We need the following three technical lemmas. The first one is an immediate corollary of the properties of Lyapunov charts.

Lemma 14.1.3. *Let x be a Lyapunov–Perron regular point for f.*

1. *If $y, z \in B(0, e^{-\chi_p^+(x)-3\varepsilon}q(x))$, then $f_x(y), f_x(z) \in B(0, q(f(x)))$ and*

$$\|f_x(y) - f_x(z)\| \le e^{\chi_p^+(x)+2\varepsilon}\|y - z\|.$$

2. *If $z, z' \in W_i'(x)$, then*

$$\|f_x^{-1}(z) - f_x^{-1}(z')\| \le e^{-\chi_i^+(x)+2\varepsilon}\|z - z'\|.$$

Lemma 14.1.4. *Fix* $x \in Q^\ell$.

1. *If* $y \in \Psi_x(W_i'(x))$ *satisfies* $\|\Psi_x^{-1}y\| \le \ell\rho_+(x)$, *then* $\rho_i(f^j(x), f^j(y)) < \delta$ *for* $0 \le j \le n_+(x)$, *and*
$$f^{n_+(x)}(y) \in \Psi_{f^{n_+(x)}(x)}(W_i'(f^{n_+(x)}(x))).$$
2. *If* $y \in \mathcal{P}_0^n(x) \cap \Psi_x(W_i'(x))$ *for some* $n \ge 0$, *then* $\rho_i(f^j(x), f^j(y)) < \delta$ *for* $0 \le j \le n$.

Proof of the lemma. It follows from Lemma 14.1.3 that $f_x^j(\Psi_x^{-1}y) \in W_i'(f^j(x))$ and
$$\|f_x^j(\Psi_x^{-1}y)\| \le \|\Psi_x^{-1}y\| e^{(\chi_p^+(x)+2\varepsilon)j}, \quad 0 < j \le n_+(x),$$
provided that
$$\|\Psi_x^{-1}y\| e^{(\chi_p^+(x)+2\varepsilon)k} \le q(f^k(x))e^{-\chi_p^+(x)-3\varepsilon} \quad \text{for all } 0 \le k < j.$$
Since $W_i'(f^j(x))$ is a graph over $B_i^u(0, q(f^j(x)))$ with slope at most 1/3 (see Section 8.4), we obtain
$$\rho_i(f^j(x), f^j(y)) < K\|\Psi_{f^j(x)}^{-1}(f^j(y))\| < \delta \quad \text{for} \quad 0 \le j \le n_+(x).$$
To prove the second statement, note that if $y \in \mathcal{P}(x)$, then $\|\Psi_x^{-1}y\| \le \ell\rho(x) \le \ell\rho_+(x)$, and we have the desired bound for $0 \le j \le n_+(x)$. Furthermore, if $n > n_+(x)$ and $y \in \mathcal{P}_0^n(x)$, then
$$f^{n_+(x)}(y) \in \mathcal{P}(f^{n_+(x)}(x)) \cap \Psi_{f^{n_+(x)}(x)}(W_i'(f^{n_+(x)}(x))),$$
and we can apply the first statement with $f^{n_+(x)}(x)$ and $f^{n_+(x)}(y)$ respectively in place of x and y. The statement follows now by an inductive argument. □

Lemma 14.1.5. *For each* $\delta > 0$, *there exists a measurable function* $n_0 \colon M \to \mathbb{N}$ *such that for* ν*-almost every* x,
$$\mathcal{P}_0^n(x) \cap \eta_i(x) \subset D_i(x, n, \delta), \quad \textit{for all } n \ge n_0(x).$$

Proof of the lemma. Set
$$n_0(x) = \inf\{n \in \mathbb{N} : f^n(x) \in Q^\ell\}. \tag{14.2}$$
Consider an arbitrary point x with the property that $f^n(x) \in Q^\ell$ infinitely often as $n \to \pm\infty$. Let $y \in \mathcal{P}_0^{n_0(x)}(x) \cap \eta_i(x)$. It suffices to show that $\rho_i(f^j(x), f^j(y)) < \delta$ for $0 \le j \le n_0(x)$ and that
$$f^{n_0(x)}(y) \in \Psi_{f^{n_0(x)}(x)}(W_i'(f^{n_0(x)}(x))). \tag{14.3}$$
Indeed, by (14.3) and the second statement of Lemma 14.1.4 applied to $f^{n_0(x)}(y)$, we obtain $\rho_i(f^j(x), f^j(y)) < \delta$ for $n_0(x) \le j \le n$. Assume that $x \notin Q^\ell$. Let $r_0 = -n_-(f^{n_0(x)}(x)) + n_0(x)$. (This is the largest negative integer such that $f^{r_0}(x) \in Q^\ell$.) Since η_i is increasing, we have that $f^{r_0}(y) \in \eta_i(f^{r_0}(x))$, which, by the choice of Q^ℓ, lies in $\Psi_{f^{r_0}(x)}(W'(f^{r_0}(x)))$. Furthermore, the choice of $\rho(x)$ guarantees that

$f^{-j}\mathcal{P}(f^{n_0(x)}(x))$ lies inside the chart at $f^{-j+n_0(x)}(x)$ for $j = 1, \dots, n_-(f^{n_0(x)}(x))$. Therefore,

$$\begin{aligned}\|\Psi^{-1}_{f^{r_0}(x)} f^{r_0}(y)\| &\leq \|\Psi^{-1}_{f^{r_0+1}(x)} f^{r_0+1}(y)\| \leq \cdots \\ &\leq \|\Psi^{-1}_{f^{n_0(x)}(x)} f^{n_0(x)}(y)\| \\ &\leq \ell\rho(f^{n_0(x)}(x)) \leq \ell\rho_+(f^{r_0}(x)).\end{aligned}$$

The desired result follows from the first statement of Lemma 14.1.4. □

We proceed with the proof of the theorem. By Lemma 14.1.5, for ν-almost every x and for all sufficiently small $\delta > 0$,

$$\begin{aligned}\overline{h}_i(x,\delta) &= \overline{\lim_{n\to\infty}} -\frac{1}{n}\log \nu_i^u(x)(D_i(x,n,\delta)) \\ &\leq \overline{\lim_{n\to\infty}} -\frac{1}{n}\log \nu_i^u(x)(\mathcal{P}_0^n(x)) \\ &\leq \overline{\lim_{n\to\infty}} -\frac{1}{n}\log \nu_i^u(x)((\mathcal{P}\vee\eta_i)_0^n(x)).\end{aligned} \tag{14.4}$$

Letting $\delta \to 0$ we obtain using Lemma 14.1.2 that

$$\int_M \overline{h}_i(x)\,\nu(x) \leq H_\nu(\eta_i | f\eta_i). \tag{14.5}$$

We conclude from (14.1) and (14.5) that

$$\int_M \underline{h}_i(x)\,\nu(x) = \int_M \overline{h}_i(x)\,\nu(x) = H_\nu(\eta_i | f\eta_i).$$

This readily implies the desired statements. □

14.1.2 Leaf Pointwise Dimensions

For each $x \in \Lambda^\ell$, set

$$\underline{d}_i^u(x) = \underline{\lim_{r\to 0}} \frac{\log \nu_i^u(x)(B_i^u(x,r))}{\log r} \quad\text{and}\quad \overline{d}_i^u(x) = \overline{\lim_{r\to 0}} \frac{\log \nu_i^u(x)(B_i^u(x,r))}{\log r}.$$

(Recall that $u(x) \leq i(x) \leq p(x)$ and $B_i^u(x,r) \subset V_i^u(x)$ is the ball at x of radius r with respect to the induced distance on $V_i^u(x)$.) Clearly, for every x,

$$\underline{d}_i^u(f(x)) = \underline{d}_i^u(x) \quad\text{and}\quad \overline{d}_i^u(f(x)) = \overline{d}_i^u(x).$$

Theorem 14.1.6. *For ν-almost every $x \in M$ the following hold:*

1. $d_i^u(x) := \underline{d}_i^u(x) = \overline{d}_i^u(x)$.
2. *For $u(x) + 1 \leq i \leq p(x)$,*

$$0 \leq d_i^u(x) - d_{i-1}^u(x) = \frac{h_i(x) - h_{i-1}(x)}{\chi_i^+(x)} \leq \dim E_i(x).$$

Proof. We first assume that ν is ergodic. In this case, for $i = u, \dots, p$,

$$h_i(x) = h_i, \quad \chi_i^+(x) = \chi_i^+ \quad \text{and} \quad \underline{d}_i^u(x) = \underline{d}_i^u, \quad \overline{d}_i^u(x) = \overline{d}_i^u$$

for some constants h_i, χ_i^+, $\underline{d}_i^u$, and $\overline{d}_i^u$.

Lemma 14.1.7. $\underline{d}_p^u = \overline{d}_p^u = h_p/\chi_p^+$.

Proof of the lemma. By Theorem 14.1.1, given $\varepsilon > 0$, there exists $\delta > 0$ such that

$$\underline{h}_p(x, K\delta) \geq h_p - \varepsilon \quad \text{for } \nu\text{-almost every } x.$$

Let $y \in B_p^u(x, e^{-n(\chi_p^+ + 3\varepsilon)}\delta q(x)^2)$ and $z = \Psi_x^{-1}(y)$. Then $z \in W_p'(x)$ and

$$\|z\| \leq e^{-n(\chi_p^+ + 3\varepsilon)}\delta q(x).$$

It follows from Lemma 14.1.3 that

$$\|f_x^k(x)\| \leq \|z\| \leq e^{(\chi_p^+ + 2\varepsilon)k}$$

for $k > 0$, provided that all iterates up to k stay inside the Lyapunov charts that is guaranteed for $k = 1, \dots, n$. Therefore, $\|f_x^k(x)\| \leq \delta$ for $0 \leq k \leq n$ and since $f_x^k(z) \in W_p'(f^k(x))$, we have

$$B_p^u(x, e^{-n(\chi_p^+ + 3\varepsilon)}\delta q(x)^2) \subset D_p(x, n, K\delta).$$

This implies that

$$\begin{aligned} \underline{d}_p^u(x) &= \varliminf_{r \to 0} \frac{\log \nu_p^u(x)(B_p^u(x, r))}{\log r} \\ &= \varliminf_{n \to \infty} \frac{\log \nu_p^u(x)(B_p^u(x, e^{-n(\chi_p^+ + 3\varepsilon)}\delta q(x)^2))}{-n(\chi_p^+ + 3\varepsilon)} \\ &\geq \frac{1}{\chi_p^+ + 3\varepsilon} \varliminf_{n \to \infty} -\frac{1}{n} \log \nu_p^u(x)(D_p(x, n, K\delta)) \geq \frac{h_p - \varepsilon}{\chi_p^+ + 3\varepsilon} \end{aligned} \tag{14.6}$$

for ν-almost every $x \in \Lambda^\ell$. Since ε can be chosen arbitrarily small, we have $\underline{d}_p^u \geq h_p/\chi_p^+$.

We now consider the partition $\mathcal{P}$ constructed in the proof of Theorem 14.1.1. It is shown in the proof of Lemma 14.1.5 that for almost every x, if $f^n(x) \in Q^\ell$ for some $n \geq 0$ then

$$f^n(\eta_p(x) \cap \mathcal{P}_0^n(x)) \subset \Psi_{f^n(x)}(W_p'(f^n(x))).$$

Fix x and $n \geq n_0(x)$ (see (14.2)). Consider $y \in \eta_p(x) \cap \mathcal{P}_0^n(x)$ and let k be the largest integer smaller than n such that $f^k(x) \in Q^\ell$. Since $f^k(y) \in \mathcal{P}(f^k(x))$,

$$\|\Psi_{f^k(x)}^{-1}(f^k(y))\| \leq \ell\rho(f^k(x)) \leq \frac{\delta}{K\ell} e^{-(\chi_p^+ + 3\varepsilon)(n-k)}.$$

Furthermore, by the definition of $n_0(x)$, we have $k \geq n_0(x)$ and $f^k(y) \in \Psi_{f^k(x)}(W_p'(f^k(x)))$. Hence by the second statement of Lemma 14.1.3,

$$\|\Psi_x^{-1}(y)\| \leq \|\Psi_{f^k(x)}^{-1}(f^k(y))\| e^{-(\chi_p^+ - 2\varepsilon)k} \leq \frac{1}{K} e^{-(\chi_p^+ - 2\varepsilon)n}.$$

Therefore,

$$\mathcal{P}_0^n(x) \cap \eta_p(x) \subset B_p^u(x, e^{-n(\chi_p^+ - 2\varepsilon)}) \quad \text{for all } n \geq n_0(x).$$

Proceeding in a similar manner to that in (14.6) and using (14.4), we obtain that

$$\begin{aligned} \overline{d}_p^u(x) &= \overline{\lim_{n\to\infty}} \frac{\log \nu_p^u(x)(B_p^u(x, e^{-n(\chi_p^+ - 2\varepsilon)}))}{-n(\chi_p^+ - 2\varepsilon)} \\ &\leq \frac{1}{\chi_p^+ - 2\varepsilon} \overline{\lim_{n\to\infty}} -\frac{1}{n} \log \nu_p^u(x)(\mathcal{P}_0^n(x)) \leq \frac{h_p}{\chi_p^+ - 2\varepsilon} \end{aligned}$$

for ν-almost every $x \in \Lambda^\ell \cap \Lambda_{p,m}$. Since ε can be chosen arbitrarily small, we have $\overline{d}_p^u \leq h_p/\chi_p^+$. □

We proceed with the proof of the theorem and establish Statements 1 and 2 for other values of i.

Lemma 14.1.8. *Given $\varepsilon > 0$, there exists a measurable function $n_0'\colon M \to \mathbb{N}$ such that for ν-almost every x and $n \geq n_0'(x)$, we have*

$$\frac{\log \nu_{i-1}^u(x)(B_{i-1}^u(x, e^{-n(\chi_i^+ - 2\varepsilon)}))}{-n(\chi_i^+ - 2\varepsilon)} \leq \overline{d}_{i-1}^u + \varepsilon, \tag{14.7}$$

$$-\frac{1}{n} \log \nu_{i-1}^u(x)(\mathcal{P}_0^n(x)) \geq h_{i-1} - \varepsilon, \tag{14.8}$$

$$\mathcal{P}_0^n(x) \cap \eta_i(x) \subset B_i^u(x, e^{-n(\chi_i^+ - 2\varepsilon)}), \tag{14.9}$$

$$-\frac{1}{n} \log \nu_i^u(x)(\mathcal{P}_0^n(x)) \leq h_i + \varepsilon. \tag{14.10}$$

Proof of the lemma. By the definition of $\overline{d}_{i-1}^u$, there exists a measurable function $n_1\colon M \to \mathbb{N}$ such that (14.7) holds for ν-almost every x and $n \geq n_1(x)$. By Theorem 14.1.1, there exist $\delta > 0$ and a measurable function $n_2\colon M \to \mathbb{N}$ such that for ν-almost every x and $n \geq n_2(x)$,

$$-\frac{1}{n} \log \nu_{i-1}^u(x)(D_{i-1}(x, n, \delta)) \geq h_{i-1} - \varepsilon.$$

By Lemma 14.1.5, there exists a measurable function $n_3\colon M \to \mathbb{N}$ such that for ν-almost every x and $n \geq n_3(x)$,

$$\mathcal{P}_0^n(x) \cap \eta_{i-1}(x) \subset D_{i-1}(x, n, \delta).$$

Repeating arguments in the proof of Lemma 14.1.7, with p replaced by i, we find that there exists a measurable function $n_4\colon M \to \mathbb{N}$ such that for ν-almost every x and $n \geq n_4(x)$,

$$\mathcal{P}_0^n(x) \cap \eta_i(x) \subset B_i^u(x, e^{-n(\chi_i^+ - 2\varepsilon)}).$$

It follows that (14.8)–(14.9) hold for $n \geq \max\{n_2(x), n_3(x), n_4(x)\}$. Finally, since $\mathcal{P}_0^n(x) \supset (\eta_i \vee \mathcal{P})_0^n$, it follows from Lemma 14.1.2 that there exists a measurable function $n_5\colon M \to \mathbb{N}$ such that (14.10) holds for ν-almost every x and $n \geq n_5(x)$. □

We can always choose N_0 sufficiently large so that the set $\Gamma = \{x \in \Lambda^\ell : n'_0(x) \le N_0\}$ has positive measure and for ν-almost every $x \in \Gamma$ there exists $n = n(x) \ge \max\{N_0, \varepsilon^{-1}\}$ for which

$$L := B^u_{i-1}(x, e^{e^{-n(\chi_i^+ - 2\varepsilon)}}) \subset \eta_{i-1}(x), \tag{14.11}$$

$$\frac{\nu^u_{i-1}(x)(L \cap \Gamma)}{\nu^u_{i-1}(x)(L)} \ge \frac{1}{2}, \tag{14.12}$$

$$\frac{\log \nu^u_i(x)(B^u_i(x, 2e^{e^{-n(\chi_i^+ - 2\varepsilon)}}))}{-n(\chi_i^+ - 2\varepsilon)} \ge \overline{d}^u_i - \varepsilon. \tag{14.13}$$

The relations (14.11)–(14.13) can be ensured due to the definition of $\overline{d}^u_i$ and the following well-known result.

Lemma 14.1.9 (Borel Density Lemma). *Let μ be a finite Borel measure and $A \subset M$ a measurable set. Then for μ-almost every $x \in A$, we have*

$$\lim_{r \to 0} \frac{\mu(B(x, r) \cap A)}{\mu(B(x, r))} = 1.$$

Furthermore, if $\mu(A) > 0$ then for each $\delta > 0$, there is a set $\Delta \subset A$ with $\mu(\Delta) > \mu(A) - \delta$ and a number $r_0 > 0$ such that for all $x \in \Delta$ and $0 < r < r_0$, we have

$$\mu(B(x, r) \cap A) \ge \frac{1}{2}\mu(B(x, r)).$$

Lemma 14.1.10. $h_i - h_{i-1} \ge \chi_i^+(\overline{d}^u_i - \overline{d}^u_{i-1})$ *for* $i = u + 1, \dots, p$.

Proof of the lemma. Fix $\varepsilon > 0$, $x \in \Gamma$, and $n = n(x) \ge \max\{N_0, \varepsilon^{-1}\}$ for which (14.7)–(14.8) hold. By (14.12) and (14.7), we have

$$\nu^u_{i-1}(x)(L \cap \Gamma) \ge \frac{1}{2}\nu^u_{i-1}(x)(L) \ge \frac{1}{2}e^{-n(\chi_i^+ - 2\varepsilon)(\overline{d}^u_{i-1} + \varepsilon)}.$$

On the other hand, by (14.8),

$$\nu^u_{i-1}(x)(\mathcal{P}^n_0(x)) \le e^{-n(h_{i-1} - \varepsilon)}.$$

It follows that the number of distinct $\mathcal{P}^n_0$-atoms intersecting $L \cap \Gamma$ is larger than $\nu^u_{i-1}(x)(L \cap \Gamma)e^{n(h_{i-1} - \varepsilon)}$.

By (14.11), $L \subset \eta_{i-1}(x) \subset \eta_i(x)$, and thus distinct $\mathcal{P}^n_0$-atoms intersecting $L \cap \Gamma$ define distinct $(\mathcal{P}^n_0 \vee \eta_i)$-atoms intersecting $L \cap \Gamma$. Note that $\mathcal{P}^n_0(y) \cap \eta_i(y) \subset B^u_i(y, e^{-n(\chi_i^+ - 2\varepsilon)})$ for $y \in L \cap \Gamma$, which guarantees that all $(\mathcal{P}^n_0 \vee \eta_i)$-atoms intersecting $L \cap \Gamma$ are contained in $B^u_i(x, 2e^{-n(\chi_i^+ - 2\varepsilon)})$. (We use here the fact that $\rho_i(x, y) \le \rho_{i-1}(x, y)$ for $y \in L$.) We conclude that the measure $\nu^u_i(x)(B^u_i(x, 2e^{-n(\chi_i^+ - 2\varepsilon)}))$ is bounded from below by the number of distinct $(\mathcal{P}^n_0 \vee \eta_i)$-atoms intersecting $L \cap \Gamma$ times the minimal measure of such an atom, and thus

$$\nu^u_i(x)(B^u_i(x, 2e^{-n(\chi_i^+ - 2\varepsilon)})) \ge \frac{1}{2}e^{-n(\chi_i^+ - 2\varepsilon)(\overline{d}^u_{i-1} + \varepsilon)}\, e^{n(h_{i-1} - \varepsilon)}\, e^{-n(h_i + \varepsilon)}.$$

Together with (14.13) this gives

$$(\overline{d}_i^u - \varepsilon)(\chi_i^+ - 2\varepsilon) \leq (\overline{d}_{i-1}^u + \varepsilon)(\chi_i^+ - 2\varepsilon) + \frac{1}{n} + h_i - h_{i-1} + 2\varepsilon,$$

and thus

$$\overline{d}_i^u - \overline{d}_{i-1}^u - 2\varepsilon \leq \frac{h_i - h_{i-1} + 3\varepsilon}{\chi_i^+ - 2\varepsilon}.$$

Since ε is arbitrary, this yields the desired result. □

We now proceed with the proof of the opposite inequality. Consider a finite-entropy partition $\mathcal{P}$ adapted to the Lyapunov charts, in the sense that for each $x \in \Lambda$ the elements of the partition $\mathcal{P}^+ = \bigvee_{n=0}^{\infty} f^n \mathcal{P}$ satisfy

$$\mathcal{P}^+(x) \subset \Psi_x(\{y \in B(0, q(x)) : \|(\Psi_{f^{-n}(x)}^{-1} \circ f^{-n} \circ \Psi_x)(y)\| \leq \delta q(f^{-n}(x))\}).$$

We assume that $\mathcal{P}$ refines the partition $\{Q^\ell, M \setminus SQ^\ell\}$. Define the measurable partition $\xi_i = \eta_i \vee \mathcal{P}^+$.

Lemma 14.1.11. *The following statements hold:*

1. *ξ is increasing and $\xi_i < \xi_{i-1}$ for $i = u+1, \dots, p$.*
2. *For ν-almost every x,*

$$\xi_i(x) \subset \Psi_x(W_i'(x) \cap R(0, r\, e^{-\chi_p^+ - 3\varepsilon} q(x))).$$

3. *$h_\nu(f, \xi_i) = H_\nu(\eta_i | f\eta_i)$.*

Proof of the lemma. The first two statements are immediate consequences of the definition of ξ_i and of the properties of Lyapunov charts (see Section 5.6). For the last statement note that since ξ_i is an increasing partition,

$$\begin{aligned} h_\nu(f, \xi_i) &= h_\nu(f, \eta_i \vee f^n \mathcal{P}^+) \\ &= H_\nu(\eta_i | f\eta_i \vee f^{n+1}\mathcal{P}^+) + H_\nu(\mathcal{P}^+ | f^{-n}\eta_i \vee f\mathcal{P}^+), \end{aligned}$$

where the first term is at most $H_\nu(\eta_i | f\eta_i)$ and the second term goes to zero as $n \to \infty$ (since $f^{-n}\eta_i \to \varepsilon$). On the other hand, since the entropy of $\mathcal{P}$ is finite, we have $h_\nu(f, \xi_i) = h_\nu(f, \eta_i \vee \mathcal{P}) \geq h_\nu(f, \eta_i)$. □

For each $i = u+1, \dots, p$, we define a metric on $\xi_i(x)/\xi_{i-1}$ for ν-almost every x. Let $E \subset Q^\ell$ be a measurable set with positive measure. We may assume that the partition $\mathcal{P}$ considered above also refines the partition $\{E, M \setminus E\}$. It follows from the properties of Lyapunov charts and Theorem 8.5.1 that there exist homeomorphisms $\psi_x : U_x \to \mathbb{R}^n$ such that (1) ψ_x and ψ_x^{-1} are Lipschitz with Lipschitz constant depending only on ℓ and (2) ψ_x takes leaves of W_i^u into parallel planes. We define the map

$$\pi : \bigcup_{n \geq 0} f^n(E) \to \mathbb{R}^{\sum_{i \geq u} \dim E_i}$$

as follows. If $x \in E$, we set $\pi(x) = \psi_y(x)$, where $y \in \Lambda^\ell \cap B(x, w)$ is such that $x \in \Psi_y(W_p'(y))$. (It can be shown that $\pi(x)$ does not depend on the choice of y.) If

$x \notin E$, we set $\pi(x) = \pi(f^{-n(x)}(x))$, where $n(x)$ is the smallest nonnegative integer such that $f^{-n(x)}(x) \in E$. Since $\xi_i > \{E, M \setminus E\}$, for each $x \in \bigcup_{n\geq 0} f^n(E)$, either $f^{-n}(\xi_i(x)) \subset E$ or $f^{-n}(\xi_i(x)) \cap E = \varnothing$. Thus for each x, the map $\pi|\xi_i(x)$ is Lipschitz with Lipschitz inverse onto a subset of $\mathbb{R}^{\sum_{j\geq i}\dim E_j}$ (see the end of Section 8.5). Furthermore, $\pi|\xi_i(x)$ takes distinct leaves of ξ_{i-1} to subsets of distinct $\mathbb{R}^{\sum_{j>i}\dim E_j}$-planes in $\mathbb{R}^{\sum_{j\geq i}\dim E_j}$.

For $x \in \bigcup_{n\geq 0} f^n(E)$, $i \geq u+1$, and $y, y' \in \xi_i(x)$, we define

$$d_i(y, y') = \|\pi_i(y) - \pi_i(y')\|,$$

where $\pi = (\pi_u, \dots, \pi_p)$. This induces a metric on $\xi_i(x)/\xi_{i-1}$ that makes $\xi_i(x)/\xi_{i-1}$ isometric to a subset of $\mathbb{R}^{\dim E_i}$. For each $i \geq u+1$, let $\mu_i(x)$ be a system of conditional measures associated with ξ_i and let

$$D_i(x, \rho) = \{y \in \xi_i(x) : d_i(x, y) < \rho\}.$$

Set

$$\delta_i(x) = \varliminf_{\rho\to 0} \frac{\log \mu_i(x)(D_i(x, \rho))}{\log \rho}.$$

This number is called the *transverse dimension* of $\xi_i(x)/\xi_{i-1}$. It is easy to see that $\delta_i(x) \leq \dim E_i$ for $i = u+1, \dots, p$.

Given $\delta > 0$, define the measurable functions $g, g_\delta, g_*\colon M \to \mathbb{R}$ by

$$\begin{aligned} g(y) &= \mu_{i-1}(y)((f^{-1}\xi_i)(y)), \\ g_\delta(y) &= \frac{1}{\mu_i(x)(D_i(y, \delta))} \int_{D_i(y,\delta)} \mu_{i-1}((f^{-1}\xi_i)(y))\, d\mu_i(y)(z), \\ g_*(y) &= \inf_{\delta>0} g_\delta(y). \end{aligned}$$

Note that for ν-almost every x and every $y \in \Lambda^\ell \cap \xi_i(x)$,

$$f^{-1}(\xi_{i-1}(y)) = \xi_{i-1}(f^{-1}(y)) \cap f^{-1}(\xi_i(x)). \tag{14.14}$$

Indeed, $f^{-1}(\xi_{i-1}(y)) \subset f^{-1}(\xi_i(y))$ since $\xi_{i-1}(y) \subset \xi_i(y)$, and

$$f^{-1}(\xi_{i-1}(y)) \subset \xi_{i-1}(f^{-1}(y))$$

since ξ_{i-1} is an increasing partition. The fact that

$$f(\xi_{i-1}(f^{-1}(y))) \cap \xi_i(x) \subset \xi_{i-1}(y)$$

is an immediate consequence of the construction of the partitions ξ_i and ξ_{i-1}. It follows from (14.14) that $g(y) = \mu_{i-1}(y)((f^{-1}\xi_{i-1})(y))$.

Lemma 14.1.12. *We have $g_\delta \to g$ ν-almost everywhere on M as $\delta \to 0$ and $\int_M -\log g_*\, d\nu < \infty$.*

Proof of the lemma. Fix a point $z \in \Lambda^\ell$ and consider the measure space $X = \xi_{i-1}(z)$ with the measure $\mu = \mu_i(z)$ and the partition $\alpha = (f^{-1}\xi_i)|X$. Fix $\lambda > 0$

and let $A = \{x \in X : \sup_{\delta>0} g_\delta(x) > \lambda\}$. For each $x \in A$, we choose $\delta(x) > 0$ such that $g_{\delta(x)}(x) > \lambda$; that is,

$$\int_{D_i(x,\delta(x))} g\,d\mu > \lambda\,\mu(D_i(x,\delta(x))).$$

Let $\mathcal{A} = \{D_i(x,\delta(x)) : x \in A\}$. We may assume that $\sup_{x\in A}\delta(x) < \infty$. It follows from the Besicovitch covering lemma (see, e.g., [149]) that there is a subcover $\mathcal{A}' \subset \mathcal{A}$ of A such that no x lies in more than a number $c = c(n)$ of elements of $\mathcal{A}'$. Therefore,

$$\mu(A) \le \sum_{B\in\mathcal{A}'} \mu(B) \le \sum_{B\in\mathcal{A}'} \frac{1}{\lambda}\int_B g\,d\mu \le \frac{c}{\lambda}\int_X g\,d\mu. \tag{14.15}$$

This implies that $g_\delta \to g$ μ-almost everywhere for every z, and thus also ν-almost everywhere. Furthermore,

$$\begin{aligned}\int_C -\log g_*\,d\mu &= \int_0^\infty \mu(\{x \in X : -\log g_*(x) > s\})\,ds\\ &= \int_0^\infty \sum_{B\in\alpha} \mu(B \cap \{x \in X : g_*(x) < e^{-s}\})\,ds.\end{aligned}$$

Arguing similarly to (14.15) we find that

$$\mu(B \cap \{x \in X : g_*(x) < e^{-s}\}) \le c\,e^{-s}.$$

Therefore,

$$\int_X -\log g_*\,d\mu \le \sum_{B\in\alpha}\int_0^\infty \min\{ce^{-s}, \mu(B)\}\,ds \le H_\mu(\alpha) + \log c + 1 < \infty.$$

Integrating over M we obtain that

$$\int_M -\log g_*\,d\nu \le H_\nu(f^{-1}\xi_i|\xi_i) + \log c + 1 < \infty.$$

This completes the proof of the lemma. □

Fix $x \in M$ and let $r_0 \le 0 < r_1 < r_2 < \dots$ be the successive times such that $f^{r_k}(x) \in E$. Note that r_0 is constant on $\xi_i(x)$. For n large enough and $0 \le k < n$, we define

$$a(x,k) = D_i(f^k(x), e^{-(\chi_i^+ + 2\varepsilon)(n-r_j)}N^j)$$

whenever $r_j \le k < r_{j+1}$, where N is some constant whose choice is given in the following lemma.

Lemma 14.1.13. *There exists a constant* $N = N(\ell)$ *such that* $a(x,k) \cap (f^{-1}\xi_i)(f^k(x)) \subset f^{-1}a(x,k+1)$.

Proof of the lemma. If $k \ne r_j - 1$ for all j, then by the definition of the distances d_i,

$$f(a(x,k)) \cap \xi_i(f^{k+1}(x)) = a(x,k+1).$$

Assume now that $k = r_j - 1$ for some j. Given $y \in E$, let $r > 0$ be the smallest integer such that $f^r(y) \in E$. By the properties of Lyapunov charts (the local unstable manifolds are graphs of functions with derivatives at most 1/3), we obtain that

$$d_i(f^r(x), f^r(z)) \leq N\, e^{r(\chi_i^+ + 3\varepsilon)} d_i(y, z)$$

for $z \in (f^{-r}\xi_i)(y)$ and some constant $N = N(\ell)$. The desired result follows. □

Lemma 14.1.14. $h_i - h_{i-1} \leq \chi_i^+ \dim E_i$ *for* $i = u+1, \dots, p$.

Proof of the lemma. Set $u_k = \mu_i(f^k(x))(a(x,k))$. We have for $m = [n(1-\varepsilon)]$ that

$$u_0 = \prod_{k=0}^{m-1} \frac{u_k}{u_{k+1}} \cdot u_m.$$

Since ν is invariant, for each $0 \leq k < n$,

$$\begin{aligned} \frac{u_k}{u_{k+1}} &= u_k \frac{\mu_i(f^k(x))(f^{-1}(\xi_i(f^{k+1}(x))))}{\mu_i(f^k(x))(a(x, k+1))} \\ &\leq \frac{\mu_i(f^k(x))(a(x,k))}{\mu_i(f^k(x))((f^{-1}\xi_i)(f^k(x)) \cap a(x,k))} \\ &\quad \times \mu_i(f^k(x))((f^{-1}\xi_i)(f^k(x))). \end{aligned} \tag{14.16}$$

The quotient in the last term is equal to $1/g_{\delta(x,n,k)}(f^k(x))$, where

$$\delta(x,n,k) = e^{-(\chi_i^+ + 3\varepsilon)(n - r_j(x))} N^j$$

and $j = \operatorname{card}\{0 < j \leq k : f^j(x) \in E\}$. If

$$I(x) = -\log \mu_i(x)((f^{-1}\xi_i)(x)),$$

the last term in (14.16) is equal to $e^{-I(f^k(x))}$. Thus,

$$\log u_0 \leq -\sum_{k=0}^{m-1} \log g_{\delta(x,n,k)}(f^k(x)) - \sum_{k=0}^{m-1} I(f^k(x)).$$

Multiplying by $-1/n$ and taking the limits as $n \to \infty$ yields

$$\begin{aligned} (\chi_i^+ + 3\varepsilon)\delta_i(x) &= (\chi_i^+ + 3\varepsilon) \varliminf_{n\to\infty} \frac{\log \mu_i(x)(D_i(x, e^{-(\chi_i^+ + 3\varepsilon)(n - r_0(x))}))}{\log e^{-(\chi_i^+ + 3\varepsilon)n}} \\ &\geq \varliminf_{n\to\infty} \frac{1}{n} \sum_{k=0}^{[n(1-\varepsilon)]} \log g_{\delta(x,n,k)}(f^k(x)) \\ &\quad + \lim_{n\to\infty} \frac{1}{n} \sum_{k=0}^{[n(1-\varepsilon)]} I(f^k(x)). \end{aligned} \tag{14.17}$$

The last limit is equal to

$$(1-\varepsilon) H_\nu(\xi_i | f\xi_i) \geq (1-\varepsilon)(h_\nu(f|\xi_i) - \varepsilon).$$

We now show that

$$\overline{\lim_{n\to\infty}} -\frac{1}{n}\sum_{k=0}^{[n(1-\varepsilon)]} \log g_{\delta(x,n,k)}(f^k(x)) \le (1-\varepsilon)[h_\nu(f,\xi_{i-1})+2\varepsilon]. \tag{14.18}$$

It follows from Lemma 14.1.12 that there exists a measurable function $\delta(x)$ such that if $\delta \le \delta(x)$, then

$$-\log g_\delta(x) \le -\log g(x) + \varepsilon.$$

Furthermore, since $\int_M -\log g_*\,d\nu < \infty$, there is a number δ' such that if $A = \{x \in M : \delta(x) > \delta'\}$ then

$$\int_{M\setminus A} -\log g_*\,d\nu \le \varepsilon.$$

We claim that for almost every x, if n is sufficiently large, then $\delta(x,n,k) \le \delta'$ for all $k \ne n(1-\varepsilon)$. We first observe that there exists $N(x)$ such that for $n \ge N(x)$,

$$\mathrm{card}\{0 \le j < n : f^j(x) \in E\} \le 2n\nu(E).$$

If $n \ge N(x)$, then

$$\delta(x,n,k) = e^{-(\chi_i^+ + 3\varepsilon)(n-r_j)} N^j \le e^{-(\chi_i^+ + 3\varepsilon)\varepsilon n} N^{2n\nu(E)}.$$

Choosing the set E in such a way that $e^{-(\chi_i^+ + 3\varepsilon)\varepsilon} N^{4\nu(E)} < 1$, we have $\delta(x,n,k) < \delta'$ for n sufficiently large. Thus,

$$\sum_{k=0}^{[n(1-\varepsilon)]} -\log g_{\delta(x,n,k)}(f^k(x)) \le \sum_{k=0,\, f^k(x)\in A}^{[n(1-\varepsilon)]} (-\log g(f^k(x)) + \varepsilon) + \sum_{k=0,\, f^k(x)\notin A}^{[n(1-\varepsilon)]} -\log g_*(f^k(x)),$$

and the limit in (14.18) can be estimated from above by

$$(1-\varepsilon)\left[\int_M -\log g\,d\nu + \varepsilon + \int_{M\setminus A} -\log g_*\,d\nu\right].$$

By the definition of g, we have

$$\int_M -\log g\,d\nu = h_\nu(f,\xi_{i-1}),$$

and thus we obtain (14.18). Combined with (14.17), this yields

$$(\chi_i^+ + 3\varepsilon)\delta_i(x) \ge (1-\varepsilon)[h_\nu(f,\xi_i) - h_\nu(f,\xi_{i-1}) - 3\varepsilon].$$

It follows from Theorem 14.1.1 and Lemma 14.1.11 that

$$h_\nu(f,\xi_i) = H_\nu(\eta_i | f\eta_i) = h_i$$

with similar identities for h_{i-1}. Since ε is arbitrary, the desired inequality follows readily from the fact that $\delta_i(x) \le \dim E_i$ for ν-almost every $x \in M$. □

We need the following technical lemma:

Lemma 14.1.15. *Let m be a probability measure on $\mathbb{R}^p \times \mathbb{R}^q$, π the projection onto $\mathbb{R}^p$, and m_t a disintegration of m with respect to π. Let $\delta \geq 0$ be such that*

$$\delta \leq \underline{\lim}_{\rho \to 0} \frac{\log m_t(B^q(s,\rho))}{\log \rho} \quad \text{for } m\text{-almost every } (s,t).$$

Then, for m-almost every (s,t),

$$\delta \leq \underline{\lim}_{\rho \to 0} \frac{\log m(B^{p+q}((s,t),\rho))}{\log \rho} - \underline{\lim}_{\rho \to 0} \frac{\log(m \circ \pi^{-1})(B^p(t,\rho))}{\log \rho}.$$

Proof of the lemma. Fix $\sigma > 0$. We can find $N_1 > 0$ and a set A_1 with $m(A_1) > 1 - \sigma$ such that for all $(s,t) \in A_1$ and $n \geq N_1$,

$$m_t(B^q(s, e^{-n})) \leq e^{-n\delta}\, e^{n\sigma}.$$

By Lemma 14.1.9, we can find $N_2 > 0$ and a set A_2 with $m(A_2) > 1 - 2\sigma$ such that for all $(s,t) \in A_2$ and $n \geq N_2$,

$$m(A_1 \cap B^{p+q}((s,t), e^{-n})) \geq \frac{1}{2} m(B^{p+q}((s,t), e^{-n})).$$

Observe that for each t, there exists $u(t)$ with $(t, u(t)) \in A_1 \cap B^q(s_0, e^{-n})$. Therefore,

$$A_1 \cap B^q(s_0, e^{-n}) \cap \pi^{-1}\{t\} \subset B^q(u(t), 2\, e^{-n}) \cap \pi^{-1}\{t\}.$$

It follows that for $(s_0, t_0) \in A_2$ and $n \geq N_2$,

$$\begin{aligned} m(B^{p+q}((s_0,t_0), e^{-n})) &\leq 2 \int_{B^p(t_0, e^{-n})} m_t(A_1 \cap B^q(s_0, e^{-n}))(m \circ \pi^{-1})(dt) \\ &\leq 2\, e^{-n\delta}\, e^{n\sigma} (m \circ \pi^{-1})(B^p(t_0, e^{-n})). \end{aligned}$$

Letting $n \to \infty$ and $\sigma \to 0$ yields the desired result. □

The following result can be obtained in a similar manner.

Lemma 14.1.16. *Let (Ω, μ) be a probability space, and let m be a probability measure on $\Omega \times \mathbb{R}^q$ such that*

$$m(dw, ds) = \int m_w(ds)\, \mu(dw).$$

If for some $\gamma \geq 0$,

$$\gamma \leq \underline{\lim}_{\rho \to 0} \frac{\log m_w(B^q(s,\rho))}{\log \rho} \quad \text{for } m\text{-almost every } (w,s),$$

then

$$\gamma \leq \underline{\lim}_{\rho \to 0} \frac{\log m(B^q(s,\rho))}{\log \rho} \quad m\text{-almost everywhere.}$$

We are now ready to establish the desired upper bound.

Lemma 14.1.17. *$h_i - h_{i-1} \le \chi_i^+(\underline{d}_i^u - \underline{d}_{i-1}^u)$ for $i = u+1, \dots, p$.*

Proof of the lemma. For each i let $\hat{\nu}_i^u(x)$ be the system of conditional measures associated with the partition ξ_i of Q^ℓ by ith local manifolds. We recall that $\eta_i = \bigvee_{n\ge 0} f^n\xi_i$. Fix $x \in Q^\ell$ and $u+1 \le i \le p$. Let

$$\hat{\pi} = \pi|_{\eta_i(x)} \colon \eta_i(x) \to \mathbb{R}^{\sum_{j\ge i}\dim E_j} = \mathbb{R}^{\sum_{j>i}\dim E_j} \times \mathbb{R}^{\dim E_i},$$

and write $\bar{\nu}_i^u(x) = \hat{\nu}_i^u(x) \circ \hat{\pi}^{-1}$. Define

$$\gamma_i(y) = \varliminf_{\rho\to 0} \frac{\log \bar{\nu}_i^u(x)(\{z \in \mathbb{R}^n : \|z_i - y_i\| < \rho\})}{\log\rho},$$

where z_i denotes the $\mathbb{R}^{\dim E_i}$-component of z. Since π is Lipschitz continuous, we have

$$\delta_i(\hat{\pi}^{-1}y) = \varliminf_{\rho\to 0} \frac{\log \hat{\nu}_i^u(\pi^{-1}x)(\{z \in \mathbb{R}^n : \|z_i - y_i\| < \rho\})}{\log\rho},$$

for $\bar{\nu}_i^u(x)$-almost every y, assuming that $\hat{\nu}_i^u(x)(\bigcup_{n\ge 0} f^n(E)) = 1$. We may also assume that

$$\delta_i(\hat{\pi}^{-1}y) \ge \frac{(1-3\varepsilon)(h_i - h_{i-1} - 3\varepsilon)}{\chi_i^+ + 3\varepsilon}$$

for $\bar{\nu}_i^u(x)$-almost every y. Applying Lemma 14.1.16 with $\Omega = \xi_i(x)/\xi_i$ and $\mathbb{R}^q = \mathbb{R}^{\dim E_i}$, we conclude that

$$\gamma_i(y) \ge \frac{(1-3\varepsilon)(h_i - h_{i-1} - 3\varepsilon)}{\chi_i^+ + 3\varepsilon}$$

for $\bar{\nu}_i^u(x)$-almost every y.

Consider now the partition of $\hat{\pi}(\xi_i(x))$ into planes of the form $z_i =$ constant. We may assume that for $\hat{\nu}_i^u(x)$-almost every y,

$$\varliminf_{\rho\to 0} \frac{\log \bar{\nu}_i^u(x)(\{z \in \mathbb{R}^n : \|z - y\| < \rho\})}{\log\rho} = \underline{d}_i^u,$$

and

$$\varliminf_{\rho\to 0} \frac{\log(\hat{\nu}_{i-1}^u(\hat{\pi}^{-1}y)\circ\hat{\pi}^{-1})(\{z \in \mathbb{R}^n : z_i = y_i, \|z - y\| < \rho\})}{\log\rho} = \underline{d}_{i-1}^u.$$

It follows from Lemma 14.1.15 that for $\bar{\nu}_i^u(x)$-almost every y,

$$\underline{d}_i^u - \underline{d}_{i-1}^u \ge \delta_i(y) \ge \frac{(1-3\varepsilon)(h_i - h_{i-1} - 3\varepsilon)}{\chi_i^+ + 3\varepsilon},$$

which gives the desired inequality. □

We conclude the proof of the theorem. By Lemma 14.1.7, we have $d_p^u := \underline{d}_p^u = \overline{d}_p^u = h_p/\chi_p^+$. Using Lemmas 14.1.10 and 14.1.17 we can show inductively that $d_i^u := \underline{d}_i^u = \overline{d}_i^u$ for $i = u, \dots, p$ and that

$$h_i - h_{i-1} = \chi_i^+(d_i^u - d_{i-1}^u) \text{ for } i = u+1, \dots, p. \tag{14.19}$$

It follows from Lemma 14.1.14 that $d_i^u - d_{i-1}^u \le \dim E_i$. This completes the proof of the theorem when ν is ergodic.

When ν is not ergodic for each $\kappa > 0$ consider the invariant sets $\Gamma(x)$ defined by

$$\Gamma(x) = \{y \in M : |h_i(y) - h_i(x)| < \kappa,\ |d_i^u(y) - d_i^u(x)| < \kappa, \\ \text{and } |\chi_i^+(y) - \chi_i^+(x)| < \kappa \text{ for } i = u, \dots, p\}.$$

The collection of these sets covers Λ^ℓ. Moreover, there exists a countable subcollection of sets $\{\Gamma^i\}_{i\in\mathbb{N}}$, which still covers Λ^ℓ. We can apply the above arguments to show that for almost every $x \in \Gamma^i$,

$$h_i(x) - h_{i-1}(x) \ge (\chi_i^+(x) - c\kappa)(\overline{d}_i^u(x) - \overline{d}_{i-1}^u(x) - c\kappa),$$
$$h_i(x) - h_{i-1}(x) \le (\chi_i^+(x) + c\kappa)(\underline{d}_i^u(x) - \underline{d}_{i-1}^u(x) + c\kappa),$$

for some constant c that does not depend on x or κ. Since $\{\Gamma^i\}_{i\in\mathbb{N}}$ is countable, letting κ go to zero yields the desired statement. □

The number $d_i^u(x)$ in Theorem 14.1.6 is called the *leaf pointwise dimension* of ν at the point x ("along" the leaf $W_i^u(x)$). The number $d_i^u(x) - d_{i-1}^u(x)$ can be interpreted as a "transverse dimension" of ν on the quotient $W_i^u(x)/W_{i-1}^u$. (Recall that each leaf $W_i^u(x)$ is foliated by leaves $W_{i-1}^u(y)$, $y \in W_i^u(x)$.) In the particular cases $i = s$ and $i = u$, the quantities

$$d^s(x) = \lim_{r\to 0} \frac{\log \nu_x^s(B^s(x,r))}{\log r} \quad \text{and} \quad d^u(x) = \lim_{r\to 0} \frac{\log \nu_x^u(B^u(x,r))}{\log r} \tag{14.20}$$

are called the *stable* and *unstable pointwise dimensions* of ν at x.

14.1.3 The Ledrappier–Young Entropy Formula

The following result by Ledrappier and L.-S. Young [125] establishes a formula for the entropy of a diffeomorphism with respect to a hyperbolic invariant measure.

Theorem 14.1.18. *If ν is a Borel measure invariant under a $C^{1+\alpha}$ diffeomorphism f, then*

$$h_\nu(f) = \int_\Lambda \sum_{i:\chi_i^+(x)>0} \chi_i^+(x)[d_i^u(x) - d_{i-1}^u(x)]\, d\nu(x).$$

Proof. By considering an ergodic decomposition we may assume without loss of generality that ν is ergodic. Summing up the equations in (14.19), we find that

$$h_p = \sum_{i=u+1}^{p} \chi_i^+(d_i^u - d_{i-1}^u).$$

It remains to show that $h_p = h_\nu(f)$.

Lemma 14.1.19. *We have*

$$h_\nu(f) = h_\nu(f, \xi_p) = H_\nu(\eta_p | f\eta_p) = h_p.$$

Proof of the lemma. The proof can be obtained by repeating arguments in the proof of Lemma 14.1.14, with ξ_{i-1} and ξ_i replaced respectively by ξ_p and $\mathcal{P}^+$ (and the exponent $\chi_i^+ + 3\varepsilon$ should be replaced by χ_p^+). Proceeding in this manner for some partition $\mathcal{P}$ as in the proof of Theorem 14.1.6 such that $h_\nu(f, \mathcal{P}^+) > h_\nu(f) - \varepsilon$, we obtain that

$$3\varepsilon \dim M \geq (1-\varepsilon)[h_\nu(f) - h_\nu(f, \xi_p) - 4\varepsilon].$$

Letting $\varepsilon \to 0$ we find that $h_\nu(f) \leq h_\nu(f, \xi_p)$. On the other hand, by Theorem 14.1.1 and Lemma 14.1.11, $h_\nu(f, \xi_p) = H_\nu(\eta_p | f\eta_p) = h_p$. □

This completes the proof of the theorem. □

14.2 Local Product Structure of Hyperbolic Measures

We establish a crucial property of an arbitrary hyperbolic measure: it has asymptotically "almost" local product structure; that is, its density can be approximated by the product of the densities on stable and unstable manifolds up to some subexponential terms. Observe that SRB measures (and, in particular, smooth hyperbolic measures) have local product structure (see Section 13.1).

Theorem 14.2.1. *Let f be a $C^{1+\alpha}$ diffeomorphism of a smooth compact Riemannian manifold M without boundary and ν an f-invariant hyperbolic measure. Then the following statements hold:*

1. *For every $\delta > 0$, there exists a set $\Lambda \subset M$ with $\nu(\Lambda) > 1 - \delta$ such that for every $x \in \Lambda$ and every sufficiently small r (depending on x),*

$$r^\delta \leq \frac{\nu(B(x,r))}{\nu_x^s(B^s(x,r))\nu_x^u(B^u(x,r))} \leq r^{-\delta}.$$

2. *For ν-almost every $x \in M$, the limit*

$$d(x) := \lim_{r\to 0} \frac{\log \nu(B(x,r))}{\log r} \tag{14.21}$$

exists, and

$$d(x) = d^s(x) + d^u(x).$$

Proof. We split the proof into several steps:

Step I: Construction of a special partition. Summarizing results in Section 14.1 we have the following. Given $0 < \varepsilon < 1$, there exists a set $\Gamma \subset M$ of measure $\nu(\Gamma) > 1 - \varepsilon/2$, an integer $n_0 \geq 1$, and a number $C > 1$ such that for every $x \in \Gamma$ and any integer $n \geq n_0$, the following properties hold:

1. For all integers $k, l \geq 1$, we have

$$C^{-1}\, e^{-(l+k)h-(l+k)\varepsilon} \leq \nu(\mathcal{P}^l_k(x)) \leq C\, e^{-(l+k)h+(l+k)\varepsilon}, \tag{14.22}$$

$$C^{-1}\, e^{-kh-k\varepsilon} \leq \nu^s_x(\mathcal{P}^0_k(x)) \leq C\, e^{-kh+k\varepsilon}, \tag{14.23}$$

$$C^{-1}\, e^{-lh-l\varepsilon} \leq \nu^u_x(\mathcal{P}^l_0(x)) \leq C\, e^{-lh+l\varepsilon}, \tag{14.24}$$

where $h = h_\nu(f)$.

2.

$$\xi^s(x) \cap \bigcap_{n\geq 0} \mathcal{P}^n_0(x) \supset B^s(x, e^{-n_0}), \tag{14.25}$$

$$\xi^u(x) \cap \bigcap_{n\geq 0} \mathcal{P}^0_n(x) \supset B^u(x, e^{-n_0}). \tag{14.26}$$

3.

$$e^{-d^s n-n\varepsilon} \leq \nu^s_x(B^s(x, e^{-n})) \leq e^{-d^s n+n\varepsilon}, \tag{14.27}$$

$$e^{-d^u n-n\varepsilon} \leq \nu^u_x(B^u(x, e^{-n})) \leq e^{-d^u n+n\varepsilon}. \tag{14.28}$$

4.

$$\mathcal{P}^{an}_{an}(x) \subset B(x, e^{-n}) \subset \mathcal{P}(x), \tag{14.29}$$

$$\mathcal{P}^0_{an}(x) \cap \xi^s(x) \subset B^s(x, e^{-n}) \subset \mathcal{P}(x) \cap \xi^s(x), \tag{14.30}$$

$$\mathcal{P}^{an}_0(x) \cap \xi^u(x) \subset B^u(x, e^{-n}) \subset \mathcal{P}(x) \cap \xi^u(x), \tag{14.31}$$

where a is the integer part of $2(1+\varepsilon)\max\{\chi^+_p, -\chi^+_1, 1\}$.

5. If $Q_n(x)$ is defined by

$$Q_n(x) = \bigcup \mathcal{P}^{an}_{an}(y) \tag{14.32}$$

where the union is taken over $y \in \Gamma$ for which

$$\mathcal{P}^{an}_0(y) \cap B^u(x, 2\, e^{-n}) \neq \varnothing \quad \text{and} \quad \mathcal{P}^0_{an}(y) \cap B^s(x, 2\, e^{-n}) \neq \varnothing,$$

then

$$B(x, e^{-n}) \cap \Gamma \subset Q_n(x) \subset B(x, 4\, e^{-n}); \tag{14.33}$$

it clearly follows from the definition of $Q_n(x)$ that for each $y \in Q_n(x)$,

$$\mathcal{P}^{an}_{an}(y) \subset Q_n(x).$$

Increasing n_0 if necessary we may also assume that

6. if $x \in \Gamma$ and $n \geq n_0$, then

$$B^s(x, e^{-n}) \cap \Gamma \subset Q_n(x) \cap \xi^s(x) \subset B^s(x, 4\, e^{-n}), \tag{14.34}$$

$$B^u(x, e^{-n}) \cap \Gamma \subset Q_n(x) \cap \xi^u(x) \subset B^u(x, 4\, e^{-n}). \tag{14.35}$$

Property (14.22) follows from Shannon–McMillan–Breiman's Theorem applied to the partition $\mathcal{P}$, while properties (14.23) and (14.24) follow from Theorem 14.1.1 and the fact that $h_p = h_\nu(f)$ (see the proof of Theorem 14.1.18). The inequalities in (14.27) and (14.28) are easy consequences of the existence of the stable and unstable pointwise dimensions d^s and d^u. Since the Lyapunov exponents at ν-almost every point are constant and equal to $\chi_1^+, \dots, \chi_p^+$, the properties (14.29), (14.30), and (14.31) follow from (14.25), (14.26), and the choice of a indicated above. The inclusions in (14.33) are based upon the continuous dependence of stable and unstable manifolds in the $C^{1+\alpha}$ topology on the base point (in each regular set).

It immediately follows from the Borel Density Lemma (Lemma 14.1.9) that one can choose an integer $n_1 \geq n_0$ and a set $\widehat{\Gamma} \subset \Gamma$ of measure $\nu(\widehat{\Gamma}) > 1 - \varepsilon$ such that for every $n \geq n_1$ and $x \in \widehat{\Gamma}$,

$$\nu(B(x, e^{-n}) \cap \Gamma) \geq \frac{1}{2}\nu(B(x, e^{-n})), \tag{14.36}$$

$$\nu_x^s(B^s(x, e^{-n}) \cap \Gamma) \geq \frac{1}{2}\nu_x^s(B^s(x, e^{-n})), \tag{14.37}$$

$$\nu_x^u(B^u(x, e^{-n}) \cap \Gamma) \geq \frac{1}{2}\nu_x^u(B^u(x, e^{-n})). \tag{14.38}$$

We establish two additional properties of the partitions $\mathcal{P}_0^k$ and $\mathcal{P}_k^0$.

Lemma 14.2.2. *There exists a positive constant $D = D(\widehat{\Gamma}) < 1$ such that for every $k \geq 1$ and $x \in \Gamma$, we have*

$$\nu_x^s(\mathcal{P}_0^k(x) \cap \Gamma) \geq D \quad \text{and} \quad \nu_x^u(\mathcal{P}_k^0(x) \cap \Gamma) \geq D.$$

Proof of the lemma. By (14.25), for every $k \geq 1$ and $x \in \Gamma$, the set $\mathcal{P}_0^k(x) \cap \Gamma$ contains the set $B^s(x, e^{-n_0}) \cap \Gamma$. It follows from (14.37) and (14.27) that

$$\nu_x^s(\mathcal{P}_0^k(x) \cap \Gamma) \geq \frac{1}{2}\mu_x^s(B^s(x, e^{-n_0})) \geq \frac{1}{2}e^{-d^s n_0 - n_0\varepsilon} =: D.$$

The second inequality in the proposition can be proved in a similar fashion using the properties (14.26), (14.28), and (14.38). □

The next statement establishes the property of the partition $\mathcal{P}$ that simulates the well-known Markov property.

Lemma 14.2.3. *For every $x \in \Gamma$ and $n \geq n_0$, we have*

$$\mathcal{P}_{an}^{an}(x) \cap \xi^s(x) = \mathcal{P}_{an}^0(x) \cap \xi^s(x),$$
$$\mathcal{P}_{an}^{an}(x) \cap \xi^u(x) = \mathcal{P}_0^{an}(x) \cap \xi^u(x).$$

Proof of the lemma. It follows from (14.30) and (14.25) that

$$\begin{aligned}\mathcal{P}_{an}^0(x) \cap \xi^s(x) \subset \mathcal{P}_{an}^0(x) \cap B^s(x, e^{-n}) \subset \mathcal{P}_{an}^0(x) \cap B^s(x, e^{-n_0})\\ \subset \mathcal{P}_{an}^0(x) \cap \mathcal{P}_0^{an}(x) \cap \xi^s(x) = \mathcal{P}_{an}^{an}(x) \cap \xi^s(x).\end{aligned}$$

Since $\mathcal{P}_{an}^{an}(x) \subset \mathcal{P}_{an}^0(x)$, this completes the proof of the first identity. The proof of the other identity is similar. □

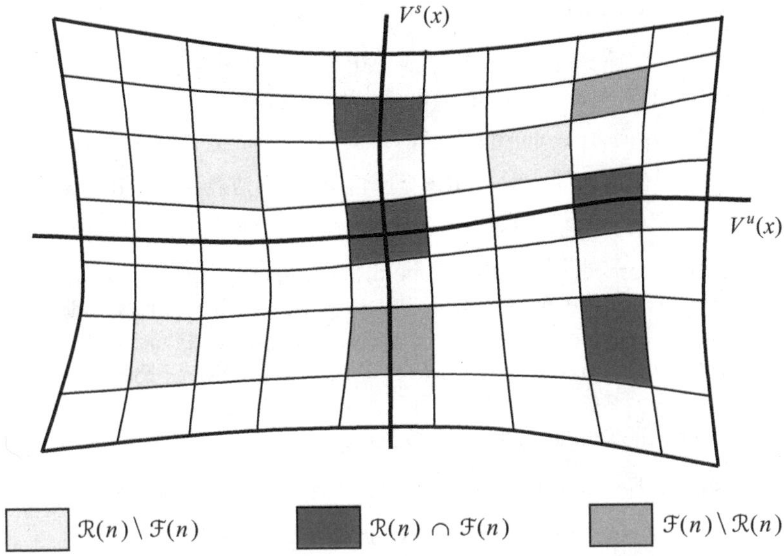

Figure 14.1. The procedure of "filling in" rectangles.

Step II: Preparatory lemmata. Fix $x \in \widehat{\Gamma}$ and an integer $n \geq n_1$. We consider the following two classes $\mathcal{R}(n)$ and $\mathcal{F}(n)$ of elements of the partition $\mathcal{P}^{an}_{an}$: (We call these elements "rectangles.")

$$\mathcal{R}(n) = \{\mathcal{P}^{an}_{an}(y) \subset \mathcal{P}(x) : \mathcal{P}^{an}_{an}(y) \cap \Gamma \neq \varnothing\};$$
$$\mathcal{F}(n) = \{\mathcal{P}^{an}_{an}(y) \subset \mathcal{P}(x) : \mathcal{P}^{0}_{an}(y) \cap \widehat{\Gamma} \neq \varnothing \text{ and } \mathcal{P}^{an}_{0}(y) \cap \widehat{\Gamma} \neq \varnothing\}.$$

The rectangles in $\mathcal{R}(n)$ carry all the measure of the set $\mathcal{P}(x) \cap \Gamma$; that is, $\sum_{R\in\mathcal{R}(n)} \nu(R \cap \Gamma) = \nu(\mathcal{P}(x) \cap \Gamma)$. Obviously, the rectangles in $\mathcal{R}(n)$ that intersect $\widehat{\Gamma}$ belong to $\mathcal{F}(n)$. If these were the only ones in $\mathcal{F}(n)$, the measure $\nu|\mathcal{P}(x) \cap \Gamma$ would have the "direct product structure" at the "level" n. One could then use the approach in [121] to estimate the measure of a ball by the product of its stable and unstable measures. In the general case, the rectangles in the class $\mathcal{F}(n)$ are obtained from the rectangles in $\mathcal{R}(n)$ (that intersect $\widehat{\Gamma}$) by "filling in" the gaps in the "product structure" (see Fig. 14.1).

We wish to compare the number of rectangles in $\mathcal{R}(n)$ and $\mathcal{F}(n)$ intersecting a given set. This allows us to evaluate the deviation of the measure ν from the direct product structure at the level n. The main observation is that for "typical" $y \in \widehat{\Gamma}$ the number of rectangles from the class $\mathcal{R}(n)$ intersecting $V^s(y)$ (respectively $V^u(y)$) is "asymptotically" the same up to a factor that grows at most subexponentially with n. However, in general, the distribution of these rectangles along $V^s(y)$ (respectively $V^u(y)$) may be different for different points y. This causes a deviation from the direct product structure. We use a simple combinatorial argument to show that this deviation grows at most subexponentially with n. One can then say that the measure μ has an "almost direct product structure."

To effect this, for each set $A \subset \mathcal{P}(x)$, we define

$$\begin{aligned} N(n, A) &= \operatorname{card}\{R \in \mathcal{R}(n) : R \cap A \neq \varnothing\}, \\ N^s(n, y, A) &= \operatorname{card}\{R \in \mathcal{R}(n) : R \cap \xi^s(y) \cap \Gamma \cap A \neq \varnothing\}, \\ N^u(n, y, A) &= \operatorname{card}\{R \in \mathcal{R}(n) : R \cap \xi^u(y) \cap \Gamma \cap A \neq \varnothing\}, \\ \widehat{N}^s(n, y, A) &= \operatorname{card}\{R \in \mathcal{F}(n) : R \cap \xi^s(y) \cap A \neq \varnothing\}, \\ \widehat{N}^u(n, y, A) &= \operatorname{card}\{R \in \mathcal{F}(n) : R \cap \xi^u(y) \cap A \neq \varnothing\}. \end{aligned}$$

Note that the number $N(n, \mathcal{P}(x))$ is the cardinality of the set $\mathcal{R}(n)$, and $N^s(n, y, \mathcal{P}(x))$ (respectively $N^u(n, y, \mathcal{P}(x))$) is the number of rectangles in $\mathcal{R}(n)$ that intersect Γ and the stable (respectively unstable) local manifold at y. The product $\widehat{N}^s(n, y, \mathcal{P}(x)) \times \widehat{N}^u(n, y, \mathcal{P}(x))$ is the cardinality of the set $\mathcal{F}(n)$ for a "typical" point $y \in \mathcal{P}(x)$.

Let $Q_n(y)$ be the set defined by (14.32).

Lemma 14.2.4. *For each $y \in \mathcal{P}(x) \cap \Gamma$ and integer $n \geq n_0$, we have*

$$\begin{aligned} N^s(n, y, Q_n(y)) &\leq \nu_y^s(B^s(y, 4e^{-n})) \cdot C\, e^{anh+an\varepsilon}, \\ N^u(n, y, Q_n(y)) &\leq \nu_y^u(B^u(y, 4e^{-n})) \cdot C\, e^{anh+an\varepsilon}. \end{aligned}$$

Proof of the lemma. It follows from (14.34) that

$$\begin{aligned} &\nu_y^s(B^s(y, 4\, e^{-n})) \geq \nu_y^s(Q_n(y)) \geq N^s(n, y, Q_n(y)) \\ &\times \min\{\nu_y^s(R) : R \in \mathcal{R}(n) \text{ and } R \cap \xi^s(y) \cap \Gamma \cap Q_n(y) \neq \varnothing\}. \end{aligned}$$

(Note that the condition $R \cap \xi^s(y) \cap Q_n(y) \neq \varnothing$ implies that $R \in \mathcal{R}(n)$.) Let $z \in R \cap \xi^s(y) \cap \Gamma \cap Q_n(y)$ for some $R \in \mathcal{R}(n)$. By Lemma 14.2.3 we obtain $\nu_y^s(R) = \mu_y^s(\mathcal{P}_{an}^0(z)) = \mu_z^s(\mathcal{P}_{an}^0(z))$. The first inequality in the lemma follows now from (14.23).

The proof of the second inequality is similar. □

Lemma 14.2.5. *For each $y \in \mathcal{P}(x) \cap \widehat{\Gamma}$ and integer $n \geq n_1$, we have*

$$\nu(B(y, e^{-n})) \leq N(n, Q_n(y)) \cdot 2C\, e^{-2anh+2an\varepsilon}.$$

Proof of the lemma. It follows from (14.36) and (14.33) that

$$\begin{aligned} &\frac{1}{2}\nu(B(y, e^{-n})) \leq \nu(B(y, e^{-n}) \cap \Gamma) \leq \nu(Q_n(y) \cap \Gamma) \\ &\leq N(n, Q_n(y)) \cdot \max\{\nu(R) : R \in \mathcal{R}(n) \text{ and } R \cap Q_n(y) \neq \varnothing\}. \end{aligned}$$

(Note that the condition $R \cap Q_n(y) \neq \varnothing$ implies that $R \in \mathcal{R}(n)$.) The desired inequality follows from (14.22). □

We now estimate the number of rectangles in $\mathcal{R}(n)$ and $\mathcal{F}(n)$.

Lemma 14.2.6. *For ν-almost every $y \in \mathcal{P}(x) \cap \widehat{\Gamma}$, there is an integer $n_2(y) \geq n_1$ such that for each $n \geq n_2(y)$, we have*

$$N(n+2, Q_{n+2}(y)) \leq \widehat{N}^s(n, y, Q_n(y)) \cdot \widehat{N}^u(n, y, Q_n(y)) \cdot 2C^2\, e^{4a(h+\varepsilon)}\, e^{4an\varepsilon}.$$

Proof of the lemma. By the Borel Density Lemma (with $A = \widehat{\Gamma}$), for ν-almost every $y \in \widehat{\Gamma}$ there is an integer $n_2(y) \geq n_1$ such that for all $n \geq n_2(y)$,

$$2\nu(B(y, e^{-n}) \cap \widehat{\Gamma}) \geq \nu(B(y, e^{-n})).$$

Since $\widehat{\Gamma} \subset \Gamma$, it follows from (14.33) that for all $n \geq n_2(y)$,

$$\begin{aligned}2\nu(Q_n(y) \cap \widehat{\Gamma}) &\geq 2\nu(B(y, e^{-n}) \cap \widehat{\Gamma}) \geq \nu(B(y, e^{-n})) \\ &\geq \nu(B(y, 4\,e^{-n-2})) \geq \nu(Q_{n+2}(y)).\end{aligned} \tag{14.39}$$

For any $m \geq n_2(y)$, by (14.22) and property (e), we have

$$\begin{aligned}\nu(Q_m(y)) &= \sum_{\mathcal{P}^{am}_{am}(z) \subset Q_m(y)} \nu(\mathcal{P}^{am}_{am}(z)) \\ &\geq N(m, Q_m(y)) \cdot C^{-1}\, e^{-2amh-2am\varepsilon}.\end{aligned}$$

Similarly, for every $n \geq n_2(y)$, we obtain

$$\nu(Q_n(y) \cap \widehat{\Gamma}) = \sum_{\mathcal{P}^{an}_{an}(z) \subset Q_n(y)} \nu(\mathcal{P}^{an}_{an}(z) \cap \widehat{\Gamma}) \leq N_n \cdot C\, e^{-2anh+2an\varepsilon},$$

where N_n is the number of rectangles $\mathcal{P}^{an}_{an}(z) \in \mathcal{R}(n)$ that have nonempty intersection with $\widehat{\Gamma}$.

Set $m = n + 2$. The last two inequalities together with (14.39) imply that

$$N(n+2, Q_{n+2}(y)) \leq N_n \cdot 2C^2\, e^{4a(h+\varepsilon)+4an\varepsilon}. \tag{14.40}$$

On the other hand, since $y \in \widehat{\Gamma}$, the intersections $\mathcal{P}^{an}_0(y) \cap \xi^u(y) \cap \widehat{\Gamma}$ and $\mathcal{P}^0_{an}(y) \cap \xi^s(y) \cap \widehat{\Gamma}$ are nonempty.

Consider a rectangle $\mathcal{P}^{an}_{an}(v) \subset Q_n(y)$ that has nonempty intersection with $\widehat{\Gamma}$. Then the rectangles $\mathcal{P}^0_{an}(v) \cap \mathcal{P}^{an}_0(y)$ and $\mathcal{P}^0_{an}(y) \cap \mathcal{P}^{an}_0(v)$ are in $\mathcal{F}(n)$ and intersect respectively the local stable and unstable manifolds at y. Hence, one can associate to any rectangle $\mathcal{P}^{an}_{an}(v) \subset Q_n(y)$ in $\mathcal{R}(n)$ that has nonempty intersection with $\widehat{\Gamma}$, the pair of rectangles $(\mathcal{P}^0_{an}(v) \cap \mathcal{P}^{an}_0(y), \mathcal{P}^0_{an}(y) \cap \mathcal{P}^{an}_0(v))$ in

$$\{R \in \mathcal{F}(n) : R \cap \xi^s(y) \cap Q_n(y) \neq \varnothing\} \times \{R \in \mathcal{F}(n) : R \cap \xi^u(y) \cap Q_n(y) \neq \varnothing\}.$$

Clearly, this correspondence is injective. Therefore,

$$\widehat{N}^s(n, y, Q_n(y)) \cdot \widehat{N}^u(n, y, Q_n(y)) \geq N_n.$$

The desired inequality follows from (14.40). □

The next goal is to compare the growth rate in n of the number of rectangles in $\mathcal{F}(n)$ with the number of rectangles in $\mathcal{R}(n)$. We start with an auxiliary result.

Lemma 14.2.7. *For each $x \in \widehat{\Gamma}$ and integer $n \geq n_1$, we have*

$$\widehat{N}^s(n, x, \mathcal{P}(x)) \leq D^{-1}C^2\, e^{anh+3an\varepsilon},$$
$$\widehat{N}^u(n, x, \mathcal{P}(x)) \leq D^{-1}C^2\, e^{anh+3an\varepsilon}.$$

Proof of the lemma. Since the partition $\mathcal{P}$ is countable, we can find points y_i such that the union of the rectangles $\mathcal{P}_0^{an}(y_i)$ is $\mathcal{P}(x)$, and these rectangles are mutually disjoint. Without loss of generality we can assume that $y_i \in \widehat{\Gamma}$ whenever $\mathcal{P}_0^{an}(y_i) \cap \widehat{\Gamma} \neq \varnothing$.

We have

$$\begin{aligned} N(n, \mathcal{P}(x)) &\geq \sum_i N^s(n, y_i, \mathcal{P}_0^{an}(y_i)) \\ &\geq \sum_{i:\mathcal{P}_0^{an}(y_i)\cap\widehat{\Gamma}\neq\varnothing} N^s(n, y_i, \mathcal{P}_0^{an}(y_i)). \end{aligned} \tag{14.41}$$

We now estimate $N^s(n, y_i, \mathcal{P}_0^{an}(y_i))$ from below for $y_i \in \widehat{\Gamma}$. By Lemmas 14.2.2 and 14.2.3, and (14.23),

$$\begin{aligned} N^s(n, y_i, \mathcal{P}_0^{an}(y_i)) &\geq \frac{\nu_{y_i}^s(\mathcal{P}_0^{an}(y_i) \cap \Gamma)}{\max\{\nu_z^s(\mathcal{P}_{an}^{an}(z)) : z \in \xi^s(y_i) \cap \mathcal{P}(x) \cap \Gamma\}} \\ &\geq \frac{D}{\max\{\nu_z^s(\mathcal{P}_{an}^{an}(z)) : z \in \xi^s(y_i) \cap \mathcal{P}(x) \cap \Gamma\}} \\ &= \frac{D}{\max\{\nu_z^s(\mathcal{P}_{an}^{0}(z)) : z \in \xi^s(y_i) \cap \mathcal{P}(x) \cap \Gamma\}} \\ &\geq DC^{-1}\, e^{anh-an\varepsilon}. \end{aligned} \tag{14.42}$$

Similarly (14.22) implies that

$$N(n, \mathcal{P}(x)) \leq \frac{\nu(\mathcal{P}(x))}{\min\{\mu(\mathcal{P}_{an}^{an}(z)) : z \in \mathcal{P}(x) \cap \Gamma\}} \leq C\, e^{2anh+2an\varepsilon}. \tag{14.43}$$

We now observe that

$$\widehat{N}^u(n, x, \mathcal{P}(x)) = \operatorname{card}\{i : \mathcal{P}_0^{an}(y_i) \cap \widehat{\Gamma} \neq \varnothing\}. \tag{14.44}$$

Putting (14.41)–(14.44) together we conclude that

$$\begin{aligned} C\, e^{2anh+2an\varepsilon} &\geq N(n, \mathcal{P}(x)) \\ &\geq \sum_{i:\mathcal{P}_0^{an}(y_i)\cap\widehat{\Gamma}\neq\varnothing} N^s(n, y_i, \mathcal{P}_0^{an}(y_i)) \\ &\geq \widehat{N}^u(n, x, \mathcal{P}(x)) \cdot DC^{-1}\, e^{anh-an\varepsilon}. \end{aligned}$$

This yields $\widehat{N}^u(n, x, \mathcal{P}(x)) \leq D^{-1}C^2\, e^{anh+3an\varepsilon}$. The other inequality can be proved in a similar way. □

We emphasize that the procedure of "filling in" rectangles to obtain the class $\mathcal{F}(n)$ may substantially increase the number of rectangles in the neighborhood of some points. However, the next lemma shows that this procedure of "filling in" does not add too many rectangles at almost every point.

Lemma 14.2.8. *For ν-almost every $y \in \mathcal{P}(x) \cap \widehat{\Gamma}$, we have*

$$\overline{\lim_{n\to+\infty}} \frac{\widehat{N}^s(n, y, Q_n(y))}{N^s(n, y, Q_n(y))} e^{-7an\varepsilon} < 1,$$

$$\overline{\lim_{n\to+\infty}} \frac{\widehat{N}^u(n, y, Q_n(y))}{N^u(n, y, Q_n(y))} e^{-7an\varepsilon} < 1.$$

Proof of the lemma. By (14.34) and (14.37), for each $n \geq n_1$ and $y \in \widehat{\Gamma}$,

$$\nu_y^s(Q_n(y)) \geq \nu_y^s(B^s(y, e^{-n}) \cap \Gamma) \geq \frac{1}{2}\nu_y^s(B^s(y, e^{-n})).$$

Since $\mathcal{P}_{an}^{an}(z) \subset \mathcal{P}_{an}^0(z)$ for every z, by virtue of (14.23) and (14.27) we obtain

$$\begin{aligned} N^s(n, y, Q_n(y)) &\geq \frac{\nu_y^s(Q_n(y))}{\max\{\nu_z^s(\mathcal{P}_{an}^{an}(z)) : z \in \xi^s(y) \cap \mathcal{P}(x) \cap \Gamma\}} \\ &\geq \frac{1}{2} \frac{\nu_y^s(B^s(y, e^{-n}))}{\max\{\nu_z^s(\mathcal{P}_{an}^0(z)) : z \in \xi^s(y) \cap \mathcal{P}(x) \cap \Gamma\}} \\ &\geq \frac{1}{2C} \frac{e^{-d^s n - n\varepsilon}}{e^{-anh + an\varepsilon}}. \end{aligned} \tag{14.45}$$

Let us consider the set

$$F = \left\{ y \in \widehat{\Gamma} : \overline{\lim_{n\to+\infty}} \frac{\widehat{N}^s(n, y, Q_n(y))}{N^s(n, y, Q_n(y))} e^{-7an\varepsilon} \geq 1 \right\}.$$

For each $y \in F$ there exists an increasing sequence of positive integers $\{m_j\}_{j=1}^\infty = \{m_j(y)\}_{j=1}^\infty$ such that

$$\begin{aligned} \widehat{N}^s(m_j, y, Q_{m_j}(y)) &\geq \frac{1}{2} N^s(m_j, y, Q_{m_j}(y)) e^{7am_j\varepsilon} \\ &\geq \frac{1}{4C} e^{-d^s m_j + am_j h + 5am_j\varepsilon} \end{aligned} \tag{14.46}$$

for all j (note that $a > 1$).

We wish to show that $\nu(F) = 0$. Assume, on the contrary, that $\nu(F) > 0$. Let $F' \subset F$ be the set of points $y \in F$ for which there exists the limit

$$\lim_{r\to 0} \frac{\log \nu_y^s(B^s(y, r))}{\log r} = d^s.$$

Clearly, $\nu(F') = \nu(F) > 0$. Then we can find $y \in F$ such that

$$\nu_y^s(F) = \nu_y^s(F') = \nu_y^s(F' \cap \mathcal{P}(y) \cap \xi^s(y)) > 0.$$

It follows from Frostman's lemma that

$$\dim_H(F' \cap \xi^s(y)) = d^s. \tag{14.47}$$

Let us consider the countable collection of balls

$$\mathcal{B} = \{B(z, e^{-m_j(z)}) : z \in F' \cap \xi^s(y); j = 1, 2, \ldots\}.$$

By the Besicovitch covering lemma one can find a subcover $\mathcal{C} \subset \mathcal{B}$ of $F' \cap \xi^s(y)$ of arbitrarily small diameter and finite multiplicity $\rho = \rho(\dim M)$. This means that for any $L > 0$ one can choose a sequence of points $\{z_i \in F' \cap \xi^s(y)\}_{i=1}^\infty$ and a sequence of integers $\{t_i\}_{i=1}^\infty$, where $t_i \in \{m_j(z_i)\}_{j=1}^\infty$ and $t_i > L$ for each i, such that the collection of balls

$$\mathcal{C} = \{B(z_i, e^{-t_i}) : i = 1, 2, \dots\}$$

comprises a cover of $F' \cap \xi^s(y)$ whose multiplicity does not exceed ρ. We write $Q(i) = Q_{t_i}(z_i)$.

The Hausdorff sum corresponding to this cover is

$$\sum_{B\in\mathcal{C}} (\operatorname{diam} B)^{d^s-\varepsilon} = (8^{d^s-\varepsilon}) \sum_{i=1}^\infty e^{-t_i(d^s-\varepsilon)}.$$

By (14.46), we obtain

$$\begin{aligned}\sum_{i=1}^\infty e^{-t_i(d^s-\varepsilon)} &\le \sum_{i=1}^\infty \widehat{N}^s(t_i, z_i, Q(i)) \cdot 4Ce^{-at_ih-4at_i\varepsilon} \\ &\le 4C \sum_{q=1}^\infty e^{-aqh-4aq\varepsilon} \sum_{i:t_i=q} \widehat{N}^s(q, z_i, Q(i)).\end{aligned}$$

Since the multiplicity of the subcover $\mathcal{C}$ is at most ρ, each set $Q(i)$ appears in the sum $\sum_{i:t_i=q} \widehat{N}^s(q, z_i, Q(i))$ at most ρ times. Hence,

$$\sum_{i:t_i=q} \widehat{N}^s(q, z_i, Q(i)) \le \rho \widehat{N}^s(q, y, \mathcal{P}(y)).$$

It follows from Lemma 14.2.7 that

$$\begin{aligned}\sum_{B\in\mathcal{C}} (\operatorname{diam} B)^{d^s-\varepsilon} &\le 4(8^{d^s-\varepsilon})C \sum_{q=1}^\infty e^{-aqh-4aq\varepsilon} \rho \widehat{N}^s(q, y, \mathcal{P}(y)) \\ &\le 4(8^{d^s-\varepsilon})D^{-1}C^3\rho \sum_{q=1}^\infty e^{-aqh-4aq\varepsilon+aqh+3aq\varepsilon} \\ &= 4(8^{d^s-\varepsilon})D^{-1}C^3\rho \sum_{q=1}^\infty e^{-aq\varepsilon} < \infty.\end{aligned}$$

Since L can be chosen arbitrarily large (and so the numbers t_i), it follows that $\dim_H(F' \cap \xi^s(y)) \le d^s - \varepsilon < d^s$. This contradicts (14.47). Hence, $\nu(F) = 0$ and this yields $\nu(F) = 0$ and this yields the first inequality in the lemma. The proof of the second inequality is similar. □

By Lemma 14.2.8, for ν-almost every $y \in \mathcal{P}(x) \cap \widehat{\Gamma}$, there exists an integer $n_3(y) \ge n_2(y)$ such that if $n \ge n_3(y)$ then

$$\widehat{N}^s(n, y, Q_n(y)) < N^s(n, y, Q_n(y))e^{7an\varepsilon}, \tag{14.48}$$

$$\widehat{N}^u(n, y, Q_n(y)) < N^u(n, y, Q_n(y))e^{7an\varepsilon}. \tag{14.49}$$

Moreover, by Luzin's Theorem, for every $\varepsilon > 0$ there exists a subset $\Gamma_\varepsilon \subset \widehat{\Gamma}$ such that

$$\nu(\Gamma_\varepsilon) > \nu(\widehat{\Gamma}) - \varepsilon \quad \text{and} \quad n_\varepsilon := \sup\{n_1, n_3(y) : y \in \Gamma_\varepsilon\} < \infty,$$

and the inequalities (14.48) and (14.49) hold for every $n \geq n_\varepsilon$.

Lemma 14.2.9. *For every $\varepsilon > 0$, if $y \in \Gamma_\varepsilon$ and $n \geq n_\varepsilon$, then*

$$\nu_y^s(B^s(y, e^{-n}))\nu_y^u(B^u(y, e^{-n})) \leq \nu(B(y, 4\, e^{-n})) \cdot 4C^3\, e^{11an\varepsilon}.$$

Proof of the lemma. Let $z \in \Gamma_\varepsilon \cap Q_n(y)$. By (14.49), if $n \geq n_\varepsilon$ then

$$\begin{aligned} N^u(n, y, Q_n(y)) &\leq \widehat{N}^u(n, y, Q_n(y)) \\ &= \widehat{N}^u(n, z, Q_n(y)) \\ &< N^u(n, z, Q_n(y))e^{7an\varepsilon} \end{aligned}$$

and

$$N^u(n, y, Q_n(y)) \leq \inf\{N^u(n, z, Q_n(y)) : z \in \Gamma_\varepsilon \cap Q_n(y)\}e^{7an\varepsilon}. \tag{14.50}$$

Since $N(n, Q_n(y))$ is equal to the number of rectangles $R \subset Q_n(y)$ such that $R \cap \Gamma \neq \varnothing$, we have

$$\widehat{N}^s(n, y, Q_n(y)) \times \inf\{N^u(n, z, Q_n(y)) : z \in Q_n(y)\} \leq N(n, Q_n(y)).$$

By (14.50), if $y \in \Gamma_\varepsilon$ and $n \geq n_\varepsilon$ then

$$N^s(n, y, Q_n(y)) \times N^u(n, y, Q_n(y)) \leq N(n, Q_n(y))e^{7an\varepsilon}.$$

As in (14.45), if $y \in \Gamma$ and $n \geq n_0$ then

$$N^s(n, y, Q_n(y)) \geq \nu_y^s(B^s(y, e^{-n})) \cdot (2C)^{-1}\, e^{anh - an\varepsilon},$$

$$N^u(n, y, Q_n(y)) \geq \nu_y^u(B^u(y, e^{-n})) \cdot (2C)^{-1}\, e^{anh - an\varepsilon}.$$

Moreover, by (14.22) and (14.33),

$$\begin{aligned} N(n, Q_n(y)) &\leq \frac{\nu(Q_n(y))}{\min\{\nu(\mathcal{P}_{an}^{an}(z)) : z \in Q_n(y) \cap \Gamma\}} \\ &\leq \nu(B(y, 4\, e^{-n})) \cdot C\, e^{2anh + 2an\varepsilon}. \end{aligned}$$

Putting together these inequalities we obtain the desired statement. □

Step III: Proof of the theorem. We can complete the proof of the theorem. Given $\varepsilon > 0$, let the set $\widehat{\Gamma}$ be as in the previous sections. By Lemmas 14.2.5 and 14.2.6, for ν-almost every $y \in \mathcal{P}(x) \cap \widehat{\Gamma}$ and $n \geq n_2(y)$, we obtain

$$\begin{aligned} \nu(B(y, e^{-n-2})) \leq\ & \widehat{N}^s(n, y, Q_n(y)) \cdot \widehat{N}^u(n, y, Q_n(y)) \\ & \times 4C^3\, e^{4a(h+\varepsilon)}\, e^{-2anh + 6an\varepsilon}. \end{aligned}$$

By (14.48), (14.49), and Lemma 14.2.4, we obtain

$$\begin{aligned}\nu(B(y, e^{-n-2})) &\le N^s(n, y, Q_n(y)) \cdot N^u(n, y, Q_n(y)) \\ &\quad \times 4C^3 e^{4a(h+\varepsilon)} e^{-2anh+20an\varepsilon} \\ &\le \nu_y^s(B^s(y, 4e^{-n}))\nu_y^u(B^u(y, 4e^{-n})) \cdot 4C^5 e^{4a(h+\varepsilon)} e^{22an\varepsilon}.\end{aligned} \tag{14.51}$$

Statement 1 of the theorem follows now immediately from (14.51) and Lemma 14.2.9.

By (14.51) and (14.20), we obtain

$$\underline{\lim}_{n\to+\infty} \frac{\log \nu(B(y, e^{-n}))}{-n} \ge d^s + d^u - 22a\varepsilon$$

for ν-almost every $y \in \widehat{\Gamma}$. Since $\nu(\widehat{\Gamma}) > 1 - \varepsilon$ and $\varepsilon > 0$ is arbitrarily small, we conclude that

$$\underline{d} := \underline{\lim}_{r\to 0} \frac{\log \nu(B(y, r))}{\log r} = \underline{\lim}_{n\to+\infty} \frac{\log \nu(B(y, e^{-n}))}{-n} \ge d^s + d^u$$

for ν-almost every $y \in M$. On the other hand, by Lemma 14.2.9 and (14.20), we obtain

$$\overline{d} := \overline{\lim}_{r\to 0} \frac{\log \nu(B(y, r))}{\log r} = \overline{\lim}_{n\to+\infty} \frac{\log \nu(B(y, e^{-n}))}{-n} \le d^s + d^u + 11a\varepsilon,$$

and it follows from the arbitrariness of ε that $\overline{d} \le d^s + d^u$. Therefore, $\underline{d} = \overline{d} = d^s + d^u$. This completes the proof of the theorem.

Step IV: The case of nonergodic measures. We show how to modify our arguments in the case when the measure ν is not ergodic. Given $\varepsilon > 0$, there is a set Γ' of measure $\ge 1 - \varepsilon$ and for every $x \in \Gamma'$, a number $h(x)$ such that Statements (14.22)–(14.35) hold if h is replaced by $h(x)$, d^s by $d^s(x)$, and d^u by $d^u(x)$. Let us fix $\kappa > 0$ and consider the invariant sets $\Gamma(x)$ defined by

$$\{y \in M : |h(y) - h(x)| < \kappa, |d^s(y) - d^s(x)| < \kappa, |d^u(y) - d^u(x)| < \kappa\}.$$

The collection of these sets covers Γ'. Moreover, there exists a countable subcollection of sets $\{\Gamma^i\}_{i\in\mathbb{N}}$, which still covers Γ'. Let ν^i be the conditional measure generated by μ on Γ^i. We can apply the above arguments to the measures μ^i and show that for almost every $x \in \Gamma^i$, the following inequality holds

$$d^s(x) + d^u(x) - c\kappa < d^i(x),$$

where $d^i(x)$ is the pointwise dimension of the measure ν^i and c does not depend on x or κ. Since $\{\Gamma^i\}_{i\in\mathbb{N}}$ is countable, letting κ go to zero yields that for ν-almost every $x \in M$,

$$d^s(x) + d^u(x) \le \underline{d}(x).$$

We show in a similar manner that $\overline{d}(x) \le d^s(x) + d^u(x)$. □

For surface diffeomorphisms our results give the following relation between the measure-theoretic entropy of a diffeomorphism, its Lyapunov exponents, and the pointwise dimension. It was first established by L.-S. Young in [236].

Theorem 14.2.10. *Let f be a $C^{1+\alpha}$ diffeomorphism of a smooth compact and ν a hyperbolic ergodic measure with Lyapunov exponents $\lambda_1 > 0 > \lambda_2$. Then*

$$\underline{d} = \overline{d} = h_\nu(f)\left(\frac{1}{\lambda_1} - \frac{1}{\lambda_2}\right).$$

14.3 Applications to Dimension Theory

Let X be a complete metric space of finite topological dimension and ν a finite Borel measure on X. Define the *Hausdorff dimension* and *lower* and *upper box dimensions* of ν by

$$\begin{aligned}
\dim_H \nu &= \inf\{\dim_H Z : \nu(Z) = 1\},\\
\underline{\dim}_B \nu &= \lim_{\delta\to 0} \inf\{\underline{\dim}_B Z : \nu(Z) \geq 1 - \delta\},\\
\overline{\dim}_B \nu &= \lim_{\delta\to 0} \inf\{\overline{\dim}_B Z : \nu(Z) \geq 1 - \delta\}
\end{aligned}$$

(see [172] for more information).

Theorem 14.3.1 ([236]). *Assume that*

$$\lim_{r\to 0} \frac{\log \nu(B(x,r))}{\log r} = d$$

for ν-almost every $x \in X$. Then

$$\dim_H \nu = \underline{\dim}_B \nu = \overline{\dim}_B \nu = d.$$

As an immediate corollary of this theorem and Theorem 14.2.1 we obtain the following result.

Theorem 14.3.2. *Let f be a $C^{1+\alpha}$ diffeomorphism of a smooth compact Riemannian manifold M without boundary preserving an ergodic Borel probability measure ν. Then for ν-almost every $x \in M$,*

$$\dim_H \nu = \underline{\dim}_B \nu = \overline{\dim}_B \nu = d.$$

14.4 Notes

The notions of local entropies and leaf pointwise dimensions were introduced and the results in Section 14.1 were obtained by Ledrappier and L.-S. Young in [124, 125]. Our presentation of these results is an elaboration of arguments presented in these papers.

Theorem 14.2.1 was proved by Barreira, Pesin, and Schmeling in [22] and it provides an affirmative solution to the conjecture commonly attributed to Eckmann and Ruelle [65]. A version of Lemma 14.2.9 was proved by Ledrappier and L.-S. Young in [125].

In [121], Ledrappier proved Theorem 14.2.1 for arbitrary SRB measures. Since SRB measures have local product structure, Statement 1 holds in this case automatically.

Let us also point out that neither of the assumptions of Theorem 14.2.1 can be omitted. Ledrappier and Misiurewicz [122] constructed an example of a smooth map of a circle preserving an ergodic measure with zero Lyapunov exponent for which there is no limit in (14.21) almost everywhere. (The problem of existence of this limit for one-dimensional maps was studied in [97].) In [177], Pesin and Weiss presented an example of a Hölder homeomorphism with Hölder constant arbitrarily close to 1 whose measure of maximal entropy is such that there is no limit in (14.21) almost everywhere (see [172] for more information).

15

Hyperbolic Measures: Topological Properties

In contrast to having just the few, albeit deep, results on ergodic properties of general hyperbolic measures described in the previous chapter, surprisingly many results on their topological properties can be established. They are generalizations to the nonuniformly hyperbolic case of well-known results on the topological behavior of uniformly hyperbolic systems, and the methods are extensions of some techniques widely used in the theory of locally maximal hyperbolic sets. These tools are not only important for applications but also help to reveal a nontrivial geometric structure of measures with nonzero exponents. We explain how one can close recurrent orbits, shadow pseudo-orbits, and determine the cohomology class of Hölder cocycles by periodic data.

The study of topological properties of hyperbolic measures originated in the work of Katok [101]. We follow the presentation in [106]. For $C^{1+\alpha}$ diffeomorphisms possessing a hyperbolic measure, we show that the combination of hyperbolicity with nontrivial recurrence reveals a rich and complicated orbit structure. The main goal is to relate the existence of nonzero Lyapunov exponents with the presence of hyperbolic periodic points, transverse homoclinic points, and hyperbolic horseshoes. One of the most striking results is that the entropy of a hyperbolic measure can be approximated by the topological entropies of (uniformly) hyperbolic invariant sets (see Theorem 15.6.1). The main tools will be the closing lemma (see Theorem 15.1.2) and the recurrence provided by the invariant measures.

15.1 The Closing Lemma

In this section we address one of the fundamental problems in dynamics:

> *Given a recurrent point x, is it possible to find a nearby periodic point y that follows the orbit of x during the period of time that the points in the orbit of x need to return very close to x?*

This sort of problem is called a *closing problem*, while the corresponding property is called a *closing property*.

We will show that for diffeomorphisms f preserving a hyperbolic measure it is possible to close certain orbits, and furthermore the periodic orbits thus obtained are hyperbolic (see Theorem 15.1.2). Similarly to the closing problem there is the *shadowing problem* or *shadowing property*:

> *Given a sequence of points $\{x_i\}$ on the manifold such that the image of x_i is very close to x_{i+1} for every i, is it possible to find a point x such that $f^i(x)$ is close to x_i for every i?*

In other words, if we have a sequence of points that resemble an orbit (a *pseudo-orbit*), then we can find a real orbit that *shadows* (or closely follows) the pseudo-orbit. These results are counterparts of the classical Closing and Shadowing Lemmas in the uniformly hyperbolic case (see, e.g., [104, Theorems 6.4.15 and 18.1.2]).

Let $f\colon M \to M$ be a $C^{1+\alpha}$ diffeomorphism of a compact smooth Riemannian manifold preserving a hyperbolic measure ν. Fix $\varepsilon > 0$, a Lyapunov–Perron regular point x, and consider the Lyapunov chart $\Psi_x\colon B(0, q(x)) \to M$ (see Theorem 5.6.1). We introduce a filtration of sets admitting uniform estimates using a simple trick based on Luzin's Theorem. This approach, although not as informative as the one in Section 2.2.2, is sufficient for the purposes of this section.

Let $\mathcal{R}$ be the set of regular points in M (see Definition 3.1.5). This set has full measure in M.

Proposition 15.1.1. *For each $\delta > 0$ and each sufficiently small $\varepsilon > 0$ (depending on δ), there is a compact set $\Lambda_\delta \subset \mathcal{R}$ such that $\nu(\Lambda_\delta) > 1 - \delta$ and the following conditions hold:*

1. *The functions q, C_ε, and A (depending on ε; see Theorem 5.6.1) and the values of the Lyapunov exponent counted with multiplicities χ_i' for $i = 1, \dots, \dim M$ are continuous on Λ_δ.*
2. *The splitting $T_x M = E^s(x) \oplus E^u(x)$ is continuous on Λ_δ.*

Proof. By Luzin's Theorem, there is a compact set $\Lambda_\delta' \subset M$ such that $\nu(\Lambda_\delta') > 1 - \delta/2$, and for each $i = 1, \dots, \dim M$, the functions $x \mapsto \chi_i(x)$ are continuous and bounded away from zero on Λ_δ'. By Theorem 5.6.1, for each sufficiently small $\varepsilon > 0$ we obtain a function q and a regular neighborhood $R(x)$ for each $x \in \Lambda_\delta'$ (see Definition 5.6.2). Applying Luzin's Theorem once again, we obtain a compact set $\Lambda_\delta \subset \Lambda_\delta'$ such that $\nu(\Lambda_\delta) > 1 - \delta$ and the functions q, C_ε, A, as well as the distributions $x \mapsto E^s(x)$ and $x \mapsto E^u(x)$ are continuous on Λ_δ. □

The sets Λ_δ play the role of regular sets (see (2.21)) and we shall use this term with respect to these sets as well. By Proposition 15.1.1, there are positive numbers

$$\begin{aligned} q_\delta &= \min\{q(x) : x \in \Lambda_\delta\}, \\ A_\delta &= \min\{A(x) : x \in \Lambda_\delta\}, \\ \chi_\delta &= \min\{|\chi_i'(x)| : x \in \Lambda_\delta \text{ and } 1 \le i \le \dim M\}. \end{aligned}$$

The following result is known as the *Closing Lemma* and was established by Katok [101].

Theorem 15.1.2. *Let f be a $C^{1+\alpha}$ diffeomorphism of a compact manifold preserving a hyperbolic Borel probability measure ν. For every $h > 0$, there exists $\beta = \beta(h, \delta) > 0$ such that if $x \in \Lambda_\delta$ and $f^m(x) \in \Lambda_\delta$ for some $m \in \mathbb{N}$ with $d(f^m(x), x) < \beta$, then there is a point $z = z(x)$ with the properties*

1. *z is a hyperbolic periodic point with $f^m(z) = z$;*
2. *there exists a constant $\lambda = \lambda(\varepsilon, \delta) \in (\exp(-\chi_\delta + \varepsilon), 1)$ such that*

$$d(f^i(z), f^i(x)) \leq 3hA_\delta \lambda^{\min\{i,m-i\}} \quad \textit{for } i = 0, \ldots, m.$$

Proof. We need a technical lemma relating two Lyapunov charts.

Lemma 15.1.3. *For each $\tau > 0$ there exists $\beta > 0$ such that if $x, y \in \Lambda_\delta$ and $d(x, y) < \beta$, then*

$$d_{C^1}({\Psi_x}^{-1} \circ \Psi_y, \mathrm{Id}_{\mathbb{R}^n}) < \tau \textit{ in } {\Psi_y}^{-1}(R(x) \cap R(y))$$

(where $\mathrm{Id}_{\mathbb{R}^n}$ is the identity map of $\mathbb{R}^n$). Furthermore, there exists $\rho > 0$ such that if $x, y \in \Lambda_\delta$, $d(x, y) < \rho$, and W is an admissible $(u, \gamma/4, 1)$-manifold near y, then the set

$$\Psi_x({\Psi_y}^{-1}(W) \cap B(0, q_\delta/2))$$

is part of an admissible $(u, \gamma, 1)$-manifold near x.

Proof of the lemma. Let $L_{yx}: T_yM \to T_xM$ be a linear map from T_yM to T_xM that preserves the hyperbolic splitting and is an isometry between the stable and unstable subspaces. Clearly, if $x, y \in \Lambda_\delta$ are sufficiently close, then ${L_{yx}}^{-1} \circ {\exp_x}^{-1} \circ \exp_y$ is close to Id_{T_yM} in the C^1 topology. By Proposition 15.1.1, the hyperbolic splitting of TM and the function $z \mapsto C_\varepsilon(z)$ are continuous at x and y. Thus, $C_\varepsilon(x)^{-1} \circ L_{yx} \circ C_\varepsilon(y)$ is close to $\mathrm{Id}_{\mathbb{R}^n}$ in the C^1 topology provided that x and y are sufficiently close. Since $\Psi_y = \exp_y \circ C_\varepsilon(y)$, we have

$${\Psi_x}^{-1} \circ \Psi_y = C_\varepsilon(x)^{-1} \circ {\exp_x}^{-1} \circ \exp_y \circ C_\varepsilon(y).$$

It follows that

$$\begin{aligned}
&{\Psi_x}^{-1} \circ \Psi_y - \mathrm{Id}_{\mathbb{R}^n} \\
&\quad = ({\Psi_x}^{-1} \circ \Psi_y - C_\varepsilon(x)^{-1} \circ L_{yx} \circ C_\varepsilon(y)) \\
&\qquad + (C_\varepsilon(x)^{-1} \circ L_{yx} \circ C_\varepsilon(y) - \mathrm{Id}_{\mathbb{R}^n}) \\
&\quad = [C_\varepsilon(x)^{-1} \circ L_{yx} \circ ({L_{yx}}^{-1} \circ {\exp_x}^{-1} \circ \exp_y - \mathrm{Id}_{T_yM}) \circ C_\varepsilon(y)] \\
&\qquad + [C_\varepsilon(x)^{-1} \circ L_{yx} \circ C_\varepsilon(y) - \mathrm{Id}_{\mathbb{R}^n}].
\end{aligned}$$

The two terms in square brackets are C^1-close to zero, and thus ${\Psi_x}^{-1} \circ \Psi_y$ is C^1-close to $\mathrm{Id}_{\mathbb{R}^n}$. The last statement is an immediate consequence of the first one. □

We continue with some auxiliary results. Given $x \in \Lambda_\delta$ and $h \in (0, 1]$, we set

$$R(x, h) = \Psi_x(B(0, hq(x))) \quad \text{and} \quad \widetilde{R}(x, h) = \Psi_x(B(0, hq_\delta)).$$

We say that the set $W \subset \widetilde{R}(x, h/2)$ is an *admissible* (u, γ, h)*-manifold* near x if $W = \Psi_x(\operatorname{graph} \varphi)$ for some γ-Lipschitz function $\varphi\colon R_k(h/2) \to R_{n-k}(h/2)$, $k = \dim E^u(x)$ with $\|\varphi(0)\| \le hq_\delta/4$. (Here we denote by $R_k(r) \subset \mathbb{R}^k$ the ball centered at zero of radius r.)

Lemma 15.1.4. *There exists $\beta = \beta(\delta, h) > 0$ with the following property. Let x, $y \in \Lambda_\delta$ with $d(f^m(x), y) < \beta$ and $f^m(x) \in \Lambda_\delta$ for some $m \in \mathbb{N}$, and let W_0 be an admissible (u, γ, h)-manifold near x. Define the set W_1 by*

$$\begin{aligned} W_0^1 &= f(W_0), \\ W_0^i &= f(W_0^{i-1} \cap R(f^{i-1}x, h)) \text{ for } i = 2, \ldots, m-1, \\ W_1 &= f(W_0^{m-1}) \cap \widetilde{R}(y, h/2). \end{aligned} \tag{15.1}$$

Then for some $\gamma = \gamma(\delta)$ the set W_1 is an admissible (u, γ, h)-manifold near y.

Proof of the lemma. By Lemma 15.1.3, W_1 is part of an admissible (u, γ, h)-manifold near y. Let $W_0 = \Psi_x(\operatorname{graph} \varphi)$ where $\varphi\colon R_k(h/2) \to R_{n-k}(h/2)$ and extend W_0 to an admissible $(u, \gamma, 2h)$-manifold $\widetilde{W}_0$ near x such that $\widetilde{W}_0 = \Psi_x(\operatorname{graph} \widetilde{\varphi})$ with $\widetilde{\varphi}\colon R_k(h) \to R_{n-k}(h)$, $\|\widetilde{\varphi}(0)\| \le hq_\delta/4$, and $\varphi = \widetilde{\varphi}|R_k(h/2)$.

Proceeding in a similar manner to that in Proposition 7.3.14 we find that for $i = 1, \ldots, m-1$,

$$f(W_0^{i-1} \cap R(f^{i-1}(x), h)) = \Psi_{f^i(x)}(\operatorname{graph} \widetilde{\varphi}_i),$$

where $\widetilde{\varphi}_i\colon R_k(h) \to R_{n-k}(h)$, $\|\widetilde{\varphi}_i(0)\| \le hq_j/(4\rho^i)$, and $\|d\widetilde{\varphi}_i\| \le \gamma\rho^i$ with $\rho = \exp(\chi_\delta + \varepsilon)$. For W_0^i as in (15.1), set

$$\widetilde{W}_1 = f(W_0^{m-1} \cap R(f^{m-1}(x), h)) \cap R(f^m(x), h).$$

We have $\widetilde{W}_1 = \Psi_{f^m(x)}(\operatorname{graph} \widetilde{\varphi}_m)$ where

$$\widetilde{\varphi}_m\colon R_k(h) \to R_{n-k}(h),\ \|\widetilde{\varphi}_m(0)\| \le hq_\delta/(4\rho^m), \text{ and } \|d\widetilde{\varphi}_m\| \le \gamma\rho^m$$

and we also have $W_1 \subset \widetilde{W}_1$, provided that β is sufficiently small. Set $z_m = \Psi_{f^m(x)}(\widetilde{\varphi}_m(0), 0)$ and $z_i = f^{i-m}(z_m)$, $i = 0, 1, \ldots, m-1$. Since $z_i \in W_0^i$, we obtain that $z_i = \Psi_{f^i(x)}(\widetilde{\varphi}_i(v_i), v_i)$ for some $v_i \in R_k(f^i(x), h)$. Clearly,

$$\|\widetilde{\varphi}_i(v_i)\| \le \|\widetilde{\varphi}_i(0)\| + \gamma\|v_i\| \le hq_\delta/(4\rho^i) + \gamma\|v_i\|.$$

Consider the manifolds $S_i = \Psi_{f^i(x)}(R_k(hq_\delta) \times \{0\})$ that are admissible (s, γ, h)-manifolds. Observe that S_m contains $f^m(x)$ and z_m. Therefore, $z_i \in S_i$ and consequently $\|v_i\| \le \gamma\|\widetilde{\varphi}_i(v_i)\|$. This yields

$$\|\widetilde{\varphi}_i(v_i)\| \le hq_\delta/(4\rho^i) + \gamma^2\|\widetilde{\varphi}_i(v_i)\|.$$

It follows that if γ is sufficiently small, then $z_i \in W_0^i \cap R(f^i(x), h)$ for all i. We have

$$W_0^i \cap R(f^i(x), h) = \Psi_{f^i(x)}(\operatorname{graph} \widetilde{\varphi}_i | D_i)$$

for some set $D_i \subset R_k(h)$. Note that D_0 contains a ball around v_0 of radius $(1-\gamma)hq_\delta/2$. Furthermore, $D_{i+1} = \bar{\varphi}_i(D_i \cap R_k(h))$ where $\bar{\varphi}_i$ denotes the unstable component of $\Psi_{f^{i+1}(x)}^{-1} \circ f \circ \Psi_{f^i(x)}$. Using the expansion of $\bar{\varphi}_i$ it is straightforward to verify that D_m contains a ball around the origin of radius

$$\min\{(1-\gamma)\lambda^{-m} hq_\delta/2, (1-\gamma)hq(f^m(x))\} \tag{15.2}$$

for some constant

$$\lambda = \lambda(\varepsilon, \delta) \in (\exp(-\chi_\delta + \varepsilon), 1).$$

Since

$$q(f^m(x)) \geq e^{\varepsilon m} q(x) \geq e^{\varepsilon m} q_\delta,$$

provided that τ (and thus, β) in Lemma 15.1.3 is sufficiently small, we have

$$(\Psi_y^{-1} \circ \Psi_{f^m(x)})(\operatorname{graph} \widetilde{\varphi}_m | D_m) = \operatorname{graph} \varphi_m$$

with the domain of φ_m covering $R_k(hq_\delta/2)$. Thus,

$$W_1 = \Psi_{f^m(x)}(\operatorname{graph} \widetilde{\varphi}_m | D_m) \cap \widetilde{R}(y, h/2)$$

is an admissible (u, γ, h)-manifold near y. □

Similarly one can define *admissible* (s, γ, h)*-manifold* near x and obtain the following result.

Lemma 15.1.5. *There exists $\beta = \beta(\delta, h) > 0$ such that if $x, y \in \Lambda_\delta$ with $d(f^{-m}(x), y) < \beta$ and $f^{-m}(x) \in \Lambda_\delta$ for some $m \in \mathbb{N}$, and W_0 is an admissible (s, γ, h)-manifold near x, then for some $\gamma = \gamma(\delta)$ the set W_1 defined by*

$$\begin{aligned} W_0^1 &= f^{-1}(W_0), \\ W_0^i &= f^{-1}(W_0^{i-1} \cap R(f^{1-i}(x), h)) \quad \textit{for } i = 2, \dots, m-1, \\ W_1 &= f^{-1}(W_0^{m-1}) \cap \widetilde{R}(y, h/2) \end{aligned}$$

is an admissible (s, γ, h)-manifold near y.

We proceed with the proof of the theorem. Let $\beta > 0$ be as in Lemmas 15.1.3–15.1.4. By Lemma 15.1.3, provided that β is sufficiently small, we can assume the distance $d_{C^1}(\Psi_x \circ \Psi_{f^m(x)}{}^{-1}, \mathrm{Id}_{\mathbb{R}^n})$ to be as small as desired. Thus by Lemma 15.1.4, if B_0 is an admissible (u, γ, h)-manifold near x for some sufficiently small $\gamma = \gamma(\delta) > 0$, then the set B_1 defined by

$$\begin{aligned} B_0^1 &= f(B_0), \\ B_0^i &= f(B_0^{i-1} \cap R(f^{i-1}(x), h)) \quad \text{for } i = 2, \dots, m-1, \\ B_1 &= f(B_0^{m-1}) \cap \widetilde{R}(x, h/2), \end{aligned}$$

is an admissible (u, γ, h)-manifold near x. Similarly, by Lemma 15.1.5, if A_0 is an admissible (s, γ, h)-manifold near x, then the set A_1 defined by

$$
\begin{aligned}
A_0^1 &= f^{-1}(A_0),\\
A_0^i &= f^{-1}(A_0^{i-1} \cap R(f^{m-i+1}(x), h)) \quad \text{for } i = 2, \ldots, m-1,\\
A_1 &= f^{-1}(A_0^{m-1}) \cap \widetilde{R}(x, h/2),
\end{aligned}
$$

is an admissible (s, γ, h)-manifold near x.

Lemma 15.1.6. *For any sufficiently small γ, each admissible (s, γ, h)-manifold near x intersects any admissible (u, γ, h)-manifold near x at exactly one point and the intersection is transverse.*

Proof of the lemma. The existence of a unique intersection is immediate from the uniform transversality of the families of admissible (s, γ, h)- and (u, γ, h)-manifolds near x, provided that γ is sufficiently small. This also implies that the intersection is transverse. □

By the lemma, A_1 and B_1 intersect at exactly one point, say z_{11}. We define inductively manifolds $A_k, A_k^1, \ldots, A_k^{m-1}$ for $k = 1, 2, \ldots$ by

$$
\begin{aligned}
A_k^1 &= f^{-1}(A_k),\\
A_k^i &= f^{-1}(A_k^{i-1} \cap R(f^{m-i+1}(x), h)) \text{ for } i = 2, \ldots, m-1,\\
A_{k+1} &= f^{-1}(A_k^{m-1}) \cap \widetilde{R}(x, h/2),
\end{aligned}
$$

and manifolds $B_k, B_k^1, \ldots, B_k^{m-1}$ for $k = 1, 2, \ldots$ by

$$
\begin{aligned}
B_k^1 &= f(B_k),\\
B_k^i &= f(B_k^{i-1} \cap R(f^{i-1}(x), h)) \text{ for } i = 2, \ldots, m-1,\\
B_{k+1} &= f(B_k^{m-1}) \cap \widetilde{R}(x, h/2).
\end{aligned}
$$

By Lemmas 15.1.4 and 15.1.5, A_k is an admissible (s, γ, h)-manifold near x and B_k is an admissible (u, γ, h)-manifold near x, for each $k \in \mathbb{N}$. By Lemma 15.1.6, A_k intersects B_j at exactly one point, say z_{kj}. We note that $z_{10} = x$ and $z_{01} = f^m(x)$.

Lemma 15.1.7. *If $k \geq 1$ and $j \geq 0$, then $f^m(z_{kj}) = z_{k-1,j+1}$.*

Proof of the lemma. Since

$$f^m(z_{kj}) \in f^m A_k \subset f^{m-1}(A_{k-1}^{m-1}) \subset \cdots \subset A_{k-1},$$

it suffices to show that $f^i(z_{kj}) \in B_j^i$ for $i = 1, \ldots, m-1$. Indeed,

$$f^m(z_{kj}) = f(f^{m-1}(z_{kj})) \in f(B_j^{m-1})$$

and

$$f^m(z_{kj}) \in A_{k-1} \subset \widetilde{R}(x, h/2)$$

imply that $f^m(z_{kj}) \in B_{j+1}$. We proceed by induction in i. Assume that $f^{i-1}(z_{kj}) \in B_j^{i-1}$. Since

$$f^{i-1}(z_{kj}) \in f^{i-1}A_k \subset f^{i-2}A_k^{m-1} \subset f^{i-3}A_k^{m-2} \subset \cdots \\ \subset fA_k^{m-i+2} = A_k^{m-i+1} \cap R(f^{i-1}(x), h),$$

we have

$$f^i(z_{kj}) \in f(B_j^{i-1} \cap R(f^{i-1}(x), h)) = B_j^i.$$

This completes the proof of the lemma. $\square$

We show that the sequence $\{z_{k,k-1}\}_{k\in\mathbb{N}}$ converges. By Lemma 15.1.3, for all k_1, k_2, $j \geq 1$,

$$\begin{aligned} c\|\Psi_x{}^{-1}(z_{k_1-1,j+1}) - \Psi_x{}^{-1}(z_{k_2-1,j+1})\| \\ \geq \|\Psi_{f^m(x)}{}^{-1}(z_{k_1-1,j+1}) - \Psi_{f^m(x)}{}^{-1}(z_{k_2-1,j+1})\| \\ \geq \lambda^{-m}\|\Psi_x{}^{-1}(z_{k_1j}) - \Psi_x{}^{-1}(z_{k_2j})\| \end{aligned}$$

for some constant $c = c(\tau) \geq 1$ with λ as in (15.2), since

$$f^i(z_{kj}) = f^{-1}(f^{i+1}(z_{kj})) \in f^{-1}(B_j^{i+1}) = B_j^i \cap R(f^i(x), h)$$

and each manifold $B_j^i \cap R(f^i(x), h)$ is part of an admissible (u, γ, h)-manifold. Arguing similarly, one can show that for each k, j_1, $j_2 \geq 1$,

$$\|\Psi_x{}^{-1}(z_{k+1,j_1-1}) - \Psi_x{}^{-1}(z_{k+1,j_2-1})\| \leq \lambda'\|\Psi_x{}^{-1}(z_{kj_1}) - \Psi_x{}^{-1}(z_{kj_2})\|,$$

where $\lambda' := c\lambda^m < 1$, provided that τ is sufficiently small. Since both z_{kk} and $z_{k,k-1}$ are in A_k, for $i = 0, \ldots, k$,

$$f^{mi}(z_{kk}) = z_{k-i,k+i} \in A_{k-i} \quad \text{and} \quad f^{mi}(z_{k,k-1}) = z_{k-i,k+i-1} \in A_{k-i}.$$

This implies that

$$\begin{aligned} \|\Psi_x{}^{-1}(z_{kk}) - \Psi_x{}^{-1}(z_{k,k-1})\| &\leq \lambda'\|\Psi_x{}^{-1}(z_{k-1,k+1}) - \Psi_x{}^{-1}(z_{k-1,k})\| \\ &\cdots \\ &\leq (\lambda')^k\|\Psi_x{}^{-1}(z_{0,2k}) - \Psi_x{}^{-1}(z_{0,2k-1})\| \\ &\leq (\lambda')^k 2\,\mathrm{diam}\, A_0. \end{aligned} \tag{15.3}$$

Similarly, since both $z_{k+i+1,k-i}$ and $z_{k+i,k-i}$ are in B_{k-i} for $k - i \geq 0$,

$$\begin{aligned} \|\Psi_x{}^{-1}(z_{k+1,k}) - \Psi_x{}^{-1}(z_{kk})\| &\leq \lambda'\|\Psi_x{}^{-1}(z_{k+2,k-1}) - \Psi_x{}^{-1}(z_{k+1,k-1})\| \\ &\cdots \\ &\leq (\lambda')^k\|\Psi_x{}^{-1}(z_{2k+1,0}) - \Psi_x{}^{-1}(z_{2k,0})\| \\ &\leq (\lambda')^k 2\,\mathrm{diam}\, B_0. \end{aligned} \tag{15.4}$$

Therefore for each $k \geq 1$,

$$\begin{aligned}\|\Psi_x{}^{-1}(z_{k+1,k}) - \Psi_x{}^{-1}(z_{k,k-1})\| &\leq \|\Psi_x{}^{-1}(z_{k+1,k}) - \Psi_x{}^{-1}(z_{kk})\| \\ &\quad + \|\Psi_x{}^{-1}(z_{kk}) - \Psi_x{}^{-1}(z_{k,k-1})\| \\ &\leq (\lambda')^k 2(\operatorname{diam} A_0 + \operatorname{diam} B_0).\end{aligned}$$

This implies that

$$\sum_{k=1}^{\infty} \|\Psi_x{}^{-1}(z_{k+1,k}) - \Psi_x{}^{-1}(z_{k,k-1})\| < \infty,$$

and hence $\{z_{k,k-1}\}_{k\geq 1}$ converges to a point $z = z(x) \in \widetilde{R}(x, h/2)$ as $k \to \infty$. It follows from (15.3) and (15.4) that the sequence $\{z_{k-1,k}\}_{k\geq 1}$ has the same limit z. Since $f^m(z_{k,k-1}) = z_{k-1,k}$,

$$f^m(z) = \lim_{k\to\infty} f^m(z_{k,k-1}) = \lim_{k\to\infty} z_{k-1,k} = z.$$

We estimate $d(f^i(z), f^i(x))$ for $i = 0, \ldots, m-1$. Let A_0^s be the image of a k-dimensional plane parallel to $R_k(hq_\delta)$ passing through z, and

$$A_i^s = f^{-1}(A_{i-1}^s) \cap R(f^{m-1}(x), h)$$

for $i = 1, \ldots, m$. By Lemma 15.1.6, the manifolds A_m^s and B_0 intersect at a unique point y. We obtain

$$\begin{aligned}&d(f^i(z), f^i(x)) \\ &\leq d(f^i(z), f^i(y)) + d(f^i(y), f^i(x)) \\ &\leq d(z, y)\lambda^i + d(f^m(y), f^m(x))\lambda^{m-i} \\ &\leq \lambda^{\min\{i,m-i\}}[d(z, y) + d(f^m(y), f^m(z)) + d(f^m(z), f^m(x))].\end{aligned}$$

Since $f^m(z) = z$, and y and z belong to the same stable manifold, using Theorem 5.6.1 we find that

$$d(f^i(z), f^i(x)) \leq 3hA_\delta\lambda^{\min\{i,m-i\}}.$$

We show that z is hyperbolic. For $y \in \mathbb{R}^n \equiv \mathbb{R}^k \times \mathbb{R}^{n-k}$ and $\gamma > 0$, we consider the cones

$$\begin{aligned}C_\gamma^u(y) &= \{(v, u) \in T_y\mathbb{R}^n : \|v\| \leq \gamma\|u\|\}, \\ C_\gamma^s(y) &= \{(v, u) \in T_y\mathbb{R}^2 : \|u\| \leq \gamma\|v\|\}.\end{aligned}$$

Proceeding as in the proof of Proposition 7.3.1, one can verify that for some γ and some $\rho < 1$, if $x \in \Lambda_\delta$ and $y \in R(x)$, then

$$df_x C_\gamma^u(y) \subset C_{\rho\gamma}^u(f_x(y)).$$

Furthermore, proceeding as in the proof of Proposition 7.3.2, we find that if $w \in C_\gamma^u(y)$, then $\|df_x w\| \geq K\|w\|$ for some constant $K = K(\lambda) > 1$. Similarly,

$$df_x^{-1} C_\gamma^s(y) \subset C_{\rho\gamma}^s(f_x^{-1}(y))$$

and if $w \in C^s_\gamma(y)$, then $\|df_x^{-1}w\| > K\|w\|$. Set

$$F^m_{x,z} = d_{\widetilde{z}_{m-1}} f_{f^{m-1}(x)} \cdots d_{\widetilde{z}_0} f_x,$$

where $\widetilde{z}_0 = \Psi_x^{-1}(z)$ and $\widetilde{z}_i = f_{f^{i-1}(x)}(\widetilde{z}_{i-1})$ for $i = 1, \ldots, m-1$. It follows that

$$F^m_{x,z}(C^u_\gamma(\widetilde{z}_0)) \subset C^u_{\rho^m\gamma}(\widetilde{z}_0),$$

and if $w \in C^u_\gamma(\widetilde{z}_0)$, then $\|F^m_{x,z}w\| \geq K^m\|w\|$. Similarly,

$$(F^m_{x,z})^{-1}(C^s_\gamma(\widetilde{z}_0)) \subset C^s_{\rho^m\gamma}(\widetilde{z}_0)$$

and if $w \in C^s_\gamma(\widetilde{z}_0)$, then $\|(F^m_{x,z})^{-1}w\| \geq K^m\|w\|$. For $0 < \gamma < 1$, we define cones in T_zM by

$$\widetilde{C}^u_\gamma(z) = d\Psi_x C^u_\gamma(\widetilde{z}_0) \quad \text{and} \quad \widetilde{C}^s_\gamma(z) = dC^s_\gamma(\widetilde{z}_0).$$

Observe that

$$\begin{aligned} d_z f^m &= d_{\widetilde{z}_m}\Psi_{f^m(x)} \circ F^m_{x,z} \circ d_z\Psi_x^{-1} \\ &= d_{\widetilde{z}_m}\Psi_{f^m(x)} \circ d_z\Psi_x^{-1} \circ d_{\widetilde{z}_0}\Psi_x \circ F^m_{x,z} \circ d_z\Psi_x^{-1} \end{aligned}$$

and that

$$d_{\widetilde{z}_0}\Psi_x \circ F^m_{x,z} \circ d_z\Psi_x^{-1} \colon T_zM \to T_zM.$$

We have $T_zM = E^s(z) \oplus E^u(z)$ with

$$E^s(z) = d_{\widetilde{z}_0}\Psi_x(\mathbb{R}^k \times \{0\}) \quad \text{and} \quad E^u(z) = d_{\widetilde{z}_0}\Psi_x(\{0\} \times \mathbb{R}^{n-k}).$$

Since $d(x, f^m(x)) < \beta$, it follows that

$$d\Psi_{f^m(x)} C^u_{\rho\gamma}(\widetilde{z}_m) \subset \widetilde{C}^u_{\rho^{1/2}\gamma}(z) \quad \text{and} \quad d\Psi_{f^m(x)} C^s_{\rho\gamma}(\widetilde{z}_m) \subset \widetilde{C}^s_{\rho^{1/2}\gamma}(z).$$

Thus,

$$\begin{aligned} df^m(\widetilde{C}^u_\gamma(z)) &= d_{\widetilde{z}_m}\Psi_{f^m(x)} \circ d_z\Psi_x^{-1} \circ d_{\widetilde{z}_0}\Psi_x \circ F^m_{x,z} \circ d_z\Psi_x^{-1}(\widetilde{C}^u_\gamma(z)) \\ &= d_{\widetilde{z}_m}\Psi_{f^m(x)} \circ d_z\Psi_x^{-1} \circ d_{\widetilde{z}_0}\Psi_x \circ F^m_{x,z}(C^u_\gamma(\widetilde{z}_0)) \\ &\subset d_{\widetilde{z}_m}\Psi_{f^m(x)} \circ d_z\Psi_x^{-1} \circ d_{\widetilde{z}_0}\Psi_x(C^u_{\rho^m\gamma}(\widetilde{z}_0)) \\ &= d_{\widetilde{z}_m}\Psi_{f^m(x)} \circ d_z\Psi_x^{-1}(\widetilde{C}^u_{\rho^m\gamma}(\widetilde{z}_0)) \subset \widetilde{C}^u_{\rho^{m/2}\gamma}(z). \end{aligned}$$

Similarly, $df^{-m}(\widetilde{C}^s_\gamma(z)) \subset \widetilde{C}^s_{\rho^{m/2}\gamma}(z)$. Taking $\rho' = \rho^{m/2}$ we obtain that

$$df^m(\widetilde{C}^u_\gamma(z)) \subset \widetilde{C}^u_{\rho'\gamma}(z) \quad \text{and} \quad df^{-m}(\widetilde{C}^s_\gamma(z)) \subset \widetilde{C}^s_{\rho'\gamma}(z).$$

Moreover, if $w \in \widetilde{C}^u_\gamma(z)$, then

$$\begin{aligned} \|df^m(w)\| &= \|d_{\widetilde{z}_m}\Psi_{f^m(x)} \circ F^m_{x,z} \circ d_z\Psi_x^{-1}(w)\| \\ &\geq c_1\|F^m_{x,z} \circ d_z\Psi_x^{-1}(w)\| \\ &\geq c_1K^m\|d_z\Psi_x^{-1}(w)\| \geq c_2K^m\|w\|, \end{aligned}$$

for some constants c_1 and c_2 depending only on δ and β. There is also a similar statement for vectors inside the stable cone. By Theorem 6.1.2, these conditions imply hyperbolicity. □

15.2 The Shadowing Lemma

We present an analogue of the shadowing lemma for nonuniformly hyperbolic diffeomorphisms. Given $a \in \mathbb{Z} \cup \{-\infty\}$ and $b \in \mathbb{Z} \cup \{\infty\}$, a sequence $\{x_n\}_{a<n<b}$ is called an *ε-pseudo-orbit* or simply an *ε-orbit* for f if $d(x_{n+1}, x_n) < \varepsilon$ for all $a < n < b$. We say that it is *β-shadowed* by the orbit of a point x if $d(x_n, f^n(x)) < \beta$ for all $a < n < b$. Given $\delta > 0$, set

$$\widetilde{\Lambda}_\delta = \bigcup_{x \in \Lambda_\delta} R(x),$$

where Λ_δ is the set constructed in Proposition 15.1.1 and $R(x)$ is a regular neighborhood of x (see Section 5.6).

The following result is known as the *Shadowing Lemma* and was established by Katok [102] (see also [106]).

Theorem 15.2.1. *Let f be a $C^{1+\alpha}$ diffeomorphism of a compact manifold preserving a hyperbolic measure ν. For every sufficiently small $\beta > 0$, there exists $\varepsilon = \varepsilon(\beta, \delta)$ such that given an ε-pseudo-orbit $\{x_m\} \subset \widetilde{\Lambda}_\delta$, there exists $y \in M$ such that its orbit β-shadows $\{x_m\}$.*

Proof. For each $m \in \mathbb{Z}$ consider the points x_{m-1}, x_m, x_{m+1}, and x_{m+2}, and choose $z_{m-1}, z_m, z_{m+1}, z_{m+2} \in \Lambda_\delta$ such that $x_{m+i} \in \widetilde{R}(z_{m+i}, 1)$, for $i = -1, 0, 1, 2$. We will produce a pseudo-orbit $\{x_m^1\}$, satisfying $d(f(x_m^1), x_{m+1}^1) < \zeta\, d(f(x_m), x_{m+1})$ for some $0 < \zeta < 1$.

Choose an admissible $(u, \gamma, 1)$-manifold V_{m-1}^u near z_{m-1} with $x_{m-1} \in V_{m-1}^u$, for some sufficiently small γ. Similarly, choose an admissible $(s, \gamma, 1)$-manifold V_m^s near z_{m+1} with $f(x_m) \in V_m^s$, and consider $f(V_{m-1}^u)$ and $f^{-1}(V_m^s)$. Arguing as in Proposition 7.3.14 for sufficiently small ε, we find that $f(V_{m-1}^u)$ is an admissible $(u, \gamma, 1)$-manifold near z_m and $f^{-1}(V_m^s)$ is an admissible $(s, \gamma, 1)$-manifold near z_m. By Lemma 15.1.6, the intersection $f(V_{m-1}^u) \cap f^{-1}(V_m^s)$ consists of a single point, say x_{m-1}^1.

Similarly, choose an admissible manifold V_m^u near z_m with $x_m \in V_m^u$ and an admissible manifold V_{m+1}^s near z_{m+2} containing $f(x_{m+1})$. Let $x_m^1 \in f^{-2}(V_{m+1}^s) \cap f(V_{m-1}^u)$. To produce x_{m+1}^1, consider an admissible manifold V_{m+2}^s and let $x_{m+1}^1 \in f^{-2}(V_{m+2}^s) \cap f(V_m^u)$. Arguing as in Theorem 15.1.2, one can show that there exists $\zeta < 1$ such that $d(f(x_m^1), x_{m+1}^1) < \zeta d(f(x_m), x_{m+1})$.

Using induction one can prove that for all $k > 0$, there exists a $\zeta^k \varepsilon$-pseudo-orbit $\{x_m^k\} \subset \widetilde{\Lambda}_\delta$. It follows that there exists the limit $y = \lim_{k\to\infty} x_0^k$, and

$$d(f^i(y), x_i) = \lim_{k\to\infty} d(f^i(y), x_i^k) \le C\varepsilon$$

for some constant $C > 0$. This gives the desired result. □

The version of the shadowing lemma presented here requires that the pseudo-orbit $\{x_n\}$ lie completely in some regular set $\widetilde{\Lambda}_\delta$. There is a slightly more general version of the shadowing lemma, which allows pseudo-orbits to visit all the regular sets. More precisely, let $\{\varepsilon_k\}_{k\geq 1}$ be a sequence of positive real numbers, $\{x_n\}_{n\in\mathbb{Z}}$ a sequence of points in Λ, and $\{s_n\}_{n\in\mathbb{Z}}$ a sequence of positive integers. We say that $\{x_n\}_{n\in\mathbb{Z}}$ is an $\{\varepsilon_k\}_{k\geq 1}$*-pseudo-orbit* if for all $n \in \mathbb{Z}$,

1. $x_n \in \widetilde{\Lambda}_{1/s_n}$;
2. $|s_n - s_{n+1}| \leq 1$;
3. $d(f(x_n), x_{n+1}) \leq \varepsilon_{s_n}$.

Given $\eta > 0$, we say that a point $x \in M$ η*-shadows* the $\{\varepsilon_k\}_{k\geq 1}$-pseudo-orbit $\{x_n\}_{n\in\mathbb{Z}}$ if for all $n \in \mathbb{Z}$,

$$d(f^n(x), x_n) \leq \eta\gamma_{s_n},$$

where $\gamma_\ell = \gamma_0\, e^{-\varepsilon\ell}$ for some $\gamma_0 > 0$ and $\varepsilon > 0$.

Theorem 15.2.2 (Hirayama [84]). *Let f be a $C^{1+\alpha}$ diffeomorphism of a compact manifold preserving a hyperbolic measure ν. Given $\eta > 0$, there exists a sequence of positive real numbers $\{\varepsilon_n\}_{n\in\mathbb{Z}}$ such that any $\{\varepsilon_k\}_{k\geq 1}$-pseudo-orbit is η-shadowed by a unique point.*

15.3 The Livshitz Theorem

The following result is a nonuniformly hyperbolic version of the Livshitz Theorem determining the cohomology class of Hölder cocycles by periodic data. It describes a class of functions φ for which the cohomological equation

$$\varphi(x) = \psi(f(x)) - \psi(x) \tag{15.5}$$

has a solution in the space of Borel measurable functions.

We recall that the *support* of a measure ν is the set $\operatorname{supp}\nu \subset M$ composed of the points $x \in M$ such that $\nu(B(x, \varepsilon)) > 0$ for any $\varepsilon > 0$, where $B(x, \varepsilon)$ denotes the ball of radius ε around x. We note that $\operatorname{supp}\nu$ is a closed set and that any other closed set of measure 1 contains $\operatorname{supp}\nu$. Let $\operatorname{Per}_h(f)$ be the set of all hyperbolic periodic points of f.

The following result was established by established by Katok [102] (see also [106]).

Theorem 15.3.1. *Let f be a $C^{1+\alpha}$ diffeomorphism of a compact manifold preserving a hyperbolic measure ν. Let $\varphi\colon M \to \mathbb{R}$ be a Hölder continuous function such that for each periodic point z with $f^m(z) = z$, we have $\sum_{i=0}^{m-1} \varphi(f^i(z)) = 0$. Then there exists a Borel measurable function ψ such that* (15.5) *holds for ν-almost every x.*

Proof. We assume that ν is ergodic. (The modifications required to treat the general nonergodic case are simple and are left to the reader.) Then there exists $x \in \operatorname{supp}\nu$

such that $\operatorname{supp} \nu \subset \overline{\mathcal{O}(x)}$, where $\mathcal{O}(x)$ denotes the orbit of x. Given $\delta > 0$, let Λ_δ be as in Proposition 15.1.1. We may assume that $\operatorname{supp} \nu \subset \overline{\mathcal{O}(x)}$ for some $x \in \Lambda_\delta$ and that for every $\delta > 0$, the intersection $\mathcal{O}(x) \cap \Lambda_\delta$ is dense in Λ_δ. Set $\psi(x) = 0$ and

$$\psi(f^n(x)) = \sum_{k=0}^{n-1} \varphi(f^k(x))$$

for each $k \in \mathbb{N}$. To extend ψ continuously to Λ_δ we show that ψ is uniformly continuous on $\mathcal{O}(x) \cap \Lambda_\delta$. Let $\varepsilon = \varepsilon(\beta, \delta)$ be as in Theorem 15.1.2 for some $\beta > 0$. If $n_2 > n_1$ are such that $d(f^{n_2}(x), f^{n_1}(x)) < \varepsilon$ then by Theorem 15.1.2, there exists a hyperbolic periodic point z, satisfying

$$d(f^i(f^{n_1}(x)), f^i(z)) \le 3\beta\, A_\delta \lambda^{\min\{i, m-i\}}$$

for $i = 0, \dots, n_2 - n_1 - 1$. Since φ is Hölder continuous,

$$|\varphi(x) - \varphi(y)| \le C\, d(x, y)^r$$

for some $C > 0$ and $0 < r \le 1$, and we obtain

$$|\varphi(f^{n_1+i}(x)) - \varphi(f^i(z))| \le C(3\beta\, A_\delta)^r \max\{\lambda^{ri}, \lambda^{r(n_2-n_1-i)}\}.$$

Thus,

$$\begin{aligned}
&|\psi(f^{n_2}(x)) - \psi(f^{n_1}(x))| \\
&= \left| \sum_{i=0}^{n_2-n_1-1} [\varphi(f^{n_1}(f^i(x))) - \varphi(f^i(x))] + \sum_{i=0}^{n_2-n_1-1} \varphi(f^i(z)) \right| \\
&\le C(3\beta\, A_\delta)^r \sum_{i=0}^{n_2-n_1-1} \max\{\lambda^{ri}, \lambda^{r(n_2-n_1-1)}\} + \left| \sum_{i=0}^{n_2-n_1-1} \varphi(f^i(z)) \right| \\
&\le C(3\beta\, A_\delta)^r N
\end{aligned}$$

for some constant $N > 0$ independent of n_1 and n_2. This shows that we can extend ψ continuously to Λ_δ. Now we extend ψ to $\bigcup_{i=0}^{\infty} f^i(\Lambda_\delta)$ as follows. If $y \in f(\Lambda_\delta) \setminus \Lambda_\delta$ then

$$\psi(y) = \varphi(f^{-1}(y)) + \psi(f^{-1}(y)),$$

and we use the same expression for all $y \in \bigcup_{i=0}^{n+1} f^i(\Lambda_\delta) \setminus \bigcup_{i=0}^{n} f^i(\Lambda_\delta)$. Since $\nu(\bigcup_{i=0}^{\infty} f^i(\Lambda_\delta)) = 1$, the function ψ is defined almost everywhere and clearly satisfies (15.5). □

15.4 Hyperbolic Periodic Orbits

Let $f\colon M \to M$ be a $C^{1+\alpha}$ diffeomorphism of a compact Riemannian manifold. We assume that f preserves a hyperbolic measure ν. In this section we study periodic orbits of diffeomorphisms possessing hyperbolic measures. The main results in this direction were obtained by Katok [101]. We follow the presentation in [106]. Recall that $\mathrm{Per}_h(f)$ denotes the set of hyperbolic periodic points for f.

Theorem 15.4.1. $\operatorname{supp} \nu \subset \overline{\mathrm{Per}_h(f)}$.

Proof. Fix $x_0 \in \operatorname{supp} \nu$ and $\varepsilon > 0$. Since $\nu(B(x_0, \varepsilon/4)) > 0$, we can find $\delta > 0$ such that $\nu(B(x_0, \varepsilon/4) \cap \Lambda_\delta) > 0$. Choose $\beta = \beta(\varepsilon/(12A_\delta), \delta)$ as in Theorem 15.1.2. Consider a set $B \subset B(x_0, \varepsilon/4) \cap \Lambda_\delta$ of diameter less than β with $\nu(B) > 0$. By Poincaré's Recurrence Theorem, for ν-almost every $x \in B$, there is a positive integer $n(x)$ such that $f^{n(x)}(x) \in B$, and hence $d(f^{n(x)}(x), x) < \beta$. By Theorem 15.1.2, we conclude that there is a hyperbolic periodic point z of period $n(x)$ such that $d(x, z) < 3(\varepsilon/(12A_\delta))A_\delta = \varepsilon/4$ and $d(x_0, z) \le d(x_0, x) + d(x, z) < \varepsilon/4$. □

When all the Lyapunov exponents have the same sign, one can characterize $\operatorname{supp} \nu$ completely.

Corollary 15.4.2. *Assume that the measure ν is ergodic. If all the Lyapunov exponents of f are negative (respectively, positive) on a set of full ν-measure, then* $\operatorname{supp} \nu$ *is an attracting (respectively, repelling) periodic orbit; that is,* $\operatorname{supp} \nu = \{p, \dots, f^{m-1}(p)\}$ *for some point $p \in M$ such that $f^m(p) = p$ and all the eigenvalues of $d_p f^m$ have absolute value less than* 1 *(respectively, greater than* 1*).*

Proof. Without loss of generality, we may assume that all the Lyapunov exponents are negative. Otherwise, all the Lyapunov exponents for f^{-1} are negative and we apply the following reasoning to obtain an attracting periodic orbit of f^{-1}, which is thus a repelling periodic orbit of f. Suppose that x is a recurrent point of f and that $x \in \Lambda_\delta$ for some $\delta > 0$. (It exists by Poincaré's Recurrence Theorem.) Then there exist a positive integer m and a number $\varepsilon > 0$ such that $f^m(\overline{B(x, \varepsilon)}) \subset \overline{B(x, \varepsilon)}$ and $f^m | B(x, \varepsilon)$ is a contraction. Therefore there exists a point $y \in M$ such that $f^m(y) = y$ and since x is recurrent, it must belong to the orbit of y. Thus, x must be periodic (and the orbits of x and y coincide). Since ν is ergodic, there exists exactly one periodic orbit in $\operatorname{supp} \nu$. □

Araújo [8] showed that Corollary 15.4.2 holds for C^1 diffeomorphisms.

It is well known that in the neighborhood of any transverse homoclinic point there exists a hyperbolic horseshoe, that is, a (uniformly) hyperbolic invariant set obtained by a horseshoelike construction (see, e.g., [104, Theorem 6.5.5]). This phenomenon persists under small perturbations. We will show that transverse homoclinic points are present whenever the diffeomorphism possesses hyperbolic measures that are nonatomic.

Theorem 15.4.3. *Let ν be a continuous and nonatomic hyperbolic measure. Then*

1. $\operatorname{supp} \nu$ *is contained in the closure of the set of hyperbolic periodic points that have transverse homoclinic points;*
2. *if ν is ergodic, then* $\operatorname{supp} \nu$ *is contained in the closure of the set of transverse homoclinic points of exactly one hyperbolic periodic point.*

Proof. By Corollary 15.4.2 applied to each ergodic component, not all Lyapunov exponents of f can have the same sign (in each ergodic component). Fix $x_0 \in \operatorname{supp} \nu$ and $\varepsilon > 0$. Choose $\delta > 0$ such that $\mu(B(x_0, \varepsilon/4) \cap \Lambda_\delta) > 0$. Pick distinct points $x_1, x_2 \in B(x_0, \varepsilon/4)$ such that $\mu(B(x_i, r) \cap \Lambda_\delta) > 0$ for every $r > 0$ and $i = 1, 2$ with

$$d(x_1, x_2) < \min\{\varepsilon/4, \rho(\delta)\}/2. \tag{15.6}$$

(The points x_1 and x_2 exist because ν is nonatomic.) Furthermore, we can consider subsets

$$B_i \subset B(x_i, d(x_1, x_2)/10) \cap \Lambda_\delta, \tag{15.7}$$

such that $\mu(B_i) > 0$ and for $i = 1, 2$,

$$\operatorname{diam} B_i < \beta = \beta(3d(x_1, x_2)/100A_\delta, k, \delta).$$

By Poincaré's Recurrence Theorem we can find points $y_1 \in B_1$ and $y_2 \in B_2$, and positive integers m_1 and m_2 such that $f^{m_i}(y_i) \in B_i$ for $i = 1, 2$. Therefore applying Theorem 15.1.2 we find hyperbolic periodic points z_1 and z_2 such that for $i = 1, 2$,

$$d(z_i, y_i) < d(x_1, x_2)/100. \tag{15.8}$$

We have

$$\begin{aligned} d(x_1, x_2) - d(x_1, y_1) - d(y_1, z_1) - d(z_2, y_2) - d(y_2, x_2) \\ \le d(z_1, z_2) \\ \le d(z_1, y_1) + d(y_1, x_1) + d(x_1, x_2) + d(x_2, y_2) + d(y_2, z_2). \end{aligned}$$

It follows from (15.7) and (15.8) that

$$\frac{1}{2} d(x_1, x_2) \le d(z_1, z_2) \le 2d(x_1, x_2). \tag{15.9}$$

The first inequality shows that $z_1 \ne z_2$.

By Theorem 15.1.2, $V^s(z_i)$ is an admissible $(s, \gamma, 1)$-manifold near y_1, and $V^u(z_i)$ is an admissible $(u, \gamma, 1)$-manifold near y_i for $i = 1, 2$. Moreover, it follows from the second inequality in (15.9) and (15.6) that $d(z_1, z_2) \le \rho(\delta)$, and hence by Lemma 15.1.6, $V^s(z_1)$ and $V^u(z_2)$ intersect transversally and $V^s(z_2)$ and $V^u(z_1)$ also intersect transversally. The existence of these transversal intersections guarantees the existence of transversal homoclinic points for z_1 and z_2.

To prove the second statement observe that since the closure of the set of transverse homoclinic points of a periodic point $x \in M$ is f-invariant, it follows from ergodicity that $\operatorname{supp} \nu$ is contained in exactly one of these closures. □

Let $P_m(f)$ be the number of m-periodic points of f, that is, the number of points $x \in M$ such that $f^m(x) = x$. We say that a compact f-invariant set $\Lambda \subset M$ is a *horseshoe* for a C^1 diffeomorphism $f : M \to M$ if there exist positive integers p and k, and disjoint closed sets $\Lambda_1, \dots, \Lambda_k$ such that $\Lambda = \Lambda_1 \cup \dots \cup \Lambda_k$, $f^k(\Lambda_i) = \Lambda_i$, $f(\Lambda_i) = \Lambda_{i+1}$ for $i = 0, \dots, k-1$, $f(\Lambda_k) = \Lambda_1$, and $f^k|\Lambda_1$ is conjugate to the full two-sided shift on p symbols. A *hyperbolic horseshoe* Λ is a (uniformly) hyperbolic set Λ, which is a horseshoe. The two-dimensional horseshoes defined in Section 6.1.1 are examples of hyperbolic horseshoes.

Corollary 15.4.4. *Let ν be a continuous and nonatomic hyperbolic measure. Then there exists a hyperbolic horseshoe $\Lambda \subset M$ for f and*

$$\overline{\lim_{m\to\infty}} \frac{1}{m} \log P_m(f) \geq h(f|\Lambda) > 0. \tag{15.10}$$

Proof. The existence of a hyperbolic horseshoe follows immediately from Theorem 15.4.3. Moreover,

$$h(f|\Lambda) = \overline{\lim_{m\to+\infty}} \frac{1}{m} \log P_n(f|\Lambda) > 0, \tag{15.11}$$

since $f|\Lambda$ is a compact locally maximal hyperbolic set for f (see, e.g., [104, Theorem 18.5.1]). The desired result follows from the inequality $P_n(f|\Lambda) \leq P_n(f)$. □

The following result is an immediate consequence of Corollary 15.4.4.

Corollary 15.4.5. *If ν is a continuous and nonatomic hyperbolic measure then $h(f) > 0$.*

For $C^{1+\alpha}$ *surface* diffeomorphisms, it follows from Corollary 10.3.4 that the converse of Corollary 15.4.5 holds. Combining this observation with Corollary 15.4.4 we obtain the following result:

Corollary 15.4.6. *Any $C^{1+\alpha}$ diffeomorphism of a compact surface with positive topological entropy possesses a hyperbolic horseshoe.*

Proof. Let $f: M \to M$ be a $C^{1+\alpha}$ diffeomorphism. By the variational principle for the topological entropy (see, e.g., [104, Theorem 4.5.3]), $h(f) = \sup_\nu h_\nu(f)$, where the supremum is taken over all f-invariant ergodic Borel probability measures. Hence, we can find an ergodic f-invariant measure ν with $h_\nu(f) > 0$. Such a measure must be nonatomic, because for ergodic measures with atoms, the measure-theoretic entropy must be zero. Let $\chi_1' \leq \chi_2'$ be the Lyapunov exponents of ν. By Corollary 10.3.3, $\chi_1' < 0$ and $\chi_2' > 0$. Hence, ν is hyperbolic and the desired result follows immediately from Corollary 15.4.4. □

This fact may be considered as a topological counterpart of the following statement (see Chapter 9): if f is a $C^{1+\alpha}$ diffeomorphism of a compact two-dimensional manifold with a smooth invariant measure ν and $h_\nu(f) > 0$, then there is a set Λ of positive measure such that $f|\Lambda$ is metrically isomorphic to an ergodic topological Markov chain.

Building upon the fact that periodic orbits are dense in the support of the hyperbolic measure ν one can approximate some characteristics of the measure, e.g., its Lyapunov exponents, by their values at periodic orbits. The following result describes approximations of the smallest (respectively, biggest) Lyapunov exponent.

Set

$$\chi(x) = \min\{|\chi_i'(x)| : 1 \le i \le \operatorname{diam}(M)\}.$$

If ν is an ergodic hyperbolic measure, then $\chi(x) = \chi_\nu$, where χ_ν is a nonzero constant.

Theorem 15.4.7. *Let $f\colon M \to M$ be a $C^{1+\alpha}$ diffeomorphism of a compact smooth manifold. If ν is an ergodic hyperbolic invariant probability measure for f and $x \in \operatorname{supp} \nu$, then for any $\rho > 0$, any neighborhoods V of x and W of $\operatorname{supp} \nu$, and any continuous functions $\varphi_1, \dots, \varphi_k$, there is a hyperbolic periodic point $z \in V$ such that*

1. *the orbit of z is contained in W;*
2. *$\chi(z) \ge \chi_\nu - \rho$;*
3. *if $m(z)$ is the period of z then for $i = 1, \dots, k$,*

$$\left| \frac{1}{m(z)} \sum_{k=0}^{m(z)-1} \varphi_i(f^k(x)) - \int_M \varphi_i \, d\nu \right| < \rho.$$

Proof. Let $\{z_m\}_{m\ge1} = \{z_m(x)\}_{m\ge1}$ be the sequence of periodic points constructed in the proof of Theorem 15.1.2 with the property that $z_m \to x \in \operatorname{supp} \nu \cap \Lambda_\delta$ as $m \to \infty$. Since the Lyapunov exponents are continuous on Λ_δ, we have that $\chi(z_m) \to \chi(x)$ as $m \to \infty$.

Suppose that $B(x, \alpha) \subset V$ for a sufficiently small $\alpha > 0$ and let $B \subset B(x, \alpha) \cap \Lambda_\delta$ be a set of diameter less than $\beta = \beta(\varepsilon/12A_\delta, \delta)$ and with $\nu(B) > 0$. Note that for ν-almost all points $x_0 \in M$ there exists a sequence of positive integers m_ℓ such that $f^{m_\ell}(x_0) \to x_0$ as $\ell \to \infty$ and we have the weak convergence

$$\frac{1}{m_\ell} \sum_{i=0}^{m_\ell-1} \delta(f^i(x_0)) \to \nu \text{ as } k \to \infty,$$

where $\delta(y)$ denotes the δ-measure at y. Since $\nu(\Lambda_\delta) > 0$, we can always assume that $x_0 \in \operatorname{supp} \nu$ and that $f^{m_\ell}(x_0) \in \Lambda_\delta$ for every $\ell \ge 1$. Let us pick a number $\ell \ge 1$ such that for $j = 1, \dots, k$,

$$\left| \frac{1}{m_\ell} \sum_{i=0}^{m_{\ell-1}} \varphi_j(f^i(x_0)) - \int_M \varphi_i \, d\nu \right| < \frac{\rho}{2}. \tag{15.12}$$

Now choose $h' > 0$ such that

$$|\varphi_j(\omega_1) - \varphi_j(\omega_2)| < \rho/2, \tag{15.13}$$

whenever $d(\omega_1, \omega_2) \le h'$ and then choose $r > 0$ such that $B(f^i(x_0), r) \subset W$ for $i = 0, \dots, m_l$. (This is always possible because $f^i(x_0) \in \operatorname{supp} \nu$ for every i.) By Theorem 15.1.2, with $h = \min\{h', r/(3h' A_\delta)\}$, there is a hyperbolic periodic point

$z = z(x_0)$ such that $f^{m_\ell}(z) = z$ and for $i = 0, \ldots, m_\ell$,

$$d(f^i(z), f^i(x_0)) \le r.$$

This implies that the orbit of z belongs to W and since $r < h'$, we conclude by (15.12) and (15.13) that

$$\left|\frac{1}{m_\ell}\sum_{i=0}^{m_\ell-1}\varphi_j(f^i(z)) - \int_M \varphi_i\,d\nu\right| \le \frac{1}{m_\ell}\sum_{i=0}^{m_\ell-1}|\varphi_j(f^i(z)) - \varphi_j(f^i x_0)|$$
$$+\left|\frac{1}{m_\ell}\sum_{i=0}^{m_\ell-1}\varphi_j(f^i(x_0)) - \int_M \varphi_i\,d\nu\right| < \rho.$$

This completes the proof. □

Corollary 15.4.8. *If $\{f_n\}_{n\ge1}$ is a sequence of $C^{1+\alpha}$ diffeomorphisms $f_n\colon M \to M$ converging to f in the C^1 topology, then for each sufficiently large n, f_n has a hyperbolic invariant probability measure ν_n such that $\{\nu_n\}_{n\ge1}$ converges weakly to ν. Furthermore, ν_n may be chosen to be supported on hyperbolic periodic points.*

Proof. The desired result follows from the separability of the space of continuous functions on M and an inductive application of Theorem 15.4.7 to the diffeomorphisms f_n. □

An application of Corollary 15.4.8 to a constant sequence of diffeomorphisms yields the following result.

Corollary 15.4.9. *For a $C^{1+\alpha}$ diffeomorphism $f\colon M \to M$ of a compact smooth manifold, one of the following mutually exclusive properties holds:*

1. *The measures supported on hyperbolic periodic points are weakly dense in the set of hyperbolic measures.*
2. *There are no hyperbolic measures (and hence there are no hyperbolic periodic points).*

Corollaries 15.4.8 and 15.4.9 suggest a "weak stability" of hyperbolic measures.

The estimate (15.10) can be strengthened in the following way to become a multiplicative estimate. The result is due to Ugarcovici [215]. We say that an ergodic measure ν is *locally maximal* in the class of f-invariant ergodic measures if there exists a neighborhood of ν in the weak* topology (in this class) such that for any measure μ in this neighborhood, we have $h_\nu(f) \ge h_\mu(f)$.

Theorem 15.4.10. *Assume that $h_\nu(f) > 0$. If ν is not a locally maximal ergodic measure in the class of f-invariant ergodic measures then there exist multiplicatively enough periodic orbits, which are equidistributed with respect to ν. In other words, for any $\rho > 0$ and any collection of continuous functions $\varphi_1, \ldots, \varphi_k$, there exist a sequence of numbers $n_m \to \infty$ and sets $P_{n_m} = P_{n_m}(\rho, \varphi_1, \ldots, \varphi_k)$ of* *periodic orbits of period n_m such that for any $z \in P_{n_m}$,*

$$\left| \frac{1}{n_m} \sum_{i=1}^{n_m} \varphi_i(f^l(z)) - \int \varphi_i \, d\nu \right| < \rho$$

and

$$\overline{\lim_{m\to\infty}} \frac{\operatorname{card} P_{n_m}}{e^{n_m h_\nu(f)}} \geq 1.$$

Proof. We recall the definition of the measure-theoretic entropy in terms of Bowen's metric d_n. For $x, y \in M$, let

$$d_n(x, y) = \max_{0\leq i\leq n-1} d(f^n(x), f^n(y)).$$

For $\varepsilon > 0$ and $\delta > 0$, let $N_n(\varepsilon, \delta)$ be the minimal number of ε-balls $B_n(x, \varepsilon)$ in the d_n-metric, which cover a set of measure $\geq 1 - \delta$. Then (see [104])

$$h_\mu(f) = \lim_{\varepsilon\to 0} \underline{\lim_{n\to\infty}} \frac{\log N_n(\varepsilon, \delta)}{n} = \lim_{\varepsilon\to 0} \overline{\lim_{n\to\infty}} \frac{\log N_n(\varepsilon, \delta)}{n}. \tag{15.14}$$

Fix a small number $r > 0$ and functions $\varphi_1, \dots, \varphi_k \in C(M)$. Since the measure ν is not locally maximal in the class of f-invariant ergodic measures, we can find a hyperbolic measure μ such that $h_\mu(f) - 3r > (1 + r)h_\nu(f)$ and

$$\left| \int_M \varphi_i \, d\mu - \int_M \varphi_i \, d\nu \right| < \frac{r}{3}, \quad i = 1, \dots, k.$$

Choose $\varepsilon > 0$ and $\delta > 0$ such that if $d(x, y) < \varepsilon$ then $|\varphi_i(x) - \varphi_i(y)| < r/3$ and

$$\overline{\lim_{n\to\infty}} \frac{\log N_n(\varepsilon, 2\delta)}{n} > h_\mu(f) - r. \tag{15.15}$$

Let ξ be a finite partition of M with $\operatorname{diam}\xi < \beta(\delta, \varepsilon/3)$ (see Theorem 15.1.2) and $\xi > \{\Lambda_\delta, M \setminus \Lambda_\delta\}$ where Λ_δ is a regular set of positive measure. Consider the set

$$\Lambda_{n,\delta} = \Bigg\{ x \in \Lambda_\delta : f^m(x) \in \xi(x) \text{ for some } m \in [n, (1 + r)n] \text{ and}$$
$$\left| \frac{1}{s} \sum_{j=0}^{s-1} \varphi_i(f^j(x)) - \int_M \varphi_i \, d\mu \right| < \frac{r}{3} \text{ for } s \geq n, \ i = 1, \dots, k \Bigg\}.$$

By Birkhoff's Ergodic Theorem, $\mu(\Lambda_{n,\delta}) \to \mu(\Lambda_\delta)$. For sufficiently large ℓ, we have that $\mu(\Lambda_{n,\delta}) > 1 - 2\delta$. Let $E_n \subset \Lambda_{n,\delta}$ be an (n, ε)-separated set of maximal cardinality; in other words, the cover by ε-balls in the d_n-metric centered at points in E_n is a minimal cover of $\Lambda_{n,\delta}$. In view of (15.15), there exist infinitely many n such that

$$\operatorname{card} E_n \geq e^{n(h_\mu(f)-2r)}.$$

By the Closing Lemma, for each $x \in E_n$, there exists a hyperbolic periodic point $z(x)$ of period $m(x)$ such that $d(f^k(z(x)), f^k(x)) < \varepsilon/3$ for every $k =$

$0, \dots, m(x) - 1$. For distinct points $x, y \in E_n$, we have $z(x) \neq z(y)$. We have that

$$\sum_{m=n}^{(1+r)n} \operatorname{card} R_m \geq \operatorname{card} E_n,$$

and hence

$$\max_{n \leq m \leq (1+r)n} \operatorname{card} P_m \geq \frac{\operatorname{card} E_n}{rn}.$$

(Recall that P_m is the number of hyperbolic periodic points of period m.) Therefore, we can find a sequence m_n, $n \leq m \leq (1+r)n$ such that

$$\operatorname{card} P_{m_n} \geq \frac{\operatorname{card} E_n}{rn} \geq \frac{e^{n(h_\mu(f)-2r)}}{rn} = e^{n(h_\mu(f)-3r)} \geq e^{n(1+r)h_\nu(f)}.$$

This implies that

$$\operatorname{card} P_{m_n} \geq e^{m_n h_\nu(f)}.$$

Moreover, for $z(x) \in P_{m_n}$ we have

$$\begin{aligned}
&\left| \frac{1}{m_n} \sum_{j=0}^{m_n-1} \varphi_i(f^j(z)) - \int_M \varphi_i \, d\nu \right| \\
&\leq \left| \frac{1}{m_n} \sum_{j=0}^{m_n-1} \varphi_i(f^j(z)) - \frac{1}{m_n} \sum_{j=0}^{m_n-1} \varphi_i(f^j(x)) \right| \\
&\quad + \left| \frac{1}{m_n} \sum_{j=0}^{m_n-1} \varphi_i(f^j(x)) - \int_M \varphi_i \, d\mu \right| + \left| \int_M \varphi_i \, d\mu - \int_M \varphi_i \, d\nu \right| \\
&\leq \frac{r}{3} + \frac{r}{3} + \frac{r}{3} = r
\end{aligned}$$

and the desired result follows. □

One can strengthen Theorem 15.4.7 and obtain an approximation of all the values of the Lyapunov exponent of an ergodic hyperbolic measure by the values of the Lyapunov exponent at a hyperbolic periodic point.

Theorem 15.4.11 (see [224]). *Let f be a $C^{1+\alpha}$ diffeomorphism of a compact p-dimensional manifold M preserving an ergodic hyperbolic measure ν with Lyapunov exponents $\chi_1 \leq \cdots \leq \chi_k < 0 < \chi_{k+1} \leq \cdots \leq \chi_p$. Then for every $\varepsilon > 0$ there exists a hyperbolic periodic point x such that for any $i = 1, \dots, p$,*

$$|\chi_i - \chi_i(x)| \leq \varepsilon;$$

where $\chi_i(x)$ are the Lyapunov exponents at x.

15.5 Topological Transitivity and Spectral Decomposition

By Theorem 15.4.3 we know that if ν is an ergodic hyperbolic measure, then its support is either a hyperbolic periodic orbit or is contained in the closure of the transverse homoclinic points of a hyperbolic periodic orbit. Building upon these results we shall prove the following Spectral Decomposition Theorem.

Theorem 15.5.1. *Let $f\colon M \to M$ be a $C^{1+\alpha}$ diffeomorphism of a compact smooth manifold possessing a hyperbolic invariant measure. For each $\delta > 0$ the set Λ_δ can be decomposed into finitely many closed f-invariant sets $\Lambda_\delta^{(i)}$ such that for each i one can find a point $x \in M$ for which $\Lambda_\delta^{(i)} \subset \overline{\{f^n x : n \in \mathbb{Z}\}}$.*

Proof. Let $\rho = \rho(\delta)$ be as in Lemma 15.1.3 and let $x_1, x_2 \in \Lambda_\delta$ be such that $\mu(B(x_i, \varepsilon) \cap \Lambda_\delta) > 0$ for some fixed $\varepsilon > 0$ and $i = 1, 2$ and $d(x_1, x_2) \le \rho/2$. Then, by Theorem 15.4.3 there are hyperbolic periodic points $z_1 = z(x_1)$ and $z_2 = z(x_2)$ such that $V^s(z_1)$ intersects $V^u(z_2)$ transversally and $V^u(z_1)$ intersects $V^s(z_2)$ transversally. This property of having transverse intersections of stable and unstable manifolds induces an equivalence relation for arbitrary hyperbolic periodic points z_1 and z_2. We define an equivalence relation on Λ_δ as follows: the points $x_1, x_2 \in \Lambda_\delta$ are equivalent if they belong to the closure of transverse homoclinic points of the same periodic point x. It remains to prove that we can have at most a finite number of such equivalence classes on Λ_δ.

We observe that if $x \in \Lambda_\delta$, then all points $y \in B(x, \rho/2) \cap \Lambda_\delta$ are in the same equivalence class. By compactness, there can exist only a finite number of equivalence classes. The desired statement follows from the topological transitivity of the closure of the transverse homoclinic points of a fixed hyperbolic periodic point. □

15.6 Entropy, Horseshoes, and Periodic Points

Set

$$\chi(\Lambda) = \inf\{\chi_\nu : \operatorname{supp} \nu \text{ is a periodic orbit on } \Lambda\}.$$

The following result was established by Katok [101, 102] (see also [106]).

Theorem 15.6.1. *Let $f\colon M \to M$ be a $C^{1+\alpha}$ diffeomorphism of a compact smooth manifold and ν an ergodic hyperbolic measure for f. If $h_\nu(f) > 0$, then for any $\varepsilon > 0$ and any finite number of continuous functions $\varphi_1, \dots, \varphi_k$ on M, there exists a hyperbolic horseshoe Λ such that*

1. $h(f|\Lambda) > h_\nu(f) - \varepsilon$;
2. *Λ is contained in a ε-neighborhood of* supp ν;
3. $\chi(\Lambda) > \chi_\nu - \varepsilon$;
4. *there is a measure ν_0 supported on Λ such that for $i = 1, \dots, k$,*

$$\left| \int_M \varphi_i \, d\nu_0 - \int_M \varphi_i \, d\nu \right| < \delta.$$

Proof. Given $\ell \geq 1$, let ζ be a finite measurable partition of M refining the partition $\{\Lambda_\delta, M \setminus \Lambda_\delta\}$. Fix $r > 0$. For each $m \geq 1$, let Λ_δ^m be the set of points $x \in \Lambda_\delta$ such that $f^q(x) \in \zeta(x)$ for some $q \in [m, (1+r)m]$ and

$$\left| \frac{1}{s} \sum_{j=0}^{s-1} \varphi_i(f^j(x)) - \int_M \varphi \, d\nu \right| < \frac{r}{2}$$

for $s \geq m$ and $i = 1, \ldots, k$. Using Birkhoff's Ergodic Theorem one can show that $\nu(\Lambda_\delta^m) \to \nu(\Lambda_\delta)$ as $m \to \infty$. From now on we choose m such that $\nu(\Lambda_\delta^m) > \nu(\Lambda_\delta) - r$. Given $\delta > 0$, there exists a cover $\{R(x_1), \ldots, R(x_t)\}$ of Λ_δ by closed rectangles (with $x_i \in \Lambda_\delta^m$) and numbers $\lambda \in (0, 1)$, satisfying $e^{-\chi_\nu - \delta} < \lambda < e^{-\chi_\nu + \delta}$, and $\gamma > 0$ such that

1. $\Lambda_\delta \subset \bigcup_{i=1}^t B(x_i, \delta)$ with $B(x_i, \delta) \subset \operatorname{int} R(x_i)$ for each i;
2. $\operatorname{diam} R(x_i) < r$ for each i;
3. if $x \in \Lambda_\delta \cap B(x_i, \delta)$ and $f^m(x) \in \Lambda_\delta \cap B(x_j, \delta)$ for some $m > 0$, then the connected component

   $$\mathcal{C}(R(x_i) \cap f^{-m}(R(x_j)), x)$$

   of $R(x_i) \cap f^{-m}(R(x_j))$ that contains x is an admissible (s, γ)-rectangle in $R(x_i)$ and

   $$f^m(\mathcal{C}(R(x_i) \cap f^{-m}(R(x_j)), x))$$

   is an admissible (u, γ)-rectangle in $R(x_j)$;
4. for $k = 0, \ldots, m$,

   $$\operatorname{diam} f^k(\mathcal{C}(R(x_i) \cap f^{-m}(R(x_j)), x)) \leq 3 \operatorname{diam} R(x_i) \lambda^{\min\{k, m-k\}}.$$

Here an *admissible* (s, γ)*-rectangle* is the set of points

$$\{(v, u) \in [-h, h]^2 : u = \theta \psi_1(v) + (1-\theta)\psi_2(v), 0 \leq \theta \leq 1\},$$

where ψ_1 and ψ_2 are two (s, γ)-curves (for some $h \leq 1$, in each Lyapunov chart). The definition of (u, γ)-rectangles is analogous. The desired cover can be easily obtained using the behavior of (s, γ)- and (u, γ)-curves under iteration by f. The last property follows from Theorem 15.1.2.

Let $E_m \subset \Lambda_\delta^m$ be an (m, ε)-separated set of maximal cardinality. Then $\Lambda_\delta^m \subset \bigcup_{x \in E_m} B_m(x, \varepsilon)$, and by (15.14) there exist infinitely many m such that $\operatorname{card} E_m \geq e^{m(h_\nu(f) - r)})$. For each $q \in [m, (1+r)m]$, let

$$V_q = \{x \in E_m : f^q(x) \in \zeta(x)\}$$

and let n be the value of q that maximizes $\operatorname{card} V_q$. Since $e^{mr} > mr$, we have $\operatorname{card} V_n \geq e^{m(h_\nu(f) - 3r)}$. Consider now the value j for which $\operatorname{card}(V_n \cap R(x_j))$ is maximal. Then

$$\operatorname{card}(V_n \cap R(x_j)) \geq \frac{1}{t} \operatorname{card} V_n \geq \frac{1}{t} e^{m(h_\nu(f) - 3r)}. \tag{15.16}$$

Each point $x \in V_n \cap R(x_j)$ returns to the rectangle $R(x_j)$ in n iterations, and thus $\mathcal{C}(R(x_j) \cap f^n(R(x_j)), f^n(x))$ is an admissible (u, γ)-rectangle in $R(x_j)$, and

$$f^{-n}(\mathcal{C}(R(x_j) \cap f^n(R(x_j)), f^n(x)))$$

is an admissible (s, γ)-rectangle in $R(x_j)$. This follows from the fact that $d(x_j, x) < \delta$ and $d(f^n(x), x_j) < \delta$ and from Property 2 of the cover. If $y \in \mathcal{C}(R(x_j) \cap f^{-n}(R(x_j)), x)$ then by the last property of the cover, $d(f^i(x), f^i(y)) \leq 3r$ for $i = 0, \dots, n$. This implies that given a point $y \in \mathcal{C}(R(x_j) \cap f^{-n}(R(x_j)), x) \setminus \{x\}$, we must have $y \notin V_n$; otherwise it would contradict the separability of V_n. Hence there exist card V_n disjoint admissible (s, γ)-rectangles, which are mapped by f^n onto card V_n admissible (u, γ)-rectangles.

Let

$$\Lambda(m) = \bigcup_{l \in \mathbb{Z}} f^{nl} \left(\bigcup_{x \in V_n \cap R(x_j)} \mathcal{C}(R(x_j) \cap f^{-n}(R(x_j)), x) \right).$$

The map $f^n|\Lambda(m)$ is conjugate to the full shift on card$(V_n \cap R(x_j))$ symbols. Now observe that for each $y \in \Lambda(m)$ its orbit remains in the union of the regular neighborhoods $R(x_j), \dots, R(f^n(x_j))$, and thus $f^n|\Lambda(m)$ is a hyperbolic horseshoe.

The entropy of $f^n|\Lambda(m)$ equals log card$(V_n \cap R(x_j))$. By (15.16),

$$h(f|\Lambda(m)) = \frac{1}{n} \log \operatorname{card}(V_n \cap R(x_j)) \geq \frac{1}{n} \log \frac{1}{t} e^{m(h_\nu(f) - 3r)}.$$

Since $m/n > 1/(1 + r)$, we obtain the desired statements. □

The following result is an immediate consequence of Theorem 15.6.1.

Corollary 15.6.2. *Assume that ν is ergodic and $h_\nu(f) > 0$. There is a sequence of f-invariant measures ν_n supported on hyperbolic horseshoes Λ_n such that*

1. *$\nu_n \to \nu$ as $n \to \infty$ in the weak* topology;*
2. *$h_{\nu_n}(f) \to h_\nu(f)$.*

Corollary 15.6.3. *Assume that ν is ergodic and $h_\nu(f) > 0$. Given $\varepsilon > 0$, we have*

$$h_\nu(f) \leq \overline{\lim_{m \to \infty}} \frac{1}{m} \log^+ \operatorname{card}\{x \in M : f^m(x) = x, \ \chi(x) \geq \chi_\nu - \varepsilon\}.$$

In particular,

$$h(f) \leq \overline{\lim_{m \to \infty}} \frac{1}{m} \log^+ P_m(f).$$

Proof. Recall that the identity (15.11) holds for any hyperbolic horseshoe $\Lambda \subset M$. Thus the first statement follows from the first and second statements of Theorem 15.6.1 and Corollary 15.6.2. The second statement follows immediately from the first and the variational principle for the topological entropy. □

In the two-dimensional case any measure with positive entropy is hyperbolic (see Corollary 10.3.3). Therefore Corollary 15.6.3 implies the following relation between periodic points and topological entropy.

Corollary 15.6.4. *For any $C^{1+\alpha}$ diffeomorphism $f: M \to M$ of a compact surface,*

$$h(f) \leq \overline{\lim_{m\to\infty}} \frac{1}{m} \log^+ P_m(f). \tag{15.17}$$

Proof. When $h(f) = 0$ there is nothing to prove. If $h(f) > 0$ then for every $\varepsilon \in (0, 1)$ we can find an ergodic f-invariant Borel probability measure ν such that $h_\nu(f) > (1 - \varepsilon)h(f) > 0$. By Corollary 10.3.3, ν is hyperbolic. It follows from Corollary 15.6.3 that

$$(1 - \varepsilon)h(f) < \overline{\lim_{m\to\infty}} \frac{1}{m} \log^+ P_m(f).$$

Since ε is arbitrary, the desired result follows. □

We point out that the inequality (15.17) may be strict. In the multidimensional case (15.17) may not hold.

The following result shows that hyperbolic measures persist under C^1 perturbations and it is a simple consequence of Theorem 15.6.1.

Corollary 15.6.5. *Let $f_n: M \to M$ be a sequence of diffeomorphisms converging to f in the C^1 topology. There exists an f_n-invariant ergodic measure ν_n satisfying the following properties:*

1. *$\nu_n \to \nu$ as $n \to \infty$, in the weak* topology.*
2. *$h_{\nu_n}(f_n) \to h_\nu(f)$ as $n \to \infty$.*
3. *$\chi_{\nu_n} \to \chi_\nu$ as $n \to \infty$.*

15.7 Continuity Properties of Entropy

We present two results describing continuity properties of topological and measure-theoretic entropies of diffeomorphisms. The first result relates to the topological entropy of a surface diffeomorphism of class $C^{1+\alpha}$ and follows easily from Corollary 15.6.5 and the structural stability of horseshoes (see, e.g., [104, Theorem 18.2.1]).

Theorem 15.7.1. *The topological entropy on the space of $C^{1+\alpha}$ diffeomorphisms of a given surface is lower semicontinuous.*

The second result concerns the measure-theoretic entropy of diffeomorphisms of class C^∞.

Theorem 15.7.2 (Newhouse [158]). *For a C^∞ map $f: M \to M$ of a compact smooth Riemannian manifold, the map $\nu \mapsto h_\nu(f)$ is upper semicontinuous on the space of f-invariant Borel probability measures on M.*

Proof. We start with some auxiliary results. Let $\Lambda \subset M$ be a compact subset. Given $n \in \mathbb{N}$, $\varepsilon > 0$, and $x \in \Lambda$, set

$$B(x, n, \varepsilon) = \{y \in M : d(f^i(x), f^i(y)) \leq \varepsilon \text{ for } i = 0, \dots, n\}.$$

Let $r(n, \delta, \varepsilon, \Lambda, x) = \max \operatorname{card} F$, where the maximum is taken over all sets $F \subset \Lambda \cap B(x, n, \varepsilon)$ such that for each $y \neq z$ in F there exists $i \in [0, n-1] \cap \mathbb{N}$ for which $d(f^i(x), f^i(y)) > \delta$. Let

$$r(n, \delta, \varepsilon, \Lambda) = \sup_{x \in \Lambda} r(n, \delta, \varepsilon, \Lambda, x)$$

and

$$r(\varepsilon, \Lambda) = \lim_{\delta \to 0} \overline{\lim_{n \to \infty}} \frac{1}{n} \log r(n, \delta, \varepsilon, \Lambda).$$

Given a probability measure ν on Λ, set

$$r_\nu(f, \varepsilon) = \lim_{\sigma \to 1} \inf_{\nu(\Lambda) > \sigma} r(\varepsilon, \Lambda).$$

Lemma 15.7.3. *If ν is an f-invariant Borel probability measure and β a finite partition with* $\operatorname{diam} \beta < \varepsilon$, *then*

$$h_\nu(f) \leq h_\nu(f, \beta) + r_\nu(f, \varepsilon).$$

Proof of the lemma. Consider a partition $\alpha = \{E_1, \dots, E_r, M \setminus \bigcup_{i=1}^r E_i\}$ with E_i compact. We have

$$H_\nu(\alpha_n) \leq H_\nu(\beta_n) + H_\nu(\alpha_n | \beta_n), \tag{15.18}$$

where $\alpha_n = \bigvee_{i=0}^{n-1} f^{-i}\alpha$, and thus we must estimate $H_\nu(\alpha_n|\beta_n)$ in terms of $r(n, \delta, \varepsilon, \Lambda)$. Given a compact set $\Lambda \subset M$, consider the partitions $\gamma = \{\Lambda, M \setminus \Lambda\}$ and $\beta'_n = \beta_n \vee \gamma$. Choose $\delta > 0$ such that

$$\delta < \frac{1}{4} \min_{i \neq j} d(E_i, E_j).$$

We consider $n = kN$ with $k, n \in \mathbb{N}$. Let $D \in \beta'_n$, and take $x_D \in D$. Let F_D be a maximal subset of $\Lambda \cap B(x_D, n, \varepsilon)$ such that for $x \neq y$ in F_D there exists $j \in [0, k) \cap \mathbb{N}$ with

$$d(f^{jN}(x), f^{jN}(y)) > 2\delta.$$

Clearly, F_D is a $(n, 2\delta)$-separated set; that is, for each $x \neq y$ in F_D there exists $j \in [0, n) \cap \mathbb{N}$ such that $d(f^j(x), f^j(y)) > 2\delta$, so that $\operatorname{card} F_D \leq r(n, 2\delta, \varepsilon, \Lambda)$. We have

$$\begin{aligned} H_\nu(\alpha_n|\beta_n) &\leq H_\nu(\alpha_n \vee \gamma | \beta_n) \\ &= H_\nu(\gamma|\beta_n) + H_\nu(\alpha_n | \beta_n \vee \gamma) \leq \log 2 + H_\nu(\alpha_n | \beta'_n) \end{aligned}$$

and

$$\begin{aligned}
H_\nu(\alpha_n|\beta'_n) &= \sum_{D\in\beta'_n} H_{\nu_D}(\alpha_n|D)\nu(D) \\
&= \sum_{D\in\beta'_n : B\subset\Lambda} H_{\nu_D}(\alpha_n|D)\nu(D) + \sum_{D\in\beta'_n : B\cap\Lambda=\varnothing} H_{\nu_D}(\alpha_n|D)\nu(D) \\
&\leq \sum_{D\in\beta'_n : B\subset\Lambda} H_{\nu_D}(\alpha_n|D)\nu(D) + n\ \log\operatorname{card}\alpha \cdot \nu(M\setminus\Lambda),
\end{aligned}$$

where $\nu_D(A) = \nu(A\cap D)/\nu(D)$. For each $D \in \beta'_n$ with $D \subset \Lambda$ and each $A \in \alpha_n|D$, take $x_A \in A$ and $y_A \in F_D$ such that

$$d(f^{jN}(x_A), f^{jN}(y_A)) \leq 2\delta$$

for all $j \in [0, k) \cap \mathbb{N}$. There are at most 2^k choices of sets A, which can correspond to a given point y_A. Hence,

$$H_{\nu_D}(\alpha_n|D) = -\sum_{A\in\alpha_n|D} \nu_D(A)\log\nu_D(A) \leq k\ \log 2 + \operatorname{card} F_D.$$

This yields

$$H_\nu(\alpha_n|\beta_n) \leq \log 2 + (k\ \log 2 + \operatorname{card} F_D)\nu(\Lambda) + n\ \log\operatorname{card}\alpha\cdot\nu(M\setminus\Lambda).$$

Given $\varepsilon' > 0$, choose N so large that $\log 2/N < \varepsilon'$ and assume that

$$\log\operatorname{card}\alpha\cdot\nu(M\setminus\Lambda) < \varepsilon'. \tag{15.19}$$

Choosing $\delta > 0$ sufficiently small (depending on α and Λ), we obtain

$$\begin{aligned}
\overline{\lim_{n\to\infty}} \frac{1}{n} H_\nu(\alpha_n|\beta_n) &\leq \overline{\lim_{k\to\infty}} \max_{0\leq j<N} \frac{1}{kN+j} H_\nu(\alpha_{kN+j}|\beta_{kN+j}) \\
&\leq \overline{\lim_{k\to\infty}} \frac{1}{kN} [H_\nu(\alpha_{kN}|\beta_{kN}) + H_\nu(\alpha_{N-1}|\beta_{kN})] \\
&= \overline{\lim_{k\to\infty}} \frac{1}{kN} H_\nu(\alpha_{kN}|\beta_{kN}) \\
&\leq 2\varepsilon' + \overline{\lim_{n\to\infty}} \frac{1}{n} \log r(n, 2\delta, \varepsilon, \Lambda).
\end{aligned}$$

It follows from (15.18) that

$$h_\nu(f,\alpha) \leq h_\nu(f,\beta) + 2\varepsilon' + r(\varepsilon,\Lambda),$$

and thus by (15.19), $h_\nu(f) \leq h_\nu(f,\beta) + 2\varepsilon' + r_\nu(f,\varepsilon)$. This completes the proof of the lemma. □

Lemma 15.7.4. *If* $\lim_{\varepsilon\to 0}\sup_\nu r_\nu(f,\varepsilon) = 0$ *with the supremum taken over all* f*-invariant Borel probability measures, then the map* $\nu \mapsto h_\nu(f)$ *is upper semi-continuous.*

Proof of the lemma. Let ν be an f-invariant Borel probability measure. Take $\delta > 0$ and choose $\varepsilon > 0$ such that $\sup_\nu r_\nu(f,\varepsilon) < \delta$. Let β be a partition such that

$\operatorname{diam}\beta < \varepsilon$, $\nu(\partial A) = 0$ for $A \in \beta$, and $h_\nu(f) < h_\nu(f, \beta) + \delta$. For each $n \in \mathbb{N}$, the map $\nu \mapsto (1/n)H_\nu(\beta_n)$ is continuous, and thus

$$\nu \mapsto h_\nu(f, \beta) = \inf_{n\in\mathbb{N}} \frac{1}{n} H_\nu(\beta_n)$$

is upper semicontinuous (since it is the infimum of continuous functions). By Lemma 15.7.3, for any f-invariant Borel probability measure μ,

$$h_\mu(f) \le h_\mu(f, \beta) + r_\mu(f, \varepsilon) \le h_\mu(f, \beta) + \delta.$$

For μ sufficiently close to ν, we have $h_\mu(f, \beta) < h_\nu(f, \beta) + \delta$ and thus

$$h_\mu(f) < h_\mu(f, \beta) + \delta < h_\nu(f, \beta) + 2\delta \le h_\nu(f) + 2\delta.$$

This establishes the desired upper semicontinuity. □

Consider a C^r map $\gamma\colon D_k \to M$, where $D_k \subset \mathbb{R}^k$ is the unit disk. Given $n \in \mathbb{N}$ and $\varepsilon > 0$, set

$$v_k(\gamma, \varepsilon, n) = \sup_{x\in M} v_k\left((f^{n-1} \circ \gamma)|\gamma^{-1}(B(x, n, \varepsilon))\right),$$

where v_k denotes the k-volume. We also set

$$\underline{v}(f, \varepsilon) = \sup_\gamma \varliminf_{n\to\infty} \frac{1}{n} \log^+ v_k(\gamma, \varepsilon, n),$$

$$\overline{v}(f, \varepsilon) = \sup_\gamma \varlimsup_{n\to\infty} \frac{1}{n} \log^+ v_k(\gamma, \varepsilon, n),$$

with the suprema taken over all C^r maps γ and all $k = 1, \dots, \dim M$.

Lemma 15.7.5. *We have $r_\nu(f, \varepsilon) \le \underline{v}(f, 2\varepsilon)$ for every f-invariant Borel probability measure ν.*

Proof of the lemma. Consider the inverse limit $\widetilde{f}\colon \widetilde{M} \to \widetilde{M}$ of f, given by

$$\widetilde{f}(x_0, x_1, \dots) = (f(x_0), x_0, x_1, \dots)$$

on the space $\widetilde{M} = \{(x_0, x_1, \dots) : f(x_{i+1}) = x_i \text{ for } i \ge 0\}$. Let $\widetilde{\nu}$ be the unique $\widetilde{f}$-invariant probability measure on $\widetilde{M}$ such that $\pi_*\widetilde{\nu} = \nu$, where $\pi : \widetilde{M} \to M$ is the natural projection $\pi(x_0, x_1, \dots) = x_0$. Consider also the pullback $T\widetilde{M}$ of TM induced by π. By Proposition 2.1.2, there exists a set $\Gamma \subset \widetilde{M}$ of full $\widetilde{\nu}$-measure on which $T\widetilde{M}$ is equivalent to the trivial bundle $\Gamma \times \mathbb{R}^m$. Let $T\widetilde{f}\colon \Gamma \times \mathbb{R}^m \to \Gamma \times \mathbb{R}^m$ be the vector bundle map defined by

$$T\widetilde{f}(x, v) = (\widetilde{f}, d_{\pi(x)}f(v)).$$

We consider the distance on $\widetilde{M}$ given by $d(x, y) = \sum_{i\ge0} 2^{-i} d(x_i, y_i)$.

Let $\widetilde{\Lambda}$ be the set of regular points in M (see Definition 3.1.5). This set has full measure in M. Using arguments similar to those in Proposition 15.1.1 for each $\tau > 0$ and each sufficiently small $\varepsilon > 0$ (depending on τ), there is a compact set $\widetilde{\Lambda}_\tau \subset \Gamma \cap \widetilde{\Lambda}$ such that $\nu(\widetilde{\Lambda}_\tau) > 1 - \tau$ and the following conditions hold:

1. The functions q, C_ε, A (see Theorem 5.6.1; these functions depend on ε), and the values of the Lyapunov exponent counted with their multiplicities χ_i' for $i = 1, \dots, \dim M$ are continuous on $\widetilde{\Lambda}_\tau$.
2. The splitting $T_x\widetilde{M} = E^s(x) \oplus E^c(x) \oplus E^u(x)$ into stable, central, and unstable subspaces is continuous on $\widetilde{\Lambda}_\tau$.

Set $\Lambda_\tau = \pi(\widetilde{\Lambda}_\tau)$, choose $x \in \Lambda_\tau$, and consider a maximal (n, δ)-separated set $E \subset \Lambda_\tau \cap B(x, n, \varepsilon)$. For each $y \in E$, let $\widetilde{y} \in \widetilde{\Lambda}_\tau$ be such that $\pi(\widetilde{y}) = y$. Then $\widetilde{E} = \{\widetilde{y} : y \in E\}$ is a (n, ε)-separated set for $\widetilde{f}$ (since $d(x, y) \geq d(\pi(x), \pi(y))$ for all $x, y \in \widetilde{M}$).

There is a neighborhood $V(\widetilde{y}, n, \varepsilon)$ of $\widetilde{y} \in \widetilde{E}$ such that

$$\pi(V(\widetilde{y}, n, \varepsilon)) \subset B(\pi(\widetilde{y}), n, \varepsilon), \tag{15.20}$$

with the property that $\widetilde{f}^j(V(\widetilde{y}, n, \varepsilon))$, $j = 1, \dots, n - j - 1$, is a regular set for $T\widetilde{f}$. Given $\eta > 0$, there is also a subset $F \subset \widetilde{E}$ with $\operatorname{card} F \geq e^{-\eta n} \operatorname{card} E$ such that all $\widetilde{y} \in F$ are in a small neighborhood V of some $z \in \widetilde{E}$. We assume that this neighborhood is sufficiently small so that the stable, central, and unstable subspaces are continuous and have constant dimensions on $\widetilde{E} \cap V$. If $W \subset M$ is a C^1-disk through $\pi(z)$, which is C^1-close to a small neighborhood Z of $\pi(z)$ in $\exp_{\pi(z)}(E^c(z) \oplus E^u(z))$, and $\widetilde{y} \in V \cap F$, then the local stable manifold $V^s(\pi(\widetilde{y}))$ has a nonempty transverse intersection with W. Consider a C^r map $\gamma\colon D_k \to M$ such that γ is C^1-close to Z. Then one can show that each set $\pi(V(\widetilde{y}, n, \varepsilon))$ cuts off a piece $\gamma_{\widetilde{y}}$ of γ for each $\widetilde{y} \in F$. Furthermore, there is a subset $\widetilde{F} \subset F$ such that

$$\operatorname{card} \widetilde{F} \geq C(\delta)e^{-\eta n} \operatorname{card} F,$$

with the property that $\gamma_{\widetilde{y}} \cap \gamma_{\widetilde{z}} = \varnothing$ for any $\widetilde{y} \neq \widetilde{z}$ in $\widetilde{F}$, where $C(\delta)$ is a positive constant depending only on δ. Using the properties of each Lyapunov chart one can easily show that

$$v_k(f^{n-1}\gamma_{\widetilde{y}}) \geq C(\widetilde{\Lambda}_\tau)e^{-\eta n}$$

for some positive constant $C(\widetilde{\Lambda}_\tau)$ depending only on $\widetilde{\Lambda}_\tau$. Since the image of $\gamma_{\widetilde{y}}$ is contained in $\pi(V(\widetilde{y}, n, \varepsilon))$, it follows from (15.20) that it is also contained in $B(x, n, 2\varepsilon)$. This establishes the desired inequality. □

We also need a result of Yomdin that we state without proof. Set

$$R(f) = \lim_{n\to\infty} \frac{1}{n} \log \max_{x \in M} \|d_x f^n\|.$$

Lemma 15.7.6 (Yomdin [234,235]). *For any C^r map $f\colon M \to M$,*

$$\lim_{\varepsilon\to 0} \overline{v}(f, \varepsilon) \leq \frac{\dim M}{r} R(f).$$

By Lemmas 15.7.5 and 15.7.6, when f is C^∞,

$$\lim_{\varepsilon\to 0} \sup_\nu r_\nu(f, \varepsilon) \leq \lim_{\varepsilon\to 0} \underline{v}(f, 2\varepsilon) \leq \lim_{\varepsilon\to 0} \overline{v}(f, 2\varepsilon) = 0.$$

It follows from Lemma 15.7.4 that the measure-theoretic entropy is upper semi-continuous. □

As an important manifestation of Theorem 15.7.2, we obtain that any C^∞ map has ergodic measures of maximal entropy.

It was established independently by Newhouse [158] and Yomdin [235] that the topological entropy is upper semicontinuous in the class of C^∞ maps of a compact manifold. Combining this with Theorem 15.7.1 we conclude that the topological entropy is continuous for C^∞ diffeomorphisms of a compact surface.

Bibliography

[1] V. Alekseev, Quasirandom dynamical systems. I: quasirandom diffeomorphisms, *Math. USSR-Sb.* **5** (1968), 73–128.

[2] V. Alekseev, Quasirandom dynamical systems. II: one-dimensional nonlinear vibrations in a periodically perturbed field, *Math. USSR-Sb.* **6** (1968), 505–560.

[3] V. Alekseev, Quasirandom dynamical systems. III: quasirandom vibrations of one-dimensional oscillators, *Math. USSR-Sb.* **7** (1969), 1–43.

[4] J. Alves, SRB measures for nonhyperbolic systems with multidimensional expansion, *Ann. Sci. École Norm. Sup.* (4) **33** (2000), 1–32.

[5] J. Alves, C. Bonnatti and M. Viana, SRB measures for partially hyperbolic systems whose central direction is mostly expanding, *Invent. Math.* **140** (2000), 351–398.

[6] D. Anosov, Geodesic flows on closed Riemann manifolds with negative curvature, *Proc. Steklov Inst. Math.* **90** (1969), 1–235.

[7] D. Anosov and Ya. Sinai, Certain smooth ergodic systems, *Russ. Math. Surv.* **22** (1967), 103–167.

[8] V. Araújo, *Non-zero Lyapunov exponents, no sign changes and Axiom A*, preprint.

[9] L. Arnold, *Random Dynamical Systems*, Monographs in Mathematics, Springer, Berlin, 1998.

[10] V. Arnold and A. Avez, *Problèmes Ergodiques de la Mécanique Classique*, Monographies Internationales de Mathématiques Modernes 9, Gauthier-Villars, Paris, 1967.

[11] A. Avila and J. Bochi, A formula with some applications to the theory of Lyapunov exponents, *Isr. J. Math.* **131** (2002), 125–137.

[12] A. Avila and M. Viana, Simplicity of Lyapunov spectra: proof of the Zorich–Kontsevich conjecture, *Acta Math.*, in press.

[13] M. Babillot, On the mixing property for hyperbolic systems, *Isr. J. Math.* **129** (2002), 61–76.

[14] W. Ballmann, *Lectures on Spaces of Nonpositive Curvature*, with an appendix by M. Brin, Birkhäuser, Boston, MA, 1995.

[15] W. Ballmann and M. Brin, On the ergodicity of geodesic flows, *Ergodic Theory Dyn. Syst.* **2** (1982), 311–315.

[16] W. Ballmann, M. Brin, and P. Eberlein, Structure of manifolds of nonpositive curvature. I, *Ann. Math.* (2) **122** (1985), 171–203.

[17] A. Baraviera and C. Bonatti, Removing zero Lyapunov exponents, *Ergodic Theory Dyn. Syst.* **23** (2003), 1655–1670.

[18] L. Barreira, *Pesin theory*, in *Encyclopaedia of Mathematics*, Supplement Volume I, edited by M. Hazewinkel, Kluwer, Boston, MA, 1997, pp. 406–411.

[19] L. Barreira and Ya. Pesin, Lectures on Lyapunov exponents and smooth ergodic theory, in *Smooth Ergodic Theory and Its Applications*, *Proc. Symp. Pure Math. 69*, edited by

A. Katok, R. de la Llave, Ya. Pesin, and H. Weiss, *American Mathematical Soceity, Providence, RI*, 2001, pp. 3–89.

[20] L. Barreira and Ya. Pesin, *Lyapunov Exponents and Smooth Ergodic Theory*, Univ. Lect. Ser. 23, American Mathematical Society, Providence RI, 2002.

[21] L. Barreira and Ya. Pesin, Smooth ergodic theory and nonuniformly hyperbolic dynamics, with appendix by O. Sarig, in *Handbook of Dynamical Systems 1B*, edited by B. Hasselblatt and A. Katok, Elsevier, New York, 2006, pp. 57–263.

[22] L. Barreira, Ya. Pesin, and J. Schmeling, Dimension and product structure of hyperbolic measures, *Ann. Math.* (2) **149** (1999), 755–783.

[23] L. Barreira and J. Schmeling, Sets of "non-typical" points have full topological entropy and full Hausdorff dimension, *Isr. J. Math.* **116** (2000), 29–70.

[24] L. Barreira and C. Silva, Lyapunov exponents for continuous transformations and dimension theory, *Discrete Contin. Dyn. Syst.* **13** (2005), 469–490.

[25] M. Benedicks and L. Carleson, The dynamics of the Hénon map, *Ann. Math.* (2) **133** (1991), 73–169.

[26] M. Benedicks and L.-S. Young, Sinai–Bowen–Ruelle measures for certain Hénon maps, *Invent. Math.* **112** (1993), 541–576.

[27] J. Bochi and M. Viana, Pisa lectures on Lyapunov exponents, in *Dynamical Systems II, Pubbl. Cent. Ric. Mat. Ennio Giorgi, Scuola Norm. Sup., Pisa*, 2003, pp. 23–47.

[28] J. Bochi and M. Viana, The Lyapunov exponents of generic volume-preserving and symplectic maps, *Ann. Math.* (2) **161** (2005), 1423–1485.

[29] C. Bonatti, L. Díaz, and R. Ures, Minimality of strong stable and unstable foliations for partially hyperbolic diffeomorphisms, *J. Inst. Math. Jussieu* **1** (2002), 513–541.

[30] C. Bonatti, L. Díaz, and M. Viana, Dynamics beyond uniform hyperbolicity. A global geometric and probabilistic perspective, *Encyclopaedia of Mathematical Sciences 102*, Mathematical Physics III, Springer, Berlin, 2005.

[31] C. Bonatti, C. Matheus, M. Viana, and A. Wilkinson, Abundance of stable ergodicity, *Comment. Math. Helv.* **79** (2004), 753–757.

[32] C. Bonnatti and M. Viana, SRB measures for partially hyperbolic systems whose central direction is mostly contracting, *Isr. J. Math.* **115** (2000), 157–193.

[33] P. Bougerol, Kalman filtering with random coefficients and contractions, *SIAM J. Control Optim.* **31** (1993), 942–959.

[34] R. Bowen, Mixing Anosov flows, *Topology* **15** (1976), 77–79.

[35] M. Brin, Topological transitivity of a certain class of dynamical systems, and flows of frames on manifolds of negative curvature, *Funct. Anal. Appl.* **9** (1975), 8–16.

[36] M. Brin, The topology of group extensions of C-systems, *Mat. Zametki* **18** (1975), 453–465.

[37] M. Brin, Bernoulli diffeomorphisms with nonzero exponents, *Ergodic Theory Dyn. Syst.* **1** (1981), 1–7.

[38] M. Brin, Hölder continuity of invariant distributions, in *Smooth Ergodic Theory and Its Applications*, *Proc. Symp. Pure Math. 69*, edited by A. Katok, R. de la Llave, Ya. Pesin, and H. Weiss, *American Mathematical Society, Providence, RI*, 2001, pp. 99–101.

[39] M. Brin, J. Feldman, and A. Katok, Bernoulli diffeomorphisms and group extensions of dynamical systems with non-zero characteristic exponents, *Ann. Math.* (2) **113** (1981), 159–179.

[40] M. Brin and G. Stuck, *Introduction to Dynamical Systems*, Cambridge University Press, Cambridge, 2002.

[41] L. Bunimovich, Mushrooms and other billiards with divided phase space, *Chaos* **11** (2001), 802–808.

[42] K. Burns, D. Dolgopyat, and Ya. Pesin, Partial hyperbolicity, Lyapunov exponents and stable ergodicity, *J. Stat. Phys.* **108** (2002), 927–942.

[43] K. Burns, D. Dolgopyat, Ya. Pesin, and M. Pollicott, Stable ergodicity of partially hyperbolic attractors with nonzero exponents, *Pure Appl. Math. Q.*, in press.

[44] K. Burns and M. Gerber, Continuous invariant cone families and ergodicity of flows in dimension three, *Ergodic Theory Dyn. Syst.* **9** (1989), 19–25.

[45] K. Burns, C. Pugh, M. Shub, and A. Wilkinson, Recent results about stable ergodicity, in *Smooth Ergodic Theory and Its Applications*, *Proc. Symp. Pure Math. 69*, edited by A. Katok, R. de la Llave, Ya. Pesin and H. Weiss, *American Mathematical Society Providence, RI*, 2001, pp. 327–366.

[46] K. Burns and R. Spatzier, Manifolds of nonpositive curvature and their buildings, *Inst. Hautes Études Sci. Publ. Math.* **65** (1987), 35–59.

[47] K. Burns and A. Wilkinson, Stable ergodicity of skew products, *Ann. Sci. École Norm. Sup.* (4) **32** (1999), 859–889.

[48] K. Burns and A. Wilkinson, On the ergodicity of partially hyperbolic systems, *Ann. Math.* (2), in press.

[49] D. Bylov, R. Vinograd, D. Grobman and V. Nemyckii, *Theory of Lyapunov Exponents and Its Application to Problems of Stability*, Izdat. "Nauka," Moscow, 1966, in Russian.

[50] M. Chaperon and F. Coudray, Invariant manifolds, conjugacies and blow-up, *Ergodic Theory Dyn. Syst.* **17** (1997), 783–791.

[51] B. Chirikov, A universal instability of many-dimensional oscillator systems, *Phys. Rep.* **52** (1979), 264–379.

[52] I. Cornfeld, S. Fomin, and Ya. Sinai, *Ergodic Theory*, Springer, 1982.

[53] R. de la Llave and C. Wayne, On Irwin's proof of the pseudostable manifold theorem, *Math. Z.* **219** (1995), 301–321.

[54] D. Dolgopyat, On dynamics of mostly contracting diffeomorphisms, *Comm. Math. Phys.* **213** (2000), 181–201.

[55] D. Dolgopyat, Limit theorems for partially hyperbolic systems, *Trans. Amer. Math. Soc.* **356** (2004), 1637–1689.

[56] D. Dolgopyat, On differentiability of SRB states for partially hyperbolic systems, *Invent. Math.* **155** (2004), 389–449.

[57] D. Dolgopyat, Averaging and invariant measures, *Mosc. Math. J.* **5** (2005), 55–66.

[58] D. Dolgopyat, H. Hu, and Ya. Pesin, An example of a smooth hyperbolic measure with countably many ergodic components, in *Smooth Ergodic Theory and Its Applications*, *Proc. Symp. Pure Math. 69*, edited by A. Katok, R. de la Llave, Ya. Pesin and H. Weiss, *American Mathematical Society, Providence, RI*, 2001, pp. 102–115.

[59] D. Dolgopyat and Ya. Pesin, Every compact manifold carries a completely hyperbolic diffeomorphism, *Ergodic Theory Dyn. Syst.* **22** (2002), 409–435.

[60] D. Dolgopyat and A. Wilkinson, Stable accessibility is C^1 dense, in Geometric methods in dynamics. II, *Astérisque* 287, 2003, pp. 33–60.

[61] P. Eberlein, Geodesic flows on certain manifolds without conjugate points, *Trans. Am. Math. Soc.* **167** (1972), 151–170.

[62] P. Eberlein, Geodesic flows on negatively curved manifolds. I, *Ann. Math.* (2) **95** (1972), 492–510.

[63] P. Eberlein, When is a geodesic flow of Anosov type? I, *J. Differ. Geom.* **8** (1973), 437–463.

[64] P. Eberlein, When is a geodesic flow of Anosov type? II, *J. Differ. Geom.* **8** (1973), 565–577.

[65] J. Eckmann and D. Ruelle, Ergodic theory of chaos and strange attractors, *Rev. Mod. Phys.* **57** (1985), 617–656.

[66] L. Eliasson, Reducibility and point spectrum for linear quasi-periodic skew-products, *Proceedings of the International Congress of Mathematicians* (Berlin, 1998), *Doc. Math.* **Extra Vol. II** (1998), 779–787.

[67] D. Epstein, Foliations with all leaves compact, *Ann. Inst. Fourier (Grenoble)* **26** (1976), 265–282.

[68] A. Fathi, M. Herman and J. Yoccoz, A proof of Pesin's stable manifold theorem, in Geometric Dynamics (Rio de Janeiro, 1981), *Lect. Notes Math. 1007*, edited by J. Palis, Springer, 1983, pp. 177–215.

[69] A. Fathi, F. Laudenbach, and V. Poénaru, Travaux de Thurston sur les surfaces, in Séminaire Orsay, *Astérisque* 66–67, 1979.

[70] N. Fenichel, Persistence and smoothness of invariant manifolds for flows, *Indiana Univ. Math. J.* **21** (1971/1972), 193–226.

[71] R. Feres and A. Katok, *Ergodic theory and dynamics of G-spaces (with special emphasis on rigidity phenomena)*, in *Handbook of Dynamical Systems 1A*, edited by B. Hasselblatt and A. Katok, Elsevier, 2002, pp. 665–763.

[72] G. Forni, On the Lyapunov exponents of the Kontsevich–Zorich cocycle, in *Handbook of Dynamical Systems 1B*, edited by B. Hasselblatt and A. Katok, Elsevier, 2006, pp. 549–580.

[73] A. Freire and R. Mañé, On the entropy of the geodesic flow in manifolds without conjugate points, *Invent. Math.* **69** (1982), 375–392.

[74] M. Gerber, Conditional stability and real analytic pseudo-Anosov maps, *Mem. Am. Math. Soc.* **54** (1985), no. 321.

[75] M. Gerber and A. Katok, Smooth models of Thurston's pseudo-Anosov maps, *Ann. Sci. École Norm. Sup.* (4) **15** (1982), 173–204.

[76] I. Gol'dsheid and G. Margulis, Lyapunov indices of a product of random matrices, *Russ. Math. Surv.* **44** (1989), 11–71.

[77] R. Gunesch, *Precise asymptotics for periodic orbits of the geodesic flow in nonpositive curvature*, preprint, 2003.

[78] R. Gunesch, *Precise volume estimates in nonpositive curvature*, preprint, 2003.

[79] J. Hadamard, Sur l'itération et les solutions asymptotiques des équations différentielles, *Bull. Soc. Math. France* **29** (1901), 224–228.

[80] H. Hasselblatt, Regularity of the Anosov splitting and of horospheric foliations, *Ergodic Theory Dyn. Syst.* **14** (1994), 645–666.

[81] B. Hasselblatt and Ya. Pesin, Partially hyperbolic dynamical systems, in *Handbook of Dynamical Systems 1B*, edited by B. Hasselblatt and A. Katok, Elsevier, 2006, pp. 1–55.

[82] M. Hénon, A two dimensional mapping with a strange attractor, *Comm. Math. Phys.* **50** (1976), 69–77.

[83] M. Herman, Une méthode pour minorer les exposants de Lyapunov et quelques examples montrant le caractére local d'un théorèm d'Arnold et de Moser sur le tore de dimension 2, *Comment. Math. Helv.* **58** (1983), 453–502.

[84] M. Hirayama, Periodic probability measures are dense in the set of invariant measures, *Discrete Contin. Dyn. Syst.* **9** (2003), 1185–1192.

[85] M. Hirayama and Ya. Pesin, Non-absolutely continuous invariant foliations, *Isr. J. Math.*, in press.

[86] M. Hirsch, C. Pugh, and M. Shub, *Invariant Manifolds*, *Lect. Notes. Math. 583*, Springer, 1977.

[87] E. Hopf, Statistik der geodätischen linien in mannigfaltigkeiten negativer krümmung, *Ber. Verh. Sächs. Akad. Wiss. Leipzig* **91** (1939), 261–304.

[88] H. Hu, Some ergodic properties of commuting diffeomorphisms, *Ergodic Theory Dyn. Syst.* **13** (1993), 73–100.

[89] H. Hu, Ya. Pesin, and A. Talitskaya, Every compact manifold carries a hyperbolic Bernoulli flow, in *Modern Dynamical Systems and Applications*, Cambridge University Press, 2004, pp. 347–358.

[90] H. Hu, Ya. Pesin, and A. Talitskaya, *Coexistence of zero and nonzero Lyapunov exponents on positive measure subsets in a transitive volume-preserving diffeomorphism*, preprint, 2006.

[91] S. Hurder and A. Katok, Ergodic theory and Weil measures for foliations, *Ann. Math.* (2) **126** (1987), 221–275.

[92] M. Irwin, A new proof of the pseudostable manifold theorem, *J. London Math. Soc.* (2) **21** (1980), 557–566.

[93] M. Jakobson, Absolutely continuous invariant measures for one-parameter families of one-dimensional maps, *Comm. Math. Phys.* **81** (1981), 39–88.

[94] M. Jakobson and G. Światek, One-dimensional maps, in *Handbook of Dynamical Systems 1A*, edited by B. Hasselblatt and A. Katok, Elsevier, 2002, pp. 599–664.

[95] R. Johnson, On a Floquet theory for almost periodic, two-dimensional linear systems, *J. Differ. Equ.* **37** (1980), 184–205.

[96] V. Kaĭmanovich, Lyapunov exponents, symmetric spaces and a multiplicative ergodic theorem for semisimple Lie groups, *J. Sov. Math.* **47** (1989), 2387–2398.

[97] B. Kalinin and V. Sadovskaya, On pointwise dimension of non-hyperbolic measures, *Ergodic Theory Dyn. Syst.* **22** (2002), 1783–1801.

[98] A. Karlsson and G. Margulis, A multiplicative ergodic theorem and nonpositively curved spaces, *Comm. Math. Phys.* **208** (1999), 107–123.

[99] A. Katok, A conjecture about entropy, in *Smooth Dynamical Systems*, edited by D. Anosov, Mir, 1977, pp. 181–203.

[100] A. Katok, Bernoulli diffeomorphisms on surfaces, *Ann. Math.* (2) **110** (1979), 529–547.

[101] A. Katok, Lyapunov exponents, entropy and periodic orbits for diffeomorphisms, *Inst. Hautes Études Sci. Publ. Math.* **51** (1980), 137–173.

[102] A. Katok, Nonuniform hyperbolicity and structure of smooth dynamical systems, in *Proceedings of the International Congress of Mathematicians* (Warsaw, 1983), PWN, 1984, pp. 1245–1253.

[103] A. Katok, with the collaboration of K. Burns, Infinitesimal Lyapunov functions, invariant cone families and stochastic properties of smooth dynamical systems, *Ergodic Theory Dyn. Syst.* **14** (1994), 757–785.

[104] A. Katok and B. Hasselblatt, *Introduction to the Modern Theory of Dynamical Systems*, Cambridge University Press, 1995.

[105] A. Katok and J. Lewis, Global rigidity results for lattice actions on tori and new examples of volume-preserving action, *Isr. J. Math.* **93** (1996), 253–280.

[106] A. Katok and L. Mendoza, Dynamical systems with nonuniformly hyperbolic behavior, supplement to *Introduction to the Modern Theory of Dynamical Systems*, by A. Katok and B. Hasselblatt, Cambridge University Press, 1995.

[107] A. Katok and L. Mendoza, *Smooth ergodic theory*, unpublished notes.

[108] A. Katok and R. Spatzier, First cohomology of Anosov actions of higher rank Abelian groups and applications to rigidity, *Inst. Hautes Études Sci. Publ. Math.* **79** (1994), 131–156.

[109] A. Katok and R. Spatzier, Subelliptic estimates of polynomial differential operators and applications to cocycle rigidity, *Math. Res. Lett.* **1** (1994), 193–202.

[110] A. Katok and J.-M. Strelcyn, with the collaboration of F. Ledrappier and F. Przytycki, Invariant Manifolds, Entropy and Billiards; Smooth Maps with Singularities, *Lect. Notes. Math. 1222*, Springer, 1986.

[111] S. Kerckhoff, H. Masur, and J. Smillie, Ergodicity of billiard flows and quadratic differentials, *Ann. Math.* (2) **124** (1986), 293–311.

[112] Yu. Kifer, *Ergodic Theory of Random Transformations*, Progress in Probability and Statistics 10, Birkhäuser, 1986.

[113] Yu. Kifer and P.-D. Liu, Random dynamics, in *Handbook of Dynamical Systems 1B*, edited by B. Hasselblatt and A. Katok, Elsevier, 2006, pp. 379–499.

[114] J. Kingman, Subadditive processes, in *École d'Été de Probabilités de Saint-Flour V–1975*, *Lect. Notes. Math. 539*, Springer, 1976, pp. 167–223.

[115] G. Knieper, Hyperbolic dynamics and Riemannian geometry, in *Handbook of Dynamical Systems 1A*, edited by B. Hasselblatt and A. Katok, Elsevier, 2002, pp. 453–545.

[116] M. Kontsevich, Lyapunov exponents and Hodge theory, in *The Mathematical Beauty of Physics* (Saclay, 1996), Adv. Ser. Math. Phys. 24, World Scientific Publishing, 1997, pp. 318–332.

[117] A. Krámli, N. Simányi, and D. Szász, A "transversal" fundamental theorem for semi-dispersing billiards, *Comm. Math. Phys.* **129** (1990), 535–560; errata in **138** (1991), 207–208.

[118] U. Krengel, On Rudolph's representation of aperiodic flows, *Ann. Inst. H. Poincaré Sect. B* **12** (1976), 319–338.

[119] R. Krikorian, Réductibilité des systèmes produits-croisés à valeurs dans des groupes compacts, *Astérisque* 259, 1999.

[120] F. Ledrappier, Propriétes ergodiques des mesures de Sinaï, *Inst. Hautes Études Sci. Publ. Math.* **59** (1984), 163–188.

[121] F. Ledrappier, Dimension of invariant measures, in *Proceedings of the Conference on Ergodic Theory and Related Topics, II* (Georgenthal, 1986), Teubner-Texte Math. 94, Teubner, 1987, pp. 116–124.

[122] F. Ledrappier and M. Misiurewicz, Dimension of invariant measures for maps with exponent zero, *Ergodic Theory Dyn. Syst.* **5** (1985), 595–610.

[123] F. Ledrappier and J.-M. Strelcyn, A proof of the estimate from below in Pesin's entropy formula, *Ergodic Theory Dyn. Syst.* **2** (1982), 203–219.

[124] F. Ledrappier and L.-S. Young, The metric entropy of diffeomorphisms. I: characterization of measures satisfying Pesin's entropy formula, *Ann. Math.* (2) **122** (1985), 509–539.

[125] F. Ledrappier and L.-S. Young, The metric entropy of diffeomorphisms. II: relations between entropy, exponents and dimension, *Ann. Math.* (2) **122** (1985), 540–574.

[126] J. Lewowicz, Lyapunov functions and topological stability, *J. Differ. Equ.* **38** (1980), 192–209.

[127] J. Lewowicz, Lyapunov functions and stability of geodesic flows, in *Geometric Dynamics* (Rio de Janeiro, 1981), *Lect. Notes. Math. 1007*, edited by J. Palis, Springer, 1983, pp. 463–479.

[128] J. Lewowicz and E. Lima de Sá, Analytic models of pseudo-Anosov maps, *Ergodic Theory Dyn. Syst.* **6** (1986), 385–392.

[129] P.-D. Liu and M. Qian, Smooth Ergodic Theory of Random Dynamical Systems, *Lect. Notes Math. 1606*, Springer, 1995.

[130] C. Liverani, Birth of an elliptic island in a chaotic sea, *Math. Phys. Electron. J.* **10** (2004), Paper 1, 13 pp.

[131] C. Liverani and M. Wojtkowski, Ergodicity in Hamiltonian systems, in *Dynamics Reported Expositions in Dynamical Systems* 4, Springer, 1995, pp. 130–202.

[132] S. Luzzatto, Stochastic-like behaviour in nonuniformly expanding maps, in *Handbook of Dynamical Systems 1B*, edited by B. Hasselblatt and A. Katok, Elsevier, 2006, pp. 265–326.

[133] S. Luzzatto and M. Viana, Parameter exclusions in Hénon-like systems, *Russ. Math. Surv.* **58** (2003), 1053–1092.

[134] A. Lyapunov, *The General Problem of the Stability of Motion*, Taylor & Francis, 1992.

[135] R. MacKay, Hyperbolic structure in classical chaos, in *Quantum Chaos* (Varenna, 1991), *Proc. Int. Sch. Phys. Enrico Fermi CXIX*, North-Holland, 1993, pp. 1–50.

[136] R. MacKay, *Renormalisation in Area-Preserving Maps*, *Adv. Ser. Nonlinear Dyn. 6*, World Scientific, 1993.

[137] R. Mañé, A proof of Pesin's formula, *Ergodic Theory Dyn. Syst.* **1** (1981), 95–102; errata in **3** (1983), 159–160.

[138] R. Mañé, Lyapunov exponents and stable manifolds for compact transformations, in *Geometric Dynamics* (Rio de Janeiro, 1981), *Lect. Notes Math. 1007*, edited by J. Palis, Springer, 1983, pp. 522–577.

[139] R. Mañé, *Ergodic Theory and Differentiable Dynamics*, Ergebnisse der Mathematik und ihrer Grenzgebiete 3. Folge·Band 8, Springer, 1987.

[140] R. Mañé, The Lyapunov exponents of generic area preserving diffeomorphisms, in *International Conference on Dynamical Systems* (Montevideo, 1995), *Pitman Res. Notes Math. Ser. 362*, Longman, 1996, pp. 110–119.

[141] R. Mañé, On the topological entropy of geodesic flows, *J. Differ. Geom.* **45** (1997), 74–93.

[142] A. Manning, Topological entropy and the first homology group, in *Dynamical Systems* (Warwick 1974), *Lect. Notes Math. 468*, edited by A. Manning, Springer, 1975, pp. 185–190.

[143] A. Manning, Topological entropy for geodesic flows, *Ann. Math.* (2) **110** (1979), 567–573.

[144] G. Margulis, Certain applications of ergodic theory to the investigation of manifolds of negative curvature, *Funct. Anal. Appl.* **3** (1969), 335–336.

[145] G. Margulis, *On some Aspects of the Theory of Anosov Systems*, with a survey by R. Sharp, *Springer Monographs in Mathematics*, Springer, 2004.

[146] R. Markarian, Non-uniformly hyperbolic billiards, *Ann. Fac. Sci. Toulouse Math.* (5) **3** (1994), 1207–1239.

[147] H. Masur, Interval exchange transformations and measured foliations, *Ann. Math.* (2) **115** (1982), 169–200.

[148] H. Masur, Ergodic theory of translation surfaces, in *Handbook of Dynamical Systems 1B*, edited by B. Hasselblatt and A. Katok, Elsevier, 2006, pp. 527–547.

[149] P. Mattila, *Geometry of Sets and Measures in Euclidean Spaces*, Camb. Stud. Adv. Math. 44, Cambridge University Press, 1995.

[150] H. McCluskey and A. Manning, Hausdorff dimension for horseshoes, *Ergodic Theory Dyn. Syst.* **3** (1983), 251–260; errata in **3** (1983), 319.

[151] C. McMullen, Dynamics on $K3$ surfaces: Salem numbers and Siegel disks, *J. Reine Angew. Math.* **545** (2002), 201–233.

[152] J. Milnor, Fubini foiled: Katok's paradoxical example in measure theory, *Math. Intelligencer* **19** (1997), 30–32.

[153] M. Misiurewicz and F. Przytycki, Entropy conjecture for tori, *Bull. Acad. Polon. Sci. Sér. Sci. Math. Astronom. Phys.* **25** (1977), 575–578.

[154] M. Misiurewicz and F. Przytycki, Topological entropy and degree of smooth mappings, *Bull. Acad. Polon. Sci. Sér. Sci. Math. Astronom. Phys.* **25** (1977), 573–574.

[155] C. Moore, Amenable subgroups of semisimple groups and proximal flows, *Isr. J. Math.* **34** (1979), 121–138.

[156] L. Mora and M. Viana, Abundance of strange attractors, *Acta Math.* **171** (1993), 1–71.

[157] J. Moser, On the volume elements on a manifold, *Trans. Am. Math. Soc.* **120** (1965), 286–294.

[158] S. Newhouse, Continuity properties of entropy, *Ann. Math.* (2) **129** (1989), 215–235.

[159] Z. Nitecki, *Differentiable Dynamics: An Introduction to the Orbit Structure of Diffeomorphisms*, M.I.T. Press, 1971.

[160] V. Niţică and A. Török, An open dense set of stably ergodic diffeomorphisms in a neighborhood of a non-ergodic one, *Topology* **40** (2001), 259–278.

[161] D. Ornstein and B. Weiss, Geodesic flows are Bernoullian, *Isr. J. Math.* **14** (1973), 184–198.

[162] V. Oseledets, A multiplicative ergodic theorem. Liapunov characteristic numbers for dynamical systems, *Trans. Mosc. Math. Soc.* **19** (1968), 197–221.

[163] R. Osserman and P. Sarnak, A new curvature invariant and entropy of geodesic flows, *Invent. Math.* **77** (1984), 455–462.

[164] O. Perron, Über stabilität und asymptotishes Verhalten der Lösungen eines Systemes endlicher Differenzengleichungen, *J. Reine Angew. Math.* **161** (1929), 41–64.

[165] O. Perron, Die stabilitätsfrage bei Differenzengleichungen, *Math. Z.* **32** (1930), 703–728.

[166] Ya. Pesin, An example of a nonergodic flow with nonzero characteristic exponents, *Func. Anal. Appl.* **8** (1974), 263–264.

[167] Ya. Pesin, Families of invariant manifolds corresponding to nonzero characteristic exponents, *Math.* USSR-Izv. **40** (1976), 1261–1305.

[168] Ya. Pesin, A description of the π-partition of a diffeomorphism with an invariant measure, *Math. Notes* **22** (1977), 506–515.

[169] Ya. Pesin, Characteristic Ljapunov exponents, and smooth ergodic theory, *Russ. Math. Surv.* **32** (1977), 55–114.

[170] Ya. Pesin, Geodesic flows on closed Riemannian manifolds without focal points, *Math. USSR-Izv.* **11** (1977), 1195–1228.

[171] Ya. Pesin, Formulas for the entropy of the geodesic flow on a compact Riemannian manifold without conjugate points, *Math. Notes* **24:4** (1978), 796–805.

[172] Ya. Pesin, *Dimension Theory in Dynamical Systems: Contemporary Views and Applications*, Chicago Lectures in Mathematics, Chicago University Press, 1997.

[173] Ya. Pesin, General theory of smooth hyperbolic dynamical systems, in *Dynamical Systems, Ergodic Theory and Applications II, Ergodic Theory of Smooth Dynamical Systems*, edited by Ya. Sinai, Encyclopedia of Mathematical Sciences 100, Springer, 2000, pp. 113–191.

[174] Ya. Pesin, *Lectures on Partial Hyperbolicity and Stable Ergodicity*, Zürich Lectures in Advanced Mathematics, European Mathematical Society, 2004.

[175] Ya. Pesin and Ya. Sinai, Hyperbolicity and stochasticity of dynamical systems, *Sov. Sci. Rev. Sect. C: Math. Phys. Rev. 2*, Harwood Academic, 1981, pp. 53–115.

[176] Ya. Pesin and Ya. Sinai, Gibbs measures for partially hyperbolic attractors, *Ergodic Theory Dyn. Syst.* **2** (1982), 417–438.

[177] Ya. Pesin and H. Weiss, On the dimension of deterministic and random Cantor-like sets, symbolic dynamics, and the Eckmann–Ruelle conjecture, *Comm. Math. Phys.* **182** (1996), 105–153.

[178] K. Petersen, *Ergodic Theory*, Camb. Stud. Adv. Math. 2, Cambridge University Press, 1989.

[179] M. Pollicott, *Lectures on Ergodic Theory and Pesin Theory on Compact Manifolds*, London Math. Soc. Lect. Note Ser. 180, Cambridge University Press, 1993.

[180] V. Potapov, The multiplicative structure of J-contractive matrix functions, *Am. Math. Soc. Transl.* (2) **15** (1960), 131–243.

[181] V. Potapov, A theorem on the modulus. I, *Am. Math. Soc. Transl.* (2) **138** (1988), 55–65.

[182] V. Potapov, Linear fractional transformation of matrices, *Am. Math. Soc. Transl.* (2) **138** (1988), 21–35.

[183] F. Przytycki, Examples of conservative diffeomorphisms of the two-dimensional torus with coexistence of elliptic and stochastic behaviour, *Ergodic Theory Dyn. Syst.* **2** (1982), 439–463.

[184] C. Pugh, The $C^{1+\alpha}$ hypothesis in Pesin theory, *Inst. Hautes Études Sci. Publ. Math.* **59** (1984), 143–161.

[185] C. Pugh and M. Shub, Ergodic attractors, *Trans. Am. Math. Soc.* **312** (1989), 1–54.

[186] C. Pugh and M. Shub, Stable ergodicity and Julienne quasi-conformality, *J. Eur. Math. Soc.* **2** (2000), 1–52.

[187] M. Ragunathan, A proof of Oseledec's multiplicative ergodic theorem, *Isr. J. Math.* **32** (1979), 356–362.

[188] S. Rasband, *Chaotic Dynamics of Nonlinear Systems*, John Wiley & Sons, 1990.

[189] A. Rodríguez Hertz, F. Rodríguez Hertz, and R. Ures, *Partially hyperbolic systems with 1D-center bundle: I. Stable ergodicity*, preprint.

[190] D. Rudolph, A two-valued step coding for ergodic flows, *Math. Z.* **150** (1976), 201–220.

[191] D. Ruelle, A measure associated with Axiom A attractors, *Am. J. Math.* **98** (1976), 619–654.

[192] D. Ruelle, An inequality for the entropy of differentiable maps, *Bol. Soc. Bras. Mat.* **9** (1978), 83–87.

[193] D. Ruelle, Ergodic theory of differentiable dynamical systems, *Inst. Hautes Études Sci. Publ. Math.* **50** (1979), 27–58.

[194] D. Ruelle, Characteristic exponents and invariant manifolds in Hilbert space, *Ann. Math.* (2) **115** (1982), 243–290.

[195] D. Ruelle, Perturbation theory for Lyapunov exponents of a toral map: extension of a result of Shub and Wilkinson, *Isr. J. Math.* **134** (2003), 345–361.

[196] D. Ruelle and A. Wilkinson, Absolutely singular dynamical foliations, *Comm. Math. Phys.* **219** (2001), 481–487.

[197] R. Sacker and G. Sell, Lifting properties in skew-product flows with applications to differential equations, *Mem. Am. Math. Soc.* **11** (1977), no. 190.

[198] R. Sacker and G. Sell, A spectral theory for linear differential systems, *J. Differ. Equ.* **27** (1978), 320–358.

[199] G. Sell, The structure of a flow in the vicinity of an almost periodic motion, *J. Differ. Equ.* **27** (1978), 359–393.

[200] M. Shub and A. Wilkinson, Pathological foliations and removable zero exponents, *Invent. Math.* **139** (2000), 495–508.

[201] Ya. Sinai, Classical dynamic systems with countably-multiple Lebesgue spectrum. II, *Izv. Akad. Nauk SSSR Ser. Mat* **30** (1966), 15–68.

[202] Ya. Sinai, Dynamical systems with elastic reflections, *Russ. Math. Surv.* **68** (1970), 137–189.

[203] Ya. Sinai, Some rigorous results on decay of correlations, supplement to the book *Statistical Irreversibility in Nonlinear Systems* by G. Zaslavskij, Nauka, 1970, pp. 124–139.

[204] Ya. Sinai, Gibbs measures in ergodic theory, *Russ. Math. Surv.* **27** (1972), 21–69.

[205] Ya. Sinai and N. Chernov, Entropy of hard spheres gas with respect to the group of space-time translations, *Tr. Semin. Im. I. G. Petrovskogo* **8** (1982), 218–238, in Russian.

[206] S. Smale, Diffeomorphisms of the 2-sphere, *Proc. Am. Math. Soc.* **10** (1959), 621–626.

[207] S. Smale, Diffeomorphisms with many periodic points, in *Differential and Combinatorial Topology* (A symposium in Honor of Marston Morse), Princeton University Press, 1965, pp. 63–80.

[208] S. Smale, Differentiable dynamical systems, *Bull. Am. Math. Soc.* **73** (1967), 747–817.

[209] A. Tahzibi, C^1-generic Pesin's entropy formula, *C. R. Math. Acad. Sci. Paris* **335** (2002), 1057–1062

[210] A. Talitskaya, *Partially Hyperbolic Phenomena in Dynamical Systems with Discrete and Continuous Time*, Ph.D. Thesis, Pennsylvania State University, 2004.

[211] Ph. Thieullen, Fibrés dynamiques asymptotiquement compacts. Exposants de Lyapunov. Entropie. Dimension, *Ann. Inst. H. Poincaré. Anal. Non Linéaire* **4** (1987), 49–97.

[212] Ph. Thieullen, Ergodic reduction of random products of two-by-two matrices, *J. Anal. Math.* **73** (1997), 19–64.

[213] W. Thurston, On the geometry and dynamics of diffeomorphisms of surfaces, *Bull. Am. Math. Soc.* **19** (1988), 417–431.

[214] M. Tsujii, Regular points for ergodic Sinai measures, *Trans. Am. Math. Soc.* **328** (1991), 747–777.

[215] I. Ugarcovici, On hyperbolic measures and periodic orbits, *Discrete Contin. Dyn. Syst.* **16** (2006), 505–512.

[216] W. Veech, Gauss measures for transformations on the space of interval exchange maps, *Ann. Math.* (2) **115** (1982), 201–242.

[217] W. Veech, The Teichmüller geodesic flow, *Ann. Math.* (2) **124** (1986), 441–530.

[218] W. Veech, Flat surfaces, *Am. J. Math.* **115** (1993), 589–689.

[219] M. Viana, Strange attractors in higher dimensions, *Bol. Soc. Bras. Mat.* **24** (1993), 13–62.

[220] M. Viana, Multidimensional nonhyperbolic attractors, *Inst. Hautes Études Sci. Publ. Math.* **85** (1997), 63–96.

[221] M. Viana, Almost all cocycles over any hyperbolic system have non-vanishing Lyapunov exponents, *Ann. Math.* (2), in press.

[222] Q. Wang and L.-S. Young, Strange attractors with one direction of instability, *Comm. Math. Phys.* **218** (2001), 1–97.

[223] Q. Wang and L.-S. Young, From invariant curves to strange attractors, *Comm. Math. Phys.* **225** (2002), 275–304.

[224] Z. Wang and W. Sun, *Liapunov exponents of hyperbolic measure and hyperbolic periodic points*, preprint.

[225] M. Wojtkowski, Invariant families of cones and Lyapunov exponents, *Ergodic Theory Dyn. Syst.* **5** (1985), 145–161.

[226] M. Wojtkowski, Measure theoretic entropy of the system of hard spheres, *Ergodic Theory Dyn. Syst.* **8** (1988), 133–153.

[227] M. Wojtkowski, Systems of classical interacting particles with nonvanishing Lyapunov exponents, in *Lyapunov Exponents* (Oberwolfach, 1990), *Lect. Notes Math. 1486*, edited by L. Arnold, H. Crauel, and J.-P. Eckmann, Springer, 1991, pp. 243–262.

[228] M. Wojtkowski, W-flows on Weyl manifolds and Gaussian thermostats, *J. Math. Pures Appl.* (9) **79** (2000), 953–974.

[229] M. Wojtkowski, Magnetic flows and Gaussian thermostats on manifolds of negative curvature, *Fundam. Math.* **163** (2000), 177–191.

[230] M. Wojtkowski, Monotonicity, J-algebra of Potapov and Lyapunov exponents, in *Smooth Ergodic Theory and Its Applications, Proc. Symp. Pure Math. 69*, edited by A. Katok, R. de la Llave, Ya. Pesin and H. Weiss, *American Mathematical Society, Providence, RI*, 2001, pp. 499–521.

[231] M. Wojtkowski and C. Liverani, Conformally symplectic dynamics and symmetry of the Lyapunov spectrum, *Comm. Math. Phys.* **194** (1998), 47–60.

[232] Z. Xia, Existence of invariant tori in volume-preserving diffeomorphisms, *Ergodic Theory Dyn. Syst.* **12** (1992), 621–631.

[233] J.-C. Yoccoz, Travaux de Herman sur les tores invariants, Séminaire Bourbaki, Vol. 1991/92, *Astérisque* 206, 1992, Exp. No. 754, 4, 311–344.

[234] Y. Yomdin, C^k-resolution of semialgebraic mappings. Addendum to: "Volume growth and entropy," *Isr. J. Math.* **57** (1987), 301–317.

[235] Y. Yomdin, Volume growth and entropy, *Isr. J. Math.* **57** (1987), 285–300.

[236] L.-S. Young, Dimension, entropy and Lyapunov exponents, *Ergodic Theory Dyn. Syst.* **2** (1982), 109–124.

[237] L.-S. Young, Lyapunov exponents for some quasi-periodic cocycles, *Ergodic Theory Dyn. Syst.* **17** (1997), 483–504.

[238] L.-S. Young, Statistical properties of dynamical systems with some hyperbolicity, *Ann. Math.* (2) **147** (1998), 585–650.

[239] R. Zimmer, Ergodic theory, group representations, and rigidity, *Bull. Am. Math. Soc.* **6** (1982), 383–416.

[240] R. Zimmer, *Ergodic Theory and Semisimple Groups*, Monographs in Mathematics 81, Birkhäuser, 1984.

[241] A. Zorich, Finite Gauss measure on the space of interval exchange transformations. Lyapunov exponents, *Ann. Inst. Fourier* (Grenoble) **46** (1996), 325–370.

[242] A. Zorich, How do the leaves of a closed 1-form wind around a surface?, in *Pseudoperiodic Topology*, edited by V. Arnold, M. Kontsevich, and A. Zorich, *Am. Math. Soc. Transl. Ser. 2, 197, American Mathematical Society, Providence, RI*, 1999, pp. 135–178.

[243] A. Zorich, Flat surfaces, in *Frontiers in Number Theory, Physics and Geometry I: On Random Matrices, Zeta Functions and Dynamical Systems*, edited by P. Cartier, B. Julia, P. Moussa, and P. Vanhove, Springer, 2006, pp. 439–586.

Index

For EU product safety concerns, contact us at Calle de José Abascal, 56–1°, 28003 Madrid, Spain or eugpsr@cambridge.org.

www.ingramcontent.com/pod-product-compliance
Lightning Source LLC
LaVergne TN
LVHW102010210926
848483LV00010B/244

* 9 7 8 0 5 2 1 8 3 2 5 8 8 *